Handbook of Vinyl Polymers

Radical Polymerization, Process, and Technology

Second Edition

PLASTICS ENGINEERING

Founding Editor

Donald E. Hudgin

Professor
Clemson University
Clemson, South Carolina

1. Plastics Waste: Recovery of Economic Value, *Jacob Leidner*
2. Polyester Molding Compounds, *Robert Burns*
3. Carbon Black-Polymer Composites: The Physics of Electrically Conducting Composites, *edited by Enid Keil Sichel*
4. The Strength and Stiffness of Polymers, *edited by Anagnostis E. Zachariades and Roger S. Porter*
5. Selecting Thermoplastics for Engineering Applications, *Charles P. MacDermott*
6. Engineering with Rigid PVC: Processability and Applications, *edited by I. Luis Gomez*
7. Computer-Aided Design of Polymers and Composites, *D. H. Kaelble*
8. Engineering Thermoplastics: Properties and Applications, *edited by James M. Margolis*
9. Structural Foam: A Purchasing and Design Guide, *Bruce C. Wendle*
10. Plastics in Architecture: A Guide to Acrylic and Polycarbonate, *Ralph Montella*
11. Metal-Filled Polymers: Properties and Applications, *edited by Swapan K. Bhattacharya*
12. Plastics Technology Handbook, *Manas Chanda and Salil K. Roy*
13. Reaction Injection Molding Machinery and Processes, *F. Melvin Sweeney*
14. Practical Thermoforming: Principles and Applications, *John Florian*
15. Injection and Compression Molding Fundamentals, *edited by Avraam I. Isayev*
16. Polymer Mixing and Extrusion Technology, *Nicholas P. Cheremisinoff*
17. High Modulus Polymers: Approaches to Design and Development, *edited by Anagnostis E. Zachariades and Roger S. Porter*
18. Corrosion-Resistant Plastic Composites in Chemical Plant Design, *John H. Mallinson*

19. Handbook of Elastomers: New Developments and Technology, *edited by Anil K. Bhowmick and Howard L. Stephens*
20. Rubber Compounding: Principles, Materials, and Techniques, *Fred W. Barlow*
21. Thermoplastic Polymer Additives: Theory and Practice, *edited by John T. Lutz, Jr.*
22. Emulsion Polymer Technology, *Robert D. Athey, Jr.*
23. Mixing in Polymer Processing, *edited by Chris Rauwendaal*
24. Handbook of Polymer Synthesis, Parts A and B, *edited by Hans R. Kricheldorf*
25. Computational Modeling of Polymers, *edited by Jozef Bicerano*
26. Plastics Technology Handbook: Second Edition, Revised and Expanded, *Manas Chanda and Salil K. Roy*
27. Prediction of Polymer Properties, *Jozef Bicerano*
28. Ferroelectric Polymers: Chemistry, Physics, and Applications, *edited by Hari Singh Nalwa*
29. Degradable Polymers, Recycling, and Plastics Waste Management, *edited by Ann-Christine Albertsson and Samuel J. Huang*
30. Polymer Toughening, *edited by Charles B. Arends*
31. Handbook of Applied Polymer Processing Technology, *edited by Nicholas P. Cheremisinoff and Paul N. Cheremisinoff*
32. Diffusion in Polymers, *edited by P. Neogi*
33. Polymer Devolatilization, *edited by Ramon J. Albalak*
34. Anionic Polymerization: Principles and Practical Applications, *Henry L. Hsieh and Roderic P. Quirk*
35. Cationic Polymerizations: Mechanisms, Synthesis, and Applications, *edited by Krzysztof Matyjaszewski*
36. Polyimides: Fundamentals and Applications, *edited by Malay K. Ghosh and K. L. Mittal*
37. Thermoplastic Melt Rheology and Processing, *A. V. Shenoy and D. R. Saini*
38. Prediction of Polymer Properties: Second Edition, Revised and Expanded, *Jozef Bicerano*
39. Practical Thermoforming: Principles and Applications, Second Edition, Revised and Expanded, *John Florian*
40. Macromolecular Design of Polymeric Materials, *edited by Koichi Hatada, Tatsuki Kitayama, and Otto Vogl*
41. Handbook of Thermoplastics, *edited by Olagoke Olabisi*
42. Selecting Thermoplastics for Engineering Applications: Second Edition, Revised and Expanded, *Charles P. MacDermott and Aroon V. Shenoy*
43. Metallized Plastics, *edited by K. L. Mittal*
44. Oligomer Technology and Applications, *Constantin V. Uglea*
45. Electrical and Optical Polymer Systems: Fundamentals, Methods, and Applications, *edited by Donald L. Wise, Gary E. Wnek, Debra J. Trantolo, Thomas M. Cooper, and Joseph D. Gresser*

46. Structure and Properties of Multiphase Polymeric Materials, *edited by Takeo Araki, Qui Tran-Cong, and Mitsuhiro Shibayama*
47. Plastics Technology Handbook: Third Edition, Revised and Expanded, *Manas Chanda and Salil K. Roy*
48. Handbook of Radical Vinyl Polymerization, *edited by Munmaya K. Mishra and Yusuf Yagci*
49. Photonic Polymer Systems: Fundamentals, Methods, and Applications, *edited by Donald L. Wise, Gary E. Wnek, Debra J. Trantolo, Thomas M. Cooper, and Joseph D. Gresser*
50. Handbook of Polymer Testing: Physical Methods, *edited by Roger Brown*
51. Handbook of Polypropylene and Polypropylene Composites, *edited by Harutun G. Karian*
52. Polymer Blends and Alloys, *edited by Gabriel O. Shonaike and George P. Simon*
53. Star and Hyperbranched Polymers, *edited by Munmaya K. Mishra and Shiro Kobayashi*
54. Practical Extrusion Blow Molding, *edited by Samuel L. Belcher*
55. Polymer Viscoelasticity: Stress and Strain in Practice, *Evaristo Riande, Ricardo Díaz-Calleja, Margarita G. Prolongo, Rosa M. Masegosa, and Catalina Salom*
56. Handbook of Polycarbonate Science and Technology, *edited by Donald G. LeGrand and John T. Bendler*
57. Handbook of Polyethylene: Structures, Properties, and Applications, *Andrew J. Peacock*
58. Polymer and Composite Rheology: Second Edition, Revised and Expanded, *Rakesh K. Gupta*
59. Handbook of Polyolefins: Second Edition, Revised and Expanded, *edited by Cornelia Vasile*
60. Polymer Modification: Principles, Techniques, and Applications, *edited by John J. Meister*
61. Handbook of Elastomers: Second Edition, Revised and Expanded, *edited by Anil K. Bhowmick and Howard L. Stephens*
62. Polymer Modifiers and Additives, *edited by John T. Lutz, Jr., and Richard F. Grossman*
63. Practical Injection Molding, *Bernie A. Olmsted and Martin E. Davis*
64. Thermosetting Polymers, *Jean-Pierre Pascault, Henry Sautereau, Jacques Verdu, and Roberto J. J. Williams*
65. Prediction of Polymer Properties: Third Edition, Revised and Expanded, *Jozef Bicerano*
66. Fundamentals of Polymer Engineering, *Anil Kumar and Rakesh K. Gupta*
67. Handbook of Polypropylene and Polymer, *Harutun Karian*
68. Handbook of Plastic Analysis, *Hubert Lobo and Jose Bonilla*
69. Computer-Aided Injection Mold Design and Manufacture, *J. Y. H. Fuh, Y. F. Zhang, A. Y. C. Nee, and M. W. Fu*
70. Handbook of Polymer Synthesis: Second Edition, *Hans R. Kricheldorf and Graham Swift*
71. Practical Guide to Injection Blow Molding, *Samuel L. Belcher*

72. Plastics Technology Handbook: Fourth Edition, *Manas Chanda and Salil K. Roy*
73. Handbook of Vinyl Polymers: Radical Polymerization, Process, and Technology, Second Edition, *edited by Munmaya K. Mishra and Yusuf Yagci*

Handbook of Vinyl Polymers

Radical Polymerization, Process, and Technology

Second Edition

Edited by

Munmaya K. Mishra
Present Affiliation:
Philip Morris USA
Richmond, Virginia

Yusuf Yagci
Istanbul Technical University
Maslak, Istanbul, Turkey

CRC Press is an imprint of the
Taylor & Francis Group, an **informa** business

CRC Press
Taylor & Francis Group
6000 Broken Sound Parkway NW, Suite 300
Boca Raton, FL 33487-2742

First issued in paperback 2019

ISBN-13: 978-0-8247-2595-2 (hbk)
ISBN-13: 978-0-367-38711-2 (pbk)

Library of Congress Cataloging-in-Publication Data

Handbook of vinyl polymers : radical polymerization, process, and technology / editors, Munmaya Mishra and Yusuf Yagci. -- 2nd ed.
p. cm.
Rev. ed. of: Handbook of radical vinyl polymerization / Munmaya K. Mishra, Yusuf Yagci. 1998.
Includes bibliographical references and index.
ISBN 978-0-8247-2595-2 (alk. paper)
1. Vinyl polymers. 2. Polymerization. I. Mishra, Munmaya K. II. Yagci, Yusuf, 1952- III. Mishra, Munmaya K. Handbook of radical vinyl polymerization. IV. Title.

QD281.P6M632 2008
668.4'236--dc22 2007020735

Visit the Taylor & Francis Web site at
http://www.taylorandfrancis.com

and the CRC Press Web site at
http://www.crcpress.com

Dedication

To my wife, Bidu

Munmaya K. Mishra

To my wife, Emine

Yusuf Yagci

Contents

Preface

The field of vinyl polymerization has grown very large indeed. The momentum of extensive investigations on radical vinyl polymerization, undertaken in many laboratories, has carried us to an advanced stage of development. Consequently, we are attempting in this *Handbook of Vinyl Polymers: Radical Polymerization, Process, and Technology* to present current knowledge of the subject in an integrated package.

The book is divided into five sections that include a total of 21 chapters. The first three chapters provide the fundamental aspects; the following 10 chapters offer a detailed description of the radical initiating systems and mechanisms, along with the technical processes. This includes comprehensive information on living polymerization, functionalization polymers, and block and graft copolymers. The book also contains a section on Vinyl Polymer Technology with seven chapters that describe the recent advances on composites, recycling, and processing of vinyl polymers. The book ends with a chapter that presents a variety of data on monomers and polymerization.

It is hoped that this presentation will prove useful to investigators in the area of vinyl polymers. The book offers much that is of value, presenting basic information in addition to providing a unified, interlocking look at recent advances in the field of vinyl polymers. Although selected parts of this discipline have been reviewed in the past, this is the first time that the entire field has been comprehensively and critically examined in a book. However, it would scarcely be possible in a single volume to do justice to all the excellent research in various branches of the subject; selection of the material to be included was difficult and an element of arbitrariness was unavoidable.

This is an interdisciplinary book written for the organic chemist/polymer scientist who wants comprehensive, up-to-date critical information about radical vinyl polymerization and technology, as well as for the industrial researcher who wants to survey the technology of vinyl polymers leading to useful products.

Specifically, this book will serve in the following ways: (1) as a reference book for researchers in vinyl polymers, (2) as a coherent picture of the field and a self-educating introductory and advanced text for the practicing chemist who has little background in vinyl polymers, and (3) as one of a group of textbooks for courses in the graduate-level curriculum devoted to polymer science and engineering.

It would not have been possible to complete a project like this without the help and participation of numerous individuals. We gratefully acknowledge all the contributors who made this book possible. Last, with love and appreciation, we acknowledge our wives Bidu Mishra and Emine Yagci for their timely encouragement, sacrifice, and support during long afternoons, weekends, early mornings, and holidays spent on this book. Without their help and support, this project would never have started or been completed.

Munmaya K. Mishra
Yusuf Yagci

Contributors

Metin H. Acar
Istanbul Technical University
Istanbul, Turkey

Rajesh D. Anandjiwala
CSIR Materials Science and Manufacturing
and
Nelson Mandela Metropolitan University
Port Elizabeth, South Africa

Nergis Arsu
Yildiz Technical University
Istanbul, Turkey

Cheol Hoon Cheong
LG Chemical Company
Yeochon, Korea

Kyu Yong Choi
University of Maryland
College Park, Maryland

Chapal K. Das
Indian Institute of Technology
Kharagpur, India

Tanya Das
Indian Institute of Technology
Kharagpur, India

Norman G. Gaylord
Drew University
New Providence, New Jersey

Gurkan Hizal
Istanbul Technical University
Istanbul, Turkey

Hanafi Ismail
University Science Malaysia
Nibong Tebal Penang, Malaysia

Maya Jacob John
CSIR Materials Science and Manufacturing
Port Elizabeth, South Africa

Sandeep Kumar
Indian Institute of Technology
Kharagpur, India

Byung-Gu Kwag
LG Chemical Company
Yeochon, Korea

Ram N. Mahaling
Indian Institute of Technology
Kharagpur, India

Munmaya K. Mishra
Present Affiliation:
Philip Morris USA Research Center
Richmond, Virginia

Ali E. Muftuoglu
Istanbul Technical University
Istanbul, Turkey

Madhumita Mukherjee
Indian Institute of Technology
Kharagpur, India

Mir Mohammad A. Nikje
Imam Khomeini International University
Qazvin, Iran

Seung Young Park
LG Chemical Company
Yeochon, Korea

Rathanasamy Rajasekar
Indian Institute of Technology
Kharagpur, India

Tanmoy Rath
Indian Institute of Technology
Kharagpur, India

Chantara T. Ratnam
Malaysian Institute for Nuclear
Technology Research
Selangor Draul Ehsan, Malaysia

Chaganti S. Reddy
Indian Institute of Technology
Kharagpur, India

Ivo Reetz
Istanbul Technical University
Istanbul, Turkey

M. Atilla Tasdelen
Istanbul Technical University
Istanbul, Turkey

Sabu Thomas
Mahatma Gandhi University
Kerala, India

Umit Tunca
Istanbul Technical University
Istanbul, Turkey

Yusuf Yagci
Istanbul Technical University
Maslak, Istanbul, Turkey

Part I

The Fundamentals of Radical Vinyl Polymerization

1 The Fundamentals

Yusuf Yagci and Munmaya K. Mishra

CONTENTS

1.1 INTRODUCTION

1.1.1 WHAT ARE RADICALS?

Organic molecules containing an unpaired electron are termed free radicals or radicals, and radicals are generally considered unstable species because of their very short lifetimes in the liquid and gaseous state. The instability of free radicals is a kinetic instead of a thermodynamic property.

Free radicals can undergo four general types of reactions: (1) transfer or abstraction, (2) elimination, (3) addition, and (4) combination or coupling. These reactions can be illustrated by the following example [1]. The pyrolysis of ethane in the gas phase is a free-radical reaction and the products are formed from the initial homolytic decomposition:

$$CH_3\text{—}CH_3 \rightarrow 2 \cdot CH_3 \tag{1.1}$$

Four basic types of reactions account for the mechanism and products of free-radical polymerization:

1. Transfer/Abstraction Reaction (hydrogen-atom transfer reaction between methyl radical and ethane):

$$\cdot CH_3 + H\text{—}CH_2CH_3 \rightarrow H_3C\text{—}H + \cdot CH_2CH_3 \tag{1.2}$$

2. Elimination Reaction (elimination of hydrogen atom from the ethyl radical):

$$CH_2\text{—}CH_2\text{—}H + \cdot CH_2CH_3 \rightarrow CH_2{=}CH_2 + H\text{—}CH_2CH_3 \tag{1.3}$$

This reaction is better known as a disproportionation reaction.

3. Combination Reaction (formation of propane by combination of methyl and ethyl radicals):

$$\cdot CH_3 + \cdot\ CH_2CH_3 \rightarrow CH_3CH_2CH_3 \tag{1.4}$$

4. Addition Reaction (addition of methyl radicals to ethylene to form propyl radicals):

$$\cdot CH_3 + CH_2{=}CH_2 \rightarrow CH_3CH_2CH_2\cdot \tag{1.5}$$

1.1.2 How Are Radicals Generated?

Virtually all free-radical chain reactions require a separate initiation step in which a radical species is generated in the reaction mixture or by adding a stable free radical (generated by a separate initiation step) directly to the reactants. Radical initiation reactions, therefore, can be divided into two general types according to the manner in which the first radical species is formed; these are (1) homolytic decomposition of covalent bonds by energy absorption or (2) electron transfer from ions or atoms containing unpaired electrons followed by bond dissociation in the acceptor molecule.

1.1.2.1 Homolytic Decomposition of Covalent Bonds

Organic compounds may decompose into two or more free-radical fragments by energy absorption. The energy includes almost any form, including thermal, electromagnetic (ultraviolet and high-energy radiation), particulate, electrical, sonic, and mechanical. The most important of these are the thermal and electromagnetic energies. For the generation of free radicals by energetic cleavage, the important parameter is the bond dissociation energy, D. The bond dissociation energy is the energy required to break a particular bond in a particular molecule. The bond dissociation energy can be used to calculate the approximate rate of free-radical formation at various temperatures according to the following reaction:

$$R_2 \xrightarrow{\Delta} 2R^{\cdot} \tag{1.6}$$

$$\frac{d[R\cdot]}{dt} = k[R_2] \tag{1.7}$$

$$K = Ae^{-D/RT} \tag{1.8}$$

Pure thermal dissociation is generally a unimolecular reaction and D is very close to the activation energy, ΔE^{+}. For a unimolecular reaction [1], the frequency factor, A, is generally of the order 10^{13}–10^{14} sec^{-1}. Most practical thermal initiators are compounds with bond dissociation energies in the range of 30 to 40 kcal mol^{-1}. This range of dissociation energies limits the types of useful compound to those containing fairly specific types of covalent bonds, notably oxygen–oxygen bonds, oxygen–nitrogen bonds, and sulfur–sulfur bonds, as well as unique bonds present in azo compounds.

1.1.3 Comparison of Free-Radical and Ionic Olefin Polymerization Reactions

The mechanism of free radical compared with ionic chain-growth polymerization has many fundamental differences. The differences involve not only the rate and manner of polymer chain growth for each type of polymerization but also the selection of monomers suitable for each type of polymerization. The variety of behaviors are listed in Tables 1.1 and 1.2.

TABLE 1.1
Polymerizability of Monomers by Different Polymerization Mechanisms

	Types of Polymerization		
Monomer	**Radical**	**Cationic**	**Anionic**
Acrylonitrile	Yes	No	Yes
Acrylamide	Yes	No	No
1-Alkyl olefins	No	Yes	No
Acrylates	Yes	No	Yes
Aldehydes	No	Yes	Yes
Butene-1	No	No	No
Butadiene-1,3	Yes	Yes	Yes
1,1-Dialkyl olefins	No	Yes	No
1,3-Dienes	Yes	Yes	Yes
Ethylene	Yes	No	Yes
Halogenated olefins	Yes	No	No
Isoprene	Yes	Yes	Yes
Isobutene	No	Yes	No
Ketones	No	Yes	Yes
Methacrylic esters	Yes	No	Yes
Methacrylamide	Yes	No	Yes
Methacrylonitrile	Yes	No	Yes
Methyl styrene	Yes	Yes	Yes
Styrene	Yes	Yes	Yes
Tetrafluoroethylene	Yes	No	No
Vinyl chloride	Yes	No	No
Vinyl fluoride	Yes	No	No
Vinyl ethers	No	Yes	No
Vinyl esters	Yes	No	No
Vinylidene chloride	Yes	No	Yes
N-Vinyl carbazole	Yes	Yes	No
N-Vinyl pyrrolidone	Yes	Yes	No

TABLE 1.2
Comparison of Free-Radical and Ionic Olefin Polymerization

Free Radical

1. End groups in growing polymer chains are truly free species.
2. It is generally felt that solvent polarity exerts no influence on free-radical propagation.
3. Radical polymerization reaction demonstrates both combination and disproportionation termination reaction steps involve two growing polymer chains.
4. Termination reactions are bimolecular.
5. Due to the high rate of bimolecular termination, the concentration of growing polymer chains must be maintained at a very low level, in order to prepare high-molecular-weight polymer.
6. Radical polymerizations are versatile and can be initiated effectively in gas, solid, and liquid phases. Polymerizations can be performed in bulk, solution, precipitation, suspension, and emulsion techniques. Each process has its own merits and special characteristics.

Ionic

1. End groups always have counterions, more or less associated.
2. The association of the counterions, their stability, and the ionic propagation depend on the polarity of the medium.
3. In cationic polymerization reaction, combination and disproportionation reactions occur between the end groups and the counterion of an active polymer chain (anion capture and proton release).
4. Termination reactions are unimolecular.
5. In ionic polymerization, a much higher concentration of growing polymer chains may be maintained without penalty to the molecular weights produced. In ionic polymerization, no tendency exists for two polymer chain end groups of like ionic charge to react. Due to much higher concentration of growing polymer chains in homogeneous polymerization reaction, the rates of ionic polymerization can be many times higher than that of a free-radical polymerization of the same monomer, even though the activation energies for propagation are comparable.
6. Ionic polymerization is limited experimentally almost entirely to solution or bulk methods, although crystalline, solid-state polymerization is observed in some cases.

REFERENCE

1. H. E. de la Mare and W. E. Vaughan, *J. Chem. Ed.*, 34, 10 (1951).

2 Chemistry and Kinetic Model of Radical Vinyl Polymerization

Yusuf Yagci and Munmaya K. Mishra

CONTENTS

2.1 CHEMISTRY

Radical vinyl polymerization is a chain reaction that consists of a sequence of three steps: initiation, propagation, and termination.

2.1.1 Chain Initiation

The chain initiation step involves two reactions. In the first step, a radical is produced by any one of a number of reactions. The most common is the homolytic decomposition of an initiator species I to yield a pair of initiator or primary radicals $R^{\cdot}$:

$$I \xrightarrow{k_d} 2R^{\cdot} \tag{2.1}$$

where k_d is the rate constant for the catalyst dissociation.

The second step involves the addition of this radical $R^{\cdot}$ to the first monomer molecule (M) to produce the chain-initiating species $M_1^{\cdot}$:

$$R^{\cdot} + M \xrightarrow{k_i} M_1^{\cdot} \tag{2.2}$$

where k_i is the rate constant for the second initiation step.

2.1.2 Chain Propagation

Propagation consists of the growth of $M^{\cdot}_1$ by the successive addition of large numbers of monomer molecules (M). The addition steps may be represented as follows:

$$M^{\cdot}_1 + M \xrightarrow{k_p} M^{\cdot}_2 \quad (2.3)$$

$$M^{\cdot}_2 + M \xrightarrow{k_p} M^{\cdot}_3 \quad (2.4)$$

$$M^{\cdot}_3 + M \xrightarrow{k_p} M^{\cdot}_4 \quad (2.5)$$

$$M^{\cdot}_4 + M \xrightarrow{k_p} M^{\cdot}_5 \quad (2.6)$$

etc. In general, the steps may be represented as

$$M^{\cdot}_n + M \xrightarrow{k_p} M^{\cdot}_{n+1} \quad (2.7)$$

where k_p is the propagation rate constant.

2.1.3 Chain Termination

Chain propagation to a high-molecular-weight polymer takes place very rapidly. At some point, the propagating polymer chain stops growing and terminates. Termination may occur by various modes:

1. Combination (coupling):

   Two propagating radicals react with each other by combination (coupling) to form a dead polymer:

   $$\sim\sim M\cdot + \cdot M \sim\sim \xrightarrow{k_{tc}} \sim\sim M\text{–}M \sim\sim \quad (2.8)$$

   where k_{tc} is the rate constant for termination by coupling.
2. Disproportionation:

   This step involves a hydrogen radical that is beta to one radical center transferred to another radical center to form two dead polymer chains (one saturated and one unsaturated):

   $$\sim\sim CH_2\text{–}\overset{H}{\underset{|}{\overset{|}{C}}}\cdot + \overset{H}{\underset{|}{\overset{|}{C}}}\text{–}\overset{H}{\underset{H}{\overset{|}{\underset{|}{C}}}}\sim\sim \longrightarrow \sim\sim CH_2\text{–}\overset{H}{\underset{|}{\overset{|}{C}}}\text{–}H + \overset{H}{\underset{|}{\overset{|}{C}}}=\overset{H}{\overset{|}{C}}\sim\sim \quad (2.9)$$

   where k_{td} is the rate constant for termination by disproportionation.

Termination can also occur by a combination of coupling and disproportionation. The two different types of termination may be represented as follows:

$$M^{\cdot}_x + M^{\cdot}_y \xrightarrow{k_{tc}} M_{x+y} \tag{2.10}$$

$$M^{\cdot}_x + M^{\cdot}_y \xrightarrow{k_{td}} M_x + M_y \tag{2.11}$$

In general, the termination step may be represented by the

$$M^{\cdot}_x + M^{\cdot}_y \xrightarrow{k_t} \text{polymer} \tag{2.12}$$

2.2 KINETIC MODEL (RATE EXPRESSIONS)

By considering Eq. (2.1), the rate of decomposition, R_d, of the initiator (I) may be expressed by the following equation in which k_d is the decay or rate constant:

$$R_d = -\frac{d[I]}{dt} = k_d[I] \tag{2.13}$$

Similarly, by considering Eq. (2.2), the rate of initiation, R_i, may be expressed as follows, where k_i is the rate constant for the initiation step:

$$R_i = \frac{d[M^{\cdot}_1]}{dt} = k_i[R^{\cdot}][M] \tag{2.14}$$

The rate of initiation, which is the rate-controlling step in free-radical polymerization, is also related to the efficiency of the production of two radicals from each molecule of initiator, as shown in the following rate equation:

$$R_i = 2k_d f[I] \tag{2.15}$$

Propagation is a bimolecular reaction which takes place by the addition of the free radical to another molecule of monomer, and by many repetitions of this step as represented in Eqs. (2.3)–(2.7). The propagation rate constant k_p is generally considered independent of the chain length.

The rate of monomer consumption (–d[M]/dt), which is synonymous with the rate of polymerization (R_p), may be defined as

$$-\frac{d[M]}{dt} = R_p[M^{\cdot}_n][M] + k_i[R^{\cdot}][M] \tag{2.16}$$

For long chains, the term k_i[R-][M] may be negligible, as the amount of monomer consumed by the initiation step (Eq. (2.2)) is very small compared with that consumed in propagation steps. The equation for R_p may be rewritten as

$$R_p = k_p[M^{\cdot}_n][M] \tag{2.17}$$

The termination of the growing free-radical chains usually takes place by coupling of two macroradicals. Thus, the kinetic chain length (ν) is equal to the half of the degree of polymerization, DP/2. The reaction for the bimolecular termination is presented in Eq. (2.8). The kinetic equation for termination by coupling is

$$R_t = -\frac{d[M^\cdot]}{dt} = 2k_t[M^\cdot]^2 \tag{2.18}$$

Termination of free-radical chain polymerization may also take place by disproportionation. The description for chain termination by disproportionation is given in Eq. (2.9). The kinetic chain length (ν) is the number of monomer molecules consumed by each primary radical and is equal to the rate of propagation divided by the rate of initiation for termination by disproportionation. The kinetic equation for the termination by disproportionation is

$$R_{td} = 2k_{td}[M^\cdot]^2 \tag{2.19}$$

The equation for the kinetic chain length for termination by disproportionation may be represented as

$$\nu_t = \frac{R_p}{R_i} = \frac{R_p}{R_{td}} = \frac{k_p[M][M^\cdot]}{2k_{td}[M^\cdot]^2} = \frac{k_p[M]}{2k_{td}[M^\cdot]} = DP = \frac{k'''[M]}{[M^\cdot]} \tag{2.20}$$

The rate of monomer–radical change can be described as the monomer–radical formed minus (monomer radical utilized), that is

$$\frac{d[M^\cdot]}{dt} = k_i[R^\cdot][M] - 2k_t[M^\cdot]^2 \tag{2.21}$$

It is experimentally found that the number of growing chains is approximately constant over a large extent of reaction. Assuming a "steady-state" condition $d[M^\cdot]/dt = 0$, and

$$k_i[R^\cdot][M] = 2k_t[M^\cdot]^2 \tag{2.22}$$

$[M^\cdot]$ can be derived by solving Eq. (2.22):

$$[M^\cdot] = \left(\frac{k_i[R^\cdot][M]}{2k_t}\right)^{1/2} \tag{2.23}$$

Similarly, assuming a "steady-state" condition for the concentration of $R^\cdot$ and taking into consideration Eqs. (2.14) and (2.15), the following equation may be derived:

$$\frac{d[R^\cdot]}{dt} = 2k_d f[I] - k_i[R^\cdot][M] = 0 \tag{2.24}$$

Solving for $[R^\cdot]$ from Eq. (2.24) gives

$$[R^\cdot] = \frac{2k_d f[I]}{k_i[M]} \tag{2.25}$$

Substituting $[R^\cdot]$ into Eq. (2.23), the expression for $[M^\cdot]$ can be represented as

$$[M^\cdot] = \left(\frac{k_d f[I]}{kt}\right)^{1/2} \tag{2.26}$$

which contains readily determinable variables. Then, by using the relationship for $[M^\cdot]$ as shown in Eqs. (2.17), (2.18), and (2.20), the equation for the rate of polymerization and kinetic chain length can be derived as follows:

$$R_p = k_p[M][M^\cdot] = k_p[M]\left(\frac{k_d f[I]}{k_t}\right)^{1/2} = [M][I]^{1/2}\left(\frac{k_p^2 k_d f}{k_t}\right)^{1/2} = k'[M][I]^{1/2} \tag{2.27}$$

where $k' = (k_p^2 k_d f / k_t)^{1/2}$,

$$R_t = 2k_t[M^\cdot]^2 = \frac{2k_t k_d f[I]}{k_t} = 2k_d f[I] \tag{2.28}$$

$$\overline{DP} = \frac{R_p}{R_i} = \frac{k_p[M](k_d f[I]/k_t)^{1/2}}{2k_d f[I]} = \frac{k_p[M]}{2(k_d k_t f[I])^{1/2}} \tag{2.29}$$

$$\overline{DP} = \frac{M}{[I]^{1/2}} \frac{k_p}{(2k_d k_t f)^{1/2}} = \frac{M}{[I]^{1/2}} k'' \tag{2.30}$$

where $k'' = k_p/(2k_d k_t f)^{1/2}$.

2.3 CONCLUSIONS

The following conclusions may be made about free-radical vinyl polymerization using a chemical initiator:

- The rate of propagation is proportional to the concentration of the monomer and the square root of the concentration of the initiator.
- The rate of termination is proportional to the concentration of the initiator.
- The average molecular weight is proportional to the concentration of the monomer and inversely proportional to the square root of the concentration of the initiator.
- The first chain that is initiated rapidly produces a high-molecular-weight polymer.
- The monomer concentration decreases steadily throughout the reaction and approaches zero at the end.

- Increasing the temperature increases the concentration of free radical and, thus, increases the rate of reactions, but decreases the average molecular weight.
- If the temperature exceeds the ceiling temperature (T_c), the polymer will decompose and no propagation will take place at temperatures above the ceiling temperature.

3 Special Characteristics of Radical Vinyl Polymerization

Yusuf Yagci and Munmaya K. Mishra

CONTENTS

3.1 INITIATOR HALF-LIFE

Depending on the structure, various radical initiators decompose in different modes and the rates of decomposition are different. The differences in the decomposition rates of various initiators can be conveniently expressed in terms of the initiator half-life $t_{1/2}$, defined as the time for the concentration of I to decrease to one-half its original value.

3.2 INITIATOR EFFICIENCY

In radical polymerization, the initiator is inefficiently used due to various side reactions. In addition, the amount of initiator that initiates the polymerization is always less than the amount of initiator that is decomposed during a polymerization. The side reactions are chain transfer to initiator (discussed later) (i.e., the induced decomposition of initiator by the attack of propagating radicals on the initiator) and the radicals' reactions to form neutral molecules instead of initiating polymerization.

The initiator efficiency (I_{eff}) is defined as the fraction of radicals formed in the primary step of initiator decomposition, which is successful in initiating polymerization.

3.3 INHIBITION AND RETARDATION

Certain substances, when added to the polymerization system, may react with the initiating and propagating radicals concerting them either to non-radical species or to less reactive radicals to undergo propagation. Such additives are classified according to their effectiveness:

Inhibitors stop every radical, and polymerization is completely ceased until they are consumed.

Retarders, on the other hand, are less efficient and halt a portion of radicals. In this case, polymerization continues at a slower rate.

For example, in the case of thermal polymerization of styrene [1], benzoquinone acts as an inhibitor. When the inhibitor has been consumed, polymerization regains its momentum and proceeds at the same rate as in the absence of the inhibitor. Nitrobenzene [1] acts as a retarder and lowers the polymerization rate, whereas nitrosobenzene [1] behaves differently. Initially, nitrosobenzene acts as an inhibitor but is apparently converted to a product that acts as a retarder after the inhibition period. Impurities present in the monomer may act as inhibitors or retarders. The inhibitors in the commercial monomers (to prevent premature thermal polymerization during storage and shipment) are usually removed before polymerization or, alternatively, an appropriate excess of initiator may be used to compensate for their presence.

The useful class of inhibitors includes molecules such as benzoquinone and chloranil (2,3,5,6-tetrachlorobenzoquinone) that react with chain radicals to yield radicals of low reactivity. The quinones behave very differently [2–5]. Depending on the attack of a propagating radical at the carbon or oxygen sites, quinone and ether are the two major products [5] formed, respectively. The mechanism may be represented as follows:

Attack on the ring carbon atom yields intermediate radical, which can undergo further reaction to form the quinone:

$$O{=}C_6H_4{=}O \xrightarrow{M_n^{\bullet}} O{=}C_6H_4(H)(M_n){-}O^{\bullet} \xrightarrow{M_n^{\bullet}} O{=}C_6H_3(M_n){=}O + M_n{-}H$$

Rearrangement

$$HO{-}C_6H_3(M_n){-}O^{\bullet} \qquad (3.1)$$

Attack of propagating radical at oxygen yields the ether type radical (aryloxy radical):

$$M_n^{\bullet} + O{=}C_6H_4{=}O \dashrightarrow M_n\text{–}O\text{–}C_6H_4\text{–}O^{\bullet} \qquad (3.2)$$

These preceding radicals, including the aryloxy radical, may undergo further reactions such as coupling or termination with other radicals.

The effect of quinones depends on the polarity of the propagating radicals. Thus, p-benzoquinone and chloranil, which are electron poor, act as inhibitors toward electron-rich propagating radicals (i.e., vinyl acetate and styrene), but only as retarders toward the electron-poor acrylonitrile and methyl methacrylate propagating radicals [6]. It is interesting to note that the inhibiting ability toward the electron-poor monomers can be increased by the addition of an electron-rich third component such as an amine (triethylamine).

Polyalkyl ring-substituted phenols, such as 2,4,6-trimethyl-phenol act as more powerful retarders than phenol toward vinyl acetate polymerization. The mechanism for retardation may involve hydrogen abstraction followed by coupling of the phenoxy radical with other polymer radicals:

$$M_n^{\bullet} + (2,4,6\text{-}R_3C_6H_2OH) \dashrightarrow M_n\text{–}H + (2,4,6\text{-}R_3C_6H_2O^{\bullet}) \qquad (3.3)$$

The presence of sufficient electron-donating alkyl groups facilities the reaction.

Dihydroxybenzences and trihydroxybenzenes such as 1,2-dihydroxy-4-t-butylbenzene, 1,2,3-trihydroxybenzene, and hydroquinone (p-dihydroxybenzene) act as inhibitors in the presence of oxygen [7, 8]. The inhibiting effect of these compounds is produced by their oxidation to quinones [9].

Aromatic nitro compounds act as inhibitors and show greater tendency toward more reactive and electron-rich radicals. Nitro compounds have very little effect on methyl acrylate and methyl methacrylate [5, 10, 11] but inhibit vinyl acetate and retard styrene polymerization. The effectiveness increases with the number of nitro groups in the ring [12, 13]. The mechanism of radical termination involves attack on both the aromatic ring and the nitro group. The reactions are represented as follows:

Attack on the ring:

$$C_6H_5\text{–}NO_2 \xrightarrow{M_n^{\bullet}} (M_n)(H)\dot{C}_6H_5\text{–}NO_2 \xrightarrow{M_n^{\bullet}} M_n\text{–}C_6H_4\text{–}NO_2 + M_n - H \qquad (3.4)$$

$$(M_n)(H)\dot{C}_6H_5\text{–}NO_2 \xrightarrow{M} M_n\text{–}C_6H_4\text{–}NO_2 + HM^{\bullet} \qquad (3.5)$$

Attack on the nitro group:

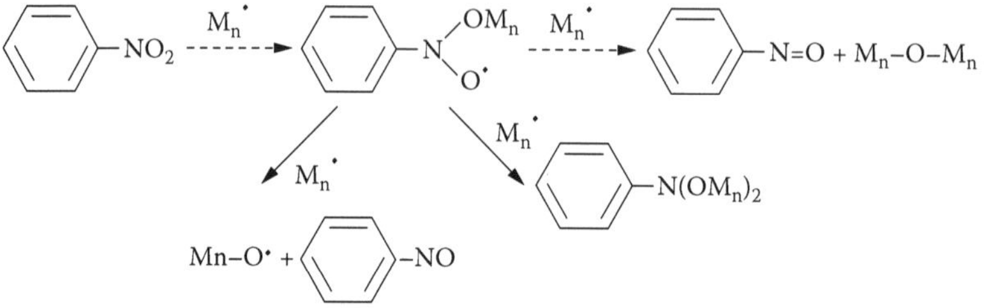

$$C_6H_5{-}NO_2 \xrightarrow{M_n^\bullet} C_6H_5{-}N{<}^{OM_n}_{O^\bullet} \xrightarrow{M_n^\bullet} C_6H_5{-}N{=}O + M_n{-}O{-}M_n \quad (3.6)$$

$$C_6H_5{-}N{<}^{OM_n}_{O^\bullet} \xrightarrow{M_n^\bullet} Mn{-}O^\bullet + C_6H_5{-}NO$$

$$C_6H_5{-}N{<}^{OM_n}_{O^\bullet} \xrightarrow{M_n^\bullet} C_6H_5{-}N(OM_n)_2 \quad (3.7)$$

where M and $M_n^\bullet$ are the monomer and the propagating radical, respectively.

Oxidants such as $FeCl_3$ and $CuCl_2$ are strong inhibitors [14–17]. The termination of growing radicals may be shown by the following reaction:

TABLE 3.1
Inhibitor Constants

Inhibitor	Monomer	Temperature (°C)	Constant (z)
Aniline	Methyl acrylate	50	0.0001
	Vinyl acetate	50	0.015
p-Benzoquinone	Acrylonitrile	50	0.91
	Methyl methacrylate	50	5.7
	Styrene	50	518.0
Chloranil	Methyl methacrylate	44	0.26
	Styrene	50	2,010
$CuCl_2$	Acrylonitrile	60	100
	Methyl methacrylate	60	1,027
	Styrene	50	11,000
DPPH	Methyl methacrylate	44	2,000
p-Dihydroxybenzene	Vinyl acetate	50	0.7
$FeCl_3$	Acrylonitrile	60	3.3
	Styrene	60	536
Nitrobenzene	Methyl acrylate	50	0.00464
	Styrene	50	0.326
	Vinyl acetate	50	11.2
Oxygen	Methyl methacrylate	50	33,000
	Styrene	50	14,600
Phenol	Methyl acrylate	50	0.0002
	Vinyl acetate	50	0.012
Sulfur	Methyl methacrylate	44	0.075
	Vinyl acetate	44	470
1,3,5-Trinitrobenzene	Methyl acrylate	50	0.204
	Styrene	50	64.2
	Vinyl acetate	50	404
1,2,3-Trihydroxybenzene	Vinyl acetate	50	5.0
2,4,6-Trimethylphenol	Vinyl acetate	50	5.0

$$\sim\!\sim CH_2\dot{C}H-C_6H_5 + FeCl_3 \longrightarrow \begin{cases} \sim\!\sim CH_2CHCl-C_6H_5 + FeCl_2 \\ \sim\!\sim CH{=}CH-C_6H_5 + HCl + FeCl_2 \end{cases} \quad (3.8)$$

Oxygen is a powerful inhibitor. It reacts with radicals to form the relatively unreactive peroxy radical, which may undergo further reaction:

$$M^{\cdot}_n + O_2 \rightarrow M_n\text{—}OO^{\cdot} \quad (3.9)$$

It may react with itself or another propagating radical to form inactive products [18–20]. It is interesting to note that oxygen is also an initiator in some cases. The inhibiting or initiating capabilities of oxygen are highly temperature dependent. Other inhibitors include chlorophosphins [21], sulfur, aromatic azo compounds [22], and carbon. The inhibitor constants of various inhibitors for different monomers are presented in Table 3.1.

3.4 CHAIN TRANSFER

Chain transfer is a chain-stopping reaction. It results in a decrease in the size of the propagating polymer chain. This effect is due to the premature termination of a growing polymer chain by the transfer of a hydrogen or other atom from some compound present in the system (i.e., monomer, solvent, initiator, etc.). These radical displacement reactions are termed chain transfer reactions and may be presented as

$$M^{\cdot}_n + CTA \xrightarrow{k_{tr}} M_n\text{—}C + TA^{\cdot} \quad (3.10)$$

where k_{tr} is the chain transfer rate constant, CTA is the chain transfer agent (may be initiator, solvent, monomer, or other substance), and C is the atom or species transferred. The rate of a chain transfer reaction may be given as

$$R_{tr} = k_{tr}[M^{\cdot}][CTA] \quad (3.11)$$

The new radical $A^{\cdot}$, which is generated by the chain transfer reaction, may reinitiate polymerization:

$$TA^{\cdot} + M \xrightarrow{k_a} M^{\cdot} \quad (3.12)$$

The effect of chain transfer on the polymerization rate is dependent on whether the rate of reinitiation is comparable to that of the original propagating radical. Table 3.2 shows the different phenomena. The rate equation for the chain transfer reaction may be represented as (also known as the Mayo equation):

$$\frac{1}{X_n} = \frac{k_t R_p}{K_p''[M]''} + C_M + C_S\frac{[S]}{[M]} + C_I\frac{k_t R_p''}{k_p'' f\, k_d [M]^3} \quad (3.13)$$

TABLE 3.2
Effect of Chain Transfer on R_p and X_n

Rate Constants[a]	Mode	Effect on Rp	Effect on Xn
1. kp << ktr ka~ kp	Telomerization	None	Large decrease
2. kp >> ktr ka~ kp	Normal chain transfer	None	Decrease
3. kp << ktr ka < kp	Degradative chain transfer	Large decrease	Large decrease
4. kp >> ktr ka < kp	Retardation	Decrease	Decrease

[a] k_p, k_{tr}, and k_a are the rate constants for propagation, transfer, and reinitiation steps, respectively.

where X_n, R_p, k_t, C, S, and M are the degree of polymerization, rate of polymerization, termination rate constant, chain transfer constant, chain transfer agent, and monomer, respectively.

3.4.1 Chain Transfer to Monomer

The chain transfer constants of various monomers at 60°C are presented in Table 3.3. The monomer chain transfer constants C_M are generally small ($10^{-5} - 10^{-4}$) for most monomers because the reaction involves breaking the strong vinyl CH–bond:

$$M_n^{\bullet} + CH_2{=}\overset{\overset{\displaystyle X}{|}}{C}H \longrightarrow M_n{-}H + CH_2{=}\overset{\overset{\displaystyle X}{|}}{C}{\cdot} \tag{3.14}$$

On the other hand, when the propagating radicals (polyvinyl acetate, ethylene, and vinyl chloride) have very high reactivity, the C_M is usually large. In the case of vinyl acetate polymerization [26], chain transfer to monomer has been generally attributed to transfer from the acetoxy methyl group:

$$M_n{\cdot} + CH_2{=}CH{-}O{-}CO{-}CH_3 \rightarrow M_n{-}H + CH_2{=}CH{-}O{-}CO{-}CH_2{\cdot} \tag{3.15}$$

TABLE 3.3
Chain Transfer Constants of Monomers

Monomer	$C_M \times 10^4$	Reference
Acrylamide	0.6	[12]
Acrylonitrile	0.26–0.3	[12]
Ethylene	0.4–4.2	[12]
Methyl acrylate	0.036–0.325	[12]
Methyl methacrylate	0.07–0.25	[12, 23]
Styrene	0.30–0.60	[24]
Vinyl acetate	1.75–2.8	[12]
Vinyl chloride	10.8–16	[12, 25]

However, a different mechanism had been suggested by Litt and Chang [27]. By using vinyl trideuteroacetate and trideuterovinyl acetate, they indicated that more than 90% of the transfer occurs at the vinyl hydrogens:

$$M_n\cdot + CH_2{=}CH{-}O{-}CO{-}CH_3 \rightarrow M_n{-}H + \dot{C}H{=}CH{-}O{-}CO{-}CH_3 \text{ and/or } CH{=}\dot{C}H{-}O{-}CO{-}CH_3 \quad (3.16)$$

The very high value of C_M for vinyl chloride may be explained by the following reactions. It is believed [28] to occur by β-scission transfer of Cl to the monomer from the propagating center or, more likely, after that center undergoes intramolecular Cl migration [29]:

$$\sim\!\sim CH_2{-}CH(Cl){-}CH(Cl){-}\dot{C}H_2 + CH_2{=}CHCl \dashrightarrow \sim\!\sim CH_2{-}CH(Cl){-}CH{=}CH_2 + ClCH_2{-}\dot{C}HCl \quad (3.17)$$

$$\sim\!\sim CH_2{-}CH(Cl){-}\dot{C}H{-}CH_2Cl + CH_2{=}CHCl \dashrightarrow \sim\!\sim CH_2{-}CH{=}CH{-}CH_2Cl + ClCH_2{-}\dot{C}HCl \quad (3.18)$$

The C_M value of vinyl chloride is high enough that the number-average molecular weight that can be achieved is 50,000–120,000.

3.4.2 Chain Transfer to Initiator

The transfer constants (C_I) for different initiators are presented in Table 3.4. The value of C_I for a particular initiator is dependent on the nature (i.e., reactivity) of

TABLE 3.4
Initiator Chain Transfer Constant

Initiator	Temperature (°C)	C_I for Polymerization STY[a]	MMA[a]	AM[a]	References
2,2′-Azobisisobutyronitrile	60	0.091–0.14	0.02	—	[23, 24]
t-Butyl peroxide	60	0.00076–0.00092	—	—	[12]
t-Butyl hydroperoxide	60	0.035	—	—	[12]
Benzoyl peroxide	60	0.048–0.10	0.02	—	[12]
Cumyl peroxide	50	0.01	—	—	[12]
Cumyl hydroperoxide	60	0.063	0.33	—	[12]
Lauroyl peroxide	70	0.024	—	—	[12]
Persulfate	40	—	—	0.0026	[30]

[a] STY = styrene; MMA = methyl methacrylate; AM = acrylamide.

the propagating radical. For example, a very large difference occurs in C_I for cumyl hydroperoxide toward the poly(methyl methacrylate) radical compared with the polystyryl radical. Peroxides usually have a significant chain transfer constant. The transfer reactions may be presented as follows:

$$M^{\cdot}_{n} + RO\text{—}OR \longrightarrow M_{n}\text{—}OR + RO^{\cdot} \tag{3.19}$$

where R is an alkyl or acyl group. The acyl peroxides have higher transfer constants than the alkyl peroxides due to the weaker bond of the former. The hydroperoxides are usually the strongest transfer agents among the initiators. The transfer reaction probably involves the hydrogen atom abstraction according to the following reaction:

$$M^{\cdot}_{n} + ROO\text{—}H \longrightarrow M_{n}\text{—}H + ROO^{\cdot} \tag{3.20}$$

The transfer reaction with azonitriles [23, 24] probably occurs by the displacement reaction, which is presented as follows:

$$M^{\cdot}_{n} + RN = NR \longrightarrow M_{n} - R + R^{\cdot} + N_{2} \tag{3.21}$$

3.4.3 Chain Transfer to Chain Transfer Agent

The chain transfer to the different substances other than the initiator and monomer (referred to as the chain transfer agent) is another special case. The example is the solvent or may be another added compound. The transfer constants for various compounds are listed in Table 3.5.

The transfer constant data presented in Table 3.5 may provide the information regarding the mechanism and the relationship between structure and reactivity in radical displacement reactions. For example, the low C_5 values for benzene and cyclohexane are due to the strong C—H bonds present. It is interesting to note that transfer to benzene does not involve hydrogen abstraction but the addition of the propagating radical to the benzene ring [31] according to

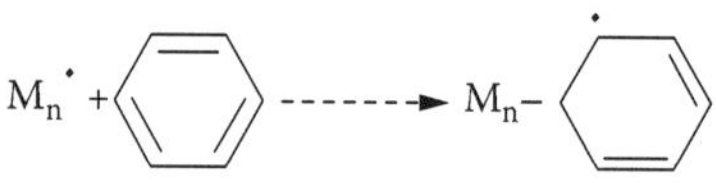

$$M_n^{\cdot} + \text{C}_6\text{H}_6 \dashrightarrow M_n\text{–C}_6\text{H}_6^{\cdot} \tag{3.22}$$

The C_S values for toluene, isopropylbenzene, and ethylbenzene are higher than benzene. This is due to the presence of the weaker benzylic hydrogens and can be abstracted easily because of the resonance stability of the resultant radical:

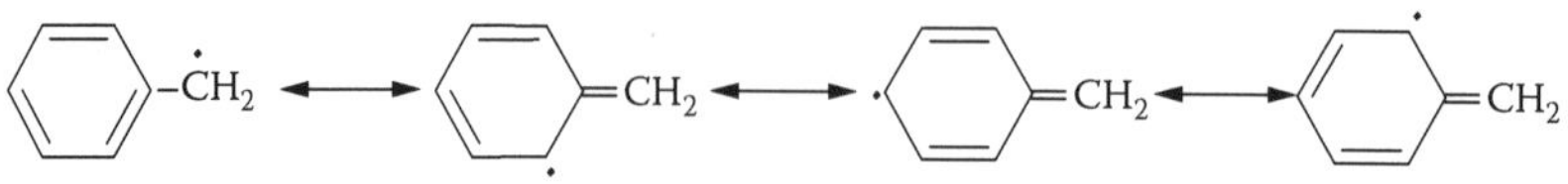

$$\text{C}_6\text{H}_5\text{–}\dot{\text{C}}\text{H}_2 \longleftrightarrow \cdot\text{C}_6\text{H}_5\text{=CH}_2 \longleftrightarrow \cdot\text{C}_6\text{H}_5\text{=CH}_2 \longleftrightarrow \cdot\text{C}_6\text{H}_5\text{=CH}_2 \tag{3.23}$$

Primary halides such as n-butyl bromide and chlorine have low transfer constants like aliphatics. This may be explained by the low stability of a primary alkyl

TABLE 3.5
Transfer Constants for Chain Transfer Agents

Transfer Agent	$C_S \times 10^4$, Polymerization at 60°C	
	Styrene	**Vinyl Acetate**
Acetic acid	2.0	1.1
Acetone	4.1	11.7
Benzene	0.023	1.2
Butylamine	7.0	—
t-Butyl benzene	0.06	3.6
n-Butyl chloride	0.04	10
n-Butyl bromide	0.06	50
n-Butyl alcohol	1.6	20
n-Butyl iodide	1.85	800
n-Butyl mercaptan	210,000	480,000
Cyclohexane	0.031	7.0
2-Chlorobutane	1.2	—
Chloroform	3.4	150
Carbon tetrachloride	110	10,700
Carbon tetrabromide	22,000	390,000
Di-n-Butyl sulfide	22	260
Di-n-Butyl disulfide	24	10,000
Ethylbenzene	0.67	55.2
Ethyl ether	5.6	45.3
Heptane	0.42	17.0 (50°C)
Isopropylbenzene	0.82	89.9
Toluene	0.125	21.6
Triethylamine	7.1	370

Source: J. Brandup and E. H. Immergut, Eds., with W. McDowell, *Polymer Handbook*, Wiley-Interscience, New York, 1975.

radical upon abstraction of Cl or Br. In contrast, n-butyl iodide shows a much higher C_S value, which transfers an iodide atom due to the weakness of the C–I bond. The high transfer constants for disulfides are due to the weak S–S bond. Amines, ethers, alcohols, acids, and carbonyl compounds have higher transfer constants than those of aliphatic hydrocarbons, due to the C–H bond breakage and stabilization of the radical by an adjacent O, N, or carbonyl group. The high C_S values for carbon tetrachloride and carbon tetrabromide are due to the weak carbon–halogen bonds. These bonds are especially weak because of the resonance stabilization of the resultant trihalocarbon radicals formed by the halogen abstraction:

$$:\ddot{\text{Cl}}-\underset{:\ddot{\text{Cl}}:}{\dot{\text{C}}}-\ddot{\text{Cl}}: \longleftrightarrow :\dot{\text{Cl}}=\underset{:\ddot{\text{Cl}}:}{\text{C}}-\ddot{\text{Cl}}: \longleftrightarrow :\ddot{\text{Cl}}-\underset{:\ddot{\text{Cl}}:}{\text{C}}=\dot{\text{Cl}}: \longleftrightarrow :\ddot{\text{Cl}}-\underset{\overset{\|}{:\dot{\text{Cl}}:}}{\text{C}}-\ddot{\text{Cl}}: \qquad (3.24)$$

3.4.4 Chain Transfer to Polymer

Chain transfer to polymer is another case of the various types of reaction described earlier. This process results in the formation of a radical site on a polymer chain that may be capable of polymerizing monomers to produce a branched polymer as follows:

$$M_n^{\bullet} + \sim\sim CH_2{-}CH_2 \sim\sim\sim \dashrightarrow M_n{-}H + \sim\sim CH_2{-}\dot{C}H \sim\sim \xrightarrow{M} \sim\sim CH_2{-}\underset{\underset{|}{M_m}}{\underset{|}{C}H} \sim\sim \tag{3.25}$$

REFERENCES

1. G. V. Schulz, *Chem. Ber.*, 80, 232 (1947).
2. P. A. Small, *Adv. Polym. Sci.*, 18, 1 (1975).
3. K. Yamamoto and M. Sugimoto, *J. Macromol. Sci.*, A13, 1067 (1979).
4. M. H. George, in *Vinyl Polymerization*, Vol. 1, G. E. Ham, Ed., Marcel Dekker, Inc., New York, 1967, Part I.
5. G. C. Eastmond, Chain transfer, inhibition and retardation, in *Comprehensive Chemical Kinetics*, Vol. 14A, C. H. Bamford and C. F. H. Tipper, Eds., American Elsevier, New York, 1976.
6. A. A. Yassin and N. A. Risk, *Polymer*, 19, 57 (1978); *J. Polym. Sci. Polym. Chem. Ed.*, 16, 1475 (1978); *Eur. Polym. J.*, 13, 441 (1977).
7. R. Prabha and U. S. Nanadi, *J. Polym. Sci. Polym. Chem. Ed.*, 15, 1973 (1977).
8. J. J. Kurland, *J. Polym. Sci. Polym. Chem. Ed.*, 18, 1139 (1980).
9. K. K. Georgieff, *J. Appl. Polym. Sci.*, 9, 2009 (1965).
10. G. V. Schulz, *Chem. Ber.*, 80, 232 (1947).
11. Y. Tabata, K. Ishigure, K. Oshima, and H. Sobue, *J. Polym Sci.*, A2, 2445 (1964).
12. J. Brandup and E. H. Immergut, Eds., with W. McDowell, *Polymer Handbook*, Wiley-Interscience, New York, 1975.
13. G. C. Eastmond, Kinetic data for homogeneous free radical polymerizations of various monomers, in *Comprehensive Kinetics*, Vol. 14A, C. H. Bamford and C. F. H. Tipper, Eds., American Elsevier, New York, 1976.
14. K. Matsuo, G. W. Nelb, R. G. Nelb, and W. H. Stockmayer, *Macromolecules*, 10, 654 (1977).
15. N. C. Billingham, A. J. Chapman, and A. D. Jenkins, *J. Polym. Sci. Polym. Chem. Ed.*, 18, 827 (1980).
16. N. N. Das and M. H. George, *Eur. Polym. J.*, 6, 897 (1970); 7, 1185 (1971).
17. P. D. Chetia and N. N. Das, *Eur. Polym. J.*, 12, 165 (1976).
18. T. Koenig and H. Fischer, Cage effects, in *Free Radicals*, Vol. I, J. K. Kochi, Ed., Wiley, New York, 1973.
19. H. Maybod and M. H. George, *J. Polym. Sci. Polym. Lett. Ed.*, 15, 693 (1977).
20. M. H. George and A. Ghosh, *J. Polym. Sci. Polym. Chem. Ed.*, 16, 981 (1978).
21. H. Uemura, T. Taninaka, and Y. Minoura, *J. Polym. Lett. Ed.*, 15, 493 (1977).
22. S. E. Nigenda, D. Cabellero, and T. Ogawa, *Makromol. Chem.*, 178, 2989 (1977).
23. G. Ayrey and A. C. Haynes, *Makromol. Chem.*, 175, 1463 (1974).
24. J. G. Braks and R. Y. M. Huang, *J. Appl. Polym. Sci.*, 22, 3111 (1978).

25. M. Carenza, G. Palma, and M. Tavan, *J. Polym Sci. Polym. Chem. Ed.*, 10, 2781, 2853 (1972).
26. S. Nozakura, Y. Morishima, and S. Murahashi, *J. Polym.Sci. Polym. Chem. Ed.*, 10, 2781, 2853 (1972).
27. M. H. Litt and K. H. S. Chang, in *Emulsion Polymers, Vinyl Acetate (Pap. Symp.)*, M. S. El-Aasser and J. W. Vanderhoff, Eds., Applied Science, London, 1981, pp. 89–171.
28. W. H. Starnes Jr., F. C. Schilling, K. B. Abbas, R. E. Cais, and F. A. Bovey, *Macromolecules*, 12, 556 (1979).
29. W. H. Starnes, Mechanistic aspects of the degradation and stabilization of poly(vinyl chloride), in *Developments in Polymer Degradation* — 3, N. Grassie, Ed., Applied Science, London, 1980.
30. S. M. Shawki and A. E. Hamielec, *J. Appl. Polym. Sci.*, 23, 334 (1979).
31. P. C. Deb and S. Ray, *Eur. Polym. J.*, 14, 607 (1978).

Part II

The Initiating Systems

4 Initiation of Vinyl Polymerization by Organic Molecules and Nonmetal Initiators

Ivo Reetz, Yusuf Yagci, and Munmaya K. Mishra

CONTENTS

4.1 INTRODUCTION

Radicals are produced in a special reaction for starting a radical polymerization. Because free radicals are reactive intermediates that possess only very limited lifetimes, radicals are generally produced in the presence of a monomer that is to be polymerized. They react very rapidly with the monomer present. The rate of the reaction of initially formed free radicals with the monomer (the initiation step) is high compared with the rate of radical formation; thus, the latter process is rate determining. Therefore, radical generation by respective initiators is a very characteristic and important feature of radical initiation.

Three main classes of reaction lead to the generation of free radicals:

- The thermally initiated homolytic rupture of atomic bonds
- The light-induced or radiation-induced rupture of atomic bonds
- The electron transfer from ions or atoms onto an acceptor molecule, which undergoes bond dissociation

Substances that deliver radicals are referred to as initiators. This chapter covers only thermal and redox initiators. Photo and high-energy polymerizations are described in other chapters. Besides polymerizations that start with the decomposition of the initiator, thermally initiated polymerizations exist, in which no initiator is present. In these cases, the monomer (styrene or methyl methacrylate) is itself able to generate initiating sites upon heating. This type of polymerization is also briefly described here.

4.2 RADICAL GENERATION BY THERMALLY INDUCED HOMOLYSIS OF ATOMIC BONDS

If one works with thermal initiators, the bond dissociation energy is introduced into the polymerization mixture in the form of thermal energy. This energy input is the necessary prerequisite for bond homolysis. Thermolabile initiators are usually employed in a temperature range between 50°C and 140°C. To have high initiation rates, the activation energy of thermal initiators has to be on the order of 120–170 kJ mol^{-1}. This activation energy brings about a strong temperature dependence of the dissociation, which is reasonable because initiators should have good storage stability at room temperature but produce radicals at slightly elevated temperatures. Only a few functional groups meet these demands, especially azo compounds and peroxides, which are of practical importance.

4.2.1 Azo Initiators

Among azo compounds (R—N=N—R′, R and R′ either alkyl or aryl) it is mostly the alkyl and alkyl-aryl derivatives that possess sufficient thermal latency for employing these substances in initiation. Simple azoalkanes decompose at temperatures above 250°C. They are not used as thermal initiators for their relative stability, but may be used for photochemical radical production when subjected to ultraviolet (UV) light of appropriate wavelengths (n–π- transition of the N=N double bond). A very great improvement in thermal reactivity is gained by introducing a nitrile group in proximity to the azo link (see Table 4.1).

The most prominent azo initiator, 2,2′-azo (bisisobutyronitrile), AIBN, is an exceptionally important initiator in industrial polymer synthesis [13, 14]. Heated AIBN decomposes, giving two 2-cyano-2propyl radicals and molecular nitrogen. Notably, the generation of the energetically favored and very stable nitrogen molecule must be an important driving force in the decomposition of azo compounds. The activation energy of reaction (4.1) is only 129 kJ mol^{-1}, whereas the bond dissociation

TABLE 4.1
Decomposition Characteristics of Selected Azo Compounds

Azo Compound	E_a, kJ mol^{-1}	k_d, s^{-1}	References
$CH_3—N{=}N—CH_3$	214	—	[1, 2]
$CH_3—CH_2$ - NÅ—$CH_2—CH_3$	202	—	[3, 4]
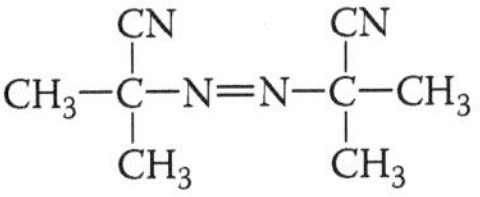 $CH_3-C(CN)(CH_3)-N{=}N-C(CN)(CH_3)-CH_3$	142	1.7×10^{-4} (80°C)	[5]
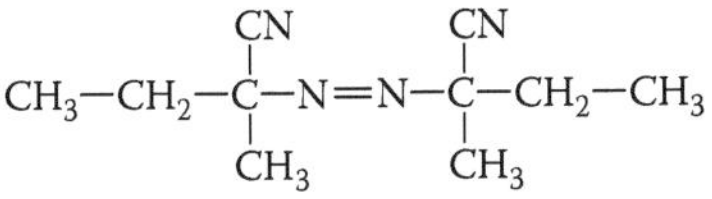 $CH_3-CH_2-C(CN)(CH_3)-N{=}N-C(CN)(CH_3)-CH_2-CH_3$	130	8.7×10^{-5} (80°C)	[5]
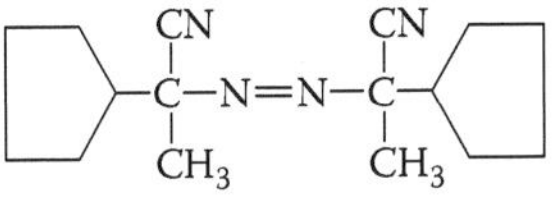 cyclobutyl$-C(CN)(CH_3)-N{=}N-C(CN)(CH_3)-$cyclobutyl	—	1.3×10^{-4} (80°C)	[6]
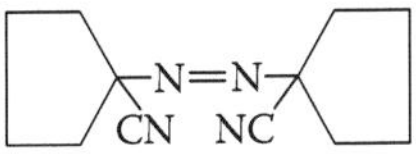 1-cyanocyclopentyl$-N{=}N-$1-cyanocyclopentyl (CN, NC)	141	7.4×10^{-5} (80°C)	[7]
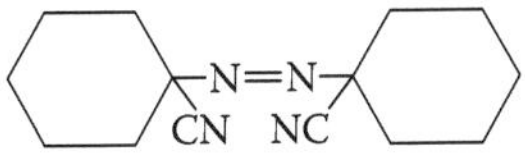 1-cyanocyclohexyl$-N{=}N-$1-cyanocyclohexyl (CN, NC)	148	6.5×10^{-6} (80°C)	[5, 7]
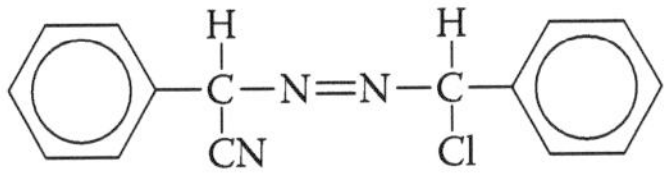 $C_6H_5-CH(CN)-N{=}N-CH(Cl)-C_6H_5$	89	5.8×10^{-2} (12°C)	[8]
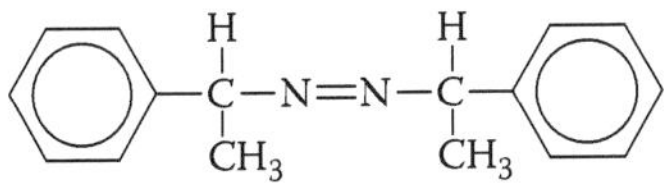 $C_6H_5-CH(CH_3)-N{=}N-CH(CH_3)-C_6H_5$	136	$5.4. \times 10^{-5}$ (100°C)	[9]
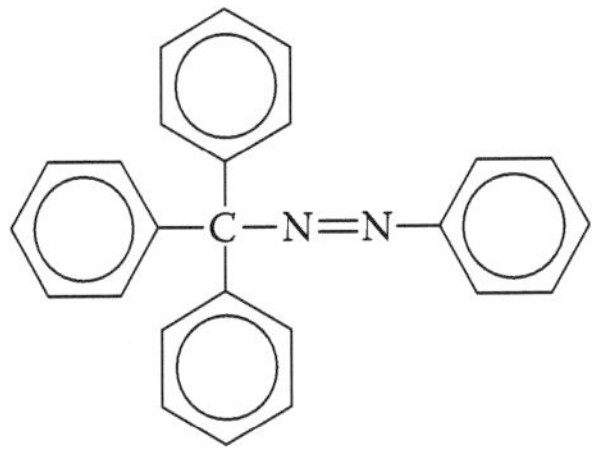 $(C_6H_5)_3C-N{=}N-C_6H_5$	122	3.5×10^{-4}(55°C)	[10, 11]

(Continued)

TABLE 4.1 (CONTINUED)
Decomposition Characteristics of Selected Azo Compounds

Azo Compound	E_a, kJ mol^{-1}	k_d, s^{-1}	References
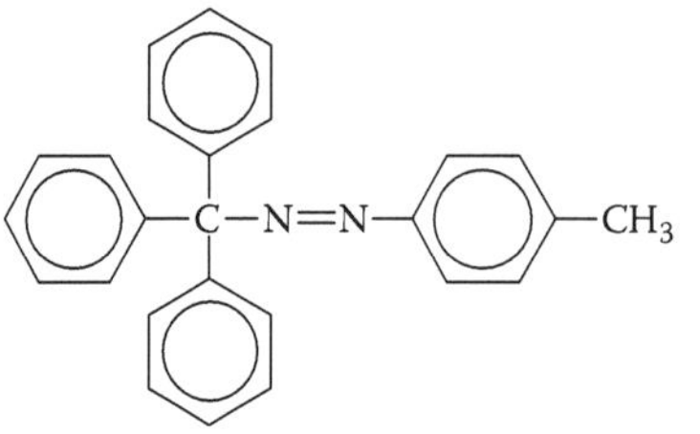 $-C-N{=}N-$ $-CH_3$	100	2.2×10^{-4} (53° C)	[11]

Note: See also R. Zand, Azo catalysts, in *Encyclopedia of Polymer Science and Technology*, Vol. 2, H. F. Mark, N. G. Gaylord, and N. M. Bikales, Eds., Interscience, New York, 1965, p. 278. E_a: activation energy for decomposition; k_d: decomposition rate constant, see Eq. (4.2).

energy of the C–N bond in AIBN ~300 kJ mol^{-1}. Usually, polymerizations initiated by AIBN are performed at 50–80°C.

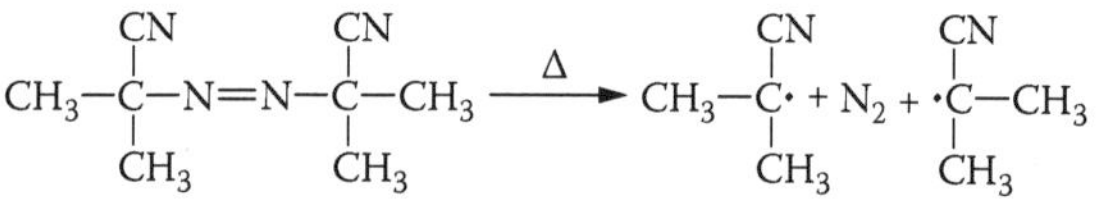

$$CH_3-\underset{CH_3}{\overset{CN}{C}}-N{=}N-\underset{CH_3}{\overset{CN}{C}}-CH_3 \xrightarrow{\Delta} CH_3-\underset{CH_3}{\overset{CN}{C}}\cdot + N_2 + \cdot\underset{CH_3}{\overset{CN}{C}}-CH_3 \quad (4.1)$$

Heated AIBN decomposes under continuous evolution of initiating radicals following strictly first-order reaction kinetics:

$$I \xrightarrow{k_d} 2R\cdot \quad (4.2)$$

The decomposition rate v_d is expressed as

$$v_d = -\frac{d[I]}{dt} = k_d[I] \quad (4.3)$$

where [I] is the initiator concentration, t is time, and k_d is the rate constant for the decomposition of the initiator.

For AIBN, k_d does not depend on the solvent. At 50°C, k_d is ~2×10^{-6} sec^{-1}, which in other terms means a half-life of 96 h.

Not all radicals formed in the decomposition are actually available for reacting with the monomer and initiating the growing of a polymer chain. A considerable loss of radicals is brought about by the so-called cage effect: After the dissociation (C—N bond rupture), the radicals formed are still very close to each other, surrounded by solvent molecules (in a solvent cage), which prevents them from diffusing apart. About 10^{-11} sec are necessary for them to move out of the solvent cage. During this short period of time, the radicals collide due to molecular motions and may recombine to give various combination products:

$$\mathrm{CH_3{-}C(CN)(CH_3){-}N{=}N{-}C(CN)(CH_3){-}CH_3} \rightarrow 2\mathrm{(CH_3)_2C{=}C{=}N\cdot} \leftrightarrow 2\mathrm{CH_3{-}\dot{C}(CH_3)(CN)}\ \textbf{(A)}$$

$$\mathrm{(CH_3)_2C{=}C{=}N\cdot} + \textbf{(A)} \rightarrow \mathrm{(CH_3)_2C{=}C{=}N{-}C(CH_3)(CN){-}CH_3}\ \textbf{(B)}$$

$$\textbf{(A)} \rightarrow \underset{\textbf{(C)}}{\mathrm{CH_3{-}C(CN)(CH_3){-}C(CN)(CH_3){-}CH_3}} + \underset{\textbf{(D)}}{\mathrm{CH_3{-}C(CN)(CH_3){-}H}} + \underset{\textbf{(E)}}{\mathrm{CH_3{-}C(CN)(CH_3){-}CH_2{-}C(CN)(CH_3){-}C(CN)(CH_3){-}CH_3}} \quad (4.4)$$

The main product of the cage recombination is a thermally unstable ketimine (B), which redecomposes, yielding 2-cyano-2-propyl radicals (A). On the other hand, the simultaneously generated tetramethylsuccidonitrile (C) is thermally stable and does not yield new radicals upon heating. An investigation of stable combination products of AIBN in toluene has shown that 84% of (C), 3.5% of isobutyronitrile (D), and 9% of 2,3,5-tricyano-2,3,5-trimethylhexane (E) are formed [15]. Thus, about half of the initially formed radicals are consumed this way. Only the remaining portion of radicals, the ones that are able to escape the solvent cage, are actually available for the polymerization. In scientific terms, one speaks of a radical yield U_r (ratio of mole radicals which react with the monomer to mole of initially formed radicals), which is 0.5 for AIBN. Introducing bulky substituents may significantly enhance the radical yield. These prevent the recombination of carbon centered radicals in the solvent cage. For example, for 1,1′-diphenyl-1,1′-diacetoxyazoethane, the radical yield amounts to 0.9 (i.e., only 10% of initially formed radicals is consumed by cage reactions):

$$\mathrm{C_6H_5{-}C(CH_3)(O{-}C({=}O){-}CH_3){-}N{=}N{-}C(CH_3)(O{-}C({=}O){-}CH_3){-}C_6H_5} \quad (4.5)$$

In general, the rate of radical formation v_r of primary radicals that are actually available for polymerization can be written as

$$v_r = 2U_r v_d = 2U_r k_d[I] \quad (4.6)$$

The initiations step is the addition of primary radicals to the olefinic double bond of the monomer:

$$\underset{R\cdot}{CH_3-\overset{CN}{\underset{CH_3}{\overset{|}{\underset{|}{C}}}}\cdot} + \underset{M}{CH_2{=}\underset{R}{\underset{|}{CH}}} \xrightarrow{k_i} \underset{P\cdot}{CH_3-\overset{CN}{\underset{CH_3}{\overset{|}{\underset{|}{C}}}}-CH_2-\underset{R}{\underset{|}{\dot{C}H}}} \tag{4.7}$$

By this reaction, the primary radical is incorporated into the growing polymer radical. After the polymerization, it may be detected as a terminal group of the polymer by suitable analytical techniques. The rate of initiation v_i can be expressed as

$$v_i = k_i[R\cdot][M] \tag{4.8}$$

If a stationary radical concentration occurs and all primary radicals really start a polymerization, the following equations hold:

$$\frac{d[R\cdot]}{dt} = v_r - v_i = 0 \tag{4.9}$$

$$v_i = 2U_r k_d[I] \tag{4.10}$$

As Eq. (4.10) implies, the initiation rate increases with the concentration of the initiator. The faster the initiation, the higher the decomposition rate and, therefore, the radical yield of the initiator. Naturally, it also rises with temperature, because at higher temperature, more radicals are formed. Notably, the molecular weight of the polymer, provided no cross-linking occurs, mostly drops with increasing concentration of primary radicals by whichever means.

As Table 4.1 shows, among azonitriles, the substitution pattern has a strong influence on reactivity. Cyclic azonitriles are somewhat less reactive, an effect which has been attributed to a decreased resonance energy attributable to angular strain in the ring of the 1-cyanocycloalkyl radical. Azotriphenylmethane initiators are extremely thermosensitive, which brings about high radical concentration, but is also connected with relatively poor storage stability. The reactivity of these compounds derives from the resonance stabilization of the triphenylmethyl radical formed. Azotriphenylmethane initiators may also be utilized as photoinitiators, as they possess chromophoric phenyl groups. For emulsion polymerization, water-soluble azo initiators were developed, the hydrophilicity of which was provided by substituents including carboxyl [16], acetate, sulfonates, amide [17], and tertiary amine [18] groups.

Azo initiators of a special type, namely macro-azo-initiators (MAIS), are of great importance for the synthesis of block and graft copolymers. MAIs are polymers or oligomers that contain azo groups in the main chain, at the end of the main chain, or at a side chain. With the first two initiators, one obtains block copolymers, whereas with the latter, graft copolymers. For the synthesis of MAIs, low-molecular-weight azo compounds are necessary, which have to possess groups that enable monomers to attach to them. Frequently, condensation or addition reactions with the azo compounds listed in Table 4.2 are used to prepare azo-containing macroinitiators.

TABLE 4.2
Bifunctional Azo Compounds Used Frequently in Polycondensation and Addition Reactions

Formula	Abbreviation	References for Synthesis	References for Block Copolymerization
$\left[HO-\overset{O}{\overset{\|}{C}}\!-\!(CH_2)_2\!-\!\underset{CN}{\overset{CH_3}{C}}\!-\!N=\right]_2$	ACPA	[19]	[20–22]
$\left[Cl-\overset{O}{\overset{\|}{C}}\!-\!(CH_2)_2\!-\!\underset{CN}{\overset{CH_3}{C}}\!-\!N=\right]_2$	ACPA	[23, 24]	[12, 21, 24–45]
$\left[HO-(CH_2)_3\!-\!\underset{CN}{\overset{CH_3}{C}}\!-\!N=\right]_2$	ACPO	[46]	[45–51]

For converting these initiators into macroinitiators, diamines, glycols, or diisocyanates are often used. Block copolymers of amide and vinyl monomer blocks are easily produced when these macroinitiators are heated in the presence of a second monomer; this polymerization is a radical vinyl polymerization:

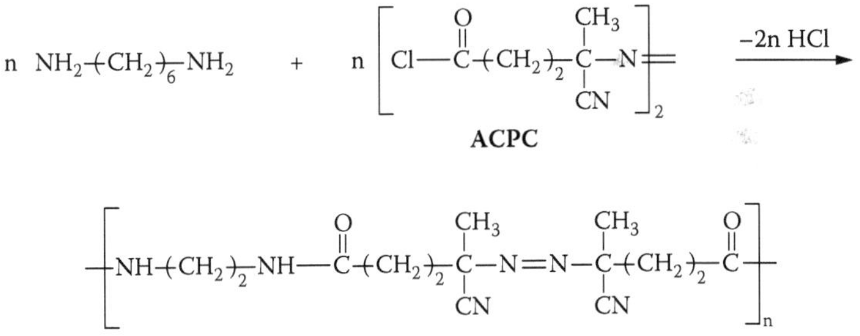

$$n\ NH_2-(CH_2)_6-NH_2 \quad + \quad n\left[Cl-\overset{O}{\overset{\|}{C}}-(CH_2)_2-\underset{CN}{\overset{CH_3}{C}}-N=\right]_2 \ \underset{\textbf{ACPC}}{} \xrightarrow{-2n\ HCl}$$

$$\left[-NH-(CH_2)_2-NH-\overset{O}{\overset{\|}{C}}-(CH_2)_2-\underset{CN}{\overset{CH_3}{C}}-N=N-\underset{CN}{\overset{CH_3}{C}}-(CH_2)_2-\overset{O}{\overset{\|}{C}}-\right]_n \tag{4.11}$$

Besides condensation reactions, the bifunctional ACPC may also be used for cationic polymerization [52–55] (e.g., of tetrahydrofuran). The polymer obtained by the method depicted in reaction (4.12) contains exactly one central azo bond [56] and is a suitable macroinitiator for the thermally induced block copolymerization of vinyl monomers. Initiators like ACPC are referred to as transformation agents because they are able to initiate polymerizations of different modes: radical vinyl polymerization (azo site) and cationic or condensation polymerization (chlorocarbonyl group) [55]. Thus, a transformation from one type of active center (e.g., cation) to another type (radical) occurs in the course of polymerization, which makes it possible to combine chemically very unlike monomers into one tailor-made block copolymer.

Several macro-azo initiators are also transformation agents (i.e., possess two different reactive sites) [57]. These may be used for the synthesis of triblock copolymers. Much work has also been done on the use of azo initiators in graft copolymerization [58]:

$$\left[\mathrm{Cl(O{=})C{-}(CH_2)_2{-}C(CH_3)(CN){-}N{=}} \right]_2 \ (\text{ACPC}) \xrightarrow{2\,Ag^+X^-} \left[\mathrm{X^-O^+{\equiv}C{-}(CH_2)_2{-}C(CH_3)(CN){-}N{=}} \right]_2 \xrightarrow{2n+2\ \text{THF}} \left[\mathrm{X^-\ \text{(THF)}O^+{-}\!\left((CH_2)_4{-}O\right)_n{-}C({=}O){-}(CH_2)_2{-}C(CH_3)(CN){-}N{=}} \right]_2 \quad (4.12)$$

4.2.2 Peroxide Initiators

Commercially, peroxides are used as oxidizing, epoxidizing, and bleaching agents, as initiators for radical polymerization, and as curing agents. As far as polymerization and curing are concerned, use is made of the propensity of peroxides for homolytic decomposition. The ease of radical formation is considerably influenced by the substituents at the peroxide group, as is demonstrated in Table 4.3. For practical applications, diacyl peroxides are used foremost; alkyl hydroperoxides and their esters, peroxyesters, and the salts of peracids are also of importance. As seen in Table 4.3, in the case of peroxy initiators, a considerable influence of solvent on the decomposition kinetics often occurs.

The decomposition of peroxide initiators, which is mostly initiated by heating, involves the rupture of the weak O—O bond, as is illustrated in the example of dibenzoyl peroxide (BPO) and di-*tert*-butyl peroxide.

$$\mathrm{C_6H_5{-}C({=}O){-}O{-}O{-}C({=}O){-}C_6H_5} \xrightarrow{k_d} 2\ \mathrm{C_6H_5{-}C({=}O){-}O^{\bullet}} \quad (4.13)$$

$$\mathrm{CH_3{-}C(CH_3)_2{-}O{-}O{-}C(CH_3)_2{-}CH_3} \xrightarrow{k_d} 2\ \mathrm{CH_3{-}C(CH_3)_2{-}O^{\bullet}} \quad (4.14)$$

As a rule, the radicals formed in this reaction start the initiation. They are, however, sometimes able to undergo further fragmentation, yielding other radicals:

$$\mathrm{C_6H_5{-}C({=}O){-}O^{\bullet}} \longrightarrow \mathrm{C_6H_5^{\bullet}} + CO_2 \quad (4.15)$$

TABLE 4.3
Selected Peroxy Compounds as Thermal Free-Radical Initiators

Peroxy Compound	Solvent	E_a, kJ mol^{-1}	k_d, s^{-1}	References
$C_6H_5-C(=O)-O-O-C(=O)-C_6H_5$	benzene	124	2.5×10^{-5} (80°C)	[59, 60]
	n-butanol	—	6.1×10^{-4} (80°C)	[61]
	cyclohexane	—	7.7×10^{-5} (80°C)	[61]
$CH_3-C(=O)-O-O-C(=O)-CH_3$	benzene	136	1.6×10^{-4} (85°C)	[62, 63]
			8.7×10^{-5} (80°C)	[64]
	tert-butanol	134	4.9×10^{-5} (80°C)	[65]
	CCl_4	—	5.5×10^{-4} (80°C)	[64]
$CH_3-CH_2-C(=O)-O-O-C(=O)-CH_2-CH_3$	benzene	129	2.4×10^{-4} (85°C)	[66]
$CH_3-(CH_2)_6-C(=O)-O-O-C(=O)-(CH_2)_6-CH_3$	benzene	—	3.8×10^{-4} (85°C)	[67]
$(CH_3)_2CH-C(=O)-O-O-C(=O)-CH(CH_3)_2$	benzene	—	2.4×10^{-4} (40°C)	[68]
$(CH_3)_3C-O-O-C(CH_3)_3$	benzene	142	7.8×10^{-8} (80°C) $\approx 2.7 \times 10^{-5}$ (130°C)	[69]
$(CH_3)_3C-O-O-H$	Benzene	171	2.0×10^{-5} (170°C)	[70]
	dodecane	128	1.3×10^{-6} (80°C)	[71]
$(CH_3)_3C-C(=O)-O-O-C(CH_3)_3$	benzene	120	7.6×10^{-4} (85°C)	[62]
$C_6H_5-C(=O)-O-O-C(CH_3)_3$	benzene	145	1.0×10^{-5} (100°C)	[62]

(Continued)

TABLE 4.3 (CONTINUED)
Selected Peroxy Compounds as Thermal Free-Radical Initiators

Peroxy Compound	Solvent	E_a, kJ mol^{-1}	k_d, s^{-1}	References
$KO{-}S(=O)_2{-}O{-}O{-}S(=O)_2{-}OK$	water		6.9×10^{-5} (80°C)	[72]

Note: E_a: activation energy for decomposition; k_d: decomposition rate constant, see Eq. (4.2).

$$(CH_3)_3C{-}O^{\bullet} \longrightarrow \dot{C}H_3 + CH_3{-}C(=O){-}CH_3 \qquad (4.16)$$

Whether or not a fragmentation according to the reactions illustrated in Equation 4.9 and Equation 4.10 takes place depends on the reactivity of the primary formed oxygen-centered radicals toward the monomer. In the case of BPO, a fragmentation occurs with phenyl radical formation (reaction in Equation 4.9), only in the absence of the monomer. In the presence of the monomer, the benzoyl oxy radicals react with monomer before decarboxylation. Aliphatic acyloxy radicals, on the other hand, undergo fragmentation already in the solvent cage whereby recombination products are produced that are not susceptible to further radical formation. As a result, the radical yield U_r for these initiator is smaller than 1:

$$CH_3{-}C(=O){-}O{-}O{-}C(=O){-}CH_3 \longrightarrow \boxed{CH_3{-}C(=O){-}\dot{O} + \dot{O}{-}C(=O){-}CH_3} \longrightarrow CH_3{-}C(=O){-}O{-}CH_3 + CO_2 \qquad (4.17)$$

The so-called *induced decomposition* of peroxides is another side reaction leading to a diminished radical yield. In the case of acylperoxides, primary formed radicals may attack the carbonylic oxygen atom of diacyl peroxides, leading to the formation of a carbon-centered radical:

$$R{-}C(=O){-}O{-}O{-}C(=O){-}R \xrightarrow{R''^{\bullet}} R{-}\dot{C}(OR'){-}O{-}O{-}C(=O){-}R \longrightarrow R{-}C(OR')=O + \dot{O}{-}C(=O){-}R \qquad (4.18)$$

The decomposition of dibenzoyl peroxide in dimethylaniline is extremely fast, which is also due to induced decomposition. The reaction mechanism involves the formation of radical cation and a subsequent transformation into radicals and stable species. In polymer synthesis, small quantities of dimethylaniline are sometimes added to BOP to promote radical generation:

$$
\begin{aligned}
&(CH_3)_2N{-}C_6H_5 + C_6H_5{-}\overset{O}{\overset{\|}{C}}{-}O{-}O{-}\overset{O}{\overset{\|}{C}}{-}C_6H_5 \\
&\downarrow \\
&\left[(CH_3)_2\dot{N}^{+}{-}C_6H_5\right]\, O^{-}{-}\overset{O}{\overset{\|}{C}}{-}C_6H_5 + C_6H_5{-}\overset{O}{\overset{\|}{C}}{-}O\cdot \\
&\downarrow \\
&\dot{C}H_2(CH_3)N{-}C_6H_5 + C_6H_5{-}\overset{O}{\overset{\|}{C}}{-}O{-}H
\end{aligned}
\qquad (4.19)
$$

Another example is the induced decomposition in the presence of butyl ether. In this case, the reaction is very likely to involve the formation of α-butoxy butyl radicals:

$$
\begin{aligned}
&C_6H_5{-}\overset{O}{\overset{\|}{C}}{-}O\cdot + C_4H_9OC_4H_9 \longrightarrow CH_3{-}(CH_2)_2{-}\dot{C}H{-}OC_4H_9 + C_6H_5{-}\overset{O}{\overset{\|}{C}}{-}O{-}H \\
&\downarrow + C_6H_5{-}\overset{O}{\overset{\|}{C}}{-}O{-}O{-}\overset{O}{\overset{\|}{C}}{-}C_6H_5 \\
&C_6H_5{-}\overset{O}{\overset{\|}{C}}{-}O\cdot + CH_3{-}(CH_2)_2{-}CH(O{-}C(=O){-}C_6H_5){-}OC_4H_9
\end{aligned}
\qquad (4.20)
$$

Usually, in the case of induced decomposition, one initiating molecule disappears without the formation of two radicals. What would be possible if the initiator species would undergo dissociation? In other words, the total number of radicals is smaller for the induced decomposition, not higher as sometimes assumed. The consumption of the initiator is, at the same time, faster than the normal dissociation. In fact, the decomposition of the initiator is faster than one would follow from first-order reaction kinetics:

$$-\frac{d[I]}{dt} = k_d[I] + k_{ind}[I]^x \tag{4.21}$$

As reactions (4.18–4.20) imply, the mechanism of induced decomposition does very much depend on the solvent. Furthermore, the extent to which induced decomposition occurs changes with the type of peroxy initiator used and the actual monomer, because induced decomposition may often be triggered by any of the radicals present in the reaction mixture.

The choice of the proper peroxy initiator largely depends on its decomposition rate at the reaction temperature of the polymerization. BPO is the major initiator for bulk polymerization of polystyrene or acrylic ester polymers, where temperatures from 90 220°C are encountered. Dilauroyl, dicaprylyl, diacecyl, and di-*tert*-butyl peroxides are also used. In the case of suspension polymerization of styrene, where temperatures between 85°C and 120°C are applied the initiators also range in activity from BPO to di-*tert*-butyl peroxide. In suspension polymerization of vinyl chloride (reaction temperatures of 45–60°C for the homopolymer), thermally very labile peroxides such as diisopropyl peroxydicarbonate and *tert*-butyl peroxypavilate are used.

As far as the handling of peroxides is concerned, it must be noted that upon heating, peroxides may explode. Special precautions have to be taken with peroxides of a low carbon content, such as diacetyl peroxide, as they are often highly explosive. In the pure state, peroxides should be handled only in very small amounts and with extreme care. Solutions of high peroxide content are also rather hazardous.

Besides low-molecular-weight peroxides, numerous works on macromolecular peroxide initiators have also been published, which are useful in the preparation of block copolymers [73]. As illustrated in Table 4.4, various compounds have been reacted via condensation or addition reaction to yield macro-peroxy-initiators.

As far as the decomposition of macro-peroxy-initiators is concerned, it has been found that the decomposition rate is about the same as for structurally similar low-molecular-weight peroxy initiators [73], despite the cage effect that obviously leads to some propensity to recombination reactions. High-molecular-weight peroxy initiators have been mostly used to combine two monomers that polymerize by radical addition polymerization. Examples are block copolymers consisting of a polyacrylamide block and a random polyacrylamide copolymer as a second block [78–82], and block copolymers of polymethyl methacrylate/poly vinyl acetate [83, 84], polystyrene/polyacrylonitrile [85], and polystyrene/polyhydroxymethyl acrylate [86]. In block copolymerization, the good solubility, especially of aliphatic macroperoxyinitiators in common monomers, is being used. Block copolymers prepared by macro-peroxy-initiators often show interesting surface activity (useful for coatings

TABLE 4.4
Macroinitiators Having Peroxy Groups Synthesized from Two Components

Structure of Peroxide Group	Reactant A	Reactant B	References
$-C(=O)-O-O-C(=O)-$	H–O–O–H	Phthaloyl chloride ($C_6H_4(COCl)_2$)	[74]
		$Cl-C(=O)-(CH_2)_n-C(=O)-Cl$	[75]
$-C(=O)-O-O-$	$H-O-O-C(CH_3)_2-CH_2-CH_2-C(CH_3)_2-O-O-H$	Cyclohexane-1,2-dicarbonyl chloride ($C_6H_{10}(COCl)_2$)	[76]
		$Cl-C(=O)-(CH_2)_n-C(=O)-Cl$	[77]
	$H-O-O-C(CH_3)_2-C{\equiv}C-C(CH_3)_2-O-O-H$	Cyclohexane-1,2-dicarbonyl chloride ($C_6H_{10}(COCl)_2$)	[76]
	$H-O-O-C(CH_3)_2-C_6H_4-C(CH_3)_2-O-O-H$	Cyclohexane-1,2-dicarbonyl chloride ($C_6H_{10}(COCl)_2$)	[76]

Note: See also A. Ueda and S. Nagai, in *Macromolecular Design: Concept and Practice*, M. K. Mishra, Ed., Polymer Frontiers Int. Inc., New York, 1994, p. 265.

and adhesives). Further, they find application as antishrinking agents [87–89] and as compatibilizers in polymer blends.

4.2.3 Persulfate Initiators

Persulfate initiators generate free radicals upon the thermally induced scission of O—O bonds, thus resembling the organic peroxides discussed previously. For potassium persulfate, the decomposition rate constant is 9.6×10^{-5} sec^{-1} at 80°C and the activation energy amounts to 140 kJ mol^{-1} (in 0.1 mol L^{-1} NaOH [90].

Interestingly, quite different reactions occur, depending on the pH of the reaction media. In alkaline and neutral media, two radical anions are formed from one persulfate molecule. In strongly acidic surrounding, however, no radicals are generated, giving rise to a suppression of polymerization with lowering the pH:

$$(S_2O_8)^{2-} \rightarrow 2SO_4^{\cdot-},\ pH \geq 7 \tag{4.22}$$

$$(S_2O_8)^{2-} + H^+ \rightarrow SO_4^{2-} + HSO_4^{\cdot},\ pH < 7 \tag{4.23}$$

The fact that the peroxydisulfate ion may initiate polymerization of certain vinyl monomers has been known for some time [91]. In most practical polymerizations, however, peroxysulfate is used together with reducing agent in redox-initiating systems.

4.3 RADICAL FORMATION VIA ELECTRON TRANSFER-REDOX-INITIATING SYSTEMS

Oxidation agents, such as hydroperoxides or halides, in conjunction with electron donors, like metal ions, may form radicals via electron transfer:

$$H—O—O—H + Fe^{2+} \rightarrow H\dot{O} + OH^- + Fe^{3+} \tag{4.24}$$

$$Fe^{2+} \rightarrow Fe^{3+} + e^- \quad \text{(oxidation)} \tag{4.25}$$

$$H—O—O—H + e^- \rightarrow H\dot{O} + OH^- \quad \text{(reduction)} \tag{4.26}$$

The initiator systems used in this type of polymerization consists, therefore, of two components: an oxidizing agent and a reducing agent. If hydrogen peroxides are used as the oxidizing agent, one hydroxyl radical and one hydroxyl ion are formed, in contrast to direct thermal initiation, where two hydroxyl radicals are generated (*vide ante*). The hydroxyl ion formed in redox systems is stabilized by salvation. As a result, the thermal activation energy is relatively low, usually 60–80 kJ mol^{-1} lower than for the direct thermal activation. Therefore, using redox systems, polymerizations can be conducted at low temperatures, which is advantageous in terms of energy saving and prevention of thermally induced termination or depolymerization. In technical synthesis, peroxide-based redox systems are used, for example, for the copolymerization of styrene and butadiene at 5°C, the so-called cold rubber process, and for the polymerization of acrylonitrile in the aqueous phase. In addition to peroxides, many other oxidizing agents may be used in radical polymerization. Table 4.5 gives an idea of the variety of systems being used.

TABLE 4.5
Initiating Systems in Redox Initiation

Oxidizing Agent	Initiating Radical	Reducing Agent	References
H–O–O–H	$HO^{\bullet}$	Fe^{2+}	[92–94]
		NO_3^-	[95]
		NO_2^-	[95]
		NH_3	[94]
		HSO_3^-/SO_3^{2-}	[91]
		$HS{-}C(=NH){-}NH_2$	[96–98], see also [28]
R–O–O–R	$RO^{\bullet}$	Fe^{2-}	[99–101]
		$HS{-}C(=NR){-}NHR$	[98, 102, 103]
$C_6H_5{-}N(CH_3)_2$	$RO^{\bullet} + {}^{\bullet}CH_2{-}N(CH_3){-}C_6H_5$		[104], see also [15]
R–O–O–H	$RO^{\bullet}$	Fe^{2-}	[105, 106]
	$R^{\bullet}$	BR'_3	[107], see also [26], [27]
O_2	$R^{\bullet}$	BR'_3	[107], see also [26], [27]
Br_2	$BR^{\bullet}$	Fe^{2+}	[108]
	$BR^{\bullet}$ and ${}^{\bullet}S{-}C(=NH){-}NH_2$	$HS{-}C(=NR){-}NHR$	[109]
$S_2O_8^{2-}$	$SO_4^{-\bullet}$	Ag^+	[95, 110]
		Fe^{2+}	[111]
		$HS{-}C(=NH){-}NH_2$	[98]
$H_2P_2O_8^{2-}$	$HPO_4^{-\bullet}$	Ag^+	[112]

(Continued)

TABLE 4.5 (CONTINUED)
Initiating Systems in Redox Initiation

Oxidizing Agent	Initiating Radical	Reducing Agent	References
		$S_2O_3^{2-}$	[113]
$C_6H_5N(CH_3)_2$	Ph-N(CH$_3$)CH$_2$$^\cdot$	$C_6H_5N(CH_3)CH_2^{\cdot}$	[114]

If metal ions are used as the reducing agent, the danger of contaminating the polymer with heavy metals exists, which may be a source of easy oxidizability of the polymer. Furthermore, high concentrations of metal ions in their higher oxidation state may lead to termination reactions according to

$$\sim\sim\sim\cdot + FeCl_3 \longrightarrow \sim\sim\sim Cl + FeCl_2 \tag{4.27}$$

To keep the concentration of metal salt small, additional reducing agents are added, which react with the metal ion, as demonstrated in the example of the system potassium peroxidesulfate, sodium sulfite, and ferrous sulfate in reactions (4.28) and (4.29):

$$S_2O_8^{2-} + Fe^{2+} \rightarrow SO_4^{-\cdot} + SO_4^{2-} + Fe^{3+} \tag{4.28}$$

$$Fe^{2+} + SO_3^{2-} \rightarrow SO_3^{-\cdot} + Fe^{3+} \tag{4.29}$$

Because the rate of reaction (4.29) is very high compared with that of reaction (28), Fe^{3}- ions formed are instantaneously reduced to Fe^{2}-, which allows the use of only catalytic quantities of iron salt for initiation. The redox system depicted in reactions (4.28) and (4.29) is used in the previously mentioned technical polymerization of acrylonitrile. In the cold rubber process, systems consisting of hydroperoxide, ferrous salts, and rongalite are used.

For initiating radical polymerization at temperatures as low as −50°C to −100°C, boroalkyls are applied as reducing agents [107]. Upon reaction with oxygen or with organic hydroperoxides, they are able to abstract alkyl radicals which act is initiating species. The alkyl radicals generated are of extraordinary reactivity, as they are usually not stabilized by resonance. Therefore, a number of side reactions, such as chain transfers, may occur:

$$BR_3 + O_2 \rightarrow R_2BOOR \tag{4.30}$$

$$R_2BOOR + BR_3 \rightarrow R_2BOBR_2 + R_2BOR + R\cdot \tag{4.31}$$

Thiourea derivatives are another often used reducing agent, which is able to produce radicals upon reaction with peroxides and other oxidizing agents. Upon this reaction, both sulfur-centered radicals and radicals stemming from the oxidizing agent are formed. In the polymerization of methylmethacrylate with the hydrogen peroxide/thiourea system, both amino and hydroxyl end groups were found, of which the latter predominate [98]. If bromine was used as the oxidizing agent, end-group analysis implied that the sulfur-centered radical is the major initiating species [109]:

$$\mathrm{H{-}S{-}C(NH_2){=}NH} + \mathrm{H{-}O{-}O{-}H} \longrightarrow \mathrm{{\cdot}S{-}C(NH_2){=}NH} + \dot{\mathrm{O}}\mathrm{H} + \mathrm{H_2O} \tag{4.32}$$

With many redox systems, coupled reactions take place that make it necessary to choose the appropriate systems in accordance with the monomer and the polymerization conditions. If the redox reaction is slow, there will be a low yield of radicals and therefore a low polymerization rate. On the other hand, if the redox reaction is fast compared with the initiation step, the majority of initiating radicals will be consumed by radical termination reactions. Therefore, redox systems are modified by further additives. For example, heavy metal ions may be complexed with substances such as citrates, which adjust their reactivity to a reduced level. Thus, redox systems for technical polymerization are complex formulations that enable one to obtain optimum results at well-defined reaction conditions.

4.4 THERMALLY INDUCED RADICAL FORMATION WITHOUT INITIATOR

In general, an initiator is added to vinyl monomers to produce initiating radicals upon the desired external stimulation. Indeed, most impurity-free vinyl polymers do not initiate upon heating, making it unavoidable to introduce free-radical initiators. However, monomers like styrene and methyl methacrylate derivatives may polymerize without any added initiator. The mechanism involves the spontaneous formation of radicals in the purified monomers.

For styrene, the conversion of monomer per hour rises from ~0.1% at 60°C to about 14% at 140°C. Thus, the effect has to be encountered, especially for polymerizations at higher temperatures. Furthermore, when a styrene-based monomer is to be purified by distillation, the addition of inhibitors and distillation at reduced pressure is advisable to avoid the distillate's becoming viscous. Another difficulty occurring during distillation is the formation of polymer in the column, which can also be prevented by distilling in vacuo. The initiation of a styrene-based monomer is assumed to involve a (4 + 2) cycloaddition of the Diels–Alder type with a subsequent hydrogen transfer from the dimer to another monomer molecule:

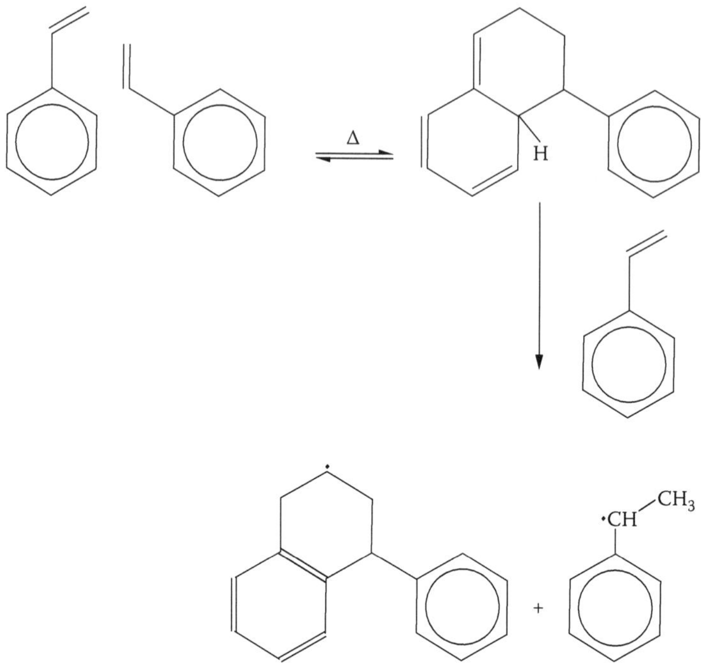

(4.33)

The radicals thus generated initiate the polymerization, provided they do not deactivate by mutual combination or disproportionation. Due to their low ceiling temperature, α-substituted styrenes hardly undergo thermal polymerization in the absence of initiator.

With methyl methacrylate, thermal self-polymerization also occurs, but with a rate about two orders of magnitude smaller than with styrene.

REFERENCES

1. H. C. Ramsperger, *J. Am. Chem. Soc.*, 49, 912 (1927).
2. C. Steel and A. F. Trotman-Dickenson, *J. Chem. Soc.*, 975 (1959).
3. R. Renaud and L. C. Leitch, *Can. J. Chem.*, 32, 545 (1954).
4. W. D. Clark, Ph.D. dissertation, University of Oregon (1959).
5. C. G. Overberger, M. T. O'Shaughnessay, and H. Thalit, *J. Am. Chem. Soc.*, 71, 2661 (1949).
6. C. G. Overberger and A. Lebovits, *J. Am. Chem. Soc.*, 76, 2722 (1954).
7. C. G. Overberger, H. Biletch, A. B. Finestone, J. Tilker, and J. Herbert, *J. Am. Chem. Soc.*, 75, 2078 (1953).
8. A. Nersasian, Ph.D. thesis, University of Michigan (1954), University Microfilms Publication No. 7697.
9. S. G. Cohen, S. J. Groszos, and D. B. Sparrow, *J. Am. Chem. Soc.*, 72, 3947 (1950).
10. S. G. Cohen, F. Cohen, and C. H. Wang, *J. Org. Chem.*, 28, 1479 (1963).
11. S. G. Cohen and C. H. Wang, *J. Am. Chem. Soc.*, 75, 5504 (1953).
12. R. Zand, Azo catalysts, in *Encyclopedia of Polymer Science and Technology*, Vol. 2, H. F. Mark, N. G. Gaylord, and N. M. Bikales, Eds., Interscience, New York, 1965, p. 278.
13. J. Ulbricht, *Grundlagen der Synthese von Polymeren*, Hüthig & Wepf, Basel, 1992.

14. H. Ulrich, *Introduction to Industrial Polymers*, Hanser, Munich, 1982.
15. A. F. Bickel and W. A. Waters, *Rec. Trav. Chim.*, 69, 1490 (1950).
16. R. W. Upson (to Du Pont), U.S. Patent 2,599,299 (1952).
17. R. W. Upson (to Du Pont), U.S. Patent 2,599,300 (1952).
18. A. L. Barney (to Du Pont), U.S. Patent 2,744,105 (1952).
19. R. M. Haines and W. A. Waters, *J. Chem. Soc.*, 4256 (1955).
20. K. Matsukawa, A. Ueda, H. Inoue, and S. Nagai, *J. Polym. Sci., Part A: Polym. Chem.*, 28, 2107 (1990).
21. O. Nuyken, J. Dauth, and W. Pekruhn, *Angew. Makromol. Chem.*, 187, 207 (1991).
22. O. Nuyken, J. Dauth, and W. Pekruhn, *Angew. Makromol. Chem.*, 190, 81 (1991).
23. D. A. Smith, *Makromol. Chem.*, 103, 301 (1967).
24. A. Ueda, Y. Shiozu, Y. Hidaka, and S. Nagai, *Kobunshi Ronbunshu*, 33, 131 (1976).
25. B. Hazer, *Makromol. Chem.*, 193, 1081 (1992).
26. O. S. Kabasakal, F. S. Güner, A. T. Erciyes, and Y. Yagci, *J. Coat. Technol.*, 67, 47 (1995).
27. Y. Kita, A. Ueda, T. Harada, M. Tanaka, and S. Nagai, *Chem. Express*, 1, 543 (1986).
28. A. Ueda and S. Nagai, *J. Polym. Sci., Part A: Polym. Chem.*, 24, 405 (1986).
29. A. Ueda and S. Nagai, *J. Polym. Sci. Polym. Chem. Ed.*, 25, 3495 (1987).
30. B. Hazer, *Macromol. Chem. Phys.*, 196, 1945 (1995).
31. A. Ueda and S. Nagai, *Kobunshi Ronbunshu*, 43, 97 (1986).
32. J. M. G. Cowie and M. Yazdani-Pedram, *Br. Polym. J.*, 16, 127 (1984).
33. T. O. Ahn, J. H. Kim, J. C. Lee, H. M. Jeong, and J.-Y. Part, *J. Polym Sci., Part A: Polym. Chem.*, 31, 435 (1993).
34. T. O. Ahn, J. J. Kim, H. M. Jeong, S. W. Lee, and L. S. Park, *J. Polym. Sci., Part B: Polym. Phys.*, 32, 21 (1994).
35. C. I. Simionescu, E. Comanita, V. Harabagiu, and B. C. Simionescu, *Eur. Polym. J.*, 23, 921 (1987).
36. C. I. Simionescu, V. Harabagiu, E. Comanita, V. Hamciuc, D. Giurgiu, and B. C. Simionescu, *Eur. Polym. J.*, 26, 565 (1990).
37. H. Terada, Y. Haneda, A Ueda, and S. Nagai, *Macromol. Rept.*, A31, 173 (1994).
38. Y. Haneda, H. Terada, M. Yoshida, A. Ueda, and S. Nagai, *J. Polym. Sci., Part A: Polym. Chem.*, 32, 2641 (1994).
39. G. Galli, E. Chiellini, M. Laus, M. C. Bignozzi, A. S. Angeloni, and O. Francescangeli, *Macromol. Chem. Phys.*, 195, 2247 (1994).
40. A. Ueda and S. Nagai, *J. Polym. Sci., Polym. Chem. Ed.*, 22, 1783 (1984).
41. A. Ueda and S. Nagai, *J. Polym. Sci., Polym. Chem. Ed.*, 22, 1611 (1984).
42. Y. Yagci, Ü. Tunca, and N. Bicak, *J. Polym. Sci., Part C: Polym. Lett.*, 24, 491 (1986).
43. Y. Yagci, Ü. Tunca, and N. Bicak, *J. Polym. Sci., Part C: Polym. Lett.*, 24, 49 (1986).
44. Ü. Tunca and Y. Yagci, *J. Polym. Sci., Part A: Chem.*, 28, 1721 (1990).
45. Ü. Tunca and Y. Yagci, *Polym. Bull.*, 26, 621 (1991).
46. J. Furukawa, S. Takamori, and S. Yamashita, *Angew. Makromol. Chem.*, 1, 92 (1967).
47. H. Yürük, A. B. Özdemir, and B. M. Baysal, *J. Appl. Polym. Sci.*, 31, 2171 (1986).
48. H. Kinoshita, M. Ooka, N. Tanaka, and T. Araki, *Kobunshi Ronbunshu*, 50, 147 (1993).
49. H. Kinoshita, N. Tanaka, and T. Araki, *Makromol. Chem.*, 194, 829 (1993).
50. H. Yürük, S. Jamil, and B. M. Baysal, *Angew. Makromol. Chem.*, 175, 99 (1990).
51. C. H. Bamford, A. D. Jenkins, and R. Wayne, *Trans. Faraday Soc.*, 56, 932 (1960).
52. Y. Yagci, G. Hizal, A. Önen, and I. E. Serhatli, *Makromol. Symp.*, 84, 127 (1994).

53. Y. Yagci, *Polym. Commun.,* 26, 8 (1985).
54. Y. Yagci, *Polym. Commun.*, 27, 21 (1986).
55. Y. Yagci and I. Reetz, Azo initiators and transformation agents for block copolymer synthesis, in *Handbook of Engineering Polymeric Materials*, N. P. Cheremisinoff, Ed., Marcel Dekker, New York, 1997, pp. 735–753.
56. G. Hizal and Y. Yagci, *Polymer*, 30, 722 (1989).
57. S. Denizligil, A. Baskan, and Y. Yagci, *Makromol. Chem. Rapid Commun.*, 16, 387 (1995).
58. O. Nuyken and B. Volt, in *Macromolecular Design, Concept and Practice*, M. K. Mishra, Ed., Polymer Frontiers International, Inc., New York, 1994, p. 313.
59. T. M. Babchinitser, K. K. Mozgova, and V. V. Korshak, *Dokl. Akad. Nauk SSSR*, 173, 575 (1967).
60. M. Matsuda and S. Fujii, *J. Polym. Sci., A-1*, 5, 2617 (1967).
61. E. A. S. Cavell, *Makromol. Chem.*, 54, 70 (1962).
62. R. A. Gregg, D. M. Alderman, and R. R. Mayo, *J. Am. Chem. Soc.*, 70, 3740 (1948).
63. G. N. Freidlin and K. A. Solop, *Vysokomolekul. Soed.*, 7, 1060 (1965).
64. C. H. Bamford, A. D. Jenkins, and R. Johnson, *Proc. Roy. Soc., Ser. A*, 241, 364 (1957).
65. G. Akazome, S. Sakai, and K. Maurai, *Kogyo Kagaky Zasshi*, 63, 592 (1960).
66. R. A. Bird and K. E. Russel, *Can. J. Chem.*, 43, 2123 (1965).
67. O. L. Mageli and J. R. Kolczynski, Peroxy compounds, in *Encyclopedia of Polymer Science and Technology*, Vol. 9, H. F. Mark, N. G. Gaylord, and N. N. Bikales, Eds., Interscience, New York, 1995, p. 814.
68. S. Imoto, J. Ukida, and T. Kominami, *Kobunshi Kagaku*, 14, 127 (1957).
69. R. A. Gregg and F. R. Mayo, *J. Am. Chem. Soc.*, 75, 3530 (1953).
70. R. N. Chadha and G. S. Misra, *Indian J. Phys.*, 28, 37 (1954).
71. T. Berezhnykh-Földes and T. Tudös, *Eur. Polym. J.*, 2, 219 (1966).
72. M. H. George and P. F. Onyon, *Trans. Faraday Soc.*, 59, 1390 (1963).
73. A. Ueda and S. Nagai, in *Macromolecular Design: Concept and Practice*, M. K. Mishra, Ed., Polymer Frontiers Int. Inc., New York, 1994, p. 265.
74. H. V. Pechmann and L. Vanino, *Ber.*, 27, 1510 (1894).
75. M. S. Tsvetkov, R. F. Markovskaya, and A. A. Sorokin, *Vysokomol. Soed., Ser. B.*, 11, 519 (1969).
76. O. Suyama, K. Taura, and M. Kato, *Jpn. Kokai*, H3-252413 (1991).
77. B. Hazer, *J. Polym. Sci., Polym. Chem.*, 25, 3349 (1987).
78. H. Ohmura and M. Nakayama, *Jpn. Kokai*, S56-93722 (1981).
79. H. Ohmura and M. Nakayama, *Jpn. Kokai*, S56-76423 (1981).
80. H. Ohmura and M. Nakayama, *Jpn. Kokai*, S56-133312 (1981).
81. H. Ohmura and M. Nakayama, *Jpn. Kokai*, S56-93723 (1981).
82. H. Ohmura and M. Nakayama, *Jpn. Kokai*, S55-142016 (1980).
83. M. Nakayama, M. Matsushima, S. Banno, and N. Kanazawa, *Jpn. Kokai*, S56-79113 (1981).
84. M. Matsushima and M. Nakayama, *Jpn. Kokai*, S55-75414 (1980).
85. B. Hazer, *Angew. Makromol. Chem.*, 129, 31 (1985).
86. H. Ohmura, H. Dohya, Y. Oshibe, and T. Yamamoto, *Kobunshi Ronbunshu*, 45, 857 (1988).
87. K. Fukushi and A. Hatachi, *Jpn. Kokai*, S58-189214 (1983).
88. M. Nakayama, M. Matsushima, S. Banno, and N. Kanazawa, *Jpn. Kokai*, S56-79133 (1981).
89. K. Fukushi, *Jpn. Kokai*, S60-217222 (1985).
90. I. M. Kolthoff and I. K. Miller, *J. Am. Chem. Soc.*, 73, 3055 (1951).
91. R. G. R. Bacon, *Trans. Faraday Soc.*, 42, 140 (1946).

92. W. J. R. Evans and J. H. Baxendale, *Trans. Faraday Soc.*, 42, 140 (1946).
93. G. G. Haber and J. M. Weiss, *Proc. Soc.*, 147A, 332 (1934).
94. E. Halfpenny and P. L. Robinson, *J. Chem. Soc.*, 928 (1952).
95. W. Kern, H. Cherdron, and R. C. Schulz, *Makromol. Chem.*, 24, 141 (1957).
96. A. Hebeish, S. H. Abdel-Fattah, and M. H. El-Rafie, *J. Appl. Polym. Sci.*, 22(8), 2253 (1978).
97. T. Sugimura, N. Yasumuto, and Y. Minoura, *J. Polym. Sci.*, A3, 2935 (1965).
98. A. R. Mukharjee, R. P. Mitra, A. M. Biswas, and S. Maity, *J. Polym. Sci., Part A-1*, 5, 135 (1967).
99. W. Kern, *Makromol. Chem.*, 2, 48 (1948).
100. H. Hasegawa, *J. Chem. Soc. Jpn.*, 31, 696 (1958).
101. W. Kern, *Monatsh. Chem.*, 88, 763 (1957).
102. T. Sugimura and Y. Minoura, *J. Polym. Sci., Part A-1*, 4, 2735 (1966).
103. T. Sugimura and Y. Minoura, *J. Polym. Sci., Part A-1*, 4, 2721 (1966).
104. S. Morsi, A. B. Zaki, and M. A. El Khyami, *Eur. Polym. J.*, 13, 851 (1977).
105. I. M. Kolthoff, *J. Am. Chem. Soc.*, 71, 3789 (1949).
106. K. Kharasch, *J. Org. Chem.*, 18, 332 (1953).
107. J. Furukawa and T. Tsurata, *J. Polym. Sci.*, 28, 227 (1958).
108. N. Uri, *Chem. Rev.*, 50, 375 (1952).
109. T. K. Sengupta, D. Parmanick, and S. R. Palit, *Indian J. Chem.*, 7, 908 (1969).
110. R. W. Rainward, *J. Polym. Sci.*, 2, 16 (1947).
111. I. M. Kolthoff, A. I. Medalia, and H. P. Raen, *J. Am. Chem. Soc.*, 73, 1733 (1951).
112. S. Lenka, P. L. Nayak, and A. K. Dhal, *Makromol. Chem., Rapid Commun.*, 1, 313 (1980).
113. P. L. Nayak, S. Lenka, and M. K. Mishra, *J. Polym. Sci., Polym. Chem. Ed.*, 19(3), 839 (1981).
114. Y. Okada, Y. Ohno, and K. Himoji, *Daigaku Kenkyu Hokoku (Jpn.)*, 30A, 122 (1977).

5 Chemical Initiation by Metals or Metal-Containing Compounds

Yusuf Yagci, Ivo Reetz, and Munmaya K. Mishra

CONTENTS

5.1 INTRODUCTION

The initiation by metals and metal-containing compounds generally takes place as a redox process [1]. In this type of initiation, free radicals responsible for polymerization are generated as transient intermediates in the course of a redox reaction. Essentially, this involves an electron transfer process followed by scission to give normally one free radical. The oxidant is generally referred to as the initiator or the catalyst, and the reducing agent is called the activator or the accelerator. Notably, depending on its oxidation state, the metal can act as reducing or oxidizing agent.

The special features of redox initiation are as follows:

1. Very short (almost negligible) induction period.
2. A relatively low energy of activation (in the range of 10–20 kcal mol^{-1}) as compared with 30 kcal mol^{-1} for thermal initiation. This enables the

polymerization to be performed at a relatively low temperature, thereby decreasing the possibility of side reactions, which may change the reaction kinetics and the properties of the resulting polymer.
3. The polymerization reaction is controlled with ease at low temperature, and comparatively high-molecular-weight polymers with high yields can be obtained in a very short time.
4. Convenient access is available to a variety of tailor-made block copolymers.
5. Redox polymerizations also provide direct experimental evidence for the existence of transient radical intermediates generated in redox reactions, which enables the identification of these radicals as terminating groups, helping to understand the mechanism of redox reactions.

A wide variety of redox reactions between metals or metal compounds and organic matter may be employed in this context. Because most of them are ionic in nature, they may be conveniently performed in aqueous solution and occur rather rapidly even at relatively low temperatures. As a consequence, redox systems with many different compositions have been developed into initiators that are very efficient and useful, particularly for suspension and emulsion polymerization in aqueous media [2], which is dealt with in detail in Chapter 6. The low-temperature (at ~5°C) copolymerization of styrene and butadiene for the production of GR-S rubber was made possible with the success of these catalytic systems.

Commonly used oxidants in redox polymerization include peroxides, cerium(IV) salts, sodium hypochloride, persulfates, peroxydiphosphate, and permanganate. Reducing agents are, for example, the salts of metals like Fe^{2+}, Cr^{2+}, V^{2+}, Ce^{2+}, Ti^{3+}, Co^{2+}, Cu^{2+}, oxoacids of sulfur, hydroxyacids, and so forth.

A typical example of a redox initiation with a metal compound as activator is the initiation by the system H_2O_2 and ferrous(II) salts [3]. In the course of this reaction, hydroxy radicals are evolved which are very reactive initiators. The reaction scheme is as follows:

$$H_2O_2 + Fe^{2+} \rightarrow Fe^{3+} + \dot{O}H + OH^- \tag{5.1}$$

In the subsequent sections, redox reactions involving metal carbonyls, metal chelates and ions, and permanganate as reducing agents will be reviewed. The other redox systems applied for suspension polymerization are the subject of Chapter 6.

5.2 TYPES OF INITIATION

5.2.1 Initiation by Metal Carbonyls

5.2.1.1 Thermal Initiation

It is well known from extensive electron-spin resonance (ESR) studies [4–6] that organic halides in conjunction with an organometallic derivative of a transition element of groups VIA, VIIA, and VIII, with the metal in a low oxidation state, give rise to free-radical species. Kinetic studies of the initiation of polymerization [7, 8] have related that in all systems containing organometallic derivatives and organic halides, the radical-producing reaction is basically an electron transfer process from transition metal to halide as presented in the following equation (M being the transition metal):

$$M^{\circ} + X_3C\text{—}R \rightarrow M^yX + X_2\dot{C}\text{—}R \tag{5.2}$$

In this process, the organic halide is split into an ion and a radical fragment. Free-radical formation by oxidation of molybdenum carbonyl with carbon tetrachloride and carbon tetrabromide has been studied in detail [6]. The overall reaction may be represented as follows:

$$Mo^{\circ} + CCl_4 \rightarrow Mo^1Cl + \dot{C}Cl_3 \tag{5.3}$$

Manganese pentacarbonyl chloride has also been used [9] as a thermal initiator for free-radical polymerization in the presence of halide and non-halide additives. At 60°C, it is 10 times as active as azobisisobutyronitrile toward methyl methacrylate polymerization. In the absence of additives, manganese pentacarbonyl chloride does not initiate the polymerization of methyl methacrylate significantly at temperatures up to 80°C; even at 100°C, initiation is very slow. Analysis of the polymers produced indicates that, with CCl_4 as the additive, initiation occurs through CCl_3 radicals and no manganese is found in the polymers. Angelici and Basolo [10] have reported measurements of the rates of ligand exchange reactions undergone by $Mn(CO)_5Cl$ and have concluded that the rate-determining step is dissociation:

$$Mn(CO)_5Cl \rightarrow Mn(CO)_4Cl + CO^{\cdot} \tag{5.4}$$

The preceding reaction is followed by the rapid combination of $Mn(CO)_4Cl$ with the ligand L so that the overall process is the replacement of CO by L.

Bamford et al. [11] have shown that $Mn(CO)_5Cl$ is a very reactive solution, even in nonpolar solvents. Thus, in benzene solution at 25°C, the dimer $(Mn(CO)_4Cl)_2$ is readily formed if carbon monoxide is removed by evacuation or a stream of nitrogen:

$$Mn(CO)_5Cl \rightarrow 1/2[Mn(CO)_4Cl]_2 + CO^{\cdot} \tag{5.5}$$

In a donor solvent such as methyl methyacrylate, ligand exchange occurs at 25°C and monomeric and dimeric complexes such as (A) and (B) are produced [11]:

```
        M                      M      Cl
        |                      |    /    \
(CO)3---Mn---Cl         (CO)3 Mn          Mn (CO)3
        |                           \    /
        M                             Cl
       (A)                            (B)
```

The radicals may be generated from thermal decomposition of (B) according to

$$(B) \rightarrow MnM(CO)_3 + MnCl_2 + M + 3CO^{\cdot} \tag{5.6}$$

In the presence of CCl_4, the reaction may be presented as follows:

$$MnM(CO)_3 + CCl_4 \xrightarrow{2CO} Mn(CO)_5Cl + M + \dot{C}Cl_3 \tag{5.7}$$

An alternative initiation mechanism starting from the intermediate (A) also involves radical generation on the halide.

$$(CO)_2MnMM_3Cl \xrightarrow{\text{rate determining}} (CO)_2MnM_2Cl + M \tag{5.8}$$

$$(CO)_2MnM_2Cl + CCl_4 \rightarrow MnCl_2 + 2M + 2CO + \dot{C}Cl_3 \tag{5.9}$$

Bamford and Mullik [12] had reported the methyl methacrylate radical polymerization initiated by the thermal reactions of methyl and acetyl manganese carbonyls. The initiating species are claimed to be methyl radicals formed from the reaction of methyl methacrylate with the transition metal derivative through an activated complex. In the presence of additives such as CCl_4 and C_2F_4, however, initiating radicals are derived from the additives as was proved by the analysis of the resulting polymers (i.e., initiation by CCl_3 radicals would introduce thee chlorine atoms into each polymer chain).

In the case of perfluoromethyl and perfluoroacetyl manganese carbonyls [13], the initiating mechanism does not involve complexation with the monomer, as illustrated in Eqs. (5.10)–(5.13) for perfluoromethyl manganese carbonyl:

$$CF_3Mn(CO)_5 \rightleftharpoons CF_3Mn(CO)_4 + CO \tag{5.10}$$

$$CF_3COMn(CO)_5 \rightleftharpoons CF_3COMn(CO)_4 + CO \tag{5.11}$$

$$CF_3Mn(CO)_4 \rightleftharpoons \dot{C}F_3 + Mn(CO)_4 \tag{5.12}$$

$$CF_3COMn(CO)_4 \rightleftharpoons CF_3CO^{\cdot} + Mn(CO)_4 \tag{5.13}$$

Tetrabis(triphenyl phosphate)nickel (NiP_4) is an interesting example of the large class of organometallic derivative which, in the presence of organic halides, initiate free-radical polymerization [5]. It was shown that the kinetics of initiation at room temperature are consistent with a mechanism in which ligand displacement by monomer leads to a reactive species readily oxidized by the halide [14–16].

The generation mechanism of the initiating radicals species is reported in Eqs. (5.14)–(5.16):

$$NiP_4 \underset{k_2}{\overset{k_1}{\rightleftharpoons}} NiP_3 + P \tag{5.14}$$

$$NiP_3 + M \overset{k}{\rightleftharpoons} M\text{—}Nip_3 \tag{5.15}$$

$$M\text{—}NiP_3 \xrightarrow{CCl_4} nR^{\cdot} \tag{5.16}$$

in which M represents the monomer and *n* is the number of radicals arising from reaction of a single molecule of complex. Notably, each complex yields approximately one free radical. Detailed studies [17] on the preceding system using methyl methacrylate and styrene as monomer revealed that both monomers behave similarly in dissociation and complexation steps. But the reaction between the M · · ·

NiP_3 complex and carbon tetrachloride shows marked kinetic differences in the two systems.

5.2.1.2 Photochemical Initiation

Two different types of photochemical initiation based on transition metal carbonyls in conjunction with a coinitiator were proposed [18]. Both systems require a "coinitiator." In the case of Type 1 initiation, the coinitiator is an organic halide, whereas Type 2 initiation is effective with a suitable olefin or acetylene.

Type 1 Initiation. The basic reaction (2), described for the metal carbonyl-initiating system, may occur thermally and photochemically. Among all the transition metal derivatives studied, manganese and rhenium carbonyls [$Mn_2(CO)_{10}$ and $Re_2(CO)_{10}$, respectively], which absorb light at rather long wavelengths, are the most inconvenient derivatives for the photoinitiation. The initiating systems $Mn_2(CO)_{10}$/CCl_4 and $Re_2(CO)_{10}$/CCl_4 were first studied by Bamford et al. [19, 20]. The principal radical-generating reaction is an electron transfer from transition metal to halide, the former assuming a low oxidation state (presumably the zero state). Whether electronically excited metal carbonyl compounds can react directly with halogen compounds has not been determined. From flash photolysis studies, it was inferred that electronically excited $Mn_2(CO)_{10}$ decomposes in cyclohexane or *n*-heptane via two routes, both being equally important:

$$Mn_2(CO)_{10} \xrightarrow{h\nu} \left[Mn_2(CO)_{10}\right]^* \longrightarrow 2\,Mn(CO)_5 \qquad (5.17)$$

$$\left[Mn_2(CO)_{10}\right]^* \longrightarrow Mn_2(CO)_9 + CO \qquad (5.18)$$

According to Bamford [21, 22], excited manganese carbonyl can react with a "coordinating compound" L in the following way:

$$[Mn_2(CO)_{10}]^* + L \rightarrow Mn(CO)_5\text{—}L\text{—}Mn(CO)_5 \qquad (5.19)$$

$$Mn(CO)_5\text{—}L\text{—}Mn(CO)_5 \rightarrow Mn(CO)_5\text{—}L + Mn(CO)_5 \qquad (5.20)$$

Both products of reaction (5.20) are capable of undergoing dissociative electron transfer with appropriate organic halides. The rate constants, however, are different:

$$Mn(CO)_5 + X_3C\text{—}R \xrightarrow{\text{fast}} Mn(CO)_5X + X_2\dot{C}\text{—}R \qquad (5.21)$$

$$Mn(CO)_5\text{—}L + X_3C\text{—}R \xrightarrow{\text{slow}} Mn(CO)_5X + X_2\dot{C}\text{—}R + L \qquad (5.22)$$

A similar mechanism [23] could hold for $Re_2(CO)_{10}$; in this case, the vinyl monomer used in the system may function as the coordinating compound L. In all these systems studied, the rate of radical generation strongly depends on the halide concentration. Apparently, no initiation occurs when no halide is present. With increasing halide concentration, the rate increases and reaches a plateau value such that the rate is not affected by halide concentration. For practical applications, it was advised to use minimum halide concentration at the plateau condition. The reactivity of halides increases with multiple substitution in the order $CH_3Cl < CH_2Cl_2 < CHCl_3 < CCl_4$ and with introduction of electron-withdrawing groups. Bromine compounds are much more reactive than the corresponding chlorine compounds and saturated F and I compounds are ineffective.

The metal carbonyl photoinitiating system has been successfully applied to the block copolymer synthesis [18]. In this case, prepolymers having terminal halide groups are irradiated in the presence of $Mn_2(CO)_{10}$ to generate initiating polymeric radicals:

$$2\sim\sim\sim CX_3 + Mn_2(CO)_{10} \xrightarrow{h\nu} 2\sim\sim\sim \dot{C}X_2 + 2Mn(CO)_5X \qquad (5.23)$$

Notably, low-molar-mass radicals are not formed and homopolymerization cannot occur. Moreover, metal atoms do not become bound to the polymer in these processes. Polymeric initiators with terminal halide groups can be prepared in different ways. Anionic polymerization [24, 25], group transfer polymerization [26], metal carbonyl initiation [27], chain transfer reaction [28], condensation reactions [29], and functional initiator [30] approaches have been successfully applied for halide functionalization and a wide range of block copolymers were prepared from the obtained polymers by using metal carbonyl photoinitiation. A similar approach [18] to obtain graft copolymers involves the use of polymer possessing side chains with photo-active halide groups:

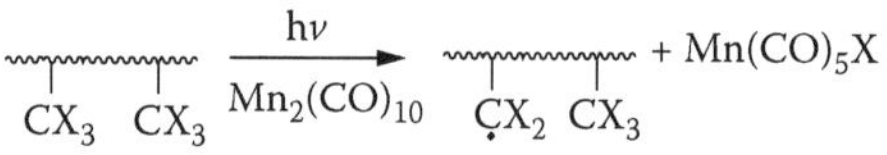

$$\underset{CX_3\quad CX_3}{\sim\sim\sim\sim} \xrightarrow[Mn_2(CO)_{10}]{h\nu} \underset{\dot{C}X_2\quad CX_3}{\sim\sim\sim\sim} + Mn(CO)_5X \qquad (5.24)$$

The grafting reaction leads to the synthesis of a network if combination of macroradicals is the predominant termination route. Network formation versus grafting of branches onto trunk polymers has been intensively studied using poly(vinyl trichloroacetate) as the trunk polymer. Styrene, methyl methacrylate, and chloroprene were grafted onto various polymers, including biopolymers [31–34]. These examples illustrate the broad versatility of the method. Actually, any blocking and grafting reactions by using this method appear feasible, provided suitable halide-containing polymers are available. In this connection, the reader's attention is also directed to previous reviews devoted to photoblocking and photografting [18, 35].

Type 2 Initiation. Bamford and Mullik [36] reported that pure tetrafluoroethylene is polymerized at –93°C upon irradiation in the presence of a low concentration of $Mn_2(CO)_{10}$ or $Re_2(CO)_{10}$. On the basis of this observation, other common vinyl monomers such as styrene and methyl methacrylate were photopolymerized at ambient temperatures with the systems $Mn_2(CO)_{10}/C_2F_4$ and $Re_2(CO)_{10}C_2F_4$. This method was also used for cross-linking and surface grafting [37, 38]. The polymers obtained this way possess metal atoms, as illustrated next for polymethylmethacrylate.

$$(CO)_5Mn-CF_2CF_2CH_2\overset{\displaystyle CH3}{\underset{\displaystyle COOCH_3}{\overset{|}{\underset{|}{C}}}}\sim\sim$$

(I)

$$(CO)_5Re-\overset{\displaystyle R_1}{\underset{\displaystyle R_2}{\overset{|}{\underset{|}{C}}}}-\overset{\displaystyle R_3}{\underset{\displaystyle R_4}{\overset{|}{\underset{|}{C}}}}-CH_2\overset{\displaystyle CH_3}{\underset{\displaystyle COOCH_3}{\overset{|}{\underset{|}{C}}}}\sim\sim$$

(II)

5.2.2 Initiation by Metal Complexes

Chelate complexes of certain transition metal ions can initiate free-radical polymerization of vinyl monomers.

Some of the important systems, such as Cu II-acetylacetonate in dimethylsulfoxide for polymerization of methyl methacrylate [39]; Cu II–chitosan for methyl methacrylate and acrylonitrile polymerization [40]; Cu II–(vinylamino–vinylacetamide) copolymer [41], Cu II–(α, ω-diaminoalkane) [42], and Cu II–imidazole [43] for acrylonitrile polymerization; and Cu II–amine in CCl_4 for acrylonitrile and methyl methacrylate polymerization [44]; Mn III–acetylacetonate for vinyl chloride polymerization [45]; Mn III, co III, and Fe III–acetylacetonate for methyl methacrylate polymerization [46]; Ni II–bis(acetylacetonate)–($Et_3Al_2Cl_3$) for isoprene polymerization [47]; vanadyl acetylacetonate–tributyl borane for methyl methacrylate polymerization [48]; Cu II–polyvinylamine for acrylonitrile and methyl methacrylate polymerization [49, 50]; and Cu II–acetylacetonate with ammonium trichloroacetate [51], Cu II–bis-ephedrine [52] in CCl_4, and Mn III–acetylacetonate [53] for the polymerization of various vinyl monomers, have already been reported.

Some of the copolymerization reactions using metal complexes were also a subject of interest for various groups of workers [54–56]. All the investigators predicted the initiation process to be essentially the scission of a ligand as free radical, with the reduction of the metal to a lower valency state. The reduction of the metal ion was confirmed by spectral and ESR measurements. It has also been illustrated that the ability of the metal chelates for the polymerization of vinyl monomers could be enhanced by the addition of various foreign substrates, particularly halogen-containing compounds [57–60] and compounds of electron-donating [61–63] or electron-accepting [53] properties. In the majority of cases reported so far, polymerization proceeded through typical radical processes.

Allen [64] reported vinyl polymerization using ammonium trichloroacetate and bis-acetylacetonate–Cu II. On the basis of the result at 80°C proposed by Bamford et al. [51] that when ammonium trichloroacetate was not in excess, the actual initiation was a 1:1 complex of two components decomposing to give the trichloromethyl radical by an internal electron-exchange reaction:

$$Cu^{II}Acac_2 + CCl_3COONH_3 \underset{}{\overset{k}{\rightleftharpoons}} I \xrightarrow{k_d} \dot{C}Cl_3 + CO_2 + II$$

where II is an unspecified CuI complex. A possible structure for formula (I) was suggested :

$$\mathrm{O{\cdots}C({\cdots}O){-}CCl_3Cu^{II}} \qquad (5.25)$$

The trichloromethyl radical was the only initiating radical proved by other workers [52, 65].

Uehara et al. [66] also reported the polymerization of methyl methacrylate initiated by bis (acetylacetonato) meta (II) and chloral, where the metal M is either Mn(II) or CO(II). It can be frequently seen that the activity of a metal complex as an initiator of radical polymerization increases in the coexistence of an organic halide. This effect was attributed to the redox reaction between the metal complex and the organic halide [67–69]. The mechanism may be presented as

$$\underset{\text{(I)}}{\mathrm{CCl_3CHO + M(acac)_2}} \rightleftharpoons \text{Complex} \longrightarrow \underset{\text{II}}{\mathrm{H_3C{-}C({=}O){-}CH{<}[C^{\bullet}(CH_3){-}O{-}M^{III}(acac){-}O{-}CH(CCl_3)]}} \qquad (5.26)$$

$$\text{II + MMA} \longrightarrow \mathrm{H_3C{-}C({=}O){-}C{<}[{=}C(CH_3){-}O{-}M^{III}(acac){-}O{-}CH(CCl_3)]} + \underset{\text{III}}{\mathrm{CH_3{-}\dot{C}(CH_3){-}C({=}O){-}OCH_3}} \qquad (5.27)$$

$$\text{III} + \text{MMA} \longrightarrow \text{propagation} \qquad (5.28)$$

The addition of chloral to the Co(II) complex indicates the transformation from octahedral to tetrahedral symmetry [70], supporting the formation of complex I.

In the polymerization of acrylonitrile of Mn(acac)$_3$, the initiation mechanism is considered to occur through the homolytic fission of the metal–oxygen bonds, as pointed out by Arnett and Mendelsohn [71]. This mechanism is also supported by Bamford and Lind [72].

The first step is the formation of activated species (I) in equilibrium with Mn(acac)$_3$. On reaction with the monomer, it yields the radical that initiates the polymerization. The reaction scheme is as:

$$\mathrm{(acac)_2Mn{<}[O{-}C(CH_3){=}CH{-}C(CH_3){=}O{\rightarrow}]} \xrightleftharpoons{K} \mathrm{(acac)_2Mn{<}[O{\cdots}C(CH_3){=}CH{-}C(CH_3){=}O{\rightarrow}]} \qquad (5.29)$$

$$I + CH_2{=}\overset{|}{\underset{|}{C}} \xrightarrow{k_i} Mn^{III}(acac)_2 + \begin{matrix} O=C \\ \quad\;\; \backslash \\ \quad\quad CH{-}CH_2{-}\overset{|}{\underset{|}{C}}^{\bullet} \\ \quad\;\; / \\ O=C \end{matrix} \tag{5.30}$$

$$R^{\bullet} + M \xrightarrow{k_i'} RM_1^{\bullet} \tag{5.31}$$

$$RM_1^{\bullet} + M \xrightarrow{k_p} RM_2^{\bullet} \tag{5.32}$$

$$RM_{n-1} + M \xrightarrow{k_p} RM_n^{\bullet} \tag{5.33}$$

$$RM_n^{\bullet} + RM_m^{\bullet} \xrightarrow{k_t} \text{dead polymer} \tag{5.34}$$

$$RM_n^{\bullet} + \text{complex} \xrightarrow{k_{t'}} \text{dead polymer} \tag{5.35}$$

5.2.3 Initiation by Metal Ions

5.2.3.1 Manganese(III)

The oxidation of various organic substrates using Mn^{3+} for the initiation of free-radical polymerization has been extensively studied [73–90]. In almost all the systems studied, initiating radicals are postulated to be formed from the decomposition of the complex between Mn^{3+} and organic substrate as depicted in Eqs. (5.36)–(5.42). This mechanism also considers the mutual termination of growing radicals.

$$Mn^{3+} + \text{substrate} \rightleftharpoons \text{complex} \tag{5.36}$$

$$\text{Complex} \rightarrow R^{\bullet} + Mn^{2+} + H^{+} \tag{5.37}$$

$$R^{\bullet} + Mn^{3+} \rightarrow Mn^{2+} + \text{product} \tag{5.38}$$

$$R^{\bullet} + M \rightarrow R{-}M_1^{\bullet} \tag{5.39}$$

$$R{-}M_1^{\bullet} + M \rightarrow R{-}M_2^{\bullet} \tag{5.40}$$

$$R{-}M_n^{\bullet}\ R{-}M_m^{\bullet} \rightarrow R{-}M_n + R{-}M_m \tag{5.41}$$

$$R{-}M_n^{\bullet} + Mn^{3+} \rightarrow R{-}M_n + Mn^{2+} + H^{+} \tag{5.42}$$

In the polymerization of acrylonitrile [77] in which organic acids were employed as the reducing agent, the order of reactivity of the acids has been found to be

citric > tartaric> ascorbic> oxalic> succinic> glutaric> adipic

Similar studies concerning the comparison of alcohol reactivity for the polymerization of methyl methacrylate were also performed by Nayak and co-workers [83].

The order of the reactivity of alcohols for the Mn^{3+}—alcohol redox couple was found to be in the following order:

1-propanol > glycerol > ethylene glycol > isobutyl alcohol

> 1-butanol > 1,2-propanediol > cycloheptanol > cyclohexanol

> cyclopentanol

Polymerization of methyl methacrylate with the $Mn(OH)_3$–hydrazine systems was investigated by two independent groups. Bond et al. [84] found that the rate of polymerization at constant $Mn(OH)_3$ concentration was independent of the monomer concentration and varied with the pH and temperature. Rehmann and Brown [85, 86] applied the same system to the emulsion polymerization of methyl methacrylate and reported that the rate of polymerization was proportional to surface area of $Mn(OH)_3$ formed in the system.

The redox polymerization of acrylonitrile initiated by dimethylsufloxide–Mn^{3+} in H_2SO_4 and $HC1O_4$ was investigated by Devi and Mahadevan [87–89]. Trivalent manganese forms a complex with dimethyl sulfoxide, followed by a reversible electron transfer. The radicals formed from the dissociation radical ion initiates the polymerization:

$$Mn^{3+} + (CH_3)_2SO \rightarrow \text{Complex} \rightarrow (CH_3)_2\overset{+}{S}—\dot{O} \tag{5.43}$$

$$(CH_3)_2\overset{+}{S}—\dot{O} + H_2O \rightarrow (CH_3)_2SO + \dot{O}H + H^+ \tag{5.44}$$

The rate of polymerization varied directly with the dimethylsufloxide concentration and was proportional to the square of the monomer and independent of the oxidant. They also investigated the polymerization of methyl methacrylate with Mn^{3+} and reducing agents such as dimethyl sulfoxide, diacetone alcohol, and malonic acid. All the reducing agents formed the complexes of varying stability with Mn^{3+}, from which initiating species are produced.

The efficiency of the cyclohexanone/Mn^{3+} redox system for the polymerization of acrylonitrile and methyl methacrylate in perchloric acid and sulfuric acid media was investigated [81]. It was found that the rate of polymerization was independent of the oxidant concentration and varied linearly with the monomer and reducing agent concentration. Complex formation and termination mechanism was found to be different in two acidic media. In perchloric media, the termination is effected by the oxidant, whereas in sulfuric acid, primary radicals terminate growing chains.

Drummund and Waters [90] employed various organic substrates as reducing agents in Mn^{3+} pyrophosphate-based redox systems.

An interesting variation of the Mn^{3+}/organic substrate redox method applies to grafting methyl methacrylate onto cellulose and polyvinylalcohol [91, 92]. The method has also been applied to graft vinyl monomers onto collagen [93]. Cakmak[94] described the use of manganese acetate redox system for block copolymerization of acrylonitrile with polyacrylamide. In this case, terminal carboxylic groups incorporated to polyacrylamide acted as the reducing agent.

5.2.3.2 Cerium(IV)

For initiating vinyl polymerizations, Ce^{4+} ions alone [95–105] or in conjunction with suitable reducing agents, which include formaldehyde [106, 107], malonic acid [108], dextrin [109], dimethyl formamide [110], starch [111], pinacol [112–115], amines [116, 117], alcohols [118–124, 132–135], carboxylic acids [142–144], amino acids [150, 151], thiourea [125–127], acetophenone [128], thiomalic acid [129], and 2-mercaptoethanol [131], may be used.

Pramanick and Sankar [99] investigated the polymerization of methyl methacrylate polymerization initiated by only ceric ions and found that the mechanism of initiation depends strongly on the acidity of the medium and is independent of the nature of anion associated with the ceric ion. In a moderately acidic medium, the primary reaction is the formation of hydroxyl radical by ceric-ion oxidation of water. When ceric sulfate is used, the hydroxyl radicals initiate the polymerization and appear as end groups in the polymer molecule. If, on the other hand, ceric ammonium sulfate or a mixture of ceric sulfate and ammonium sulfate are used, some of the hydroxyl radicals react with the ammonium ion, producing ammonium radicals, and both radicals act as initiators, giving polymers with both hydroxyl and amino end groups. In the polymerization of acrylamide by ceric salt, the infrared (IR) spectra suggests the formation of monomer–ceric salt complexes in aqueous solution [98]. This coordination bond presumably consists of both σ- and π-type bonds. It is likely that for acrylamide, the reaction mechanism is not the redox type, but based on complex formation.

Various alcohols such as benzyl alcohol [24], ethanol [120], ethylene glycol [119], and 3-chloro-1 propanol [132] have been employed with ceric ions to form redox systems for homopolymerization or graft copolymerization. Regarding block copolymerization [36], the alcohol used is generally a dialcohol or multifunctional oligomeric or even a high-molecular-weight alcohol. A typical mechanism, based on the oxidation of a special, azo-containing polyethyleneglycol, is illustrated in Eq. (5.45) [133–135].

$$\mathrm{HO}\!\left[\mathrm{CH_2\text{-}CH_2\text{-}O}\right]_n\!\mathrm{-\overset{O}{\overset{\|}{C}}-\underset{CH_3}{\overset{CH_3}{C}}-N{=}N-\underset{CH_3}{\overset{CH_3}{C}}-\overset{O}{\overset{\|}{C}}}\!\left[\mathrm{O\text{-}CH_2\text{-}CH_2}\right]_n\!\mathrm{OH} + \mathrm{Ce(IV)} \tag{5.45}$$

$$\downarrow$$

$$\dot{\mathrm{O}}\!\left[\mathrm{CH_2\text{-}CH_2\text{-}O}\right]_n\!\mathrm{-\overset{O}{\overset{\|}{C}}-\underset{CH_3}{\overset{CH_3}{C}}-N{=}N-\underset{CH_3}{\overset{CH_3}{C}}-\overset{O}{\overset{\|}{C}}}\!\left[\mathrm{O\text{-}CH_2\text{-}CH_2}\right]_n\!\dot{\mathrm{O}} + \mathrm{Ce(III)} + \mathrm{H^+}$$

$$\Big\downarrow\ m\ \mathrm{CH_2{=}\underset{CONH_2}{CH}}$$

$$\left[\mathrm{\underset{CONH_2}{CH}\text{-}CH_2}\right]_m\!\mathrm{O}\!\left[\mathrm{CH_2\text{-}CH_2\text{-}O}\right]_n\!\mathrm{-\overset{O}{\overset{\|}{C}}-\underset{CH_3}{\overset{CH_3}{C}}-N{=}N-\underset{CH_3}{\overset{CH_3}{C}}-\overset{O}{\overset{\|}{C}}}\!\left[\mathrm{O\text{-}CH_2\text{-}CH_2}\right]_n\!\mathrm{O}\!\left[\mathrm{CH_2\text{-}\underset{CONH_2}{CH}}\right]_m \tag{5.46}$$

The mechanism depicted in Eqs. (5.45) and (5.46) involves the production of one proton and of oxygen-centered radicals, which initiate vinyl polymerization in the presence of the monomer. As a result, a polymer with one central azo bond is formed. When heated in the presence of a second monomer, this macro-initiator is split at the azo side, giving rise to two initiating macro-radicals (Eq. (5.46)). The final result is tailor-made multiblock copolymers [136]. Other hydroxyl-groups containing high-molecular-weight compounds used in conjunction with Ce(IV) salts include methyl cellulose and methyl hydroxypropyl cellulose [137].

In addition to alcohols, pyrroles have also been found to be suitable activators in the Ce(IV)-initiated polymerization. In some recent work, polypyrrole was synthesized by an oxidation of pyrrole with Ce(IV) [138, 139].

The polymerization of acrylonitrile [116] by ceric ions was found to be accelerated by secondary and tertiary amines, but not by primary amines. This phenomenon may be because the acceleration is due to a redox reaction between ceric ions and amines and, therefore, depends on the electron-donating ability of the substituents. The order of reactivity of amines is triethanolamine > triethylamine > diethanolamine > diethylamine. Pramanick [117] polymerized methyl methacrylate in the presence of $Ce(ClO_4)_4$ and monoamines and reported the formation of poly(methyl methacrylate) containing amine end groups. With ethanolamines, products containing reactive OH groups were obtained.

Various amino acids, such as serine, glucine, or phenylalanine, have been employed in conjunction with Ce(IV) for the radical polymerization of acrylamide [140, 141]. Polymerizations were conducted in sulfuric acid solution. It was found that the resulting polymers contained carboxylic end groups. The mechanism of initiation is illustrated in the example of phenylalanine:

$$C_6H_5{-}CH_2{-}CH(COOH)(NH_2) + Ce^{4+} \longrightarrow C_6H_5{-}CH_2{-}\dot{C}(COOH)(NH_2) + Ce^{3+} + H^+ \quad (5.47)$$

$$\xrightarrow{n\,CH_2{=}CH{-}C(=O){-}NH_2} \text{Polymerization}$$

In the polymerization of acrylamide and of acrylonitrile, carboxylic acids also have been used in conjunction with Ce(IV) in diluted sulfuric acid solution [150–154]. The carboxylic acids that turned out to be useful in this respect include malonic acid, tartaric acid, and citric acid. In all cases, the polymers were found to be equipped with carboxylic end groups. In one work [145], polymerization and electrolysis were performed simultaneously. This allows Ce(III) to be converted to Ce(IV) in the course of polymerization. The highest polymerization rates were obtained when stainless-steel electrodes were used for Ce(III) oxidation.

Another system for the polymerization of acrylamide are chelating polyaminocarboxylic acids with Ce(IV) [146–148]. In these systems, the redox reaction is followed by a decarboxylation to yield the initiating carbon-centered radical. It was found that diethylenetriamine pentaacetic acid (DTPA) is slightly more effective than ethylenediamine tetraacetic acid (EDTA). The efficiency of nitrilotriacetic acid (NDA) (see Eq. (5.48)) is smaller than that of EDTA.

$$\text{HCOO}-\text{CH}_2-\underset{\underset{\text{O}=\text{C}-\text{OH}}{|}}{\underset{\text{CH}_2}{|}}{\overset{\overset{\text{COOH}}{|}}{\overset{\text{CH}_2}{|}}{\text{N}}} + \text{Ce}^{4+} \longrightarrow \text{HCOO}-\text{CH}_2-\underset{\dot{\text{C}}\text{H}_2}{\overset{\overset{\text{COOH}}{|}}{\overset{\text{CH}_2}{|}}{\text{N}}} + \text{CO}_2 + \text{H}^+ + \text{Ce}^{3+} \xrightarrow{n\,\text{CH}_2=\text{CH}-\text{C(=O)NH}_2} \text{Polymerization} \quad (5.48)$$

5.2.3.3 Vanadium(V)

Vanadium(V) in the presence of various organic reducing agents has been used as an effective initiator in the polymerization of vinyl monomers [149]. In this redox system, the initiating radicals are also generated from the reducing agent by the decomposition of the intermediate complex formed between oxidant and reductant. Vanadium(V) with a large number of organic substrates, namely cyclohexanol [150–152], lactic [152] and tartaric acid [153], cyclohexanone [154], cyclohexane [155], ethylene glycol [156], thiourea, ethylene thiourea [157–159], and propane 1,2-diol [160], has been used in free-radical polymerization processes.

Based on the systematic investigation [156] of the V^{5+}/alcohol redox system for the polymerization of acrylonitrile, the order of the activity of the alcoholic compounds was found to be

ethane 1,2-diol > propane 1,3-diol > cyclohexanol > butane 1,4-diol

> pinacol > 1-butanol > iso-propyl alcohol > sec-butyl alcohol

A vanadium (V)-based redox system has been applied to grafting of vinyl monomers onto various polymeric substrates (Table 5.1). Besides graft copolymers, homopolymers were also formed.

TABLE 5.1
Grafting of Vinyl Monomers onto Polymers by Using Vanadium(V)-Based Redox System

Monomer	Trunk Polymer	Reference
Acrylonitrile	Collagen	[161]
Vinyl acetate	Collagen	[161]
Methyl methacrylate	Cellulose	[162]
Acrylonitrile	Dialdehyde cellulose	[163]
Acrylonitrile	Polymethacrolein	[163]
Vinyl pyridine	Polyacrylamide	[163]
Vinyl monomers	Wool	[159]

5.2.3.4 Cobalt(III)

Cobalt (II) invariably exists as an octahedrally coordinated ion, and has *d* electrons which can become involved both in electron transfer reaction and ligand bonding [164]. The powerful oxidizing capacity of trivalent cobalt has been shown by several investigators [165–183]. A wide variety of organic compounds — aromatic as well as aliphatic aldehydes, olefins, ketones, hydrocarbons, and alcohols — have been found to be susceptible to oxidation by cobaltic ions, and the kinetics of these reactions have been reported in detail. That Co^{3+} could initiate the vinyl polymerization was suggested by Baxendale et al. [184]. Later, Santappa and co-workers [185–187] investigated the polymerization of methyl methacrylate, methyl acrylate, acrylonitrile, and acrylamide initiated by a redox system involving Co^{3+}. From the experimental results, the following general mechanism was proposed:

$$Co^{3+} + M \rightarrow Co^{2+} + R^{\bullet} \tag{5.49}$$

$$Co(OH)^{2+} + M \rightarrow Co(OH)^{+} + R^{\bullet} \tag{5.50}$$

$$R^{\bullet} + M \rightarrow R—M^{\bullet} \tag{5.51}$$

$$R—M^{\bullet} + (n-1)M \rightarrow R—M_n^{\bullet} \tag{5.52}$$

$$R—M_n^{\bullet} + Co^{3+} \rightarrow R—M_n + Co^{2+} + H^{+} \tag{5.53}$$

$$R—M_n^{\bullet} + Co(OH)^{2+} \rightarrow R—M_n + Co(OH)^{+} + H^{+} \tag{5.54}$$

Notably, cobaltic ions participate in both initiation and termination processes. These authors [188] also investigated the polymerization of methyl methacrylate initiated by Co^{3+}/*tert*-butyl alcohol and found that the redox system is operative only at high concentration. The cobaltous chloride/dimethyl aniline redox system for the polymerization of acrylamide was also reported [189].

The aqueous polymerization of methyl methacrylate initiated by the potassium trioxalate cobaltate (II) complex was studied by Guha and Palit [190]. At a relatively higher concentration (>0.001 mol L^{-1}), this compound can initiate aqueous polymerization of methyl methacrylate in the dark at room temperature. The complex is highly photosensitive, which can photoinitiate polymerization. A detailed end-group analysis of the obtained polymers indicated that carboxyl and hydroxyl radicals, which are from the decomposition of the photoexcited complex, are the initiating species.

5.2.3.5 Chromium(VI)

Chromic acid is one of the most versatile oxidizing agents [191]. Viswanathan and Santappa [192] investigated chromic acid/reducing agent (*n*-butanol, ethylene glycol, cyclohexanone, and acetaldehyde) initiated polymerization of acrylonitrile. These authors [193] also observed that the percentage of conversion to polymer was more with acrylonitrile monomer and much less with monomers such as methyl acrylate and acrylamide under similar experimental conditions. This difference of reactivities of monomers could be explained by assuming that Cr^{6+} species terminated the chain radicals more effectively with respect to the latter monomers than with polyacrylonitrile radicals.

The Cr^{6+}/1-propanol, Cr^{6+}/1,2-propane diol, Cr^{6+}/phenyl *tert*-butyl alcohol, Cr^{6+}/thiourea, and Cr^{6+}/ethylene thiourea systems have been studied in the polymerization of acrylonitrile [194, 195]. These studies furnished information on polymerization kinetics and the general mechanism of chromic acid oxidations. The mechanism involves the formation of unstable species such as Cr^{6+} and Cr^{5+}.

The following reaction scheme involving the initiation by Cr^{4+} or $R^{\bullet}$ and termination by Cr^{6+}, which was in line with the experimental results, was proposed:

$$HCrO_4^- + R + 2H^+ \rightarrow Cr^{4+} + \text{product} \tag{5.55}$$

$$R + Cr^{4+} \rightarrow R^{\bullet}Cr^{3+} + H^+ \tag{5.56}$$

$$R^{\bullet} + Cr^{6+} \rightarrow Cr^{5+} + \text{product} \tag{5.57}$$

$$R + Cr^{5+} \rightarrow Cr^{3+} + \text{product} \tag{5.58}$$

$$R^{\bullet} + M \rightarrow R\text{—}M^{\bullet} \tag{5.59}$$

$$Cr^{4+} + M \rightarrow M^{\bullet} + Cr^{3+} + H^+ \tag{5.60}$$

$$R\text{—}M_n^{\bullet} + Cr^{6+} \rightarrow R\text{—}M_n + Cr^{5+} + H^+ \tag{5.61}$$

Potassium chromate in conjunction with a variety of reducing agents was used to initiate emulsion copolymerization of styrene and butadiene [196]. Arsenic oxide was fuond to be the most powerful reducing agent. Here, again, the formation of unstable species Cr^{6+} and Cr^{5+} was responsible for the initiation.

$$Cr^{6+} + As^{3+} \rightarrow Cr^{4+} + As^{5+} \tag{5.62}$$

$$Cr^{4+} + M \rightarrow M^{\bullet} + Cr^{3+} + H^+ \tag{5.63}$$

The Cr_2O_3/$NaHSO_3$ redox system for the aqueous polymerization of methyl methacrylate was also described [197]. Nayak et al. [198] reported grafting methyl methacrylate onto wool by using a hexavalent chromium ion. In this case, macroradicals were produced by reaction of Cr^{6+} with wool in the presence of perchloric acid.

5.2.3.6 Copper(II)

The Cu^{2+}/potassium disulfide [199] and Cu^{2+}/metabisulfide [200] redox systems have been used in the polymerization of acrylonitrile and acrylamide, respectively. The cupric sulfate-hydrazine redox system in which hydrazyl radicals are responsible for the initiation was studied in the absence and presence of molecular oxygen. The Cu^{2-}/hydrazine hydrate [201–204] and Cu^{2+}/2-aminoethanol [205] systems were used for the polymerization of vinyl monomers. Misra [206] demonstrated that the polymerization of acrylamide could be initiated by the Cu^{2+} /metabisulfide redox system. Initiating systems of cupric(II) ions in conjunction with dimethyl aniline [207] and α-amylase [208] were also reported. Cupric-ion-based redox reactions were successfully applied [209] to graft vinyl monomers onto wool and Nylon-6.

5.2.3.7 Iron(III)

Bamford et al. [210] and Bengough and Ross [211] reported that ferric salt acts as an electron transfer agent. Cavell et al. [212] showed that the rate of polymerization is proportional to the reciprocal of the concentration of the ferric salt. The role of ferric salt in the polymerization of acrylamide initiated by ceric salt was studied by Narita et al. [113].

The polymerization of methyl methacrylate in acidic solution by iron metal was reported earlier [213]. Saha and co-workers [214] studied the mechanism of methyl methacrylate in the presence of ferric chloride. They proposed that the hydroxyl radical formed by the chemical decomposition of the system containing ferric salt is the active species for initiating polymerization. Narita et al. reported [215] the polymerization of acrylamide initiated by ferric nitrate and suggested that a complex of monomer and metallic salt generates an active monomer radical capable of initiating vinyl polymerization.

The reaction between Fe^{3+} and monomercaptides was studied extensively [216–218]. It was shown that complexes formed between Fe^{3+} and monomercaptides such as thioglycolate or cysteinate invariably undergo redox reaction in which the monomercaptides oxidized to disulfide, and Fe^{3+} is reduced to Fe^{2+}. The formation of the intermediate thiol radical by the interaction of iron(III) with mercaptans, which can initiate vinyl polymerization, was reported by Wallace [219]. The Fe^{3+}/thiourea redox pair was investigated for the initiation of polymerization of methyl methacrylate, styrene, and acrylonitrile by several research groups [220–227]. In general, the initiating species is formed by the abstraction of the hydrogen atom of the –SH group of the isothiourea form in the presence of the ferric ion. It was also found that the rate of polymerization was effected by the substitution of the amino group of the thiourea.

Brown and Longbottom [228] reported the redox system of hydrazine and ferric ammonium sulfate for the polymerization of methyl methacrylate. *N*-halsoamines in conjunction with Fe^{2+} were found [229] to be efficient redox initiators for the polymerization of methyl methacrylate. Amino radicals formed according to the following reaction initiate the polymerization:

$$R_2NCl + Fe^{2+} \rightarrow R_2N^{\bullet} + [FeCl]^{2+} \tag{5.64}$$

The trimethyl amine oxide/Fe^{2+} system in aqueous medium initiates the polymerization of methyl methacrylate in a similar electron transfer process [230]. Interestingly, acrylonitrile and acrylamide were not polymerizable with this system. On the other hand, acrylamide was polymerized by iron(III) with bisulfite [231] and 4,4′-azobis(cyanopentanoic acid) [232] redox couples.

Narita et al. [233] investigated the polymerization of methyl methacrylate in the presence of ferric nitrate. The ferric nitrate in dilute solutions was found to initiate the polymerization. At a comparatively higher concentration, the ferric salt reacts as an electron transfer agent, and the rate of polymerization is decreased with increasing concentration. The following reaction mechanism was proposed:

$$Fe^{3+} + M \rightleftharpoons Fe^{2+} + R^{\bullet} + H^{+} \tag{5.65}$$

Fe^{3+}-induced redox reactions were used in grafting methyl methacrylate onto cellulose [234] and acrylonitrile and acrylamide onto polyamides such as Nylon 6,6 and 6,10 [235].

5.2.4 Initiation by Permanganate-Containing Systems

The permanganate ion is known [236] to be a versatile oxidizing agent, because of its ability to react with almost all types of organic substrates. Its reaction is most interesting because of the several oxidation states to which it can be reduced, the fate of manganese ion being largely determined by the reaction conditions; in particular, the acidity of the medium. Considerable work has been done in elucidating the mechanism of permanganate oxidations of both organic and inorganic substrates and many of these are well understood. The permanganate ion coupled with simple water-soluble organic compounds acts as an efficient redox system for the initiation of vinyl polymerization.

Palit and co-workers [237, 238] used a large number of redox initiators containing permanganate as the oxidizing agent. The reducing agents are oxalic acid, citric acid, tartaric acid, isobutyric acid, glycerol, bisulfite (in the presence of dilute H_2SO_4), hydrosulfite (in the presence of dilute H_2SO_4), and so forth. The peculiarity of the permanganate system is that two consecutive redox systems exist in the presence of monomer:

1. The monomer (reductant) and permanganate (oxidant)
2. Added reducing agent (reductant) and separated manganese dioxide (oxidant)

Konar and Palit [239] studied the aqueous polymerization of acrylonitrile and methyl methacrylate initiated by the permanganate oxalic acid redox system. The rate of polymerization is independent over a small range. The molecular weight of polymers is independent of oxalic acid concentration in the range where the rate of polymerization is independent of the oxalic acid concentration. However, the molecular weight decreased at a higher concentration of oxalic acid with an increasing concentration of catalyst and temperature. The addition of salts, such as Na_2SO_4, and complexing agents, such as fluoride ions and ethylene diaminetetracetic acid, decreased the rate of polymerization, whereas the addition of detergents and salts, such as $MnSO_4$, at low concentrations increased the rate.

Weiss [240] reported the activation of oxalic acid and observed that it acquires an increased reducing power when treated with an insufficient amount of an oxidizing agent ($KMnO_4$). The action of $KMnO_4$ on oxalic acid at room temperature is a relatively slow process that occurs in steps. A possible mechanism was given by Launer and Yost [241] for the generation of carboxyl radicals ($C_2O_4^{\cdot-}$ or $COO^{\cdot-}$), which appear to be the initiating radicals in this system:

$$Mn^{4+} + C_2O_4^{2-} \xrightarrow{\text{measurable}} Mn^{3+} + CO_2 + {}^{\cdot}COO^{-} \tag{5.66}$$

$$Mn^{4+} + {}^{\cdot}COO^{-} \xrightarrow{\text{rapid}} Mn^{3+} + CO_2 \tag{5.67}$$

$$Mn^{3+} + 2C_2O_4^{2-} \xrightarrow{\text{rapid}} [Mn(C_2O_4)_2]^{-} \tag{5.68}$$

$$Mn^{3+} + C_2O_4^{2-} \xrightarrow{\text{measurable}} Mn^{2+} + {}^{\cdot}COO^{-} + CO_2 \tag{5.69}$$

$$Mn^{3+} + {}^{\cdot}COO^{-} \xrightarrow{\text{rapid}} Mn^{2+} + CO_2 \tag{5.70}$$

Weiss [240] suggested that the continuous production of active oxalic acid ion radical ($C_2O_4^{\cdot-}$) in this system is governed by the reaction

$$Mn^{3+} + C_2O_4^{2-} \rightarrow \dot{C}_2O_4^{-} + Mn^{2+} \tag{5.71}$$

At room temperature, this active oxalic acid ion radical has a life of about $\frac{1}{2}$ hr. Therefore, the system behaves in such a manner that the aqueous polymerization caused by the reaction of monomer with carboxyl radicals tends toward its completion within $\frac{1}{2}$ hr or so after initiation.

The aqueous polymerization of acrylic acid, methacrylic acid, acrylamide, and methacrylamide using potassium permanganate coupled with a large number of organic substrates as the reducing agent was studied by Misra et al. [242–246]. The rate of polymerization of acrylic and methacrylic acid initiated by the permanganate/oxalic acid redox system was investigated in the presence of certain natural salts and water-soluble organic solvents, all of which depress the rate of polymerization, whereas Mn^{2+} ions have been found to increase the initial rate but to depress the maximum conversion [242].

The rate of acrylamide polymerization initiated by the permanganate/tartaric acid [245] and the permanganate/citric acid [246] redox system increase with increasing catalyst and monomer concentration. The initial rate increased with increasing temperature, but the conversion decreased beyond 35°C. The addition of neutral salts like $Co(NO_3)_2$ and $Ni(NO_3)_2$, organic solvents, and complexing agents reduced the rate and percentage of conversion. However, the addition of $MnSO_4$ or the injection of more catalyst at intermediate stages increased both initial rate and the maximum conversion.

The redox reaction of tartaric acid and manganic pyrophosphate was studied by Levesley and Waters [247]. They suggested the formation of a cyclic complex between the two components that dissociate with loss of carbon dioxide and formation of free-radical $R\dot{C}H(OH)$, capable of initiating vinyl polymerization. The distinguishing feature of the permanganate system is that two consecutive redox systems operate in the presence of the monomer (i.e., permanganate [oxidant] and monomer [reductant], separated manganese dioxide [oxidant], and the added reducing agent [reductant]).

In the aqueous polymerization of acrylamide initiated by the permanganate/tartaric acid system, the permanganate first reacts with tartaric acid to generate the highly reactive Mn^{3+} ions and the active free radical, capable of initiating the polymerization. The detailed mechanism of the latter reaction could be presented by Eqs. (5.72)–(5.81):

$$\begin{array}{l} CH(OH)COOH \\ | \\ CH(OH)COOH \end{array} + Mn^{4+} \xrightarrow{slow} \begin{array}{l} \dot{C}H(OH) \\ | \\ CH(OH)COOH \end{array} + Mn^{3+} + CO_2 + H^+ \quad (5.72)$$

$$\begin{array}{l} \dot{C}H(OH) \\ | \\ CH(OH)COOH \end{array} + Mn^{4+} \xrightarrow{fast} \begin{array}{l} CHO \\ | \\ CH(OH)COOH \end{array} + Mn^{3+} + H^+ \quad (5.73)$$

$$2\begin{array}{l} CH(OH)COOH \\ | \\ CH(OH)COOH \end{array} + Mn^{3+} \xrightarrow{rapid} \left[Mn \left\{ \begin{array}{l} \dot{C}H(OH)\ COO \\ | \\ CH(OH)COOH \end{array} \right\}_2 \right] + 4H^+ \quad (5.74)$$

$$\begin{array}{l} COOH \\ | \\ CH(OH) \\ | \\ CH(OH)COOH \end{array} + Mn^{3+} \xrightarrow{fast} \begin{array}{l} \dot{C}HO \\ | \\ CH(OH)COOH \end{array} + Mn^{2+} + H^+ \quad (5.75)$$

$$\begin{array}{l} \dot{C}HO \\ | \\ CH(OH)COOH \end{array} \xrightarrow{Mn^{4+}/Mn^{3+}} \begin{array}{l} COOH \\ | \\ CH(OH)COOH \\ \text{(tartronic acid)} \end{array} \quad (5.76)$$

$$\begin{array}{l} COOH \\ | \\ CH(OH)COOH \end{array} \begin{cases} \xrightarrow[slow]{Mn^{3+}} \underset{B}{\dot{C}H(OH)COOH} + CO_2 + Mn^{2+} + H^+ \\ \xrightarrow[slow]{Mn^{3+}} \begin{array}{l} COOH \\ | \\ \dot{C}(OH)COOH \\ \quad C \end{array} + Mn^{2+} + H^+ \\ \xrightarrow[slow]{Mn^{3+}} \begin{array}{l} COOH \\ | \\ \dot{C}H(OH) \\ \quad D \end{array} + CO_2 + Mn^{2+} + H^+ \end{cases} \quad (5.77)$$

$$\mathrm{B} \xrightarrow[\text{fast}]{Mn^{3+}} \begin{array}{l} CHO \\ | \\ COOH \end{array} + Mn^{2+} + H^{+} \tag{5.78}$$

$$\mathrm{C} \xrightarrow[\text{fast}]{Mn^{3+}} \begin{array}{l} COOH \\ | \\ CHO \end{array} + CO_2 + H^{+} + Mn^{2+} \tag{5.79}$$

$$\mathrm{D} \xrightarrow[\text{fast}]{Mn^{3+}} \begin{array}{l} COOH \\ | \\ CHO \end{array} + H^{+} + Mn^{2+} \tag{5.80}$$

$$\begin{array}{l} CHO \\ | \\ COOH \end{array} + Mn^{3+} \longrightarrow \begin{array}{l} COOH \\ | \\ COOH \end{array} + Mn^{2+} \quad \begin{array}{l} COOH \\ | \\ COOH \end{array} + Mn^{3+} \xrightarrow{\text{slow}} \begin{array}{l} \dot{C}OO \\ | \\ COO^{-} \end{array} + 2H^{+} + Mn^{2+}$$

$$\mathrm{E} \longrightarrow \dot{C}O_2^{-} \underset{\mathrm{F}}{+} CO_2 \tag{5.81}$$

The free radicals A, B, C, EE, and F are all capable of initiating the polymerization of acrylamide.

In the case of acrylamide polymerization initiated by the citric acid/permanganate system, the oxidation of citric acid leads to a keto-dicarboxylic acid, which upon drastic oxidation, transformed into acetone and carbon dioxide [246]. The mechanism of the redox system is as follows:

At low concentration of $KMnO_4$:

$$\underset{\mathrm{I}}{\begin{array}{c} CH_2\text{–}COOH \\ | \\ HOOC\text{–}C\text{–}OH \\ | \\ CH_2\text{–}COOH \end{array}} + Mn^{4+} \xrightarrow{\text{slow}} \underset{\mathrm{II}}{\begin{array}{c} CH_2\text{–}COOH \\ | \\ \dot{C}\text{–}OH \\ | \\ CH_2\text{–}COOH \end{array}} + Mn^{3+} + CO_2 + H^{+} \tag{5.82}$$

$$\mathrm{I} + Mn^{3+} \xrightarrow{\text{slow}} \mathrm{II} + Mn^{2+} + CO_2 + H^{+} \tag{5.83}$$

The free radicals II initiate polymerization and the reaction (Eq. (5.82)) is the main rate-determining step.

At high concentrations of $KMnO_4$:

$$\mathrm{II} + Mn^{3+} \xrightarrow{\text{fast}} \underset{\mathrm{III}}{\begin{array}{c} H_2C\text{-}COOH \\ | \\ C{=}O \\ | \\ H_2C-COOH \end{array}} + Mn^{3+} + H^{+} \tag{5.84}$$

$$II + Mn^{3+} \xrightarrow{\text{fast}} III + Mn^{2+} + H^{+} \tag{5.85}$$

Shukla and Mishra [248] studied the aqueous polymerization of acrylamide initiated by the potassium permanganate/ascorbic acid redox system. Ascorbic acid has been used in a reducing agent with several oxidants (i.e., H_2O_2 [249], $K_2S_2O_8$ [250], and *tert*-butyl peroxybenzoate [251]) to produce free radicals capable of initiating polymerization in the aqueous media. The initial rate of polymerization was proportional to the first power of the oxidant and monomer concentration and independent of ascorbic acid concentration in the lower concentration ranges. At higher concentrations of ascorbic acid, the rate of polymerization and the maximum conversion decreased as the temperature was increased from 20°C to 35°C. The overall activation energy was 10.8 kcal mol^{-1}. The rate of polymerization decreased by the addition of water-miscible organic solvents or salts such as methyl alcohol, ethyl alcohol, isopropyl alcohol, potassium chloride, and sodium sulfate, whereas the rate increased by the addition of Mn^{2+} salts and complexing agents such as NaF.

Permanganate oxidizes ascorbic acid to form threonic acid and oxalic acid as presented next. The permanganate reacts with oxalic acid to produce the .COO– radical, which initiates polymerization.

O=C–(O)–
O=C
O=C
H–C–
HO–C–H
H_2COH

Ascorbic Acid, keto form; Dehydroascorbic Acid

or

O=C
HO–C–H
OH–C–H
H–C–OH
HO–C–H
H_2COH

Ascorbic Acid, hydrated form

$\xrightarrow[\text{H}_2\text{SO}_4\text{, KMnO}_4]{\text{Oxidation}}$

COOH
H–C–OH
OH–C–H
CH_2OH

Threonic Acid

+ COOH–COOH (5.86)

The effect of some additives on aqueous polymerization of acrylamide initiated by the permanganate/oxalic acid redox system was studied by Hussain and Gupta [252]. The rate of polymerization was increased in the presence of alkali metal chlorides. However, the rate was decreased in the presence of cupric chloride and ferric chloride. Anionic and cationic detergents showed a marked influence on the rate of polymerization.

Permanganate based redox systems were used to graft vinyl monomers onto various natural and synthetic polymers (Table 5.2). In these systems, macroradicals were formed by a redox reaction between the manganese(IV) ion and the polymer to be grafted, according to the following general reaction:

~~(H)(H)~~ + Mn^{4+} ⟶ ~~(•)(H)~~ + Mn^{3+} + H^+

↓ Monomer (5.87)

Graft Copolymer

TABLE 5.2
Grafting of Vinyl Monomers onto Polymers by Using Permanganate-Based Redox System

Monomer	Trunk Polymer	Redox System	References
Methyl methacrylate	Silk	$KMnO_4 - H_2SO_4$	[253]
Methyl methacrylate	Silk	$KMnO_4$ – oxalic acid	[254]
Methyl methacrylate	Nylon-6	$KMnO_4$ – various acids	[255, 256]
Acrylonitrile	Nylon-6	$KMnO_4$ – various acids	[255, 256]
Acrylic acid	Nylon-6	$KMnO_4$ – various acids	[255, 256]
Butyl methacrylate	Cellulose	$KMnO_4$	[257]
Acrylonitrile	Starch	$KMnO_4$	[258]

REFERENCES

1. G. S. Mishra and U. D. N. Bajpai, *Prog. Polym. Sci.*, 8, 61 (1982).
2. F. A. Bovey, I. M. Kolthoff, A. J. Medalia, and E. J. Meehan, *Emulsion Polymerization*, Interscience Publishers, New York, 1955.
3. F. Haber and J. Weiss, *Proc. R. Soc. London*, A147, 332 (1934).
4. C. H. Bamford and F. J. T. Fildes, *Am. Chem. Soc., Polym. Prepr.*, 11(2), 927 (1970).
5. C. H. Bamford, *Eur. Polym. J.*, 7 (Suppl. 1) (1969).
6. C. H. Bamford, G. C. Eastmond, and F. J. T. Fildes, *Proc. R. Soc. London*, A326(1567), 431 (1972).
7. C. H. Bamford and I. J. Sakamoto, *J. Chem. Soc., Faraday Trans. I*, 70, 330 (1974).
8. C. H. Bamford and I. J. Sakamoto, *J. Chem. Soc., Faraday Trans. I*, 70, 334 (1974).
9. C. H. Bamford and S. U. Mullik, *J. Chem. Soc., Faraday Trans. I*, 72(10), 2218 (1976).
10. R. J. Angelici and F. F. Basolo, *J. Am. Chem. Soc.*, 84, 2495 (1962).
11. C. H. Bamford, J. W. Burley, and M. Coldbeck, *J. Chem. Soc., Dalton*, 1846 (1972).
12. C. H. Bamford and S. U. Mullik, *J. Chem Soc., Faraday Trans. I*, 74(7), 1634 (1978).
13. C. H. Bamford, *J. Chem. Soc., Faraday Trans. I*, 74(7), 1648 (1978).
14. C. H. Bamford and K. Hargraves, *Nature* (London), 209, 292 (1966).
15. C. H. Bamford and K. Hargraves, *Proc. R. Soc. London*, A297, 425 (1967).
16. C. H. Bamford and E. O. Hughes, *Proc. R. Soc. London*, A326, 469 (1972).
17. C. H. Bamford and E. O. Hughes, *Proc. R. Soc. London*, A326, 489 (1972).
18. C. H. Bamford, in *New Trends in the Polymer Photochemistry*, N. S. Allen and J. F. Rabek, Eds., Elsevier Applied Science, New York, 1985, pp. 129–145.
19. C. H. Bamford, P. A. Crowe, and R. P. Wayne, *Proc. R. Soc. London*, A284, 455 (1965).
20. C. H. Bamford, P. A. Crowe, J. Hobbs, and R. P. Wayne, *Proc. R. Soc. London*, A292, 153 (1965).
21. P. S. Church, H. Hermann, F. W. Grevels, and K. Schaffner, *J. Chem. Soc., Chem. Commun.*, 785 (1994).

22. C. H. Bamford, in *Reactivity Mechanism and Structure in Polymer Chemistry,* A. D. Jenkins and A. Ledwith, Eds., Wiley, New York, 1974, p. 52.
23. C. H. Bamford and S. U. Mullik, *J. Chem. Soc., Faraday Trans. I,* 71, 625 (1975).
24. C. H. Bamford, G. C. Eastmond, J. Woo, and D. H. Richards, *Polymer,* 23, 643 (1982).
25. G. C. Eastmond, K. J. Parr, and J. Woo, *Polymer,* 29, 950 (1988).
26. C. H. Bamford and X. Z. Han, *Polymer,* 22, 1299 (1981).
27. G. C. Eastmond and J. Grigor, *Makromol. Chem., Rapid Commun.,* 7, 325 (1986).
28. A. K. Alimoglu, Ph.D. thesis, University of Liverpool (1979).
29. C. H. Bamford, J. P. Middleton, K. D. Al-Lamee, and J. Papronty, *Br. Polym. J.,* 19, 269 (1987).
30. Y. Yagci, M. Muller, and W. Schanabel, *Macromol. Rep.,* A28 (Suppl. 1), 37 (1991).
31. C. H. Bamford, G. C. Eastmond, and D. Whittle, *Polymer,* 10, 771 (1969).
32. C. H. Bamford, G. C. Eastmond, and D. Whittle, *Polymer,* 12, 241 (1971).
33. J. Asworth, C. H. Bamford, and E. G. Smith, *Pure Appl. Chem.,* 30, 25 (1972).
34. C. H. Bamford, J. P. Middleton, and K. D. Al-Lamee, *Polymer,* 27, 1981 (1986).
35. Y. Yagci and W. Schnabel, *Prog. Polym. Sci.,* 15, 551 (1990).
36. C. H. Bamford and S. U. Mullik, *Polymer,* 14, 38 (1973).
37. C. H. Bamford and S. U. Mullik, *Polymer,* 17, 225 (1976).
38. C. H. Bamford and S. U. Mullik, *Polymer,* 19, 948 (1979).
39. C. J. Shahani and N. Indictor, *J. Polym. Sci.,* 16, 2997 (1978).
40. Y. Inaki, M. Otsuru, and K. Takemoto, *J. Macromol. Sci.-Chem.,* A12, 953 (1978).
41. K. Kimura, K. Hanabusha, Y. Inaki, and K. Takemoto, *Angew. Makromol. Chem.,* 52, 129 (1976).
42. Y. Inaki, S. Nakagawa, K. Kimura, and K. Takemoto, *Angew. Makromol. Chem.,* 48, 29 (1975).
43. Y. Inaki, H. Shirai, and K. Takemoto, *Angew. Makromol. Chem.,* 45, 51 (1975).
44. Y. Inaki, M. Ishiyana, K. Hibino, and K. Takemoto, *Makromol. Chem.,* 176, 3135 (1975).
45. S. Hoering, J. Schnellenberg, K. V. Blogrodskaya, G. Optiz, and J. Ulbricht, *Z. Wiss, Techn. Hochsch "Carl Schorlemmer" Leuno-Merseburg,* 20271 (1978); *Chem. Abstr.,* 90, 23792g (1979).
46. S. Lenka and P. L. Nayak, *J. Macromol. Sci.-Chem.,* A18(5), 695 (1982).
47. M. Nagata, H. Nishiki, M. Sahai, V. Sakakibora, and U. N. Yasumasa, *Kyoto Kogei Sen'i Daigaku, Sen'igakubu Gakujutsu Hokoku,* 161, 8 (1977); *Chem. Abstr.,* 87, 185031y (1977).
48. K. Kojima, M. Yoshikuni, J. Ishizu, and S. Umeda, *Nippon Kogaku Kaishi,* 12, 222 (1975); *Chem. Abstr.,* 84, 106135x (1976).
49. K. Kimura, Y. Inaki, and K. Takemoto, *Angew. Makromol. Chem.,* 49, 103 (1976).
50. K. Kimura, Y. Inaki, and K. Takemoto, *J. Macromol. Sci.-Chem.,* A9, 1399 (1975).
51. C. H. Bamford, G. C. Eastmond, and J. A. Rippon, *Trans. Faraday. Soc.,* 59, 2548 (1963).
52. Y. Amono and T. Uno, *J. Chem. Soc. Jpn., Pure Chem. Sec.,* 86, 1105 (1965).
53. C. H. Bamford and D. J. Lin, *J. Chem. Soc., Chem. Commun.,* 792 (1966).
54. M. L. Eritsyan, B. V. Zolotukhina, and G. F. Zolotukhina, *Arm. Khim. Zh.,* 29, 784 (1976); *Chem. Abstr.,* 86, 73184y (1977).
55. K. Kimura, Y. Inaki, and K. Takemoto, *J. Macromol. Sci.-Chem.,* A10, 1223 (1976).
56. K. Belogorodshcaya, L. I. Ginzburg, and A. F. Nikolaev, *Vyskomol. Soedin., Ser. B.,* 17, 115 (1975).
57. E. G. Kasting, H. Naarmann, H. Reis, and G. Berding, *Angew. Chem.,* 77, 313 (1965).
58. K. Uehara, M. Tanaka, and N. Murata, *Kogyo Kagaku Zasshi,* 70, 1564 (1967).

59. J. Barton and M. Lazar, *Makromol. Chem.*, 124, 38 (1969).
60. J. Barton, M. Lazar, J. Nemcek, and Z. Manasek, *Makromol. Chem.*, 124, 50 (1969).
61. K. Uehara, Y. Kataoka, M. Tanaka, and N. Murata, *J. Chem. Soc. Jpn., Ind. Chem. Sec.*, 72, 754 (1969).
62. C. H. Bamford and A. N. Ferrar, *Proc. R. Soc. London, Ser. A*, 321, 425 (1971).
63. J. Barton, P. Werner, and J. Vickova, *Makromol. Chem.*, 172, 77 (1973).
64. P. E. M. Allen, *Eur. Polym. J.*, 5, 335 (1969).
65. C. H. Bamford and V. J. Robinson, *Polymer*, 7, 573 (1966).
66. K. Uehara, Y. Ohasi, and M. Tanaka, *J. Polym. Sci., Polym. Chem. Ed.*, 15, 707 (1977).
67. C. H. Bamford and C. A. Fineh, *Trans. Faraday Soc.*, 59, 548 (1963).
68. G. G. Olive and S. Olive, *Makromol. Chem.*, 88, 117 (1965).
69. K. Uehara, T. Nishi, T. Matsumura, F. Tamura, and N. Murata, *Kogyo Kagaku Zasshi*, 70, 755 (1967).
70. S. Buffangi and T. M. Dunn, *J. Chem. Soc.*, 5105 (1961).
71. E. M. Arnett and M. A. Mendelsohn, *J. Am. Chem. Soc.*, 84, 3821 (1962).
72. C. H. Bamford and D. J. Lind, *Chem. Ind.*, 1627 (1965).
73. A. Y. Drummond and W. A. Waters, *J. Chem. Soc.*, 497 (1955).
74. J. S. Litter, *J. Chem. Soc.*, 827 (1962).
75. W. A. Waters and J. S. Littler, in *Oxidation in Organic Chemistry,* K. B. Wilberg, Ed., Academic, London, 1965, Chap. 3.
76. N. Ganga Devi and V. Mahadevan, *Makromol. Chem.*, 152, 177 (1972).
77. P. L. Nayak, R. K. Samal, and M. C. Nayak, *Eur. Polym. J.*, 14, 287 (1978).
78. P. L. Nayak, R. K. Samal, and M. C. Nayak, *J. Macromol. Sci.-Chem.*, A11, 827 (1977).
79. P. L. Nayak, R. K. Small, and M. C. Nayak, *J. Macromol. Sci.-Chem.*, A12, 1815 (1978).
80. P. L. Nayak, R. K. Small, and H. Baral, *J. Macromol. Sci.-Chem.*, A11, 1071 (1977).
81. N. Ganga Devi and V. Mahadevan, *Makromol. Chem.*, 166, 209 (1973).
82. N. Ganga Devi and V. Mahadevan, *J. Polym. Sci.*, A1(10), 903 (1972).
83. P. L. Nayak, R. K. Samal, and M. C. Nayak, *J. Polym. Sci.*, A1(7), 1 (1979).
84. J. Bond and H. M. Longbottom, *J. Appl. Polym. Sci.*, 13, 2333 (1969).
85. A. Rehmann and C. W. Brown, *J. Appl. Polym. Sci.*, 23, 2019 (1979).
86. A. Rehmann and C. W. Brown, *J. Appl. Polym. Sci.*, 23, 3027 (1979).
87. N. Ganga Devi and V. Mahadevan, *J. Chem. Soc.*, D13, 797 (1970).
88. N. Ganga Devi and V. Mahadevan, *Curr. Sci.*, 39, 37 (1970).
89. N. Ganga Devi and V. Mahadevan, *J. Polym. Sci., Part* A, 11, 1553 (1973).
90. A. Y. Drummund and W. A. Waters, *J. Chem. Soc.*, 2836 (1953).
91. H. Singh, R. T. Thumpy, and V. B. Chipalkatty, *J. Polym. Sci.*, A1, 1247 (1965).
92. D. Namasivaya, B. K. Patnaik, and R. T. Thumpy, *Makromol. Chem.*, 205, 147 (1967).
93. K. Satish Baba, K. Panduranga Rao, K. T. Joseph, M. Santappa, and Y. Nayudamma, *Leath. Sci.* (*Madras India*), 21, 261 (1974).
94. I. Cakmak, *Macromol. Rep.*, A31(Suppl. 1 & 2), 85 (1994).
95. R. G. R. Bacon, *Quart. Rev.* (*London*), 9, 288 (1937).
96. J. Saldick, *J. Polym. Sci.*, 19, 73 (1956).
97. P. Cremoneshi, *Ric. Doc. Tessile*, 4, 59 (1967).
98. T. Toru, N. Masanori, H. Yashuhiko, and S. Ichiro, *J. Polym. Sci., B*, 5, 509 (1967).
99. D. Pramanick and S. K. Sarkar, *Colloid Polymer Sci.,* 254, 989 (1976).
100. B. C. Singh, B. K. Mishra, A. Rout, N. Mullick, and M. K. Rout, *Makromol. Chem.*, 180, 953 (1979).

101. D. Sudhakar, K. S. V. Srinivasan, K. T. Joseph, and M. Santappa, *J. Appl. Polym. Sci.*, 23, 2923 (1979).
102. B. N. Mishra, I. Kaur, and R. Dogra, *J. Appl. Polym. Sci.*, 24, 1595 (1979).
103. A. Moce and S. Lapanze, *Makromol. Chem.*, 180, 1599 (1579).
104. A. Habeish, A. T. El-Aref, and M. H. El-Rafie, *Angew. Makromol. Chem.*, 79, 195 (1979).
105. A. E. El-Ashmawy, A. A. Abrahim, and H. El-Salied, *Cell. Chem. Technol.*, 13, 153 (1979).
106. S. V. Subramanium and M. Santappa, *Makromol. Chem.*, 112, 1 (1968).
107. M. Santappa and V. Ananthanarayanan, *Proc. Ind. Acad. Sci., Sect. A,* 62, 150 (1965).
108. S. V. Subramanium and M. Santappa, *J. Polym. Sci., A-1*, 6, 493 (1968).
109. R. A. Wallace and D. G. Young, *J. Polym. Sci., A-1*, 4, 1179 (1966).
110. G. M. Guzman, *Anales Real. Soc. Espan. Fiz. Quin. (Madrid), Ser. B*, 60, 307 (1964).
111. S. Kimura and M. Imoto, *Chem. Abstr.*, 55, 14972c (1961).
112. G. Mino, S. Kaizerman, and E. Rasmussen, *J. Am. Chem. Soc.*, 81, 1494 (1959).
113. H. Narita, S. Okamoto, and S. Machida, *Makromol. Chem.*, 125, 15 (1969).
114. H. Narita, S. Sakata, T. Oda, and S. Machida, *Angew. Makromol. Chem.*, 32, 91 (1973).
115. H. Narita, T. Okimoto, and S. Machida, *Kyoto Kogei Sen't Daigaku Sen'igakubu Gakujutsu Hokuku*, 18, 63 (1976).
116. S. K. Saha and A. K. Choudhary, *J. Polym. Sci., A-1*, 10, 797 (1972).
117. D. Pramanick, *Colloid Polym. Sci.*, 257, 41 (1979).
118. J. Lalitha and M. Santappa, *Vigyan Parishad Anusaudhan Patrika*, 4, 139 (1961).
119. A. A. Katai, V. K. Kulshrestha, and R. H. Marchessantt, *J. Polym. Sci., C-1*, 403 (1963).
120. K. Kaesiyama, *Bull. Chem. Soc. (Japan)*, 42, 1342 (1962).
121. A. Rout, S. P. Rout, B. C. Singh, and M. Santappa, *Makromol. Chem.*, 178, 639 (1977).
122. N. Mohanty, B. Pradhan, and M. C. Mohantay, *Eur. Polym. J.*, 15, 743 (1979).
123. K. Sajjad, Q. Anwarruddin, and L. V. Natrajan, *Curr. Sci. (India)*, 48, 156 (1979).
124. K. N. Rao, S. Sondu, B. Senthuran, and T. N. Rao, *Polym. Bull.*, 2, 43 (1980).
125. A. Rout, S. P. Rout, B. C. Singh, and M. Santappa, *Eur. Polym. J.*, 13, 497 (1977).
126. D. Pramanick, A. K. Chatterjee, and S. K. Sarkar, *Makromol. Chem.*, 180, 1085 (1979).
127. D. Pramanick and A. K. Chatterjee, *J. Polym. Sci., A-1*, 18, 311 (1980).
128. A. R. Swayam, P. Rout, M. Mullick, and B. C. Singh, *J. Chem. Sci., A-1,* 16, 391 (1978).
129. G. S. Misra and G. P. Dubey, *Polym. Bull.*, 1, 671 (1979).
130. M. M. Hussain and A. Gupta, *J. Makromol. Sci.-Chem.*, A11, 2177 (1977).
131. J. Ulbritch and W. Seidel, *Plaster Kantscuk*, 20, 6 (1973).
132. G. Mino, S. Kaizerman, and E. Rasmussen, *J. Polym. Sci.*, 38, 393 (1959).
133. I. Cakmak, B. Hazer, and Y. Yagci, *Eur. Polym. J.*, 27, 101 (1990).
134. A. T. Erciyes, M. Erim, B. Hazer, and Y. Yagci, *Angew. Makromol. Chem.* 200, 163 (1992).
135. Ü. Tunca, E. Serhatli, and Y. Yagci, *Polym. Bull.*, 22, 483 (1983).
136. I. Cakmak, *Macromol. Rep.*, A32, 197 (1995).
137. O. Galioglu, A. B. Soydan, A. Akar, and A. S. Saraç, *Angew. Makromol Chem.*, 214, 19 (1994).
138. A. S. Sarac, C. Erbil, and B. Ustamehmetoglu, *Polym. Bull.*, 33, 535 (1994).

139. A. S. Sarac, B. Ustamehmetoglu, M. I. Mustafaev, and C. Erbil, *J. Polym. Sci., Part A: Polym. Chem.*, 33, 1581 (1995).
140. C. Erbil, A. B. Soydan, A. Z. Aroguz, and A. S. Sarac, *Angew. Makromol. Chem.*, 213, 55 (1993).
141. C. Özeroglu, O. Güney, A. S. Sarac, and M. I. Mustafaev, *J. Appl. Polym. Sci.*, 60, 759 (1996).
142. A. S. Sarac, A. Göcmen, and B. Basaran, *J. Sol. Chem.*, 19, 7 (1990).
143. C. Erbil, B. Ustamehmetoglu, G. Uzelli, and A. S. Sarac, *Eur. Polym. J.*, 30, 2 (1994).
144. A. S. Sarac, H. Basak, A. B. Soydan, and A. Akar, *Angew. Makromol. Chem.*, 198, 191 (1992).
145. A. S. Sarac, B. Ustamehmetoglu, and C. Erbil. *Polym. Bull*, 32, 91 (1994).
146. A. S. Sarac, C. Erbil, and A. B. Soydan, *J. Appl. Polym. Sci.*, 44, 877 (1992).
147. C. Erbil, C. Cin, A. B. Soydan, and A. S. Sarac, *J. Appl. Polym. Sci.*, 47, 1643 (1993).
148. A. S. Sarac, C. Erbil, and F. Durap, *Polym. Int.*, 40, 179 (1996).
149. M. K. Mishra, *J. Macromol. Sci. —Rev. Macromol. Chem.*, C19, 193 (1981).
150. S. Saccubai, K. Jiji, and M. Santappa, *Ind. J. Chem.*, 4, 493 (1966).
151. S. Saccubai and M. Santappa, *Macromol. Chem.*, 117, 50 (1968).
152. S. Saccubai and M. Santappa, *J. Polym. Sci., A-1*, 7, 643 (1969).
153. T. R. Mohanty, B. C. Singh, and P. L. Nayak, *J. Polym. Sci*, 13, 2075 (1975).
154. T. R. Mohanty, B. C. Singh, and P. L. Nayak, *Makromol. Chem.*, 175, 2345 (1974).
155. R. K. Samal, P. L. Nayak, and T. R. Mohanty, *Macromolecules*, 10, 489 (1977).
156. P. L. Nayak, T. R. Mohanty, and R. K. Samal, *Makromol. Chem.*, 178, 2975 (1977).
157. B. C. Singh, T. R. Mohanty, and P. L. Nayak, *Eur. Polym. J.*, 12, 371 (1976).
158. B. C. Singh, T. R. Mohanty, and P. L. Nayak, *J. Macromol. Sci.-Chem.*, A9, 1149 (1975).
159. P. L. Nayak, S. Lenka, and N. C. Pati, *J. Appl. Polym. Sci.*, 22, 3301 (1978).
160. P. L. Nayak, S. Lenka, and N. C. Pati, *Angew. Makromol. Chem.*, 68, 117 (1978).
161. R. G. Grigoryan, *Arm. Chim. Zh.*, 32, 239 (1979).
162. S. Lenka, P. L. Lenka, and M. K. Mishra, *J. Appl. Polym. Sci.*, 25, 1323 (1980).
163. R. M. Livshits and Z. A. Rogovin, *Vysokomol. Soed.,* 4, 784 (1962).
164. B. R. James, J. R. Lyons, and R. P. Williams, *Biochemistry*, 1, 379 (1962).
165. W. A. Waters and J. S. Littler, in *Oxidation in Organic Chemistry*, K. B. Wilberg, Ed., Academic, London, 1965, p. 185.
166. D. G. Hoare and W. A. Waters, *J. Chem. Soc.,* 965 (1962).
167. D. G. Hoare and W. A. Waters, *J. Chem.* Soc., 971 (1962).
168. T. A. Cooper and W. A. Waters, *J. Chem. Soc.*, 1538 (1964).
169. D. G. Hoare and W. A. Waters, *J. Chem. Soc.*, 2552 (1964).
170. W. A. Waters and J. Kemp, *Proc. R. Soc. London*, A274, 480 (1963).
171. C. E. H. Bawn and A. G. White, *J. Chem. Soc.*, 331 (1951).
172. C. E. H. Bawn and A. G. White, *J. Chem. Soc.*, 339 (1951).
173. C. E. H. Bawn and A. G. White, *J. Chem. Soc.*, 343 (1951).
174. C. E. H. Bawn, *Discuss. Faraday Soc.*, 14, 181 (1953).
175. C. E. H. Bawn and M. A. Sharp, *J. Chem. Soc.*, 1856 (1957).
176. C. E. H. Bawn and J. E. Jolley, *Proc. R. Soc. London*, A237, 297 (1957).
177. L. H. Sutcliffe and J. R. Weber, *Trans. Faraday Soc.*, 52, 1225 (1956).
178. L. H. Sutcliffe and J. R. Weber, *Trans. Faraday Soc.*, 52, 1225 (1956).
178a. L. H. Sutcliffe and J. R. Weber, *Trans. Faraday Soc.,* 55, 1892 (1959).
179. L. H. Sutcliffe and J. R. Weber, *Trans. Faraday Soc.*, 57, 91 (1961).
180. L. H. Sutcliffe and G. Hargreaves, *Trans. Faraday Soc.*, 51, 786 (1955).

181. J. B. Kirwin, F. D. Peat, P. J. Proll, and L. H. Sutcliffe, *J. Phys. Chem.*, 67, 2288 (1963).
182. K. G. Ashurst and W. C. E. Higginson, *J. Chem. Soc.*, 343 (1956).
183. D. R. Rosseinsky and W. C. E. Higginson, *J. Chem. Soc.*, 31 (1960).
184. J. H. Baxendale and C. F. Wells, *Trans. Faraday. Soc.*, 53, 800 (1957).
185. K. Jijie, M. Santappa, and M. Mahadevan, *J. Polym. Sci., A1*, 4, 377 (1966).
186. K. Jijie, M. Santappa, and M. Mahadevan, *J. Polym. Sci., A1*, 4, 393 (1966).
187. M. Santappa, M. Mahadevan, and K. Jijie, *Proc. Indian Acad. Sci.*, 64 (1966).
188. K. Jijie and M. Santappa, *Proc. Indian Acad. Sci.*, 65, 124 (1967).
189. S. Gil Soo, *Yongu, Pogo Yongnam Taehakkyo Kongop Kisul Yonguso*, 7, 43 (1979).
190. T. Guha and S. R. Palit, *J. Polym. Sci., A,* 2, 1731 (1963).
191. W. A. Waters and J. S. Littler, in *Oxidation in Organic Chemistry*, K. B. Wilberg, Ed., Academic, London, 1965, p. 69.
192. S. Viswanathan and M. Santappa, *J. Polym. Sci., A1,* 9, 1685 (1971).
193. S. Viswanathan and M. Santappa, *Makromol. Chem.*, 126, 234 (1970).
194. P. L. Nayak, T. R. Mohanty, and R. K. Samal, *J. Macromol. Sci.-Chem.*, A10, 1239 (1976).
195. A. Rout, S. P. Rout, B. C. Singh, and M. Santappa, *J. Macromol. Sci.-Chem.* A11, 957 (1977).
196. I. M. Kolthof and E. J. Meehan, *J. Polym. Sci.*, 9, 327 (1952).
197. A. S. Risk and M. H. Nossair, *Ind. J. Chem., Sec.* A, 16, 564 (1978).
198. P. L. Nayak, S. Lenka, and N. C. Pati, *J. Polym. Sci., A1*, 17, 345 (1979).
199. S. Yu., L. Paikachev, and N. Mizerovski, *Tr. Ivannov Khim., Tekhnol. Inst.*, 12, 125 (1970).
200. G. S. Misra and S. L. Dubey, *J. Macromol. Sci.-Chem.*, A13, 31 (1979).
201. J. Bond and P. I. Lee, *J. Polym. Sci., A1*, 6, 2621 (1968).
202. J. Bond and P. I. Lee, *J. Polym. Sci., A1,* 7, 379 (1969).
203. J. Bond and P. I. Lee, *J. Appl. Polym. Sci.*, 13, 1215 (1969).
204. M. H. El-Rafie, S. H. Abdel-Fatteh, E. M. Khalil, and A. Habeish, *Angew. Makromol. Chem.*, 87, 63 (1980).
205. J. Barton and J. M. Vicekova, *Makromol. Chem.*, 178, 513 (1977).
206. G. S. Misra and S. L. Dubey, *J. Macromol. Sci.-Chem.*, A13, 31 (1979).
207. T. Sato, M. Takada, and T. Otsu, *Makromol. Chem.*, 148, 239 (1971).
208. M. Imoto, N. Sakade, and T. Ouchi, *J. Polym. Sci., A1*, 15 (1977).
209. M. H. El-Rafie and A. Habeish, *J. Appl. Polym. Sci.*, 19, 1815 (1975).
210. C. H. Bamford, A. D. Jenkins, and R. Johnstone, *Proc. R. Soc. London*, A239, 214 (1957).
211. W. I. Bengough and I. C. Ross, *Trans. Faraday Soc.*, 62, 2251 (1966).
212. E. A. Cavell, I. T. Gilson, and A. C. Meeks, *Makromol. Chem.*, 73, 145 (1964).
213. R. Inoue and T. Yamauchi, *Bull. Chem. Soc. Jpn.*, 26, 135 (1953).
214. M. K. Saha, A. R. Mukherjee, P. Ghosh, and S. R. Palit, *J. Polym. Sci, Part C*, 16, 159 (1967).
215. H. Narita, Y. Sakumoto, and S. Machida, *Makromol. Chem.*, 143, 279 (1971).
216. M. P. Schubert, *J. Am. Chem. Soc.*, 54, 4977 (1932).
217. D. L. Leussing and I. M. Kolthoff, *J. Am. Chem. Soc.*, 75, 3904 (1953).
218. N. Tanaka, I. M. Kolthoff, and W. Stricks, *J. Am. Chem. Soc.*, 77, 1996 (1955).
219. T. J. Wallace, *J. Org. Chem.*, 31, 3071 (1966).
220. B. M. Mandal, U. S. Nandi, and S. R. Palit, *J. Polym. Sci., A1,* 7, 1407 (1969).
221. J. C. Milco and L. Nicolas, *J. Polym. Sci., A1*, 4, 713 (1966).
222. J. C. Milco and L. Nicolas, *J. Polym. Sci., A1,* 7, 1407 (1969).
223. J. C. Milco and L. Nicolas, *J. Polym. Sci., A1*, 8, 67 (1970).

224. H. Narita, A. Ostaki, and S. Machida, *Makromol. Chem.*, 178, 3217 (1977).
225. H. Narita, Y. Kazuse, and M. Araki, *Kyoto Kogei Sen't Daigaku Sen'igakubu Gakujutsu Hokuku*, 8, 89 (1978).
226. V. A. Laprev, M. G. Voronkov, E. N. Baiborodina, N. N. Shagleaeva, and T. N. Rakhmatulina, *J. Polym. Sci., A1*, 17, 34411 (1979).
227. A. Habeish, S. H. Abdel-Fatteh, and A. Bendak, *Angew. Makromol. Chem.*, 37, 911 (1974).
228. C. W. Brown and H. M. Longbottom, *J. Appl. Polym. Sci.*, 17, 1787 (1973).
229. A. K. Bentia, B. M. Mandal, and S. R. Palit, *Makromol. Chem.*, 175, 413 (1974).
230. A. K. Bentia, *J. Ind. Chem. Soc.*, 54, 1148 (1977).
231. G. Talamani, A. Turalla, and E. Vianello, *Chem. Ind. (Milan)*, 45, 335 (1963).
232. E. A. S. Cavell, I. T. Gilson, and A. C. Meeks, *Makromol. Chem.*, 73, 145 (1964).
233. H. Narita et al., *Makromol. Chem.,* 152, 143 (1972).
234. J. C. Milco and Nicolas, *J. Polym. Sci, A1*, 4, 713 (1966).
235. V. P. Kien and R. C. Schulz, *Makromol. Chem.*, 180, 1825 (1979).
236. R. Stewart, in *Oxidation in Organic Chemistry*, K. B. Wilberg, Ed., Academic, London, 1965, p. 1.
237. S. R. Palit and R. S. Konar, *J. Polym. Sci.*, 57, 609 (1962).
238. R. S. Konar and S. R. Palit, *J. Polym. Sci.*, 58, 85 (1963).
239. R. S. Konar and S. R. Palit, *J. Polym. Sci., A2,* 2, 1731 (1964).
240. J. Weiss, *Discuss. Faraday Soc.*, 2, 188 (1947).
241. H. F. Launer and D. M. Yost, *J. Am. Chem. Soc.*, 56, 2571 (1934).
242. G. S. Misra, J. S. Shukla, and H. Narain, *Makromol. Chem.*, 119, 74 (1968).
243. G. S. Misra and H. Narain, *Makromol. Chem.*, 113, 85 (1968).
244. G. S. Misra and C. V. Gupta, *Makromol. Chem.*, 168, 105 (1973).
245. G. S. Misra and J. J. Rebello, *Makromol. Chem.*, 175, 3117 (1974).
246. G. S. Misra and J. J. Rebello, *Makromol. Chem.*, 176, 21 (1976).
247. P. Levesley and W. A. Waters, *J. Chem. Soc.*, 217 (1953).
248. J. S. Shukla and D. C. Mishra, *J. Polym. Sci., Polym. Chem. Ed.*, 11, 751 (1973).
249. Kureha Chemical Works Ltd., British Patent 895,153 (May 2, 1962).
250. Z. Csuros, M. Gara, and I. Cyurkovics, *Acta Chem. Acad. Sci. Hung.*, 29, 207 (1961).
251. Kennoro and Hiroshi-Takida, Japan Synthetic Chemical Industry Co., Jpn. Patent 1345 (Feb. 2, 1962).
252. M. Hussain and A. Gupta, *Makromol. Chem.*, 178, 29 (1977).
253. N. C. Pati, S. Lenka, and P. L. Nayak, *J. Macromol. Sci.-Chem.*, A13, 1157 (1979).
254. G. Pand, N. C. Pati, and P. L. Nayak, *J. Appl. Polym. Sci.*, 25, 1479 (1980).
255. M. I. Khalil, S. H. Abdel-Fatteh, and A. Kantouch, *J. Appl. Polym. Sci.*, 19, 2699 (1975).
256. S. H. Abdel-Fateh, E. Allam, and M. A. Mohharom. *J. Appl. Polym. Sci.*, 20, 1049 (1976).
257. R. Teichmann, *Acta Polym.*, 30, 60 (1979).
258. A. Habeish, I. El-Thalouth, M. A. El-Kashouti, and S. H. Abdel-Fatteh, *Angew. Makromol. Chem.*, 78, 101 (1979).

6 Suspension Polymerization Redox Initiators

Munmaya K. Mishra, Norman G. Gaylord, and Yusuf Yagci

CONTENTS

6.1 INTRODUCTION

The conditions under which radical polymerizations are performed are both of the homogeneous and heterogeneous types. This classification is usually based on whether the initial reaction mixture is homogeneous or heterogeneous. Some homogeneous systems may become heterogeneous as polymerization proceeds due to insolubility of the polymer in the reaction media. Heterogeneous polymerization is extensively used as a means to control the thermal and viscosity problems. Three types of heterogeneous polymerization are used: precipitation, suspension, and dispersion.

The term suspension polymerization (also referred to as bead or pearl polymerization) refers to polymerization in an aqueous system with a monomer as a dispersed phase, resulting in a polymer as a dispersed solid phase. The suspension polymerization is performed by suspending the monomer as droplets (0.001–1 cm in diameter) in water (continuous phase). In a typical suspension polymerization, the initiator is dissolved in the monomer phase. Such initiators are often referred to as oil-soluble initiators. Each monomer droplet in a suspension is considered to be a small bulk polymerization system and the kinetics is the same as that of bulk polymerization. The suspension of a monomer is maintained by agitation and the use of stabilizers. The suspension polymerization method is not used with monomers, which are highly soluble in water or where a polymer has too high a glass transition temperature. The method is used commercially to prepare vinyl polymers such as polystyrene, poly(methyl methacrylate), poly(vinyl chloride), poly(vinyl acetate), poly(vinylidene chloride), and poly(acrylonitrile). Various types of redox initiator are used to prepare such polymers by suspension polymerization. The following examples describe the various types of initiating systems for suspension polymerization.

Suspension polymerization is essentially equivalent to bulk polymerization but is performed in a reaction medium in which the monomer is insoluble and dispersed as a discrete phase (e.g., droplets), with a catalyst system that generates or permits the entry of radical species within the suspended monomer phase or droplets. The following review presents examples of initiators for bulk polymerization as well as suspension polymerization, as initiating systems suitable for bulk polymerization due to monomer-soluble catalysts are potentially useful in suspension polymerization.

6.2 ACYL PEROXIDE

Acyl peroxides may be defined as substances of the type

$$\mathrm{R\overset{\overset{O}{\|}}{C}OO\overset{\overset{O}{\|}}{C}R'}$$

where R and R′ are either alkyl or aryl. Acyl peroxides have been one of the most frequently used sources of free radicals, and interest in their various modes of decomposition has been keen. Acyl peroxides (i.e., benzoyl peroxide [Bz_2O_2] and lauroyl peroxide [LPO]) have been used extensively as the initiator for suspension polymerization of styrene [1–4], vinyl chloride [5–7], and vinyl acetate [8].

6.2.1 Fe^{2+} as Reductant

Kern and other investigators [9, 10] found Bz_2O_2 to be very effective in both aqueous and nonaqueous media with or without heavy metals as a component, Kern [9] based his theory of reaction on Haber's earlier suggestions and formulated the production of radicals as an electron transfer process. He proposed a Haber–Weiss type of mechanism for two-component systems:

$$Fe^{2+} + (RCOO)_2 \rightarrow Fe^{3+} + RCOO^{\cdot} + RCOO^{-} \tag{6.1}$$

where $RCOO^{\cdot}$ is the active species.

In the presence of a third component, a reducing agent (YH_2), the reaction continues as follows:

$$Fe^{3+} + YH_2 \rightarrow Fe^{2+} + YH^{\cdot} + H^{+} \tag{6.2}$$

$$Fe^{3+} + YH^{\cdot} \rightarrow Fe^{2+} + Y + H^{+} \tag{6.3}$$

$$RCOO^{\cdot} + YH_2 \rightarrow RCOOH + YH^{\cdot} \tag{6.4}$$

The effect of activators like $FeSO_4$ [11, 12] for emulsion polymerization and ferric stearate [13] for bulk polymerization of vinyl monomers in combination with acyl peroxide has been studied. The ferrous ion catalyzed decomposition of Bz_2O_2 in ethanol has been studied in some detail by Hasegawa and co-workers [14, 15]. The cycle, which requires reducing of Fe^3- by solvent-derived radicals, yields a steady-state concentration of Fe^2- after a few minutes, shown spectroscopically to be proportional to the initial concentration of the ferrous ion [14]. The second-order rate constant for the following reaction was found to be 8.4 L mol^{-1} sec^{-1} at 25°C, with an activation energy of 14.2 kcal mol^{-1}:

$$Bz_2O_2 + Fe^{2+} \rightarrow BzO^- + BzO^\cdot + Fe^{3+} \tag{6.5}$$

$$BzO^\cdot + EtOH \rightarrow BzOH + MeC^\cdot HOH \tag{6.6}$$

$$MeC^\cdot HOH + Fe^{3+} \rightarrow Fe^{2+} + AcH + H^+ \tag{6.7}$$

The suspension polymerization of vinyl chloride using lauroyl peroxide (LPO) and a water-soluble Fe^{2-} salt [16, 17] and monomer-soluble [18–20] Fe^{2-} salt as the reducing agent has been studied. In the case of a monomer-soluble reducing agent like ferrous caproate, the mechanism of initiation of the polymerization is considered to be a one-electron transfer reaction in the monomer phase as follows:

$$\begin{aligned} &C_{11}H_{23}COO-OOCC_{11}H_{23} + (C_5H_{11}COO-)_2Fe \rightarrow C_{11}H_{23}COO^\cdot \\ &+C_{11}H_{23}COO-Fe(-OOCC_5H_{11})_2 \end{aligned} \tag{6.8}$$

$$C_{11}H_{23}COO^\cdot + (C_5H_{11}COO-)_2Fe \rightarrow C_{11}H_{23}COO-Fe(-OOC_5H_{11})_2 \tag{6.9}$$

Das and Krishnan [21] had reported the suspension polymerization of vinyl acetate and vinyl alcohol using a redox pair of Bz_2O_2 and ferrous octoate (reducing agent). A high degree of polymerization was achieved using this redox-pair initiating system.

6.2.1.1 Suspension Polymerization of Vinyl Chloride

In the suspension polymerization of vinyl chloride using LPO and a water-soluble reducing agent [16, 17], $Fe(OH)_2$ (produced by in situ reaction of a ferrous salt and an alkali metal hydroxide), the conversion was 80% and 65% by using a Na maleate–styrene copolymer and poly(vinyl alcohol) as the dispersing agent, respectively. The reaction was performed according to the recipe presented in Table 6.1.

The suspension polymerization of vinyl chloride was also performed at –15°C using a monomer-soluble reducing agent like ferrous caproate [18, 19]. The molecular weight of the poly(vinyl chloride) decreased as the concentration of the iron(II) system increased, because of chain termination reactions. Konishi and Nambu [20] also reported low-temperature polymerization of vinyl chloride using the LPO–ferrous caproate redox system. The reaction was studied by varying the temperature from –30°C to +30°C with a molar ratio of oxidant to reductant of 1:1. The activation energy of the overall rate of polymerization was 6.5 kcal mol^{-1}. The initial rate increased, and the degree of polymerization decreased, with increasing ratio of

TABLE 6.1
Typical Recipe: Suspension Polymerization of Vinyl Chloride[a]

Ingredients	Amount (ppm)
0.03% Aqueous dispersing agent	200
$FeSO_4$	0.15
Vinyl chloride	100
Lauroyl peroxide	0.2
$HCCl–CCl_2$	40
0.5% Aqueous NaOH	1.7

[a] Polymerization for 5 h at 20°C.

ferrous caproate to LPO. The relative efficiencies of the peroxide with the reducing agent ferrous caproate were measured and are presented in Table 6.2.

A moderate rate of polymerization and a maximum yield were obtained by appropriate, continuous charging of the catalyst ingredients instead of the one-time addition. The syndiotacticity was increased as the polymerization temperature decreased. The initial rate was increased with the increasing rate of ferrous caproate to LPO, but after passing through the ratio of unity, the maximum yield of the polymer suddenly became lower. This could be attributed to the decrease in the number of initiating radicals as shown in reaction (6.9). The oxidation–reduction reaction initiates and the polymerization can proceed readily in the monomer phase by using a monomer-soluble reducing agent.

6.2.2 Sn^{2+} as Reductant

Organic peroxides may decompose in a number of different ways when treated with ions of variable oxidation number. The reaction can be rationalized on the basis of the following general reaction:

$$+2e\ M^{n+} + R-\overset{\overset{\displaystyle O}{\|}}{C}-O:O-\overset{\overset{\displaystyle O}{\|}}{C}-R+ \rightarrow M^{(n+2)} + 2R-\overset{\overset{\displaystyle O}{\|}}{C}-O^{-} \tag{6.10}$$

TABLE 6.2
Relative Efficiency of the Peroxides

Peroxides	Temperature (°C)	Rate of Polymerization (% h)
Lauroyl peroxide	−15	4.5
2,4-Dichlorobenzoyl peroxide	−15	1.5
Benzoyl peroxide	−15	1.4
Cumene hydroperoxide	−15	0.8
Di-tert-butyl hydroperoxide	−15	0.7

Source: A. Konishi and K. Nambu, *J. Polym. Sci.*, 54, 209 (1961).

The reaction of diacyl peroxide with stannous chloride in acid solution in room temperature or at a slightly elevated temperature is used in the quantitative analysis of the peroxygen compounds [22]. The reaction of the peroxygen compound with stannous chloride in the acid medium is apparently rapid and complete enough at room temperature to serve as a quantitative assay method. However, no information is available as to the nature of the decomposition products (i.e., radical or ionic). In the absence of other evidence, the most reasonable mechanism would appear to be a heterolytic process as shown in reaction (6.11):

$$\mathrm{R{-}\overset{\overset{O}{\|}}{C}{-}O{-}O{-}\overset{\overset{O}{\|}}{C}{-}R + Sn^{2+} \rightleftharpoons R{-}\overset{\overset{O}{\|}}{C}{-}O{-}\underset{\underset{R-C=O}{|}}{O} \longrightarrow Sn^{2+} \longrightarrow R{-}\overset{\overset{O}{\|}}{C}{-}O^{-} + R{-}\overset{\overset{O}{\|}}{C}{-}O{-}Sn^{3+}}$$

$$\longrightarrow \mathrm{2R{-}\overset{\overset{O}{\|}}{C}{-}O^{-} + Sn^{4+}} \tag{6.11}$$

Some evidence of the free-radical mechanism of polymerization using a peroxygen compound and Sn^{2-} halides exists. The effective polymerization of vinyl chloride in the presence of the peroxyester $-SnCl_2$ catalyst system confirms the generation of free radicals [23]. This contrasts with the reported rapid decomposition of diacyl peroxides in solution at room temperature in the presence of various metal halides, to nonradical species through ionic intermediates. Thus, a polar carbonyl inversion mechanism is proposed in the decomposition of benzoyl peroxide and/or other diacyl peroxide in the presence of aluminum chloride [24–27], antimony pentachloride [26–28], and boron trifluoride [25–27].

However, radical generation has been confirmed in the polymerization of various monomers in the presence of a catalyst system consisting of an aluminum alkyl and either a diacyl peroxide or a peroxyester (i.e., peroxygen compounds containing carbonyl groups) [29–31]. The proposed mechanism of decomposition involves complexation of the AIR_3 with the carbonyl group of the peroxide as well as with the monomer, resulting in an electron shift, which weakens the peroxy linkage:

$$\begin{array}{c} \mathrm{AlR_3} \\ \diagup \qquad \diagdown \\ \mathrm{R{-}\overset{\overset{O}{\|}}{C}{-}O{\vdots}O{-}} \quad \mathrm{M} \end{array} \tag{6.12}$$

Although this mechanism may be operative to some extent, a redox mechanism analogous to that normally invoked in redox catalyst systems containing a peroxygen compound for the initiation of polymerization, considered to be a two-electron transfer, probably plays a major role:

$$\mathrm{R{-}\overset{\overset{O}{\|}}{C}{-}OO{-}\overset{\overset{O}{\|}}{C}{-}R + 2e \longrightarrow R{-}\overset{\overset{O}{\|}}{C}{-}O^{-} + R{-}\overset{\overset{O}{\|}}{C}{-}O^{\cdot}} \tag{6.13}$$

$$\mathrm{Sn^{2+} \longrightarrow Sn^{4+} + 2e} \tag{6.14}$$

$$2R-\overset{\overset{\displaystyle O}{\|}}{C}-OO-\overset{\overset{\displaystyle O}{\|}}{C}-R \; + Sn^{2+} \longrightarrow 2\,R-\overset{\overset{\displaystyle O}{\|}}{C}-O^{-} \; + \; 2R-\overset{\overset{\displaystyle O}{\|}}{C}-O^{\bullet} + Sn^{4+} \tag{6.15}$$

Another — a one-electron, transfer mechanism — may be suggested for the formation of free radicals as follows:

$$R-\overset{\overset{\displaystyle O}{\|}}{C}-OO-\overset{\overset{\displaystyle O}{\|}}{C}-R + Sn^{2+} \longrightarrow R-\overset{\overset{\displaystyle O}{\|}}{C}-O^{-} \; + \; R-\overset{\overset{\displaystyle O}{\|}}{C}-O^{\bullet} + Sn^{3+} \tag{6.16}$$

As Sn^{3+} is very unstable after formation, it may undergo reaction in two ways in which it may again be reduced to Sn^{2+} or oxidized to a Sn^{4+} state. The reactions are as follows:

$$R-\overset{\overset{\displaystyle O}{\|}}{C}-OO-\overset{\overset{\displaystyle O}{\|}}{C}-R + Sn^{3+} \longrightarrow R-\overset{\overset{\displaystyle O}{\|}}{C}-O^{+} + R-\overset{\overset{\displaystyle O}{\|}}{C}-\dot{O} + Sn^{2+} \tag{6.17}$$

The $R-\overset{\overset{\displaystyle O}{\|}}{C}-O^{\bullet}$ radical, reaction (6.16), and $R-\overset{\overset{\displaystyle O}{\|}}{C}-OO^{\bullet}$ radical, reaction (6.17), may react with acyl peroxide as follows:

$$^{\bullet}O-\overset{\overset{\displaystyle O}{\|}}{C}-R + R-\overset{\overset{\displaystyle O}{\|}}{C}-OO-\overset{\overset{\displaystyle O}{\|}}{C}-R \dashrightarrow (R-\overset{\overset{\displaystyle O}{\|}}{C})_2O + R-\overset{\overset{\displaystyle O}{\|}}{C}-OO^{\bullet} \tag{6.18}$$

$$^{\bullet}O-O-\overset{\overset{\displaystyle O}{\|}}{C}-R + R-\overset{\overset{\displaystyle O}{\|}}{C}-OO-\overset{\overset{\displaystyle O}{\|}}{C}-R \dashrightarrow (R-\overset{\overset{\displaystyle O}{\|}}{C}-)_2O + R-\overset{\overset{\displaystyle O}{\|}}{C}-O^{\bullet} + O_2 \tag{6.19}$$

or, in the other step, Sn^{3+} produced in reaction (6.16) may be oxidized to the Sn^{4+} state as follows:

$$2R-\overset{\overset{\displaystyle O}{\|}}{C}-OO-\overset{\overset{\displaystyle O}{\|}}{C} + Sn^{3+} \dashrightarrow 2\,R-\overset{\overset{\displaystyle O}{\|}}{C}-O^{-} + 2R-\overset{\overset{\displaystyle O}{\|}}{C}-O^{\bullet} + Sn^{4+} \tag{6.20}$$

The mechanism of polymerization may be represented as follows:

Initiation

$$M + R^{\bullet} \xrightarrow{k_i} M^{\bullet} \tag{6.21}$$

Propagation

$$M^{\bullet} + M \xrightarrow{k_p} M^{\bullet} \tag{6.22}$$

$$M^{\bullet}_{n-1} + M \xrightarrow{k_p} M^{\bullet}_{n} \tag{6.22a}$$

Termination

$$M^{\bullet}_x + M^{\bullet}_y \rightarrow \text{Dead polymer (mutual)} \quad (6.23)$$

$$M^{\bullet}_x + R^{\bullet} \rightarrow \text{Dead polymer (linear)} \quad (6.23a)$$

where M is the monomer, $R^{\bullet}$ is the initiating radical, and k_i and k_p are the rate constants.

6.2.2.1 Suspension Copolymerization of Acrylonitrile with Methyl Acrylate and with Styrene

Kido et al. reported the suspension copolymerization of acrylonitrile–methyl acrylate [32] and acrylonitrile–styrene [33] using dilauroyl peroxide and the $SnCl_2$ redox system.

In the case of suspension copolymerization of acrylonitrile and methyl acrylate, mixtures of 40–85% acrylonitrile and 15–60% methyl acrylate were polymerized in an H_2O suspension using inorganic dispersants according to the typical recipe presented in Table 6.3 to produce 100-μ spherical copolymer beads.

In the case of suspension copolymerization of acrylonitrile–styrene, mixtures of 10–40 wt% acrylonitrile and 40–90 wt% styrene are polymerized in H_2O in the presence of inorganic dispersing agents according to the typical recipe presented in Table 6.4, to produce transparent copolymer beads containing >90% 100–400-μ mesh particles.

6.2.3 Cu^{2+} as Reductant: Suspension Polymerization of Vinyl Chloride

Recently, Cozens [34] has reported the suspension polymerization of vinyl chloride using a LPO–Cu^{2+} metal chelate redox pair system. The suspension polymerization of vinyl chloride [35] was also studied using a diacyl peroxide such as Bz_2O_2–Cu^{2+} as the redox initiator. The microsuspension polymerization of vinyl chloride was performed at 40–60°C. The conversion of 85% was obtained after 10 h of polymerization according to the typical recipe presented in Table 6.5.

TABLE 6.3
Typical Recipe: Suspension Copolymerization of Acrylonitrile and Methyl Acrylate[a]

Ingredients	Amount (ppm)
Water	200
$SnCl_2 \cdot 2H_2O$	0.02
75:25 Acrylonitrile-methyl acrylate	250
Dilauroyl peroxide	0.5
HCC– CCl_2	40
0.5% Aqueous NaOH	1.7

[a] Polymerization for 1 h at 250°C followed by 15 h at 60°C at stirring rate of 1000 rpm.

TABLE 6.4
Typical Recipe: Suspension Copolymerization of Acrylonitrile and Styrene[a]

Ingredients	Amount (ppm)
Water	150
Hydroxylapatite	2
Polyethylene glycol alkyl aryl ether phosphate	0.01
$SnCl_2 \times 2H_2O$	0.02
Acrylonitrile	25
Styrene	75
tert-Dodecyl mercaptan	0.5
Dilauroyl peroxide	0.71
HCC– CCl_2	240

[a] Polymerization for 1 h at 25°C followed by 15 h at 60°C at stirring rate of 400 rpm.

6.2.4 Tertiary Amine as Reductant

The use of tertiary amines as cocatalysts with metal ions in aqueous polymerization has been the subject of study of various workers [36]. No nucleophilic displacement in peroxidic oxygen has received more attention than that by amines [37]. Extensive studies with acyl peroxide were performed by several workers [38–51].

The amine–peroxide combination as an initiator for vinyl polymerization has been investigated extensively by various workers. Solution polymerization of vinyl chloride [52] and styrene and methyl methacrylate [53], bulk polymerization of styrene [54], and dead-end polymerization of styrene and methyl methacrylate [55] were performed using the benzoyl peroxide–dimethylaniline initiating system. Lal and Green [56] have reported the effect of various amine accelerators on the bulk polymerization of methyl methacrylate with benzoyl peroxide. At about the same time, Imoto and Takemoto [57] had reported the solution polymerization of acrylonitrile in the presence of a substituted benzoyl peroxide–dimethylaniline redox system. In another article, Takemoto et al. [58] have reported the solution polymerization of

TABLE 6.5
Typical Recipe: Suspension Polymerization of Vinyl Chloride[a]

Ingredients	Amount (g)
Water	700
Vinyl chloride	675
Benzoyl peroxide	0.675
$CuSO_4 \cdot 5H_2O$	45 mg

[a] Polymerization for 10 h at 50°C.

styrene using benzoyl peroxide and various di-n-alkylaniline redox systems. In a series of articles, O'Driscoll and McArdle reported on the bulk polymerization of styrene at 0°C [59] and higher temperatures [60] using benzoyl peroxide–dimethylaniline, and the bulk polymerization of styrene [61] using substituted diethylaniline and benzoyl peroxide. The efficiencies of free-radical production by various substituted benzoyl peroxides and substituted di-n-alkylanilines have also been studied [59–65]. Recently, the feasibility of the triethylamine–benzoyl peroxide [55] redox system to induce photopolymerization in solution has been reported.

The presence of free radicals in the reaction of tertiary amines and benzoyl peroxide has been observed by electron spin resonance (ESR) spectroscopy [67–69]. The reaction of amines with acyl peroxide is much more rapid than the thermal decomposition of the peroxide alone [70]. For example, benzoyl peroxide [53] with dimethylaniline at 0°C in styrene or chloroform exhibits an apparent second-order rate constant of $2.3 \times 10^{-4}\ \mathrm{sec}^{-1}$. However, acetyl [41] and lauroyl peroxide [71, 72] react somewhat slower.

Recently, Morsi et al. [73] have studied the rate of charge transfer interactions in the decomposition of organic peroxides. O'Driscoll and Richezza [74] have also reported the ultraviolet absorbance study of the complex formation between benzoyl peroxide and dimethylaniline. According to Horner and Schwenk [45], the mechanism for the polymerization of vinyl monomers by benzoyl peroxide and dimethylaniline is as follows:

$$\text{Ø}-\underset{\substack{\|\\ \text{O}}}{\text{C}}-\text{O}-\text{O}-\underset{\substack{\|\\ \text{O}}}{\text{C}}-\text{Ø} + \text{Ø}-\text{N}\begin{matrix}\diagup \text{CH}_3\\ \diagdown \text{CH}_3\end{matrix} \longrightarrow \left[\text{Ø}-\overset{\bullet +}{\text{N}}\begin{matrix}\diagup \text{CH}_3\\ \diagdown \text{CH}_3\end{matrix}\right]\text{ØCOO}^- + \text{ØCO}\dot{\text{O}} \tag{6.24}$$

$$\left[\text{Ø}-\overset{\bullet +}{\text{N}}\begin{matrix}\diagup \text{CH}_3\\ \diagdown \text{CH}_3\end{matrix}\right]\text{ØCOO}^- \longrightarrow \dot{\text{Ø}}-\text{N}\begin{matrix}\diagup \text{CH}_3\\ \diagdown \text{CH}_3\end{matrix} + \text{ØCOOH} \tag{6.25}$$

$$\text{Ø}-\overset{\bullet +}{\text{N}}\begin{matrix}\diagup \text{CH}_3\\ \diagdown \text{CH}_3\end{matrix} + \text{M} \longrightarrow \text{Ø}-\overset{\overset{\text{CH}_3}{|}}{\underset{\underset{\text{CH}_3}{|}}{\overset{+}{\text{N}}}}-\dot{\text{M}} \tag{6.26}$$

$$\text{Ø}-\overset{\overset{\text{CH}_3}{|}}{\underset{\underset{\text{CH}_3}{|}}{\overset{+}{\text{N}}}}-\dot{\text{M}} + n\text{M} \longrightarrow \text{Ø}-\overset{\overset{\text{CH}_3}{|}}{\underset{\underset{\text{CH}_3}{|}}{\overset{+}{\text{N}}}}-\dot{\text{M}}_{n+1} \tag{6.27}$$

$$2\text{Ø}-\overset{\overset{\text{CH}_3}{|}}{\underset{\underset{\text{CH}_3}{|}}{\overset{+}{\text{N}}}}-\dot{\text{M}}_{n+1} \xrightarrow[\text{Combination}]{\text{Disprop. or}} 1 \text{ to } 2 \text{ Polymers} \tag{6.28}$$

where steps (6.24) and (6.25) represent the formation of free radicals, step (6.26) the initiation of the monomer, step (6.27) the chain propagation, and step (6.28) the termination by combination or disproportionation.

They suggested that the dimethylaniline radical is the initiator. However, the mechanism was later questioned by Imoto et al. [52]. They suggested that the active radical (benzoate radical) produced by the interaction between benzoyl peroxide and dimethylaniline initiates the vinyl chloride polymerization.

In a later study, Horner [38] postulated the detailed reaction mechanism of tertiary amine with benzoyl peroxide and pictured the initiation of polymerization by benzoate radical. Mechanistically speaking, the first stage of the amine–peroxide reaction is, unquestionably, nucleophilic attack on the O–O bond. Imoto and Choe [75] have studied the detailed aspects of the mechanism of the reaction between substituted benzoyl peroxide in the presence of dimethylaniline (DMA). The mechanism of the reaction of Bz_2O_2 with substituted dimethylaniline was studied by Horner et al. [44, 45]. They have indicated that the higher the electron density of the lone pair on the nitrogen atom of substituted dimethylaniline, the stronger the promoting effect of the amine on the decomposition rate of Bz_2O_2. It was shown that the more abundant the quantity of DMA, the faster the decomposition velocity of Bz_2O_2.

In their study, Imoto and Choe [75] suggested the reversible formation of a complex intermediate III, which subsequently decomposes into free radicals as follows:

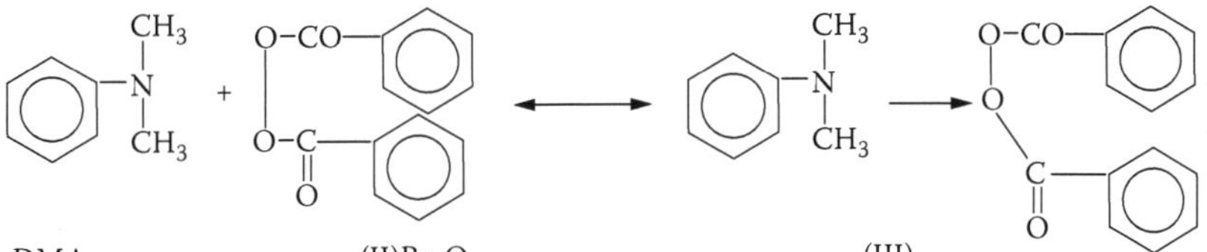

DMA + (II)Bz_2O_2 ⟷ (III) (6.29)

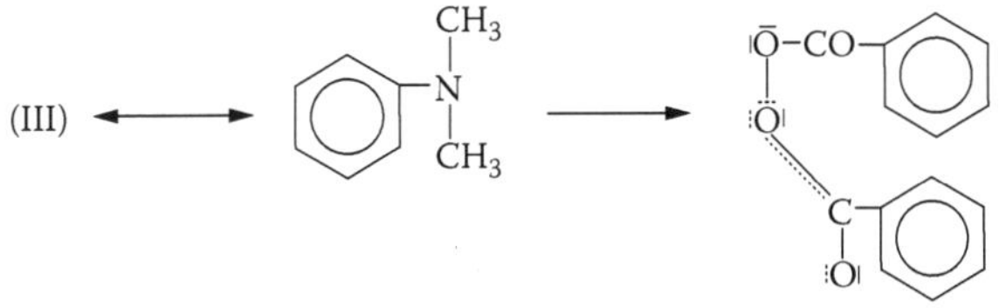

(III) ⟷ ⟶ (6.30)

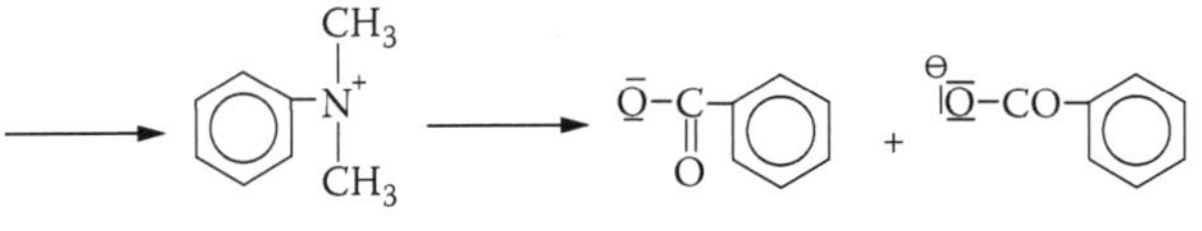

⟶ (IV) ⟶ + (V) (6.31)

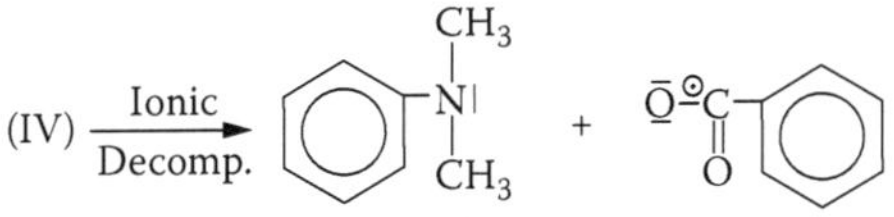

(IV) $\xrightarrow[\text{Decomp.}]{\text{Ionic}}$ (I, DMA) + (VI) (6.32)

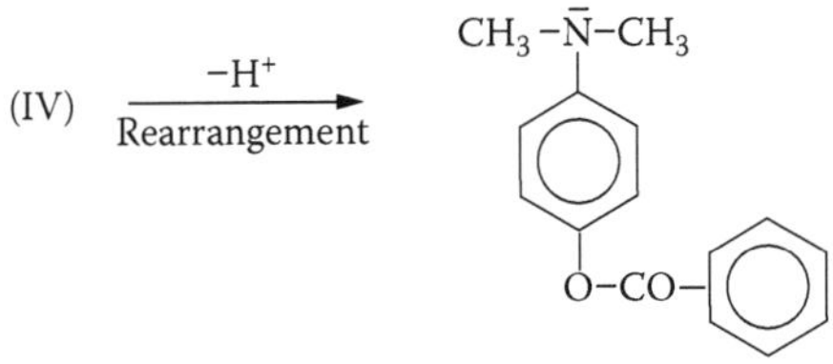

(IV) $\xrightarrow[\text{Rearrangement}]{-H^+}$ (VII) (6.33)

$$\text{(IV)} \xrightarrow[\text{Decomposition}]{\text{Radical}} \underset{\text{(VIII)}}{C_6H_5\overset{\oplus}{\dot{N}}(CH_3)_2} + \underset{\text{(IX)}}{\dot{O}-\overset{O}{\overset{\|}{C}}-C_6H_5} \quad (6.34)$$

Although it appears clear that Bz_2O_2 and DMA undergo a bimolecular reaction that gives rise to free radicals, the exact nature of the process is controversial. Thus, Horner [38] has proposed the formation of a "complex" in reaction (6.35) as the rate-determining step, which subsequently gives rise to the observed products.

$$\left[C_6H_5-\overset{\cdot +}{N}(CH_3)_2 \quad \begin{matrix} C_6H_5C(=O)-O \\ C_6H_5C(=O)-O^- \end{matrix} \right] \quad (6.35)$$

Imoto and Choe [75] have suggested the reversible formation of a complex, which subsequently decomposes into free radicals. However, Walling and Indictor [53] have suggested a new approach toward the free-radical mechanism between benzoyl peroxide and dimethylamine. They suggested that the rate-controlling step is a nucleophilic displacement on the peroxide by DMA to yield a quaternary hydroxylamine derivative. The reaction is as follows:

$$Bz_2O_2 + DMA \longrightarrow \left[C_6H_5-\underset{CH_3}{\overset{CH_3}{N}}OCOC_6H_5 \right]^+ C_6H_5COO^- \quad (6.36)$$

Such a formulation parallels that proposed for the bimolecular reaction [76] of peroxides and phenols, and, as it leads to an ionic product, should have a considerable negative entropy of activation. As has been pointed out previously [77], it also accounts for the accelerating effects of electron-supplying groups on the amine and electron-withdrawing groups on the peroxide, and parallels a plausible formulation of three other reactions: the reaction of peroxides with secondary amines, the formation of amine oxides in the presence of hydrogen peroxides, and the initiation of polymerization by amine oxides in the presence of acylating agents [78]. The product of reaction (36) has only a transient existence and decomposes by at least two possible paths:

$$\left[C_6H_5-\underset{CH_3}{\overset{CH_3}{N}}OCOC_6H_5 \right]^+ \longrightarrow C_6H_5\underset{CH_3}{\overset{CH_3}{N^{\cdot +}}} + C_6H_5COO^{\cdot} \quad (6.37)$$

$$\left[C_6H_5-\underset{CH_3}{\overset{CH_3}{N}}OCOC_6H_5 \right]^+ \longrightarrow \left[C_6H_5\overset{CH_3}{N}{=}CH_2 \right]^+ + C_6H_5COOH \quad (6.38)$$

Reaction (6.37), which gives Horner's [38] intermediate, represents a free-radical path and would account for the initiation of polymerization. As no significant amount of nitrogen is found in the resulting polymers, the amine fragment may well disappear by reacting with peroxide. Reaction (6.38) represents a nonradical breakdown and would account for the low efficiency of the system as a polymerization initiator. Admittedly, the same products could arise from a radical disproportionation closer to that suggested by Horner, but in the latter case, the reaction would have to occur in the same solvent "cage" as reaction (6.37), because otherwise, reaction (6.39) would compete with the initiation of polymerization and the efficiency of the latter would not show the independence of Bz_2O_2 and DMA concentration actually observed.

$$C_6H_5\overset{\overset{\displaystyle CH_3}{|}}{\underset{\underset{\displaystyle CH_3}{|}}{N^{\dot{+}}}} + C_6H_4COO^{\cdot} \longrightarrow \left[C_6H_5\ \overset{\overset{\displaystyle CH_3}{|}}{N} = CH_2 \right]^{+} + C_6H_5COOH \quad (6.39)$$

6.2.4.1 Suspension Polymerization of Vinyl Chloride

No induction period existed in the solution polymerization of vinyl chloride [52] initiated by the benzoyl peroxide–dimethylaniline system in various solvents such as tetrahydrofuran, ethylene dichloride, dioxane, cyclohexanone, methylethyl ketone, and so forth. The initial rate of polymerization and the conversion was directly and inversely proportional to the temperature, respectively. The polymerization was restricted to only 20% conversion, probably due to the complete consumption of benzoyl peroxide. Without the monomer, the extent of decomposition on benzoyl peroxide reaches a constant value regardless of the temperature and amount of dimethylaniline. It was seen that the greater the amount of dimethylaniline, the faster the initial rate of polymerization and the lower the conversion. The degree of polymerization of vinyl chloride obtained by the redox system benzoyl peroxide–dimethylaniline was generally lower than the polymer obtained by the benzoyl peroxide system alone. The activation energy of the polymerization by the redox system was lower than that of the benzoyl-peroxide-alone initiated polymerization and found to be 12.5 kcal mol^{-1}. The initial rate of polymerization could be expressed as

$$\left(\frac{d(PVC)}{dt}\right)_{t\to 0} = k(Bz_2O_2)^{1/2}(DMA)^{1/2} \quad (6.40)$$

The results in the solution polymerization of vinyl chloride are summarized in Table 6.6.

6.2.4.2 Suspension Polymerization of Acrylonitrile

The solution polymerization of acrylonitrile [57] has been studied in benzene at 40°C by a dilatometer using dimethylaniline and various substituted benzoyl peroxides. It was found that the initial rate of polymerization increased with increasing the molar ratio of DMA/Bz_2O_2 from 0–5 by keeping the Bz_2O_2 concentration at 5.57×10^{-5} mol L^{-1}. On the other hand, after a considerable polymerization time has elapsed,

TABLE 6.6
Solution Polymerization of Vinyl Chloride in Tetrahydrofuran; Bz_2O_2 = 0.52 mol (%)

DMA Bz_2O_2	Temperature (°C)	Initial Rate (% min)	Maximum Conversion (%)	DP
0.16	50	0.430	11.5	75
1.00	50	0.509	9.5	110
1.20	50	0.500	8.6	70
1.61	50	0.590	7.8	—
0.80	20	0.037	>26	85
0.80	30	0.120	<25	—
0.80	40	0.200	20.5	60
0.80	50	0.400	17.5	75
0.80	60	0.600	15.5	—

Source: M. Imoto, T. Otsu, and K. Kimura, *J. Polym. Sci.*, 15, 475 (1955).

the polymer yields in the presence of a large quantity of DMA frequently became smaller than the yield obtained in the presence of smaller quantity of DMA. The relation between the initial rates of polymerization R_p^0, and concentration of Bz_2O_2 and DMA may be expressed as

$$R_P^0 = k(Bz_2O_2)^{1/2}(DMA)^{1/2} \tag{6.41}$$

The initial rate was also found to be directly proportional to the monomer concentration. On the basis of the kinetic data, a rate equation may be derived as follows:

$$Bz_2O_2 + DMA \xrightarrow{k_1} BzO^{\cdot} \tag{6.42}$$

$$BzO^{\cdot} + M \xrightarrow{k_1} P_1^{\cdot} \text{ (initiation)} \tag{6.43}$$

$$P_n^{\cdot} + M \xrightarrow{k_p} P_{n+1}^{\cdot} \quad \text{(propagation)} \tag{6.44}$$

$$P_n^{\cdot} + P_n^{\cdot} \xrightarrow{k_t} P_n + P_{n'} \quad (\text{or } P_{n+n'}) \quad \text{(temination)} \tag{6.45}$$

Introduction of the steady state leads to

$$R = k_p(P_n^{\cdot})(M) = k_p\left(\frac{k_i(M)}{k_t}\right)^{1/2}(BzO^{\cdot})^{1/2}(M) \tag{6.46}$$

Again, assuming the steady state for $(BzO^{\cdot})$, the following equation will be drawn:

$$\frac{d(BzO^{\cdot})}{dt} = k_1(Bz_2O_2)(DMA) - k_i(BzO^{\cdot})(M) = 0 \tag{6.47}$$

From the previously mentioned equation, the following expression is readily obtained:

$$R = \frac{k_p k_1^{1/2}}{k_t^{1/2}} (M)(Bz_2O_2)^{1/2}(DMA)^{1/2} \tag{6.48}$$

6.2.4.3 Suspension Polymerization of Styrene

The polymerization of styrene in solution [53, 58] and bulk [54, 55, 59–61] by the redox system benzoyl peroxide–di-n-alkylaniline has been studied considerably by many researchers. Different dialkylanilines (DAAs) such as dimethylaniline (DMA), diethylaniline (DEA), di-n-butylanilines, di-n-octylaniline, and di-n-decylaniline combined with benzoyl peroxide have been studied for the solution polymerization of styrene [58] in benzene at 30°C. It was found that the initial rate of polymerization increased with a decrease of the molar ratio of Bz_2O_2. The degree of polymerization decreased with the decrease of molar ratio of Bz_2O_2/DAA. The initiator efficiency seemed to increase gradually with the number of carbons of the alkyl groups in the DAAs, with the exception of di-n-octylaniline.

The kinetics of the bulk polymerization of styrene has been studied in detail at 30°C and 60°C by a dilatometer [60] using the benzoyl peroxide–dimethylaniline redox system. Also, the initiating efficiency of the ring-substituted diethylanilines–benzoyl peroxide system [61] at 30°C for styrene polymerization has been reported. A mathematical treatment for the free-radical production by Bz_2O_2–DMA has been derived for the styrene polymerization [55, 59] at 0°C. The initial rates of polymerization R_p^0 for 30°C and 60°C are as follows:

At 30°C,

$$R_p^0 = 1.67 \times 10^{-3}([Bz_2O_2][DMA])^{0.497} \tag{6.49}$$

At 60°C,

$$R_p^0 = 5.25 \times 10^{-3}([Bz_2O_2][DMA])^{0.418} \tag{6.50}$$

The results of O'Driscoll and Schmidt [60] were different from those of Meltzer and Tobolsky [54] for R_p^0 (initial rate of polymerization) as a function of the catalyst concentration. In the latter work, it was shown that the rate law $R_p^0 = K([Bz_2O_2][DMA])^a$ held over a wide range of catalyst concentrations and temperatures. The value of exponent a was 0.5 at low temperatures, as expected for a bimolecular reaction between amine and peroxide. At 30°C, 45°C, and 60°C, the values found by Melzer and Tobolsky [54] were 0.39, 0.38, and 0.33, respectively. However, according to O'Driscoll and Schmidt [60], these values were 0.5 and 0.42 at 30°C and 60°C, respectively. The lower value at 60°C may be attributed to the existence of an induction period at the lower catalyst concentration.

In conclusion, it was demonstrated that the kinetics of polymerization are the same at higher and lower temperature. The efficiency of the reaction in initiating polymerization appears to fall slightly with increasing temperature.

6.2.4.4 Suspension Polymerization of Methyl Methacrylate

Lal and Green [56] have studied extensively the bulk polymerization of methyl methacrylate at 25°C using various amines, mostly tertiary amines. The total yield of polymer depends on the heat developed during polymerization as well as the production of free radicals. The rate of polymerization increased or decreased with the substitution at the para position of dimethylaniline by electron-donating groups or electron-withdrawing groups, respectively. Aliphatic tertiary amines are much less reactive than aromatic tertiary amines for accelerating polymerization, whereas primary amines, aliphatic as well as aromatic, act as inhibitors. Substitution of the methyl groups in dimethylaniline by ethyl groups does not change the reactivity of the amine for accelerating the polymerization; however, when propyl groups are substituted for methyl groups, the reactivity is somewhat reduced. Replacement of methyl groups in dimethylaniline by droxy ethyl groups does not materially affect the reactivity of the amine for accelerating polymerization. Tribenzylamine decomposes benzoyl peroxide very rapidly (less than 5 min), but no polymer is obtained in the bulk polymerization of methyl methacrylate. The amine may function as its own inhibitor. The molecular weights of the polymers obtained are in the neighborhood of 120,000 ± 10,000 in the case of trialkylamines.

Very recently, the feasibility of aliphatic tertiary amine like the triethylamine–benzoyl peroxide redox-initiating system in photopolymerization of methyl methacrylate [66] has been reported. In the dilatometric study of methyl methacrylate polymerization at 35°C with various solvents, the initiator exponent was 0.34. The monomer exponent depends on the solvents used. In acetonitrile, pyridine, and bromobenzene, the monomer exponent was 0.5, 0.67, and 1.1, respectively, within the concentration range studied. Benzene and chloroform give first-order dependence of rate on [monomer] and behave as normal (inert) diluents. The activation energy is 3.2 kcal mol^{-1}.

6.2.5 Quaternary Ammonium Salts as Reductants

Quaternary salts in combination with benzoyl peroxide are known to induce vinyl polymerization in emulsion systems [79, 80]. Quaternary salts are also potential photoinitiators for vinyl polymerization [81]. The use of quaternary salts in combination with peroxides as redox initiators for suspension polymerization of styrene [82] and polymerization of methyl methacrylate [83, 84] in bulk or in solution have been explored.

The polymerization of MMA with the cetyltrimethyl ammonium bromide (CTAB)–Bz_2O_2 redox system [84] and the cetylpyridinium bromide (CPB)–benzoyl peroxide redox system [83] was strongly inhibited by hydroquinone, whereas the inhibitory effect of air or oxygen was marginal. A radical mechanism is thus indicated. End-group analysis for amino end groups by the dye technique [85] clearly indicated the incorporation of basic (amino) end groups. When a dilute solution of Bz_2O_2 was mixed with an equal volume of a dilute solution of quaternary salt, for example, CPB, the UV absorption spectrum of the mixture was not the average of the spectra of the two solutions. The absorbance difference may be attributed to the rapid equilibrium between the formation of a complex and the components.

Thus, the species effective for initiating polymerization appears to be the complex of the peroxide and CPB, which subsequently decomposes by a radical mechanism. The concentration of the initiating species [I] in the polymerization may be expressed as

$$[I] = K_c\,[CPB][Bz_2O_2] \tag{6.51}$$

where K_c is the equilibrium constant for complexation:

$$CPB + Bz_2O_2 \overset{K_c}{\rightleftharpoons} \underset{\text{Initiating Complex (I)}}{[CPB \cdots Bz_2O_2]} \tag{6.52}$$

The mechanism is similar to that of the cetyltrimethyl ammonium bromide–benzoyl peroxide redox system. The radical generation process may be considered to include the following steps:

1. Complexation

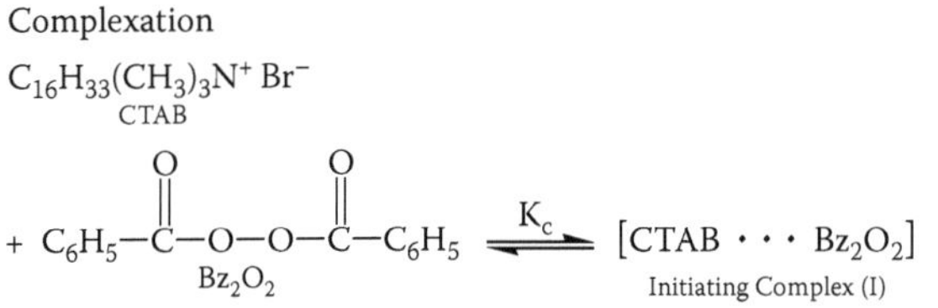

Complexation

$$\underset{\text{CTAB}}{C_{16}H_{33}(CH_3)_3N^+\,Br^-} + \underset{Bz_2O_2}{C_6H_5-\overset{\overset{O}{\|}}{C}-O-O-\overset{\overset{O}{\|}}{C}-C_6H_5} \overset{K_c}{\rightleftharpoons} \underset{\text{Initiating Complex (I)}}{[CTAB \cdots Bz_2O_2]} \tag{6.53}$$

2. Radical generation

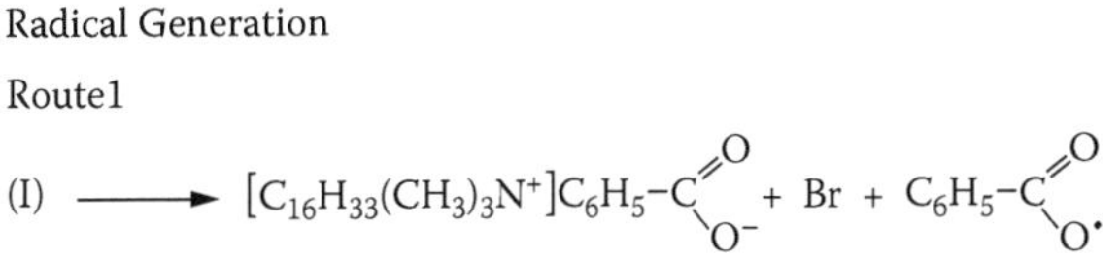

Radical Generation

Route1

$$(I) \longrightarrow [C_{16}H_{33}(CH_3)_3N^+]C_6H_5-C(=O)O^- + Br + C_6H_5-C(=O)O^{\bullet} \tag{6.54}$$

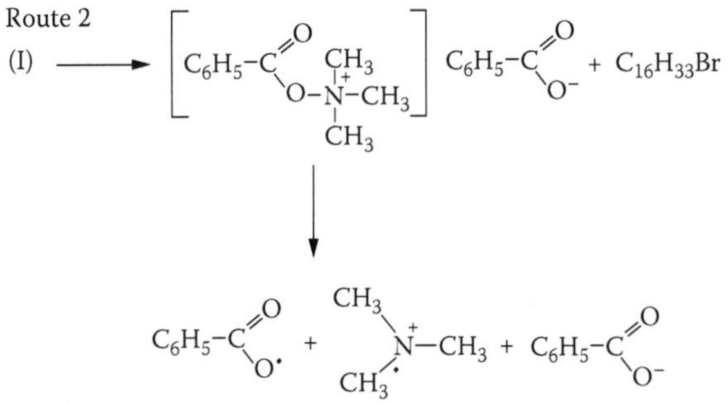

Route 2

$$(I) \longrightarrow \left[C_6H_5-C(=O)-O-\overset{+}{N}(CH_3)_3\right] C_6H_5-C(=O)O^- + C_{16}H_{33}Br \tag{6.55}$$

$$\downarrow$$

$$C_6H_5-C(=O)O^{\bullet} + (CH_3)_2\overset{+\bullet}{N}-CH_3 + C_6H_5-C(=O)O^-$$

$$(CH_3)_2\overset{+\bullet}{N}-CH_3 \longrightarrow CH_3-N(CH_3)-\dot{C}H_2 \tag{6.56}$$

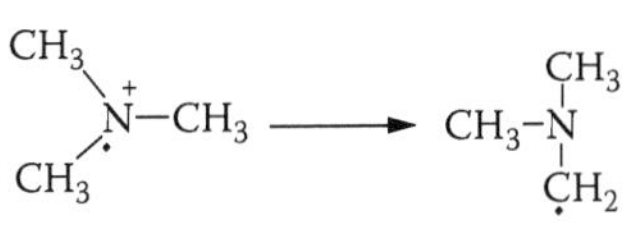

The radical generation step is apparently influenced by monomer (M) and solvent molecules (S), which possibly compete in reaction with the initiating complex (I). The radical generation reactions influenced by monomer and solvent may then be expressed as

$$(I) + M \xrightarrow{k_1} \text{radicals} \tag{6.57}$$

$$(I) + S \xrightarrow{k_2} \text{radicals} \tag{6.58}$$

The rate of initiation R_i may then be written as

$$R_i = K_c\,[CATB][Bz_2O_2]\,(k_1[M] + k_2[S]) \tag{6.59}$$

6.2.5.1 Suspension Polymerization of Methyl Methacrylate

The polymerization of methyl methacrylate was studied dilatometrically at 40°C under bulk and high-dilution conditions using CPB–Bz_2O_2 [83] and CTAB–Bz_2O_2 [84] redox systems in polar solvents such as alcohol, acetone, or dimethyl formamide. The effect of several solvents/additives on the polymerization revealed that dimethyl formamide (DMF), acetonitrile, and pyridine acted as rate-enhancing solvents; benzene, methanol, chloroform, and acetone acted as inert diluents; formamide and acetamide cause pronounced retardation. In the case of the CTAB–Bz_2O_2 system under bulk condition (using DMF 10% of the total), the rate was practically independent of $[Bz_2O_2]$ up to 0.025 M, whereas the kinetic order with respect to CTAB was about 0.16 for a concentration up to 0.001 M. At high dilution (DMF 50% of the total), the rate of polymerization was proportional to $[Bz_2O_2]^{0.5}$ and $[CTAB]^{0.5}$. At the high-dilution condition in DMF (50% v/v), R_p increased with $[Bz_2O_2]$ up to 0.025 M and remained constant with a further increase in $[Bz_2O_2]$. R_p increased with increasing [CTAB] up to 0.001 M and then decreased with a further increase in [CTAB]. It was found that R_p increased with increasing DMF content up to about 30%. This accelerating effect of DMF was not apparent with further dilution and the usual effect of monomer concentration was found, the order with respect to the monomer being unity. The overall activation energy was 11.2 kcal mol^{-1}.

However, in the case of CPB–Bz_2O_2 initiating system, R_p was proportional to $([CPB][Bz_2O_2])^{0.18}$ both in near-bulk and high-dilution conditions. The [CPB] was between 0.1×10^{-3} and 8×10^{-3} M and $[Bz_2O_2]$ with between 3×10^{-3} and 100×10^{-3} M. The activation energy for polymerization was 13.6 kcal mol^{-1}. DMF, acetonitrile, and pyridine acted as rate-enhancing solvents in the redox polymerization, whereas formamide and acetamide behaved as retarding additives.

6.2.5.2 Suspension Polymerization of Styrene

In the case of suspension polymerization of styrene [82] using the Bz_2O_2–lauryl puridinium chloride redox system, about 100% conversion with 1-mm-diameter polystyrene beads was obtained using the typical recipe presented in Table 6.7.

TABLE 6.7
Typical Recipe: Suspension Polymerization of Styrene[a]

Ingredients	Amount (g)
Water	1000
Styrene	900
3 μ Mg silicate	0.5
Benzoyl peroxide	2.7
Lauryl pyridinium chloride	0.03
tert-Butyl perbenzoate	3.6

[a] Polymerization for 10 h at 80–120°C.

6.2.6 Nitrite as Reductant

The reaction between hydrogen peroxide and sodium nitrite was studied in detail by Halfpenny and Robinson [86] in 1952. The characteristics of the reaction, particularly in the presence of a bromide ion, and the evolution of oxygen with certain concentration of peroxides suggested the possible formation of free radicals. They demonstrated their occurrence by observing the polymerization of methyl methacrylate. In the early 1950s, Schulz et al. [87] had reported the acrolein polymerization by H_2O_2–$NaNO_2$ as a redox initiator. The reports of the use of nitrites as a reducing agent in polymerization are very few. Patent literature reports the suspension polymerization of vinyl pyridine [88] and vinyl chloride [89] using $NaNO_2$ as the reducing agent in combination with acyl peroxides like Bz_2O_2 or lauroyl peroxide. The mechanism may be written in one step as follows:

$$\mathrm{R-\overset{\overset{O}{\|}}{C}-OO-\overset{\overset{O}{\|}}{C}-R+NO_2^- \longrightarrow R-\overset{\overset{O}{\|}}{C}-O^{\cdot}+NO_2^{\cdot}+R-\overset{\overset{O}{\|}}{C}-O^-} \quad (6.60)$$

According to Halfpenny and Robinson [86] in the light of the mechanism for the H_2O_2–nitrite redox system, the various steps of the reaction system may be written as follows:

$$\mathrm{NO_2^- + H^+ \longrightarrow HNO_2} \quad (6.61)$$

$$\mathrm{HNO_2+R-\overset{\overset{O}{\|}}{C}-OO-\overset{\overset{O}{\|}}{C}-R \longrightarrow R-\overset{\overset{O}{\|}}{C}-O_2NO+R-\overset{\overset{O}{\|}}{C}-OH} \quad (6.62)$$

$$\mathrm{R-\overset{\overset{O}{\|}}{C}-O_2NO \longrightarrow R-\overset{\overset{O}{\|}}{C}-O^{\cdot}+{}^{\cdot}NO_2} \quad (6.63)$$

$$\mathrm{R-\overset{\overset{O}{\|}}{C}-O^{\cdot}+{}^{\cdot}NO_2 \longrightarrow R-\overset{\overset{O}{\|}}{C}NO_3} \quad (6.64)$$

When peroxide is abundant, the acyl radical provides a means of oxygen liberation according to the following reactions:

$$\mathrm{R-\overset{\overset{\displaystyle O}{\|}}{C}-O^{\cdot}+R-\overset{\overset{\displaystyle O}{\|}}{C}-OO-\overset{\overset{\displaystyle O}{\|}}{C}-R\longrightarrow R-\overset{\overset{\displaystyle O}{\|}}{C}-O-\overset{\overset{\displaystyle O}{\|}}{C}-R+R-\overset{\overset{\displaystyle O}{\|}}{C}-O-O^{\cdot}} \quad (6.65)$$

$$\mathrm{R-\overset{\overset{\displaystyle O}{\|}}{C}-OO-\overset{\overset{\displaystyle O}{\|}}{C}-R+R-\overset{\overset{\displaystyle O}{\|}}{C}-O-O^{\cdot}\longrightarrow R-\overset{\overset{\displaystyle O}{\|}}{C}-O^{\cdot}+R-\overset{\overset{\displaystyle O}{\|}}{C}-O-\overset{\overset{\displaystyle O}{\|}}{C}+O_2} \quad (6.66)$$

The preceding free radicals take part in the initiation and the termination processes in polymerization.

6.2.6.1 Suspension Polymerization of Vinyl Chloride

Vinyl chloride [89] with or without a comonomer has been suspension polymerized using a mixed-catalyst system (i.e., lauroyl peroxide and 2-ethylhexyl peroxydicarbonate) with a reducing agent $NaNO_2$ in two reactors maintained at different temperatures. The polymerization was performed according to the typical recipe presented in Table 6.8. After 50 h, the conversion values in the first and second reactors were 15% and 90%, respectively, and no side-wall deposition was noted. The polymer had a weight-average degree of polymerization (DP) of 1020, a plasticizer absorbability of 29.2%, a thermal stability of 75 min, and gel time of 2.5 min. The product prepared by polymerization at 58°C in both reactors had a weight-average DP of 1010, a plasticizer absorption of 24.8%, a thermal stability of 65 min, and a gel time of 4.0 min.

6.2.6.2 Suspension Polymerization of Vinyl Pyridine

Vinyl pyridine with or without comonomers was polymerized by suspension polymerization [88] in H_2O in the presence of fatty acid esters or phthalates to give polymers in spherical powder form. The polymerization was performed according to the

TABLE 6.8
Typical Recipe: Suspension Polymerization of Vinyl Chloride[a]

Ingredients	Amount (ppm)
Water	140
Vinyl chloride	100
2-Ethylhexyl peroxydicarbonate	0.04
Lauroyl peroxide	0.01
80% Saponified poly(vinyl acetate)	0.07
$NaNO_2$	0.002

[a] Polymerization at 61°C in the first reactor and at 57°C in the second reactor for 50 h.

TABLE 6.9
Typical Recipe: Suspension Polymerization of Vinyl Pyridine[a]

Ingredients	Amount (ppm)
Water	800
4-Vinyl pyridine	247
Styrene	62
Benzoyl peroxide	4
Dioctyl phthalates	50
NaCl	234
$NaNO_2$	2.9
Hydroxyethyl cellulose	4.5

[a] Polymerization for 9 h at 80°C at a stirring rate of 250 rpm.

recipe provided in Table 6.9 to yield a final product of 850 ml of yellow transparent spherical copolymer beads. When dioctyl phthalate was omitted, a similar composition yielded large lumps.

6.3 ALKYL PEROXIDE

Alkyl peroxides are extensively used for the suspension polymerization of styrenic monomers [90–93] and vinyl chloride [94, 95].

6.3.1 Alkyl Boron as Reductant

Alkyl boron compounds can initiate the polymerization of vinyl monomers in the presence of a suitable cocatalyst. A common feature of the cocatalyst (i.e., peroxides [96–98], hydroperoxides [99], amines [100], and organic halides [101]) is that it can be considered an "electron-donating" compound. In most of the systems investigated, it has been established that the reaction is a free-radical polymerization [101–104], but the rate equation is not simple, suggesting a complex mechanism in which coordination of the organometallic compound is a rate-determining step [100, 101, 105]. Furthermore, the reaction order changes when the organometallic compound to cocatalyst ratio changes for peroxides [97], oxygen [102–104, 106, 107], hydroperoxides [99], hydrogen peroxide [108], and organic halides [101]. This change of order, which may be a consequence of complex formation, is attributed to various causes [101, 106, 108], but in most cases, no satisfactory explanation is given.

The oxidation of trialkylborons by molecular oxygen generally produces alkoxy boron compounds via intermediate peroxides [109, 110], although Mirviss [111] has reported hydrocarbons among the products. In addition, vinyl monomers polymerize at room temperature in the presence of trialkylborons and air [112]. Free radicals are evidently produced at some stage in the reaction. Free radicals have been assumed to arise from the hemolytic decomposition of peroxidic intermediates [111, 113] even though these peroxides are very stable at room temperature [110]. Others have suggested that the free radicals are produced in a reaction between the peroxide and unoxidized trialkylboron [98, 114].

The high rate of peroxide formation in the oxidation of triethylboron seems to rule out the possibility of a long-lived oxygen–triethylboron complex that rearranges to the peroxide. This has been stated by various authors [109–111]. Only Zutty and Welch [109] have provided experimental evidence in the case of tri-n-butylboron. A transient intermediate cannot be excluded. No indication exists that free radicals arise during the oxidation of triethylboron. Both triethylboron and peroxide were required to initiate vinyl polymerization of methyl methacrylate in agreement with Bawn and co-workers [98]. The results [104] indicate that the ethyl radical was produced in a reaction between triethylboron and the peroxide, wherein the peroxide was reduced. No evidence existed for the presence of the ethoxy radical arising from hemolytic decomposition of the peroxide. Iodide was an efficient trap for the ethyl radical. From the work of Hansen and Hamman [104], it is indicated with some uncertainty that the reduction is a 1:1 reaction. It is unlikely that the ethyl radical was the only one produced, but the structure of a companion radical could not be ascertained. The amount of iodine consumed indicates that the radicals were not produced in each reactive act. A possible explanation [115] is that the reduction is a "cage reaction":

$$\begin{array}{c} >B{-}C_2H_5 \\ \uparrow \\ C_2H_5O{\cdots}OB \end{array} \longrightarrow\ >B{-}OC_2H_5 + C_2H_5^{\cdot} + >BO^{\cdot} \tag{6.67}$$

The formation of a cage should be especially favored by coordination of one of the oxygen atoms to boron. Radical recombination would lead to the alkoxy compounds commonly observed, whereas diffusion from the cage could lead to the products derived from free radicals.

Recently, Abuin et al. [96] investigated the kinetic features of bulk polymerization of methyl methacrylate using triethyl boron (TEB)–di-*tert*-butylperoxide mixture as the radical initiator. From their data, it can be seen that, working at a constant di-*tert*-butylperoxide concentration, the reaction rate increases as the TEB concentration increases, reaching a maximum and then decreasing with further TEB addition. If the TEB only modifies the initiation rate, a simple free-radical mechanism would predict that at a given temperature, the following equation should hold true:

$$R\lambda = \text{constant} \tag{6.68}$$

where R is the measured polymerization rate and λ is the mean chain length.

However, their data show that, at a high TEB concentration, the product ($R\lambda$) decreases when TEB increases. This effect can be related to the occurrence of chain transfer to the organometallic compound. The data can be treated according to the following reaction scheme:

$$\text{TEB} + X \rightarrow 2R^{\cdot} \tag{6.69}$$

$$R^{\cdot} + M \rightarrow M^{\cdot} \tag{6.70}$$

$$M^{\cdot} + M \rightarrow M^{\cdot} \tag{6.71}$$

$$M^{\cdot} + M^{\cdot} \rightarrow \text{polymer} \tag{6.72}$$

$$M^{\cdot} + M \rightarrow Q^{\cdot} + \text{polymer} \tag{6.73}$$

$$M^{\cdot} + TEB \rightarrow P^{\cdot} + \text{polymer} \tag{6.74}$$

$$P^{\cdot} + M \rightarrow M^{\cdot} \tag{6.75}$$

$$Q^{\cdot} + M \rightarrow M^{\cdot} \tag{6.76}$$

$$P^{\cdot} + M^{\cdot} \rightarrow \text{polymer} \tag{6.77}$$

where M, X, and $R^{\cdot}$ represent monomer, peroxide, and radical, respectively.

The chain transfer reactions are considered to be

$$M^{\cdot} + TEB \rightarrow MH + P^{\cdot} \tag{6.78}$$

$$M^{\cdot} + TEB \rightarrow C_2H_5^{\cdot} + MB(C_2H_5)_2 \tag{6.79}$$

A reaction similar to reaction (6.79) has been reported as being extremely fast for several radicals conjugated to a carbonyl group [116].

Initiation by TEB–di-*tert*-butyl peroxide (DTP) shows [97] the following main characteristics:

1. At low (TEB/DTP), the initiation step follows a rate law represented by

$$R_i \rightarrow k_{80}(TEB)(DTP) \tag{6.80}$$

2. When the peroxide concentration is kept constant, the rate of initiation increases, reaching a maximum and then decreasing when the TEB concentration increases.

The following mechanism is consistent with these findings:

$$TEB + MMA \rightarrow C_1 \tag{6.81}$$

$$2TEB + DTP \rightarrow C_2 \tag{6.82}$$

$$C_1 + DTP \rightarrow \text{radicals} \tag{6.83}$$

where C_1 and C_2 represent complexed forms of the TEB. The proposed mechanism gives the following expression for the rate of initiation:

$$R_i = \frac{k_{83}K_{81}(MMA)(DTP)_0(TEB)}{1 + K_{82}(TEB)^2} \tag{6.84}$$

where $(DTP)_0$ is the total peroxide concentration, (TEB) is the concentration of TEB uncomplexed, and K_{81} and K_{82} are the equilibrium constants of reactions (6.81) and (6.82), respectively. The concentration of uncomplexed TEB can be obtained from the total concentration $(TEB)_0$ by solving

$$(TEB)_0 = \frac{(TEB)[1 + K_{81}(MMA)] + 2K_{82}(DTP)_0(TEB)^2}{1 + K_{82}(TEB)^2} \tag{6.85}$$

Similarly, for a rate at low $(TEB)_0$, Eq. (6.84) reduces to

$$R_i = \frac{k_{83}K_{81}(MMA)(DTP)_0(TEB)_0}{1 + K_{81}(MMA)} \tag{6.86}$$

Similarly, the maximum rate for a given $(TEB)_0$ could be derived to be

$$(R_i)_{max} = \frac{k_{83}K_{81}(MMA)(DTP)_0}{2K_{82}^{1/2}} \tag{6.87}$$

6.3.1.1 Bulk Polymerization of Methyl Methacrylate

Methyl methacrylate [97] was bulk polymerized at 50°C using t-butyl peroxide–triethylboron (TEB) as the initiator. The rate of initiation by the mixture of triethylboron and t-butyl peroxide was first order with respect to peroxide. The order in triethylboron changes from 1 at a low triethylboron/peroxide ratio to nearly 0 at a high triethylboron/peroxide ratio. The results [97] are given in Table 6.10.

Abuin et al. [96] also reported the bulk polymerization of methyl methacrylate at 20°C using triethylboron–di-t-butylperoxide at various triethyl-boron concentrations. The amount of polymer produced was proportional to the reaction time. The results are presented in Table 6.11.

Abuin et al. [96] have also compared the rate of polymerization initiated by the mixture containing different peroxides and it is found that the rate with dimethyl peroxide is nearly 32 times faster than with di-*tert*-butylperoxide as a cocatalyst. This difference can be attributed to the steric hindrance introduced by the bulky *tert*-butyl groups. Similarly, it is interesting to note the difference between TEB and triethyl aluminum (TEA) as the cocatalyst with peroxides. With TEB, alkyl and acylic peroxides show similar cocatalytic activities [97]. On the other hand, it has been reported that TEA is only active when acyl peroxides are employed [117].

TABLE 6.10
Bulk Polymerization of Methyl Methacrylate at 50°C

[t-Butyl Peroxide] (mol L^{-1})	[TEB] (mol L^{-1})	Time (min)	Conversion % (% L^{-1})	E (kcal mol^{-1})
0.062	0.034	60	3.2	
		6	4.1	
		135	9.7	10.0
0.062	0.063	60	6.0	
		60	5.5	
		100	10.3	
		120	10.9	

TABLE 6.11
Bulk Polymerization of Methyl Methacrylate at 20°C

[TEB] (10^{-2} M)	[DTP][a] (M)	Rate of Polymerization)[b] ($\times 10^{-6}$ m sec^{-1})	Mean Chain Length
1.18	0.062	3.65	5000
6.80	0.062	4.50	4000
15.40	0.062	4.14	3800
24.50	0.062	2.74	5500
38.80	0.062	2.50	5000

[a] DTP = di-t-butyl peroxide.
[b] Averaged over a 185-min reaction time.

This difference can be related to the monomeric state of TEB; TEA is mainly present as a dimer [117].

6.4 PERESTERS (PEROXYESTERS OF CARBONIC ACID)

Although peroxydicarbonates are useful low-temperature initiators for vinyl polymerization [118, 119], little has been published about the characteristics of their thermal decomposition. The rate of decomposition was determined for several of these compounds ($[ROC(O)O]_2$, R = Et, i-Pr, $PhCH_2$, $NO_2CMe_2CH_2$ [120]; R = i-Pr [104]) in the early 1950s. Razuvaev et al. have since added others to the list (R = Me, Bu, i-Bu, t-Bu, amyl, cyclohexyl [122], and Ph [123, 124]). A belief exists that peroxydicarbonates are particularly sensitive to radical-induced decomposition [119–121]). This is undoubtedly true for the pure substances [119, 120]. The addition of 1% of iodine to pure diisopropyl peroxydicarbonate reduces [120] the rate of decomposition by a factor of 60. Among all the percarbonates, phenyl peroxydicarbonate [123, 124] may prove to act differently because decarboxylation yields the resonance-stabilized phenoxy radical. This peroxide is said to be more labile than other peroxydicarbonates and inhibits [123] instead of initiates polymerization [124]. Peroxydicarbonates are very efficient initiators for the suspension polymerization of vinyl chloride [125–139], vinylidiene fluoride [16], and styrene [17], and the copolymerization of vinyl acetate [18, 19] with other monomers. Peroxydicarbonates [140–143] are also proved to be efficient-radical initiators in conjunction with various reducing agents for vinyl polymerization.

6.4.1 Mercaptans as Reductant

Mercaptans have been proved to be an efficient reducing agent with H_2O_2 [144, 145] to initiate vinyl polymerization. It has also been used with Bz_2O_2 [146] for emulsion polymerization of vinyl monomers. The activation of persulfate by reducing agents such as thiols [147–155] has been extensively studied and the combination has been used for vinyl polymerization. Starkweather et al. [156] and Kolthoff et al. [157, 158] have demonstrated the catalytic effect of thiols in persulfate-initiated emulsion polymerization of styrene with or without butadiene.

The use of 2-mercaptoethanol as reducing agent in conjunction with peroxydicarbonate for the suspension polymerization of vinyl chloride [159, 160] has been reported in the patent literature. During the redox reaction, hydrogen is extracted from thiol by the homolysis of the –S–H group to give a sulfur radical. The mechanism may be proposed as follows:

$$\mathrm{R{-}O{-}\overset{\overset{O}{\|}}{C}{-}O{-}O{-}\overset{\overset{O}{\|}}{C}{-}O{-}R + R{-}S{-}H \longrightarrow RS^{\cdot} + R{-}O{-}\overset{\overset{O}{\|}}{C}{-}OH + R{-}O{-}\overset{\overset{O}{\|}}{C}{-}O^{\cdot}} \tag{6.88}$$

6.4.1.1 Suspension Polymerization of Vinyl Chloride

Suspension polymerization of vinyl chloride has been reported using 2-mercaptoethanol as a reductant with bis (2-ethylhexyl) peroxycarbonate [160] and diisopropyl peroxydicarbonate [159]. Thus, in the case of (2-ethylhexyl) peroxycarbonate [160], mixtures of vinyl chloride with or without comonomers 100, C_{2-6} compounds, having –SH and –OH groups 0.001–0.1 ppm, C_{4-18} alkyl vinyl ether 0.01–1.0 ppm, and benzyl alcohol with or without C_{1-4} alkyl substituents 0.01–1.0 ppm are stirred to give PVC or copolymers having equally good porosity, heat stability, and processability.

In the case of diisopropyl peroxydicarbonate [159], an 80–20:20–80 mixture of partially saponified poly(vinyl acetate) and a cellulose ether was used as the dispersing agent. The suspension polymerization or copolymerization of vinyl chloride was performed in the presence of a compound having a –SH, –OH, or $–CO_2H$ groups to reduce the amount of chain transfer agent required.

6.4.2 Sulfide and Dithionate as Reductant

The oxyacids of sulfur such as sulfite, bisulfite, bisulfate, thiosulfate, metabisulfite, and dithionate proved to be efficient reducing agents in the redox-initiated polymerization of vinyl monomers. Numerous articles in these areas have been reported in the literature. Palit et al. [161–164] and Roskin et al. [165–167] have reported the polymerization of vinyl monomers using the persulfate–dithionate redox system. Chaddha et al. [168] also reported the persulfate–sulfide redox system to initiate polymerization. The use of sulfide [169, 170] and dithionate [171] as reducing agents in conjunction with organic hydroperoxide, like cumene hydroperoxide and iron salt in emulsion polymerization, has been described. Tadsa and Kakitani have reported the suspension polymerization of vinyl chloride by percarbonate–sodium sulfide [172] and percarbonate–sodium dithionate [173] systems.

The general initiation reaction in these systems can be schematically represented as

$$\mathrm{R{-}O{-}\overset{\overset{O}{\|}}{C}{-}O{-}O{-}\overset{\overset{O}{\|}}{C}{-}O{-}R \; + A^{-2} \longrightarrow R{-}O{-}\overset{\overset{O}{\|}}{C}{-}O^{-} + R{-}O{-}\overset{\overset{O}{\|}}{C}{-}O^{\cdot} + A^{\cdot -}} \tag{6.89}$$

TABLE 6.12
Typical Recipe: Suspension Polymerization of Vinyl Chloride by Dioctyl Peroxydicarbonate–Sodium Sulfide System[a]

Ingredients	Amount (ppm)
Water	150
Vinyl chloride	100
Partially saponified poly(vinyl acetate)	0.1
Dioctyl peroxydicarbonate	0.05
Na_2Sx	0.01

[a] Polymerization under stirring for 5.8 h at 58°C.

where A is sulfide or dithionate;

$$\mathrm{R{-}O{-}\overset{\overset{\displaystyle O}{\|}}{C}{-}O{-}O{-}\overset{\overset{\displaystyle O}{\|}}{C}{-}O{-}R} + S^{-2} + S_3O_6^{-2} \longrightarrow \mathrm{R{-}O{-}\overset{\overset{\displaystyle O}{\|}}{C}{-}O^-} + \mathrm{R{-}O{-}\overset{\overset{\displaystyle O}{\|}}{C}{-}O^{\cdot}} + {}^{\cdot}S^- + S_3O_6^{\cdot-} \quad (6.90)$$

These indicated radicals initiate polymerization.

6.4.2.1 Suspension Polymerization of Vinyl Chloride

The presence of sulfide or dithionate also prevents scale formation during polymerization. Thus, in the system dioctyl peroxydicarbonate–sodium sulfide [172], vinyl chloride with or without vinyl comonomers was polymerized in the presence of 0.1–1000 ppm (based on monomers) inorganic sulfides according to the recipe presented in Table 6.12 to give PVC with good heat stability, with no scale formation, compared with 450 g m^{-2} for a similar run without Na_2S_x.

Similarly, in the case of dioctyl peroxydicarbonate–$Na_2S_3O_6$ [173], the polymerization was performed according to the recipe in Table 6.13 to give PVC with good heat stability. Scale formation in the preceding polymerization was 5 g m^{-2}, compared with 550 g m^{-2} for a similar run without $Na_2S_3O_6$.

TABLE 6.13
Typical Recipe: Suspension Polymerization of Vinyl Chloride by Dioctyl Peroxydicarbonate –$Na_2S_3O_6$ System[a]

Ingredients	Amount (ppm)
Water	150
Vinyl chloride	100
Partially saponified poly(vinyl acetate)	0.1
Dioctyl peroxydicarbonate	0.04
$Na_2S_3O_6$	0.001

[a] Polymerization under stirring for 6.4 h at 58°C.

6.4.3 Alkyl Borane as Reductant

Despite the great number of investigations [106, 112, 114, 174–177] in which alkyl boron compounds were used as initiators of vinyl polymerization, most of the main features of the mechanism involved for the initiating system, such as alkyl boron compounds in the absence of air [106, 113, 114, 175, 178–181], peroxides or hydroperoxides in conjunction with trialkylboron (A_3B) compounds [98, 99, 182] have not yet been demonstrated. The reports on a percarbonate–alkyl boron redox system for vinyl polymerization are very few. The bulk polymerization of vinyl chloride by the redox system consisting of diisopropyl peroxydicarbonate–triethylboron has been reported by Ryuichi and Isao [183]. Ryabov et al. [184] also reported the low-temperature polymerization of vinyl chloride by the dicyclohexyl peroxydicarbonate–tri-n-butylboron redox system. In light of the mechanism described by Contreras et al. [97], the following mechanism may be suggested for the percarbonate–alkyl boron system:

$$A_3B + R-O-\overset{\overset{\displaystyle O}{\|}}{C}-O-O-\overset{\overset{\displaystyle O}{\|}}{C}-O-R \longrightarrow A_2BO-C-O-R + A^{\cdot} + R-O-\overset{\overset{\displaystyle O}{\|}}{C}-O^{\cdot} \quad (6.91)$$

$$R-O-\overset{\overset{\displaystyle O}{\|}}{C}-O^{\cdot} \longrightarrow R-O^{\cdot} + CO_2 \quad (6.92)$$

6.4.3.1 Bulk Polymerization of Vinyl Chloride

In the case of the diisopropyl peroxydicarbonate–triethylboron redox system [183], 26.4% di-butyl phthalate solution containing 0.01624 g of diisopropyl peroxydicarbonate was chilled to –78°C in a pressure vessel, 15 g of vinyl chloride was added followed by 6.928×10^{-5} mol Et_3B in hexane under nitrogen, and the mass was kept at –20°C for 7 h to give 4.4% polymerization. Similar polymerization at 10°C with 0.02706 g of Et_3B produced 16.64% polymer, but no polymer was obtained with the use of diisopropyl peroxydicarbonate alone.

6.5 PERESTERS (PEROXYESTERS OF CARBOXYLIC ACID)

Peroxyesters of carboxylic acids have been extensively used for the suspension polymerization of vinyl monomers (Table 6.14). Several patents have appeared on the suspension polymerization of vinyl chloride [185–188], styrenic monomers [189, 190], and methyl methacrylate [191].

6.5.1 Sn^{2+} Salts as Reductant

A decomposition of a peroxyester by a stannous salt involves 2 mol of perester because the oxidation of stannous ion to stannic is a two-electron transfer; that is,

$$Sn^{2+} \; 2R'-\overset{\overset{\displaystyle O}{\|}}{C}-O-O-R \longrightarrow Sn^{4+} \; 2R'-\overset{\overset{\displaystyle O}{\|}}{C}-O^{-} + 2^{\cdot}O-R \quad (6.93)$$

TABLE 6.14
Peroxyesters for Vinyl Monomers Polymerization

Initiators	Monomers
tert-Bu peroxyneodecanoate	Vinyl chloride
2,4,4-Trimethylpentyl peroxyphenoxyacetate	Vinyl chloride
Diphenyl peroxyoxalate	Vinyl chloride
Di-*tert*-Bu diperoxyazelate	Styrene
Di-*tert*-Bu peroxyhexahydroterephthalate	α-Methyl styrene/styrene/acrylonitrile
tert-Bu 2-ethylhexane peroxoate	Styrene
tert-Bu peroxy-2-ethylhexanoate	Methyl methacrylate

The stoichiometry shown indicates that a 2:1 perester/Sn^{2-} mole ratio should result in complete perester decomposition. However, this is not in accord with the experimental observation in the reaction [192, 193] between t-butyl peroctoate (TBPO) and stannous octoate (SnOct); that is, the decomposition occurs rapidly to the extent of approximately 40% and then the TBPO concentration remains unchanged. This may be attributed to the requirement for the availability of stannous ion and the failure of stannous octoate to undergo ionic dissociation; that is, stannous octoate may possess some covalent character.

It is noteworthy that the analytical procedure for the quantitative determination of TBPO involves reaction with excess stannous chloride in an aqueous medium, followed by back titration of excess stannous ions. The aqueous medium results in the hydrolysis of stannous chloride to produce a solution of stannous hydroxide in aqueous hydrochloric acid. Stannous octoate may not hydrolyze in a neutral aqueous medium. Thus, the absence of complete dissociation and/or hydrolysis prevents the stoichiometric interaction of stannous octoate and TBPO. The possible presence of a TBPO–stannous octoate complex may also play a role in the failure to complete the reaction, as suggested by the observed presence of residual peroxide and residual stannous ions after the decomposition of TBPO has proceeded to the maximum extent.

The presence of a vinyl chloride monomer (VCM) has been shown to reduce even the limited extent of BPO decomposition by stannous octoate. This may be attributed to a VCM–stannous octoate complex, the presence of which has been experimentally confirmed [192]. Apparently, the stannous octoate in this complex, which contains VCM and stannous octoate in a 1:2 molar ratio, dissociates or hydrolyzes or interacts in some other manner with TBPO (e.g., by complexation with the carbonyl group) to an even lesser extent than stannous octoate in the absence of VCM.

It would appear that the failure to achieve the theoretical complete decomposition of TBPO at a 2:1 TBPO/Sn^{2+} ratio is due to the unavailability of Sn^{2+} in the concentration necessary to achieve the indicated stoichiometry. Thus, the TBPO–stannous octoate complex and the VCM–stannous octoate complex reduce the availability of stannous octoate for TBPO decomposition. Further, if the decomposition of TBPO requires the presence of a stannous ion, the incomplete hydrolysis or dissociation of stannous octoate per se or completed with TBPO and/or VCM, under the decomposition condition, reduces the availability of a stannous ion.

It is obvious that a route to effective, stoichiometric decomposition of TBPO in the presence of a stannous salt requires complete dissociation or hydrolysis of the latter through a change in reaction condition (e.g., an acidic pH) or the use of a more rapidly hydrolyzed stannous salt. It should be noted that stannous chloride, which generates an acidic medium on hydrolysis, quantitatively decomposes TBPO. Further, stannous lauroate, which contains the lauroate moiety, in the presence of emulsifiers such as sodium lauryl sulfate or dodecylbenzene sulfonate results in a more rapid polymerization rate and a higher conversion of VCM than stannous octoate, indicative of a greater availability of the effective reductant (i.e., a stannous ion). The stannous laurate may be solubilized in the aqueous phase and the resultant microenvironment promotes hydrolysis and/or dissociation, in contrast to the situation with water-insoluble stannous octoate and stearate.

Because it is necessary to increase the availability of the stannous salt for hydrolysis and/or dissociation, it is desirable to utilize an additive which competes with TBPO and VCM in complex formation with stannous salts. In this connection, it has been noted that the decomposition of TBPO in the presence of stannous octoate proceeds to a greater extent when an ester such as ethyl acetate is present. Further, the suspension system contains esters such as sorbate esters and an ethereal compound (i.e., methylcellulose) and promotes a more complete decomposition of TBPO and polymerization of VCM. Both ester groups and ethereal oxygen have the ability to complex with metal compounds, including stannic and stannous derivatives, and can therefore effectively compete with TBPO and/or VCM in complex formation.

An important point to be considered is the necessity to generate radicals from TBPO on a continuous basis to achieve effective polymerization. If the hydrolysis and/or dissociation of a stannous salt occurs rapidly and promotes rapid TBPO decomposition, the resultant radicals may be generated too rapidly for effective initiation of VCM polymerization to high conversions to high-molecular-weight PVC. Thus, it is necessary to promote the formation or release of reactive reductant at the desired rate throughout the polymerization period. It is also necessary to provide for complete decomposition of TBPO to yield PVC that does not contain residual TBPO.

The mechanism for the polymerization of VCM in the presence of TBPO–stannous octoate may be described as follows. The t-butoxy radical adds to VCM and initiates polymerization, in lieu of hydrogen abstraction. The reaction between the substrate radical (i.e., a VCM radical or the propagating chain radical) and the stannic ion results in the termination of the radical chain reaction and regeneration of the stannous ion. Thus, the latter is available for decomposition of BPO to generate additional chain-initiating t-butoxy radicals. However, the termination of chain growth results in an inefficient consumption of radicals:

$$2R-\overset{\overset{\displaystyle O}{\|}}{C}-O-O-tBu+Sn^{2+}\longrightarrow 2R-\overset{\overset{\displaystyle O}{\|}}{C}-O^{-}+Sn^{4+}+2tBuO^{\cdot} \tag{6.94}$$

$$tBuO^{\cdot}+CH_2=\underset{\underset{\displaystyle Cl}{|}}{C}H_2\longrightarrow 2tBuO-CH_2\,\underset{\underset{\displaystyle Cl}{|}}{C}H^{\cdot} \tag{6.95}$$

$$\underset{\substack{|\\ \mathrm{Cl}}}{\mathrm{tBuO-CH_2\,CH^{\cdot}}} + \mathrm{nCH_2=}\underset{\substack{|\\ \mathrm{Cl}}}{\mathrm{CH_2}} \longrightarrow \mathrm{tBuO-[CH_2\,}\underset{\substack{|\\ \mathrm{Cl}}}{\mathrm{CH}}]_n\mathrm{CH_2-}\underset{\substack{|\\ \mathrm{Cl}}}{\mathrm{CH^{\cdot}}} \quad (6.96)$$

$$\mathrm{R-\overset{O}{\overset{\|}{C}}-O^-}\ \ \mathrm{Sn^{4+}}\ \ \mathrm{^-O-C-R} + \mathrm{\sim CH_2\underset{Cl}{CH^{\cdot}}}\ \ \mathrm{^{\cdot}\underset{Cl}{CH}CH_2\sim} \longrightarrow 2\mathrm{\sim CH_2\underset{Cl}{CH}-O-\overset{O}{\overset{\|}{C}}-R} + \mathrm{Sn^{2+}} \quad (6.97)$$

$$\mathrm{\sim CH_2-\underset{Cl}{CH^{\cdot}}}\ \ \mathrm{^{\cdot}CH_2-\underset{Cl}{CH}\sim} \longrightarrow \mathrm{\sim CH_2\underset{Cl}{CH_2}} + \mathrm{\underset{Cl}{CH}=CH\sim} \quad (6.98)$$

6.5.1.1 Suspension Polymerization of Vinyl Chloride

Gaylord et al. have reported the suspension polymerization of vinyl chloride using the redox system, such as t-butyl peroxyoctoate–$SnCl_2$ [23, 194] and t-butyl peroxyoctoate–stannous carboxylate [192, 193]. The polymerization of VCM in the presence of the redox system has several unusual characteristics that can be explained on the basis of the previous description.

1. The redox polymerization of VCM requires considerably higher TBPO concentration than the conventional thermal polymerization. The decomposition of TBPO at 50°C requires the presence of stannous octoate (SnOct) in some specific and reactive form, presumably stannous ions. In view of the unavailability of stannous octoate in this form, due to its complexation with TBPO and VCM as well as its failure to hydrolyze and/or dissociate, the concentration of active reductant is much lower than the amount of stannous octoate charged. Because the TBPO/SnOct ratio is maintained constant, it is necessary to increase the amount of TBPO so that a sufficient amount of active reductant is available. It is also possible that the stannic species generated by the BPO oxidation of stannous octoate participate in the termination of propagating chains or interact with radicals generated from TBPO. It is well known that metals in the higher valance state (e.g., ferric and stannic compounds) react with free radicals and, as a result of electron transfer, convert the latter to cationic species that cannot add monomers such as VCM. It is, therefore, necessary to increase the TBPO concentration to provide additional radicals and the propagating chains therefrom, to compensate for those lost by electron transfer. The coordination or complexation of stannous octoate with the chlorine atoms appended to the PVC chains may also reduce the concentration of stannous octoate available for TBPO reduction, as the PVC particles are insoluble in VCM and therefore remove the appended stannous octoate from the active locus of polymerization.
2. The redox polymerization generally does not go to completion except after extremely long reaction times and even then, the reaction mixture contains a large amount of undecomposed TBPO. Based on the decomposition

studies, it also contains residual stannous species. The presence of undecomposed TBPO and stannous species at the leveling off or termination of VCM polymerization is consistent with the presence of TBPO–SnOct complex and/or the unavailability of reactive stannous species to complete the stoichiometric decomposition. Unhydrolyzed or PVC-bonded stannous octoate may be in the system, but not capable of reducing TBPO.

3. Although decomposition studies show that the decomposition of TBPO in the presence of stannous octoate proceeds rapidly to about 40% during the first 2 h and then remains essentially unchanged over the next 20 h, the polymerization continues for more than 10 h. The indicated rapid decomposition of TBPO in the presence of stannous octoate does not occur when VCM is present. In fact, the decomposition rate is greatly reduced. This is actually desirable because the rapid decomposition generates radicals at a faster rate than VCM can add to it. Further, the presence of the suspending agents results in interference with the VCM–stannous octoate complex, possibly by forming a suspending agent–stannous octoate complex which slowly makes active stannous reductant available and therefore extends the time for radical generation, analogous to the behavior of peresters in thermal decomposition.

The suspension polymerization of VCM [23, 194] in a bottle was performed with the suspension recipe given in Table 6.15. The attempted polymerization of VCM in the presence of 0.055 ml (0.23 mmol) TBPO (0.5% by weight of VCM), in the absence of stannous chloride dehydrate, failed to yield any polymer after 20 h at 50°C. This is consistent with the TBPO half-life of 133 h at 50°C (10 h half-life 74°C). The suspension polymerization of VCM in the presence of 0.5 wt% peroxyoctoate (POT), stannous chloride dehydrate 0.052 g (TBPO/$SnCl_2$ molar ratio = 1), and glacial acetic acid 2 ml resulted in a 5% yield of polymer after 13 h at 50°C. The low yield of PVC indicated that the TBPO–$SnCl_2$ interaction either yielded predominantly nonradical products or proceeded so rapidly in the VCM droplet as to preclude effective polymerization. When the suspension polymerization of VCM in the presence of 0.5 wt% TBPO and VCM-insoluble stannous chloride dehydrate (POT/$SnCl_2$ molar ratio = 1) was conducted in the absence of acetic acid, the yield of PVC was 82% after 13 h

TABLE 6.15
Typical Recipe Suspension Polymerization of Vinyl Chloride[a]

Ingredients	Amount (ml)
Water	21
Polyoxyethylene sorbitan monostearate (1% aqueous solution)	1
Sorbitan monostearate (1% aqueous solution)	2
Methocel A15 (15 cps viscosity grade methylcellulose) (1% aqueous solution)	2
Vinyl chloride	10 g
TBPO	0.055

[a] Polymerization for 20 h at 50°C.

at 50°C. This may be attributed to the interaction of the monomer-insoluble $SnCl_2$ or the hydrated ions thereof with the TBPO in the VCM at the water–monomer droplet interface to generate radicals at a slow useful rate.

In the case of perester–stannous carboxylate redox system [192, 193] with or without a complexing agent for vinyl chloride polymerization, the polymerization recipe was the same as described earlier for the $SnCl_2$ as reductant. In each case, 10 g of VCM was taken in the experiment for polymerization at 50°C. The conversion was increased in the presence of complexing agents. Some of the results are presented in Table 6.16.

6.5.2 Mercaptans as Reductant

The use of mercaptans as the reducing agent in the emulsion or aqueous polymerization of vinyl polymerization is not new. Its efficiency as the reducing agent in

TABLE 6.16
Vinyl Chloride Polymerization at 50°C with Perester–Stannous Carboxylate Redox System with or without Complexing Agent

Stannous Carboxylate (mmol)	TBPO[a] (mmol)	Complexing Agent[b] Type (mmol)	Time (h)	Conversion %
Octoate 0.23	0.46	—	9	60
Octoate 0.23	0.46	—	11	80
Octoate 0.23	0.46	—	20	92
——	0.46	—	20	0
Octoate 0.115	0.23	—	12	45
Stearate 0.115	0.23	—	18	30
Laurate 0.23	0.46	—	2	10
Laurate 0.23	0.46	—	6	40
Laurate 0.23	0.46	—	8	65
Laurate 0.23	0.46	—	10	90
Laurate 0.23	0.46	—	15	96
Laurate 0.115	0.23	—	9	45
Laurate 0.115	0.23	—	16	51
Laurate 0.115	0.23	DOP (0.23)	9	68
Laurate 0.115	0.23	DOP (0.23)	16	73
Laurate 0.046	0.092	—	7	15
Laurate 0.046	0.092	DOP (0.092)	7	23
Laurate 0.115	0.23	—	9	60[c]
Laurate 0.115	0.23	DOP (0.23)	9	83[c]
Laurate 0.115	0.23	DOA (0.23)	9	65
Laurate 0.115	0.23	2-EHA (0.23)	9	64
Laurate 0.115	0.23	TEP (0.23)	9	65

[a] TBPO = t-butyl peroxyoctoate.
[b] DOP = dioctyl phthalate; EHA = ethylhexanoic acid; TEP = triethylphosphate.
[c] Polymerization at 55°C.

conjunction with oxidants like H_2O_2 [144, 145], Bz_2O_2 [146], and $K_2S_2O_8$ [147–158] for vinyl polymerization has been reported. The feasibility of mercaptans [195–197] as the reductant with peroxyesters of carboxylic acid has been described for suspension copolymerization or graft copolymerization. The general mechanism may be written as

$$\mathrm{R-\overset{\overset{\displaystyle O}{\|}}{C}-O-O-R'} + \mathrm{R-S-H} \longrightarrow \mathrm{R^{\bullet}S} + \mathrm{R-\overset{\overset{\displaystyle O}{\|}}{C}-OH} + \mathrm{R-O^{\bullet}} \quad (6.99)$$

The preceding radicals take part in the initiation of polymerization.

6.5.2.1 Suspension Graft Copolymerization of Styrene and Acrylonitrile to Polybutadiene Latex

In the suspension graft copolymerization [195], ABS polymers having enhanced physical properties like high impact strength and low polybutadiene content are prepared by graft copolymerization of styrene and acrylonitrile onto a polybutadine polymer latex (particle size 1000–3000 Å, gel content 20–85%, and swell index ≈18–150) in an aqueous medium in the presence of a suspending agent and a catalyst. The suspension graft copolymerization was performed according to the typical recipe presented in Table 6.17 to yield ABS resin beads. The dried beads were blended with antioxidant and extruded to form pellets, which were molded to form a sample having Izod impact strength >13 ft-lb/in., tensile stress at yield 4500, elastic modulus 2.25×10^5, and shear–Izod ratio <4.2.

TABLE 6.17
Typical Recipe: Suspension Graft Polymerization of Styrene and Acrylonitrile to Poly(butadiene) Latex[a]

Ingredients	Amount (g)
Part A	
Styrene	521
Acrylonitrile	211
Antioxidant (as a 10% solution in acrylonitrile)	3.5
tert-Butyl peroxypivalate	1.8
tert-Butyl peroctoate	0.54
tert-Dodecyl mercaptan	2.4
Part B	
Water	3050
Pliolite	2104
Latex	281
2.5% Solution of poly(vinyl alcohol)	700
0.5% Solution of poly(ethylene oxide)	100

[a] Part A was added to Part B and the mixture was maintained at 68°C for 4 h and at 100°C for 1 h.

TABLE 6.18
Typical Recipe: Suspension Copolymerization of Styrene and Acrylonitrile[a]

Ingredients	Amount (g)
Water	25,000
$Ca_3(PO4)_2$	150
Styrene	11,000
Actylonitrile	6000
tert-Butyl 3,5,5-trimethyl perhexanoate	25
tert-Butyl peracetate	15
tert-$C_{12}H_{25}SH$	50
Styrene	1600

[a] Polymerization for 5 h at 100°C and 2 h at 125°C.

6.5.2.2 Suspension Copolymerization of Acrylonitrile and Styrene

The acrylonitrile–styrene copolymer [196] was prepared by suspension polymerization in the presence of 0.005–0.05% (based on monomers) *tert*-Bu, 3,5,5-trimethyl perhexanoate, and *tert*-butyl peracetate at 110–140°C according to a typical recipe presented in Table 6.18 to give a copolymer (unreacted monomer 0.1%) with lower yellow neon and haze than a control (unreacted monomer) prepared without *tert*-Bu peracetate.

6.5.3 Alkyl Borane as Reductant

Alkyl borane as the reductant in redox polymerization is well known. It has been used previously in conjunction with alkyl peroxide and peroxyester of carbonic acid. The mechanism of alkyl borane–peroxyester of carboxylic acid is similar to that previously described. Suspension polymerization or copolymerization of vinyl chloride by the redox system such as monotertiary butyl permaleate–Et_3B or Bu_3B or iso-Bu_3B has been reported in patent literature [198].

6.5.3.1 Suspension Polymerization of Vinyl Chloride or Its Mixture

Vinyl chloride by itself or mixed with C_2H_4, propylene, or isobutylene is polymerized [198] in the presence of Et_3B, Bu_3B or iso-Bu_3B, and mono-*tert*-Bu permaleate at –30°C to +80°C. Thus, 500 g of vinyl chloride was polymerized at 20°C for 10 h with 1200 cm^3 of double-distilled H_2O, and 1.5 g of H_2O-soluble suspension stabilizer was introduced. Bu_3B (3.14 g) was introduced with the exclusion of oxygen and 4.0 g of mono-*tert*-Bu permaleate was added; 75.6% polymer conversion was achieved. Much lower yields of polymer were obtained when a peracetate or perhenzoate was used instead of the permaleate.

6.5.4 Bisulfite as Reductant

Bisulfite is one of the oldest reducing agents used in polymerization. Vinyl polymerizations using bisulfite as the reductant in conjunction with $K_2S_2O_8$ as the oxidant have been reported [155, 199–215] as early as 1946. The use of bisulfite with H_2O_2 [199, 216, 217] as well as with organic peroxide like Bz_2O_2 [218] to initiate polymerization is also well known. The persulfate-bisulfite system has been used for the polymerization of acrylonitrile [204, 219–222], methyl acrylate [223], styrene [211], chlorotrifluoro ethylene [201], and so forth. The bisulfite–persulfate combination along with Fe^{2-} has also been used to polymerize acrylonitrile [207, 224]. Bisulfite is also used with other oxidants like peroxydicarbonate [225], Cr_2O_3 [226], oxygen [227], and $KBrO_3$ [227] for redox initiation. These initiating systems are only restricted to emulsion or aqueous polymerization. Very few reports have been published about suspension polymerization of vinyl chloride [228] and suspension graft polymerization of vinyl pyridine to polyolefins [229] using a perester of the carboxylic acid –$NaHSO_3$ redox system.

In the light of the mechanism of persulfate–bisulfite redox initiation, the mechanism for the perester–bisulfite redox system may be suggested as follows:

$$R-\overset{\overset{\displaystyle O}{\|}}{C}-O-O^{-}-R' + HSO_3^{-} \longrightarrow HSO_3^{\cdot} + R-\overset{\overset{\displaystyle O}{\|}}{C}-O + R'O^{\cdot} \quad (6.100)$$

The preceding radicals take part in the initiation process.

6.5.4.1 Suspension Polymerization and Copolymerization of Vinyl Chloride

Vinyl chloride [228] was polymerized and copolymerized in suspension at low temperature in the presence of a peroxide, a reducing agent, and a copper accelerator. Thus, the vinyl chloride-2-ethylhexyl acrylate copolymer was prepared in 100% yield by using the recipe given in Table 6.19.

6.5.4.2 Suspension Graft Copolymerization of Vinyl Pyridine

Vinyl pyridine-grafted polyolefins [229] having improved dyeability were prepared with >0.02 wt% based on the monomer of a perester catalyst and >0.1 wt% based

TABLE 6.19
Typical Recipe: Suspension Polymerization of Vinyl Chloride-2-ethylhexyl Acrylate Copolymer[a]

Ingredients	Amount (g)
Water	70
Vinyl chloride	28.5
2-Ethylhexyl acrylate	1.5
Fluoronic F-68	0.3
Lupersol-11 (t-butyl peroxypivalate)	0.06
$NaHSO_3$	0.5
$CuCl_2 \cdot 2H_2O$ (0.00039% with respect to the monomer)	

[a] Polymerization at 61°F.

on the monomer of a reducing agent promoter selected from lower-valent salts of multivalent metals, hydrosulfite, or alkali metal formaldehyde sulfoxylate. Thus, the polypropylene–styrene–vinylpyridine-graft copolymer prepared in the presence of 1 wt% sodium hydrosulfite and 0.5 wt% *tert*-butyl 2-ethyl perhexanoate at 90°C was melt-spun into fibers which were dyed to a light-fast wash-resistant deep red shade with Capracyl Red G.

6.5.5 Monosaccharide as the Reductant

6.5.5.1 Suspension Polymerization of Vinyl Chloride

A process for the bulk or suspension polymerization of vinyl chloride in the presence of a redox catalyst system consisting of a peroxyester and a monosaccharide or carboxylic acid esters of monosaccharide was described by Gaylord [230]. The monosaccharides which were used as reductants include pentoses and hexoses wherein the carbonyl group is either an aldehyde or ketone; that is, polyhydroxy aldehydes commonly referred to as aldoses and polyhydroxy ketones commonly referred to as ketoses.

Representative monosaccharides or reducing sugars include arabinose, xylose, lyxose, ribose, glucose, mannose, allose, galactose, tallose, altrose, idose, fructose, and sorbose. The preferred concentration of peroxyester is generally between 0.5% and 1% by weight of the vinyl chloride. The peroxyester/reductant mole ratio is generally 1/0.1–1. The preferred temperature for the suspension polymerization was in the 20–60°C range and the weight ratio of monomer and water was about 2:1.

Although the peroxyester–monosaccharide or peroxyester–monosaccharide–carboxylic acid ester catalyst system is useful in the bulk and suspension polymerization of vinyl chloride, the redox system may also be used in the copolymerization of vinyl chloride with vinylidene chloride, vinyl acetate, and other monomers which undergo copolymerization with vinyl chloride.

6.5.6 Metal Mercaptides as Reductant

Gaylord et al. [231] described the bulk or suspension polymerization of ethylenically unsaturated monomers, particularly vinyl chloride, using a catalyst system consisting of a monomer-soluble peroxyester or diacyl peroxide and a reducing agent which is a stannous or antimony(III) mercaptide.

The peroxygen compound/reductant mole ratio was about 1:0.1–1. The concentration of peroxyester was about 0.05–1% by weight of the vinyl halide monomer. The concentration of both peroxygen compound and reductant may be reduced by the addition of complexing agents that contain suitable functional groups. Alternatively, the addition of complexing agents increases the rate of polymerization at a given concentration of peroxygen compound and reductant.

The rate of decomposition of a peroxygen compound such as t-butyl peroxyoctoate in the presence of a stannous or antimony(III) mercaptide is decreased in the presence of vinyl chloride, presumably due to the formation of a complex between the reductant and the monomer. However, when a complexing agent containing carbonyl functionality (e.g., a ketone, lactone, carboxylic acid, or carboxylic ester) is present, the complex formation is decreased and the rate and extent of decomposition of the peroxygen compound increases, even in the presence of the monomer. The increased rate and extent of decomposition of a peroxyester or diacyl peroxide in the presence of the complexing agent is accompanied by an increase in the rate and extent of polymerization of vinyl chloride.

The complexing agents which may be used in the process of the present invention are organo-soluble and contain carbonyl groups or phosphorus–oxygen linkages. Thus, ketones, carboxylic acids and esters, and phosphate esters are effective complexing agents. The latter may be saturated or unsaturated, cyclic or acyclic, branched or linear, substituted or unsubstituted.

6.5.7 Ascorbic/Isoascorbic Acid or Esters as Reductant

Ascorbic acid has been used extensively as a sole reducing agent or in combination with cupric, ferrous, or ferric salts for the polymerization of vinyl chloride in the presence of water-soluble catalysts including hydrogen peroxide [232–235], potassium persulfate [236], cumene hydroperoxide [237], acetyl cyclohexanesulfonyl peroxide [238], and a mixture of hydrogen peroxide and acetyl cyclohexanesulfonyl peroxide [239].

Ascorbic acid has also been used as a complexing agent in the polymerization of vinyl chloride [240] in the presence of a diacyl peroxide and various water-soluble metal salts. Similarly, 6-O-polmitoyl-l-ascorbic acid has been used as a reducing agent in the polymerization of vinyl chloride in the presence of hydrogen peroxide [241] and methyl ethyl ketone peroxide [242].

6.5.7.1 Suspension Polymerization of Vinyl Chloride

Gaylord [243] has described the bulk or suspension polymerization of vinyl chloride using a catalyst system consisting of a monomer-soluble peroxyester or diactyl peroxide as oxidant and a 6-O-alkanoyl-l-ascorbic acid as a reducing agent.

Bulk or suspension polymerization may be performed at temperatures in the 20–60°C range. Gaylord [244] has also described the use of isoascorbic acid as the reducing agent in combination with a peroxygen compound as the catalyst system for the suspension polymerization of vinyl chloride.

A comparison of the results obtained with ascorbic acid and isoascorbic acid, in the suspension polymerization of vinyl chloride at 50°C, in the presence of t-butyl peroxyoctoate (t-BPOT) at a peroxyester/reductant mole ratio of 2:1 is given in Table 6.20. The use of isomeric 6-O-alkanoy-d-ascorbic acid has been found to

TABLE 6.20
Polymerization of Vinyl Chloride at 50°C

Reductant	t-BPOT (wt%)	Time (h)	Conversion %
Ascorbic acid	0.3	8.5	40.5
	0.1	8.5	20.0
Isoascorbic acid	0.1	8.5	70.5
	0.1	16.0	70.5
	0.05	16.0	40.0

result in a significantly higher rate of polymerization, permitting the use of lower concentrations of peroxyester to achieve faster reaction.

6.6 HYDROPEROXIDES

Generally, hydroperoxides are derivatives of hydrogen peroxide, with one hydrogen replaced by an organic radical:

$$\underset{\text{hydrogen peroxide}}{H-O-OH} \qquad \underset{\text{hydroperoxide}}{R-O-O-H}$$

Hydroperoxide chemistry had its heyday in the decade 1950–1960, following the firm establishment of these compounds as reactive intermediates in the autoxidation of olefins. Afterward, many reports regarding vinyl polymerization involving hydroperoxide alone or coupled with a suitable reducing agent have appeared in the literature.

6.6.1 Sulfur Dioxide as Reductant

Many reports have been published on the use of SO_2 as the reductant to initiate the polymerization. For example, Polish workers [245] studied the emulsion polymerization of the styrene $-SO_2$ system using cumene and pinene hydroperoxide. Gomes and Lourdes [246] investigated the liquid SO_2–cumene hydroperoxide system. Ghosh et al. used the SO_2 in combination with hetero-cyclic compounds by pyridine, tetrahydrofuran, and N-N′-dimethylformamide for photopolymerization [247–249] as well as aqueous polymerization [250]. Mazzolini et al. [251–254] reported the organic hydroperoxide–SO_2 redox pair and a nucleophilic agent to polymerize vinyl chloride in bulk at subzero temperatures. Patron and Moretti [255] have also reported on the bulk polymerization of vinyl chloride using the same type of system at 20°C.

The decomposition of organic hydroperoxides by the action of SO_2 depends on the reaction medium; for example, cumyl hydroperoxide (CHP) is quantitatively decomposed into phenol and acetone if the reaction is performed in an anhydrous weakly nucleophilic or non-nucleophilic medium (e.g., CCl_4, CH_3CN, CH_3CH_2Cl, CH_2-CHCl). This type of decomposition, which proceeds through an ionic mechanism without formation of radicals, can also be obtained [256, 257] with perchloric acid, ferric chloride in benzene, and sulfuric acid. It is, therefore, inferred that SO_2 behaves as a strong acid toward the decomposition of CHP in anhydrous, weakly nucleophilic or non-nucleophilic solvents.

For a redox reaction to take place, according to the Lewis theory of acids and bases, it is necessary that the reductant (SO_2) acts as a base toward the oxidant (hydroperoxide), to allow the transfer of electrons from the former to the latter [258]. The condition is fulfilled by the addition to the reaction medium of a strongly nucleophilic agent N- (e.g., OH-) to transform the SO_2 into the conjugate base NSO_2^-. When water is added to the system hydroperoxide–SO_2 and the concentration of the former increases, the absorbance at 272 mμ, characteristic of phenol, diminishes, whereas a new absorbance maximum, ranging between 237 and 255 mμ, emerges due to a mixture of 1-methylstyrene (3%), acetophenone (60%), and cumyl alcohol (37%). When water is added to the system hydroperoxide–SO_2 in an organic medium, a situation analogous to the emulsion polymerization by hydroperoxide and SO_2 is induced [259]. This demonstrates the possibility of switching the mechanism of reaction between hydroperoxide and SO_2 in an essentially organic medium from an ionic mechanism to a radical one, thus offering a way for the initiation of vinyl polymerization at low temperature.

According to Mazzolini et al. [251], the kinetic expressions for the continuous bulk polymerization of vinyl chloride by the hydroperoxide–SO_2 nucleophilic agent may be as follows:

Production of radicals:

$$ROOH + CH_3OSO_2^- \rightarrow CH_3OH + R^{\cdot} + {}^-SO^{\cdot}_3 \tag{6.101}$$

and $d(R^{\cdot})/dt = 2K_d(ROOH)(CH_3SO_2^-)$, where K_d is the velocity constant for the radical production.

Initiation of polymerization:

$$R^{\cdot} + M \rightarrow M^{\cdot} \tag{6.102}$$

where M is vinyl chloride monomer and

$$\frac{d(M^{\cdot})}{dt} = K_a(R^{\cdot})(M) \tag{6.103}$$

where K_a is the velocity constant for monomer addition to primary radicals.

Propagation:

$$M^{\cdot} + M \rightarrow M^{\cdot} \tag{6.104}$$

and

$$\frac{d(M)}{dt} = K_p(M^{\cdot})(M) \tag{6.105}$$

where K_p is the velocity constant for the propagation reaction.

Termination:

$$M^{\cdot} + M^{\cdot} \rightarrow P \tag{6.106}$$

$$\frac{d(M^{\cdot})}{dt} = 2K_r(M^{\cdot})^2 \tag{6.107}$$

where K_r is the velocity constant for the combination reaction.

Under stationary conditions input = output + reaction amount, the balance for the catalyst will be

$$F_0(C)_0 = F(C) + K_d(C)(S)V \tag{6.108}$$

where

F_0 = feed rate of all liquid streams to reactor, volume per unit time
F = output rate of the liquid fraction at overflow from reactor, volume per unit time
$(C)_0$ = hydroperoxide concentration in liquid feed
(C) = hydroperoxide concentration in reactor (or in reactor overflow)
(S) = concentration of compound $CH_3OSO_2^-$ in reactor (or in reactor overflow)
V = volume occupied by liquid phase in reactor

At sufficient dwell time and $(ROSO_2^-)/(CHP)$ molar ratios, F(C) is negligible if compared with $F_0(C)_0$ and $K_d(C)(S)V$. As the catalyst decomposition approaches completion, an expression for (C) can thus be assumed:

$$(C) = \frac{F_0(C)_0}{K_d(S)V} \tag{6.109}$$

The balance for the monomer is

$$F_0(M)_0 = F(M) + K_p(M^\cdot)(M)V \tag{6.110}$$

$(M)_0$ (monomer concentration in feed) and (M) (monomer concentration in liquid phase of overflow) being equal for a bulk polymerization, the monomer conversion can be expressed as

$$c = \frac{F_0 - F}{F_0} \tag{6.111}$$

or from the previous equation $c = K_p(M^\cdot)V/F_0$.

The balance for $(M^\cdot)$ is

$$2fK_d(C)(S)V = F(M^\cdot) + 2K_r(M^\cdot)^2V \tag{6.112}$$

where f is the efficiency of initiating radicals (i.e., the fraction of radicals taking part in polymer chain initiation).

The term $2fK_d(C)(S)V$ can be assumed to be equal to $2f(C)_0F_0$. $2K_r(M^\cdot)^2V$ is equal to twice the number of macromolecules formed per unit time. Both can be experimentally estimated. $F(M^\cdot)$ appears to be negligible if compared with $2fK_d(C)(S)V$ and $2K_r(M^\cdot)^2V$. Then the preceding equation becomes

$$2K_r(M^\cdot)^2 = 2fK_d(C)(S) \tag{6.113}$$

or substituting the value of (C),

$$2K_r(M^\cdot)^2 = \frac{V}{2fF_0(C)_0} \tag{6.114}$$

Then, the conversion can be expressed as

$$C = \left(\frac{K_p}{K_r^{1/2}}\right) f^{1/2}(C)_0^{1/2} V^{1/2} F_0^{-1/2} \tag{6.115}$$

for conversion, not exceeding about 20%, V can be assumed as

$$V = V_0(1 - c) \tag{6.116}$$

thus V_0/F_0 being the conventional dwell time, O, a final equation may be written as

$$\frac{c}{(1-c)^{1/2}} = \left(\frac{K_p}{K_r^{1/2}}\right) f^{1/2}(c)_0^{1/2} Q^{1/2} \tag{6.117}$$

In other words, the conversion is proportional to the square root of the hydroperoxide concentration and the dwell time.

The following mechanism can be proposed for the radical decomposition of the hydroperoxide by SO_2 and nucleophilic agent:

$$SO_2 + CH_3O^- \rightarrow CH_3OSO_2^- \tag{6.118}$$

$$\begin{aligned} &ROOH + CH_3OSO_2^- \rightarrow ROOSO_2^- + CH_2OH \rightarrow RO^{\cdot} \\ &+ \ {}^{-}SO_3^{\cdot} + CH_3OH \end{aligned} \tag{6.119}$$

or

$$\mathrm{Ph{-}C(CH_3)_2{-}O{-}OH} + \left[\mathrm{CH_3{-}O{-}S^{+}(O^{(-)})(O^{(-)})}\right]^- \longrightarrow CH_3OH + \left[\mathrm{Ph{-}C(CH_3)_2{-}OO{-}S(O)(O)}\right] \longrightarrow \mathrm{Ph{-}C(CH_3)_2{-}O^{\cdot}} + {}^{\cdot}SO_3^- \tag{6.120}$$

The oxycumyl radical may further decompose into 1-methylstyrene, acetophenone, and cumyl alcohol, or the radical itself, its fragments $(CO_3^{\cdot}, OH^{\cdot})$, or radicals derived from chain transfer reactions may initiate polymerization. The ${}^{-}SO_3^{\cdot}$ radical is easily identified as an end group in the polymer chain. The rate-determining step for the whole catalytic reaction appears to be the formation of the complex (I), as indicated by the fact that an asymptotic limit for the polymerization rate is reached only when the $(CH_3OSO_2^{\cdot})/(CHP)$ ratio is in considerable excess over the stoichiometric ratio of 1.

6.6.1.1 Bulk Polymerization of Vinyl Chloride

During the bulk polymerization of vinyl chloride [251], when cumyl or *tert*-Bu hydroperoxides and SO_2 are used with ethers, ketones, or alcohols, sulfone groups

TABLE 6.21
Influence of Nucleophilic Agents on Bulk Polymerization of Vinyl Chloride[a]

Nucleophilic Agents	% (OMW)	Conversion %
None	0.00	0.00
Acetophenone	0.60	1.90
Cyclehexanone	0.49	5.10
Acetone	0.29	1.20
Methyl ethyl ether	0.36	3.70
Ethyl ether	0.52	3.20
Methanol	0.16	6.80
Methanol	5.00	10.50
Butanol	0.36	6.00
Dimethyl amine	0.22	0.40
Dimethyl formamide	0.36	0.51

[a] Temperature = –30°C; CHP = 0.15% (OMW); SO_2 = 1.6% (OMW); nucleophilic agent as specified; addition time of catalyst components into monomer = 1 h. Total reaction time = 2 h. (OMW = on monomer weight.)

are incorporated in the polymer chain because of copolymerization of SO_2. When the hydroperoxides and SO_2 are used with MeO or EtO (from Na or Mg alkoxides), SO_2 copolymerization is completely suppressed, provided the MeO^-/SO_2 or EtO^-/SO_2 ratio is at least 1:1. When the feed rate of hydroperoxide is constant, the maximum monomer conversion in continuous bulk polymerization is reached when the SO_2/hydroperoxide ratio is ≥1.5:1. The percentage conversion for the various nucleophilic agents used are presented in Table 6.21.

The hydroperoxide –SO_2 system reacted in the vinyl chloride monomer at –30°C. Without any nucleophilic agent, the reaction proceeds via the usual ionic path and no polymerization is detected. With the addition of alcohols, ketones, and ethers, the redox reaction is promoted and substantial quantities of polymer are formed. When weak nucleophilic agents, like ethers and ketones, are used, polymerization yields are low. High conversions were obtained with alcohols. The best yield was obtained by the addition of 5% methanol on the monomer weight. The polymerization rate, at constant CHP and SO_2 concentrations, approaches the maximum when the $(CH_3O^-)/(SO_2)$ ratio is at least 1, employing either sodium or magnesium methoxide at a $(CH_3O^-)/(SO_2)$ ratio of 1. The SO_2 is completely transformed into the salt of methyl sulfurous acid. The systematic polymerization study was performed using a $(CH_3O\text{-})/SO_2$ ratio of 1:1 to assure complete neutralization of SO_2 and avoid its copolymerization. The syndiotacticity index was 2.1–2.2 for polymers prepared at –30°C, and 2.4–2.5 for polymers prepared at –50°C. The glass transition temperature T_g was 100°C for polymers obtained at –30°C, and 104°C for polymers obtained at –50°C.

The previously described catalytic system was also effective [251] with other vinyl monomers over wide temperature ranges. The results are given in Table 6.22.

In two other patents reported by Mazzolini et al. [252, 254] for bulk polymerization of vinyl chloride, they used the same type of catalytic system as discussed previously.

TABLE 6.22
Polymerization of Vinyl Monomers by the CHP/SO_2/$Mg(OCH_3)_2$ Catalytic System[a]

Monomer	Temperature (C)	Conversion %
Vinyl acetate	−30	22.0
	−60	6.0
Vinyl formate	−30	19.0
Acrylonitrile	−30	23.0
Styrene	50	15.5
Acrylamide (30% in methanol)	−30	21.0
2-Hydroxyethyl acrylate	20	56.5
	−30	27.5
tert-Butylaminoethyl methacrylate	20	46.5

[a] CHP = 0.25% (OMW); SO_2 = 0.2% (OMW); $Mg(OCH_3)$ = 0.14% (on moles). Catalyst addition time = 5 h. Total reaction time = 5 h. (OMW = on monomer weight).

Thus, vinyl chloride was polymerized at −30°C in the presence of a mixture of cumene hydroperoxide or *tert*-Bu hydroperoxide, a methanolic solution of SO_2, and a methanolic solution of NaOMe, NaOEt, or KOMe.

In another German patent, Mazzolini et al. [253] reported the low-temperature bulk polymerization of vinyl chloride in the presence of a catalyst system consisting of an organic hydroperoxide, SO_2, and at least one alkali metal alcoholate at a [ROX]–[SO_2]/[R′OOH] mole ratio of 0–0.5 and 0.005–1% mercapto compound which gave a degree of conversion >18% and a polymer with outstanding physical and chemical properties. The typical recipe for the polymerization is presented in Table 6.23.

Patron and Moretti [255] also reported the bulk polymerization of vinyl chloride at >0°C in the presence of a catalyst system consisting of an organic hydroperoxide, SO_2, and an alcohol or metal alcoholate. A 25% conversion was obtained at 25°C. The PVC recovered had an intrinsic viscosity of 1.3 and bulk density of 0.41 g cm^{-3}.

TABLE 6.23
Typical Recipe: Bulk Polymerization of Vinyl Chloride[a]

Ingredients	Amount(g hr^{-1})
Liquid vinyl chloride (at (30°C)	200,000
Cumene hydroperoxide	240
SO_2	150
MeONa	136
Mercaptoethanol	60

[a] 36.2 kg h^{-1} (22.5% conversion); PVC has intrinsic viscosity 1.38 dl g^{-1}.

They [260] also reported the bulk polymerization of vinyl chloride by taking a mixture containing liquid vinyl chloride at –30°C and a catalyst composition containing cumene hydroperoxide, SO_2, Na methylate, and 2-mercaptoethanol that was continuously fed to a reactor. The molar weight concentration ratio of the catalyst composition was (NaOME)(SO_2)/cumene hydroperoxide = 0.1. The polymerization yielded PVC with an intrinsic viscosity of 1.3 dl g^{-1}.

6.6.2 Sulfite as Reductant

The oxyacids of sulfur such as sulfite [155, 199, 220, 253–265] form an efficient redox system in conjunction with persulfates to initiate vinyl polymerization. Sully [266] examined the $Cu^{2+} - SO_3^{2-}$ system in air. The $ClO_3^- - SO_3^{2-}$ system has been used in the polymerization of acrylonitrile [267, 268] and acrylamide [268, 269]. The $KBrO_3$–Na_2SO_3–H_2SO_4 system is also an effective redox initiator [270], giving rise to polymers containing strong acid end groups. All the preceding initiating systems have been employed in aqueous or emulsion polymerization. Reports of the use of sulfite as the reductant with organic peroxides or hydroperoxides are very few. Melacini et al. [271] have reported the bulk polymerization of acrylonitrile by redox system such as cumene hydroperoxide–dimethyl sulfite. The mechanism of initiation may be described as

$$ROOH + SO_3^{2-} \rightarrow RO^{\cdot} + OH^- + SO_3^{\cdot -} \tag{6.121}$$

These radicals take part in the initiation step. t-Butyl hydroperoxide (*tert*-BHP) forms free radicals with $SOCI_2$ in the presence of methanol which initiates the polymerization of vinyl chloride successfully [272]. It was proposed that as a first step, $SOCI_2$ reacts with methanol to yield methyl chlorosulfite with which *tert*-BHP reacts to form methyl *tert*-butyl peroxysulfite, which decomposes to give free radicals.

6.6.2.1 Bulk Polymerization of Acrylonitrile

Acrylonitrile [271] polymers were prepared in bulk in high yields under controlled conditions at room temperature to 60°C in 30–90 min using radical catalysts with decomposition rate constants >1 hr^{-1}. Thus, 1600 g of acrylonitrile containing 300 ppm of water was kept at 50°C, and 3.2 g of cumene hydroperoxide, 23.2 g of dimethyl sulfite, and 18.1 g of magnesium methylate in 150 cm^3 of MeOH were added. The conversion achieved in 15 min represented a final conversion of 77% in a continuous polymerization system.

REFERENCES

1. R. Vicari, Eur. Patent Application EP 3,43,986 (1989); *Chem. Abstr.*, 112(20), 180085u (1989).
2. S. Zalwert, M. Jendraszek, M. Durak, R. Jurczak, J. Muranski, and B. Preising, *Chem. Abstr.*, 111(2), 8327t (1988).
3. Y. Akasaki, N. Yabuchi, and T. Oki, Jpn. Kokai Tokkyo Koho, JP 63,191,805; *Chem. Abstr.*, 110(26), 232571e (1988).

4. T. Takeda and K. Kono, Jpn. Kokai Tokkyo Koho, JP 63,066,209 (1988); *Chem. Abstr.*, 109(18), 150296j (1987).
5. V. Kuhlwilm, H. Kaltwasser, M. Ernst, R. D. Klodt, S. Kreissl, and L. Nosske, German Patent DD 254,946 (1988).
6. P. Osanu, F. Georgescu, E. Georgescu, S. Curcaneanu, I. Ichim, R. Negretu, G. Arnautu, and M. Georgescu, Romanian Patent RO 85,539 B1 (1985); *Chem. Abstr.*, 104(20), 16904w (1985).
7. R. A. Marshall, J. W. Hershberger, and S. A. Hershberger, Eur. Patent Application EP 3,06,433 A2 (1989); *Chem. Abstr.*, 111(8), 58532b (1989).
8. H. Meissner and G. Heublein, *Acta Polym.*, 36(6), 343 (1985).
9. W. Kern, *Makromol. Chem.*, 1, 209, 249 (1948); 2, 48 (1948).
10. E. S. Raskin, *Zh. Prikl. Khim.*, 30, 1030 (1957); *Chem. Abstr.*, 51, 18692 (1957).
11. I. M. Kolthoff and M. Youse, *J. Am. Chem. Soc.*, 72, 3431 (1950).
12. S.-S. Chein, Y. Pinkang, and M. Chao, *Res. Bull. Taiwan Fertilizer Co.*, 4, 1–29 (1951).
13. A. D. Stepukhovich, M. D. Gol'Dfein, and V. G. Marinin, *Vysokomol. Soyed.*, 8(7), 1185 (1966).
14. S. Hasegawa and N. Nishimura, *Bull. Chem. Soc. Jpn.*, 33, 775 (1960).
15. S. Hasegawa, N. Nishimura, S. Mitsumoto, and K. Yokoyama, *Bull. Chem. Soc. Jpn.*, 36, 522 (1963).
16. A. M. Sharetskii, S. V. Svetozarskii, E. N. Zil'berman, and I. B. Kotlyar, U.S. Patent 3,594,359.
17. A. M. Sharetskii, S. V. Svetozarskii, E. N. Zil'berman, and I. B. Kotlyar, Br. Patent 1,164,250.
18. J. Ulbricht and V. N. Thanh, *Plaste Kaut.*, 21(3), 186, 190 (1974).
19. J. Ulbricht and G. Mueller, *Plaste Kaut.*, 21(6), 410 (1974).
20. A Konishi and K. Nambu, *J. Polym. Sci.*, 54, 209 (1961).
21. J. Das and V. Krishnan, *Proc. Sem. Polym. Surf. Coat.: Recent Dev.*, 25-6, Oil Technol. Assoc. India South. (1982); *Chem. Abstr.*, 100(8), 52116v (1982).
22. R. M. Johnson and I. W. Siddigi, in *The Determination of Organic Peroxides*, Pergamon Press, Oxford, 1970, p. 40.
23. N. G. Gaylord, U.S. Patent 4,269,958 (1981).
24. J. T. Edward, H. S. Chang, and S. A. Samad, *Can. J. Chem.*, 40, 804 (1962).
25. R. Huisgen and W. Edl, *Angew. Chem.*, 74, 588 (1962); *Angew. Chem. Int. Ed.*, 1, 458 (1962).
26. D. Z. Denney, T. M. Valega, and D. B. Denney, *J. Am. Chem. Soc.*, 86, 46 (1964).
27. S. Sivaram, R. K. Singhal, and I. S. Bhardwaj, *Polym. Bull.*, 3, 27 (1980).
28. D. B. Denney and D. Z. Denney, *J. Am. Chem. Soc.*, 84, 2455 (1962).
29. Ye. L. Kopp and Ye. B. Milovskaya, *Vysokomol. Soyed.*, A11, 750 (1969); *Polym. Sci. USSR*, 11, 848 (1969).
30. L. V. Zamoishaya, S. I. Vinogradeva, and Ye. B. Molovskaya, *Vysokomol. Soyed.*, A13, 1484 (1971); *Polym. Sci. USSR*, 13, 1670 (1971).
31. Ye. L. Kopp, O. S. Mikhaiycheva, and Ye. B. Milovskaya, *Vysokomol. Soyed.*, A14, 2653 (1972); *Polym. Sci. USSR*, 14, 3087 (1972).
32. K. Kido, H. Hakamori, G. Asai, and K. Kushida, Jpn. Kokai Tokkyo Koho, JP 8,002,206 (1976).
33. K. Kido, H. Wakamori, G. Asai, and K. Kushida, Jpn. Kokai Tokkyo Koho, JP 78,126,094 (1978).
34. R. J. Cozens, Eur. Patent Application EP 94,160 A1 (1983).
35. T. Dittrich, M. C. Kindelberger, H. Schirge, and R. Schlege, German Patent DD 215,790 A1 (1984); *Chem. Abstr.*, 103(12), 88347x (1984).
36. M. K. Mishra, *J. Macromol. Sci.—Rev. Macromol. Chem.*, C20(1), 149 (1981).

37. E. J. Behrman and J. O. Edwards, *Prog. Phys. Org. Chem.*, 4, 93 (1967).
38. L. Horner, *J. Polym. Sci.*, 18, 438 (1955).
39. L. Horner and B. Anders, *Chem. Ber.*, 95, 2470 (1962).
40. L. Horner and C. Betzel, *Ann. Chem.*, 579, 175 (1953).
41. L. Horner, H. Brüggemann, and K. H. Knappe, *Ann. Chem.*, 626, 1 (1959).
42. L. Horner and H. Junkermann, *Ann. Chem.*, 591, 53 (1955).
43. L. Horner and W. Kirmse, *Ann. Chem.*, 597, 48 (1958); 597, 66 (1958).
44. L. Horner and K. Scherf, *Ann. Chem.*, 573, 35 (1951).
45. L. Horner and E. Schwenk, *Ann. Chem.*, 566, 69 (1950).
46. L. Horner and H. Steppan, *Ann. Chem.*, 606, 47 (1957).
47. J. T. Edward, *J. Chem. Soc.*, 1464 (1954); 222 (1965).
48. J. T. Edward and S. A. Samad, *Can. J. Chem.*, 41, 1027 (1963); 41, 1638 (1963).
49. C. W. Capp and E. G. E. Hawkins, *J. Chem. Soc.*, 4106 (1953).
50. S. Gambarjan, *Ber.*, 42, 4003 (1909); 58B, 1775 (1925).
51. S. Gambarjan and O. A. Chaltykyan, *Ber.*, 60B, 390 (1927).
52. M. Imoto, T. Otsu, and K. Kimura, *J. Polym. Sci.*, 15, 475 (1955).
53. C. Walling and N. Indicator, *J. Am. Chem. Soc.*, 80, 5814 (1958).
54. T. H. Meltzer and A. V. Tobolsky, *J. Am. Chem. Soc.*, 76, 5178 (1954).
55. N. Nishimura, *J. Polym. Sci., A1*, 7, 2015 (1969).
56. J. Lal and R. Green, *J. Polym. Sci.*, 18, 403 (1955).
57. M. Imoto and K. Takemoto, *J. Polym. Sci.*, 18, 377 (1955).
58. K. Takemoto, A. Nishio, Y. Iikubo, and M. Imoto, *Makromol. Chem.*, 42, 97 (1960).
59. K. F. O'Driscoll and S. A. McArdle, *J. Polym. Sci.*, 40, 557 (1959).
60. K. F. O'Driscoll and J. F. Schmidt, *J. Polym. Sci.*, 45, 189 (1960).
61. K. F. O'Driscoll and E. N. Riccheza, *Makromol. Chem.*, 47, 15 (1961).
62. K. F. O'Driscoll, T. P. Konen, and K. M. Connolly, *J. Polym. Sci., A-1*, 5, 1789 (1967).
63. K. F. O'Driscoll, P. F. Lyons, and R. Patisga, *J. Polym. Sci.*, 3, 1567 (1965).
64. T. Sato, K. Takemoto, and M. Imoto, *Makromol. Chem.*, 104, 297 (1967).
65. H. Yano, R. Takemoto, and M. Imoto, *J. Macromol. Sci. Chem.*, 2. 81 (1968); 2, 739 (1968).
66. P. Ghosh and N. Mukherji, *Eur. Polym. J.*, 15, 797 (1979).
67. F. N. Mazitova and Yu. M. Ryzhmanov, *Dokl. Akad. Nauk SSSR*, 161, 1346 (1965); F. N. Mazitova, Yu. M. Ryzhmanov, R. R. Shagidullin, and I. A. Lamanova, *Neftekhim*, 5, 904 (1965); F. N. Manitova, Yu. M. Ryshmanova, Yu. V. Yabloka, and O. S. Durova, *Dokl. Akad. Nauk SSSR*, 153, 354 (1963).
68. K. Somono and O. Kikuchi, *Kogyo Kagaku Zasshi*, 68, 1527 (1965).
69. S. D. Stayrova, G. V. Peregudev, and M. F. Margritova, *Dokl. Akad. Nauk SSSR*, 157, 636 (1964).
70. P. D. Bartlett and K. Nozaki, *J. Am. Chem. Soc.*, 68, 2299 (1947).
71. G. Favini, *Gazz. Chim. Ital.*, 89, 2121 (1959).
72. Y. Okada, *Kogyo Kagaku Zasshi*, 65, 1085 (1962); *Chem. Abstr.*, 58, 6664.
73. S. E. Morsi, A. B. Zaki, and M. A. El-Khyami, *Eur. Polym. J.*, 13, 851 (1977).
74. K. F. O'Driscoll and E. N. Richezza, *J. Polym. Sci.*, 46, 211 (1960).
75. M. Imoto and S. Choe, *J. Polym. Sci.*, 15, 485 (1955).
76. C. Walling and R. B. Hodgson Jr., *J. Am. Chem. Soc.*, 80, 228 (1957).
77. C. Walling, in *Free Radicals in Solution*, John Wiley & Sons, New York, 1957, pp. 590–594.
78. V. Boekelheide and D. L. Harrington, *Chem. Ind.*, 1423 (1955).
79. Yu. V. Lebedev, S. N. Trubitsyna, M. M. Askrov, and M. F. Margaritova, *Vyskomol. Soyed.*, 15, 612 (1973).
80. S. N. Trubitsyna, M. F. Margaritva, and S. S. Medvedev, *Vysokomol. Soyed.*, 7, 1968 (1965); *Chem. Abstr.*, 55, 2524i.

81. T. Sato and T. Otsu, *J. Polym. Sci., Polym. Chem. Ed.*, 12, 2943 (1974).
82. T. Taniguchi and N. Hosokawa, Jpn. Kokai Tokkyo Koho, JP 8,034,247 (1978).
83. P. Ghosh and S. N. Maity, *Eur. Polym. J.*, 15, 787 (1979).
84. P. Ghosh and S. N. Maity, *Eur. Polym. J.*, 14, 855 (1978).
85. S. Maiti and M. K. Saha, *J. Polym. Sci., A-1,* 5, 151 (1967).
86. E. Halfpenny and P. L. Robinson, *J. Chem. Soc.*, 928 (1952).
87. R. C. Schulz, H. Cherdon, and W. Kern, *Makromol. Chem.*, 24, 141 (1947).
88. Koei Chemical Co., Ltd., Jpn. Kokai Tokkyo Koho, JP 8,111,915; *Chem Abstr.*, 95, 0824t.
89. Toa Gosei Chemical Industry Co. Ltd., Jpn. Kokai Tokkyo Koho, JP 81,118,407; *Chem. Abstr.*, 96, 7536v.
90. J. G. Murray, Eur. Patent Application EP 61,890; *Chem. Abstr.*, 98(2), 49,186.
91. J. G. Murray, U.S. Patent 4,367,320.
92. T. Nimura and T. Aoyanagi, Jpn. Kokai Tokkyo Koho, JP 60,245,612; *Chem. Abstr.*, 105(2), 6975e.
93. S. Mishima, T. Suzuki, and H. Shibata, Jpn. Kokai Tokkyo Koho, JP 61,171,705; *Chem. Abstr.*, 106(8), 50820v.
94. K. Satomi and K. Kamimura, Jpn. Kokai Tokkyo Koho, JP 01,110,511; *Chem. Abstr.*, 111(24), 215099.
95. R. Negretu, P. Osamu, S. Curcaneanu, I. Dimitriu, N. Chiroiu, P. Gluck, and M. Patrascu, Romania Patent RO 91,473; *Chem. Abstr.*, 108(2), 6654m.
96. E. Abuin, J. Carnejo, and E. A. Lissi, *Eur. Polym. J.*, 11, 779 (1975).
97. L. Contreras, J. Crotewold, E. A. Lissi, and R. Rozas, *J. Polym. Sci., A-1*, 7, 2341 (1969).
98. C. E. H. Bawn, D. Margerison, and M. M. Richardson, *Proc. Chem. Soc.*, 397 (1959).
99. A. Misono, Y. Vchida, and K. Yamada, *Bull. Chem. Soc. Jpn.*, 39, 2458 (1966).
100. K. Kojima, Y. Iwata, M. Nagayama, and S. Iwabuchi, *Polym. Lett.*, 8, 541 (1970).
101. M. Yoshikuni, M. Asami, S. Iwabuchi, and K. Kajima, *J. Polym. Sci., Polym. Chem. Ed.*, 11, 3115 (1973).
102. J. Furukawa, T. Tsuruta, T. Fueno, R. Sakata, and K. Ito, *Makromol. Chem.*, 30, 109 (1959).
103. B. Borsini and M. Cipolla, *J. Polym. Sci. B*, 2, 291 (1964).
104. R. L. Hansen and R. R. Hamman, *J. Phys. Chem.*, 67, 2868 (1963).
105. S. Iwabuchi, M. Ueda, M. Kobayashi, and K. Kojima, *Polym. J.*, 6, 185 (1974).
106. R. L. Hansen, *J. Polym. Sci. A*, 2, 4215 (1964).
107. J. Furkawa, T. Tsuruta, and S. Shiotani, *J. Polym. Sci.*, 40, 237 (1959).
108. E. Abyin, E. A. Lassi, and A. Yañez, *Polym. Lett.*, 7, 515 (1970).
109. N. L. Zutty and F. J. Welch, *J. Org. Chem.*, 25, 861 (1960).
110. A. G. Davies and D. G. Hare, *J. Chem. Soc.*, 438 (1959).
111. S. B. Mirviss, *J. Am. Chem. Soc.*, 83, 3051 (1961).
112. J. Furukawa, T. Tsuruta, and S. Inoue, *J. Polym. Sci.*, 26, 234 (1957).
113. S. Inoue, T. Tsuruta, and J. Furukawa, *Makromol. Chem.*, 40, 13 (1961).
114. F. J. Welch, *J. Polym. Sci.*, 61, 243 (1962).
115. J. Frank and E. Rabinowitch, *Trans. Faraday Soc.*, 30, 120 (1934).
116. H. C. Brown and G. W. Kabalka, *J. Am. Chem. Soc.*, 92, 710 (1970).
117. E. B. Milovskaya, T. G. Shuravleva, and L. V. Zamoyskaya. *J. Polym. Sci. C*, 16, 899 (1967).
118. F. Strain, U.S. Patent 2,464,062 (1949).
119. W. A. Strong, *Ind. Eng. Chem.*, 56, 33 (1964).
120. F. Strain, W. E. Bissinger, W. R. Dial, H. Rudolph, B. J. Dewitt, H. C. Strevens, and J. H. Longston, *J. Am. Chem. Soc.*, 72, 1254 (1950).

121. S. G. Cohen and D. B. Sparrow, *J. Am. Chem. Soc.*, 72, 611 (1950).
122. G. A. Razuvaev, L. M. Terman, and D. M. Yanovskii, *Vysokimol. Soyed.*, B9, 208 (1967).
123. G. A. Razuvaev, L. M. Terman, and D. M. Yanovskii, *Zh. Org. Khim.*, 1, 274 (1965).
124. G. A. Razuvaev, L. M. Terman, and K. F. Bol'shakova, *Vysokomol. Soyed.*, B10, 38 (1968); *Chem. Abstr.*, 68, 69359.
125. V. Ya. Kolesnikov, V. A. Popov, Yu. A. Zvereva, V. I. Zegel'man, and E. P. Shvarev, *Plast. Massy*, 8, 7 (1983); *Chem. Abstr.*, 99(16), 123046w.
126. T. Xie, Z. Yu, Q. Cai, and Z. Pan, *Huagong Xuebao*, 2, 93 (1984); *Chem. Abstr.*, 101(26), 231062s.
127. G. P. Osanu, E. I. Georgescu, F. Georgescu, S. Curcaneanu, R. Negretie, G. Arnautu, I. Ichim, N. Chiroiu, and A. Szakacs, Romanian Patent RO 83251; *Chem Abstr.*, 102(26), 221340.
128. K. H. Graul, J. Weber, V. Steinert, W. Thuemmler, C. Schulz, and H. Kaltwasser, German Patent DD 223716; *Chem. Abstr.*, 104(14), 110373x.
129. N. Satomi, H. Hagii, and K. Aoshima, Japanese Patent JP 60258212; *Chem. Abstr.*, 105(6), 43552k.
130. M. Stanescu, M. R. Creanga, I. Deaconescu, A. Stane, N. Chiroiu, I. Rus, and G. Zanescu, Romanian Patent RO 86,797; *Chem. Abstr.*, 105(14), 115561n.
131. N. Satomi, H. Hagai, H. Nagai, Jpn. Kokai Tokkyo Koho, JP 61130315; *Chem. Abstr.*, 106(4), 19182q.
132. H. Nilsson and C. Silvergren, Eur. Patent Application EP 209,504; *Chem. Abstr.*, 106(4), 103241r.
133. R. B. Hawrylko, U. S. Patent 4,668,707; *Chem. Abstr.*, 107(14), 116432z.
134. H. Ishimi, Y. Hirai, and S. Toyonishi, Jpn. Kokai Tokkyo Koho, JP 62,086,005; *Chem. Abstr.*, 107(26), 237501e.
135. A. A. Matyakubov, S. Masharipov, and K. A. Nazhimov, *Uzb. Khim. Zh.*, 4, 48 (1987); *Chem. Abstr.*, 108(6), 38492y.
136. G. P. Osanu, R. Negretu, N. Chiroiu, E. Georgescu, G. X. Arnautu, I. Ichim, E. Molovan, D. Vaduva, M. Monu, and M. Georgescu, Romanian Patent RO 89406; *Chem. Abstr.*, 108(8), 56799h.
137. S. Schnabel, J. Weber, and W. Thuemmler, German Patent DD 251,755; *Chem. Abstr.*, 109(12), 938261.
138. V. P. Malhotra, V. K. Tandon, R. K. Diwan, V. K. Saroop, and M. K. Behi, Indian Patent IN 164417; *Chem. Abstr.*, 112(14), 119634f.
139. T. Amano and S. Hashida, Jpn. Kokai Tokkyo Koho, JP 01,275,610; *Chem. Abstr.*, 112(16), 140049d.
140. K. Ihara, Y. Noda, and T. Amano, Jpn. Kokai Tokkyo Koho, JP 01,129,005; *Chem. Abstr.*, 112(6), 36761u.
141. H. Iwami, Jpn. Kokai Tokkyo Koho, JP 01,161,004; *Chem. Abstr.*, 11(26), 233888w.
142. M. Stanescu, M. R. Creanga, I. Deaconescu, A. Stane, N. Chiroiu, I. Ros, and G. Zanescu, Romanian Patent RO 90,603; *Chem. Abstr.*, 108(14), 113164y.
143. T. L. Latypov, A. A. Yul'chibaev, Kh. Yu. Kuzieva, *Uzb. Khim. Zh.*, 4, 62 (1983); *Chem. Abstr.*, 99(22), 176337y.
144. S. Maiti and S. R. Palit, *J. Polym. Sci.*, B8(7), 515 (1970).
145. N. Shevit and M. Konigsbuch, *J. Polym. Sci. C*, 16, 43 (1966).
146. V. F. Kazanskaya and S. V. Smirnova, *Vysokomol. Soyed. Ser. B*, 12(7), 523 (1970).
147. G. S. Misra and S. L. Dubey, *Colloid Polym. J.*, 257, 156 (1979); *J. Polym. Sci. A-1*, 17, 1393 (1979).
148. G. S. Misra and N. M. Bhasilal, *Eur. Polym. J.*, 14, 901 (1978); *J. Macromol. Sci.-Chem.*, A12(9), 1275 (1978).

149. M. M. Hussain, S. N. Misra, and A. Gupta, *Makromol. Chem.*, 176, 2861 (1975); 177, 2991 (1976).
150. M. M. Hussain, S. N. Misra, and R. D. Singh, *Makromol. Chem.*, 179, 295 (1978).
151. M. M. Hussain and S. N. Misra, *Makromol. Chem.*, 179, 41 (1978); *Ind. J. Chem.*, 17A, 118 (1978).
152. F. T. Wall, F. W. Banes, and G. D. Sando, *J. Am. Chem. Soc.*, 68, 1429 (1946).
153. R. Synder, J. Stewart, R. Allen, and R. Dearborn, *J. Am. Chem. Soc.*, 68, 1422 (1946).
154. R. L. Eager and C. A. Winkler, *Can. J. Res.*, 26B, 527 (1948).
155. I. M. Kolthoff, *J. Am. Chem. Soc.*, 74, 4419 (1952); *J. Polym. Sci.*, 15, 459 (1955).
156. H. W. Starkweather, P. O. Bare, A. S. Carter, F. B. Hill, V. R. Hurka, C. J. Mighton, P. Sanders, H. Yonker, and M. Yanker, *Ind. Eng. Chem.*, 39, 210 (1947).
157. I. M. Kolthoff and W. J. Dale, *J. Am. Chem. Soc.*, 61, 1672 (1945).
158. I. M. Kolthoff and W. H. Harris, *J. Polym. Sci.*, 2, 41 (1947).
159. R. Azuma and K. Kurimoto, Jpn. Kokai Tokkyo Koho, JP 7,791,089 (1976).
160. H. Kawakado, Y. Isobe, S. Imaizumi, T. Morita, and H. Hayashi, Jpn. Kokai Tokkyo Koho, JP 79,158,491 (1979).
161. S. R. Palit and M. Biswas, *J. Sci. Ind. Res.*, 20B, 279 (1961).
162. S. R. Palit and T. Guha, *J. Polym. Sci.*, A-1, 1877 (1963).
163. S. R. Palit and T. Guha, *J. Polym. Sci.*, 34, 243 (1959).
164. M. Biswas, T. Guha, and S. R. Palit, *J. Ind. Chem. Soc.*, 42(8), 509 (1965).
165. E. S. Roskin and G. B. Karpenko, *Izv. Vysshikh. Uchele. Zaved. Khim. Khim. Tekhnol.*, 4, 280 (1961).
166. E. S. Roskin and G. B. Karpenko, *Izv. Vysshikh. Uchele. Zaved. Khim. Khim. Tekhnol.*, 7(3), 523 (1964).
167. E. S. Roskin, *Zh. prikl. Khim.*, 30, 1030 (1957); *Chem. Abstr.*, 51, 18692 (1957).
168. S. C. Chaddha, P. Ghosh, A. R. Mukherjee, and S. R. Palit, *J. Polym. Sci.*, A-2, 4441 (1964).
169. I. M. Kolthoff and E. J. Meehan, *J. Polym. Sci.*, 9, 327 (1952).
170. I. M. Kolthoff and E. J. Meehan, *J. Polym. Sci.*, 9, 433 (1952).
171. I. M. Kolthoff and E. J. Meehan, *J. Appl. Polym. Sci.*, 1, 200 (1959).
172. T. Tadasa and H. Kakitani, Jpn. Kokai Tokkyo Koho, JP 78,114,892; *Chem. Abstr.*, 90, 39608n.
173. T. Tadasa and H. Kakitani, Jpn. Kokai Tokkyo Koho, JP 78,113,879; *Chem. Abstr.*, 90, 39614m.
174. N. Ashikari, *J. Polym. Sci.*, 28, 641 (1958).
175. J. Furkawa and T. Tsuruta, *J. Polym. Sci.*, 28, 227 (1958).
176. J. W. Fordham and C. L. Sturm, *J. Polym. Sci.*, 33, 504 (1958).
177. N. L. Zutty and F. J. Welch, *J. Polym. Sci.*, 43, 445 (1960).
178. R. D. Burkhalt and N. L. Zutty, *J. Polym. Sci. A*, 1, 1137 (1963).
179. F. S. Arimoto, *J. Polym. Sci. A-1*, 4, 725 (1966).
180. G. S. Lolesnikow and N. V. Klimentova, *J. Polym. Sci.*, 39, 560 (1959).
181. M. Bednarek, M. Olinsky, and D. Lim, *Czechoslovak. Chem. Commun.*, 32, 1575 (1967).
182. K. Noro, H. Kawazura, and E. Uemura, *Kogyo Kagaku Zasshi*, 65, 973 (1962).
183. K. Ryuichi and S. Isao, Japanese Patent 5498 (1965).
184. A. V. Ryabov, V. A. Dodonov, and Yu. A. Ivanova, *Tr. Khim. Tekhnol.*, 1, 238 (1970).
185. T. Iwashita, I. Takahasi, M. Kuso, K. Kikuchi, T. Nahamura, and S. Sakai, Jpn. Kokai Tokkyo Koho, JP 60,163,906; *Chem. Abstr.*, 104(18), 150044x.
186. H. P. Chowanitz, K. Hoehne, K. D. Weissenborn, J. Fischmann, R. D. Klodt, R. Madla, G. Weickert, G. Henschel, and G. Noll, German (East) Patent DD 237664; *Chem. Abstr.*, 106(22), 177069n.

187. M. Fukuda, S. Ishibashi, T. Ikeda, K. Kamimura, and K. Kazama, Jpn. Kokai Tokkyo Koho, JP 63,230,713; *Chem. Abstr.*, 109(26), 231747j.
188. S. Masuko, T. Kunimura, H. Takahara, and K. Fukuda, Jpn. Kokai Tokkyo Koho; *Chem. Abstr.*, 111(8), 58546j.
189. N. Mitrea, M. Stanescu, C. Casadjicov, and A. Grigorescu, Romanian Patent RO 92624; *Chem. Abstr.*, 109(12), 93829j.
190. M. Nakagawa, K. Mori, and T. Sugita, Jpn. Kokai Tokkyo Koho, JP 60,206,811; *Chem. Abstr.*, 104(20), 169454c.
191. K. Maeda and S. Aihara, Jpn. Kokai Tokkyo Koho, JP 60,231,716; *Chem. Abstr.*, 105(2), 7007c.
192. N. G. Gaylord, M. Nagler, and M. M. Fein, U.S. Patent 4,269,957 (1981); N. G. Gaylord and M. Nagler, in *Proc. IUPAC, 28th Makromol. Symp.*, p. 267; *Chem. Abstr.*, 99(12), 88643q.
193. N. G. Gaylord and M. Nagler, U.S. Patent 4,269,956 (1981).
194. N. G. Gaylord and M. Nagler, *Polym. Bull.*, 8, 395 (1983).
195. S. Papetti, U.S. Patent 4,046,839; *Chem. Abstr.*, 89, 152909n.
196. S. Kato and M. Momoka, Jpn. Kokai Tokkyo Koho, JP 7,882,892; *Chem. Abstr.*, 89,180616z.
197. S. Pappetti, German Offen. 223,811; *Chem. Abstr.,* 78, 125218 (1973).
198. Stockholms Superfosfat Fabriks A/B, Br. Patent 961,254; *Chem. Abstr.*, 61, 71364.
199. R. G. R. Bacon, *Trans. Faraday* Soc., 42, 140 (1946).
200. K. Hattori and Y. Komeda, *Kogyo Kagaku Zasshi*, 68(9), 1729 (1965).
201. R. Kojima, S. Nagase, H. Muramatsu, and H. Baba, Kogyo *Kagaku Zasshi*, 60, 499 (1957).
202. D. Campbell, *J. Polym., Sci.*, 32, 413 (1958).
203. S. Yughuchi and M. Watanabe, *Kobunshi Kagaku*, 17, 465 (1906); 18, 368 (1961).
204. Y. Tsuda, *J. Appl. Polym. Sci.*, 5, 104 (1961).
205. S. Yuguchi and M. Hosina, *Kobunshi Kagaku*, 18, 381 (1961).
206. G. M. Guzman and F. Arranz, *Real. Soc. Espan. Fis. Quim.*, B59(6), 445 (1963).
207. J. Ulbritch and P. Fritzsche, *Fascherforsch. Tentiltech.*, 15(3), 93 (1964); 14((8), 320 (1963); 14(12), 517 (1963).
208. A. N. Akopova and N. G. Korolnik, *Dokl. Akad. Nauk USSR*, 20(8), 34 (1963).
209. W. K. Wilkinson, *Macromol. Synth.*, 2, 78 (1966).
210. D. Feldman and F. Sandru, *Mater. Plast.*, 3(1), 25 (1966).
211. M. Narkis and D. H. Kohn, *J. Polym. Sci.*, 5(5), 1033 (1967).
212. T. Matsuda, T. Higushimura, and S. Okamura, *J. Macromol. Sci.-Chem.*, 2(1), 43 (1968).
213. H. Prochess and F. Patal, *Macromol. Chem.*, 114, 11 (1968).
214. R. M. Fitch and Tsang-Jan Chen, *Polym. Prepr., Am. Chem. Soc., Div. Polym. Chem.*, 10(1), 424 (1969).
215. P. C. Mark and J. Ugelstad, *Makromol. Chem.*, 128, 83 (1969).
216. R. G. R. Bacon, *Quart. Rev.* (*Lond.*). 9, 288 (1937).
217. M. Nagano and Y. Kuroda, *Sen-i-Gakkaishi*, 22(11), 479 (1966); 21(10), 541 (1965).
218. P.-K. Shen and W.-Syaliang, *Res. Bull. Taiwan Fertilizer Co.*, 11, 1–22 (1953).
219. K. L. Berry and J. H. Peterson, *J. Am. Chem. Soc.*, 73, 5195 (1951).
220. M. F. Hoover, U. S. Patent 3,832,992 (1967).
221. M. Katayama and T. Ogoshi, *Chem. High Polym.* (*Tokyo*), 13, 6 (1956).
222. Monsanto Co., Br. Patent 1,215,320 (Dec. 9, 1970).
223. A. Nakajima, N. Takaya, and M. Hoten, *Chem. Abstr.*, 68, 105640 (1968).
224. G. Talamani, A. Turolla, and E. Vianello, *Chim. Ind.* (*Milan*), 47(6), 581 (1965).
225. J. Carno, E. K. Flemminy, and W. A. Kein, *Ind. Eng. Chem. Prod.–Res. Dev.*, 8(1), 93 (1969).

226. A. S. Risk and M. H. Nossair, *Ind. J. Chem. A*, 16A(7), 564 (1978).
227. R. S. Konar and S. R. Palit, *J. Ind. Chem. Soc.*, 38, 481 (1961).
228. K. Shen, U.S. Patent 3,668,194.
229. D. F. Knaack, U.S. Patent 3,664,582.
230. N. G. Gaylord, U.S. Patent 4,261,870 (1981).
231. N. G. Gaylord, M. Nagler, and M. M. Fein, U.S. Patent 4,242,482 (1980).
232. H. I. Roll, J. Wergau, and W. Dockhorn, German Offen. 2,208,442 (1973).
233. J. A. Cornell, U.S. Patent 3,534,010 (1970).
234. K. Okamura, K. Suzuki, Y. Nojima, and H. Tanaka, Japanese Patent 18,945 ('64) (1964).
235. H. Watnabe, S. Yamanaka, and Y. Amagi, Japanese Patent 16,591 ('60) (1960).
236. K. H. Prell, E. Plaschil, and H. Germanus, German (East) Patent 75,395 (1970).
237. R. J. S. Mathews, Br. Patent 931,628 (1963).
238. A. G. Dynamite Nobel, Netherlands Application No. 6,408,790 (1965).
239. R. Buning, K. H. Diessel, and G. Bier, Br. Patent 1,180,363 (1970).
240. N. Fischer, J. Boissel, T. Kemp, and H. Eyer, U.S. Patent 4,091,197 (1978).
241. K. Kamio, T. Tadasa, and K. Nakanishi, Japanese Patent 7,107,261 (1971).
242. K. Kamio, T. Tadasa, and K. Nakanishi, Japanese Patent 7,025,513 (1970).
243. N. G. Gaylord, U.S. Patent 4,269,960 (1981).
244. N. G. Gaylord, U.S. Patent 4,382,133 (1983); 4,543,401 (1985).
245. Z. Jedlinski and A. Grycz, *Rocznicki Chem.*, 37(10), 1177 (1963).
246. A. D. Gomes and M. D. Lourdes, *J. Polym. Sci.*, A17(8), 2633 (1979).
247. P. Ghosh and S. Biswas, *Makromol. Chem.*, 182, 1985 (1981).
248. P. Ghosh, S. Jana, and S. Biswas, *Eur. Polym. J.*, 16, 89 (1980).
249. P. Ghosh, S. Biswas, and S. Jana, *Bull. Chem. Soc. Jpn.*, 54, 595 (1981).
250. P. Ghosh and S. Biswas, *J. Macromol. Sci. Chem.*, A16(5), 1033 (1981).
251. C. Mazzolini, L. Patron, A. Moretti, and M. Campanelli, *Ind. Eng. Chem. Prod. Res. Dev.*, 9(4), 504 (1970).
252. C. Mazzolini and L. Patron, *Kinet. Mech. Polyreactions, Int. Symp. Macromol. Chem. Prepr.*, 3, 65 (1969).
253. S. L. Monaco, C. Mazzolini, L. Patron, and A. Moretti, German Offen. 2,046,143 (1971).
254. C. Mazzolini, L. Patron, and A. Moretti, German Offen. 1,962,638 (1970).
255. L. Patron and A. Moretti (Chatillon Societa Anon.), Italian Patent 896 (1971); *Chem. Abstr.*, 86, 107254.
256. F. H. Seubold and W. E. Vaughan, *J. Am. Chem. Soc.*, 75, 3790 (1953).
257. A. V. Tobolsky and R. B. Mesrobian, in *Organic Peroxides*, Interscience, New York, 1954, pp. 99–100, 117–120.
258. H. Gilman, in *Organic Chemistry, Advanced Treatise*, 2nd ed., John Wiley & Sons, New York, 1948, Vol. II, p. 1858.
259. B. E. Kutsenok, M. N. Kulakova, E. I. Tinylkava, and B. A. Dolgoplosk, *Dokl. Akad. Nauk SSAR*, 125, 1076 (1959).
260. L. Patron and A. Moretti (Chatillon Societa Anon. Italiana per le Fibre Tessili Artificiali S.P.A.), Italian Patent 924,513; *Chem. Abstr.*, 86, 107253.
261. S. Ponratnam and S. L. Kapur, *Curr. Sci.*, 45, 295 (1976).
262. T. A. Mal'tseva, D. L. Snezhko, A. D. Virnic, and Z. A. Rogovin, *Izv. Vysshikh. Ucheb. Zaved. Khim. Tekhnol.*, 8(4), 651 (1965).
263. R. G. R. Bacon, *Chem. Ind.*, 897 (1953).
264. A. Nikolaev, W. Larinova, and M. Tereshchenko, *Zh. Prikl. Khim.*, 38(10), 2287 (1965).
265. J. M. Willis, *Ind. Eng. Chem.*, 41, 2272 (1949).
266. B. D. Sully, *J. Chem. Soc.*, 1948 (1950).

267. T. J. Suen, Y. Jen, and J. Lockwood, *J. Polym. Sci.*, 31, 481 (1958).
268. W. H. Thomas, E. Gleason, and G. Mino, *J. Polym. Sci.*, 24, 43 (1957).
269. N. A. Dobrynin, N. P. Dymarchuk, and K. P. Mischenko, *Zh.Obsheh. Khim.*, 40(6), 1186 (1970).
270. R. S. Konar and S. R. Palit, *J. Ind. Chem. Soc.*, 38, 481 (1961).
271. P. Melacini, L. Patron, A. Moretti, and R. Tedesco (Chatillon Societa Anon. Italiana per le Fibre Tessili Artificiali S.P.A.), Italian Patent 903,309.
272. H. Minato, H. Iwai, K. Hashimoto, and T. Yusai, *J. Polym. Sci.*, C23(2), 761 (1966).

7 Vinyl Polymerization Initiated by High-Energy Radiation

Ivo Reetz, Yusuf Yagci, and Munmaya K. Mishra

CONTENTS

7.1 INTRODUCTION

The availability of radioactive sources and particle accelerators has stimulated studies on their use for initiating chain polymerizations. These refer mainly to the radiation-induced production of free radicals, which are able to initiate vinyl polymerization. First evidence of vinyl polymerization by high-energy radiation was found before World War II [1–3], but it was in the 1950s and 1960s that numerous data on radiation-induced polymerization of many monomers were accumulated. Special attention has also been devoted to the exposure of polymeric substrates to high-energy radiation. The polymer-bound radicals and ions generated under these circumstances in the presence of a monomer may initiate graft copolymerization. Tailor-made polymers with an interesting combination of properties are thus accessible.

High-energy radiation includes electromagnetic x-rays and γ-rays and energy-rich particle rays, such as fast neutrons and α-and β-rays. As far as γ-rays are concerned, 1.25-MeV rays emitted by ^{60}Co and 0.66-MeV rays generated by ^{137}Cs have been most frequently used for polymerizations. Electrons (β-rays) produced by electron accelerators were also often applied. High-energy electrons (several MeV) were mainly employed for investigating dose-rate effects. Relatively low-energy electrons (0.2–0.5 MeV) are used successfully for industrial curing of various coatings. Electromagnetic or particle rays other than γ-rays or electrons are seldom used due to several disadvantages such as high cost, lack of penetration, and, in the case of neutrons, residual reactivity.

As far as the absorption of energy by the monomer or polymer is concerned, the predominant effect as γ-rays enter organic substrates is the Compton effect. It involves an electron ejected from an atom after collision. The ejected electron interacts with other atoms to raise their energy level to an excited state. If the electron possesses sufficient energy, another electron is ejected, leaving behind a positive ion. The excited atoms and ions can take part in further reactions in the substrate and transfer their energy or decompose into radicals that give rise to polymer formation in the presence of vinyl monomer:

$$C + \text{radiation} \rightarrow C^{+} + e^{-} \tag{7.1}$$

The cation may then form a radical by dissociation:

$$C^{+} \rightarrow A^{\cdot} + B^{+} \tag{7.2}$$

The initially ejected electron may be attracted to the cation B^{+} forming another radical:

$$B^{+} + e^{-} \rightarrow B^{\cdot} \tag{7.3}$$

Radicals may also be produced upon a sequence of reactions initiated by the capture of an ejected electron by C:

$$C + e^{-} \rightarrow C^{-} \tag{7.4}$$

$$C^{-} \rightarrow B^{\cdot} + A^{-} \tag{7.5}$$

$$A^{-} \rightarrow A^{\cdot} + e^{-} \tag{7.6}$$

Another pathway includes the direct hemolytic bond rupture upon irradiation with high-energy rays, a process involving the formation of electronically excited C particles:

$$C + \text{radiation} \rightarrow A^{\cdot} + B^{\cdot} \tag{7.7}$$

As described previously, the radiolysis of olefinic monomers results in the formation of cations, anions, and free radicals. It is possible for these species to initiate chain polymerizations. Whether radiation-induced polymerization is initiated by radicals, cations, or anions depends on the monomer and the reaction conditions. However, in most radiation-initiated polymerizations, initiating species are radicals [4]. It is usually only at low temperatures that ions are stable enough to react with a monomer [5]. At ambient temperatures or upon heating, ions are usually not stable and dissociate to yield radicals. Furthermore, the absence of moisture is crucial if one aims at ionic polymerizations [6, 7]. Thus, for styrene polymerization at room temperature, polymerization rates are by a factor of 100–1000 times higher for "super-dry" styrene than for "wet" styrene, the difference being due to the contribution of cationic polymerization [8, 9]. Radiolytic initiation can also be performed using additional initiators that are prone to undergo decomposition upon irradiation with high yields.

The reactive intermediates generated when organic matter is exposed to fast electrons generally do not differ much from those obtained by γ-irradiation. The electrons are slowed down by interactions with atoms of the absorber leading to ionizations and excitations. In monomers, ionizations and excitations are produced in a sphere of a radius of ~2-20 nm, a zone referred to as spur.

7.2 RADICAL POLYMERIZATION

As discussed previously, the interaction of high-energy rays with monomers results in the generation of free radicals. In radiation chemistry, the yield of a reaction is generally expressed in terms *G* values, that is, the number of radiolytically produced or consumed species per 100 eV absorbed. As far as radical vinyl polymerization is concerned, *G* (radical) values depend on the proneness of a monomer to form radicals. Thus, for styrene, *G* (radical) values of 0.7 are found; for vinyl acetate, the *G*(radical) value amounts to ~ 12 (see Table 7.1) [10].

Radiation-induced polymerizations may be performed in bulk, in solution, or even as emulsion polymerization [11, 12]. For solution polymerization, the possibility of generating radicals stemming from solvent also has to be taken into account. Notably, in contrast to photopolymerization, where solvents transparent to incident light are used, usually high-energy radiation is absorbed by all components of the polymerization mixture, including solvent. As is seen in Table 7.1, for a monomer with a low *G* (radical) value, the overall radical yield and, therefore, the polymerization rate may be enhanced by using solvents that easily produce radicals (e.g., halogen-containing solvents).

In some cases of solution polymerization, efficient energy transfer occurs from excited solvent molecules to monomer molecules or vice versa [13, 14]. For example, in the case of styrene polymerization in *n*-dibutyl disulfide (DBD), an energy transfer

TABLE 7.1
Free Radical Yields *G* (Radical) Values for a Few Polymerization Mixtures, Bulk and Solution Polymerization

Monomer	Solvent	G(radical)
Styrene	None (bulk)	0.69
	Benzene	0.76
	Toluene	1.15
	Chlorobenzene	8.0
	Ethyl bromide	11.8
Methyl methacrylate	None (bulk)	11.5
	Methyl acetate	10.9
Vinyl acetate	None (bulk)	12.0
	Ethyl acetate	12.0

Source: H. F. Mark, N. G. Gaylord, and N. M. Bikales, Eds., *Encyclopedia of Polymer Science and Technology,* Vol. 11, Interscience, New York, 1969, p. 702.

from styrene to DBD has been observed [15]. Very strong sensitization occurs in mixtures of carbon tetrachloride. By the addition of ~ 3% of this substance, the polymerization rate of styrene rises by a factor of 3 [16]. The acceleration of styrene polymerization by addition of small amounts of methanol has also been reported [17]. This effect has been explained in terms of a reaction of protons, stemming from methanol, with radiolytically formed styrene-based anion radicals, transforming the latter into initiating radicals. Furthermore, higher polymerization rates may be brought about by small concentrations of typical radical initiators, such as hydrogen peroxide [18, 19], that readily generate radicals upon irradiation.

7.3 GRAFTING AND CURING

Radiolytical grafting is an often-performed method for the production of specialty polymers with interesting surface and bulk properties. Important areas of radiation-induced grafting onto solid polymers include the development and production of hydrophilic surfaces and membranes, which find application in separation technology and in medicine [20–29]. Radiation is used to activate the base polymer, onto which the monomer present is grafted. Radicals and/or ions produced adjacent to the polymer backbone act as initiating sites for free-radical or ionic polymerization. The striking advantage of this method is its universality. In fact, virtually all polymers may be activated for grafting by high-energy radiation to an extent that compares well with chemical initiation. Furthermore, no requirement exists for heating the backbone trunk as would be the case if thermal activation was applied.

Upon exposing a polymer to high-energy radiation, radiolytically induced chain scission or cross-linking also has to be taken into account [30]. If the irradiation takes place in the presence of oxygen, chain scission is often observed. Oxygen acts as a radical scavenger and forms reactive peroxides when reacting with the polymer-bound radicals, giving rise to degradation processes referred to as autoxidation.

Three experimental approaches of radiolytical grafting have to be distinguished.

1. The *preirradiation technique* consists of two distinct steps: irradiation of the backbone polymer *in vacuo* or an inert gas in the absence of monomer and, subsequently, addition of a monomer. Obviously, sufficient lifetimes of the reactive species and high reactivities toward the monomer is necessary. For trapped radicals, this generally requires some degree of crystallinity or a glassy state in the polymer and storage at low temperatures [20, 31].

   In the second, the actual grafting step, the monomer has to diffuse to the active centers of the polymer. In some applications, heating is applied for increasing the mobility of both the monomer and irradiated polymer. However, a disadvantage of heating in this step is that recombination of polymer-bound radicals becomes more likely, owing to the high overall particle mobility. Solvents or swelling agents also lead to a faster diffusion of monomer to the reactive site of the backbone trunk.
2. In many cases, higher grafting yields are obtained when the preirradiation is performed in oxygen or in air [32, 33]. This phenomenon, which is ascribed to the formation of peroxides at the base polymer, is utilized for

the so-called *peroxide technique* [32–36]. Oxygen present during irradiation reacts with the reactive polymer-bound radicals generated upon irradiation. In a second step, the polymeric hydroperoxides are decomposed upon heating or ultraviolet (UV) irradiation in the presence of a monomer. The oxygen-centered, polymer-bound radicals generated give rise to a graft copolymer. Not being attached to the backbone trunk, hydroxyl radicals initiate homopolymerization of the monomer present:

$$C + \text{radiation} \rightarrow A^{\cdot} + B^{\cdot} \tag{7.8}$$

If peroxy radicals react together, peroxides are produced, which yield only polymer-bound radicals upon irradiation. Therefore, homopolymer formation does not take place. However, these peroxides are harder to activate than hydroperoxides:

$$\begin{array}{l} \vdash\!\text{O—OH} \xrightarrow{\Delta \text{ or } h\nu} \vdash\!\dot{\text{O}} + \dot{\text{O}}\text{H} \\ \qquad\qquad\qquad\quad \downarrow \text{Monomer} \downarrow \\ \qquad\qquad\qquad \text{Grafting} \quad \text{Homopolymer} \\ \vdash\!\text{O—O}\!\dashv \xrightarrow{\Delta \text{ or } h\nu} 2\vdash\!\dot{\text{O}} \xrightarrow{\text{Monomer}} \text{Grafting} \end{array} \tag{7.9}$$

Another possibility of preventing homopolymer formation is the addition of reducing agents, such as the Fe^{2+} containing Mohr's salt [22–24, 33, 36], to the monomer.

As discussed previously, peroxides are also precursors of autoxidation. To avoid excessive degradation of the trunk polymer, control of irradiation doses is necessary.

3. The simultaneous irradiation of a monomer and a base polymer is referred to as *mutual radiation grafting technique.* In this method, the monomer may be present as vapor, liquid, or in solution. To avoid degradation, the polymerization mixtures are mostly freed from oxygen.

Following this technique, reactive sites are produced on both the backbone trunk and the monomer, the latter giving rise to a sometimes appreciable yield of undesired homopolymer. In fact, the radical formation yield of the trunk polymer has to be high in comparison with that of the monomer to have little homopolymer formation. Trunk polymers that are very suitable in this respect are poly(vinyl chloride) abstraction of Cl), wool, cellulose [37], poly(amides) [38–68], or aliphatic-type polymers, such as poly(ethylene) (facile C—H bond scission) [69]. Polymers with aromatic rings in the backbone are unsuitable because they are generally quite radiation resistant. The efficiency of grafting is usually high because the reactive species produced react immediately with monomer. Another advantage is that relatively low radiation doses are sufficient for grafting, which is particularly important for radiation-sensitive base polymers, such as poly(vinyl chloride). In many cases, radiation protection by the monomer may be observed. Vinyl compounds often protect aliphatic substances from undergoing radiolytically induced reactions, a phenomenon ascribed to scavenging reactions, particularly involving hydrogen atoms [70, 71].

The presence of solvents or swelling agents exerts a significant influence on the copolymerization and sometimes on the properties of the copolymer [72–79]. For example, if styrene is grafted onto cellulose in the presence of *n*-butanol as a swelling agent, grafting is observed only at the surface of the cellulose sample. On the other hand, if methanol is used instead of butanol, grafting occurs at the surface as well as in the cellulose bulk [80]. For many backbone polymers, the yield of grafting may be significantly improved by using accelerating additives such as mineral acids [81–84]. Furthermore, higher grafting rates may be obtained by means of thermal radical initiators, such as 2, 2′-azobisisobutyronitrile [85].

A wide variety of trunk polymers have been used for grafting reactions [13]. As an example, grafting onto nylon was of interest because by grafting, surface and certain bulk properties of this important synthetic fiber may be modified. Using γ-rays for activation, various vinyl monomers were grafted onto nylon-6 backbone. The grafting of acrylic acid [86–89] on nylon was performed using a ^{60}Co source at room temperature. The amount of acrylic acid grafted on the fiber increased linearly with monomer concentration [89]. The radiation-induced graft copolymerization of styrene, acrylonitrile [38–53], methyl methacrylate [54–56], methacrylic acid [57–59], and acryl amide [60–64] and its derivatives, such as N-methylol acrylamide [60, 65–69] onto nylon-6 was studied by various workers using a ^{60}Co source. The kinetics of the process was studied by measuring radical destruction rates and the weight increase. Usually, no homopolymer was obtained [44]. The graft copolymer is evenly distributed in the amorphous area of the fiber. The fiber's crystallinity remained unchanged [51].

Sumitomo and Hachihama [32, 90], Skyes and Thomas [60], Okamura et al. [91], and Armstrong and Rutherford [92–94] compared various procedures of grafting and concluded that preirradiation of nylon in air followed by heating it in monomer (ethyl acrylate) at –100°C yielded higher grafting yields than those obtained by mutual irradiation technique or by preirradiation *in vacuo.*

A commercially important technique that is in some respect similar to radiolytically induced grafting is the *curing* of coatings by high-energy rays [95]. For curing, performed polymers, oigomers, or sometimes monomers are irradiated mainly by means of electron beam machines, whereby reactive sites, mainly radicals, are generated. The radicals are adjacent to the polymer backbone and react together, leading to cross-linking in the irradiated part of the coating. As prepolymers, mostly acrylate-based polymers prepared by monomer polymerization or the acrylation of a backbone polymer such as a poly(urethane), poly(ester), poly(ether), or epoxy polymers with relatively low molecular weight (≈ 300) are used [96–99]. Recent studies suggest that upon curing at room temperature, ions are produced that have to be taken into account for explaining the curing mechanism [100, 101].

REFERENCES

1. F. L. Hopwood and J. T. Phillips, *Proc. Phys. Soc.* (*London*), 50, 438 (1938).
2. F. Joliot, French Patent 996, 760 (1940).
3. E. Rexer, *Reichsber. Phys.* (*Suppl. Phys. Z.*) 1, 111 (1944).

4. V. T. Stannett, J. Silverman, and J. L. Garnett, in *Comprehensive Polymer Science*, Vol. 4, G. Allen and J. C. Bevington, Eds., Pergamon Press, Oxford, 1989, p. 317.
5. K. B. Wood, V. T. Stannet, and P. Sigwalt, *J. Polym. Sci., Part A: Polym. Chem.,* 33, 2909 (1995).
6. E. Collinson, F. S. Dainton, and H. A. Gillis, *J. Phys. Chem.*, 63, 909 (1959).
7. J. V. F. Best, T. H. Bates, and F. Williams, *Trans. Faraday Soc.*, 58, 192 (1962).
8. R. C. Potter, C. L. Johnson, D. J. Metz, and R. H. Bretton, *J. Polym. Sci Part A-1*, 4, 419 (1966).
9. K. Ueno, K. Hayashi, and S. Okamura, *Polymer (London)*, 7, 431 (1966).
10. H. F. Mark, N. G. Gaylord, and N. M. Bikales, Eds., *Encyclopedia of Polymer Science and Technology,* Vol. 11, Interscience, New York, 1969, p. 702.
11. E. M. Verdurmen and A. L. German, *Macromol. Chem. Phys.,* 195, 6.35 (1994).
12. V. T. Stannett and E. P. Stahel, *Prog. Polym. Process.*, 289, 3 (1992).
13. A. Chapiro, *The Radiation Chemistry of Polymeric System*, Wiley-Interscience, New York, 1962.
14. V. T. Stannet, J. D. Wellons, and H. Yasuda, *J. Polym. Sci., Part C*, 4, 551 (1963).
15. L. A. Miller and V. T. Stannet, *J. Polym. Sci., Part A-1*, 7, 3159 (1969).
16. A. Chapiro, *The Radiation Chemistry of Polymeric Science*, Wiley-Interscience, New York, 1962, p. 270.
17. F. W. Tang, M. Al-Sheikly, and J. Silverman, *Proc. 7th Tihany Symp. Radiat. Chem.*, 1991, p. 270.
18. J. Zhang, Z. Zhiping, and S. Ying, *Radiat. Phys. Chem.*, 37, 263 (1991).
19. J. Zhang, Z. P. Zhang, and S. K. Ying, *Radiat. Phys. Chem.,* 36, 393 (1990).
20. K. Uezu, K. Saito, S. Furusaki, T. Sugo, and I. Ishigaki, *Radiat. Phys. Chem.*, 40, 31 (1992).
21. I. Ishigaki, T. Sugo, T. Takayama, T. Okada, J. Okamoto, and S. Machi, *J. Appl. Polym. Sci.*, 27, 1043 (1982).
22. K. Kaji, Y. Abe, M. Murai, N. Nishioka, and K. Kosai, *J. Appl. Polym. Sci.*, 47, 1427 (1993).
23. E. A. Hegazy, M. M. El-Dessouky, and S. A. El-Sharabasy, *Radiat. Phys. Chem.*, 27, 323 (1986).
24. E. A. Hegazy, *Polymer,* 33, 96 (1992).
25. Y. Haruvy, L. A. Rajbenbach, and J. Jagur-Grodzinski, *J. Appl. Polym. Sci.*, 27, 2711 (1982).
26. J. Fuehrer and G. Ellinghorst, *Angew. Makromol. Chem.*, 93, 175 (1981).
27. D. Vierkotten and G. Ellinghorst, *Angew. Makromol. Chem.*, 113, 153 (1983).
28. M. Suzuki, Y. Tamada, H. Iwata, and Y. Ikada, *Int. Symp. Physicochem. Aspects Polym. Surf.*, 2, 923 (1983).
29. Y. Ikada, M. Zuzuki, M. Taniguchi, H. Iwata, W. Taki, H. Miyake, Y. Yonekawa, and H. Handa, *Radiat. Phys. Chem.*, 18, 1207 (1981).
30. L. W. Dickson, Report AECL-9556, Energy Can. Ltd., 1988, p. 12.
31. D. R. Johnsson, W. J. Yen, and M. Dole, *J. Phys. Chem.*, 77, 2174 (1973).
32. Y. Hachihama and H. Sumitomo, *Technol. Rep. Osaka Univ.*, 8, 481 (1958).
33. A. Wirsen and A.-C. Albertsson, *J. Polym. Sci., Part A: Polym. Chem.*, 33, 2039 (1995).
34. I. L. J. Dogué, N. Mermilliod, and F. Genoud, *J. Polym. Sci., Part A: Polym. Chem.*, 32, 2193 (1994).
35. B. Gupta, F. N. Büchi, and G. G. Scherer, *J. Polym. Sci., Part A: Polym. Chem.,* 32, 1931 (1994).
36. A. Wirsen and A.-C. Albertsson, *J. Polym. Sci., Part A: Polym. Chem.,* 33, 2049 (1995).
37. S. Basu, A. Bhattacharyya, P. Ch. Mondal, and S. N. Bhattacharya, *J. Polym. Sci., Part A: Polym. Chem.*, 32, 2251 (1994).

38. T. O'Neill, *J. Polym. Sci., Polym. Chem. Ed.*, 10, 569 (1972).
39. A. I. Bessenov, M. I. Vitushkin, P. Glazunov, Sh. A. Karapetyan, B. N. Parfanovich, G. G. Ryabchikova, and A. A. Yakubovich, *Plast. Massy,* 5, 3 (1965).
40. A. I. Kurilenko, L. V. Smetaniana, L. B. Aleksandrova, and V. L. Karpova, *Vysokomol. Soedin,* 11, 1935 (1965).
41. E. E. Magat, I. K. Miller, D. Tanner, and J. Zummerman, *J. Polym. Sci., Part C,* 4, 615 (1963).
42. A. I. Kurilenko and V. I. Glukhov, *Dokl. Akad. Nauk. SSSR,* 166(4), 901 (1966).
43. V. B. Tikhomirov, V. E. Gusev, and A. I. Kurilenko, *Teknol. Tekstil'n. Prom.*, 2, 105 (1966).
44. V. M. Goryaev, G. G. Ryabchikoa, Z. N. Tarasova, and L. G. Tokarova, in *Radiat. Khim. Polim. Mater. Simp. Moscow,* 1964, p. 171.
45. A. A. Kachen, *Vysokomol. Soedin,* 8(12), 2144 (1966).
46. A. G. Davies, *Text. Inst. Ind.*, 4(1), 11 (1966).
47. A. I. Kurilenko, V. I. Glukhov, E. P. Danilov, E. R. Klinshpont, and V. L. Karpova, in *Radiat. Khim. Polim., Mater. Simp. Moscow,* 1964, p. 143.
48. A. I. Kurilenko, L. V Aleksandrova, and L. B. Smetanina, in *Sb. Rab. Konf. Moskow,* 1965, p. 90.
49. E. F. Kertvichenko, A. A. Kachan, V. A. Vonsyatskii, and A. M. Kalinichenko, *Vysokomol. Soedin, Ser. A*, 9(6), 1382 (1967).
50. Dasgupta, Report AECL-3511, Canadian Atomic Energy Commission, 1969, p. 29.
51. E. Schamberg and J. Hoigne, German Offen. 2004494 (1976).
52. I. Molnov, *Proc. Hung. Text. Conf.*, 1, 211 (1971).
53. M. A El-Azmirly, A. H. Zahran, and M. F. Barkat, *Eur. Polym. J.,* 11, 19 (1975).
54. B. L. Testlin, in *Radiat. Khim. Polim. Mater. Simp. Moskow,* 1964, p. 131.
55. Dasgupta, Canadian Patent 855, 678 (1970), to Canadian Atomic Energy Commission.
56. K. Matsuzaki, T. Kanai, and N. Norita, *J. Appl. Polym. Sci.*, 16, 15 (1972).
57. G. J. Howard, S. R. Kim, and R. H. Peters, *J. Soc. Dyers Colour.*, 85, 468 (1969).
58. R. Roberts and J. K. Thomas, in *International Atomic Energy Conference, Applications of Large Radiation Sources in Industry*, 1959.
59. R. Roberts and J. K. Thomas, *J. Soc. Dyers Colour,* 76, 342 (1960).
60. J. A. N. Skyes and J. K. Thomas, *J. Polym. Sci.*, 55, 721 (1961).
61. I. M. Trivedi, P. C. Mehta, K. N. Rao, and M. H. Rao, *J. Appl. Polym. Sci.*, 19, 1 (1975).
62. E. Collinson, *Discuss. Faraday Soc.,* 29, 188 (1960).
63. A. I. Brodski, *Vysokomol. Soedin,* 7, 16 (1965).
64. A. Hegev, *Dtsch. Textiltech.*, 17, 311 (1967).
65. Nippon Rayon Co., Japanese Patent 4250 (1961).
66. A. S. Hoffmann and G. R. Berbeco, *Text. Res. J.*, 40 (11), 975 (1970).
67. L. Jansco, *Magy. Textiltech.,* 27(7), 333 (1974).
68. E. Schamberg and J. Hoigne, *J. Polym. Sci., Part A-1,* 8, 693 (1970).
69. K. Mori, K. Koshiishi, and Masuhara, *Kobunshu Ronbunshu,* 48, 1 (1991).
70. A. Ekstrom and J. L. Garnett, *J. Chem. Soc. A*, 2416 (1968).
71. A. Chapiro, A. M. Jendruchowska-Bonamour, and G. Lelievre, *Faraday Discuss. Chem. Soc.*, 63, 134 (1977).
72. G. Odian, *Am. Chem. Soc., Div. Polym. Prepr.*, 1(2), 327 (1960).
73. G. Odian, T. Ackev, R. Elliot, M. Sobel, and R. Klein, Report RAI-301, U.S. Atomic Energy Commission, 1962, p. 54.
74. G. Odian, M. Sobel, R. Klein, and T. Ackev, Report NYO-2530, U.S. Atomic Energy Commission, 1961, p. 7.

75. G. Odian and T. Ackev, Report TID-7643, U.S. Atomic Energy Commission, 1962, p. 233.
76. G. Odian, T. Ackev, and M. Sobel, *J. Appl. Polym. Sci.*, 7, 245 (1963).
77. G. Odian, M. Sobel, A. Rossi, R. Klein, and T. Ackev, *J. Polym. Sci., Part A-1*, 1, 639 (1963).
78. S. Machi, I. Kamel, and J. Silverman, *J. Polym. Sci., Part A-1*, 8, 3329 (1970).
79. I. Kamel, S. Machi, and J. Silverman, *J. Polym. Sci., Part A-1,* 10, 1019 (1972).
80. J. L. Garnett, *ACS Symp. Ser.*, 48, 334 (1977).
81. J. L. Garnett, *Radiat. Phys. Chem.*, 14, 79 (1979).
82. J. L. Garnett and J. D. Leeder, *ACS Symp. Ser.*, 49, 197 (1977).
83. J. L. Garnett, S. V. Jankiewicz, R. Levot, and D. F. Sangster, *Radiat. Phys. Chem.*, 25, 509 (1985).
84. C. H. Ang, J. L. Garnett, R. Levot, and M. A. Long, *J. Polym. Sci., Polym. Lett. Ed.*, 21, 257 (1983).
85. R. P. Chaplin, N. J. W. Gamage, and J. L. Garnett, *Radiat. Phys. Chem.*, 46, 949 (1995).
86. M. B. Huglin and B. L. Johnson, *J. Polym. Sci., Part A-1*, 7, 1379 (1969).
87. M. B. Huglin and B. L. Johnson, *Kolloid-Z. Z. Polym.*, 249, 1080 (1971).
88. M. B. Huglin and B. L. Johnson, *J. Appl. Polym. Sci.,* 16., 921 (1972).
89. J. H. Choi and C. Lee, *J. Korean Nucl. Soc.*, 8(3), 159 (1976).
90. H. Sumitomo and Y. Hachihama, *Kogya Kogyku Zasshi, I*, 62, 132 (1968).
91. S. Okamura, T. Iwaski, Y. Kobayashi, and K. Hayashi, in *Large Radiation Sources in Industrial Processes Conference*, Warsaw, 1960, Vol. 1, p. 459.
92. A. A. Armstrong, Jr. and H. A. Rutherford, *Text Res. J.*, 33, 264 (1960).
93. A. A. Armstrong Jr. and H. A. Rutherford, Rep. TID-7643, U.S. Atomic Energy Commission, 1962, p. 268.
94. A. A. Armstrong Jr. and H. A. Rutherford, Report NCSC-2477, U.S. Atomic Energy Commission, 1962, p. 699.
95. P. A. Dworjanyn and J. L. Garnett, *Radiat. Curing Polym. Sci. Technol.,* 1, 263 (1963).
96. J. L. Gordon and J. W. Prane, Eds., *Nonpolluting Coatings and Coating Processes,* Plenum Press, New York, 1973.
97. G. A. Senich and R. E. Florin, *J. Macromol. Sci–Rev. Macromol. Chem.*, C24(2), 239 (1984).
98. J. L. Garnett, *J. Oil Colour Chem. Assoc.,* 65, 383 (1982).
99. S. Joensso, P.-E. Sundell, J. Hultgren, D. Sheng, and C. E. Hoyle, *Prog. Org. Coat.,* 27, 107 (1996).
100. S. J. Bett, G. Fletcher, and J. L. Garnett, *Radiat. Phys. Chem.*, 28, 207 (1986).
101. S. V. Nablo and A. S. Denholm. *J. Radiat. Curing,* 7(3), 11 (1980).

8 Photoinitiated Radical Vinyl Polymerization

Nergis Arsu, Ivo Reetz, Yusuf Yagci, and Munmaya K. Mishra

CONTENTS

8.1 INTRODUCTION

When polymerizations are initiated by light and both the initiating species and the growing chain ends are radicals, we speak of *radical photopolymerization*. As for other polymerizations, molecules of appreciably high molecular weight can be formed in the course of the chain reaction. Playing the predominant role in technical polymer synthesis, vinyl monomers can be mostly polymerized by a radical mechanism. Exceptions are vinyl ethers, which have to be polymerized in an ionic mode. Light-induced *ionic* polymerization has been reviewed elsewhere [1–4] and is beyond the scope of this book.

Regarding initiation by light, it must be pointed out that the *absorption* of incident light by one or several components of the polymerization mixture is the crucial prerequisite. If the photon energy is absorbed directly by a photosensitive compound, being a monomer itself or an added initiator, this photosensitive substance undergoes a homolytic bond rupture forming radicals, which may initiate the polymerization. In some cases, however, the photon energy is absorbed by a compound that itself is not prone to radical formation. These so called sensitizers transfer their electronic excitation energy to reactive constituents of the polymerization mixture, which finally generate radicals. The radicals evolved react with intact vinyl monomer starting a chain polymerization. Under favorable conditions, a single free radical can initiate the polymerization of a thousand molecules. The spatial distribution of initiating species may be arranged in any desired manner.

Light-induced free radical polymerization is of enormous commercial use. Techniques such as curing of coatings on wood, metal and paper, adhesives, printing inks and photoresists are based on photoinitiated radical vinyl polymerization. Some other interesting applications are available, including production of laser video discs and curing of acrylate dental fillings.

In contrast to thermally initiated polymerizations, photopolymerization can be performed at room temperature. This is a striking advantage for both classical polymerization of monofunctional monomers and modern curing applications.

Photopolymerization of monofunctional monomers takes place without side reactions such as chain transfer. In thermal polymerization, the probability of chain transfer is high which brings about a high amount of branched macromolecules. Thus, low-energy stereospecific polymeric species, namely of syndiotactic configuration, may be obtained by photopolymerization. Another important use refers to monomers with low ceiling temperature. They can only be polymerized at moderate temperatures, otherwise depolymerization dominates over polymerization. By means of photopolymerization, these monomers are often easily polymerizable. Furthermore, biochemical applications, such as immobilization of enzymes by polymerization, do also usually require low temperatures. As far as curing of coatings or surfaces is concerned, it has to be noted that thermal initiation is often not practical, especially if large areas or fine structures are to be cured or if the curing formulation is placed in an environment or structure that should not be heated, such as dental fillings.

Radical photopolymerization of vinyl monomers played an important role in the early development of polymerization. One of the first procedures for polymerizing vinyl monomers was the exposure of monomer to sunlight. Blyth and Hoffmann [5] reported on the polymerization of styrene by sunlight more than 150 years ago.

Photocurable formulations are mostly free of additional organic solvents; the monomer, which serves as reactive diluent, is converted to solid, environmentally safe resin without any air pollution. UV curing is often a very fast process, taking place as described previously without heating. If the polymerization mixture absorbs solar light and the efficiency of radical formation is high, photocuring can be performed with no light source but sunlight. These features make photopolymerization an ecologically friendly and economical technology, which has high potential for further development.

8.2 PHOTOINITIATION

Photoinitiated free radical polymerization consists of four distinct steps:

1. *Photoinitiation*: Absorption of light by a photosensitive compound or transfer of electronic excitation energy from a light absorbing sensitizer to the photosensitive compound. Homolytic bond rupture leads to the formation of a radical that reacts with one monomer unit.
2. *Propagation*: Repeated addition of monomer units to the chain radical produces the polymer backbone.
3. *Chain transfer*: Termination of growing chains by hydrogen abstraction from various species (e.g., from solvent) and concomitant production of a new radical capable of initiating another chain reaction.
4. *Termination*: Chain radicals are consumed by disproportionation or recombination reactions. Termination can also occur by recombination or disproportionation with any other radical including primary radicals produced by the photoreaction. These four steps are summarized in Scheme 8.1.

Notably, the role that light plays in photopolymerization is restricted to the very first step, namely the absorption and generation of initiating radicals. The reactions

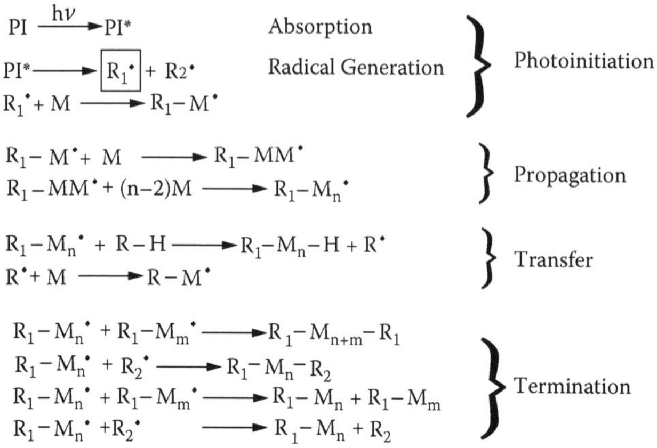

$$\mathrm{PI} \xrightarrow{h\nu} \mathrm{PI}^* \quad \text{Absorption}$$
$$\mathrm{PI}^* \longrightarrow \boxed{R_1^\bullet} + R_2^\bullet \quad \text{Radical Generation}$$
$$R_1^\bullet + M \longrightarrow R_1{-}M^\bullet$$

Photoinitiation

$$R_1{-}M^\bullet + M \longrightarrow R_1{-}MM^\bullet$$
$$R_1{-}MM^\bullet + (n{-}2)M \longrightarrow R_1{-}M_n^\bullet$$

Propagation

$$R_1{-}M_n^\bullet + R{-}H \longrightarrow R_1{-}M_n{-}H + R^\bullet$$
$$R^\bullet + M \longrightarrow R{-}M^\bullet$$

Transfer

$$R_1{-}M_n^\bullet + R_1{-}M_m^\bullet \longrightarrow R_1{-}M_{n+m}{-}R_1$$
$$R_1{-}M_n^\bullet + R_2^\bullet \longrightarrow R_1{-}M_n{-}R_2$$
$$R_1{-}M_n^\bullet + R_1{-}M_m^\bullet \longrightarrow R_1{-}M_n + R_1{-}M_m$$
$$R_1{-}M_n^\bullet + R_2^\bullet \longrightarrow R_1{-}M_n + R_2$$

Termination

SCHEME 8.1 General photopolymerization steps.

of these radicals with monomer, propagation, transfer, and termination are purely thermal processes; they are not affected by light. Because in this chapter the genuine photochemical aspects are to be discussed, propagation, transfer, and termination reactions are not depicted as long as it is not necessary for the understanding of a reaction mechanism. Instead, the photochemically produced initiating species are highlighted by a frame, as illustrated in Scheme 8.1.

8.2.1 Absorption of Light

The absorption of light excites the electrons of a molecule, which lessens the stability of a bond and can, under favorable circumstances, lead to its dissociation. Functional groups that have high absorbency, like phenyl rings or carbonyl groups, are referred to as chromophoric groups. Naturally, photoinduced bond dissociations do often take place in the proximity of the light absorbing chromophoric groups. In some examples, however, electronic excitation energy may be transferred intramolecularly to fairly distant, but easily cleavable bonds to cause their rupture.

The intensity I_a of radiation absorbed by the system is governed by the Beer Lambert law, where I_0 is the intensity of light falling on the system, l is the optical path length, and [S] is the concentration of the absorbing molecule having the molar extinction coefficient ε.

$$I_a = I_0\,(1 - e^{-\varepsilon l[S]}) \tag{8.1}$$

If the monomer possesses chromophoric groups and is sensitive toward light (i.e., it undergoes photoinduced chemical reactions with high quantum yields) one can perform photopolymerization by just irradiating the monomer. In many cases, however, monomers are not efficiently decomposed into radicals upon irradiation. Furthermore, monomers are often transparent to light at λ >320 nm, where commercial lamps emit. In these cases, photoinitiators are used. These compounds absorb light and bring about the generation of initiating radicals.

8.2.2 Radical Generation

8.2.2.1 Radical Generation by Monomer Irradiation

Some monomers are able to produce radical species upon absorption of light. Studies on various vinyl compounds show that a monomer biradical is formed.

$$M \xrightarrow{h\nu} \boxed{\cdot M \cdot} \tag{8.2}$$

These species are able to react with intact monomer molecules thus leading to growing chains. Readily available monomers, which to some extent undergo polymerization and copolymerization upon UV irradiation, are listed in Table 8.1.

However, regarding technical applications, radical generation by irradiation of vinyl monomer does not play a role due to the very low efficiency of radical formation and the usually unsatisfactory absorption characteristics.

8.2.2.2 Radical Generation by Initiators

In most cases of photoinduced polymerization, initiators are used to generate radicals. One has to distinguish between two different types of photoinitiators:

8.2.2.2.1 Type I Photoinitiators: Unimolecular Photoinitiators

These substances undergo homolytic bond cleavage upon absorption of light. The fragmentation that leads to the formation of radicals is, from the point of view of

TABLE 8.1
Photosensitive Monomers

Allyl methacrylate
Barium acrylate
Cinnamyl methacrylate
Diallyl phthatlate
Diallyl isophtalate
Diallyl terephthalate
2-Ethylhexyl acrylate
2-Hydroxyethyl methacrylate
2-Hydroxypropyl acrylate
N,N′-Methylenebisacrylamide
Methyl methacrylate
Pentaerythritol tetramethacrylate
Styrene
Tetraethylene glycol dimethacrylate
Tetrafluoroethylene
N-Vinylcarbazole
Vinyl cinnamate
Vinyl 2-fuorate
Vinyl 2-furylacrylate

chemical kinetics, a unimolecular reaction.

$$\mathrm{Pl} \xrightarrow{h\nu} \mathrm{Pl}^{*} \xrightarrow{k} \dot{\mathrm{R}}_1 + \dot{\mathrm{R}}_2 \tag{8.3}$$

$$\frac{d[\dot{\mathrm{R}}_1]}{dt} = \frac{d[\dot{\mathrm{R}}_2]}{dt} = k\ [\mathrm{Pl}^{*}] \tag{8.4}$$

The number of initiating radicals formed upon absorption of one photon is termed as quantum yield of radical formation ($\phi_{R\bullet}$)

$$\Phi_{R\bullet} = \frac{\text{Number of initiating radicals formed}}{\text{Number of photons absorbed by the photoinitiator}} \tag{8.5}$$

Theoretically, cleavage type photoinitiators should have a $\phi_{R\bullet}$ value of two because two radicals are formed by the photochemical reaction. The values observed, however, are much lower because of various deactivation routes of the photoexcited initiator other than radical generation. These routes include physical deactivation such as fluorescence or non-radiative decay and energy transfer from the excited state to other, ground state molecules, a process referred to as quenching. The reactivity of photogenerated radicals with polymerizable monomers is also to be taken into consideration. In most initiating systems, only one in two radicals formed adds to monomer thus initiating polymerization. The other radical usually undergoes either combination or disproportionation. The initiation efficiency of photogenerated radicals (f_P) can be calculated by the following formula

$$f_p = \frac{\text{number of chain radicals formed}}{\text{number of primary radicals formed}} \tag{8.6}$$

The overall photoinitiation efficiency is expressed by the quantum yield of photoinitiation (Φ_P) according to the following equation:

$$\Phi_P = \Phi_{R\bullet} \times f_P \tag{8.7}$$

Regarding the energy necessary, it has to be said that the excitation energy of the photoinitiator has to be higher than the dissociation energy of the bond to be ruptured. The bond dissociation energy, on the other hand, has to be high enough to guarantee long-term storage stability. The majority of Type I photoinitiators are aromatic carbonyl compounds with appropriate substituents, which spontaneously undergo "α-cleavage," generating free radicals according to reaction (8.8). The benzoyl radical formed by the reaction depicted is very reactive toward the unsaturations of vinyl monomers [6].

O R' OR'' —hν→ O + R' R''O (8.8)

R'= H, Alkyl, subst.Alkyl
R'' = H, Alkyl, subst.Alkyl

The α-cleavage, often referred to as Norrish Type I reaction [7] of carbonyl compounds, starts from the initiator's triplet state, which is populated via intersystem crossing. Notably, the excited triplet states are usually relatively short-lived, which prevents excited molecules from undergoing side reactions with constituents of the polymerization mixture. Although triplet quenching by oxygen can, in most cases, be neglected due to the short lifetime of the triplet states, quenching by monomer sometimes plays a role. However, this refers exclusively to monomers with low triplet energies such as styrene ($E_T = 259$ kJ mol^{-1} [8]).

If the absorption characteristics of a cleavable compound are not meeting the requirements (i.e., the compound absorbs at too low wavelengths), the use of sensitizers (S) with matching absorption spectra is recommendable. Sensitizers absorb the incident light and are excited to their triplet state. The triplet excitation energy is subsequently transferred to the photoinitiator that forms initiating radicals. This process has to be exothermic (i.e., the sensitizers' triplet energy has to be higher than the triplet energy level of the initiator). Through energy transfer, the initiator is excited and undergoes the same reactions of radical formation as if it were excited by direct absorption of light. The sensitizer molecules return to their ground state upon energy transfer; they are therefore not consumed in the process of initiation.

$$S \xrightarrow{h\nu} {}^3S^* \quad (8.9)$$

$$^3S^* + PI \rightarrow S + {}^3PI^* \quad (8.10)$$

8.2.2.2.2 Type II Photoinitiators: Bimolecular Photoinitiators

The excited states of certain compounds do not undergo Type I reactions because their excitation energy is not high enough for fragmentation (i.e., their excitation energy is lower than the bond dissociation energy). The excited molecule can, however, react with another constituent of the polymerization mixture, the so-called coinitiator (COI), to produce initiating radicals. In this case, radical generation follows second-order kinetics.

$$PI \xrightarrow{h\nu} PI^* \quad (8.11)$$

$$PI^* + COI \xrightarrow{k} \dot{R}_1 + \dot{R}_2 \quad (8.12)$$

$$\frac{d[\dot{R}_1]}{dt} = \frac{d[\dot{R}_2]}{dt} = k\,[PI^*]\,[COI] \quad (8.13)$$

Radical generation by Type II initiating systems has two distinct pathways:

1. *Hydrogen abstraction* from a suitable hydrogen donor. As a typical example, the photoreduction of benzophenone by isopropanol has been given next. Bimolecular hydrogen abstraction is limited to diaryl ketones [7]. From the point of view of thermodynamics, hydrogen abstraction is to be expected if the diaryl ketone's triplet energy is higher than the bond

dissociation energy of the hydrogen atom to be abstracted.

$$(C_6H_5)_2C{=}O \xrightarrow{h\nu} {}^3[(C_6H_5)_2C{=}O]^* \xrightarrow{(H_3C)_2CH{-}OH} (C_6H_5)_2\dot{C}{-}OH + (H_3C)_2\dot{C}{-}OH \quad (8.14)$$

2. Photoinduced *electron transfer reactions* and subsequent fragmentation. In electron transfer reactions, the photoexcited molecule, termed as sensitizer for the convenience, can act either as electron donor or electron acceptor according to the nature of the sensitizer and coinitiator. Fragmentation yields radical anions and radical cations, which are often not directly acting as initiating species themselves but undergo further reactions, by which initiating free radicals are produced.

$$S \xrightarrow{h\nu} S^* \quad (8.15)$$

$$S^* + A \longrightarrow S^{+\bullet} + A^{-\bullet} \longrightarrow \text{further reactions} \quad (8.16)$$

$$S^* + D \longrightarrow S^{-\bullet} + D^{+\bullet} \longrightarrow \text{further reactions} \quad (8.17)$$

The electron transfer is thermodynamically allowed, if ΔG calculated by the Rehm-Weller equation (Eq. (8.18)) [9] is negative.

$$\Delta G = F\,[E_{1/2}^{ox}\,(D/D^{+\cdot}) - E_{1/2}^{red}\,(A/A^{-\cdot})\,] - E_S + \Delta E_c$$

where

F: Faraday constant

$E_{1/2}^{ox}\,(D/D^{+\cdot})$, $E_{1/2}^{red}\,(A/A^{-\cdot})$: oxidation and reduction potential of donor and acceptor, respectively

E_S: singlet state energy of the sensitizer

ΔE_c: coulombic stabilization energy (8.18)

Electron transfer is often observed for aromatic ketone/amine pairs and always with dye/coinitiator systems. The photosensitization by dyes is dealt with in detail later in this chapter.

8.3 TYPE I PHOTOINITIATORS

8.3.1 Aromatic Carbonyl Compounds

8.3.1.1 Benzoin Derivatives

Benzoin and its derivatives are the most widely used photoinitiators for radical polymerization of vinyl monomers. As depicted in Reaction 8.8, they undergo α-cleavage to produce benzoyl and α-substituted benzyl radicals upon photolysis.

TABLE 8.2
Various Benzoin Derivatives: Quantum Yields of α-Scission (Φ_α), Triplet Energies (E_T), and Triplet Lifetimes (τ_T)

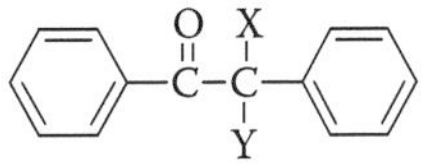

X	Y	Φ_α	E_T (kJ mol^{-1})	τ_T (10^{-9} s)	References
H	H	—	302	125	[12]
OH	H	0.87	308	0.83	[13–15]
OCH_3	H	0.44	300	<0.1	[13–15]
$OCH(CH_3)_2$	H	0.33	—	—	[16]
$OCH(CH_3)C_2H_5$	H	0.30	—	—	[16]
OC_6H_5	H	0.39	304	0.17	[13,14]
$OCOCH_3$	H	0.33	—	20	[13,14]
OH	C_6H_5	0.10	—	—	[16]
CH_3	CH_3	0.44	306	6	[12,17]
OCH_3	OCH_3	0.57	278	<0.1	[16,18–20]
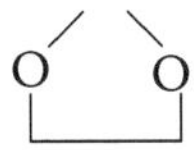 O O (cyclic)		0.11			[16]

The importance of these photoinitiators derives from the following: they possess high absorptions in the far UV region (λ_{max} = 300–400 nm, ε_{max} ≥100–200 l mol^{-1} cm^{-1}), high quantum efficiencies for radical generation [10], and a relatively short-lived triplet state [11] (Table 8.2).

Regarding the photochemistry of benzoin derivatives, starting from excited triplet states populated after intersystem crossing, Norrish Type I bond scission is the main chemical reaction occurring under various experimental conditions [21–25]. In the consequence of this bond cleavage, benzoyl and ether radicals are formed. In the absence of monomer, hydrogen abstraction takes place leading to benzaldehyde, benzil, and pinacol derivatives [21–23]. The reactivity of benzoyl and benzyl ether radicals were found to be almost the same provided the concentration of radicals is low and that of monomer high. On the other hand, if the concentration of radicals is high and that of monomer low, benzoyl radicals are more reactive toward monomer molecules present than the ether radicals [10, 24, 26]

The photoinduced α-cleavage reaction is not or only very little affected by triplet quenchers including styrene, owing to the short lifetime of the excited triplet state [27]. This circumstance makes benzoin photoinitiators particularly useful for industrial applications involving styrene monomer.

Regarding practical applications, it has to be mentioned that benzoin derivatives are only storable for limited time at ambient temperature (i.e., they slowly but steadily decompose thermally during storage). Thermally, benzylic hydrogen atoms are abstracted giving rise to benzyl radicals and various subsequent decomposition products. Several benzoin derivatives, such as benzil ketals or metholyl benzoin derivatives (*vide ante*), are more thermally stable.

A few benzoin derivatives decompose by photofragmentation mechanisms other than Norrish Type I. For example, α-halogeno acetophenones [28], oxysulphonyl ketones [29], and sulphonyl ketones [30] sufficiently undergo β-cleavage upon UV irradiation and may be used for initiation. α-Dimethylamino substituted benzoin undergoes both α-cleavage and β-photoelimination, the latter being the dominant process [14]. Because the radicals formed according to Eq. (8.19) are not reactive toward monomer and the quantum yield of α-cleavage Eq. (8.20) is low (ca. 0.04), these derivatives are not suitable photoinitiators.

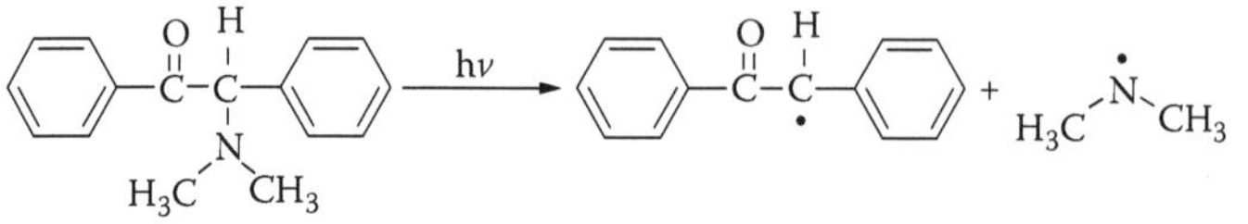

(8.19)

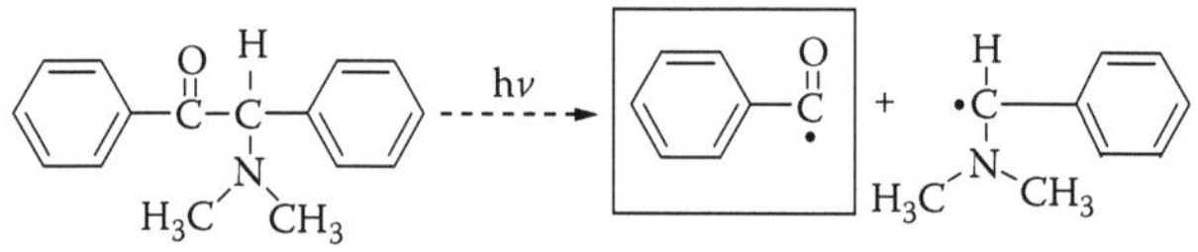

(8.20)

Another benzoin derivative, desoxybenzoin, undergoes α-scission with rather low quantum yields, too. However, it becomes an efficient initiator when utilized together with a hydogen donor, such as tertiary amine or tetrahydrofuran. In this case, initiating radicals are generated by hydrogen abstraction with comparatively high yields [31, 32].

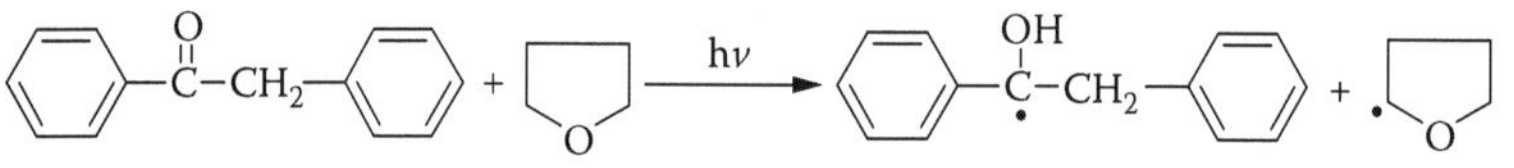

(8.21)

When used in conjunction with onium salts, various methylol functional benzoin ethers are efficient cationic photoinitiators [33,34]. One important advantage of methylol benzoin derivatives is their relatively high thermal stability (Table 8.3).

In the case of methylolybenzyl sulphonic acid esters, the initiation efficiency could be considerably enhanced by adding lithium salt to the formulations [38]. This effect has been explained in terms of formation of the sulphonic acid ester's lithium salt. This salt migrates toward the surface of the coating, and forms thereby a shield preventing oxygen to diffuse into the inner zones of the coating.

TABLE 8.3
Methylol Benzoin Derivatives

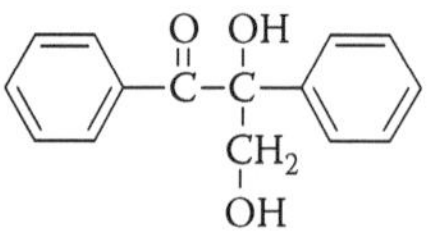 Methylolbenzoin [35,36]	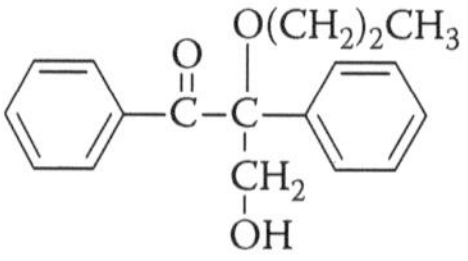 Methylolbenzoin propyl ether [35]	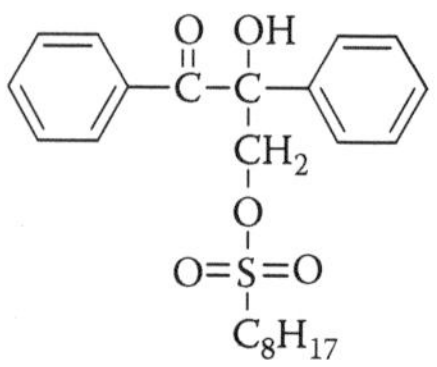 Methylolbenzoin, 1-octosulphonic ester [37–39]

TABLE 8.4
Benzilketals for Photopolymerization

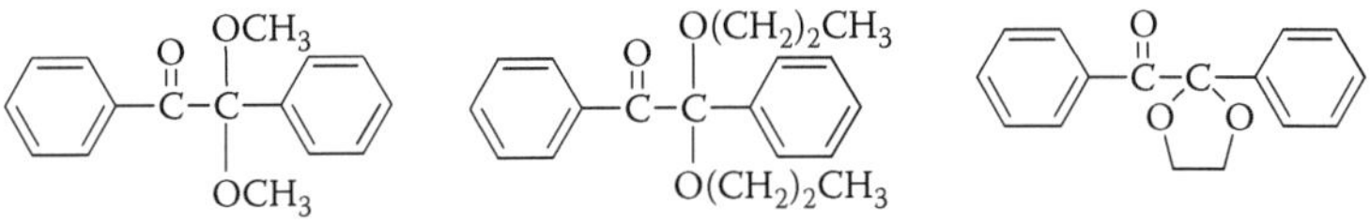

2,2-dimethoxy-2-phenylacetophenone (DMPA) [16,18–20]	2,2-diethoxy-2-phenylacethophenone	benzilglycolketal (BGK) [16,39a,40]

8.3.1.2 Benzilketals

Benzilketals are another important class of photoinitiators developed for free radical vinyl polymerization. Benzilketals exhibit higher thermal stability than benzoin compounds due to the absence of thermally labile benzylic hydrogen. The most prominent member of this class is the commercially used DMPA. Indeed, this initiator shows an excellent efficiency in photopolymerizations and is at the same time easy to synthesize. Other benzilketals are also suitable initiators but do not reach the price performance ratio of DMPA (Table 8.4).

Like benzoin ethers, benzilketals undergo α-cleavage whereby a benzoyl radical and a dialkoxybenzyl radical are formed. Although the benzoyl radicals are, as explained earlier, vigorously reacting with olefinic bonds of vinyl monomers, dialkoxybenzyl radicals were found to be of low reactivity. Actually, one of seven dialkoxy benzyl radicals formed is found to be incorporated into the polymer chain during the photopolymerization of methylmethacrylate initiated by DMPA [41]. However, to what extent this portion of dialkoxy benzyl groups is caused by termination instead of initiation remains unclear.

Dimethoxybenzyl radicals undergo a fragmentation yielding methyl radicals [42–44], which act as additional initiating species in radical vinyl polymerization [45, 18]

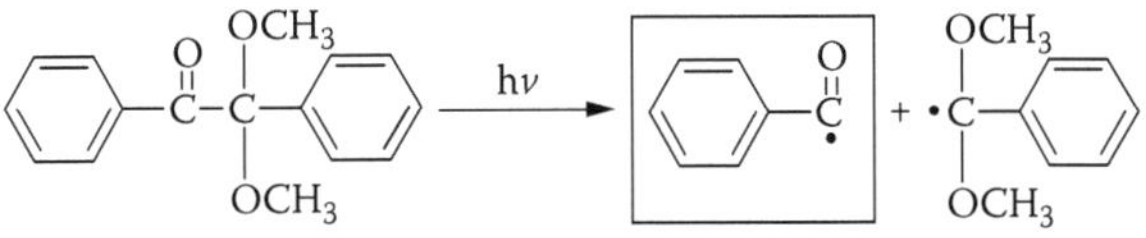

(8.22)

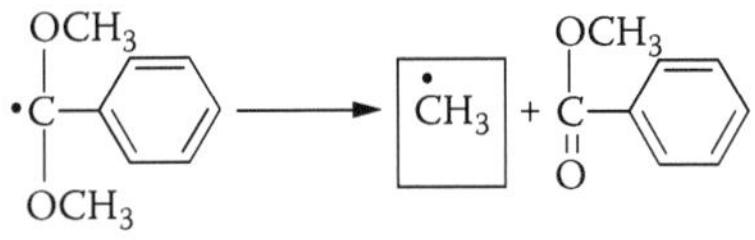

(8.23)

Mechanistic studies [18,46–48,48a] based on ESR and ENDOR analysis revealed that the fragmentation reaction depicted in Eq. 8.23 is a fast two-photon process, provided high intensity light sources (pulsed lasers) are used for photolysis. On the other hand, relatively low intensity light gives rise to a slower, thermal fragmentation.

8.3.1.3 Acetophenones

α-Substituted acetophenones are another important class of photoinitiators used in various applications of free radical polymerizations [42,49–51]. These initiators exhibit excellent initiator properties especially in micellar solutions [52]. The most prominent example of this class of photoinitiators is the commercially available α,α-diethoxyacetophenone (DEAP); furthermore 1-benzoylcyclohexanol and 2-hydroxy-2-methyl-1-phenylpropanone are initiators with good properties. Besides high efficiency the pros of acetophenones include high storage stability and little tendency toward yellowing. Regarding photochemistry, both Norrish Type I and Norrish Type II bond ruptures were evidenced [45]. However, only the α-cleavage (Norrish Type I) gives initiating radicals: benzoyl radicals directly formed upon the light-induced α-cleavage and ethylradicals, generated in a subsequent thermal fragmentation reaction.

$$C_6H_5{-}C(=O){-}CH(OCH_2CH_3)_2 \xrightarrow{h\nu} \boxed{C_6H_5{-}\dot{C}{=}O} + \boxed{\cdot CH(OCH_2CH_3)_2} \longrightarrow O{=}CH{-}OCH_2CH_3 + \boxed{\dot{C}H_2CH_3} \quad (8.24)$$

8.3.1.4 α-Aminoalkylphenones

α-Aminoalkylphenones have recently been developed for the use in pigmented photopolymerizations (Table 8.5).

These compounds possess better absorption characteristics than many other aromatic ketone photoinitators and are, therefore, quite amenable to practical applications where irradiation at longer wavelengths is desired. There is no doubt that α-aminoalkylphenones undergo α-cleavage to yield initiating benzoyl radicals and other carbon centered radicals [54,55,55a]. By means of thioxanthone as triplet sensitizer the sensitivity of the initiating formulation can be extended to the near UV or even visible region of the spectrum [56–58]. Recently, ammonium group containing benzoin ethers have turned out to be efficient, water-soluble photoinitiators in the polymerization of trimethylolpropane triacrylate [59].

8.3.1.5 O-Acyl-α-oximino Ketones

O-Acyl-α-oximino ketones are known to undergo cleavage with high quantum efficiency [60] and have been used as photoinitiators for acrylates and unsaturated polyesters [28,60–62]. Besides benzoyl radicals, phenyl radicals are produced in a secondary reaction. Both radical types are reactive in initiation. The most prominent

TABLE 8.5
α-Aminoalkylphenones for Photopolymerization: Quantum Yields of α Scission (Φ_α), Triplet Energies (E_T), and Triplet Lifetimes (τ_T)

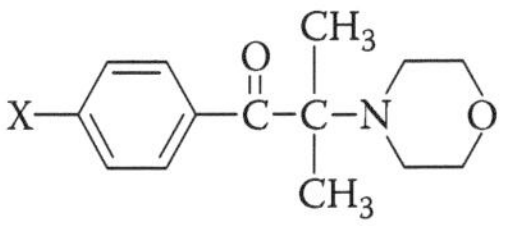

$$X\text{-}C_6H_4\text{-}C(=O)\text{-}C(CH_3)_2\text{-}N(CH_2CH_2)_2O$$

X	Φ_a	E_T (kJ mol^{-1})	τ_T (10^{-9} s)
H	1	—	1
CH_3O	1	273	0.4
CH_3S	0.88	256	10
$(CH_3)_2N$	0.014	265	2000

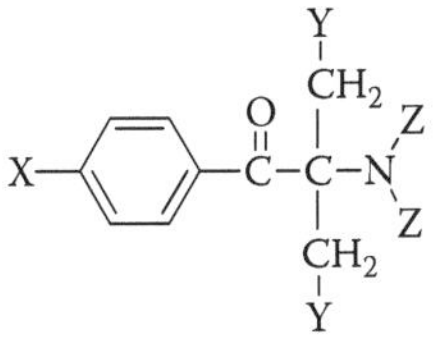

$$X\text{-}C_6H_4\text{-}C(=O)\text{-}C(CH_2Y)_2\text{-}NZ_2$$

X	Y	Z	Φ_α	E_T (kJ mol^{-1})	τ_T (10^{-9}s)
H	C_6H_5	CH_3	1	—	1
Morpholine	C_6H_5	CH_3	0.22	—	1700
Morpholine	CH_2=CH–	CH_3	>0.9	—	1500
Morpholine	CH_2=CH–	$-C_2H_4OC_2H_4-$	0.07	—	80

Source: K. Dietliker, *Chemistry & Technology of UV&EB Formulations for Coating, Inks & Paints*, Vol. 3, K. Dietliker, Ed., SITA Technology Ltd., London, (1991) p. 157.

example of these initiators is O-benzoyl-α-oximino-1-phenyl-propane-1-one, the reaction of which is illustrated in Eq. (8.25).

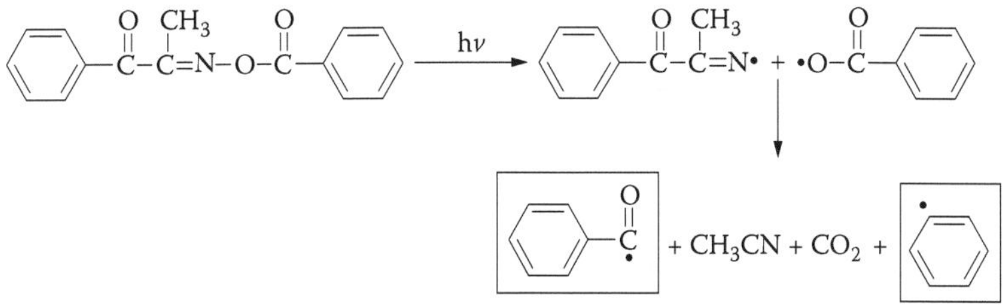

$$C_6H_5\text{-}C(=O)\text{-}C(CH_3)\text{=N-O-}C(=O)\text{-}C_6H_5 \xrightarrow{h\nu} C_6H_5\text{-}C(=O)\text{-}C(CH_3)\text{=N}\bullet + \bullet O\text{-}C(=O)\text{-}C_6H_5 \rightarrow \boxed{C_6H_5\text{-}\dot{C}=O} + CH_3CN + CO_2 + \boxed{C_6H_5\bullet} \quad (8.25)$$

Although these compounds absorb more strongly in the near UV than most of the other aromatic photoinitiators, their use as photoinitiators is limited, because they are thermally not very stable. The relatively weak N-O bond dissociates both photochemically and thermally at moderate temperatures.

8.3.1.6 Acylphosphine Oxide and Its Derivatives

Acylphosphine oxides and acylphosphonates with different structures have been used as photoinitiators for free-radical initiated photopolymerization (see Table 8.6).

TABLE 8.6
Phosphonates for Photopolymerization: Quantum Yields of Intersystem Crossing (Φ_T), α-Scission (Φ_α), and Triplet Lifetimes (τ_T)

Structure	Φ_T	Φ_α	$\tau_T(10^{-9}s)$	References
C_6H_5–C(=O)–P(=O)(OC_2H_5)–OC_2H_5	0.9	0.03	24	[63, 64]
2,4,6-(CH_3)$_3$$C_6H_2$–C(=O)–P(=O)($OC_2H_5$)–$OC_2H_5$	0.6	0.3	—	[65]
2,4,6-(CH_3)$_3$$C_6H_2$–C(=O)–P(=O)($OCH_3$)–$OCH_3$	0.6	0.3	—	[65]
2,6-$Cl_2$$C_6H_3$–C(=O)–P(=O)($OCH_3$)–$OCH_3$	0.5	0.2	—	[63]
H_3C–C(CH_3)$_2$–C(=O)–P(=O)(OC_2H_5)–OC_2H_5	0.6	0.3	30	[64, 66, 67]

Long wavelength absorption characteristics make these compounds particularly useful for the polymerization of TiO_2 pigmented formulations containing acrylate or styrene type monomers and of glass fiber reinforced polyester laminates with reduced transparency [68–73,73a]. These initiators are thermally stable up to c. 180°C and no polymerization takes place when the fully formulated systems are stored in the dark. Moreover, very little yellowing occurs in coatings cured with acylphosphine oxides. With respect to the storage of curing formulations and the actual curing, it has to be taken into account that acylphosphine oxides may react with water, alcohols, or amines, as well as what leads to the cleavage of the C-P bond [73]. By introducing bulky groups in *ortho*-position of the benzoyl group, the solvolysis is significantly slowed down. Furthermore, these substituents seem also to be able to increase the tendency for α-scission (see Table 8.6).

Extensive investigations on the photochemistry of acylphosphine oxides revealed that they do undergo α-cleavage with fairly high quantum yields [64]. Furthermore, it was found that the phosphonyl radicals formed are highly reactive toward vinyl monomers, as can be seen in Table 8.7, where rate constants of radicals generated

TABLE 8.7
Bimolecular Rate Constants (l mol^{-1} s^{-1}) of the Reaction of Phosphonyl and Benzoyl Radicals with Various Monomers in Cyclohexane at 20°C

	$\bullet P(=O)(OCH_3)-OCH_3$	$\bullet P(=O)(OCH_2CH_3)-OCH_2CH_3$	$\bullet P(=O)(OCH_3)-CH_3$	$\bullet P(=O)(OCH_2(CH_3)_2)-C_6H_5$	$C_6H_5-C(=O)\bullet$
St	2.2×10^8	2.5×10^8	8×10^7	4.5×10^7	$<2 \times 10^5$
AN	5.8×10^6	2.6×10^6	1.8×10^6	2.0×10^7	$<2 \times 10^4$
MA	1.7×10^7	1.6×10^7	1.3×10^7	2.1×10^7	—

Source: T. Sumiyoshi and W. Schnabel, *Makromol. Chem.*, 186, 1811 (1986).
St: styrene, AN: acrylonitrile, MA: methylacrylate.

from photoinitiators with various monomers are compiled.

$$2,4,6\text{-}(CH_3)_3C_6H_2-C(=O)-P(=O)(C_6H_5)-C_6H_5 \xrightarrow{h\nu} \boxed{2,4,6\text{-}(CH_3)_3C_6H_2-\dot{C}=O} + \boxed{\bullet P(=O)(C_6H_5)-C_6H_5} \quad (8.26)$$

Notably, dialkoxyphosphonyl radicals are highly reactive toward monomers. For carbon centered benzoyl radicals significantly lower rate constants are detected. The excellent reaction efficiency of phosphonyl radicals is attributed to the high electron density at the phosphorous atom and the pyramidal structure of the radicals providing more favorable streric conditions for the unpaired radical site to react with monomers.

8.3.1.7 α-Hydroxy Alkylphenones

α-Hydroxy alkylphenone is another photoinitiator containing benzoyl groups that has found practical application in many vinyl polymerizations [74–76]. This initiator has both a high light sensitivity and good thermal stability. Furthermore, coatings prepared using α-hydroxy alkylphenone do show only very little yellowing, which makes these compounds particularly suitable for clear coatings [19]. Another striking advantage is that α,α′-dialkyl hydroxyphenones are liquid at room temperature and are of relatively low polarity. Therefore, they are easy to dissolve in non-polar curing formulations [77–81].

Regarding the photochemistry of α-hydroxy alkylphenones, α-scission is the dominating reaction starting from the first excited triplet state (see Table 8.8). Although the reactivity of benzoyl radicals toward monomers is of no doubt (*vide ante*), the question whether the hydroxyalkyl radical is able to initiate polymerization is not entirely elucidated. However, for 1-hydroxycyclohexylphenylketone (see reaction (8.27) initiated polymerization of methyl 2-*tert*-butyl acrylate it has been shown by analysis of photolysis

TABLE 8.8
Various α Hydroxy Alkylphenones: Quantum Yields of α-Scission (Φ_α), Triplet Energies (E_T) and Triplet Lifetimes (τ_T)

X–C6H4–C(=O)–C(Y)(Y)–OH

X	Y	Φ_α	E_T (kJ mol⁻¹)	τ_T (10^{-9} s)	References
H	CH_3	0.2—0.3	298	30	[82, 83]
$CH(CH_3)_2$	CH_3	0.2—0.3	—	50	[82]
H	(cyclohexane ring)		281	—	[19, 41]
OCH_3	CH_3	0.38	298	12	[84]
OCH_2COOH	CH_3	0.1	—	55	[85]
OCH_2CH_2OH	CH_3	0.06	—	120	[85]
SCH_2CH_2OH	CH_3	0.4	—	10,000	[85]
$N(CH_3)_2$	CH_3	0.011	264	3300	[84]

products that hydroxyalkyl radicals add to the double bond of the monomer [53].

$$C_6H_5-\overset{O}{\overset{\|}{C}}-C(OH)(C_5H_{10}) \xrightarrow{h\nu} C_6H_5-\overset{O}{\overset{\|}{C}}\bullet + \bullet C(OH)(C_5H_{10}) \quad (8.27)$$

Recently, various derivatives of dibenzoyldihydroxy methane have been used for free radical polymerization of acrylic monomers [86]. Photocalorimetric and real-time IR investigations gave evidence that these compounds are excellent photoinitiators.

$$R-C_6H_4-C(=O)-C(OH)_2-C(=O)-C_6H_5 \xrightarrow{h\nu} R-C_6H_4-\overset{O}{\overset{\|}{C}}\bullet + \bullet C(OH)_2-C(=O)-C_6H_5 \quad (8.28)$$

8.4 PEROXY COMPOUNDS

The use of peroxides as initiators for vinyl polymerization is not new, and papers describing their ability have been published and reviewed [87, 88]. Peroxides contain two adjacent oxygen atoms with overlapping lone-pair orbitals. The average bond energy of the O-O linkage is about 143 kJ mol^{-1} (i.e., this bond is relatively weak).

In early investigations, hydrogen peroxide has been used as an initiator for the photoinduced free radical polymerization of acrylonitrile [89–91]. However, hydrogen peroxide absorbs light only weakly and solubility problems are unavoidable, especially if apolar monomers are to be polymerized.

Organic peroxides, such as benzoyl peroxide, are well-known thermal initiators. However, their use as photoinitiators in free radical polymerization is hampered by their low absorption at wavelengths above 300 nm and their thermal instability.

For photopolymerization, chromophoric groups absorbing at λ >320 nm were attached to either benzoyl peroxide [92] or perbenzoic acid ester [93–97]. In these systems, the excitation energy is transferred from the light absorbing aromatic carbonyl moieties to the perester moieties, which undergo homolytic bond scission. Although organic peroxy compounds are quite efficient in initiating vinyl polymerization, they are only scarcely used as photoinitiators because of their high thermal instability. Formulations containing peroxides may be stored for a very short time only.

$$C_6H_5\text{-}\overset{O}{\overset{\|}{C}}\text{-}C_6H_4\text{-}\overset{O}{\overset{\|}{C}}\text{-}O\text{-}O\text{-}C(CH_3)_2CH_3 \xrightarrow{h\nu} \boxed{C_6H_5\text{-}\overset{O}{\overset{\|}{C}}\text{-}C_6H_4\text{-}\overset{O}{\overset{\|}{C}}\text{-}O^{\bullet}} + \dot{O}\text{-}C(CH_3)_2CH_3$$

$$\downarrow$$

$$\boxed{C_6H_5\text{-}\overset{O}{\overset{\|}{C}}\text{-}C_6H_4^{\bullet}} + CO_2 \qquad (8.29)$$

8.5 AZO COMPOUNDS

Although diarylazo compounds are stable toward light, alkylazo compounds readily dissociate upon irradiation to give free radicals. The use of a perfluoro derivative of azomethane, namely hexafluoro azomethane as a photoinitiator of vinyl polymerization is well known [98–100]. This compound can be photolyzed to give $F_3C^{\cdot}$ radicals that do, in the presence of excess vinyl monomer, initiate polymerizations.

Regarding azonitriles, the famous thermal initiator 2,2′-azobisisobutyronitrile (AIBN) has also shown some potential for photochemical initiation. It was used to polymerize vinyl acetate [102–104], styrene [103,105], vinyl chloride [106], methyl methacrylate [103], and acrylonitrile [107]. Also of certain interest as photoinitiators for vinyl polymerization are α-azobis-1-cyclohexane-carbonitrile [108–110] and 1′,1′-azodicyclohexanecarbonitrile [111].

However, the relatively low absorbency of azo compounds has prevented their widespread use as photoinitiators. Furthermore, they are easily thermally decomposable, which reduces the lifetime of fully formulated curing mixtures.

8.6 HALOGENS AND HALOGEN-CONTAINING COMPOUNDS

In a number of early investigations, halogens such as chlorine [112–114], bromine [112,115,116], and iodine [112,117] have been employed as photoinitators for vinyl polymerization. It is assumed that the initiation involves the formation of a complex of halogen with the monomer. This complex absorbs the incident light and decomposes to yield initiating free radicals [116]. Moreover, the direct addition of photolytically formed halogen radicals could account for the initiation. In practical applications halogens are not in use as photoinitators, which is certainly due to inconvenience arising from handling, storage, and disposal.

Because the bonds between carbon and halogen atoms are relatively weak (except for C-F), halogenated compounds have some potential to be photoinitiators. Halogenated acetophenone derivatives, such as α,α,α,-trichloro-4-*tert*-butylacetophenone, were shown to undergo both β and α scission, the first process was found to dominate [28].

$$H_3C{-}C(CH_3)_2{-}C_6H_4{-}\overset{O}{\overset{\|}{C}}{-}CCl_3 \xrightarrow{h\nu} H_3C{-}C(CH_3)_2{-}C_6H_4{-}\overset{O}{\overset{\|}{C}}{-}\dot{C}Cl_2 + \boxed{\dot{Cl}} \tag{8.30}$$

$$H_3C{-}C(CH_3)_2{-}C_6H_4{-}\overset{O}{\overset{\|}{C}}{-}CCl_3 \xrightarrow{h\nu} H_3C{-}C(CH_3)_2{-}C_6H_4{-}\overset{O}{\overset{\|}{\dot{C}}} + \boxed{\dot{C}Cl_3} \tag{8.31}$$

The chlorine radicals formed according to Eq. (8.30) are very reactive: they either react with monomer molecules, thus initiating growing of a chain, or abstract hydrogen from various components of the polymerization mixture.

8.7 PHENACYL-TYPE SALTS

Phenacyl onium salts, namely N-phenacyl-N,N-dimethyl anilinium N,N-diethyldithiocarbamate (Ia), phenacyl-triphenylphosphonium-N,N diethyldithiocarbamate(II), 1-phenacyl pyridinium-N,N-diethyldithiocarbamate(III) (Eq. (8.32)), are shown to be efficient photoinitiators for polymerization of methyl methacrylate (MMA) [118].

$$C_6H_5{-}\overset{O}{\overset{\|}{C}}{-}CH_2{-}N^{+}(CH_3)_2{-}C_6H_5 \; X^{-}$$

Ia,b
a: $X^{\bullet} = SC(S)N(Et)_2$
B: $X^{\bullet} = SbF_6^{\bullet}$

$$C_6H_5{-}\overset{O}{\overset{\|}{C}}{-}CH_2{-}P^{+}(C_6H_5)_3 \; X^{-} \tag{8.32}$$

II, $X^{\bullet} = SC(S)N(Et)_2$

$$C_6H_5{-}\overset{O}{\overset{\|}{C}}{-}CH_2{-}N^{+}C_5H_5 \; X^{-}$$

III, $X^{\bullet} = SC(S)N(Et)_2$

It was considered that the same mechanism implies the initiation by all the salts, because they are structurally similar (i.e., positively charged heteroatom is located adjacent to phenacyl moiety). Plausible mechanism of the photoinitiation involves both free-radical and zwitterionic processes (Eqs. (8.33–8.35)).

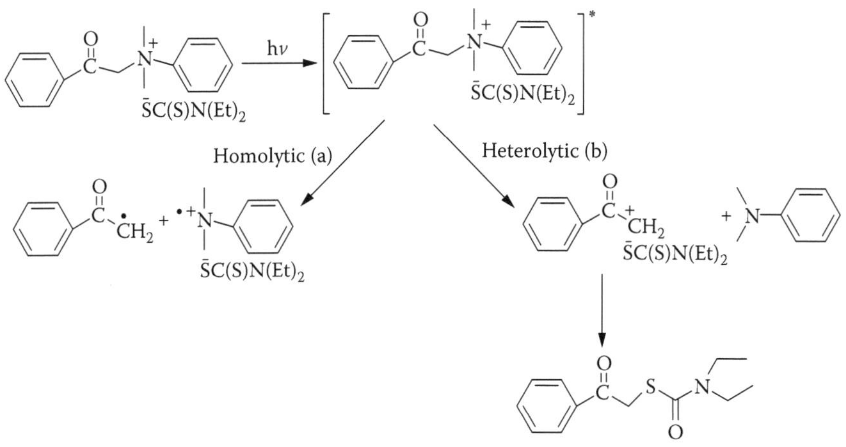

hν

Homolytic (a) Heterolytic (b)

(8.33)

Free Radical Polymerization

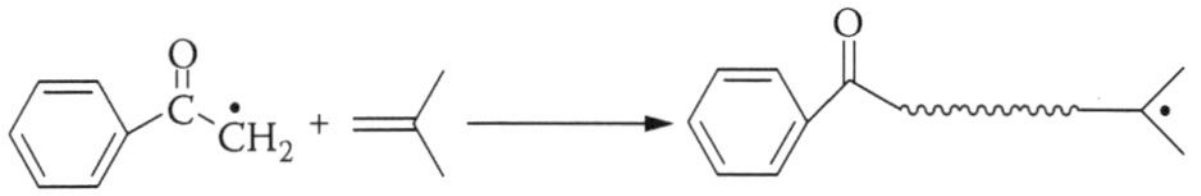

(8.34)

Zwitterionic Polymerization

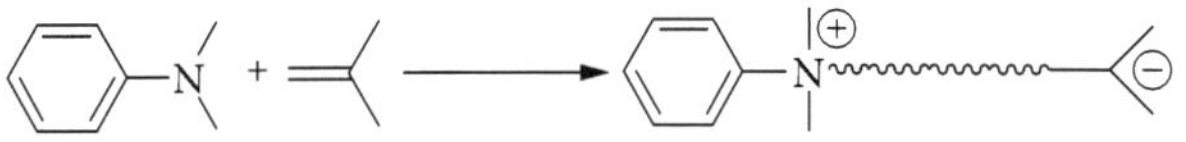

(8.35)

Electronically excited salt may undergo heterolytic cleavage resulting in the formation of phenacylium cations. Similar photodecomposition was proposed by Tachi and co-workers [119], who used phenacyl ammonium salts in photoinitiated crosslinking of epoxy compounds. Reaction (8.33) path b offers an alternative pathway in which homolytic cleavage followed by electron transfer essentially yields the same species.

Phenacyl radicals induce free radical polymerization, whereas dimethylaniline initiates zwitterionic polymerization according to the reactions (8.34) and (8.35).

8.8 TYPE II PHOTOINITIATORS

8.8.1 Aromatic Ketone/Coinitiator System

Photolysis of aromatic ketones, such as benzophenone, in the presence of hydrogen donors, such as alcohols, amines, or thiols, leads to the formation of a radical stemming from the carbonyl compound (ketyl type radical in the case of benzophenone) and another radical derived from the hydrogen donor (see Eq. (8.14)). Provided vinyl monomer is present the latter may initiate a chain polymerization. The radicals stemming from the carbonyl compound are usually not reactive toward vinyl monomers due to bulkiness and/or the delocalization of the unpaired electron.

Apart from benzophenones, thioxanthone, and anthraquinones, ketocoumarins and some 1,2 diketones are used in conjunction with coinitiators for initiating vinyl polymerizations. As has been explained earlier, both electron and hydrogen transfer

TABLE 8.9
Rate Constants for the Quenching of Benzophenone Triplet States by Monomer

Monomer	kq (l mol^{-1} s^{-1})
Styrene	3.3×10^9
N-vinyl pyrrolidone	3.6×10^8
Methyl methacrylate	6.9×10^7
Methyl acrylate	7.5×10^6

Source: R. Kuhlmann and W. Schnabel, *Polymer*, 17, 419 (1976).

can bring about radical formation in the case of type II photoinitiators. In many systems, both processes occur.

Because the initiation is based on a bimolecular reaction, type II photoinitiators initiate generally slower than type I photoinitiators. These systems are, therefore, more sensitive to the quenching of excited triplet states, which are the reactive precursors of light-induced chemical changes for carbonyl compounds. Indeed, quenching by monomers with a low triplet energy (e.g., styrene or N-vinyl carbazol) or by oxygen is often observed and may lead to relatively low curing rates (see Table 8.9).

In view of applications, the selection of the coinitiator is undoubtedly of great importance. Amines are used most of the time because of their high efficiency and the relatively low price. Excited carbonyl triplet states are usually by two to three orders of magnitude more reactive toward tertiary amines than toward alcohols or ethers. Trialkylamines are very efficient but are sometimes disregarded for their strong odor. Alkanolamines are efficient and less smelly, but they lack good thermal stability and may also cause yellowing of the final polymer material. Aromatic amines have good initiating properties but are slightly more expensive.

8.8.1.1 Benzophenones

Hydrogen abstraction by the excited triplet manifold of benzophenone, which is populated with quantum yields close to unity, from tertiary amines (N-methyl diethanolamine) is depicted in Eqs. (8.36) and (8.37) [120]. The carbon centered radical stemming from the amine is able to initiate free radical polymerizations of suitable monomers. α-Aminoalkylradicals are especially suitable for the polymerization of acrylates [121] and are less efficient for styrene polymerization, which is explainable in terms of triplet quenching by styrene. In Table 8.9, rate constants for triplet quenching by various monomers are compiled. The ketyl radicals scarcely add to olefinic double bonds due to resonance stabilization and for steric reasons, but instead undergo recombination and disproportionation reactions, as shown in Eq. (8.38). Furthermore, they may act as chain terminators in the polymerization leading to ketyl moieties incorporated into polymer chains and relatively short chains [122].

To avoid chain termination by ketyl radicals, additives such as onium salts [123–125] or certain bromocompounds [126] have turned out to be useful. These additives react with the ketyl radicals thus suppressing chain termination. In the case of onium salts, phenyl radicals, which initiate polymerizations instead of terminating growing chains, are produced by the interaction of ketyl radicals with salt entities. Thus, the overall effect of these additives is an enhancement in polymerization rate.

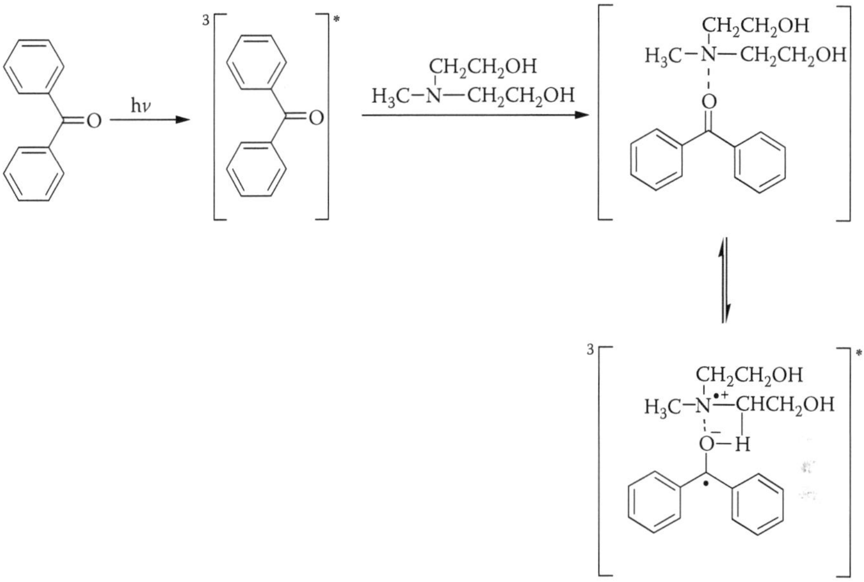

(8.36)

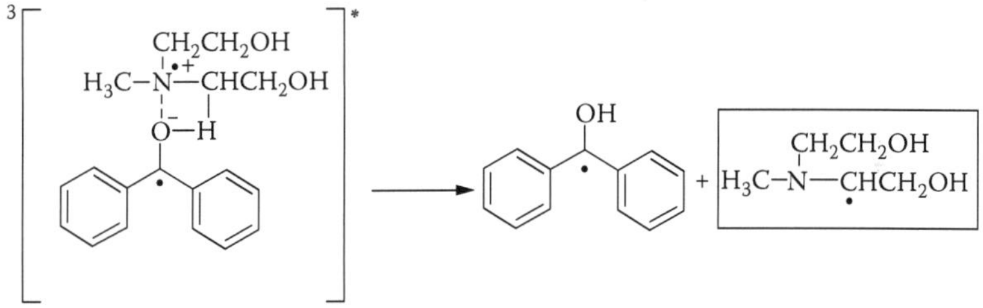

(8.37)

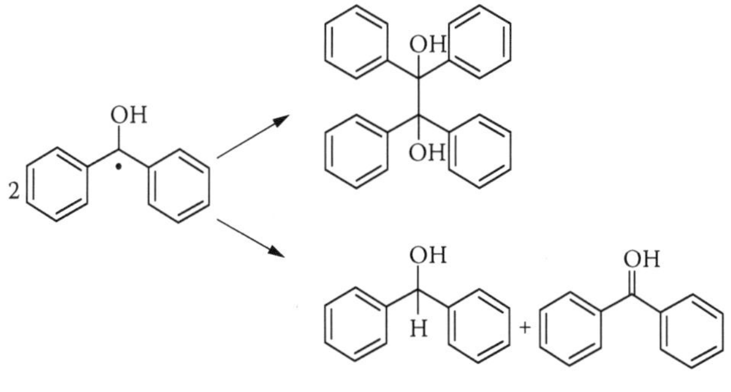

(8.38)

Recently, benzophenone based initiators with hydrogen donating amine moieties covalently attached via an alkyl spacer were introduced as photoinitiators for vinyl polymerization [101,128–132], (see Table 8.10). Though also following the general

TABLE 8.10
Miscellaneous Benzophenone Type Photoinitiators

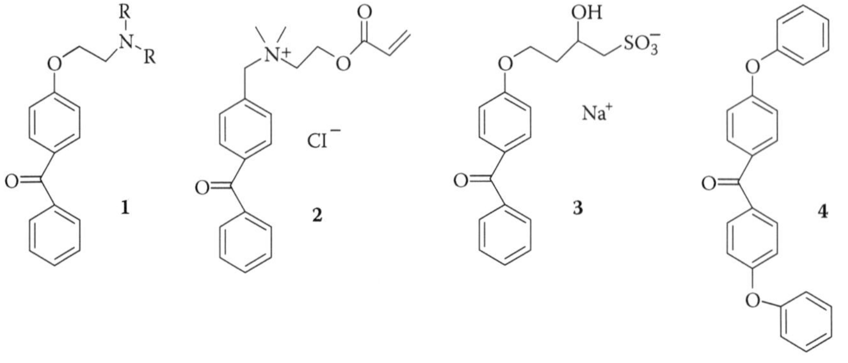

1	2	3	4
[131,138]	[133–135,137]		[136,139]

scheme of Type II initiators, the initiation is a monomolecular reaction, because both reactive sites are at the same molecule. Hydrogen transfer is suspected to be an intramolecular reaction. The ionic derivatives (2 and 3) in Table 8.9 are used for polymerization in aqueous phase [133–135]. With 4,4′-diphenoxybenzophenone (4 in Table 8.10) in conjunction with tertiary amines, much higher polymerization rates (by a factor of 8 higher) were obtained than benzophenone [136].

8.8.1.2 Michler's Ketone

Michler's ketone, 4,4′-bis(dimethylamino)benzophenone, is another efficient hydrogen abstraction type photoinitiator that possesses both chromophoric aromatic ketone and tertiary amine groups in its structure. It absorbs much stronger light of 365 nm than does benzophenone. Michler's ketone may undergo photoinduced hydrogen abstraction from ground state molecules, but with a relatively low efficiency [140]. However, in most cases it is used in conjunction with benzophenone and serves as a hydrogen donor. The mechanism involves electron transfer in the exciplex formed and subsequent hydrogen abstraction (Eqs. (8.39–8.40)) [141].

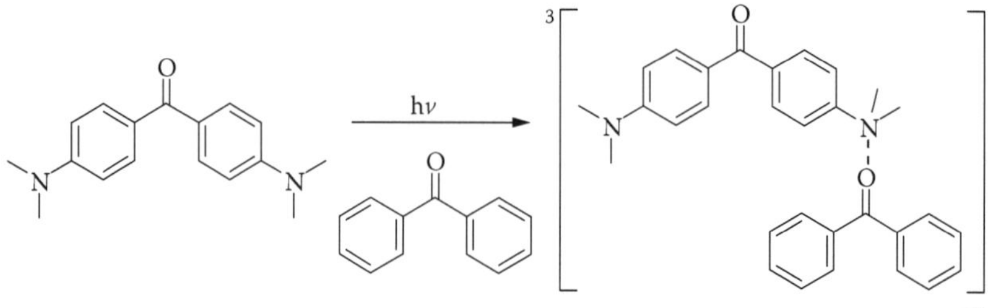

$$\xrightarrow{h\nu} \quad (8.39)$$

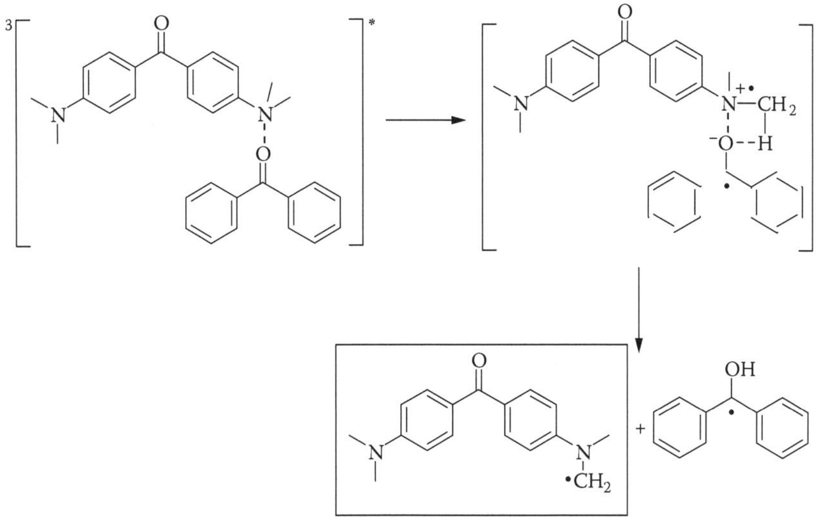

(8.40)

It is noteworthy that the combination of Michler's ketone and benzophenone gives a synergistic effect: this system is more efficient in forming initiating radicals in conjunction with amines than that of separate components. A disadvantage of Michler's ketone is the yellow color that coatings cured with this ketone possess. It prevents the utilization of the highly efficient benzophenone/Michler's ketone system in white pigmented formulations. Moreover, regarding Michler's ketone, a suspicion of carcinogenity exists.

8.8.1.3 Thioxanthones

Thioxanthones in conjunction with tertiary amines are efficient photoinitiators [142] with absorption characteristics that compare favorably with benzophenones; absorption maxima are in the range between 380 to 420 nm ($\varepsilon = 10^4$ l mol^{-1} cm^{-1}) depending on the substitution pattern. The reaction mechanism has been extensively investigated by spectroscopic and laser flash photolysis techniques [64,143–145]. It was found that the efficiency of thioxanthones in conjunction with tertiary amines are similar to that of benzophenone/amine systems (Table 8.11).

The most widely used commercial derivatives are 2-chlorothioxanthone and 2-isopropylthioxanthone. Furthermore, ionic thioxanthone derivatives have been developed, which may be employed for water based curing formulations [147,148]. A great advantage is that thioxanthones are virtually colorless and do not cause yellowing in the final products.

As for other type II initiating systems, quenching by monomer has to be taken into account, provided monomers with low triplet energies are used. Thus, the bimolecular rate constants of the reaction of various thioxanthones with styrene are between 3 to 6×10^9 l mol^{-1} s^{-1}. For acrylonitrile, the values range between 4×10^5 and 4×10^6 l mol^{-1} s^{-1}, indicating very little quenching [64].

Interestingly, when *N*-ethoxy-2-methylpyridinium salt is added to the mixture consisting of monomer (methyl methacrylate) and thioxanthone, a significant

TABLE 8.11
Thioxanthone Derivatives for Photoinitiation

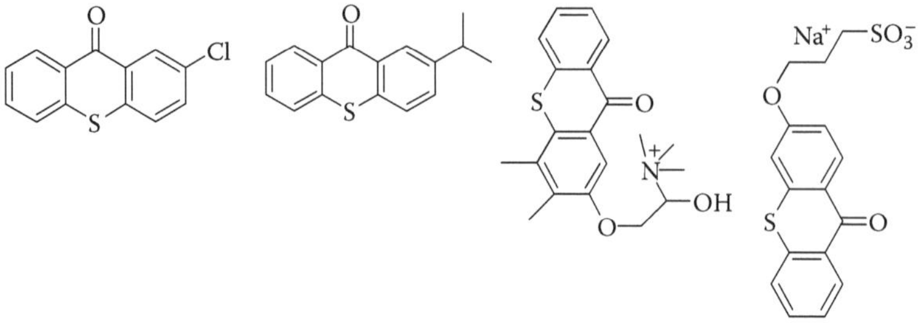

[144]	[146]	[147,148]

enhancement of the polymerization rate is detected [149]. This effect has been attributed to a reaction of ketyl radicals stemming from thioxanthone with the pyridinium salt, which leads to the generation of initiating ethoxy radicals.

8.8.1.4 2-Mercapto-thioxanthone

As an alternative hydrogen donor, Cokbaglan et al. [150] reported the synthesis and the use of a thiol derivative of thioxanthone as a photoinitiator for free radical polymerization. A great advantage is that this photoinitiator does not require an additional co-initiator.

Its capability to act as an initiator for the polymerization of methylmethacrylate, styrene, and multifunctional monomers were examined.

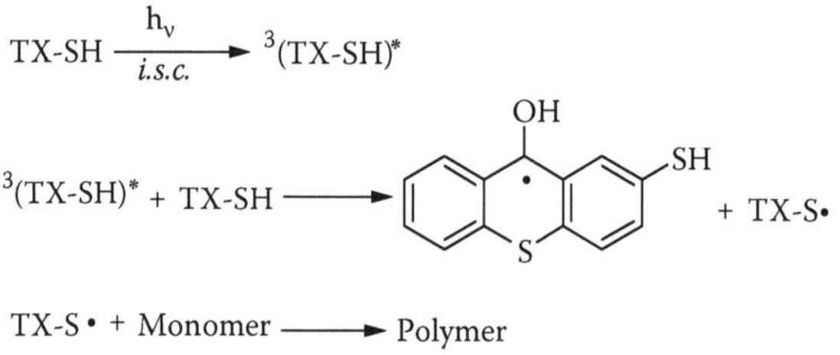

$$\text{TX-SH} \xrightarrow[i.s.c.]{h_\nu} {}^3(\text{TX-SH})^*$$

$${}^3(\text{TX-SH})^* + \text{TX-SH} \longrightarrow [\text{ketyl radical of TX-SH (OH, SH)}] + \text{TX-S}\bullet \quad (8.41)$$

$$\text{TX-S}\bullet + \text{Monomer} \longrightarrow \text{Polymer}$$

TX-SH: 2-Mercapto-thioxanthone

The mechanism is based on the intermolecular reaction of triplet, ^{3}TX-SH*, with the thiol moiety of ground state TX-SH. When TX-SH is irradiated in the presence of a monomer, it can serve as both a triplet photosensitizer and a hydrogen donor. The resulting thiyl radicals initiate the polymerization (Eq. (8.41)).

Intramolecular hydrogen abstraction was excluded by laser flash photolysis experiments. Flash photolysis (355-nm excitation) of a highly diluted solution of TX-SH affords a readily detectable transient absorption spectrum, which decayed in a first-order kinetic with a lifetime of 20 μs (Figure 8.1).

Transient absorption and lifetime is similar to the triplet-triplet absorption of the parent TX [151]. Therefore, it was concluded that the transient absorption

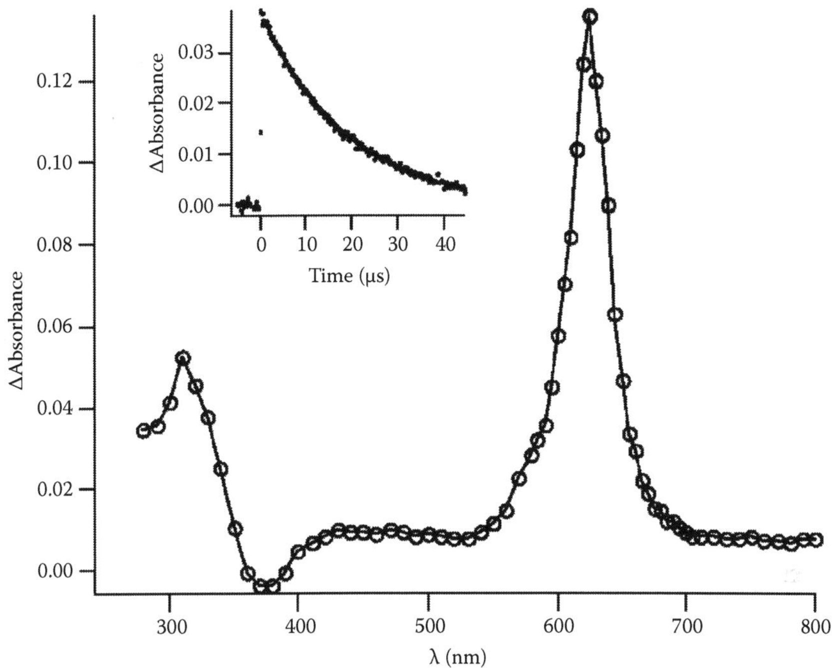

FIGURE 8.1 Transient optical absorption spectrum recorded 1-5 μs following laser excitation (355 nm, 5 ns) of 2-mercapto-thioxanthone (TX-SH) (1×10^{-4} M) in argon-saturated acetonitrile solution at 23°C. Insert: Transient absorption kinetic observed at 625 nm.

corresponds to the triplet-triplet absorption of TX-SH. If intramolecular hydrogen abstraction would dominate, then the transient decay kinetic should be significantly faster. This is consistent with an unfavorable interaction of the excited carbonyl group with the thiol moiety caused by the rigidity of the linked aromatic groups. Similar limitations were accounted for Michler's ketone [152] as a photoinitiator, which also possesses both chromophoric and hydrogen donating sites.

Phosphorescence spectra of TX covalently attached polymers may also provide further evidence for the initiation mechanism. As can be seen from Figure 8.2, phosphorescence spectra in 2-methyltetrahydrofuran at 77 K of TX-SH (spectrum a) and PMMA obtained from photopolymerization initiated by TX-SH (spectrum b) are very similar.

The excitation spectra for the emission signal are in good agreement with the absorption spectra of TX-SH (see also Figure 8.3).

The phosphorescence lifetimes at 77 K are also very similar: 147 ms and 145 ms, respectively. Thus, various spectroscopic investigations reveal that thioxanthone groups are incorporated into the polymers.

Moreover, phosphorescence measurements are useful to gain information on the triplet configuration of TX-SH. Phosphorescence spectra of ketones with n-π* nature of the lowest triplet state are usually structured, due to the vibrational progression of the C=O vibration, and π-π* triplets are mostly unstructured [7, 153)].

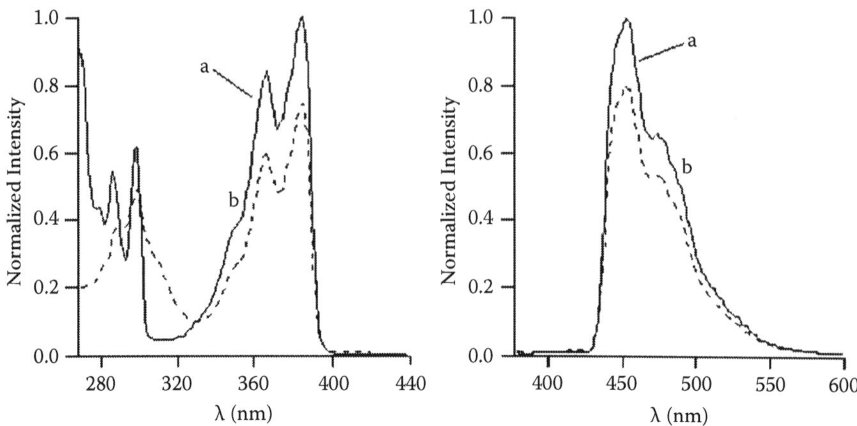

FIGURE 8.2 Phosphorescence excitation (left) and emission (right) spectra of 2-mercapto-thioxanthone (TX-SH) (a) and poly(methyl methacrylate) obtained by photoinitiated polymerization by using TX-SH (b) in 2-methyltetrahydrofuran at 77 K; $\lambda_{exc} = 375$ nm.

In addition, the phosphorescence lifetimes for n-π* triplets are significantly shorter (in the order of several milliseconds) compared with π-π* triplets (more than 100 ms) [7, 151]. Thus, the broad structureless phosphorescence of TX-SH, together with the long phosphorescence lifetime in a matrix at 77 K, indicates a π-π* nature of the lowest triplet state. This is in agreement with the π-π* nature of the lowest triplet state of unsubstituted TX. TX-SH is an efficient photoinitiator for free radical

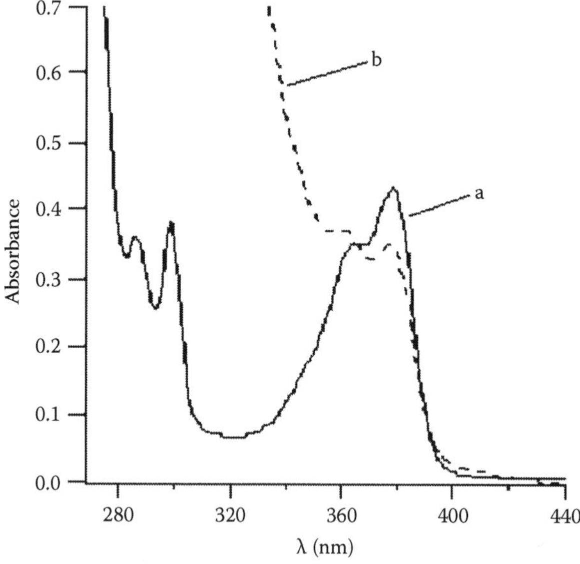

FIGURE 8.3 Absorption spectra of 2-mercapto-thioxanthone (TX-SH) (a) and poly(methyl methacrylate) obtained by photoinitiated polymerization by using TX-SH (b) in 2-methyltetrahydrofuran in the presence of air.

polymerization. This odorless new photoiniator is very attractive, because it does not require an additional hydrogen donor and initiates the polymerization of both acrylate and styrene monomers [155] in the presence and absence of air. In addition, TX-SH possesses excellent optical absorption properties in the near UV spectral region, ensuring efficient light absorption from most UV-curing tools [150].

8.8.1.5 Ketocoumarins

In conjunction with tertiary amines, ketocoumarines act as highly efficient Type II photoinitiating systems [156–159]. The spectral sensitivity of this system can be tuned to various wavelengths of the visible part of the spectrum by selection of suitable substituents. Moreover, the substitution pattern determines whether the coumarin acts as electron donor or as electron acceptor. 3-Ketocoumarins with alkoxy substituents in the 5- and 7-position show good absorption in the near UV and are excellent electron acceptors. Regarding coinitiators, alkylaryl amines are most suitable.

8.9 PHOTOINITIATION BY DECARBOXYLATION

It was found that irradiations of phenyl acetic acid [160] and 1-naphtylacetic acid in aqueous solution lose carbon dioxide (Eq. (8.42)).

$$PhCH_2CO_2^- \xrightarrow[H_2O]{h\nu} PhCH_2CO_2^{\bullet} + e^{\bullet}_{aq} \longrightarrow PhCH_2^{\bullet} + CO_2 \tag{8.42}$$

In this mechanism, electron ejection from anion gives a carboxylate radical that breaks down, giving carbon dioxide and a benzyl radical. Although irradiation of nitrophenyl acetate [161] resulted in a similar decarboxylation process, the electron ejection mechanism was discounted because of the high quantum yield for decarboxylation of the meta- and para-substituted compounds. Flash photolysis studies confirmed that nitrobenzylanions were intermediates, and it was suggested that the excited nitrophenylacetate ions lose CO_2 to give these in a one-step reaction. When α-(2,4-dinitroanilino)alkanoic acids and α-(2,4-dinitrophenoxy)alkanoic acids [161] were irradiated in both acid and basic solution, decarboxylation was also observed. Intramolecular hydrogen abstraction by the excited nitro group from the α-C-H bonds or from the carboxyl group were responsible for the decarboxylation.

It was also found that carboxylic acids can undergo decarboxylation by irradiation of either their cerium [162] or lead (IV) salts [163]. However, these two reactions occur by different mechanisms; excited cerium (IV) salts decompose to give a cerium (III) salt plus a carboxylate radical which then loses carbon dioxide, whereas the excited lead (IV) salts decompose by a multibond cleavage reaction to yield simultaneously, alkyl radicals, carbon dioxide, and a lead salt.

$$PhCO_2Pb(O_2CR)_3 \xrightarrow{h\nu} R^{\bullet} + CO_2 + P^{\bullet}b^{\bullet}(O_2CR)_3 \tag{8.43}$$

ESR spectroscopy has been employed for the confirmation of alkyl radicals produced by these reactions.

Heterocyclic compounds [164], such as acridine, quinoline, isoquinoline, and riboflavin, were found to be efficient "sensitizer" for the decarboxylation of carboxylic acids.

Pyridiyl acetic acids are also decarboxylated on irradiation and give the corresponding picoline [165]. Irradiation of α-amino acids in the presence of purines leads to decarboxylation [166].

It was reported that carboxylic acids of the type $RXCH_2CO_2H$, where X = O, S, or NH) are decarboxylated on irradiation in the presence of biacetyl [167], aromatic ketones [161,168], quinines [161], and aromatic hydro compounds [169]. A mechanism was suggested by Baum and Norman [168] who found that the decarboxylation of phenylacetic acid and phenoxyacetic acid can be phosensitized by biacetyl. They proposed that the excited triplet ketone reacts with the acid to give a biradical (Eq. (8.44)), which undergoes an acid catalyzed fragmentation reaction to give eventually carbon dioxide and radicals derived from the ketone and acid.

$$\left[CH_3COCOCH_3\right]^* + C_6H_5CH_2COOH \longrightarrow \text{biradical (}H^+\text{)} \longrightarrow CH_3\dot{C}OHCOCH_3 + CO_2 + Ph\dot{C}H_2 \qquad (8.44)$$

The alternative mechanism was suggested by Davidson et al. [162,169,170] for the decarboxylation reactions of two components that involves an exciplex intermediate. The reactions are considered as occurring via exciplex formation between the carbonyl compound and acid in which electron transfer from the acid to the excited keton occurs. Excited aromatic carbonyl compounds such as benzophenone can undergo electron transfer with sulfur or oxygen-containing carboxylic acids to give carboxylate radicals. Subsequent decarboxylation of this intermediate radical produces an alkyl radical.

$$ArXCH_2COOH + Ar_2CO_{T_1} \longrightarrow ArX^{\bullet+}CH_2COOH + Ar_2\dot{C}O^- \qquad (8.45)$$

$$\text{A: } Ar_2CO_{S_0} + ArXCH_2CO_2H \qquad \text{B: } ArX\dot{C}H_2 + CO_2 + Ar_2\dot{C}OH$$

More recently, one-component bimolecular photoinitiator systems based on the decarboxylation process were reported by Aydin et al. [171]. 2-Thioxanthone-thioacetic acid and 2-(carboxymethoxy)thioxanthone were used as photoinitiators for free radical polymerization. These one-component initiators contain light absorbing and electron donating and consequently hydrogen donating sites in one molecule, as opposed to two-component systems in which light absorbing and electron donating acidic sites are composed in independent molecules. The suggested mechanism involves redox reactions. The intermolecular electron-transfer reaction may occur between an excited thioxanthone moiety and the carboxylic acid group of another molecule (Eq. (8.46)).

$$TX\text{-}SCH_2COOH \xrightarrow{h\nu} TX\text{-}SCH_2COOH^* \xrightarrow{TX\text{-}SCH_2COOH} \ldots \tag{8.46}$$

Monomer → Polymer; + CO_2 + ... → Products

TX-SCH_2COOH: 2-Thioxanthone-thioacetic Acid

The intramolecular electron transfer should also be considered. Although a favorable interaction of the excited carbonyl group with carboxylic hydrogen is facilitated with flexible S-CH_2 or O-CH_2 spacer groups, the actual initiation involves electron transfer from a sulfur or oxygen atom. It appears that the interaction of excited carbonyl with the sulfur or oxygen atoms of the same molecule separated with a rigid aromatic ring is unlikely to occur [171].

8.10 BENZIL AND QUINONES

Benzil and quinones, such as 9,10-phenanthrene quinone and camphor quinone in combination with hydrogen donors, can be used as photoinitiators both in the UV and visible region [122,172,173]. Photopolymerization of methyl methacrylate using benzil was elaborately studied by Hutchison et al. [122]. They have observed a threefold increase in the polymerization rate when a hydrogen-donating solvent, such as THF, was used in the system indicating the importance of hydrogen abstraction.

$$\text{(camphorquinone)} + 2\,\text{THF} \xrightarrow{h\nu} \text{(ketyl radical: } CH_3, H_3C, CH_3, O, OH, H) + 2\,\text{THF}^{\bullet} \tag{8.47}$$

Amines, such as dimethylaniline and triethylamine, are also used as coinitiators for free radical polymerizations [174,175]. In these cases, initiating radicals are supposedly generated through exciplex formation followed by proton transfer. The low order of toxicity of camphorquinone and its curability by visible light makes such systems particularly useful for dental applications [172,176,177]. It is noteworthy that the reactivity is relatively low, owing to a low efficiency in hydrogen abstraction reactions. This circumstance has prevented the use of quinones in other applications.

8.11 MALEIMIDES

In the past, the polymerization reactions of maleimide (MI)-vinyl ether (VE) have received a revitalized interest in the field of UV-induced curing when these systems were presented as being able to be used in photoinitiator free formulation. A critical component in all photopolymerization reactions is the photoinitiator, which absorbs light efficiently and generates reactive species capable of initiating chain growth polymerization. The maleimide group, which is the basic UV-absorbing chromophore, is consumed during the polymerization process and absorbs little light at wavelength greater than 300 nm [178,179].

Maleimide derivatives are copolymerizable and self-bleaching photoinitiators. This prospect is particularly attractive because photoinitiators have several disadvantages. Although maleimides have been believed to be hydrogen abstracting photoinitiators it was shown that maleimides can also undergo electron transfer reactions with unsaturated monomers. The high rate (>10^9 s^{-1} in neat resin) of initiating radical formation is, however, accompanied by a low overall quantum yield associated with the electron transfer reaction. Although hydrogen back donation does not take place usually, electron back transfer within the radical ion pair is a highly favored process, especially in apolar solvents [180a].

The lower susceptibility of the maleimide systems to oxygen inhibition may also be due in part to the ability of maleimides upon excitation to abstract a hydrogen atom from a suitable C-H bond thereby generating another initiating radical. The ability of excited maleimides to hydrogen abstract from polyethers means that they can be used to initiate the polymerization of acrylates that contain these groups [180b].

Maleimides are alternative to traditional photoinitiators and they have the ability to function both as a photoinitiator and a polymerizable monomer with vinyl ethers, acrylates and styrene oxides [181–185].

$$(M)_{S0} \xrightarrow{h\nu} (M)_{S1} \longrightarrow (M)_{T1} \longrightarrow \text{polyether radical} + \text{hydroxy maleimide radical} \quad (8.48)$$

$$\text{hydroxy maleimide radical} \xrightarrow{^3O_2} \text{maleimide} + HOO^\bullet$$

Hydrogen abstraction by maleimides

It should also be noted that the presence of oxygen leads to the regeneration of the maleimide, which means that the hydrogen abstraction process does not diminish to the extent to which the alternating co-polymerization process occurs.

8.11.1 Sensitization of Maleimides

De Schryver et al. [186–189] have described the sensitization of maleimides using benzophenone to induce photoallowed [2+2] homocycloadditions, which also occur in the absence of a sensitizer. These reactions yield the cyclobutane adducts of aliphatic as well as aromatic N-substituted maleimides. De Schryver reported that these reactions may occur intermolecularly or intramolecularly, in the case of difunctional aliphatic maleimides, with the yield of intramolecular cycloadduct varying as a function of spacer length between reactive groups [186–189].

The photopolymerization of a diacrylate monomer with an N-aliphatic maleimide as the photoinitiator has been investigated by photodifferential scanning calorimetry (Photo-DSC), real-time Infrared (RT-IR) spectroscopy, and real-time UV (RT-UV) spectroscopy by Clark et al. [179]. The polymerization rates obtained for direct excitation of *N*-aliphaticmaleimides are much lower than what can be achieved with conventional cleavage type photoinitiators [179].

8.12 DYE SENSITIZED INITIATION

Most of the photoinitiators described up to now are sensitive to light at wavelengths below 400 nm. This enables an easy processability, because the sun and many artificial sources of light do overwhelmingly emit light of higher wavelengths and therefore photoinduced reactions before the intended initiation by UV light may be kept at low level. However, if the strong emission of mercury lamps or the sun in the visible region of the spectrum is to be used, photoinitiating systems that absorb visible light are required. Such systems often involve dyes as light absorbing chromophores. Numerous photoinitiated free radical polymerizations using dyes have been described and reviewed by several authors [190–194]. Initiating radicals are generated by photoinduced electron transfer. Energy transfer is not thermodynamically favorable in these systems due to low excitation energies of dyes. Depending on the nature of the dye involved, two distinct mechanisms are to be considered:

1. Electron transfer from the coinitiator to the excited, *photoreducable dye* molecule yields radical cations of the coinitiator and radical dye anions. The former can initiate the polymerization. In many cases, however, initiating radicals are formed in subsequent thermal reactions. Species deriving from the dye molecule do not react with monomer molecules.

$$D^* + COI\!: \longrightarrow (D...COI\!:)^* \longrightarrow D^{-\bullet} + \boxed{COI^{+\bullet}} \quad (8.49)$$

2. Electron transfer from the excited, *photooxidizable dye* to the coinitiator. In this case too, the initiating radicals stem mostly from coinitiator.

$$:D^* + COI \longrightarrow (:D...COI)^* \longrightarrow D^{\overset{+}{\bullet}} + COI^{-\bullet} \quad (8.50)$$

For initiating the radical polymerizations, the first depicted reaction is commonly followed.

A disadvantage of many dye containing formulations is that they lack good storage stability. Mostly this phenomenon is due to the basicity of the coinitiators that can abstract hydrogen from the dye thus leading to depletion.

8.12.1 Photoreducable Dye/Coinitiator Systems

As can be seen in Table 8.12, differently colored photoreducable dyes are used for sensitizing cationic polymerizations.

As coinitiators, the following substances have found application in conjunction with photoreducable dyes (Table 8.13).

Among amine coinitiators, phenylglycidine has been reported to be particularly efficient. As depicted in reaction (8.51), the formation of initiating radicals owes to a thermal fragmentation reaction.

D* + [phenyl]–NH–CH₂–C(=O)OH → [phenyl]–$\dot{N}H^{+}$–CH₂–C(=O)OH + $\dot{D}^{-}$ → [phenyl]–NH–$\dot{C}H_2$ + CO_2 + H^+ (8.51)

Regarding organotin compounds it is noteworthy that although systems containing these coinitiators are not above average as far as reactivity is concerned, they are superior to many other systems based on dye sensitization for their high storage stability.

D* + [phenyl]–CH_2–Sn(CH_3)₃ → [phenyl]–C–$\dot{S}n^{+}$(CH_3)₃ + $\dot{D}^{-}$ → [phenyl]–$\dot{C}H_2$ + Sn^{+}(CH_3)₃ (8.52)

Borate salts are especially useful in combination with cyanine dyes. Depending on the cyanine used, different absorption maxima exist in the visible region with usually high molar extinction coefficients (ε = ca. 10^5 l mol^{-1} cm^{-1}). Radical formation by borate is illustrated in Eq. (8.53). In contrast to many other initiating systems based on dye sensitization, cyanine/borate complexes are ionic before the electron transfer

TABLE 8.12
Absorption Characteristics of Typical Photoreducable Dyes

Type	Example	λ_{max} (nm)
Acridinium	H_2N, NH_2, N, CH_3, Cl^- Acriflavine	460
Xanthene	ONa, I, I, O, I, I, Cl, COONa, Cl, Cl, Cl Rose Bengale	565
Fluorone	I, I, O, O, OH, I, I	536
Thiazene	H_3C, N, CH_3, N, +, S, Cl^-, N, CH_3, CH_3 Polymethylene Blue	645
Polymethylene	O, N+, n, I, O, N, Cl^-, C_6H_{13}, C_6H_{13} Cyanine Dye	490...700, depending on n

TABLE 8.13
Coinitiators for Dye-Sensitized Free-Radical Polymerization

Type	Formula	Name	References
Amines	$N(CH_2CH_2OH)_3$	Triethanolamine	[195]
	C_6H_5–NH–CH_2–COOH	N-phenylglycine	[196,197]
		N,N-dimethyl-2,6-diisopropyl aniline	[198,199]
	$H_2N{-}NH_2$	Hydrazine	[200–202]
Phosphines and arsines	$(C_6H_5)_3X$ (X = P, As)	Triphenyl-phosphine, triphenylarsine	[203]
Sulphinates	$R{-}S(=O)_2{-}R'$		[204–207]
Heterocycles		Imidazole	[208]
		Oxazole	[208]
		Thiazole	[208]
Enolates		Dimedone enolate	[209]

TABLE 8.13 (CONTINUED)
Coinitiators for Dye-Sensitized Free-Radical Polymerization

Type	Formula	Name	References
Organotin comp.	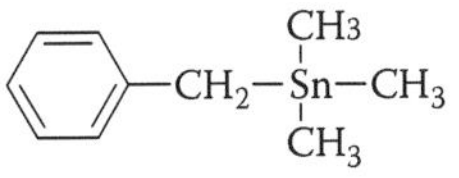 $C_6H_5-CH_2-Sn(CH_3)_3$	Benzyltrimethyl stannate	[197]
Borates	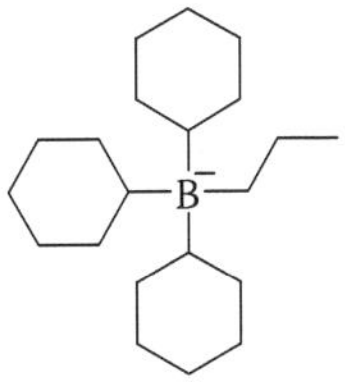	Triphenylmethyl borate	[210–212]

and then transformed into neutral species. Other systems are neutral before electron transfer and get ionic thereafter.

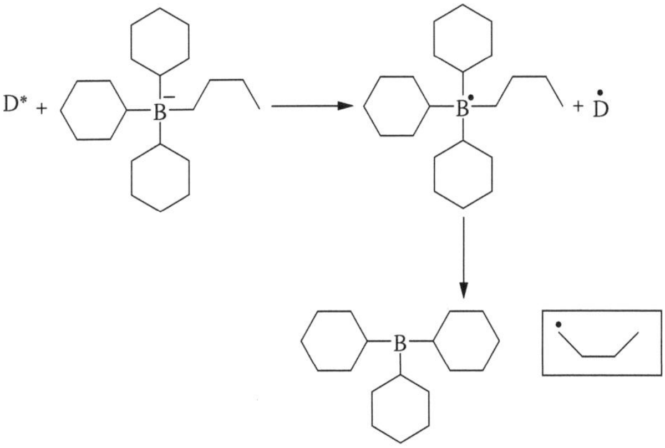

D* + (8.53)

In some studies, photoreducable dyes, such as acridine and xanthene, were used without adding any coinitiator for the photopolymerization of styrene, α-methylstyrene, and methyl methacrylate [213]. The initiation has in these cases been explained in terms of electron transfer reactions between excited dye and monomer molecules.

8.12.2 Photooxidizable Dye/Coinitiator Systems

As far as photooxidizable dyes are concerned, the oxidation of dyes such as acridine or xanthene by onium salts is to be mentioned. Onium salts, such as aryldiazonium, diaryliodonium, phosphonium, and sulphonium salts, are able to oxidize certain dyes. In Table 8.14 reduction potentials of various onium salts are depicted. Notably, the higher (more positive) the value of $E_{1/2}^{red}$ (On^+) is, the more capable is the salt to oxidize a dye (Table 8.14).

TABLE 8.14
Reduction Potentials of Onium Salt Cations, $E_{1/2}^{red}$ (On^+) in V

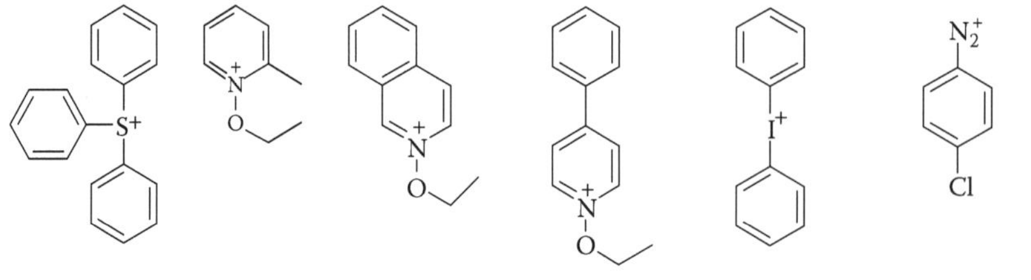

−1.1 [214]	−0.7 [215]	−0.5 [216]	−0.5 [216]	−0.2 [217]	0.35 [218]

The decomposition of diazonium salts by excited xanthene dyes (e.g., eosin, erythrosin, and rhodamin B) in ethanol solution has been attributed to the oxidation of the dye [219]. These systems were employed for a photopolymerization process in which vinyl monomers (e.g., vinylpyrrolidone, bis(acrylamide)) were crosslinked by visible light [220]. The initiation is depicted next:

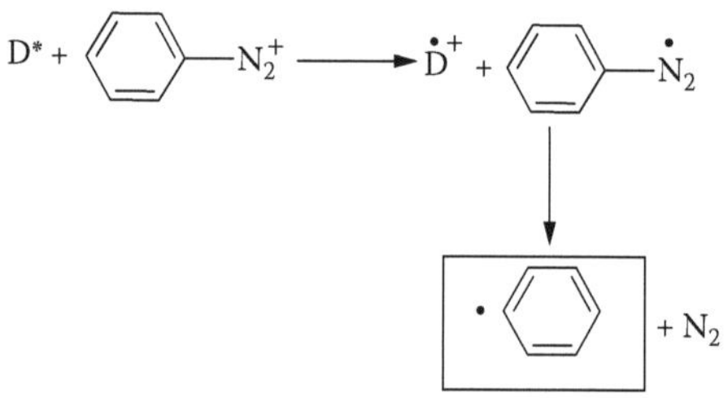

$$D^* + C_6H_5N_2^+ \longrightarrow D^{\bullet+} + C_6H_5\dot{N}_2 \longrightarrow C_6H_5^{\bullet} + N_2 \qquad (8.54)$$

The oxidation of various dyes by iodonium salts and the use of these systems for both free radical and ionic polymerization have been reported by several authors [193, 221–223]. Although the radical initiating species derive from the onium salt, dye radical cations are able to initiate cationic polymerizations.

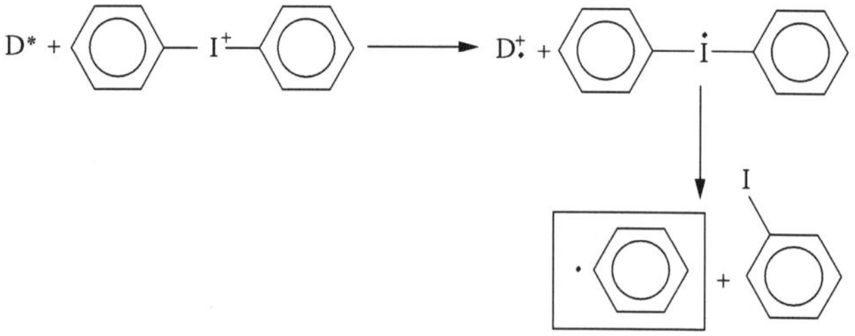

$$D^* + C_6H_5I^+C_6H_5 \longrightarrow D^{\bullet+} + C_6H_5\dot{I}C_6H_5 \longrightarrow C_6H_5^{\bullet} + C_6H_5I \qquad (8.55)$$

Pyridinium salt has been reduced by anthracene. This reaction was utilized for the light-induced polymerization of methyl methacrylate. In this case, ethoxy radicals have been found to react with monomer molecules [149].

$$\mathrm{D^* + (\text{2-methyl-}N\text{-ethoxypyridinium}) \longrightarrow D^{\cdot+} + (\text{2-methyl-}N\text{-ethoxypyridinyl radical}) \longrightarrow \text{2-methylpyridine} + \boxed{\dot{O}CH_2CH_3}} \quad (8.56)$$

Furthermore, the photoreduction of cyanine dyes by polyhalogen containing hydrocarbons was used for light-induced vinyl polymerization [224, 225].

8.13 THIOL-ENE POLYMERIZATION

Thiol-ene polymerizations are reactions between multifunctional thiol and ene (vinyl) monomers that proceed via a step growth radical addition mechanism. Initiation is achieved through generation of radical centers, the most common method being photoinitiation of the radical centers. Thiol-ene addition reaction was discovered as early as 1905 by Posner [226], and since then the reaction mechanism, polymerization kinetics, and monomer reactivities were extensively explored, and two excellent reviews were published by Jacobine [227] and Woods [228].

The generally accepted mechanism of the thiol-ene system was proposed by Kharasch et al. [229]. In this regard, it is worth mentioning the pioneering work of Morgan et al. [230], who used benzophenone to absorb light and initiate the polymerization of the radical chain process via a hydrogen abstraction reaction involving the excited benzophenone and a ground state thiol (see Scheme 8.2).

Excited benzophenone abstracts a hydrogen from a thiol monomer, generating thiyl radical that can either propagate or terminate. It is important to note that after generation of an initial thiyl radical species the thiol-ene propagation basically involves a free radical addition step followed by a chain transfer step and effectively produces addition of a thiol across a carbon-carbon double bond. The photocrosslinking of two component systems, such as thiols/acrylates, have rapidly expanded starting from the 1970s until now [231].

The reactions between a multifunctional monomer and multifunctional thiol are part of an attractive and versatile method that can be used for the preparation of adhesives, sealants, and coatings by using UV light.

When multifunctional monomers with an average functionality greater than two are utilized, highly crosslinked polymer networks are formed via a step-growth mechanism. A thermoplastic elastomer, polystyrene-*block*-polybutadiene-*block*-polystyrene (SBS) with a high vinyl content (59%), was crosslinked within a fraction of a second by UV-irradiation in the presence of a trifunctional thiol and an acylphosphine oxide photoinitiator. The formation of the polymer network in the

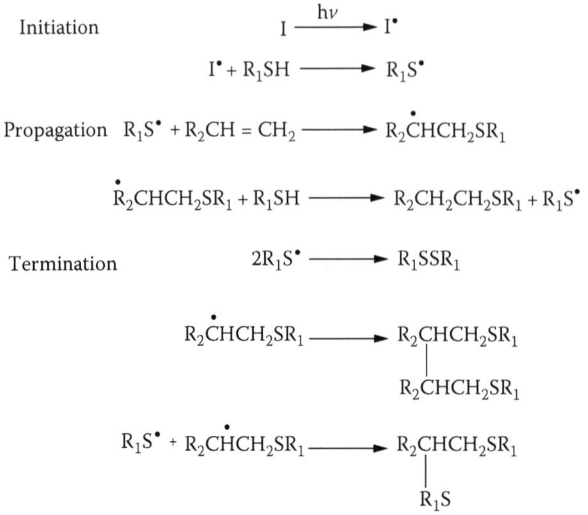

$$\text{Initiation} \qquad I \xrightarrow{h\nu} I^{\bullet}$$

$$I^{\bullet} + R_1SH \longrightarrow R_1S^{\bullet}$$

$$\text{Propagation} \quad R_1S^{\bullet} + R_2CH = CH_2 \longrightarrow R_2\dot{C}HCH_2SR_1$$

$$R_2\dot{C}HCH_2SR_1 + R_1SH \longrightarrow R_2CH_2CH_2SR_1 + R_1S^{\bullet}$$

$$\text{Termination} \qquad 2R_1S^{\bullet} \longrightarrow R_1SSR_1$$

$$R_2\dot{C}HCH_2SR_1 \longrightarrow \begin{array}{c} R_2CHCH_2SR_1 \\ | \\ R_2CHCH_2SR_1 \end{array}$$

$$R_1S^{\bullet} + R_2\dot{C}HCH_2SR_1 \longrightarrow \begin{array}{c} R_2CHCH_2SR_1 \\ | \\ R_1S \end{array}$$

SCHEME 8.2 General thiol-ene photopolymerization process.

elastomeric phase upon irradiation was followed through the vinyl group disappearance ($\nu = 1826\ cm^{-1}$) [232–234].

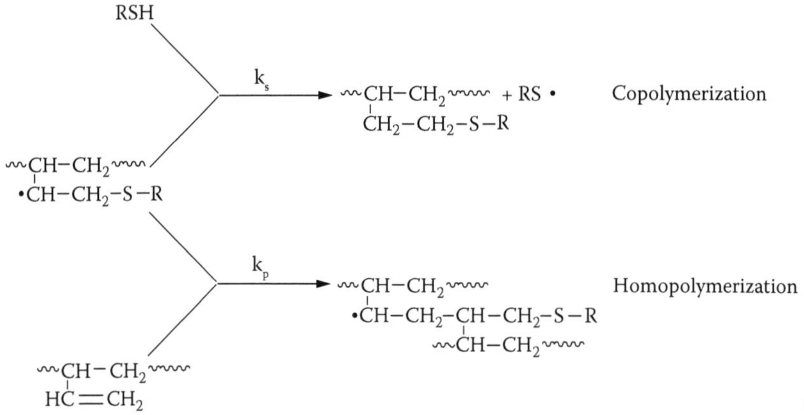

$$\begin{array}{l} RSH + \sim CH(\dot{C}H-CH_2-S-R)-CH_2\sim \xrightarrow{k_s} \sim CH(CH_2-CH_2-S-R)-CH_2\sim + RS\bullet \quad \text{Copolymerization} \\ \sim CH(\dot{C}H-CH_2-S-R)-CH_2\sim + \sim CH(HC{=}CH_2)-CH_2\sim \xrightarrow{k_p} \sim CH(\dot{C}H-CH_2-CH(\sim CH-CH_2\sim)-CH_2-S-R)-CH_2\sim \quad \text{Homopolymerization} \end{array} \tag{8.57}$$

Benzophenone and its derivatives have typically been employed as photoinitiators for thiol-ene systems [230]. Cleavage type photoinitiators such as 2,2-dimethoxy-2-phenylacetophenone initiate thiol-ene polymerization [235]. Another important aspect of thiol-ene polymerizations is that the reaction proceeds readily in the absence of an initiator. A thiol monomer (pentaerythritol tetrakis(3-mercaptopropionate) was shown to copolymerize with vinyl ether, allyl, acrylate, methacrylate, and vinylbenzene monomers. These thiol-ene polymerizations are photoinitiated without the use of photoinitiator molecules. The thiol-vinyl ether and thiol-allyl polymerizations were shown to be true stoichiometric step-growth reactions, whereas the thiol-acrylate, methacrylate, and vinyl benzene polymerizations proceeded via a combination of step growth and vinyl homopolymerization reactions. Although the polymerizations proceed with irradiated light centered around 365 nm, the reaction was significantly faster when a 254-nm light was utilized [236].

8.14 ORGANOMETALLIC PHOTOINITIATORS

Organometallic compounds have great potential as photoinitiators—many of them have satisfactory absorption characteristics and undergo photoinduced chemical reactions, which may be utilized for initiation of polymerizations. However, the use of many organometallic compounds is hampered by either their lack of thermal stability or their relatively high order of toxicity.

Titanocene initiators turned out to be the most attractive organometallic photoinitiators for visible light curing [237–240, 240a]. Fluorinated titanocenes (see Eq. (8.58)) are thermally stable (decomposition at c. 230°C) and do absorb strongly in the range between 400 and 500 nm. The mechanism by which they initiate vinyl polymerizations has been the subject of extensive investigations [241–244]. It is assumed that the initial titanocene species undergo a photoinduced isomerization yielding a coordinated unsaturated and therefore highly reactive isomer with a quantum yield of nearly 1. In the presence of acrylates, initiating biradicals and pentafluorophenyl-cyclopentenyl radicals are formed. The latter radical species are not prone to reactions with monomer but instead dimerize.

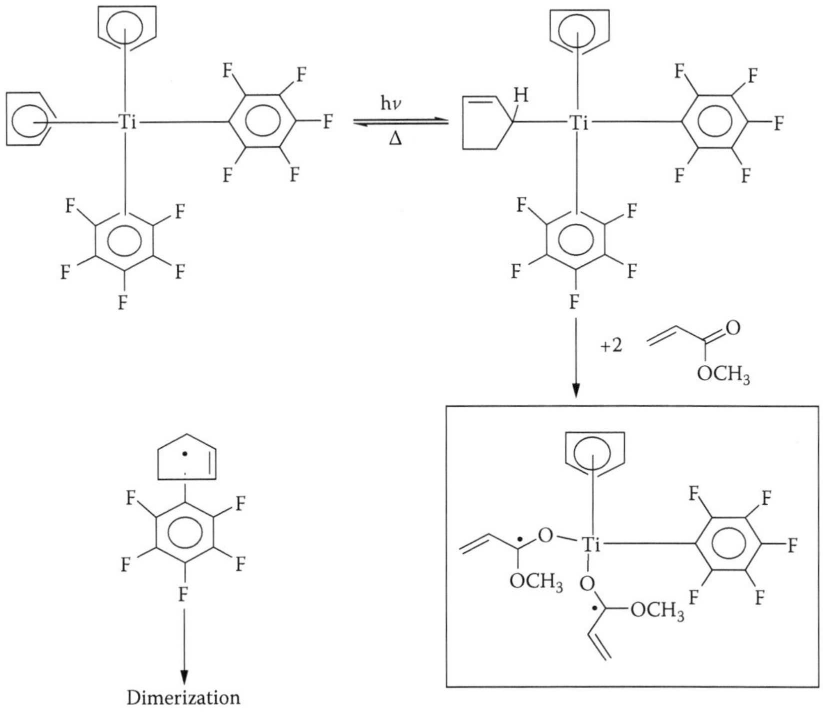

hν
Δ
H
Ti
F F F F F
+2
O
OCH_3
OCH_3
O
Dimerization

(8.58)

Apart from titanocene initiators, iron arene complexes have also been applied for light-induced radical polymerization, namely of acrylates [245–247]. Iron arene complexes were originally developed for cationic polymerization, because they release acid upon irradiation. However, light-induced reactions include the formation of alkyl radicals, which may be utilized for radical initiation.

8.15 MACROPHOTOINITIATORS

Macromolecular photoinitiators have attracted much attention in the past years, for their combined properties of polymers and those of low molecular weight photoinitiators [248–251]. Solubility or miscibility problems are often observed with coatings containing low molecular weight photoinitators. However, it is not the case with the macromolecular photoinitiators because polymers are easily miscible with the resin to be cured as well as with the final, cured film. Moreover, odor and toxicity problems do not occur with macroinitiators owing to the low volatility of large molecules. The low migration tendency of polymeric photoinitators and of photoproducts brings about a reduced proneness to yellowing of cured coatings. In many cases, macroinitiators were used to make tailor made block or graft copolymers [252]. Besides that, photoreactive polymers are of outstanding importance in photolithography.

8.15.1 Type I Macrophotoinitiators

As for low molecular weight photoinitiators, polymers that with a high yield undergo homolytic bond dissociation upon irradiation are referred to as Type I photoinitiators. In many cases of Type I macrophotoinitiators, scissile moieties like benzoin derivatives or *O*–acyl-α–oximino ketones are functional groups attached to a host polymer backbone in a polymer analog reaction. Another possibility is the attachment of photosensitive groups to monomers. In this case, macrophotoinitiators are formed by polymerization or copolymerization with another monomer. A few polymers, like polysilanes, are prone to light-induced main chain scission to an extent, which enables this reaction to be used for initiating vinyl polymerization. Polymers containing terminal photoactive benzoin groups have been synthesized using azobenzoin initiators [253, 254]. The thermal treatment of these initiators in the presence of styrene leads to benzoin groups at both ends of the polystyrene chain, because polystyryl radicals tend to terminate via recombination. Upon irradiation of the styrene based photoinitiators, benzoyl and alkoxy-benzyl radicals are produced, both capable of initiating polymerizations to give mixtures of homopolymers and block copolymers [255]. Obviously, one could also use the benzoin site of the azo benzoin initiator for a photopolymerization in the first step and activate the azo sites in a subsequent, thermal reaction [256]. However, regarding homopolymer formation the method depicted in Eq. (8.59) has to be given priority because benzoin groups are thermally stable whereas azo groups may be photolytically ruptured. Benzoin azo initiators are, furthermore, suitable for the synthesis of block copolymers with one of two monomers being cationically polymerizable [257,258].

$$\left[C_6H_5-\overset{O}{\overset{\|}{C}}-\overset{OCH_3}{\underset{C_6H_5}{C}}-CH_2-O-\overset{O}{\overset{\|}{C}}-(CH_2)_2-\overset{CH_3}{\underset{CN}{C}}-N= \right]_2 \xrightarrow[\Delta]{N_2} 2\, C_6H_5-\overset{O}{\overset{\|}{C}}-\overset{OCH_3}{\underset{C_6H_5}{C}}-CH_2-O-\overset{O}{\overset{\|}{C}}-(CH_2)_2-\overset{CH_3}{\underset{CN}{C}}\bullet$$

$$\xrightarrow{2\ CH_2=CH-C_6H_5} C_6H_5-\overset{O}{\overset{\|}{C}}-\overset{OCH_3}{\underset{C_6H_5}{C}}-CH_2-O-\overset{O}{\overset{\|}{C}}-(CH_2)_2-\overset{CH_3}{\underset{CN}{C}}-(CH_2-\underset{C_6H_5}{CH})_n-(CH_2)_2-\overset{O}{\overset{\|}{C}}-O-CH_2-\overset{H_3CO}{\underset{C_6H_5}{C}}-\overset{O}{\overset{\|}{C}}-C_6H_5$$

$$\xrightarrow{h\nu} 2\, \boxed{C_6H_5-\overset{O}{\overset{\|}{C}}\bullet} + \boxed{\bullet\overset{OCH_3}{\underset{C_6H_5}{C}}-CH_2-O-\overset{O}{\overset{\|}{C}}-(CH_2)_2-\overset{CH_3}{\underset{CN}{C}}-(CH_2-\underset{C_6H_5}{CH})_n-(CH_2)_2-\overset{O}{\overset{\|}{C}}-O-CH_2-\overset{H_3CO}{\underset{C_6H_5}{C}}\bullet}$$

$$\text{(radical 1)} \xrightarrow{MMA} \text{HomoPMMA} \qquad \text{(radical 2)} \longrightarrow \text{PSt-}b\text{-PMMA} \tag{8.59}$$

Degirmenci and co-workers [259] had applied atom transfer radical polymerization (ATRP) technique to incorporate photoinitiator moieties at the end and at the middle of the polymer chain. Thus, polystyrenes with precise functionalities and narrow molecular weight distributions were prepared by using respective halide functional initiators in the ATRP system according to the following reactions:

$$\text{C}_6\text{H}_5\text{-C(=O)-C(CH}_3)_2\text{-O-C(=O)-CH(CH}_3)\text{-Br} \xrightarrow[\text{CuBr/bpy, 110°C}]{n\ \text{CH}_2\text{=CH-C}_6\text{H}_5} \text{C}_6\text{H}_5\text{-C(=O)-C(CH}_3)_2\text{-O-C(=O)-CH(CH}_3)\text{-(CH}_2\text{-CH(C}_6\text{H}_5))_n\text{-Br}$$

$$\text{BrHC(CH}_3)\text{-C(=O)-O-H}_2\text{C-H}_2\text{C-O-C}_6\text{H}_4\text{-C(=O)-C(CH}_3)_2\text{-O-C(=O)-CH(CH}_3)\text{-Br} \xrightarrow[\text{CuBr/bpy, 110°C}]{n\ \text{CH}_2\text{=CH-C}_6\text{H}_5}$$

$$\text{Br-(HC(C}_6\text{H}_5)\text{-H}_2\text{C)}_{n/2}\text{-HC(CH}_3)\text{-C(=O)-O-H}_2\text{C-H}_2\text{C-O-C}_6\text{H}_4\text{-C(=O)-C(CH}_3)_2\text{-O-C(=O)-CH(CH}_3)\text{-(CH}_2\text{-CH(C}_6\text{H}_5))_{n/2}\text{-Br}$$

(8.60)

Similar synthetic strategy was applied [260] to macrophotoinitiators of poly (ε-caprolactone) and involved the reaction of photoinitiators, namely benzoin (B), 2-hydroxy-2-methyl-1-phenyl propan-1-one (HMPP), and 2-hydroxy-1-[4-(2-hydroxyethoxy)phenyl]-2-methyl propan-1-one (HE-HMPP), with ε-caprolactone (ε-CL) in the presence of stannous octoate catalyst. In view of the reported role of hydroxyl groups as initiators of the ring-opening polymerization, this reaction produced polymers containing a photoiniator group on one end or on the middle of the chain, derived from a single or two terminal units of the photoinitiators, respectively (Eq. (8.61)).

$$\text{C}_6\text{H}_5\text{-C(=O)-C(CH}_3)_2\text{-OH} \ \text{or} \ \text{C}_6\text{H}_5\text{-C(=O)-CH(C}_6\text{H}_5)\text{-OH} \ \text{or} \ \text{HO-H}_2\text{C-H}_2\text{C-O-C}_6\text{H}_4\text{-C(=O)-C(CH}_3)_2\text{-OH}$$

$$\downarrow n\ \varepsilon\text{-CL},\ \text{Sn (Oct)}_2\text{, 110°C, Bulk}$$

$$\text{C}_6\text{H}_5\text{-C(=O)-C(CH}_3)_2\text{-O-(C(=O)-CH}_2\text{CH}_2\text{CH}_2\text{CH}_2\text{CH}_2\text{-O)}_n\text{-H} \ \text{or} \ \text{C}_6\text{H}_5\text{-C(=O)-CH(C}_6\text{H}_5)\text{-O-(C(=O)-CH}_2\text{CH}_2\text{CH}_2\text{CH}_2\text{CH}_2\text{-O)}_n\text{-H}$$

or

$$\text{H-(O-H}_2\text{CH}_2\text{CH}_2\text{CH}_2\text{CH}_2\text{C-C(=O))}_{n/2}\text{-O-H}_2\text{C-H}_2\text{C-O-C}_6\text{H}_4\text{-C(=O)-C(CH}_3)_2\text{-O-(C(=O)-CH}_2\text{CH}_2\text{CH}_2\text{CH}_2\text{CH}_2\text{-O)}_{n/2}\text{-H}$$

(8.61)

Such prepared narrowly distributed macrophotoiniators, both polystyrene and poly(ε-caprolactone) can be used in photopolymerization of vinyl monomers [261]. The type of macrophotoiniator influences the polymerization products. Although both homo and block copolymers are formed with end-chain functional photoinitiators, mid-chain functional photoiniators yield purely block copolymers. In addition, poly(ε-caprolactone) photoinitiators represent a class of potentially useful materials in biomedical applications due to their potential benefits of biocompatablity and non-toxicity of PCL backbone and the polymeric nature of the photoinitiator.

Benzoin methyl ethers have been incorporated into a polycarbonate chain. The synthesis was achieved by polycondensing bisphenol A with phosgene in the presence of 4,4′-dihydrobenzoin methylether [262]. Irradiation of the resulting polycondensate in the presence of methyl methacrylate resulted in block copolymer formation, as illustrated in Eq. (8.62). Apart from methyl methacrylate, other vinyl monomers, such as ethyl methacrylate and acrylonitrile have been polymerized by the macroinitiator [263].

hν

MMA

Block Copolymer

(8.62)

Macroinitiators containing groups of benzoin type in the side chain were synthesized by several authors [13,264–268,271]. Some examples are summarized in Table 8.14. Notably, a spacer between the polymer backbone and the benzoin moiety enhances the activity of the initiator in vinyl polymerization considerably [36]. This phenomenon is attributed to an easier accessibility of reactive benzoin sites (Table 8.15).

Several polymerizable monomers based on either acrylic acid or styrene containing photodissociable groups have been developed (Table 8.16).

These monomers may be homopolymerized or copolymerized with various vinyl monomers via radical polymerization technique. By choosing the appropriate comonomer one can design either hydrophobic or hydrophilic macroinitiators [271]. Macrophotoinitiators prepared from the two styrene based monomers depicted were found to possess an activity comparable to that of efficient low molecular weight photoinitiators.

Polymers that possess carbonyl groups in the side chains are able to generate lateral macroradicals upon UV irradiation. The polymers depicted in Table 8.17 have

TABLE 8.15
Side Chain Benzoin Type Macroinitiators

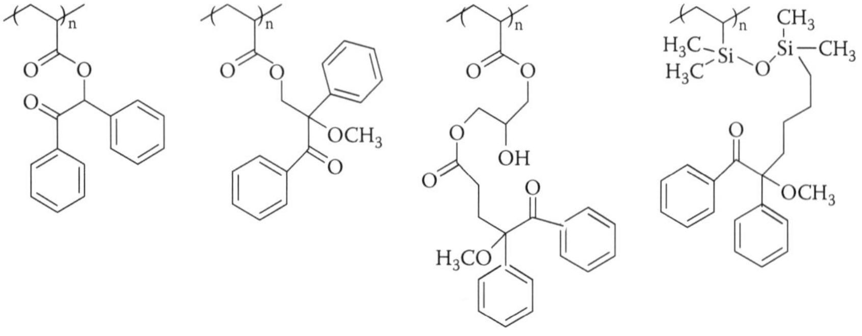

[264]	[13]	[265–267]	[268]

been used as photoinitiators. Photolysis of these initiators in the presence of vinyl monomers gives both graft copolymer and homopolymer, because radicals bound to the backbone and low molecular weight carbonyl radicals are formed.

Copolymers containing carbonyl groups in the main chain have been used in early works as photoinitiators. These copolymers undergo main chain scission upon irradiation. In the presence of vinyl monomer, the terminal radicals react with monomer initiating its polymerization. Following this method, various block copolymers (second comonomer in the block copolymer: methyl methacrylate) have been obtained (Table 8.18).

Polymer bound acyl oxime esters have also been reported to be suitable radical sources [262]. Polyesters containing acyl oxime oxides were prepared by a

TABLE 8.16
Monomers Containing Scissile Groups

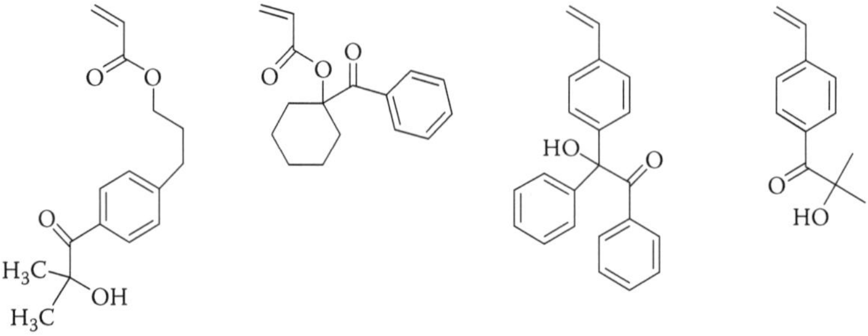

[269]	[125]	[270]	[270]

TABLE 8.17
Polymers with Side Chain Carbonyl Groups

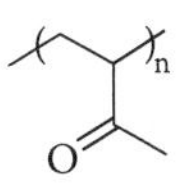 [272]	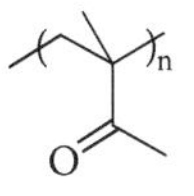 [273]	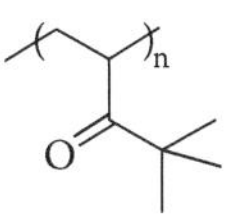 [273]

polycondensation reaction [278]. Irradiation of these compounds with light of 365 nm in the presence of vinyl monomers (styrene) gave block copolymers.

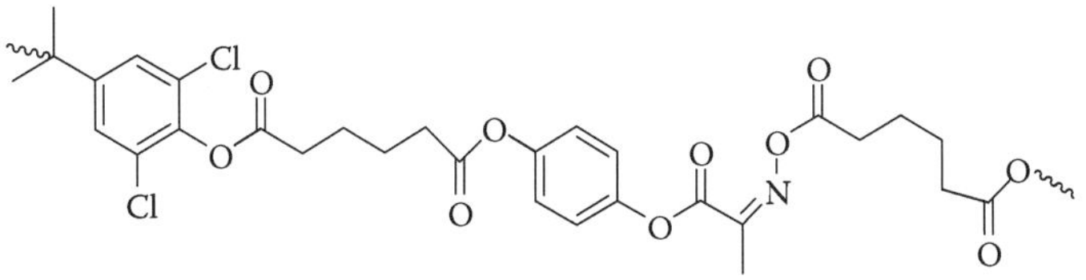

(8.63)

Polystyrene containing acyl oxime ester groups was synthesized using an azo-acyloxime ester bifunctional initiator (Eq. (8.63)) [279,280]. In the first stage, the azo-acyloxime ester initiator was heated in the presence of styrene, whereby styrene polymerization initiated by the azo sites. In a second step, the acyloxime ester sites were photochemically activated to start the polymerization of a second monomer. The striking advantage of acyloxime ester groups in macrophotoinitiators derives from the fact that upon photolysis apart from the macroradicals two low molecular weight products (CH_3CN and CO_2) are formed. These photolysis products prevent the macroradicals from recombining, which leads to relatively high initiation efficiencies.

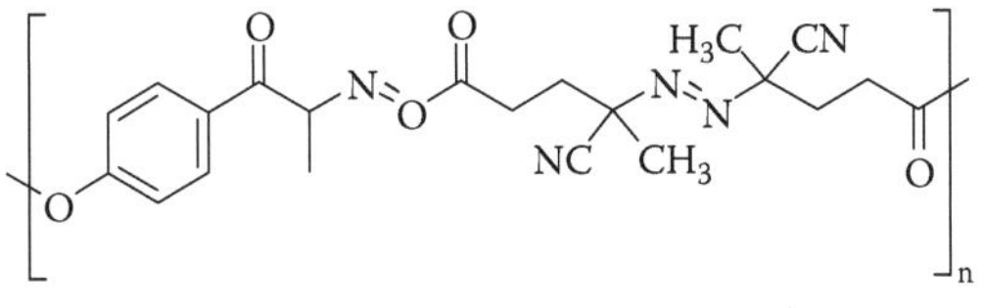

(8.64)

TABLE 8.18
Photoinitiators with In-Chain Carbonyl Groups

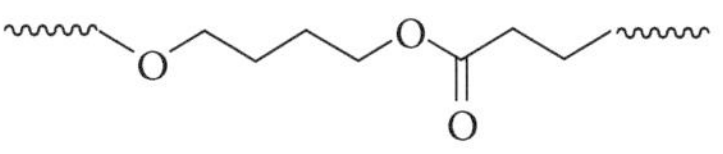 [274]	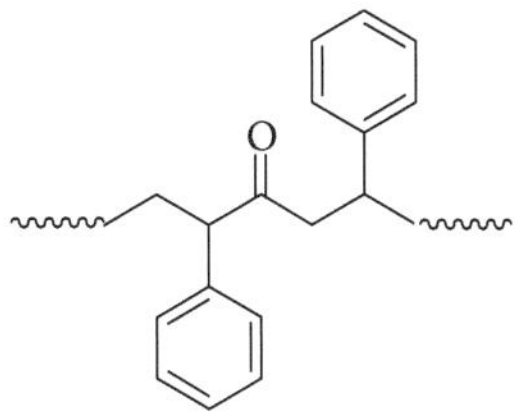 [275–277]

A similar effect was observed for polymeric initiators containing nitroso groups in their main chain [281]. Upon absorption of two photons, nitrogen is released and initiating macroradicals are formed, as depicted in Eq. (8.65). Notably, care must be taken while working with nitroso-containing macroinitiators because they may decompose violently under the influence of UV light or of heat [282].

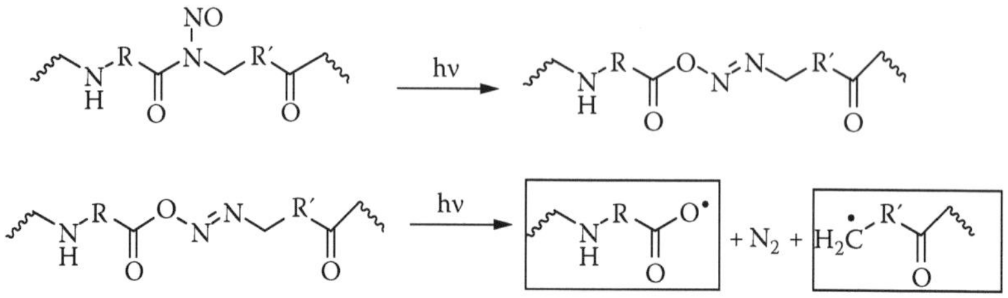

(8.65)

Photosensitive polymers containing azo groups [283,284] or triazene groups [285] in the main chain also undergo main chain scission and evolve nitrogen upon UV irradiation. With respect to photochemistry, alkylaryl initiators are of special interest, because they have better absorption characteristics than dialkyl initiators and are, on the other hand, more reactive than the respective diaryl substituted compounds.

Various organic disulfides including thioram disulfides are also capable of acting as thermal and photochemical initiators for free radical polymerization. For example, when the polymerization of styrene was initiated by tetraethylthioram disulfide, the polymer was found to contain terminal Et_2N-CS-S groups. Photolyzed in the presence of vinyl monomer these photosensitive polymers give block copolymers [286].

Peroxyesters with triplet photosensitizer functionalities are efficient initiators for free radical polymerization of monomer provided the monomers do not quench the polymer's excited triplet state. For example, methyl methacrylate was graft copolymerized photochemically (photografted) onto polystyrene containing a low fraction of peroxybenzoate Eq. (8.66) [95,138]. Upon irradiation in the presence of methyl methacrylate both homopolymer and graft copolymer were formed, as can be easily understood on the basis of Eq. (8.66).

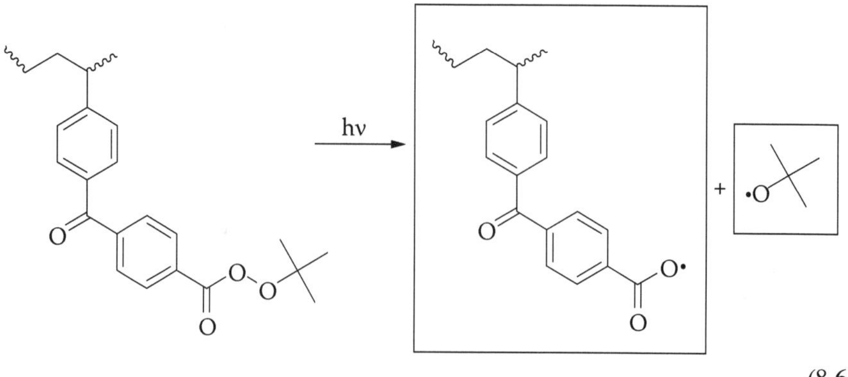

(8.66)

Polymers containing halogen atoms, especially brominated polymers, have also shown some potential in free radical polymerization. Because the bond energy of C-X is relatively low for X being chlorine or bromine, these bonds are easily photolytically ruptured. Thus, brominated polystyrene was used to initiate the polymerization of methyl methacrylate [287,288]. Furthermore, brominated polyacrylamide and polyacrylonitrile served as radical generating photoactive polymer for the grafting of acrylamide and ethyleneglycol oligomers [289, 290].

The curing of polysiloxane formulations by chemically attached groups of Type I initiating species is of great interest. Because the cured coatings possess interesting features, such as high stability toward heat and chemicals and good flexibility, they are attractive for a number of applications [291,293]. Prepared by various synthetic strategies, polysilanes containing benzoin ether [294], benzoin acrylate [295], benzilketals [296], α,α-dialkoxyacetophenones, α-hydroxyalkylphenones [296,297] and α,α-dialkylacethophenone [298] were used as macrophotoinitiators.

Due to their high photosensitivity, polysilanes are suitable sources of free radicals [299]. These silicon based polymers decompose upon absorption of UV light with quantum yields ranging between 0.2 and 0.97 [300] producing silyl radicals and silylene biradicals (Eqs. (8.67) and (8.68)). Polysilanes usually absorb light of wavelengths below 350 nm; changes of the organic substituents R_1 and R_2 as well as the number of silica atoms per chain however cause considerable alterations in the absorption characteristics [300–302]. Although radicals are formed with high yields, the initiation efficiency of polysilanes is not very high owing presumably to disproportionation reactions between the two kinds of radicals formed. However, polysilanes showed good performance in photocuring of vinyl functionalized polysiloxanes [303] and in the polymerization of several vinyl monomers [304,305]. In the latter study, upon irradiation of polysilane—vinyl copolymers in the presence of a further monomer block copolymers were synthesized [305].

$$\sim\!\!\sim Si(R_1)(R_2)-Si(R_1)(R_2)\sim\!\!\sim \xrightarrow{h\nu} \boxed{\sim\!\!\sim Si^{\bullet}(R_1)(R_2)} + \boxed{{}^{\bullet}Si(R_1)(R_2)\sim\!\!\sim} \tag{8.67}$$

$$\sim\!\!\sim Si(R_1)(R_2)-\left[Si(R_1)(R_2)\right]_n-Si(R_1)(R_2)\sim\!\!\sim \xrightarrow{h\nu} \sim\!\!\sim Si(R_1)(R_2)-\left[Si(R_1)(R_2)\right]_{n-1}-Si(R_1)(R_2)\sim\!\!\sim + \boxed{{}^{\bullet}Si(R_1)(R_2)^{\bullet}} \tag{8.68}$$

8.15.2 Type II Macrophotoinitiators

As explained earlier, the mechanism of Type II photoinitiators is based on the phenomenon that triplet states of aromatic ketones readily abstract hydrogen atoms from hydrogen donors, such as tertiary amines. Regarding macrophotoinitiators, examples with the ketone and the amine bound to polymer molecules can be cited.

Polystyrene terminated with amine functions (1 in Table 8.19) has been synthesized using an azo initiator containing terminal amine groups [306]. Heating of this

TABLE 8.19
Type II Macrophotoinitiators

1	2	3
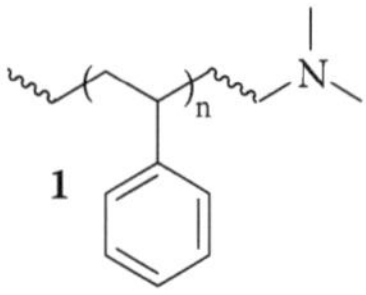 [306]	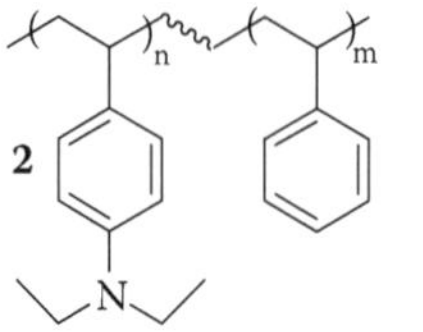 [308]	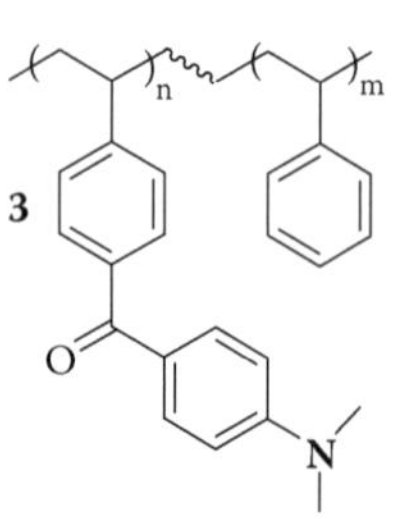 [307, 308]
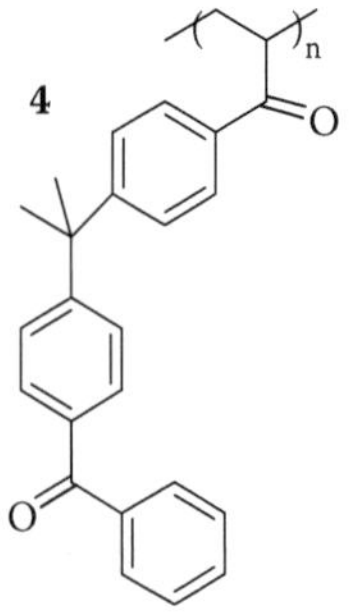 4	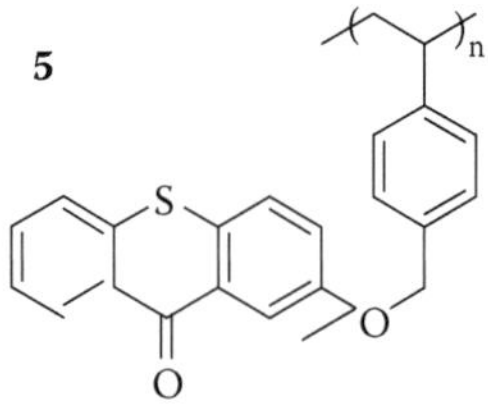 5	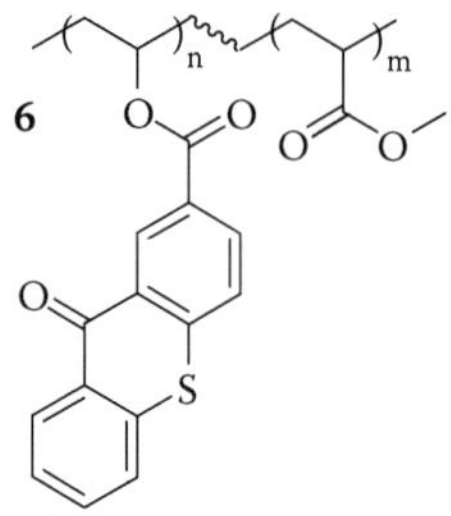 6
[309]	[310]	[311, 312]

initiator in the presence of styrene yields the desired initiator, because the amine groups are left unaltered by the thermal treatment. In a second reaction step, methyl methacrylate was block copolymerized by making use of the amine sites in a Type II photoinitiating process (Eqs. (8.69–8.71)).

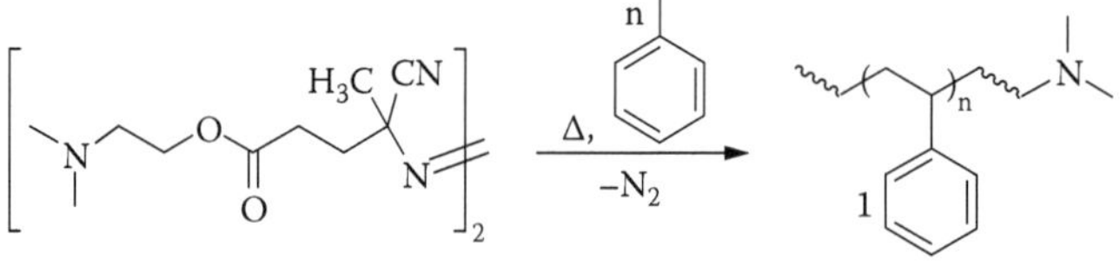

H_3C CN, N, O, $]_2$ —(n styrene; Δ, $-N_2$)→ 1 (8.69)

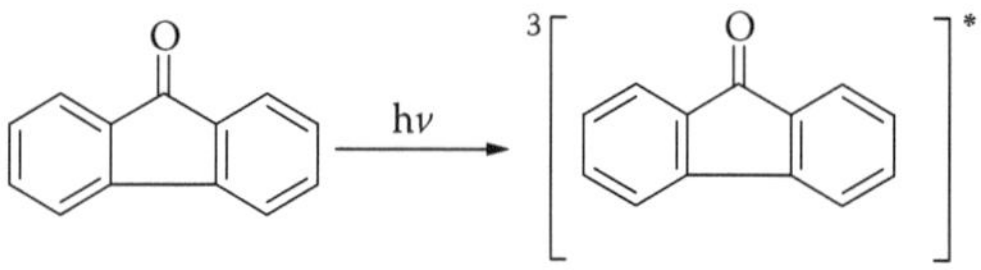

O —(hν)→ $^3[$ O $]^*$ (8.70)

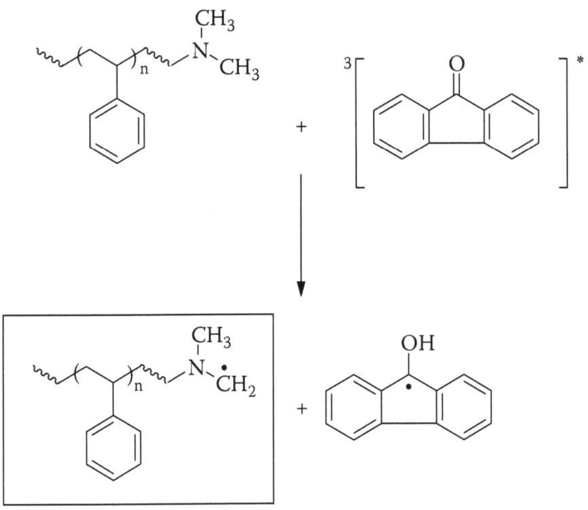

(8.71)

The macroinitiator 2 (see Table 8.19) has been obtained by copolymerization of styrene and N-N′-diethylamino styrene [307]. The initiator was used for the photopolymerization ($\lambda = 350$ nm) of 2-ethoxyethyl methacrylate with benzoin serving as coinitiator. This procedure resulted in both homopolymer (poly(2-ethoxyethyl methacrylate, initiated most probably by ketyl radicals) and the respective graft copolymer. The graft efficiency was pretty high: about 77% of the monomer converted was grafted onto the backbone polymer demonstrating the low initiation efficiency of ketyl radicals.

Various examples of attaching benzophenones to polymers can be cited [307–309, 313,314]. Containing both benzophenone and alkylarylcarbonyl type chromophoric groups, the macroinitiators 4 (Table 8.18) can be regarded as photoinitiators of Type I and Type II at the same time. They undergo main chain scission upon photolysis thus producing terminal radicals [309]. The mechanism, although not entirely elucidated, presumably involves the absorption of incident light by the benzophenone residue and a Norrish Type II reaction leading to chain scission (Type I initiator). Furthermore, a reaction of excited triplet benzophenone sites with the main chain (hydrogen abstraction) has been evidenced (Type II initiator).

The polymer bound thixanthone derivatives 5 and 6 (Table 8.18) have been used in conjunction with amines for free radical vinyl polymerization. Their initiation efficiency is similar to that of low molecular weight analogs.

8.15.3 Macrophotoinitiators with Halogen-Containing Groups

When polymers with functional groups like $-CX_3$, $-CHX_2$, or $-CH_2X$ are irradiated ($\lambda = 350\ldots450$ nm) in the presence of $Mn_2(CO)_{10}$, terminal carbon centered macroradicals are formed [315]. Upon absorption of light, $Mn_2(CO)_{10}$ decomposes yielding $Mn(CO)_5$, a compound that reacts with terminal halogen groups according to Eq. (8.74).

$$Mn_2(CO)_{10} \xrightarrow{h\nu} 2\ Mn(CO)_5 \qquad (8.72)$$

$$-(A)_n-CCl_3 + M_n(CO)_5 \longrightarrow \boxed{-(A)_n-\dot{C}Cl_2} + M_n(CO)_5Cl \qquad (8.73)$$

These terminal radicals may be used to initiate the polymerization of a second monomer, whereby block copolymers are formed. Notably, homopolymerization is not observed in this process because no low molecular weight radicals are generated upon the reaction of $Mn(CO)_5$ with the macroinitiator.

Naturally, this polymerization procedure may only be applied if the polymer end functionalized with halogen containing groups is available. Methods are available for converting terminal –OH, $-NH_2$, and –COOH groups into $-CCl_3$ groups [316]. Furthermore, polymers containing terminal $-CCl_3$ functionalities may be obtained using a bifunctional azo initiator [317].

Besides for block copolymers, the reaction of $Mn_2(CO)_{10}$ with halogen groups attached to polymer backbones has been utilized for the synthesis of graft copolymers [318–323]. A drawback of this method is that in many cases combination of macroradicals is observed giving rise to crosslinking instead of grafting. Network formation versus grafting has been studied for a number of trunk polymer/monomer combinations.

8.16 TECHNIQUES

8.16.1 RT-FTIR (Real-Time Infrared Spectroscopy)

As stated in the Section 8.1 of this chapter, the light-induced polymerization of multifunctional monomers has found widespread applications in various industrial sectors, in particular for the surface protection of organic materials by UV curable coatings, the manufacture of optical lenses and video discs as well as the production of fast-hardening composites and adhesives. Various analytical methods have been employed to study the kinetics of these ultrafast reacting systems, such as DSC, IR spectroscopy, laser nephelometry, dilatometry, etc.

The RT-FTIR method [324–327] has proved to be extremely valuable for measuring the polymerization rates and quantum yields of reactions that develop in the millisecond time scale. The basic principle of this method of kinetic analysis consists in exposing the sample simultaneously to the UV beam, which monitors the resulting decrease in the absorbance of the reactive double bond. The IR spectrophotometer must be set in the absorbance mode and the detection wavelength fixed at a value where the monomer double bond exhibits a discrete and intense absorption for acrylic monomers ($CH = CH_2$ twisting). During UV exposure the intensity of the absorption at 810 cm^{-1} is continuously monitored and this reflects the fast decrease of this band. This will accurately reflect the extent of the polymerization process because the absorbance increment $(A_{810})_0 - (A_{810})_t$ is directly proportional to the number of acrylate functions that undergo polymerization.

$$\text{Degree of Conversion} = A(_{810})_0 - A(_{810})_t / A(_{810})_0 \times 100 \tag{8.74}$$

The rate of polymerization (R_p) can be determined at any time of the reaction from the slope of the RT-FTIR kinetic curve.

$$R_p = [M]_0 \frac{(A_{810})_{t1} - (A_{810})_{t2}}{(A_{810})_{t0} - (t_2 - t_1)}$$

$$t_1, t_2 = \text{UV exposure time} \tag{8.75}$$

$[M]_0$ = The concentration of acrylate double bonds before UV exposure

The quantum yield of polymerization (ϕ_p), which corresponds to the number of functional groups polymerized per photon absorbed, is calculated from the ratio of R_p to the absorbed light intensity.

$$Q_p = \frac{R_p x l}{10^3 \times (1 - e^{-2.3A}) \times I_0} \quad (8.76)$$

where
A = absorbance of sample
l = the film thickness

The kinetic chain length (Λ) of the polymerization can be evaluated when the initiation quantum yield (ϕ_i) is known

$$\Lambda = \phi_p/\phi_i \quad (8.77)$$

In the photopolymerization of a multifunctional monomer where it is difficult to evaluate the molecular weight of the crosslinked polymer formed, quantum yield measurements appear to be the best method of evaluating the kinetic chain length of the reaction.

The induction period is observed only for experiments performed in the presence of air. It shows interference of oxygen during the polymerization process.

The residual unsaturation content (τ) of photopolymers can be quite important especially for multifunctional monomer systems where polymerization stops at an early stage as a consequence of the network formation and the resulting drop in the segmental mobility of the growing polymer chain.

$$\tau = [(A_{810})_t/(A_{810})_0]\ M_0 \quad (8.78)$$

The value of τ has been found to be highly dependent on both the monomer functionality and the chemical structure of the functionalized oligomer chain.

8.16.1.1 Advantages and Limitations of RT-FTIR Spectroscopy

This technique offers several advantages: Cure reactions occurring in a fraction of a second can now be followed continuously from the very beginning of the process to its end, thus providing a precise analysis of the quasi-instant liquid-solid phase change. RT-FTIR spectroscopy permits the study of kinetics of cure reactions over a very broad range of light intensity, in particular those performed under intense illumination and in the presence of air (i.e., experimental conditions very similar to those found in most industrial UV curing operations).

RT-FTIR spectroscopy is the first technique to provide in real-time quantitative information about the important kinetic parameters: induction period, polymerization rate, photosensitivity, and residual unsaturation of tack-free and scratch-free coatings.

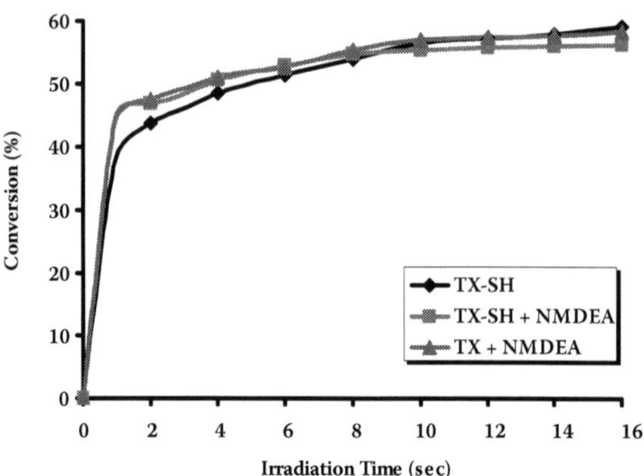

FIGURE 8.4 Real-time FTIR kinetic profiles demonstrating the photopolymerization of trimethylol propanetriacrylate containing different photoinitiators with polychromatic light. Photoinitiator (1%): (♦) TX-SH, (■) TX-SH and NMDEA, and (▲) TX and NMDEA. TX-SH: 2-mercapto-thioxanthone, TX: thioxanthone, NMDEA: N-methyldiethanolamine.

Because it is based on IR spectroscopy, the RT-FTIR technique proved to be both reliable and highly sensitive, leading to accurate and reproducible results. In addition, it offers some valuable advantages such as speed of analysis, ease of use, and low cost, for it requires only a routine IR spectrophotometer.

The cured coating can be examined at any moment of the reaction for evaluation of some of its physical characteristics, such as hardness, abrasion resistance, flexibility, solvent resistance, gloss, heat resistance, etc.

A wide range of applications can be examined by this novel technique of kinetic investigation, because essentially any fast polymerizing system, clear or pigmented, can be analyzed in real time. It can be extended to other types of radiation such as lasers, microwaves, and electron beams.

This technique has some disadvantages (e.g., vulnerable to phase/baseline changes, sample must be either viscous or enclosed, and the response speed is limited by mechanics and electronics of the spectrophotometer).

A typical example of the use of this technique is given in Figure 8.4, where the conversion of a multifunctional monomer, namely trimethylol propantriacrylate, was monitored during photopolymerization using a one-component hydrogen abstraction initiator.

8.16.1.2 Comparison with Other Analytical Methods

The performance of RTIR spectroscopy for real time analysis of UV curing reactions has been compared to the widely used differential photocalorimetry, IR radiometry, and laser interferometry (Table 8.20).

TABLE 8.20
Comparison of Monitoring UV Curing Reactions by Different Analytical Methods

Techniques	DSC	IR-R	LASER	RTIR	CURING
Response time(s)	>2	5×10^{-1}	1×10^{-3}	2×10^{-2}	—
Light intesity(mW/cm^2)	<4	<50	<1000	<500	500
Exposure time(s)	>10	>2	$>1 \times 10^{-2}$	$>1 \times 10^{-1}$	$>1 \times 10^{-1}$
Atmosphere	N_2	Air	Air	Air	Air
Sample temperature	<30°	<50°	<100°	<70°	<70°
Thickness control	Poor	Fair	Fair	Good	Fair
Rate evaluation	Yes	No	No	Yes	—
Unsaturation content	Poor	No	No	Yes	—
Properties measurement	No	Yes	No	Yes	Yes

8.16.2 Calorimetric Methods

Photopolymerization reactions that occur via chain reactions are generally exothermic because double bonds of monomers are converted into single bonds whenever the monomers add to polymer radicals. For this reason, calorimetric studies of photopolymerization have been performed using a variety of calorimeter designs. Many examples of the application of calorimetry to UV curing have employed commercial differential scanning calorimeters adapted to allow the sample to be irradiated with a UV lamp.

8.16.2.1 Differential Scanning Calorimetry

A differential scanning calorimetry (DSC) measures the rate of heat produced as well as the total amount of heat produced from a chemical reaction. In a polymerization reaction this can easily be related to the rate of conversion of monomer to polymer and to the total amount of monomer polymerized [328–332].

The DSC has been used to study polymerization in bulk, solution, and emulsion systems. Both isothermal and temperature scanning modes have been employed. Photo-DSC is a unique method to obtain fast and accurate indication on the photoinitiator performance.

The basic data obtained from calorimetric analysis of a photoinitiated polymerization are in the form of an exotherm trace as shown in Figure 8.5.

Here the rate of heat released by the sample (dH/dT) is plotted versus the elapsed time of reaction. Initial rates of polymerization (R_p) can be calculated from the maximum polymerization heat (t_{max}) of the plots of dH/dT (J/g) versus irradiation time (s) following Eq. (8.79).

$$R_p = dC/dT = dH/dT.dH_0 \tag{8.79}$$

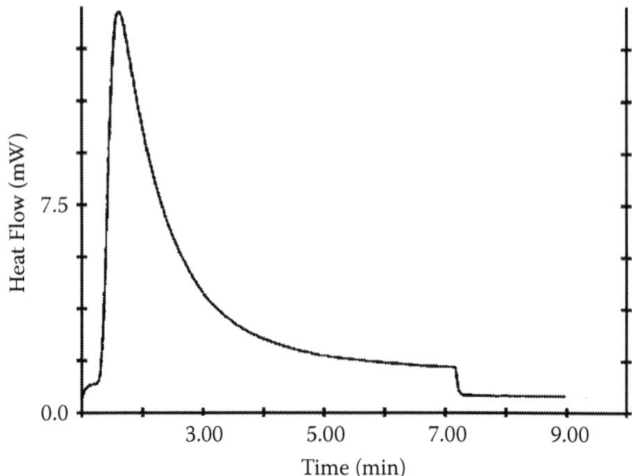

FIGURE 8.5 2-Methyl-1-[4-(methylthio)phenyl-2-morpholino]propane-1-one (Irg-907)/lauryl acrylate (LA) in air atmosphere.

The time to reach the maximum polymerization heat (t_{max}) and the peak height (h) give information about the photoinitiator activity. The area below the trace (ΔH_p) corresponds to the amount of double bonds converted (DBC).

$$DBC = \frac{\Delta H_p \times M_{LA}}{\Delta H_T}$$

where

ΔH_P = heat of polymerization (J/g)
M_{LA} = molecular weight of monomer (lauryl acrylate: 240.39 $g.mol^{-1}$)
ΔH_T = theoretical polymerization enthalpy of monomer (lauryl acrylate: 80,500 J mol^{-1}) (8.80)

The rate of polymerization R_p can also be calculated according to Eq. 8.81.

$$R_p = \frac{d \times h}{\Delta H_T} \quad (8.81)$$

where

R_p = rate of polymerization [$mol.L^{-1}.s^{-1}$]
h = height of the DSC signal [$mW.mg^{-1}$]
d = density of monomer (lauryl acrylate: 884 $g.L^{-1}$)

REFERENCES

1. Y. Yagci and I. Reetz, *Prog. Polym. Sci.*, 23, 1485 (1998).
2. J.V. Crivello, *Chemistry and Technology of UV/EB Formulations for Coating, Inks and Paints*, Vol. 3, K. Dietliker, Ed., SITA Technology Ltd., London (1991), p. 329.

3. W. Schnabel, *Macromolecular Engineering: Recent Advances*, M.K. Mishra, O. Nuyken, S. Kobayashi, Y. Yagci, and B. Sar, Eds., Plenum Press, New York (1995), p. 67.
4. J.-P. Fouassier, *Photoinitiation, Photopolymerization and Photocuring. Fundamentals and Applications*, Hanser Publishers, Munich, Vienna, New York (1995).
5. J. Blyth and A.W. Hoffmann, *Ann.*, 53, 292 (1845).
6. C. Walling, *Free Radicals in Solution*, Wiley, New York, 273 (1957).
7. N.J. Turro, *Modern Molecular Photochemistry*, The Benjamin Cummings Publishing Co., Inc., Menlo Park, CA, 368 (1978).
8. S.L. Murov, *Handbook of Photochemistry*, Marcel Dekker, Inc., New York (1973).
9. D. Rehm and A. Weller, *Isr. J. Chem.*, 8, 259 (1970).
10. L.H. Carlblom and S.P. Pappas, *J. Polym. Sci.*, 15, 1381 (1977).
11. R. Kuhlmann and W. Schnabel, *Polymer*, 18, 1163 (1977).
12. H.-G. Heine, *Tetrahedron Lett.*, 3411 (1972).
13. H. Rudolph, H.J. Rosenkranz, and H.-J. Heine, *Appl. Polym. Symp.*, 26, 157 (1975).
14. F.D. Lewis, R.T. Lauterbach, H.-G. Heine, W. Hartmann, and H.J. Rudolph, *J. Am. Chem. Soc.*, 97, 1519 (1975).
15. J. Hutchison and A. Ledwith, *Polymer*, 14, 405 (1973).
16. U. Müller and C. Vallejos, *Angew. Makromol. Chem.*, 206, 171 (1993).
17. H.-G. Heine and H.-J. Traencker, *Progr. Org. Coatings*, 3, 115 (1975).
18. C.J. Groeneboom, H.J. Hageman, T. Overeem, and A.J.M. Weber, *Makromol. Chem.*, 183, 281 (1982).
19. V. Desobry, K. Dietliker, R. Huesler, W. Rusch, and H. Loelinger, *Polym. Paint. Colour J.*, 178, 913 (1988).
20. H. Fischer, R. Baer, R. Hany, I. Verhoolen, and M. Walbiner, *J. Chem. Soc., Perkin II*, 787 (1990).
21. H. Baumann and H.-J. Timpe, *Zeitschr. Chem.*, 23, 197 (1983).
22. S.P. Pappas, *Encyclopedia of Polymer Science and Technology*, Vol. 11, John Wiley, New York, 186 (1988).
23. H. Baumann, U. Müller, D. Pfeiffer, and H.-J. Timpe, *J. Prakt. Chem.*, 324, 217 (1982).
24. H.J. Hagemann and T. Overeem, *Makromol. Chem. Rapid Commun.*, 2, 719 (1981).
25. S.P. Pappas and A.K. Chattopadhyay, *Polym. Lett. Ed.*, 13, 483 (1975).
26. S.P. Pappas and R.A. Asmus, *J. Polym. Sci., Polym. Chem. Ed.*, 20, 2643 (1982).
27. R. Kuhlmann and W. Schnabel, *Angew. Makromol. Chem.*, 70, 145 (1978).
28. G. Berner, R. Kirchmayr, and G. Rist, *J. Oil Chem. Assoc.*, 61, 105 (1978).
29. G. Berner, G. Rist, W. Rutsch, and R. Kirchmayr, *Radcure Basel*, Technical Paper FC85-446, SME Ed., Dearborn, MI (1985).
30. G. Li Bassi, L. Cadona, and F. Broggi, *Radcure 86*, Technical Paper 4-27, SME Ed., Dearborn, MI (1986).
31. C.E. Hoyle, M. Cranford, M. Trapp, Y.G. No, and K. Kim, *Polymer*, 29, 2033 (1988).
32. J.P. Fouassier and A. Merlin, *Makromol. Chem.*, 181, 1307 (1980).
33. Y. Yagci and A. Ledwith, *J. Polym. Sci., Part A: Polym. Chem.*, 26, 1911 (1988).
34. Y. Yagci and W. Schnabel, *Makromol. Chem., Macromol. Symp.*, 13/14, 161 (1988).
35. W.R. Adams, German Patent Application 2357866 (Prior 6.12.72) to Sun Chemical Corp.
36. K.-D. Ahn, K.J. Ihn, and I.C. Known, *J. Macromol. Sci. Chem.*, A23, 355 (1986).
37. H.-G. Heine, H. Rudolph, and H.J. Kreuder, German Patent Application 1919678 (1970).
38. H.J. Hagemann and L.G. Jansen, *J. Makromol. Chem.*, 189, 2781 (1988).
39. H. Angad Gaur, C.J. Groeneboom, H.J. Hagemann, G.T.M. Hakvoort, P. Oosterhoff, T. Overeem, R.J. Polman, and S. Van der Werf, *Makromol. Chem.*, 185, 1795 (1984).

39a. H. Baumann, H.-J. Timpe, V.E. Zubarev, N.V. Fok, M.Y. Mel'nikov, and M.Y. Raskazovskii, *Dokl. Akad. Nauk SSSR*, 284, 367 (1985).
40. R. Kirchmayr, G. Berner, and G. Rist, *Farbe u. Lack*, 86, 224 (1980).
41. X.T.P. Phan, *J. Radiat. Curing*, 13, 18 (1986).
42. M.R. Sander and C.L. Osborn, *Tetrahedron Lett.*, 5, 415 (1974).
43. P. Jeagermann, F. Lendzian, G. Rist, and K. Möbius, *Chem. Phys. Lett.*, 140, 615 (1987).
44. J. Huang, D. Zhu, and Y. Feng, *Gaofenzi Cailiao Kexue Yu Gongcheng*, 10, 36 (1994); C.A. 122, 134080x (1995).
45. A. Borer, R. Kirchmayr, and G. Rist, *Helv. Chim. Acta*, 61, 305 (1978).
46. F. Jent, H. Paul, and H. Fischer, *Chem. Phys. Lett.*, 146, 315 (1988).
47. H. Kurrek, B. Kriste, and W. Lubitz, *Angew. Chem., Int. Ed.*, 23, 173 (1987).
48. F. Lendzian, P. Jaegermann, and K. Möbius, *Chem. Phys. Lett.*, 120, 195 (1985).
48a. P. Jaegermann, F. Lendzian, G. Rist, and K. Möbius, *Chem. Phys. Lett.*, 140, 487 (1987).
49. J. Christensen, A.F. Jacobine, and C.J. Scanio, *J. Radiat. Curing*, 8, 12 (1981).
50. G. Berner, R. Kirchmayr, and G. Rist, *J. Radiat. Curing*, 6(2), 2 (1979).
51. C.L. Osborn, *J. Radiat. Curing*, 3, 2 (1976).
52. J.P. Fouassier and J.P. Lougnout, *J. Chem. Soc., Faraday Trans.* I, 83, 2953 (1987).
53. K. Dietliker, *Chemistry & Technology of UV&EB Formulations for Coating, Inks & Paints*, Vol. 3, K. Dietliker, Ed., SITA Technology Ltd., London, (1991) p. 145.
54. L. Misev, V. Desobry, K. Dietliker, R. Husler, M. Rembold, G. Rist, and W. Rutsch, *Proc. Conf. on Radtech*, Florence, Radtech, Fribourg, (1989) p. 359.
55. D. Leopold and H. Fischer, *J. Chem. Soc., Perkin Trans. II*, 513 (1992).
55a. N. Arsu and R.S. Davidson, *J. Photochem. Photobiol. A: Chem.*, 84, 291 (1994).
56. J.P. Fouassier, D.J. Lougnot, A. Paverne, and F. Wieder, *Chem. Phys. Lett.*, 135, 30 (1987).
57. G. Rist, A. Borer, K. Dietliker, V. Desobry, J.P. Fouassier, and D. Ruhlmann, *Macromolecules*, 25, 4182 (1992).
58. N. Arsu, R. Bowser, R.S. Davidson, N. Khan, and P.M. Moran, in *Photochemistry and Photopolymerisation*, J. M. Kelly, C.B. McArdle, M.J. De, F. Maunder, Eds., Royal Society of Chemistry, London (1993) p.15.
59. H. Tomioka, M. Harada, and I. Sumiyoshi, in *RadTech Asia* '93, *UV/EB Conf. Expo., Conf. Proc.*, 250 (1993).
60. G.A. Delzenne, U. Laridon, and H. Peeters, *Eur. Polym. J.*, 6, 933 (1970).
61. G. Berner, J. Puglisi, R. Kirchmayr, and G. Rist, *J. Radiat. Curing*, 6(2), 2 (1979).
62. V.D. McGinnis, *J. Radiat. Curing*, 2(1), 3 (1975).
63. T. Majima and W. Schnabel, *Bull. Soc. Chim. Belg.*, 99, 911 (1990).
64. W. Schnabel, *J. Radiat. Curing*, 13, 26 (1986).
65. T. Sumiyoshi, W. Schnabel, and A. Henne, *J. Photochem.*, 32, 119 (1986).
66. T. Sumiyoshi and W. Schnabel, *Makromol. Chem.*, 186, 1811 (1986).
67. T. Sumiyoshi, W. Schnabel, and A. Henne, *J. Photochem.*, 32, 191 (1986).
68. P. Lechtken, I. Buethe, and A. Hesse, DOS 2830927 (1980); assigned to BASF AG, C.A. 93, 46823u (1980).
69. H. Heine, H. Rosenkranz, and H. Rudolph, DOS 3023486 (1980); assigned to BASF AG.
70. A. Hesse, P. Lechtken, W. Nicolaus, and D. Scholz, DOS 2909993 (1980); assigned to BASF AG.
71. M. Jakobi and A. Henne, *J. Radiat. Curing*, 19 (4), 16 (1983).
72. M. Jacobi and A. Henne, *Radcure 83*, Technical Paper Nr. FC83-256, Lausanne (1983).
73. M. Jacobi, A. Henne, and A. Böttcher, *Polym. Paint Colour J.*, 175, 636 (1986).
73a. W. Rutsch, K. Dietliker, D. Leppard, M. Koehler, L. Misev, U. Kolczak, and G. Rist, *Proc. 20th Int. Conf. Org. Coat. Sci. Technol.*, 467 (1994).

74. J. Ohngemach, K.H. Neisius, J. Eichler, and C.P. Herz, *Kontakte*, 3, 15 (1980).
75. J.P. Baxter, R.S. Davidson, H.J. Hagemann, G.T.M. Havkoort, and T. Overeem, *Polymer*, 29, 1575 (1988).
76. C.P. Herz, J. Eichler, *Farbe u. Lack*, 85, 983 (1979).
77. C.L. Lee and M.A. Lutz, U.S. Patent 4,780,486 (Prior 15.5.86) to Dow Corning Corp.
78. L.J. Cottington and A. Revis, U.S. Patent 4,973,612 (Prior 30.11.89) to Dow Corning Corp.
79. M. Takamizawa, F. Okada, Y. Hara, and H. Aoki, U.S. Patent 4,273,907 (Prior 19.7.79) to Shin-Etsu Chemical Co. Ltd.
80. B. Bouvy, J. Cavezzan, and J.-M. Frances, Eur. Patent Application, 402274 (Prior 2.6.89) to Rhone-Poulenc Chimi.
81. P.A. Manis and E.R. Martin, U.S. Patent 46,069,933 (Prior 25.11.85) to SWS Silicon Corp.
82. J. Eichler, C.P. Herz, I. Naito, and W. Schnabel, *J. Photochem.*, 12, 225 (1980).
83. H. Baumann, H.-J. Timpe, V.E. Zubarev, N.V. Fok, and M.Y. Mel'nikov, *J. Photochem.*, 30, 487 (1985).
84. K. Dietliker, M. Rembold, G. Rist, W. Rutsch, and F. Sitek, *Radcure Europe 87, Conf. Proc. 3rd; 3/37*, Assoc. Finish Process SME, Dearborn, MI (1987).
85. J.-P. Fouassier, D. Burr, and F. Wieder, *J. Polym. Sci., Part A: Polym. Chem.*, 29, 1319 (1991).
86. P. Bosch, F. del Monte, J.L. Mateo, and R.S. Davidson, *J. Photochem. Photobiol., A: Chem.*, 78, 79 (1994).
87. M.K. Mishra, *J. Macromol. Sci.— Rev. Makromol. Chem.*, C20, 149 (1981).
88. M.K. Mishra, *J. Macromol. Sci.— Rev. Macromol. Chem.*, C22, 471 (1983).
89. H. Miyama, N. Harumiya, and A. Takeda, *J. Polym. Sci., Polym. Chem. Ed.*, 16, 943 (1972).
90. F.S. Daiton, *J. Phys. Colloid. Chem.*, 52, 490 (1948).
91. M. Yoshida and M. Taniyama, *Kobunshi Kagaku*, 19, 627 (1962).
92. J.E. Leffler and J.W. Miley, *J. Am. Chem. Soc.*, 93, 7005 (1971).
93. S.N. Gupta, I. Gupta, and D.C. Neckers, *J. Polym. Sci., Polym. Chem. Ed.*, 19, 103 (1981).
94. S.N. Gupta, L. Thijs, and D.C. Neckers, *J. Polym. Sci., Polym. Chem. Ed.*, 19, 855 (1981).
95. I. Gupta, S.N. Gupta, and D.C. Neckers, *J. Polym. Sci., Polym. Chem. Ed.*, 20, 147 (1982).
96. I.I. Abu-Abdoun, L. Thijs, and D.C. Neckers, *J. Polym. Sci., Polym. Chem. Ed.*, 21, 3129 (1983).
97. D.C. Neckers, *J. Radiat. Curing*, 10(2), 19 (1982).
98. J.R. Dacey and D.M. Young, *J. Chem. Phys.*, 23, 1302 (1955).
99. G.O. Pritchard, H.O. Pritchard, and A.F. Trotman-Dickenson, *Chem. Ind. (London)*, 564 (1955).
100. G.O. Pritchard, H.O. Pritchard, H.I. Schiff, and A.F. Trotmann-Dickenson, *Trans. Faraday Soc.*, 52, 849 (1956).
101. C. Carlini, *Brit. Polym. J.*, 18, 236 (1986).
102. M.S. Matheson, E.E. Auer, E.B. Bevilacqua, and E.J. Hart, *J. Am. Chem. Soc.*, 71, 2610 (1949).
103. H. Miyama, *Bull. Chem. Soc. Jpn.*, 74, 2027 (1952).
104. A. Dannil, W.W. Graessley, and J.S. Dranoff, *Chem. React. Proc. 5th Eur. Symp.*, Elsevier, Amsterdam, p. 49 (1972).
105. M.S. Matheson, E.E. Auer, E.B. Bevilacqua, and E.J. Hart, *J. Am. Chem. Soc.*, 73, 1700 (1951).
106. G.M. Burmett and W.W. Wright, *Proc. R. Soc. (London)*, 225, 37 (1954).

107. H. Miyama, N. Harumiya, and A. Takeda, *J. Polym Sci., Polym. Chem. Ed.*, 16, 943 (1972).
108. W.I. Bengough, *Nature*, 180, 1120 (1957).
109. J.A. Hicks and H.W. Melville, *Nature*, 171, 300 (1953).
110. H. Miyama, *Bull. Chem. Soc. Jpn.*, 29, 720 (1956).
111. H. Miyama, N. Harumiya, and A. Takeda, *J. Polym. Sci., Polym. Chem. Ed.*, 10, 1543 (1977).
112. S.S. Labana, *J. Macromol. Sci.— Rev. Macromol. Chem.*, C11(2), 299 (1974).
113. P. Ghosh and S. Chakraborty, *J. Polym. Sci., Polym. Chem. Ed.*, 13, 1531 (1975).
114. S.R. Rfifkov, V.A. Sechkovskaya, and G.P. Gladyshev, *Vysokomol. Soedin.*, 3, 1137 (1961).
115. P. Ghosh, P.S. Mitra, and A.N. Banerjjee, *J. Polym. Sci., Polym. Chem. Ed.*, 11, 2021 (1973).
116. N. Sakota, T. Tanigaki, and K. Tabuchi, *Kogyo Kagaku Zasshi*, 72, 975 (1969).
117. P. Ghosh and A.N. Banerjee, *J. Polym. Sci., Polym. Chem. Ed.*, 12, 375 (1974).
118. F. Kasapoglu, M. Aydın, N. Arsu, and Y. Yagcı., *J. Photochem. Photobiol. A: Chem.*, 159, 151 (2003).
119. H. Tachi, T. Yamamoto, M. Shirai, and M. Tsunooka, *J. Polym. Sci., Polym. Chem. Ed.,* 39, 1329 (2001).
120. N.S. Allen, F. Catalina, P.N. Green, and W.A. Green, *Eur. Polym. J.*, 22, 49 (1986).
121. M.R. Sander, C.L. Osborn, and D.J. Trecker, *J. Polym. Sci., Polym. Chem. Ed.*, 10, 3173 (1972).
122. J. Hutchison, M.C. Lambert, and A. Ledwith, *Polymer*, 14, 250 (1973).
123. H. Baumann and H.-J. Timpe, *Acta Polymerica*, 37, 309 (1986).
124. H. Baumann, U. Oertel, and H.-J. Timpe, *Eur. Polym. J.*, 22, 313 (1986).
125. J.-P. Fouassier, D. Ruhlmann, Y. Takimoto, M. Harada, and M. Kawabata, *J. Polym. Sci., Part A: Polym. Chem.*, 31, 2245 (1993).
126. J.P. Fouassier, A. Erddalane, F. Morlet-Savary, I. Sumiyoshi, M. Harada, and M. Kawabata, *Macromolecules*, 27, 3349 (1994).
127. R. Kuhlmann and W. Schnabel, *Polymer*, 17, 419 (1976).
128. N.S. Allen, E. Lam, J.L. Kotecha, W.A. Green, A. Timms, S. Navaratnam, and B.J. Parsons, *J. Photochem. Photobiol. A: Chemistry*, 54, 367 (1990).
129. F. Ciardelli, G. Ruggeri, M. Aglietto, D. Angiolini, C. Carlini, G. Bianchi, G. Siccardi, G. Bigogno, and L.J. Cioni, *J. Coat. Technol.*, 61, 77 (1989).
130. L. Mateo, P. Bosch, F. Catalina, and R. Sastre, *J. Polym. Sci., Part A: Polym. Chem.*, 29, 1955 (1991).
131. N.S. Allen, E. Lam, E.M. Howells, P.N. Green, A. Green, F. Catalina, and C. Peinado, *Eur. Polym. J.*, 26, 1345 (1990).
132. A. Costela, J. Dabrio, J.M. Figuera, I. Garcia-Moreno, H. Gsponer, and R. Sastre, *J. Photochem. Photobiol. A: Chem.*, 92, 213 (1995).
133. N.S. Allen, F. Catalina, J.L. Mateo, R. Sastre, W. Chen, P.N. Green, and W.A. Green, in *Radiation Curing of Polymeric Materials*, C.E. Hoyle, J.F. Kinstle, Eds., American Chemical Society, Washington D.C. (1990) p. 72.
134. J.P. Fouassier, D.J. Lougnot, and I. Zuchowicz, *Eur. Polym. J.*, 22, 933 (1986).
135. J.P. Fouassier and D.J. Lougnot, *J. Appl. Polym. Sci.*, 32, 6209 (1986).
136. C. Decker and K. Moussa, *J. Polym. Sci., Part C: Polym. Lett.*, 27, 347 (1989).
137. J.P. Fouassier, D.L. Lougnot, I. Zuchowicz, P.N. Green, H.-J. Timpe, K.P. Kronfield, and U.J. Mueller, *J. Photochem.*, 36, 347 (1987).
138. D.C. Neckers, *J. Radiat. Curing*, 10, 19 (1983).
139. C. Decker and K. Moussa. *Polym. Mater. Sci. Eng.*, 60, 547 (1989).
140. T.H. Koch and A.H. Jones, *J. Am. Chem. Soc.*, 92, 7503 (1970).

141. C.C. Wamser, G.S. Hammond, C.T. Chang, and C. Baylor Jr., *J. Am. Chem. Soc.*, 92, 6362 (1970).
142. F. Catalina, J.M. Tercero, C. Peinado, R. Sastre, J.L. Mateo, and N.S. Allen, *J. Photochem. Photobiol. A: Chem.*, 50, 249 (1989).
143. G. Amirzadeh and W. Schnabel, *Makromol. Chem.*, 182, 2821 (1981).
144. N.S. Allen, F. Catalina, P.N. Green, and W.A. Green, *Eur. Polym. J.*, 22, 793 (1986).
145. S. Yates and G.B. Schuster, *J. Org. Chem.*, 49, 3349 (1984).
146. K. Meier and H. Zweifel, *J. Photochem.*, 35, 353 (1986).
147. P.N. Green, *Polym. Paint. Resins*, 175, 246 (1985).
148. M.J. Davis, G. Gawne, P.N. Green, and W.A. Green, *Polym. Paint. Colour J.*, 176, 536 (1986).
149. N. Kayaman, A. Önen, Y. Yagci, and W. Schnabel, *Polym. Bull.*, 32, 589 (1994).
150. L. Cokbaglan, N. Arsu, Y. Yagci, S. Jockusch, and N. J. Turro, *Macromolecules*, 36, 2649 (2003).
151. I. Carmichael and G.L. Hug, in *CRC Handbook of Organic Photochemistry*, J.C. Scaiano (Ed.) CRC Press, Boca Raton, FL, Vol. 1, pp. 369–403, (1989).
152. N.J. Turro, *Modern Molecular Photochemistry*; University Science Books, Sausalito, CA (1991).
153. S.K. Lower and M.A. El-Sayed, *Chem. Rev.*, 66, 199–241 (1966).
154. J.C. Dalton and F.C. Montgomery, *J. Am. Chem. Soc.*, 96, 6230 (1974).
155. S.P. Pappas, *J. Radiat. Curing*, 14, 6 (1987).
156. S. Wu, J. Zhang, J.P. Fouassier, and D. Burr, *Ganguang Kexue Yu Kuang Juaxue*, 2, 47 (1989), C.A. 113, 61293w (1989).
157. R.S. Davidson, *J. Photochem. Photobiol. A: Chem.*, 73, 81 (1993).
158. J.L.R. Williams, D.P. Specht, and S. Farid, *Polym. Eng. Sci.*, 23, 1022 (1983).
159. D.P. Specht, P.A. Martic, and S. Farid, *Tetrahedron*, 38, 1203 (1982).
160. R.S. Davidson, *J. Chem. Soc.*, (C), 247 (1983).
161. H.I. Joschek and L.I. Grossweiner, *J. Am. Chem. Soc.,* 88, 3261 (1966).
162. R.S. Davidson and P.R. Steiner, *J. Chem. Soc.*, (C), 1682 (1983).
163. R.A. Sheldon and J.K. Kochi, *J. Am. Chem. Soc.*, 90, 6688 (1968).
164. J.K. Kochi, R.A. Sheldon, and S.S. Lande, *Tetrahedron,* 25, 1197 (1969).
165. R. Noyori, M. Kato, H. Kawanisi, and H. Nozaki, *Tetrahedron*, 25, 1125 (1969).
166. F.R. Stermitz and W.H. Huang, *J. Am. Chem. Soc.*, 92, 1446 (1970).
167. D. Elad and I. Rosenthal, *Chem. Commun.*, 905 (1969).
168. E.J. Baum and R.O.C. Norman, *J. Chem. Soc.*, (B), 227 (1968).
169. R.S. Davidson, K. Harrison, and P.R. Steiner, *J. Chem. Soc.*, (C), 3480 (1971).
170. R.S. Davidson and P.R. Steiner, *J. Chem. Soc.*, *Perkin II*, 1358 (1972).
171. M. Aydın, N. Arsu, and Y. Yagci, *Macromol. Rapid. Commun.*, 24, 718 (2003).
172. N. Mukai, H. Ige, T. Makino, and J. Atarashi, Eur. Patent Application 301516 (Prior 31.7.87) to Mitsubishi Rayon Co.
173. N.S. Allen, G. Pullen, M. Edge, I. Weddell, and F. Catalina, *Eur. Polym. J.*, 31, 15 (1995).
174. M.V. Encinas, J. Garrido, and E.A. Lissi, *J. Polym. Sci., Part A: Polym. Chem.*, 27, 139 (1989).
175. P.K. Ghosh, I.A. Weddel, and J.E. Yates, *PCT Int. Appl.* WO 94 24,170; C.A. 123, 144874u (1995).
176. H. Pfannenstiel and H. Huebner, German Patent Application 3240907 (Prior 5.11.82) to ESPE Fabrik Pharmazeutischer Präparate.
177. K.J. Goehrlich, Eur. Patent Application 166009 (Prior 22.6.84) to Dentsply GmbH.
178. D. Burget, C. Mallein, and J.P. Fouassier, *Polymer,* 44, 7671 (2003).
179. S.C. Clark, C.E. Hoyle, S. Jonsson, F. Morel, and C. Decker, *Polymer*, 40, 5063 (1999).

180a. J. von Sonntag, D. Beckert, W. Knolle, and R. Mehnert, *Radiat. Phys. Chem.*, 55, 609 (1999).
180b. R.S. Davidson, *Exploring the Science, Technology and Applications of UV and EB Curing*, SITA Technology Ltd., London (1999).
181. C.E. Hoyle, S.C. Clark, S. Jonsson, and M. Shimose, *Polymer*, 38, 5695 (1997).
182. C. Decker, F. Morel, S.C. Clark, S. Jonsson, and C.E. Hoyle, *Polym. Mater. Sci. Eng. Preprints*, 75, 198 (1996).
183. S.C. Clark, S. Jonsson, and C.E. Hoyle, *Polym. Mater. Sci. Eng. Preprints*, 76, 62 (1996).
184. S.C. Clark, S. Jonsson, and C.E. Hoyle, *Polymer Preprints*, 38(2), 363 (1997).
185. S.C. Clark, S. Jonsson, and C.E. Hoyle, *Polymer Preprints*, 37, 348 (1996).
186. F.C. De Schryver, N. Boens, and G. Smets, *J. Am. Chem. Soc.*, 96(20), 6463 (1974).
187. J. Put and F.C. De Schryver, *J. Am. Chem. Soc.*, 95(1),137 (1973).
188. F.C.De Schryver, N. Boens, and G. Smets, *J. Polym. Sci.: A*, 10,1687 (1972).
189. F.C. De Schryver, W.J. Feast, and G. Smets, *J. Polym. Sci.: A*, 8,1939 (1970).
190. D.F. Eaton, *Adv. Photochem.*, 13, 427 (1986).
191. H.-J. Timpe and S. Neuenfeld, *Kontakte*, 2, 28 (1990).
192. D. Ketley, *J. Radiat. Curing*, 9, 35 (1982).
193. H. Küstermann, H.-J. Timpe, K. Gabert, and H. Schülert, *Wiss. Zeitschr. TH Merseburg*, 29, 287 (1987).
194. H.-J. Timpe, S. Jockusch, and K. Koerner, *Radiat. Curing Polym. Sci. Technol.*, 2, 575 (1993).
195. C. Chen, *J. Polym. Sci., Part A*, 3, 1107 (1965).
196. T. Yamaoka, Y.C. Zhang, and K. Koseki, *J. Appl. Polym. Sci.*, 38, 1271 (1989).
197. D.F. Eaton, *Photogr. Sci. Eng.*, 23, 150 (1979).
198. T.L. Marino, D. Martin, and D.C. Neckers, *RadTech '94 North Am. UV/EB Conf. Exhib. Proc.*, 1, 169 (1994).
199. T. Tanabe, A. Torres-Filho, and D.C. Neckers, *J. Polym. Sci., Part A: Polym. Chem.*, 33, 1691 (1995).
200. G. Oster, *Nature*, 173, 300 (1954).
201. G. Oster, U.S. Patent 2,850,445 (1958).
202. G. Oster, U.S. Patent 3,097,096 (1963).
203. J.B. Rust, U.S. Patent 3,649,495 (1972) to Hughes Aircraft.
204. J.B. Rust, U.S. Patent 6,573,922 (1969) to Hughes Aircraft.
205. S. Levinos, U.S. Patent 3,650,927 (1972) to K & E Company.
206. U.L. Laridon, G.A. Delzenne, and H.K. Peeters, U.S. Patent 3,847,610 (1974) to Agfa Gevaert.
207. G.A. Delzenne, H.K. Peeters, and U.L. Laridon, *J. Photog. Sci.*, 22, 23 (1974).
208. G.A. Delzenne and U.L. Laridon, U.S. Patent 3,597,343 (1971) to Agfa Gaevert.
209. R.J. Allen and S. Chaberek, U.S. Patent 3,488,269 (1970) to Technical Operations, Inc.
210. S. Chatterjee, P.D. Davis, P. Gottschalk, M.E. Kurz, B. Sauerwein, X. Yang, and G.B. Schuster, *J. Am. Chem. Soc.*, 112, 6329 (1990).
211. G.B. Schuster, *Pure Appl. Chem.*, 62, 1565 (1990).
212. S. Chatterjee, P. Gottschalk, P.D. Davis, and G.B. Schuster, *J. Am. Chem. Soc.*, 110, 2326 (1988).
213. F. Takemura, *Bull. Chem. Soc. Jpn.*, 35, 1073 (1962).
214. J. Grimshaw, *The Chemistry of Sulphonium Groups*, C.J.M. Stirling and S. Patai, Eds., Wiley, New York (1981).
215. A. Böttcher, K. Hasebe, G. Hizal, Y. Yagci, P. Stellberg, and W. Schnabel, *Polymer*, 32, 2289 (1991).

216. Y. Yagci, A. Kornowski, and W. Schnabel, *J. Polym. Sci., Part A: Polym. Chem.*, 30, 1987 (1992).
217. O.A. Ptitsyna, T.W. Levashova, and K.P. Butin, *Dokl. Akad. Nauk.*, 201, 372 (1971).
218. F.M. Elofson and F.F. Gadallah, *J. Org. Chem.*, 94, 854 (1969).
219. P.E. Macrae and T.R. Wright, *J. Chem. Soc., Chem. Commun.*, 898 (1974).
220. E.J. Cerwonka, U.S. Patent 3,615,452 (1971) to GAF Corp.
221. H.-J. Timpe and S. Neuenfeld, *Kontakte*, 2, 28 (1990).
222. M. Li, J. He, J. Li, H. Song, and E. Wang, *Fushe Yanjiu Yu Fushe Gongyi Xuebao*, 58, 13 (1995), C.A. 123, 84107j (1995).
223. H.-J. Timpe, S. Ulrich, C. Decker, and J.-P.Fouassier, *Eur. Polym. J.*, 30, 1301 (1994).
224. E.M. Moore, U.S. Patent 3,495,987 (1970) to Du Pont.
225. T. Tani, *Photogr. Sci. Eng.*, 17, 11 (1973).
226. T. Posner, *Berichte*, 38, 646 (1905).
227. A.F. Jacobine, Thiol-ene photopolymers, in *Radiation Curing in Polymer Science and Technology*, Vol. III, J.P. Fouassier and J. Rabek, Eds., 219 (1993).
228. J.G. Woods, *Radiation Curable Adhesives in Radiation Curing: Science and Technology*, S.P. Pappas, Ed., Plenum Press, New York, 333 (1992).
229. M.S. Kharasch, J. Read, and F.R. Mayo, *Chem. Ind.*, 57,752 (1938).
230. C.R. Morgan, F. Magnotta, and A.D. Ketley, *J. Polym. Sci., Polym. Chem. Ed.*, 15, 627 (1977).
231. G. Eisele, J.P. Fouassier, and R. Reeb, *J. Polym. Sci., Polym. Chem. Ed.*, 35, 2333 (1997).
232. C. Decker, T. Nguyen, and T. Viet, *Macromol. Chem. Phys.*, 200, 1965 (1999).
233. C. Decker, T. Nguyen, and T. Viet, *J. Appl. Polym. Sci.*, 77, 1902 (2000).
234. C. Decker, T. Nguyen, and T. Viet, *Polymer*, 41, 3905 (2000).
235. N.B. Cramer and C.N. Bowman, *J. Polym. Sci., Polym. Chem.*, 39, 3311, (2001).
236. N.B. Cramer, J.P. Scott, and C.N. Bowman, *Macromolecules*, 35, 5361 (2002).
237. H.M. Wagner and M.D. Purbrick, *J. Photogr. Sci.*, 29, 230 (1981).
238. E. Steiner, H. Beyeler, M. Riediker, V. Desobry, K. Dietliker, and R. Heusler, Eur. Patent Application 401167 (1990).
239. C. Giannoti, *NATO Adv. Stud. Inst. Ser.*, Ser. C, 257 (1989).
240. J. Finter, M. Riediker, O. Rohde, and B. Rotzinger, *Macromol. Chem., Makromol. Symp.*, 24, 177 (1989).
240a. J.E. Beecher, J.M.J. Frechet, C.S. Willand, D.R. Robello, and D.J. Williams, *J. Am. Chem. Soc.*, 115, 12216 (1993).
241. B. Klingert, A. Roloff, B. Urwyler, and J. Wirz, *Helv. Chem. Acta*, 71, 1858 (1988).
242. A. Roloff, K. Meier, and M. Riediker, *Pure Appl. Chem.*, 58, 1267 (1986).
243. J. Finter, M. Riediker, O. Rohde, and B. Rotzinger, *Makromol.Chem., Makromol. Symp.*, 24, 177 (1989).
244. J. Finter, M. Riediker, O. Rohde, and B. Rotzinger, *Proc. 1st Meeting Eur. Polym. Fed. Eur. Symp. Polym. Mater.*, Lyon, France, paper ED03 (1987).
245. R. Bowser and R.S. Davidson, *J. Photochem. Photobiol. A: Chem.*, 77, 269 (1994).
246. J.F. Rabek, J. Lucki, M. Zuber, B.J. Qu, and W.F. Shi, *J. Macromol. Sci., Pure Appl. Chem.*, 29, 297 (1992).
247. K. Yamashita and S. Imahashi, Jpn. Kokai Tokkyo Koho 02,305,806 (1990), C.A. 114, 209330z (1990).
248. Y. Yagci and M.K. Mishra, in *Macromolecular Design: Concepts and Practice*, M.K. Mishra, Ed., Polymer Frontiers Int. Inc., (1994), p. 229.
249. R.S. Davidson, *J. Photochem. Photobiol. A: Chem.*, 69, 263 (1993).
250. C. Carlini and L. Angiolini, *Radiat. Curing Polym. Sci. Technol.*, 2, 283 (1993).

251. N.S. Allen, *Photochemistry*, 24, 405 (1993).
252. Y. Yagci and W. Schnabel, *Prog. Polym. Sci.*, 15, 551 (1990).
253. A. Önen and Y. Yagci, *J. Macromol. Sci. Chem.*, A27, 743 (1990).
254. A. Önen and Y. Yagci, *Angew. Makromol. Chem.*, 181, 191 (1990).
255. Y. Yagci and A. Önen, *J. Macromol. Sci. Chem.*, A28, 129 (1991).
256. Y. Hepuzer, M. Bektas, S. Denizligil, A. Önen, and Y. Yagci, *Macromol. Reports*, A30 (Suppl. 1&2), 111 (1993).
257. Y. Yagci, A. Önen, and W. Schnabel, *Macromolecules*, 24, 4620 (1991).
258. A. Önen and Y. Yagci, *Eur. Polym. J.*, 7, 721 (1992).
259. M. Degirmenci, I. Cianga, and Y. Yagci, *Macromol. Chem., Phys.*, 203, 1279, (2002).
260. M. Degirmenci, G. Hizal, and Y. Yagci, *Macromolecules*, 35, 8265 (2002).
261. Y. Yagci and M. Degirmenci, *ACS Symposium Series 854, Advances in Controlled Radical Polymerization*, K. Matyjaszewski, Ed., Washington, D.C., 383 2001.
262. G. Smets, *Polym. J.*, 17, 153 (1985).
263. T. Doi and G. Smets, *Macromolecules*, 22, 25 (1989).
264. Y. Kurusu, H.N. Shiyama, and M. Okawara, *J. Chem. Soc. Jap., Ind. Chem. Sec.*, 70, 593 (1967).
265. J.S. Shim, N.G. Park, U.Y. Kim, and K.D. Ahn, *Polymer (Korea)*, 8, 34 (1984).
266. K.B. Hatton, E. Irving, J.M.A. Walshe, and A. Mallaband, Eur. Patent Application 302831 (1987); assigned to Ciba Geigy.
267. H. Hageman, R.S. Davidson, and S. Lewis, *RadTech Europe*, Edinburgh 1991, Paper 54, p. 691 (1991).
268. Q.S. Lien and R.W.R. Humphreys, U.S. Patent 4,587,276 (1983); assigned to Loctite Corp.
269. M. Köhler and J. Ohngemach, *Polym. Paint Colour J.*, 178, 203 (1988).
270. R. Klos, H. Gruber, and G. Greber, *J. Macromol. Sci. Chem.*, A28 (9), 925 (1991).
271. K.-D. Ahn, *Processes in Photoreactive Polymers,* V.V. Krongauz and A.D. Trifunac, Eds., Chapman & Hall, New York (1995), p. 260.
272. J.E. Guillet and R.G.W. Norrish, *Proc. R. Soc. London*, 239, 172 (1956).
273. I. Naito, T. Ueki, S. Tabara, T. Tomiki, and A. Kinoshita, *J. Polym. Sci., Polym. Chem. Ed.*, 24, 875 (1986).
274. J. Dhanras and J.E. Guillet, *J. Polym. Sci. C*, 23, 433 (1968).
275. W. Kawai, *J. Polym. Sci., Polym. Chem. Ed.*, 15, 1479 (1974).
276. W. Kawai, *J. Macromol. Chem.*, A11, 1027 (1974).
277. W. Kawai and T. Ichiashi, *J. Polym. Sci., Polym. Chem. Ed.*, 12, 1041 (1974).
278. E. Lanza, H. Bergmans, and G. Smets, *J. Polym. Sci., Polym. Phys. Ed.*, 11, 95 (1973).
279. A. Önen, S. Denizligil, and Y. Yagci, *Angew. Makromol. Chem.*, 217, 79 (1994).
280. T. Imamoglu, A. Önen, and Y. Yagci, *Angew. Makromol. Chem.*, 224, 145 (1995).
281. H. Craubner, *J. Polym. Sci., Polym. Chem. Ed.*, 20, 1935 (1982).
282. H. Graubner, *J. Polym. Sci., Polym. Chem. Ed.*, 18, 2011 (1980).
283. O. Nuyken, J. Dauth, and J. Stebani, *Angew. Makromol. Chem.*, 207, 81 (1993).
284. O. Nuyken, J. Dauth, and J. Stebani, *Angew. Makromol. Chem.*, 207, 65 (1993).
285. J. Stebani, O. Nuyken, T. Lippert, and A. Wokaun, *Makromol. Chem., Rap. Commun.*, 14, 365 (1993).
286. M. Imoto, T. Otsu, and J. Yonezawa, *Makromol. Chem.*, 36, 93 (1960).
287. M.H. Jones, H.W. Melville, and W.H. Robertson, *Nature*, 174, 78 (1954).
288. M.H. Jones, *Can. J. Chem.*, 34, 948 (1958).
289. M.L. Miller, *Can. J. Chem.*, 36, 309 (1958).
290. H. Miyama, N. Fujii, K. Ikeda, and A. Kuwano, *Polym. Photochem.*, 6, 247 (1985).

291. G. Preiner and K. Matejcek, German Patent Application DE 3433654 (1984); assigned to Wacker Chemie.
292. C.L. Lee and M.A. Lutz, U.S. Patent 4,780,486 (1986); assigned to Dow Corning Co.
293. S. Huy and Y. Masumoto, U.S. Patent 4,935,455 (1989); assigned to Toshiba Silicone Co., Ltd.
294. A. Shirata, U.S. Patent 4,391,963 (1982); assigned to Toray Silicone Co. Ltd.
295. Y. Yagci, A. Önen, V. Harabagiu, M. Pinteala, C. Cotzur, and B.C. Simionescu, *Turk. J. Chem.*, 18, 101 (1994).
296. S.Q.S. Lin and A.F. Jacobine, U.S. Patent 4,538,388 (1984); assigned to Loctite Corp.
297. L.F. Fabrizio, S.Q.S. Lin, and A.F. Jacobine, U.S. Patent 4,536,265 (1984); assigned to Loctite Corp.
298. N.S. Allen, S.J. Hardy, A.F. Jacobine, D.M. Glaser, F. Catalina, S. Navaratnam, and B.J. Parsons, *J. Photochem. Photobiol. A: Chem.*, 62, 125 (1991).
299. R. West, A.R. Wolf, and P.J. Peterson, *J. Rad. Curing*, 13, 35 (1986).
300. P. Trefonas, R. Miller, and R.J. West, *J. Am. Chem. Soc.*, 107, 2737 (1985).
301. P. Trefonas, *Encyclopedia of Polymer Science and Technology*, Vol. 13, (1988) p. 173.
302. X.H. Zhang and R. West, *J. Polym. Sci, Chem.*, 22, 225 (1984).
303. M.A. Lutz, Eur. Patent Application 372566 (1988); assigned to Dow Corning Co.
304. T.C. Chang, Y.S. Chiu, H.B. Chen, and S.Y. Ho, *J. Chin. Chem. Soc. (Taipei)*, 41, 843 (1994).
305. D. Yücesan, H. Hostoygar, S. Denizligil, and Y. Yagci, *Angew. Makromol. Chem.*, 221, 207 (1994).
306. Y. Yagci, G. Hizal, and U. Tunca, *Polym. Commun.*, 31, 7 (1990).
307. J.F. Knistle and S.L. Watson Jr., *J. Radiat. Curing*, 2(2), 7 (1975).
308. J.L. Mateo, P. Bosch, E. Vazquez, and R. Sastre, *Makromol. Chem.*, 189, 1219 (1988).
309. I. Lukaç, C.H. Evans, J.C. Scaiano, and P. Hrdlovic, *J. Polym. Sci., Polym. Chem. Ed.*, 28, 595 (1990).
310. N.S. Allen, F. Catalina, C. Peinado, R. Sastre, J.L. Mateo, and P.N. Green, *Eur. Polym. J.*, 23, 985 (1987).
311. F. Catalina, C. Peinado, J.L. Mateo, P. Bosch, and N.S. Allen, *Eur. Polym. J.*, 28, 1533 (1992).
312. F. Catalina, C. Peinado, R. Sastre, and J.L. Mateo, *J. Photochem. Photobiol. A: Chem.*, 47, 365 (1989).
313. R.S. Davidson, A.A. Dias, and D. Illsley, *J. Photochem. Photobiol. A: Chem.*, 89, 75 (1995).
314. R.S. Davidson, A.A. Dias, and D.R. Illsley, *J. Photochem. Photobiol. A: Chem.*, 91, 153 (1991).
315. C.H. Bamford, in *New Trends in the Photochemsitry of Polymers*, N.A. Allen, and J.F. Rabek, Eds., Elsevier Applied Science Publishers, London, (1985) p. 129.
316. C.H. Bamford, J.P. Middleton, K.D. Al-Lamee, and J. Paprotny, *Br. Polym. J.*, 19, 269 (1987).
317. Y. Yagci, M. Müller, and W. Schnabel, *Macromol. Reports*, A28 (Suppl. 1), 37 (1991).
318. C.H. Bamford, in *Reactivity, Mechanism and Structure in Polymer Chemistry*, A.D. Jenkins and A. Ledwith, Eds., J. Wiley, New York, (1974) p. 52.
319. C.H. Bamford, G.C. Eastmond, and D. White, *Polymer*, 10, 771 (1969).
320. C.H. Bamford, G.C. Eastmond, and D. White, *Polymer*, 12, 241 (1972).
321. J. Ashworth, C.H. Bamford, and E.G. Smith, *Pure Appl. Chem.*, 30, 25 (1972).

322. C.H. Bamford, I.P. Middleton, and K.G. Al-Lamee, *Polymer*, 27, 1981 (1986).
323. C.H. Bamford and S.U. Mullik, *Polymer*, 17, 225 (1976).
324. C. Decker and K. Moussa, *Makromol. Chem.*, 189, 2381 (1988).
325. C. Decker and K. Moussa, *Macromolecules*, 22, 4455 (1989).
326. C. Decker and K. Moussa, *Polym. Prep.,* 29(1), 516 (1988).
327. C. Decker and K. Moussa, *Makromol. Chem.*, 191, 963 (1999).
328. V. Desorby, K.Dietliker, and R. Hüsler, *Polym. Mater. Sci. Eng.*, 60, 26 (1983).
329. R. Sastre, M. Gonde, and J.L. Mateo, *J. Photochem. Photobiol. A: Chem.*, 44, 111, (1988).
330. G.R. Tryson and A.R. Shultz, *J. Polym. Sci. Polym. Phys. Ed.*, 17, 2059, (1979).
331. M. Ikeda, Y. Teramoto, and M. Yasutake, *J. Polym. Sci: Polym. Chem. Ed.*, 16, 1175 (1978).
332. R. Liska, *J. Polym. Sci., Polym. Chem.*, 42(9), 2285 (2004).

9 Functionalization of Polymers

Yusuf Yagci and Munmaya K. Mishra

CONTENTS

9.1 INTRODUCTION

Design and synthesis of materials with novel properties is becoming an important aspect of polymer chemistry [1, 2]. Quite often, desired properties are not attainable by the properties of a single homopolymer. The synthesis of block structures is one way to adjust properties of polymers. Blocking reactions are generally accomplished in two ways; (1) successive addition of appropriate monomers in living polymerization [3], which allows preparation of block copolymers containing two or more different segments and (2) polymers with functional end groups (telechelics) can be converted to initiating species by external stimulation such as heat, light, or chemical reaction [4, 5]. Alternatively, these polymers, based on the reaction of the functional groups with other suitable low-molar-mass compounds or polymers, can be used in block condensation (Scheme 9.1). In the latter case, the number and the location of a functional group are quite important for the overall structure of block condensate. Polymers must possess exactly two functional end groups to yield polycondensates with high molecular weight and functional end groups located at the end of one chain can undergo similar condensation reactions. In this case, graft copolymers are formed (Scheme 9.2). In addition to their use in block and graft copolymerizations,

X～～～X + Y～～～Y ⟶ ～～～X'Y'～～～X'Y'～～～

X～～～X + Y—Y ⟶ ～～～X'Y'—X'Y'～～～

SCHEME 9.1

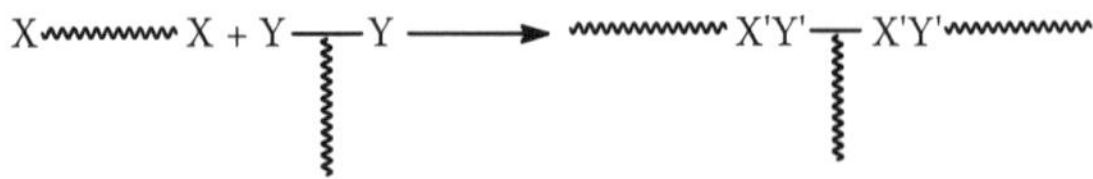

SCHEME 9.2

end groups have various effects on the properties of the incorporated polymer such as dyeability and hydrolytic and thermal stability.

Functionalization of polymers is usually achieved by either anionic or cationic [4] polymerization mechanisms in which chain-breaking reactions are of minor importance. On the other hand, radical vinyl polymerization, although easy to handle and which can be applied to most vinyl monomers, possesses some disadvantages. Noninstantaneous initiation and the random nature of the termination yield polymers of much higher polydispersities than those obtained by ionic processes. To have true telechelics (i.e., each polymer possesses two functional groups), termination should occur exclusively by combination. This requirement is not fulfilled by many monomers. Whereas styrene and its derivatives, and acrylates terminate mainly by combination, methacrylates undergo disproportionation. The latter would result with monofunctional polymers. Providing one type of primary radicals is formed (i.e., primary radicals do not participate in hydrogen abstraction reactions), this problem may be overcome by working at higher initiator concentrations. Chain transfer reactions were also frequently used to functionalize polymers. With system consisting of a functional chain transfer agent reacts not only with the propagating radical but also with the primary radicals derived from the initiator (Scheme 9.3). Although it can be reduced to very low levels, nonfunctional groups are also formed.

$$I \longrightarrow 2R\bullet$$

$$R\bullet + nM \longrightarrow RMn\bullet$$

$$RM_n + (F)\text{—S—H} \longrightarrow R\text{–}M_n\text{–}H + (F)\text{–S}\bullet$$

$$R\bullet + (F)\text{—S—H} \longrightarrow R\text{–}H + (F)\text{–S}\bullet$$

$$(F)\text{–S}\bullet + nM \longrightarrow (F)\text{–SMn}\bullet$$

SCHEME 9.3

9.2 FUNCTIONALIZATION TECHNIQUES

9.2.1 Initiator Techniques

9.2.1.1 Azo Initiators

Azo initiators are, by far, the most widely employed initiators for polymer functionalization. The well-known azo initiator 2,2′-azobisisobutyronitrile (AIBN) is highly efficient to introduce cyano end groups to polymers. Ethylene [6] and styrene [7] were polymerized with AIBN to yield the corresponding polymers with the functionalization values of 1.7 and 2, respectively. High functionalization in the case of polystyrene was due to the primary radical combination as confirmed by nuclear magnetic resonance (NMR) studies [8, 9]. Gas chromatography-mass spectroscopy (GC-MS) studies revealed that the relatively lower functionalization in the case of polyethylene results in part from the occurrence of some termination by disproportionation. The following product stemming from disproportionation was identified (Scheme 9.4).

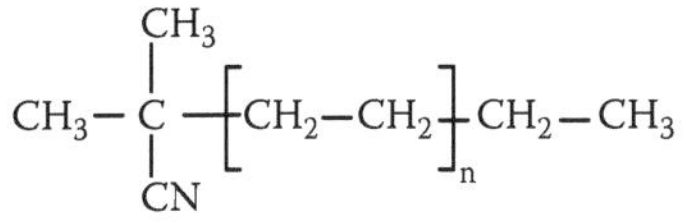

$$CH_3-\underset{CN}{\overset{CH_3}{\underset{|}{\overset{|}{C}}}}-\left[CH_2-CH_2\right]_n-CH_2-CH_3$$

SCHEME 9.4

Primary amine- and carboxy-terminated polymers may be obtained [7] from AIBN-initiated telechelics by subsequent hydrogenation and hydrolysis, respectively (Scheme 9.5).

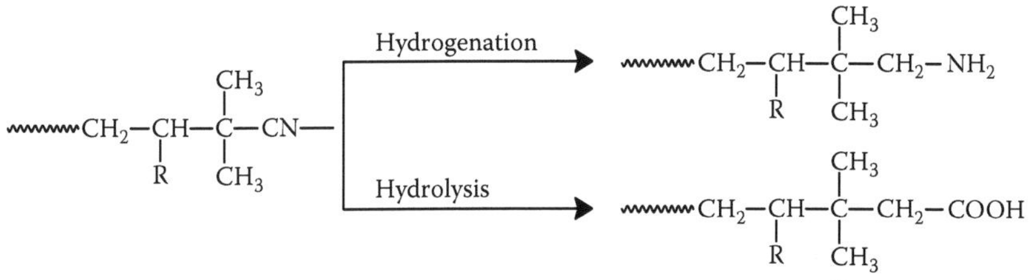

$$\sim\sim CH_2-\underset{R}{\underset{|}{CH}}-\underset{CH_3}{\overset{CH_3}{\underset{|}{\overset{|}{C}}}}-CN \xrightarrow{\text{Hydrogenation}} \sim\sim CH_2-\underset{R}{\underset{|}{CH}}-\underset{CH_3}{\overset{CH_3}{\underset{|}{\overset{|}{C}}}}-CH_2-NH_2$$

$$\sim\sim CH_2-\underset{R}{\underset{|}{CH}}-\underset{CH_3}{\overset{CH_3}{\underset{|}{\overset{|}{C}}}}-CN \xrightarrow{\text{Hydrolysis}} \sim\sim CH_2-\underset{R}{\underset{|}{CH}}-\underset{CH_3}{\overset{CH_3}{\underset{|}{\overset{|}{C}}}}-CH_2-COOH$$

SCHEME 9.5

Isocyanate functional groups can be introduced [7] by two methods as illustrated in Scheme 9.6. The former method, which requires less reaction steps, appears to be more suitable. To obtain polymers with NH_2 terminal groups, an alternative method was developed [10]. An azo-initiator containing (acyloxy)imino groups was heated

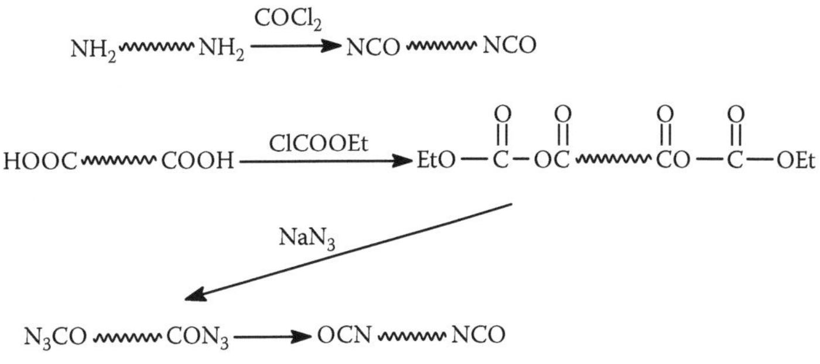

$$NH_2\sim\sim NH_2 \xrightarrow{COCl_2} NCO\sim\sim NCO$$

$$HOOC\sim\sim COOH \xrightarrow{ClCOOEt} EtO-\overset{O}{\overset{\|}{C}}-O\overset{O}{\overset{\|}{C}}\sim\sim \overset{O}{\overset{\|}{C}}O-\overset{O}{\overset{\|}{C}}-OEt$$

$$\xrightarrow{NaN_3} N_3CO\sim\sim CON_3 \longrightarrow OCN\sim\sim NCO$$

SCHEME 9.6

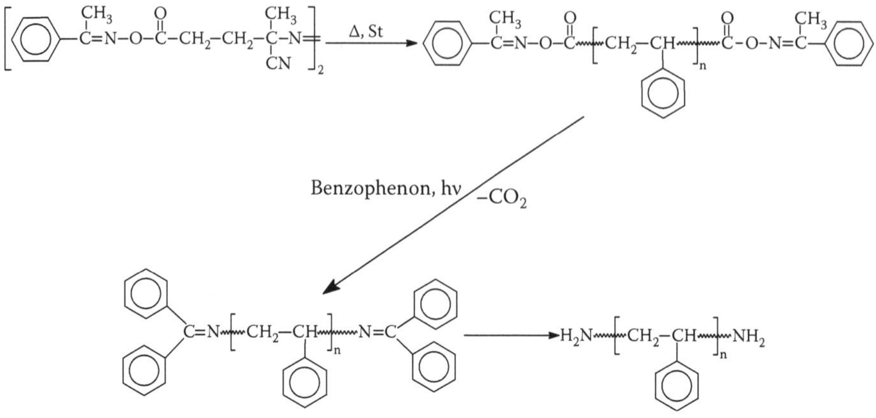

SCHEME 9.7

in the presence of styrene. Irradiation of the resulting polymers in the presence of benzophenone sensitizer and subsequent hydrolysis yielded polymers with primary amino groups (Scheme 9.7). Successful functionalization was achieved by polyamidation with bifunctional acid chlorides. Regarding photoactive group functionalization, the azo initiators listed in Table 9.1 have been successfully applied in free-radical polymerization.

Pyrene-containing polymers have attracted much attention due to their potential use as semiconductors, photoresist materials, and fluorescent probes [18]. Various methods have been developed to attach pyrene moieties to polymers [19–23]. Most recently, a new azo initiator, 4,4′-azobis(4-cyanopyrenylmethyl pentanoate) (ACPMP) was used in the synthesis of polystyrene with pyrene terminal groups by using conventional free-radical polymerization (FRP) and stable free-radical polymerization (SFRP) techniques (Scheme 9.8) [24].

Incorporation of terminal groups was confirmed by spectral measurements. Crown ether moieties, capable of binding certain ions, have also been introduced on polymers by using the functional initiator approach [25, 26].

A variety of other functional initiators has been synthesized by modifying AIBN's methyl [27] or nitrile [28] groups. Azo compounds containing hydroxyl, carboxyl, chloride, and isocyanate groups are the most frequently used initiators for the functionalization of polymers from a wide range of monomers [29–33].

Polydienes possessing these groups are of particular interest because they find application as polymeric binders [34] and can modify properties of polyamides and polyesters by introducing soft segments. Polybutadiene [35] with CN or COOMe groups, and polyisoprene and polychloroprene [36–38] with OH or COOH groups have been successfully prepared via the functional initiator approach. Hydroxy-terminated polyacrylonitrile [31] and poly(acrylonitrile-*co*-styrene) [39] were also prepared. Polystyrenes with monofunctional α-imidazole groups were synthesized by using the azo initiator as shown in Scheme 9.9.

Azo initiators possessing monomeric functional sites other than the actual azo function can be used successfully to obtain macromonomers. By using the furan [40] and tiophene [41] azo initiators shown in Scheme 9.10, macromonomers that

TABLE 9.1
Azoinitiators (I)[a] Used for Photoactive Group Functionalization and Applications

Azoinitiator(R=)	End Group	Application	References
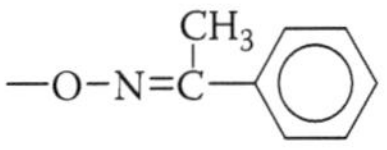 $-O-N=C(CH_3)-C_6H_5$	Amino	Polyamidation	[10]
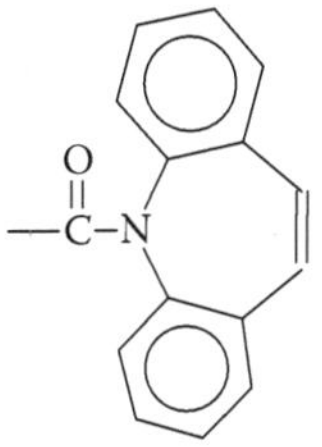 $-C(=O)-N$ (dibenzazepine)	N-Acyl dibenzazepine	Chain extension,block copolymerization	[11]
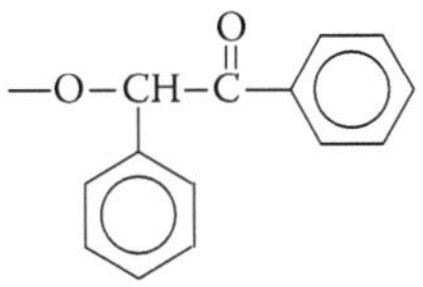 $-O-CH(C_6H_5)-C(=O)-C_6H_5$	Benzoin	Block copolymerization	[12–15]
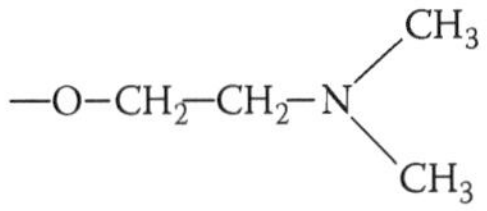 $-O-CH_2-CH_2-N(CH_3)_2$	Dimethyl amino	Block copolymerization	[16]
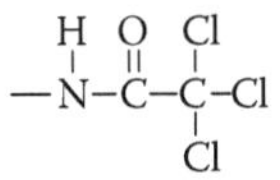 $-N(H)-C(=O)-C(Cl)_2-Cl$	Trichloroacetyl	Block copolymerization	[17]

[a] $I = \left[RO-\overset{O}{\overset{\|}{C}}-CH_2-CH_2-\underset{CN}{\overset{CH_3}{C}}-N=\right]_2$

(ACPMP)

65°C

130°C

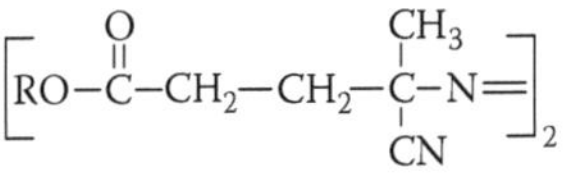

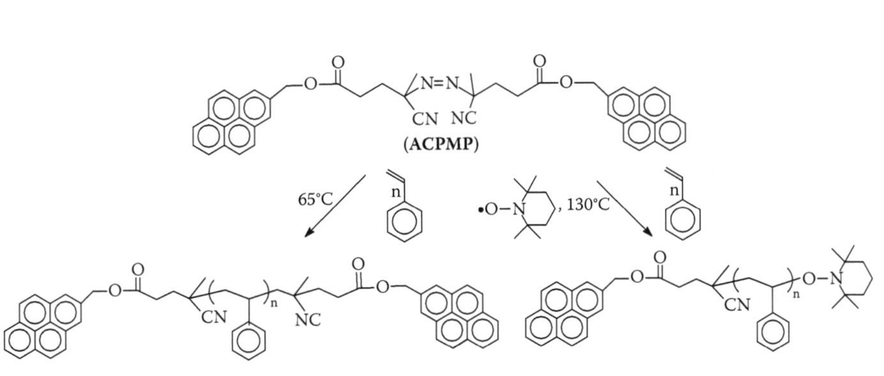

SCHEME 9.8

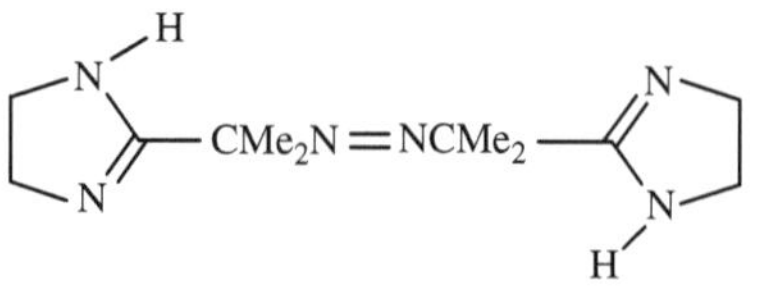

SCHEME 9.9

can act as partners for [4 + 2] cycloaddition and electropolymerization reactions, respectively, are obtained.

Initiation of styrene polymerization by means of these initiators is expected to yield monomeric sites at each end because termination is by radical–radical combination as illustrated for the thiopehene-azo initiator (Scheme 9.11). Notably, the polydispersity of polystyrene macromonomer thus obtained was rather high (1.2–1.8) [41].

In general, the termination mode of the particular monomer determines the number of functionalities per macromolecular chain. Most monomers undergo both unimolecular and bimolecular termination reactions. It is often observed that both respective monofunctional and bifunctional polymers are formed and well-defined functional polymers cannot be prepared. The use of allylmalonic acid diethylester in free-radical polymerization has been proposed to overcome the problems associated with the aforementioned functionality. In the presence of the allyl compound, the

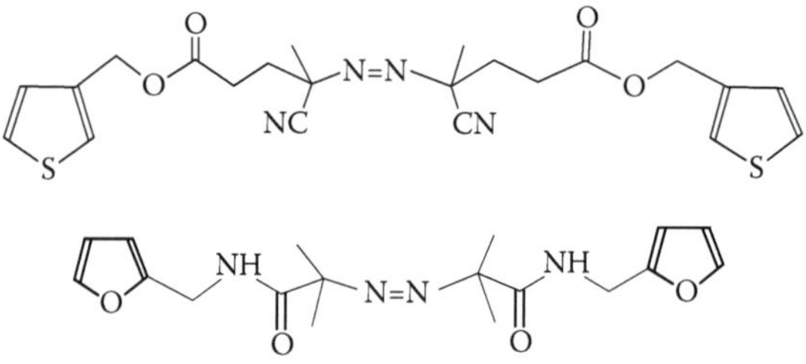

SCHEME 9.10

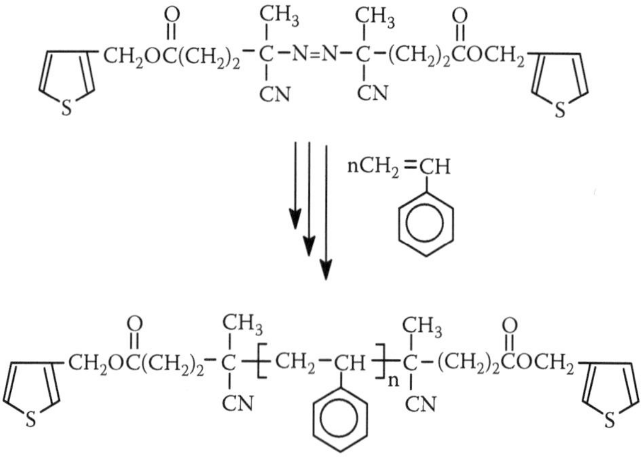

SCHEME 9.11

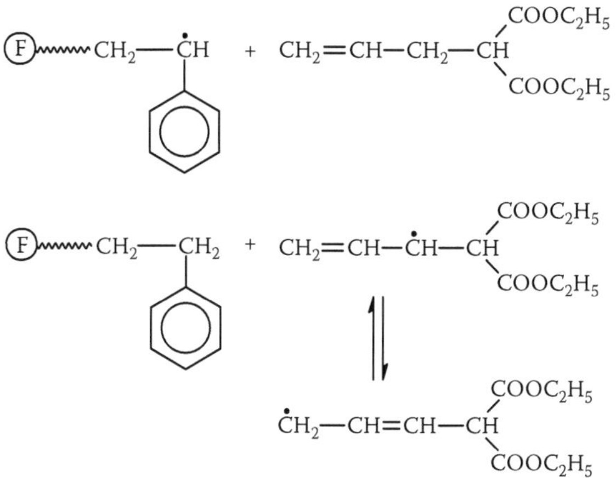

$$\text{(F)}\sim\sim CH_2-\dot{C}H(C_6H_5) + CH_2{=}CH-CH_2-CH(COOC_2H_5)_2$$

$$\text{(F)}\sim\sim CH_2-CH_2(C_6H_5) + CH_2{=}CH-\dot{C}H-CH(COOC_2H_5)_2$$

$$\updownarrow$$

$$\dot{C}H_2-CH{=}CH-CH(COOC_2H_5)_2$$

SCHEME 9.12

free-radical polymerization of monomers, regardless of their termination mode, proceeds entirely with the unimolecular termination mechanism, as shown in Scheme 9.12. Because allyl compounds lead to degradative chain transfer, the resulting allyl radical is quite stable due to the allyl resonance. Monofunctional polystyrene, polyvinylacetate, and poly (*t*-butyl methacrylate) were prepared by using this approach [42]. Subsequently, various macromonomers were derived from these polymers.

9.2.1.2 Peroxides

Acylperoxides are less frequently used in polymer functionalization due to the fact that upon heating, two different types of radicals are formed (Scheme 9.13).

On the other hand, aryloxy peroxides [43], although less reactive, have led to polystyrene telechelics. For example, benzyl chloride-and benzaldehyde-terminated polymers were obtained directly from the following benzoyl peroxides without further purification (Scheme 9.14).

Dialkylperoxycarbonate initiators were used for ester functionalization. Monomers such as ethylene [44], methyl methacrylate [45], and styrene [46] were reported to be polymerized by these initiators. The terminal ester groups can easily be transformed into hydroxyl groups by hydrolysis [34].

Redox reactions involving peroxides are another efficient way to prepare telechelics. Hydroxyl-terminated polydienes [46] and carboxylic acid-terminated

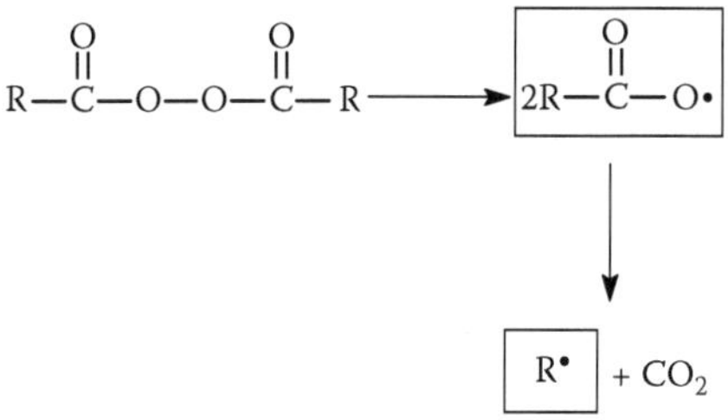

$$R-\overset{O}{\overset{\|}{C}}-O-O-\overset{O}{\overset{\|}{C}}-R \longrightarrow \boxed{2R-\overset{O}{\overset{\|}{C}}-O\bullet} \longrightarrow \boxed{R^\bullet} + CO_2$$

SCHEME 9.13

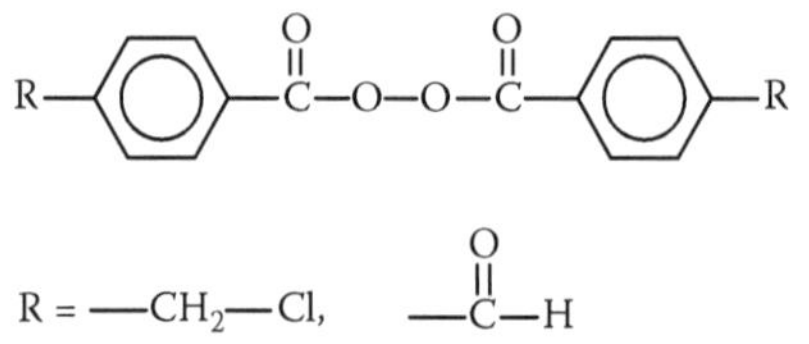

$$R-C_6H_4-\overset{O}{\overset{\|}{C}}-O-O-\overset{O}{\overset{\|}{C}}-C_6H_4-R$$

$$R = -CH_2-Cl, \quad -\overset{O}{\overset{\|}{C}}-H$$

SCHEME 9.14

polytetrafluoroethylenes [47] were synthesized by using H_2O_2 and potassium peroxy disulfate, respectively, via redox reactions.

9.2.2 Transfer Reactions

9.2.2.1 Transfer Agents

In addition to the initiator-controlled polymer functionalization, transfer reactions may add to the synthesis of functional polymers. In free-radical polymerization chain transfer, reactions involving thiols proceed via atom abstraction, as illustrated in Scheme 9.3. Consequently, these molecules do not offer any scope for introducing functionalities at both ends. However, monofunctional telechelics have been successfully prepared by using thiols. For example, Boutevin and co-workers [48, 49] introduced polymerizable vinyl groups to polyvinylchloride according to that strategy. 3-Mercaptopropionic acid has been used as a functional chain transfer agent, and the carboxylic acid group has then been reacted with glycidyl methacrylate.

Polystyrene macromonomers [50] with molecular weights of 10^3–10^4 were also prepared. Similar experiments [51, 52] using 2-mercaptoethanol, and subsequent treatment of the resulting polymer with methacrolyl chloride led to methacryl end-capped polymethacrylates and polyvinyl chloride. Typical reactions in the case of polyvinyl chloride are illustrated in Scheme 9.15.

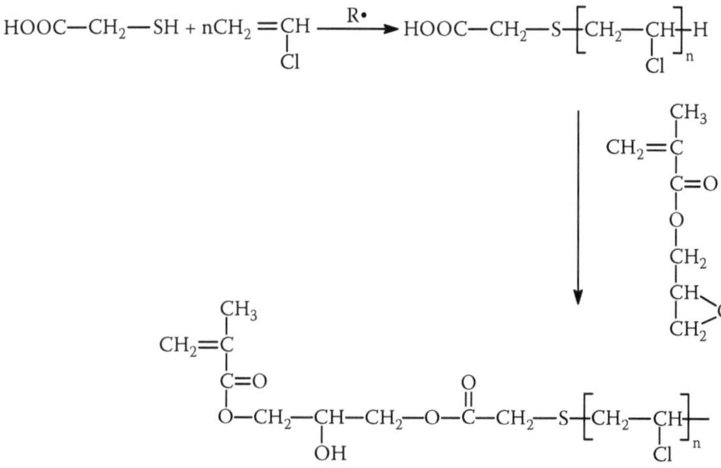

$$HOOC-CH_2-SH + nCH_2=\underset{Cl}{\underset{|}{CH}} \xrightarrow{R\bullet} HOOC-CH_2-S\left[CH_2-\underset{Cl}{\underset{|}{CH}}\right]_n H$$

$$\downarrow \; CH_2=C(CH_3)-C(=O)-O-CH_2-\underset{O}{CH-CH_2}$$

$$CH_2=C(CH_3)-C(=O)-O-CH_2-\underset{OH}{\underset{|}{CH}}-CH_2-O-\overset{O}{\overset{\|}{C}}-CH_2-S\left[CH_2-\underset{Cl}{\underset{|}{CH}}\right]_n$$

SCHEME 9.15

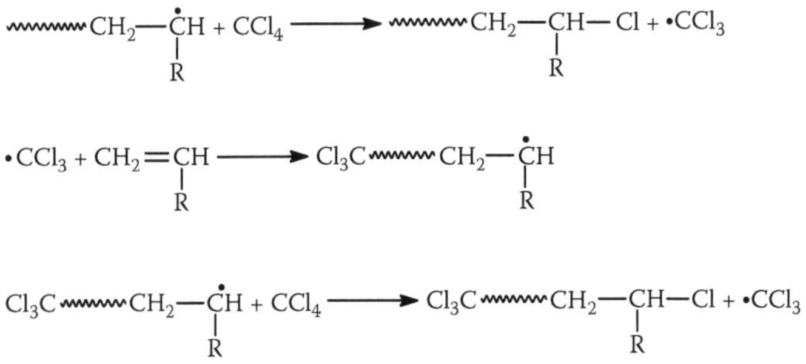

$$\sim\sim CH_2-\dot{C}H(R) + CCl_4 \longrightarrow \sim\sim CH_2-CH(R)-Cl + \bullet CCl_3$$

$$\bullet CCl_3 + CH_2{=}CH(R) \longrightarrow Cl_3C\sim\sim CH_2-\dot{C}H(R)$$

$$Cl_3C\sim\sim CH_2-\dot{C}H(R) + CCl_4 \longrightarrow Cl_3C\sim\sim CH_2-CH(R)-Cl + \bullet CCl_3$$

SCHEME 9.16

Free-radical polymerization in the presence of a chain transfer agent (telogen) is often called a telomerization reaction [53]. The most studied and well-documented telogen is CCl_4. Other analogous halogen compounds, namely CBr_4 and CHI_3, are not suitable due to their very high transfer constants, which yield only oligomeric materials. General reactions involving telomerization are represented for the case of CCl_4 in Scheme 9.16.

9.2.2.2 Iniferters

Disulfides are useful compounds in free-radical polymerization of vinyl compounds. The S-S bond present in the molecule readily decomposes to form thiyl radicals, which act as both initiators and terminators. The disulfides are also used as chain transfer agents. These compounds are called "iniferters" in short for their roles **ini**tiation, trans**fer**, and **ter**mination reactions. Polymers prepared with disulfides possess terminal weak and easily dissociable carbon-sulfur bonds, which allow further addition of monomer on a terminated polymer. When the functional group is incorporated to the disulfides, the obvious benefit would be the possibility of the preparation of functional polymers. For example, carboxylic acid and amino functionalities were introduced to polystyrene using the corresponding disulfides [54]. Diamino functional poly (*t*-butyl acrylate) was also prepared [55]. In this case, polymers were readily hydrolyzed to polyacrylic acid possessing amino terminal groups, which is a useful material for the application of polyelectrolytes. The photochemically and thermally induced iniferter properties of the tetraalkyl thiuram disulfides during free-radical vinyl polymerization were also exploited to end functionalize poly (methyl methacrylate) and polystyrene (Scheme 9.17) [56, 57]. Table 9.2 summarizes the functional iniferters used for obtaining telechelic oligomers.

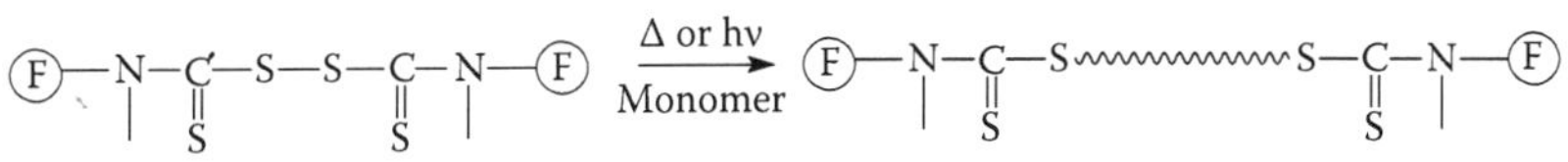

$$(F)-N(-)-C(=S)-S-S-C(=S)-N(-)-(F) \xrightarrow[\text{Monomer}]{\Delta \text{ or } h\nu} (F)-N(-)-C(=S)-S\sim\sim\sim S-C(=S)-N(-)-(F)$$

SCHEME 9.17

TABLE 9.2
Functional Polymers Prepared by Iniferters

Iniferter	Monomer	Functional Group	References
$(HO\text{–}CH_2CH_2(C_2H_5)N\text{–}C(=S)\text{–}S\text{–})_2$	MMA	OH	[58]
$((C_2H_5O)_2P(=O)\text{–}N(CH_3)\text{–}CH_2CH_2\text{–}N(C_2H_5)\text{–}C(=S)\text{–}S\text{–})_2$	MMA, St	phosphorylamide	[59]
(indane-1,3-dione–CH_2CH_2–N(piperazine)N–C(=S)–S–$)_2$	Isoprene	NH_2	[60]
$\text{—}O\text{–}CH_2\text{–}CH_2\text{–}N(CH_3)_2$	MMA, St	phenyl	[61]
$(H_2N\text{–}C_6H_4\text{–}S\text{–})_2$	St, *t*-BA	NH_2	[54]
$(HOOC\text{–}C_6H_4\text{–}S\text{–})_2$	St	COOH	[54]
$((H_3C)_3Si\text{–}O\text{–}C(C_6H_5)_2\text{–})_2$	MMA	OH	[62]
$((H_3C)_3Si\text{–}O\text{–}C(C_6H_5)(\text{furyl})\text{–})_2$	MMA, BA, STA, St,	Furanyl	[63]
$(NC\text{–}C(C_6H_4R)_2\text{–})_2$	MMA	Cl, $C(CH_3)_3$	[64]

MMA: Methyl methacrylate, St: styrene, *t*-BA: *tert*-butyl acrylate, STA: stearyl acrylate.

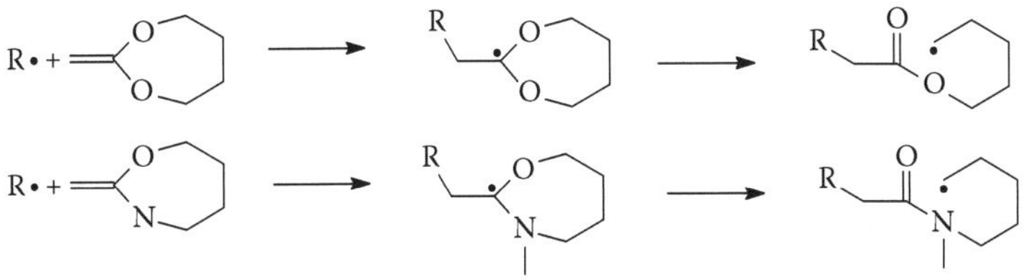

SCHEME 9.18

9.2.2.3 Addition-Fragmentation Reactions

9.2.2.3.1 Free-Radical Copolymerization of Vinyl Compounds with Unsaturated Heterocyclic Monomers

Free-radical ring-opening polymerization of cyclic ketenacetals and their nitrogen analogs occurs via formation of carbonyl bond at the expense of a less stable carbon-carbon double bond as depicted in Scheme 9.18 [65–72].

Functional polystyrene and polyethylene possessing hydroxy and carboxylic end groups in the same molecule were prepared by taking advantage of this type of so-called addition fragmentation reaction. For this purpose, 2-methylene-1,3-dioxepane was polymerized with excess styrene and ethylene. Hydrolysis of resulting polymers yielded desired telechelics (Scheme 9.19).

Similar reactions of the nitrogen derivative gave polystyrene with an aminomethyl group at one end and a carboxy group at the other end of the molecule (Scheme 9.20).

9.2.2.3.2 Open-Chain Ketenacetals

Open-chain ketenacetals undergo addition-fragmentation reactions as the cyclic ketenacetals to form an ester bond [72]. These molecules do not yield high-molecular-weight homopolymers due to the degradative chain transfer reaction. The radical copolymerization of equimolar amounts of styrene and diethyl ketenacetal, however, gives oligomeric products possessing ethoxy carbonyl groups at the chain end (Scheme 9.21).

In this case, diethyl ketenacetal acts as both monomer and moderately effective chain transfer agent. Introduction of a phenyl group increased the extent of cleavage and the efficiency of the chain transfer reaction. Thus, benzyl and ester functional polymers

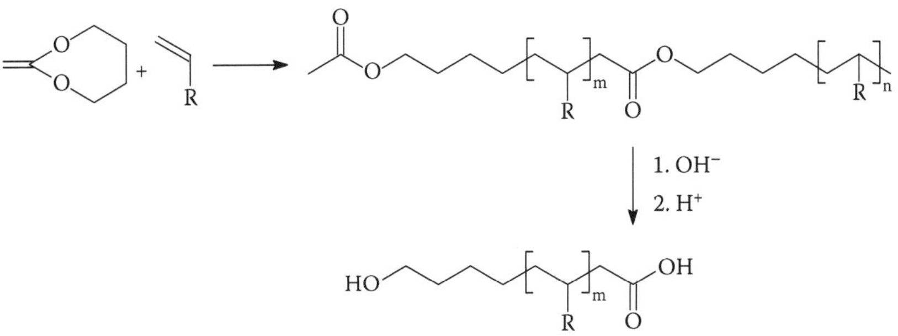

1. OH^-
2. H^+

SCHEME 9.19

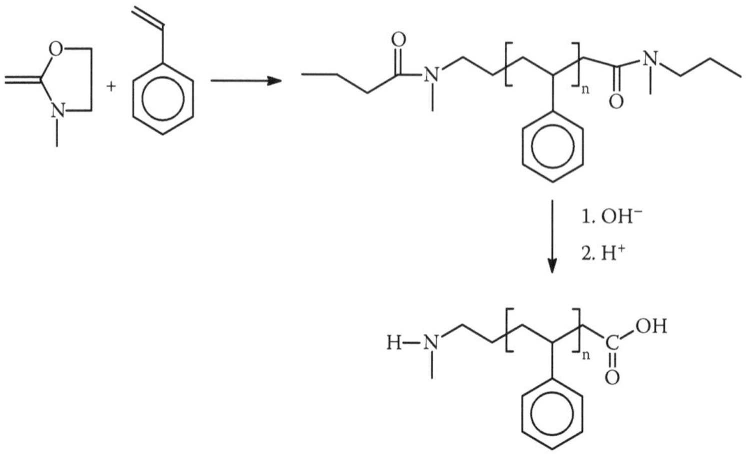

SCHEME 9.20

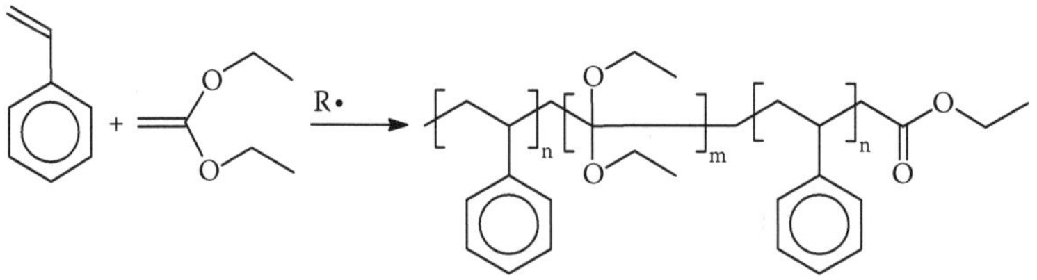

SCHEME 9.21

are essentially obtained. Ketenacetal that possesses benzylalcohol (Scheme 9.22) is able to yield polymers with two hydroxy groups at the chain ends [72].

9.2.2.3.3 Allylic Sulfides

Among many other compounds, which undergo addition-fragmentation reactions, appropriately substituted allylic sulfides have been shown to be efficient transfer agents in free-radical polymerization [73–75]. The average chain transfer constants of these compounds are in the range 0.3–3.9. Thus, molecular weight can be controlled by efficient transfer reactions. These molecules can be used to prepare a wide range of macromonomers and monofunctional and bifunctional polymers when functionality is introduced into one or both of the substituents R_1 and R_2 (Scheme 9.23).

Other allylic compounds, which participate in addition-fragmentation reactions analogous to allylic sulfides, include allylic bromides, allylic phosphanate, and allylic

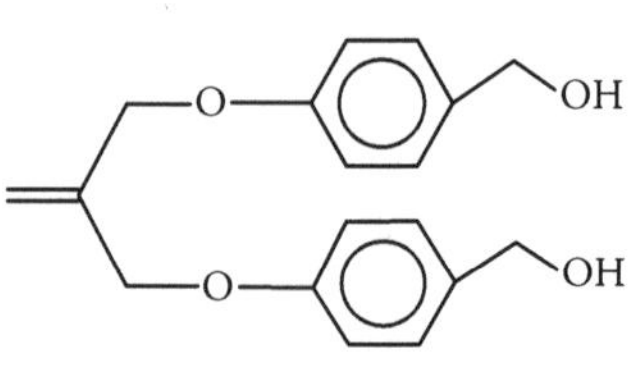

SCHEME 9.22

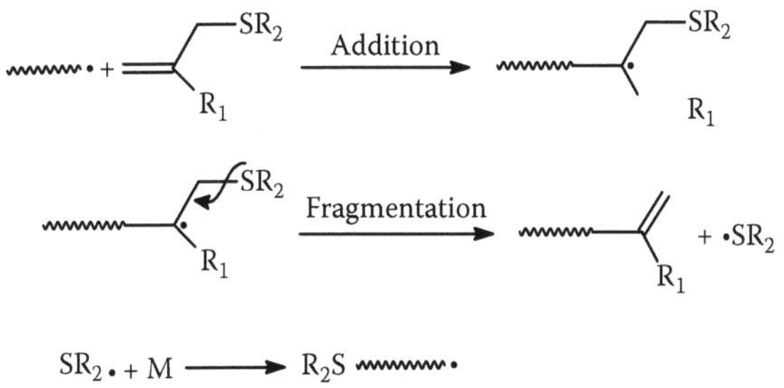

SCHEME 9.23

stannane [76, 77]. The high chain transfer constants indicate that these compounds can also be used to prepare functional polymers.

9.2.2.3.4 Allylic Peroxides

Allylic peroxides are another class of compounds, which undergo addition-fragmentation reactions [78, 79]. The striking advantage of allylic peroxides derivatives is that the epoxy functional polymers can be prepared by the addition of propagating radical to the double bond followed by γ-scission according to the following reactions (Scheme 9.24).

Epoxy functionalization of polystyrene via the previously described mechanism was evidenced by ^{1}H-NMR and ^{13}C-NMR analysis of the resulting polymer.

9.2.2.3.5 Vinyl Ethers

Chain transfer reactions involving suitably constituted vinyl ethers [80, 81] also proceeded via addition-fragmentation reactions, as illustrated in Scheme 9.25. Benzyl and benzoyl groups were thus incorporated into polystyrene and poly (methyl methacrylate).

9.3 CONTROLLED RADICAL POLYMERIZATION

Accurate control of polymerization process is an important aspect for the preparation of well-defined telechelics and end-functionalized macromolecules [82]. Such control of chain ends was traditionally accomplished using living ionic polymerization techniques. Nevertheless, it is well known that the ionic processes suffer from

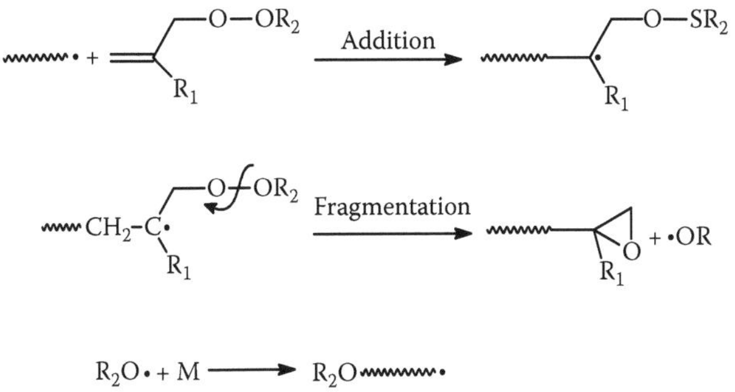

SCHEME 9.24

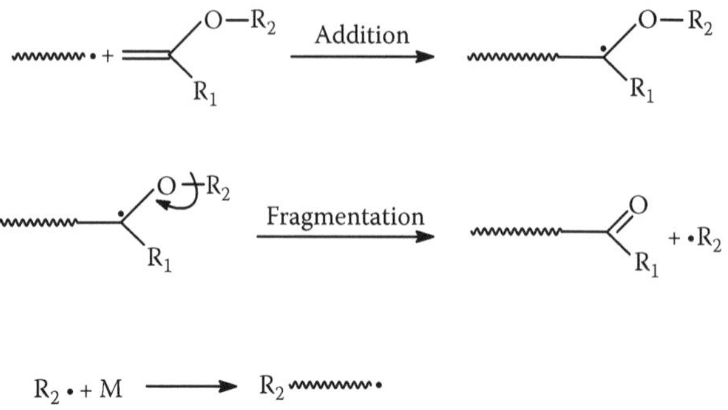

$$\sim\sim\sim\bullet + CH_2{=}C(R_1)(O{-}R_2) \xrightarrow{\text{Addition}} \sim\sim\sim CH_2{-}\dot{C}(R_1)(O{-}R_2)$$

$$\sim\sim\sim CH_2{-}\dot{C}(R_1)(O{-}R_2) \xrightarrow{\text{Fragmentation}} \sim\sim\sim CH_2{-}C(R_1){=}O + \bullet R_2$$

$$R_2\bullet + M \longrightarrow R_2\sim\sim\sim\bullet$$

SCHEME 9.25

rigorous synthetic requirements and in some cases they are sensitive to the functional groups to be incorporated. As described previously in detail, free-radical polymerization is flexible and less sensitive to the polymerization conditions and functional groups. However, conventional free-radical processes yield polydisperse polymers without molecular weight and chain end control. Competing coupling and disproportionation steps and the inefficiency of the initiation steps lead to functionalities less than or greater than theoretically expected. For example, number of initiator fragments incorporated per polymer chain by using conventional radical polymerization varied from 1.4–2.3. Recent developments in controlled/living radical polymerization provided possibility to synthesize well-defined telechelic polymers with desired functionality [83]. As it will be shown next, all three standard methods for controlled/living radical polymerization, namely, atom transfer radical polymerization (ATRP) [84, 85], stable free-radical mediated polymerization (SFRP), also called nitroxide mediated polymerization (NMP) [86], and reversible addition-fragmentation chain transfer polymerization (RAFT) [87] were used for the preparation of telechelic polymers.

9.3.1 Atom Transfer Radical Polymerization (ATRP)

ATRP [88, 89] involves reversible homolytic cleavage of a carbon-halogen bond by a redox reaction between an organic halide (R-X) and a transition metal, such as copper (I) salts, as illustrated in Scheme 9.26.

Polymer functionalization by ATRP can be achieved by using functional initiators (a) and monomers (b) and the chemical transformation of the halogen end groups (c). These routes are summarized in Scheme 9.27.

Quite a number of functional initiators were successfully used in ATRP to prepare functional styrene and acrylate type polymers [90]. For this purpose, initiators

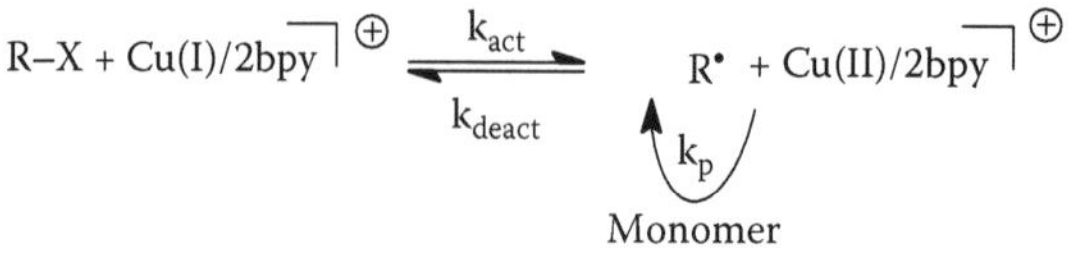

$$R{-}X + [Cu(I)/2bpy]^{\oplus} \underset{k_{deact}}{\overset{k_{act}}{\rightleftharpoons}} R^{\bullet} + [Cu(II)/2bpy]^{\oplus}$$

(Monomer, k_p)

SCHEME 9.26

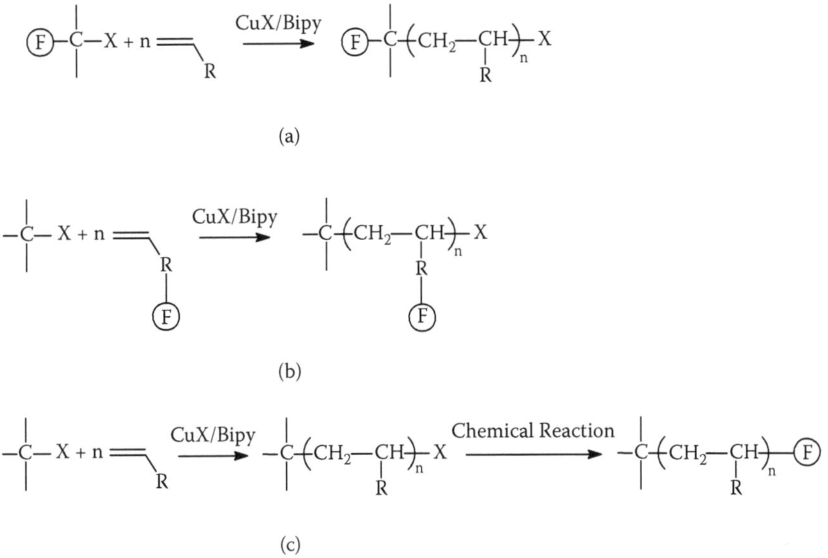

$$\text{(F)}-\overset{|}{\underset{|}{C}}-X + n\,CH_2{=}CH{-}R \xrightarrow{CuX/Bipy} \text{(F)}-\overset{|}{\underset{|}{C}}\left(CH_2-\underset{R}{\underset{|}{CH}}\right)_n X$$

(a)

$$-\overset{|}{\underset{|}{C}}-X + n\,CH_2{=}CH{-}R{-}\text{(F)} \xrightarrow{CuX/Bipy} -\overset{|}{\underset{|}{C}}\left(CH_2-\underset{R-\text{(F)}}{\underset{|}{CH}}\right)_n X$$

(b)

$$-\overset{|}{\underset{|}{C}}-X + n\,CH_2{=}CH{-}R \xrightarrow{CuX/Bipy} -\overset{|}{\underset{|}{C}}\left(CH_2-\underset{R}{\underset{|}{CH}}\right)_n X \xrightarrow{\text{Chemical Reaction}} -\overset{|}{\underset{|}{C}}\left(CH_2-\underset{R}{\underset{|}{CH}}\right)_n \text{(F)}$$

(c)

SCHEME 9.27

should be equipped not only with the desired functional groups but also with a radical stabilizing group on the α-carbon atom such as aryl, carbonyl, nitrile, and multiple halogens to ensure successful ATRP. Notably, direct bonding of halogen to aryl or carbonyl group does not facilitate radical generation. In this connection, it should also be pointed out that any functionalities in the initiator should not interfere with ATRP (i.e., should be inert to catalyst). The telechelic polymers prepared by using functional initiator approach in ATRP are presented in Table 9.3, along with the functional groups. Carboxylic acid functionalization by ATRP is rather difficult because the acid functionality poisons the catalyst. However, ATRP of MMA by using 2-bromoisobutyric acid was reported to proceed. Various protected initiators were also reported for carboxylic acid functionalization [91, 92]. Hydrolysis of the protecting groups yields polymers with the desired carboxylic acid functionalities.

Some other functionalities, including biofunctionalities [123–125], were introduced to polymers by ATRP. Substituted aromatic and aliphatic sulphonyl chlorides were shown to be efficient initiators for ATRP and used as initiators for the incorporation of functionalities to polystyrenes and polyacrylates [126]. In these applications heterogeneous CuCl $(bpy)_3$ systems were utilized.

Obviously, ATRP leads to the formation of monofunctional telechelics because the other chain always contains halogen due to the fast deactivation process. Therefore, α-ω-telechelics can only be prepared by transformation of the halide end group by means of nucleophilic substitution, free-radical chemistry, or electrophilic addition catalyzed by Lewis acids [127]. A typical example of such displacement is represented in Scheme 9.28 which illustrates the reactions used to replace the halogens with azides and consequently leading to amino-functional telechelics [128–130].

Other examples of halogen atom displacement to produce alcohol are C_{60}, epoxy, maleic anhydride, triphenylphosophine, and ketone functionalized polymers [131–133].

TABLE 9.3
Monofunctional Polymers Prepared by Using Functional Initiators in ATRP

Functional Group	Polymer	References
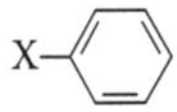 $X = H, CH_3, Br, CN, NH_2, CHO, NO_2, OCH_3$	PSt, PMA	[93–97]
CN	PSt, PMA	[98, 99]
	PSt, PMA, PtBA	[100]
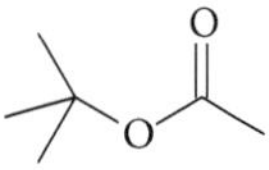	PSt, PMA	[95]
OH	PSt, PMA, PBA PMMA	[97, 101–108]
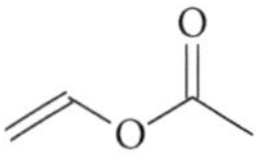	PSt	[109]
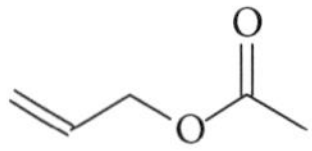	PSt, PMA	[110]
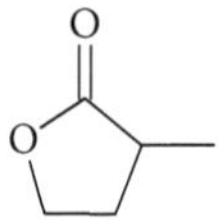	PSt, PMA	[95]
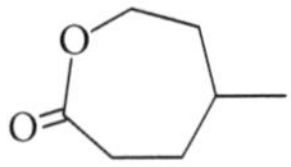	PSt	[97]
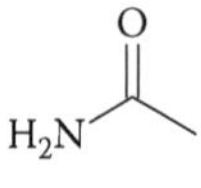 H_2N	PSt	[95]
COOH	PMMA	[111]
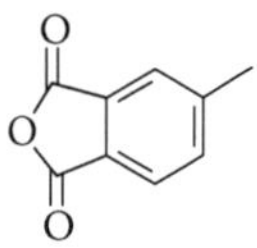	PSt	[112]
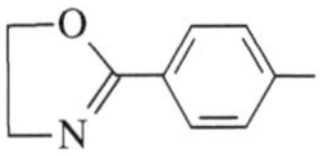	PSt	[112]
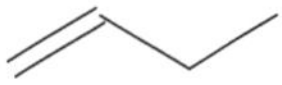	PMA	[113–115]
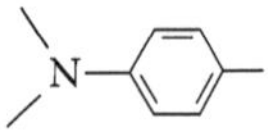	PSt	[116]

TABLE 9.3 (CONTINUED)
Monofunctional Polymers Prepared by Using Functional Initiators in ATRP

Functional Group	Polymer	References
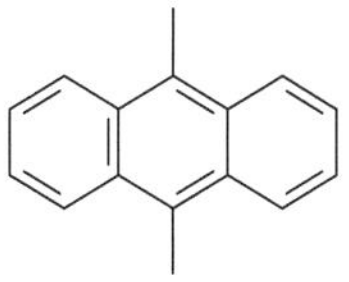	PSt	[117]
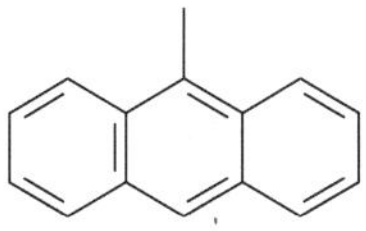	PMMA	[118]
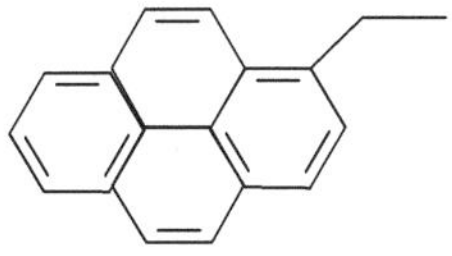	PSt	[119, 120]
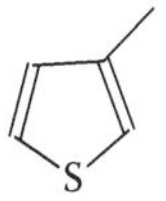 S	PMMA	[121]
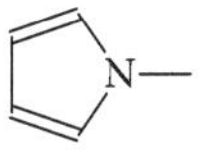 N—	PSt, PMMA	[122]

PMMA: poly (methyl methacrylate), PSt: polystyrene, PBA: poly (butyl acrylate), P*t*BA: poly (*tert*-butyl acrylate), PMA: poly (methyl acrylate).

The halide displacement is particularly important to prepare bifunctional hydroxy telechelics, which find application in the preparation of segmented polyester and polyurethanes [134]. In such applications the first hydroxyl group can be incorporated by using hydroxy-functional initiator derivatives [135]. The second hydroxyl group functionalization can then be achieved by direct displacement of a halogen group with an amino alcohol or using allyl alcohol. Diols can be prepared also by coupling of monohydroxy functional polymeric halides. In this connection, it is noteworthy to mention the recent work [116] regarding atom transfer coupling process for the synthesis of bifunctional telechelics. Polymers with monofunctional groups such as aldehyde, aromatic hydroxyl, and dimethyl amino groups were obtained by ATRP of styrene using functional initiators in the presence of the CuBr/bpy catalytic complex. The bifunctional telechelics were prepared by coupling of monofunctional

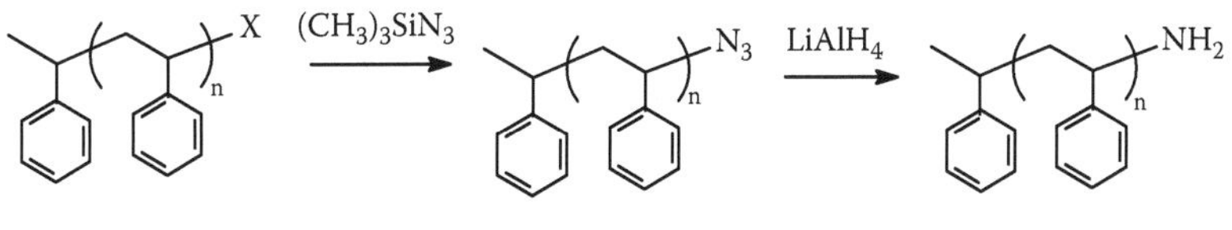

X $\xrightarrow{(CH_3)_3SiN_3}$ N_3 $\xrightarrow{LiAlH_4}$ NH_2

SCHEME 9.28

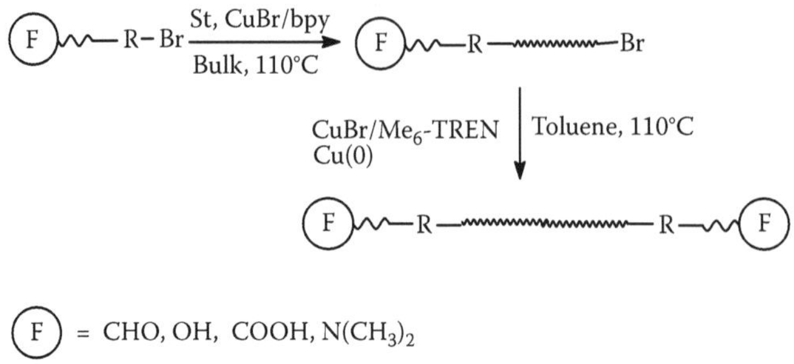

St, CuBr/bpy; Bulk, 110°C

$CuBr/Me_6$-TREN, Cu(0); Toluene, 110°C

F = CHO, OH, COOH, $N(CH_3)_2$

SCHEME 9.29

polymers in atom transfer radical generation conditions, in the absence of monomer, using CuBr as catalyst, tris[2-(dimethylamino)ethyl]amine (Me_6TREN) as ligand, Cu(0) as reducing agent and toluene as solvent. The overall process is depicted in Scheme 9.29.

This strategy was further followed by Otazaghine et al. [136] who prepared polymers with hydroxy, acid, or ester end groups and performed model studies.

Incorporation of polymerizable end groups to polymers by ATRP is limited to certain groups. As has been shown in Table 9.3, among the olefinic groups only ally and vinyl acetate groups were successfully incorporated. Obviously, the other polymerizable groups such as epoxides and oxazolines are not reactive toward radicals. To produce polymers with more reactive unsaturated end groups such as

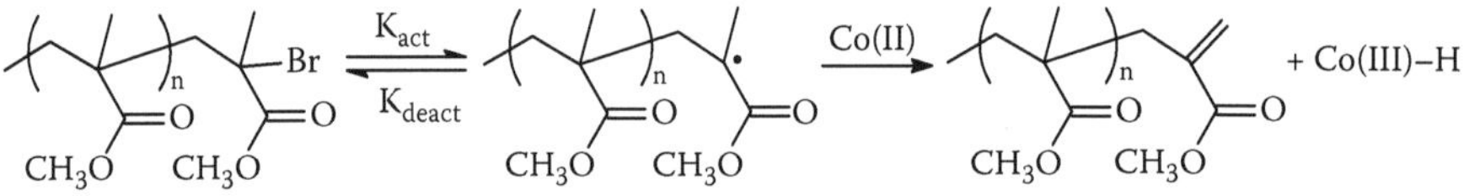

SCHEME 9.30

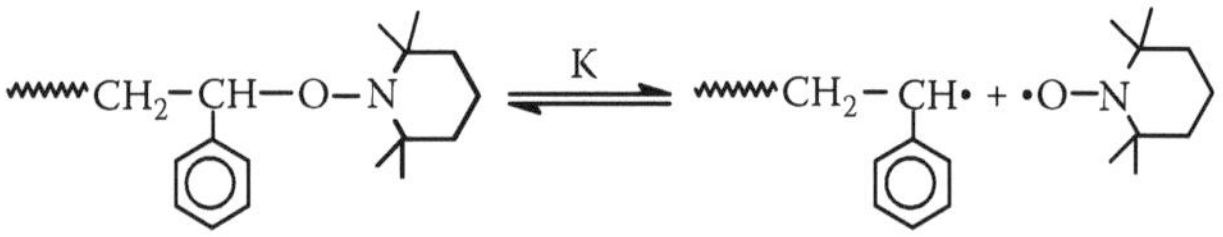

SCHEME 9.31

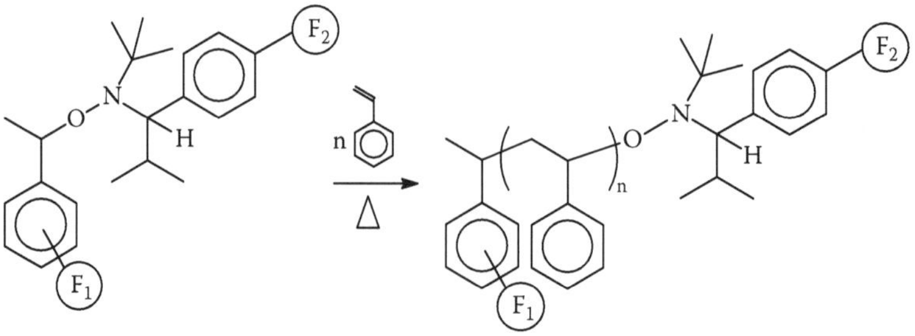

SCHEME 9.32

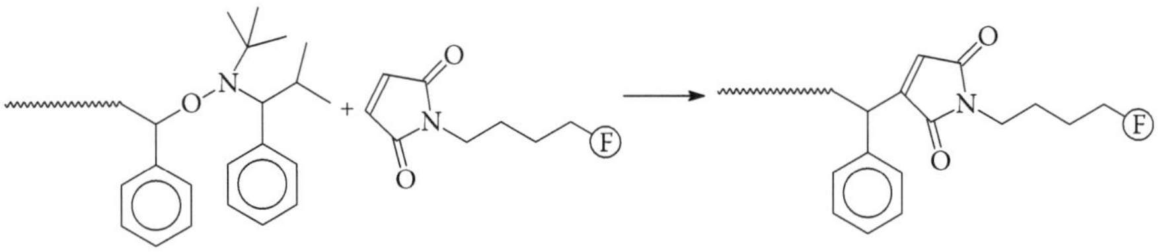

SCHEME 9.33

methacrylates, a combined ATRP and a catalytic chain transfer (CCT) process was proposed [137]. In this methodology, the catalytic chain transfer agents were added to the ATRP of MMA near to the end of polymerization leading to the formation of ω-unsaturated PMMA macromonomer with low polydispersity and controlled molecular weight.

Here, the CTC agent acts as a chain transfer terminator but does not initiate new chain in the classical manner. This process was equally applicable to the well-defined diblock macromonomers.

Similarly, Bon et al. [138] reported the use of methyl 2-bromometylacrylate as means of introducing methacrylate end groups on poly(methyl methacrylate). In addition, in this case, the functional group containing halide compound is added at the conclusion of the polymerization.

More recently, a convenient, one-pot synthesis of telechelic polymers with unsaturated end-groups was developed [139]. Addition of excess ethyl 2-bromomethacrylate to ATRP of acrylate monomers after an 80–90% conversion resulted in the formation of mono and bifunctional polymers. The average degree of end-functionality was almost quantitative.

9.3.2 Nitroxide Mediated Living Radical Polymerization (NMP)

Another controlled radical polymerization developed in recent years is "stable free-radical mediated polymerization" (SFRP), also called "nitroxide mediated radical polymerization" (NMP) [86, 140]. This type of polymerization can be realized through reversible deactivation of growing radicals by stable radicals such as 2, 2, 6, 6-tetramethyl-1-piperidinyloxy (TEMPO).

It is also possible to prepare telechelic polymers by NMP procedure because it tolerates a wide variety of functional groups [86]. For the synthesis of telechelics by NMP, two general methods are used: functional groups can be placed at the initiating chain end, F_1, or the nitroxide mediated chain end, F_2.

Telechelics with a variety of functional groups can essentially be prepared by using functional nitroxides (Table 9.4).

It was reported that high degree of functionalization (i.e., greater than 95%) is possible even at molecular weights up to 50,000–75,000 by NMP method [161].

A wide variety of functional groups including polynuclear aromatic pyrene group can also be introduced by taking advantage of monoadditioin of maleic anhydrides and maleimide derivatives to alkoxy amine end followed by elimination of mediating nitroxide radical [162]. The thermal stability of the telechelics was increased as the alkoxyamine group was removed.

TABLE 9.4
Functional Groups Attached to Polymers by Nitroxide-Mediated Living Free-Radical Procedures

Functional Group	References
OH	[141–149]
C_6H_5–	[150, 151]
F_3C–C_6H_4–	[149]
$C_{10}H_7$–	[152, 153]
—C(=O)—	[146, 153–156]
CN	[150, 157]
$PO(OR)_2$	[157, 158–160]
$P(O)_2OH$	[149]
NH_2	[86]
COOH	[86]

Telechelics can be prepared by RAFT process by selecting suitable RAFT agents [87] as explained in the previous section.

Among the controlled radical polymerization methods discussed here, ATRP is the most applied route for the preparation of telechelics because besides the initiator functionalization, the terminal halogen produced in ATRP can easily be converted to many useful functionalities (e.g., by nucleophilic substitution) [90]. Displacement of nitoxides and dihioesters is more difficult in respective NMP and RAFT processes.

9.4 CONCLUSION

Free-radical polymerization is still the most important process in view of its wide utility in terms of functionalization via various synthetic routes as described in this chapter. Functionalization of polymers can be achieved via different pathways such as initiation techniques, transfer reactions, and controlled radical polymerizations. Very rapid developments made in controlled radical polymerizations in recent years certainly revitalized its potentiality for polymer functionalization.

REFERENCES

1. M. K. Mishra, Ed., *Macromolecular Design: Concept and Practice*, Polymer Frontiers International, Inc., New York, 1994.
2. J. R. Ebdon, Ed., *New Methods of Polymer Synthesis*, Blackie and Sons, New York, 1991.

3. M. Van Beylen and M. Schwarz, *Ionic Polymerization and Living Systems*, Chapman & Hall, New York, 1993.
4. J. P. Kennedy and B. Ivan, *Designed Polymers by Carbocationic Macromolecular Engineering*, Hanser, Munich, 1991.
5. Y. Yagci and M. K. Mishra, in *Macromolecular Design: Concept and Practice*, M. K. Mishra, Ed., Polymer Frontiers International, Inc., New York, 1994, Chapters 6 and 10.
6. W. Guth and W. Heitz, *Makromol. Chem.,* 177, 1835 (1976).
7. W. Konter, B. Boemer, K. H. Koehler, and W. Heitz, *Makromol. Chem.,* 182, 2619 (1984).
8. G. Moad, D. H. Solomon, S. R. Johns, R. Johns, and R. I. Willing, *Macromolecules,* 17, 1094 (1984).
9. G. Moad, E. Rizzardo, D. H. Solomon, and S.R. Johns, *Macromol. Chem., Rapid Commun.,* 5, 793 (1984).
10. A. Onen, S. Denizligil, and Y. Yagci, *Macromolecules,* 28, 5375 (1995).
11. C. H. Bamford, A. Ledwith, and Y. Yagci, *Polymer,* 19, 354 (1978).
12. A. Onen and Y. Yagci, *J. Macromol. Sci.,* A-27, 755 (1990).
13. A. Onen and Yusuf Yagci, *Angew. Macromol. Chem.,* 181, 191 (1990).
14. Y. Yagci and A. Onen, *J. Macromol. Sci.,* A-28, 129 (1991).
15. Y. Yagci, A. Onen, and W. Schnabel, *Macromolecules,* 24, 4620 (1991).
16. Y. Yagci, G. Hizal, and U. Tunca, *Polym. Commun.,* 7, 31 (1990).
17. Y. Yagci, M. Muller, and W. Schanabel, *J.Macromol.Sci., Macromol. Report.*, A-28 (Suppl. 1), 37 (1991).
18. W. Retting, B. Strechmel, S. Schrader, and H. Seifert, *Applied Fluorescence in Chemistry, Biology and Medicine*, Springer, Berlin, 1999.
19. M.A. Winnik, A.E.C. Redpath, and K. Paton, *Polymer* 25, 91(1984).
20. S. Slomkowski, M.A. Winnik, P. Furlog, and W.F. Reynolds, *Macromolecules*, 22, 503 (1989).
21. M. Strukelj, J.M.G. Martinho, and M.A. Winnik, *Macromolecules* 24: 2488 (1991).
22. M.R. Korn and M.R. Gagne, *Chem. Commun.*, 1711 (2000).
23. M. Erdogan, Y. Hepuzer, I. Cianga, Y. Yagci, and O. Pekcan, *J. Phys. Chem. A*, 107, 8363 (2003).
24. Y.Hepuzer, U.Guvener, and Y.Yagci, *Polym. Bull.*, in press.
25. Y. Yagci, U. Tunca, and N. Bicak, *J. Polym. Sci., Polym. Lett. Ed.,* 24, 49 (1986).
26. Y. Yagci, U. Tunca, and N. Bicak, *J. Polym. Sci., Polym. Lett. Ed.,* 24, 491 (1986).
27. D. Ghatge, S. P. Vernekar, and P. O. Wadgaonkar, *Macromol. Chem., Rapid Commun.,* 4, 307 (1977).
28. R. Walz, B. Boemer, and W. Heitz, *Macromol. Chem.,* 178, 2527 (1977).
29. C. H. Bamford and A. D. Jenkins, *Nature (London),* 176, 78 (1955).
30. C. H. Bamford, A. D. Jenkins, and R. P. Wayne, *Trans. Faraday Soc.,* 56, 932 (1960).
31. C. H. Bamford, A. D. Jenkins, and R. Johnson, *Trans. Faraday Soc.,* 55, 179 (1959).
32. C. H. Bamford, A. D. Jenkins, and R. Johnson, *Trans. Faraday Soc.,* 55, 1451 (1959).
33. Y. Yagci, *Polym. Commun.,* 26, 7 (1985).
34. H. Schenko, G. Degler, H. Dongowski, R. Caspary, G. Angerer, and S. Ng, *Angew. Macromol. Chem.,* 70, 9 (1978).
35. W. Heitz, P. Ball, and M. Lattekamp, *Kautch. Bummi, Kunstst,* 34, 459 (1981).
36. S. F. Reed, *J. Polym. Sci.,* Part A-1, 9, 2029 (1971).
37. S. F. Reed, *J. Polym. Sci.,* Part A-1, 10, 649 (1972).
38. S. F. Reed, *J. Polym. Sci.,* Part A-1, 11, 55, (1973).
39. S. F. Reed, *J. Polym. Sci.,* Part A-1, 9, 2147 (1971).

40. D. Edelmann and H.Ritter, *Makromol. Chem.*, 194, 1183 (1993).
41. S. Alkan, L. Toppare, and Y. Yagci, Y. Hepuzer, *Synth. Met.*, 119, 133 (2001).
42. K. Ishize, *J. Polym. Sci., Chem. Ed.,* 28, 1887 (1990).
43. H. C. Haas, N. W. Schuler, and H. S. Kolesinski, *J. Polym. Sci., Part A-1,* 5, 2964 (1967).
44. H. N. Friendlander, *J. Polym. Sci.,* 58, 455 (1965).
45. G. A. Razuvaev, L. M. Terman, and D. M. Yanovski, *Dokl. Akad. Nauk SSSR,* 161, 614 (1965); *Chem Abstr.,* 63, 1869h (1965).
46. J. A. Verdal, P. W. Ryan, D. J. Carrow, and K. L. Kunci, *Rubber Age,* 98, 57 (1966).
47. I. M. Robinson and J. K. Kochi, *Macromolecules,* 16, 526 (1983).
48. B. Boutevin, Y. Pietrasanta, M. Taha, and T. Elsarraf, *Polym. Bull.,* 10, 157 (1983).
49. B. Boutevin, Y. Pietrasanta, M. Taha, and T. Elsarraf, *Makromol. Chem.,* 184, 2401 (1983).
50. T. Tsuda, Y. Yasuda, and T. Azuma, Japanese Patent No. 6164705; *Chem, Abstr.,* 105, 209574 (1986).
51. K. F. Gillman and E. Scogles, *J. Polym. Sci., Polym. Lett. Ed.,* 5, 477 (1967).
52. C. Bonardi, B. Boutevin, Y. Pietrasanta, and M. Taha, *Macromol. Chem.,* 186, 261 (1985).
53. H. A. Nguyen and E. Marechal, *J. Macromol. Sci.-Rev. Macromol. Chem. Phys.,* C28, 187 (1988).
54. R. M. Pierson, A. J. Constanza, and A. H. Weienstein, *J. Polym. Sci.,*17, 221 (1995).
55. A. Schefer, A. J. Grodzinsky, K. L. Prime, and J. P. Busnel, *Macromolecules,* 26, 2240 (1993).
56. T. Otsu and M. Yoshida, *Macromol. Chem., Rapid Commun.,* 3, 127 (1993).
57. G. Clouet, *Polymer Prep.,* 33(1), 895 (1992).
58. C.P.R. Nair, G. Clouet, and P. Chaumont, *J. Polym. Sci., Polym. Chem. Ed.,*27, 1795 (1989).
59. C.P.R. Nair and G. Clouet, *Makromol. Chem.*, 190, 1243 (1989).
60. G. Clouet and H.J. Juhl, *Makromol. Chem.*, 195, 243 (1994).
61. S.A. Haque and G. Clouet, *Makromol. Chem.*, 194, 315 (1994).
62. A. Bledzki, D. Braun, and K. Titzschkau, *Makromol. Chem.* 184, 745 (1983).
63. D. Edelmann and H. Ritter, *Makromol. Chem.*, 194, 2375 (1994).
64. D. Braun, T. Skrzek, S. Steinhauserbeiser, H. Tretner, and H.J. Lindner, *Makromol. Chem.*, 196, 573 (1995).
65. W. J. Bailey, P. Y. Chen, W. B. Chiao, T. Endo, L. Sidney, N. Yamamoto, N. Yamazaki, and K. Yonezawa, in *Contemporary Topics in Polymer Science,* M. Shen, Ed., Plenum Press, New York, 1979, Vol. 3, p. 29.
66. W. J. Bailey, T. Endo, B. Gabud, Y. N. Lin, Z. Ni, C. Y. Pan, S. E. Shaffer, S. R. Wu, N. Yamazaki, and K. Yanezawa, *J. Macromol., Sci.: Chem,* A-21, 979 (1984).
67. W. J. Bailey, P. Y. Chen, S. C. Chen, W. B. Chiao, T. Endo, B. Gabud, Y. N. Lin, Z. Ni, C. Y. Pan, S. E. Shaffer, L. Sidney, S. R. Wu, N. Yamamoto, N. Yamazaki, and K. Yonezawa, *J. Macromol, Sci.: Chem,* A-21, 1611 (1984).
68. W. J. Bailey, *Polym. J.,* 17, 85 (1985).
69. W. J. Bailey, *ACS Symp. Ser,* 286, 47 (1985).
70. W. J. Bailey, *Macromol. Chem., Suppl.,* 13, 171 (1985).
71. W. J. Bailey, B. Gabud, Y. N. Lin, Z. Ni, and S. R. Wu, *ACS Symp. Ser.,* 282, 147 (1985).
72. W. J. Bailey, P. Y. Chen, S. C. Chen, W. B. Chiao, T. Endo, B, B. Gabud, Y. Kuruganti, Y. N. Lin, Z. Ni, C. Y. Pan, S. E. Shaffer, L. Sidney, S. R. Wu, N. Yamamoto, K. Yonezawa, and L. L. Zhou, *Macromol. Chem., Macromol. Symp.,* 6, 81 (1986).
73. D. H. R. Barton and D. Crich, *Tetrahedron Lett.,* 25, 2787 (1984).

74. D. H. R. Barton and D. Crich, *Tetrahedron Lett.*, 26, 757 (1985).
75. E. Rizzardo, G. F. Meijs, and S. H. Thang, *Macromol. Symp.*, 98, 101 (1995).
76. E. Rizzardo, G. F. Meijs, and S. H. Thang, PCT Int.Appl.WO 88/4304A1 (16 Jun 1988); *Chem Abstr.*, 110, 7629k (1989).
77. G. F. Meijis, E. Rizzardo, and S. H. Thang, *Polym. Bull.*, 24, 501 (1990).
78. G. F. Meijis, E. Rizzardo, and S. H. Thang, *Polym. Prepr.*, 33, 893 (1992).
79. K. S. Murthy and K. Kishore, *J. Polym. Sci., Polym. Chem. Ed.*, 34, 1415 (1996).
80. G. F. Meijis and E. Rizzardo, *Macromol. Chem., Rapid Commun.*, 9, 547 (1988).
81. G. F. Meijis and E. Rizzardo, *Macromol. Chem.*, 191, 1545 (1990).
82. Y. Yagci, O. Nuyken, and V. Graubner, in *Encylopedia of Polymer Science, 3rd ed.*, (J. I. Kroschwitz, Ed.),Wiley, New York, 12, 57–130 (2005).
83. K. Matyjaszewski, Ed., *Controlled Radical Polymerization*, ACS Symp. Ser. Washington, D.C., 685, 1997.
84. K. Matyjaszewski, *Chem. Rev.*, 101, 2921 (2001).
85. M. Kamigaito and M.Sawamoto, *Chem. Rev.*, 101, 3689 (2001).
86. C. J. Hawker, A. W. Bosman, and E.Harth, *Chem. Rev.*, 101, 3661 (2001).
87. J. Chiefari and E. Rizzardo, in K. Matyjaszewski and T. P. Davis, Eds., *Handbook of Radical Polymerization*, Wiley-Interscience, New York, 2002, Chapter 12.
88. J. S. Wang and K. Matyjaszewski, *Macromolecules*, 28, 7901 (1995).
89. J. S. Wang and K. Matyjaszewski, *J. Am. Chem. Soc.*, 117, 5614 (1995).
90. X. Zhang, J. Xia, S. Gaynor, and K. Matyjaszewski, *Polym. Mat. Sci. Eng.*, 79, 409 (1998).
91. D. M. Haddleton and C. Waterson, *Macromolecules*, 32, 8732 (1999).
92. K. Matyjaszewski, S. Coca, Y. Nakawa, and J. Xia, *Polym. Mat. Sci. Eng.*, 76, 147 (1997).
93. K. Matyjaszewski, V. Coessens, V. Nakawa, Y. Xia, J. Qiu, S. G. Gaynor, S. Coca, and C. Jaseczek, *Am. Chem. Soc., Symp. Ser.*, 704, 16 (1998).
94. S. G. Gaynor and K. Matyjaszewski, *Am. Chem. Soc., Symp. Ser.*, 685, 396 (1997).
95. X. Zhang, J. Xia, and K. Matyjaszewski, *Macromolecules*, 33, 2340 (2000).
96. K. Matyjaszewski, S. M. Jo, and H.-J. Paik, *Macromolecules*, 30, 6398 (1997).
97. K. Matyjaszewski, M. Wei, J. Xia, and N. E. McDermott, *Macromolecules*, 30, 8161 (1997).
98. D. Mecerreyes, B. Athoff, K. A. Boduch, M. Trollsas, and J. L. Hedrick, *Macromolecules*, 32, 5175 (1999).
99. G. Moineau, C. Granel, P. Dubois, P. Teyssie, and R. Jerome, *Macromolecules*, 32, 27 (1997).
100. C. J. Hawker, J. L. Hedrick, E. E. Malmstorm, M. Trollsas, D. Mecerreyes, G. Moineau, P. Dubois, and R. Jerome, *Macromolecules*, 31, 213 (1998).
101. M. Desterac, K. Matyjaszewski, and B. Boutevin, *Macromol. Chem. Phys.*, 201, 265 (2000).
102. M. Desterac, K. Matyjaszewski, and B. Boutevin, in *Controlled/Living Radical Polymerization*, K. Matyjaszewski, Ed., ACS Symp. Ser., American Chemical Society, Washington D.C., 2000, Chapter 17.
103. D. M. Haddleton, C. Waterson, P. J. Derrick, C. B. Jasieczek, and A. J. Shooter, *Chem. Commun.*, 683 (1997).
104. H. Keul, A. Neumann, B. Reinung, and H. Hocker, *Macromol. Symp.*, 161, 63 (2000).
105. V. Cosessens and K. Matyjaszewski, *Macromol. Rapid Commun.*, 20, 66 (1999).
106. Y. Xu, C. Pan, and L.Tao, *J. Polym. Sci., Polym. Chem. Ed.*, 38, 436 (2000).
107. K. Matyjaszewski, K. L. Beers, A. Kern, and S. G. Gaynor, *J. Polym. Sci., Polym. Ed.*, 36, 823 (1998).
108. Y. Nakagawa and K. Matyjaszewski, *Polym. J.*, 30, 138 (1998).

109. D. M. Haddleton, A. M. Heming, D. Kukulj, D. J. Duncalf, and A. J. Shooter, *Macromolecules*, 31, 2016 (1998).
110. H. Malz, H. Comber, D. Voigt, I. Hopfe, and J. Pionteck, *Macromol. Chem. Phys.*, 200, 642 (1999).
111. A. Kajiwara and K. Matyjaszewski, *Polymer J.*, 31, 70 (1999).
112. F. Zeng, Y. Shen, and F. Pelton, *Macromolecules*, 33, 1628 (2000).
113. Y. Shen, S. Zhu, and F. Pelton, *Macromol. Chem., Phys.,* 201, 1387 (2000).
114. S. Yurteri, I. Cianga, and Y. Yagci, *Macromol. Chem. Phys.*, 204, 1171 (2003).
115. K. Ohno, K. Fujimoto, Y. Tsujii, and T. Fukuda, *Polymer*, 40, 759 (1999).
116. M. Erdogan, G. Hizal, U. Tunca, D. Hayrabetyan, and O. Pekcan, *Polymer*, 43, 1925 (2002).
117. M. R. Korn, and M. R. Gagne, *Chem. Commun.*, 1711 (2000).
118. M. Erdogan, Y. Hepuzer, I. Cianga, Y. Yagci, and O. Pekcan, *J. Phys. Chem.*, 107, 8363 (2003).
119. S. Alkan, L. Toppare, Y. Hepuzer, and Y. Yagci, *J. Polym. Sci., Polym. Chem. Ed.*, 37, 4218 (1999).
120. D. Mecerreyes, J.A. Pompose, M. Bengoetxea, and H. Grande, *Macromolecules*, 33, 5846 (2000).
121. D. M. Haddleton and K. Ohno, *Biomacromolecules*, 1, 152 (2000).
122. D. M. Haddleton, R. Edmonds, A.M. Heming, E. Keely, and E.J. Kukulj, *New J. Chem.*, 23, 477 (1999).
123. A. Marsh, A. Khan, D. M. Haddleton, and M. J. Hannon, *Macromolecules*, 32, 8725 (1999).
124. V. Percec, H. J. Kim, and B. Barboi, *Macromolecules*, 30, 8526 (1997).
125. K. Matyjaszewski, Y. Nakagawa, and S. G. Gaynor, *Macromol. Rapid Commun.*, 18, 1057 (1997).
126. V. Cosessens, Y. Nakagawa, and K. Matyjaszewski, *Polym. Bull.*, 40, 135 (1998).
127. V. Cosessens and K. Matyjaszewski, *J. Macromol. Chem. Sci., Pure Appl. Chem.*, A36, 667 (1999).
128. L. Li, C. Wang, Z. Long, and S. Fu, *J. Polym. Sci., Polym. Chem. Ed.*, 38, 4518 (2000).
129. V. Cosessens, J. Pyun, P.J. Miller, S.G. Gaynor, and K. Matyjaszewski, *Macromol. Rapid Commun.*, 21, 103 (2000).
130. E. G. Koulouri, J. K. Kallitsis, and G. Hadziioannou, *Macromolecules*, 32, 6242 (1999).
131. T. Ando, M. Kamigatio, and M. Sawamoto, *Macromolecules*, 31, 6708 (1998).
132. A. K. Shim, V. Cosessens, T. Pintauer, S. G Aynor, and K. Matyjaszewski, *Polym. Prep.* (*Am. Chem. Soc. Div. Polym. Chem.*) 37(1), 577 (1996).
133. V. Cosessens and K. Matyjaszewski, *Macromol. Rapid Commun.*, 20, 127 (1999).
134. B. Otazaghine, G. David, B. Boutevin, J. J. Robin, and K. Matyjaszewski, *Macromol. Chem. Phys.*, 205, 154 (2004).
135. J. Norman, S. C. Moratti, A. T. Slark, D. J. Irvine, and A. T. Jackson, *Macromolecules*, 35, 8954 (2002).
136. S. A. F. Bon, S. R. Morsley, C. Waterson, and D. M. Haddleton, *Macromolecules*, 33, 5819 (2000).
137. C. W. Bielawski, J. M. Jethmalani, and R. H. Grubbs, *Polymer*, 44, 3721 (2003).
138. M. K. Georges, R. P. N. Veregin, P. M. Kazmaier, G. K. Hamer, and G. Hamer, *Macromolecules*, 26, 2987 (1993).
139. E. Yoshida and T. Fujii, *J. Polym. Sci., Polym. Chem.*, 36, 269 (1998).
140. W. G. Skene, S. T. Belt, T. J. Connolly, P. Hahn, and C. Scaiano, *Macromolecules*, 31, 9103 (1998).
141. C. J. Hawker, G. G. Barclay, and J. Dao, *J. Am. Chem. Soc.*, 118, 11467 (1996).

142. E. Yoshida and A. Sugita, *J. Polym. Sci., Polym. Chem.*, 36, 2059 (1998).
143. W. G. Skene, J. C. Sciaino, N. Listigovers, P. M. Kazmaier, and M. K. Georges, *Macromolecules*, 33, 5065 (2000).
144. Y. Miura, N. Nakamura, and I. Taniguchi, *Macromolecules*, 34, 447 (2001).
145. S. Marque, H. Fischer, E. Baier, and A. Studer, *J. Org. Chem.*, 66, 1146 (2001).
146. E. Harth, B. van Horn, and C. J. Hawker, *Chem. Commun.*, 823 (2001).
147. K. Matyjaszewski, S. G. Gaynor, D. Greszta, D. Mardare, and T. J. S. Shigemoto Wang, *Macromol. Symp.*, 95, 217 (1995).
148. D. Benoit, V. Chaplinski, R. Braslau, and C. J. Hawker, *J. Am. Chem. Soc.*, 121, 3904 (1999).
149. S. Jousset and J. M. Catala, *Macromolecules*, 33, 4705 (2000).
150. T. Shigemoto and K. Matyjaszewski, *Macromol. Rapid Commun.*, 17, 347 (1996).
151. N. J. Turro, G. Lem, and I. S. Zavarine, *Macromolecules*, 33, 9782 (2000).
152. C. H. Han, M. Drache, H. Baethge, and G. Schmidt-Naake, *Macromol. Chem.*, 200, 1779 (1999).
153. R. P. N. Veregin, M. K. Georges, P. M. Kazmaier, and G. K. Hamer, *Macromolecules*, 26, 5316 (1993).
154. Y. K. Chong, F. Ercole, G. Moad, E. Rizzardo, S. H. Thang, and A. G. Anderson, *Macromolecules*, 32, 6895 (1999).
155. M. O. Zink, A. Kramer, and P. Nesvadba, *Macromolecules*, 33, 8106 (2000).
156. E. Yoshida and Y. Okada, *Bull. Chem. Soc. Jpn.*, 70, 275 (1997).
157. M. K. Georges, R. P. N. Veregin, P. M. Kazmaier, G. K. Hamer, and M. Saban, *Macromolecules*, 27, 7228 (1994).
158. D. Benoit, S. Grimaldi, S. Robin, J.P. Finet, P. Tordo, and Y. Gnanou, *J. Am. Chem. Soc.*, 122, 5929 (2000).
159. M. Radlert, E. Harth, I. Rees, and C.J. Hawker, *J. Polym. Sci., Polym. Chem. Ed.*, 38, 4749 (2000).
160. E. Harth, C. J. Hawker, W. Fan, and R. M. Waymouth, *Macromolecules*, 34, 3856 (2001).

10 Controlled/Living Radical Polymerization

Umit Tunca, Gurkan Hizal, Metin H. Acar, M. Atilla Tasdelen, Yusuf Yagci, and Munmaya K. Mishra

CONTENTS

10.1 INTRODUCTION

Today, conventional free-radical polymerization (FRP) is still one of the most widely applied processes for the preparation of polymeric materials as nearly 50% of all commercial synthetic polymers are produced by this method. The main reason for this fact is that a wide range of monomers can be polymerized and copolymerized via radical chemistry, which provides a spectrum of materials for various markets. Moreover, the polymerization does not require rigorous process conditions. On the other hand, some important elements of the polymerization process that would lead to the well-defined polymers with controlled molecular weight, polydispersity, composition, structural architecture, and functionality are poorly controlled. The importance of the synthesis of polymers with such control has been augmented due to the rising demand for the specialty polymers. Obviously, living polymerization is an essential technique for synthesizing polymers with controlled structures. Moreover, living polymerization techniques allow preparation of macromonomers, macro initiators, functional polymers, block and graft copolymers, and star polymers. In this way, the need for specialty polymers having a desired combination of properties can be fulfilled. Control of complex architectures by living polymerization has largely been achieved using living anionic and cationic as well as group transfer polymerization techniques. From the practical point of view, however, these techniques are less attractive than free-radical polymerization, because the latter can be performed much more easily. Moreover, ionic techniques are limited to a very few vinyl monomers, whereas practically all vinyl monomers can be homo and copolymerized by a free-radical mechanism.

A long-lasting goal has been the development of controlled/living radical polymerization methods. As mentioned previously, radical polymerization suffers from some defects (i.e., the control of the reactivity of the polymerizing monomers and, in turn, the control of the structure of the resultant polymer). In ionic living systems, however, the chain ends do not react with one another due to the electrostatic repulsion. On the other hand, the growing radicals very easily react with each other at diffusion controlled rates via combination and/or disproportionation. Therefore, controlled/living radical polymerization has long been considered impossible. Following the discovery of living anionic and cationic polymerization, many attempts have been made to find controlled/living radical polymerization systems to achieve a high level of control over molar mass, polydispersity, end groups, and architecture. Despite considerable progress, a truly controlled/living radical polymerization has not been developed until a little more than a decade ago. In the past decade, a number of controlled/living radical polymerization methods have been developed and the three most promising types are: atom transfer radical polymerization (ATRP) (also known as transition metal catalyzed radical polymerization), stable free-radical polymerization (SFRP) (also known as nitroxide-mediated polymerization, NMP), and reversible addition-fragmentation chain transfer (RAFT) polymerization. This chapter focuses on the recent progress on these three methods and the earlier attempts will not be considered here as they were discussed in detail in the first edition of the book.

10.2 METAL CATALYZED ATOM TRANSFER RADICAL POLYMERIZATION

Metal-catalyzed controlled/living radical polymerization (C/LRP), mediated by Cu, Ru, Ni, and Fe metal complexes, is one of the most efficient methods to produce polymers in the field of C/LRP.[1–4] Among aforementioned systems, copper-catalyzed LRP in conjunction with organic halide initiator and amine ligand, often called atom transfer radical polymerization (ATRP), received more interest. The name of atom transfer radical polymerization (ATRP) comes from the atom transfer step, which is the key elementary reaction responsible for the uniform growth of the polymeric chains. ATRP, which is the most versatile method of the controlled free-radical polymerization systems, uses a wide variety of monomers, catalysts, solvents, and reaction temperatures. This approach enables researchers to combine very different types of monomers into one polymeric structure by a one-pot or sequential two-step method. It has been used for the synthesis of polymers with well-defined compositions (e.g., block, gradient, or alternating), topologies (e.g., star, comb, or branched), and chain functionalities.[5–8] As a multicomponent system, ATRP includes the monomer, an initiator with a transferable halogen, and a catalyst (composed of a transition metal species with any suitable ligand). Both activating and deactivating components of the catalytic system must be simultaneously present. For a successful ATRP, other factors, such as solvent, temperature, concentrations, and solubility of all components (and sometimes the order of their addition) must also be taken into consideration. However, the main purpose of this chapter is to provide the reader with an overview of the atom transfer radical polymerization (ATRP). Thus, no attempt will be made to give a detailed account of the experimentation.

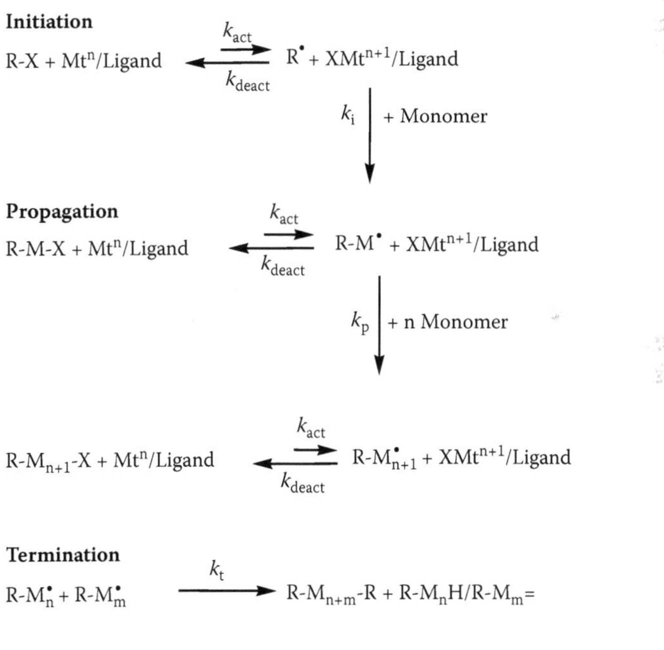

SCHEME 10.1

The general mechanism of ATRP, which is presented in Scheme 10.1, involves the abstraction of a halogen from the dormant chain by a metal center, such as complexes of Cu^I, in a redox process. Upon halogen abstraction, the free radical formed (the transient radical) can undergo propagation. However, the free radical is also able to abstract the halogen back from the metal, reproducing the dormant species. These processes are rapid, and the dynamic equilibrium that is established favors the dormant species. By this way, all chains can begin growth at the same time, and the concentration of free radicals is quite low, resulting in a reduced amount of irreversible radical-radical termination.

The rates of polymerization and polydispersity in ATRP, assuming steady-state kinetics, are given in Eqs. 10.1 and 10.2, respectively.[6, 9, 10]

$$R_p = k_{\mathrm{p}}.K_{\mathrm{eq}}\frac{[\mathrm{R\text{-}X}][\mathrm{Mt^n}]}{[\mathrm{Mt^{n+1}}]}[\mathrm{M}] \quad \text{or} \quad \ln\left(\frac{[\mathrm{M_0}]}{[\mathrm{M}]}\right) = \frac{k_{\mathrm{p}}.k_{\mathrm{act}}[\mathrm{R\text{-}X}][\mathrm{Mt^n}]}{k_{\mathrm{deact}}[\mathrm{Mt^{n+1}}]}t = k_{\mathrm{app}}t \tag{10.1}$$

$$\frac{M_w}{M_n} = 1 + \left(\frac{k_p[\mathrm{R\text{-}X}]}{k_{\mathrm{deact}}[\mathrm{Mt^{n+1}}]}\right)\left(\frac{2}{\mathrm{p}} - 1\right) = 1 + \frac{2}{k_{\mathrm{act}}[\mathrm{Mt^n}]\,t} \tag{10.2}$$

From Eq. (10.1), the rate of polymerization, R_{p}, is directly proportional to the equilibrium constant, K_{eq}, and the propagation rate constant. The proper selection of the reaction components of an ATRP process led to establishment of an appropriate equilibrium between activation and deactivation processes. The equilibrium constant ($K_{\mathrm{eq}} = k_{\mathrm{act}}/k_{\mathrm{deact}}$) plays an important role in the fate of ATRP because it determines the concentration of radicals and, therefore, the rates of polymerization and termination. K_{eq} must be low to maintain a low stationary concentration of radicals; thus the termination reaction is suppressed. However, at the early stage of polymerization, the transient radicals can terminate either by coupling or by disproportionation (k_{t}), resulting in a higher concentration of deactivator species in the system. Consequently, the equilibrium is strongly shifted toward the dormant species ($k_{\mathrm{act}} << k_{\mathrm{deact}}$); this phenomenon is known as persistent radical effect (PRE).[11,12] For the ATRP system, the rate of polymerization, R_{p}, is first order with respect to the monomer [M] and the activator [$\mathrm{Mt^n}$] concentrations and increases with the concentrations of activator, monomer, and initiator [R-X] and decreases with the increasing deactivator [Mt^{n+1}] concentration. Equation (10.2) shows that lower polydispersities are obtained at higher conversion, higher k_{deact} relative to k_{p}, higher concentration of deactivator, and higher monomer to initiator ratio, $[\mathrm{M}]_0/[\mathrm{I}]_0$. It should be noticed that Eq. (10.1) was derived for a simplified version of the reaction in Scheme 10.1. Therefore, the effect of termination reactions on the overall kinetics of polymerization was excluded. Both Fischer[12] and Fukuda et al.[13] have separately shown that under ideal conditions the persistent radical effect should result in apparent orders of reactants of less than unity for monomer, initiator, and activator. Indeed, termination reactions occur continuously throughout the polymerization. Thus, the concentration of the deactivator species increases and monomer consumption should not be simply first-order. Fischer has shown that the monomer consumption should follow Eq. (10.3).[12]

$$\ln\left(\frac{[\mathrm{M_0}]}{[\mathrm{M}]}\right) = \frac{3}{2}k_p([\mathrm{RX}]_0[\mathrm{Mt^n}]_0)^{1/3}\left(\frac{k_{\mathrm{act}}}{3k_{\mathrm{deact}}.2k_{\mathrm{t}}}\right)^{1/2} t^{2/3} \tag{10.3}$$

However, most published ATRP kinetic data represent a linear semilogarithmic plot of monomer conversion versus time. The observed deviations from ideal PRE behavior resulted from a number of sources, such as the potential of a significant amount of deactivator being present in the activator at the beginning of the polymerization.

Because the ATRP is a C/LRP, the theoretical number-average molecular weight of the polymer can be calculated as follows:

$$\text{DPn} = [\text{M}]0/[\text{I}]0 \cdot \text{Conversion}, \quad \text{DP} = \text{degree of polymerization} \qquad (10.4)$$

Well-defined polymers with molecular weights ranging from 1000 to 150,000 have been successfully synthesized by ATRP with a relatively narrow molecular weight distribution ($1.0 < M_w/M_n < 1.5$). Equation (10.2) shows how the molecular weight distribution, M_w/M_n, decreases with conversion, p. A low propagation rate constant, k_p, and a low value for the ratio of k_{act}/k_{deact} are required to obtain a polymer with narrow molecular weight distribution.[14] Higher M_w/M_n values are generally found for polyacrylates than for polystyrene or polymethacrylates due to a much higher k_p for the acrylate monomers.[6]

10.2.1 Initiators

Organic halides having a labile carbon-halogen bond are the most successfully employed initiators in ATRP. In general, these organic halides possess electron withdrawing groups and/or atoms such as carbonyl, aryl, cyano, or halogens at α-carbon to stabilize the generated free radicals. Moreover, the polymerization rates are found to be first order with respect to the concentration of initiator and the molecular weights change reciprocally with initial initiator concentration in ATRP.[15] The best control over molecular weight was achieved with bromine and chlorine atoms, whereas iodine was used for acrylate polymerization with limited success.[16] The carbon–fluorine bond strength is too strong for the fast activation–deactivation cycle with atom transfer. For the successful ATRP, initiation should be quantitative and faster than the propagation. To obtain similar reactivity of the carbon-halogen bond in the initiator and the dormant polymer end, the structure of the alkyl group, R, of the initiator should be similar to the structure of the dormant polymer end. The initiators used in ATRP such as α-haloesters (I-1 to I-7), (haloalkyl) benzenes (I-8 to I-11), sulfonyl halides (I-12 to I-18), *N*-halogenated amides, lactams, imides (I-19 to I-23), allyl halides (I-24), α-haloketons (I-25 to I-27), α-halonitriles (I-28), *N*-chlorosulfonamide (I-29), and haloalkanes (I-30 to I-32) are summarized in Table 10.1.

10.2.2 Monomers

Many types of monomers are polymerized by ATRP to produce controlled molecular weight and narrow molecular weight distribution (Table 10.2). Most of them are conjugated monomers such as styrenes, acrylates, methacrylates, acrylamides, and acrylonitrile, which possess substituents that can stabilize the propagating radicals. Each monomer has its own equilibrium constant, K_{eq}, which determines the polymerization rate in ATRP according to Eq. (10.1), even if the same catalyst is used. In this regard, methacrylates in general, particularly MMA, were found to be more easily polymerizable than styrenes and acrylates, because of the lower values of K_{eq} for the latter.

TABLE 10.1
Organic Halides Used as Initiators in ATRP[6, 7]

Initiator		Monomer
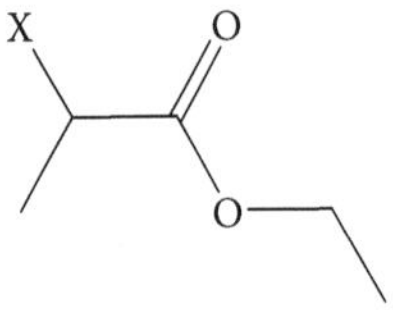 X= Cl, Br, I	I-1	Acrylates Methacrylates Acrylamides
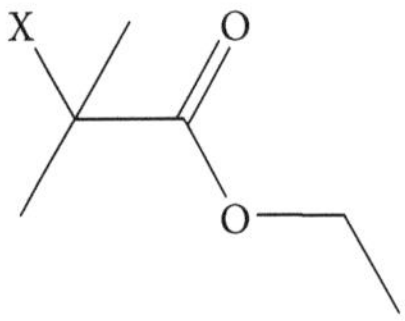 X = Br, I	I-2	Methacrylates Styrenes (St)
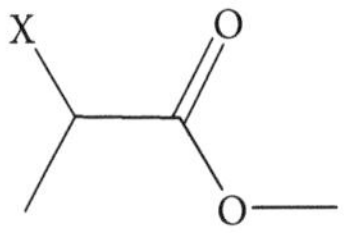 X = Cl, Br	I-3	Acrylates
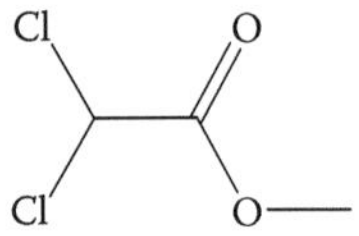	I-4	Methyl methacrylate (MMA)
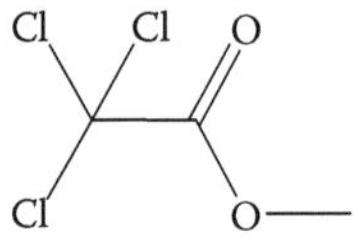	I-5	MMA
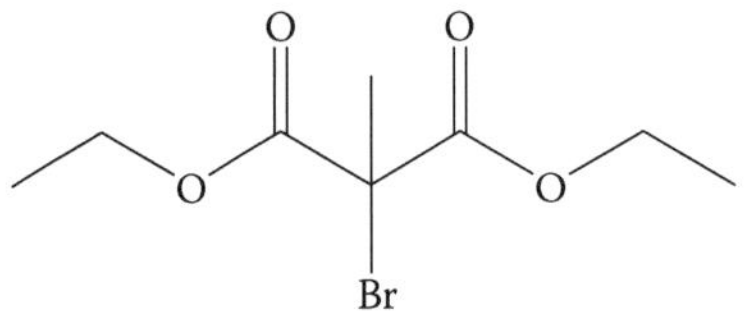	I-7	MMA
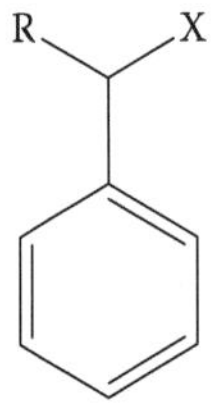 R = –H, $-CH_3$, –Phenyl X = Cl, Br	I-8	Styrenes (St), MMA

TABLE 10.1 (CONTINUED)
Organic Halides Used as Initiators in ATRP[6, 7]

Initiator		Monomer
	I-9	MMA
	I-10	MMA
X = Cl, Br	I-11	MMA[125]
R = $-CH_3$, $-OCH_3$, –F, –naphtyl, $-NO_2$, –H, –OH, –COOH	I-12	Methacrylates, acrylates, styrenes (St)
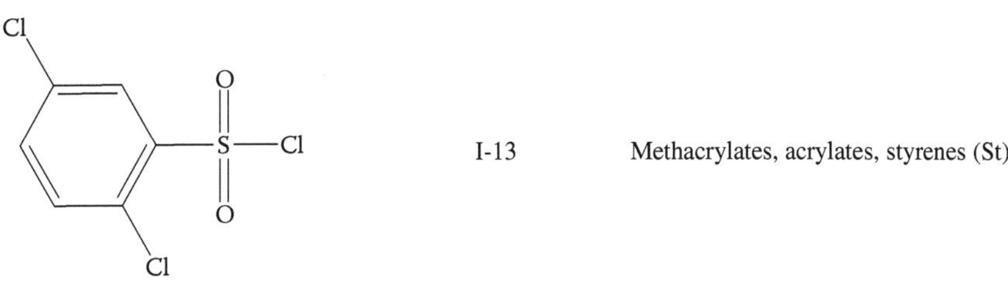	I-13	Methacrylates, acrylates, styrenes (St)

(Continued)

TABLE 10.1 (CONTINUED)
Organic Halides Used as Initiators in ATRP[6, 7]

Initiator		Monomer
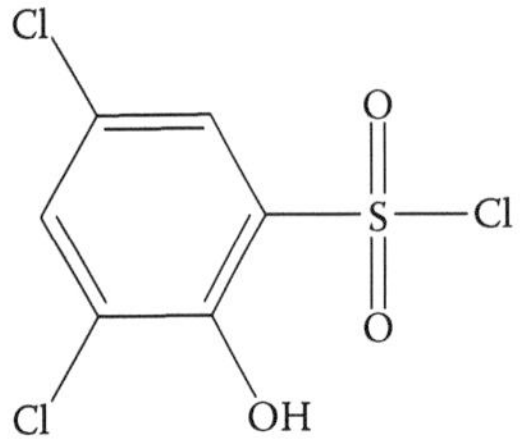	I-14	Methacrylates, acrylates, styrenes (St)
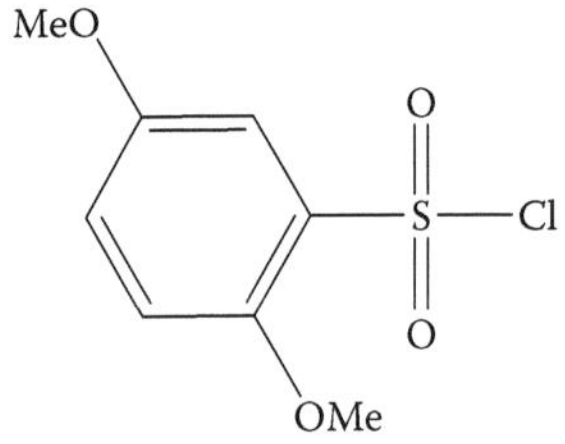	I-15	Methacrylates, acrylates, styrenes (St)
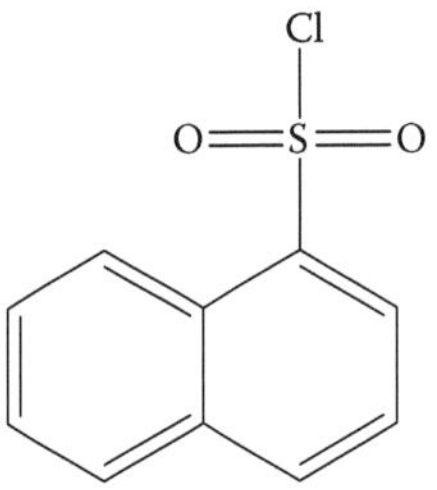	I-16	Methacrylates, acrylates, styrenes (St)
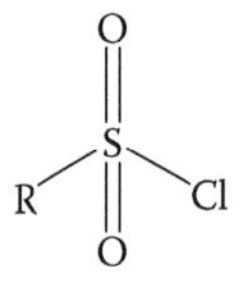 R = $-CH_3$, $-CCl_3$, $-C_4F_9$, $-C_8H_{17}$,	I-17	Methacrylates, acrylates, styrenes (St)
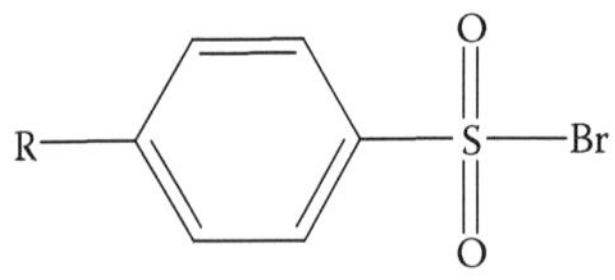 R = $-CH_3$, $-OPh-SO_2Br$	I-18	Methacrylates, acrylates, styrenes (St)[126]

TABLE 10.1 (CONTINUED)
Organic Halides Used as Initiators in ATRP[6, 7]

Initiator		Monomer
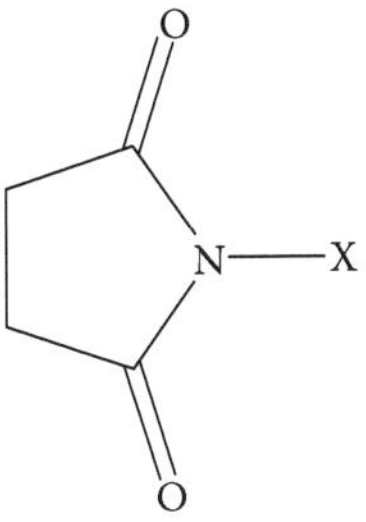 X = Cl, Br	I-19	MMA[127]
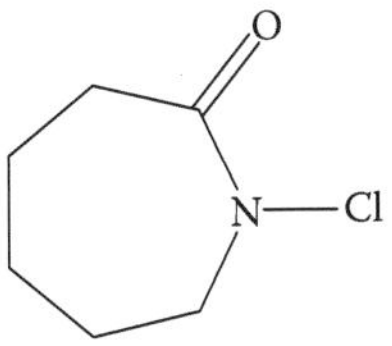	I-20	MMA[127]
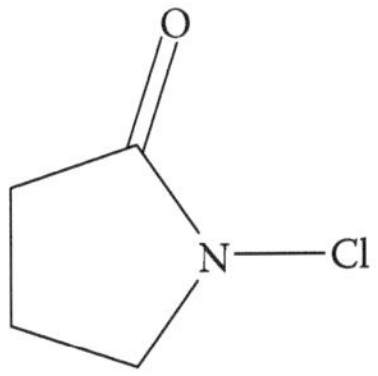	I-21	MMA[127]
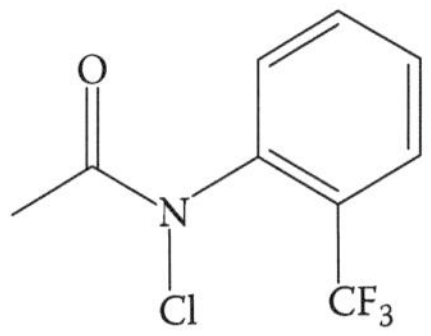	I-22	MMA[127]
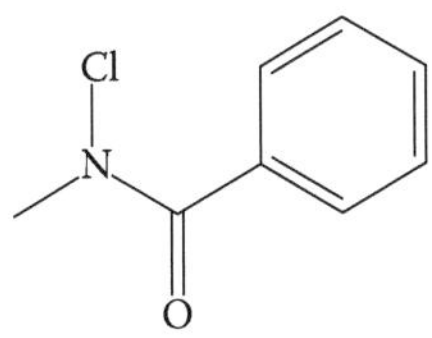	I-23	MMA[127]
 X = Cl, Br	I-24	Styrene (St)

(*Continued*)

TABLE 10.1 (CONTINUED)
Organic Halides Used as Initiators in ATRP[6, 7]

Initiator		Monomer
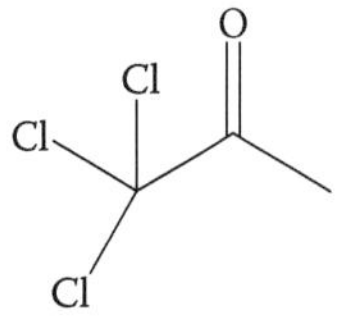	I-25	MMA
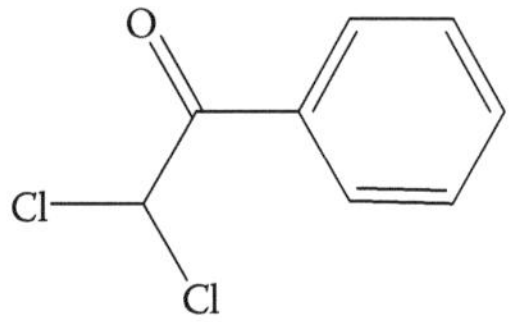	I-26	MMA
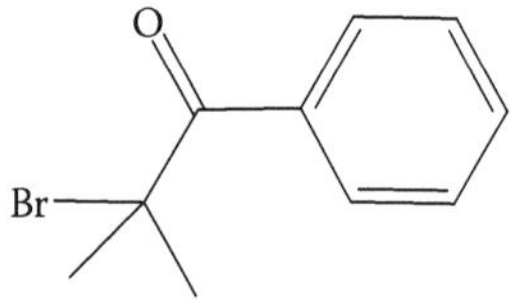	I-27	MMA
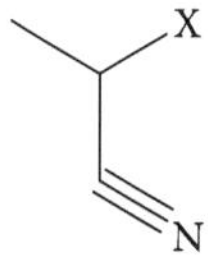 X = Cl, Br	I-28	Acrylonitrile (AN) MMA
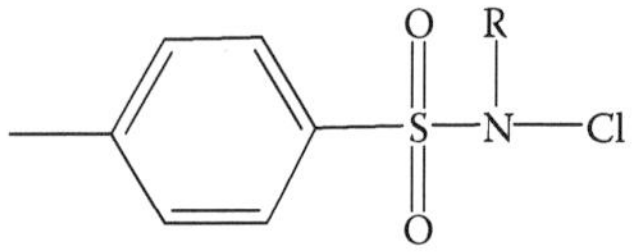 R = $-C_3H_7$	I-29	MMA, styrene (St)[128]
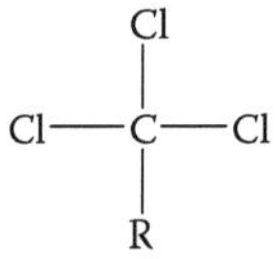	I-30	MMA, styrene (St), methylacrylate (MA)
R = –H, –Cl, –Br, –Phenyl, $-CF_3$, $-C_8H_{17}$	I-31	Styrene (St), MMA
CHI_3	I-32	
$F(CF_2)_4I$		
$F(CF_2)_6I$	I-33	Styrene (St), MMA

TABLE 10.2
Monomers Polymerized by ATRP

Monomer	References
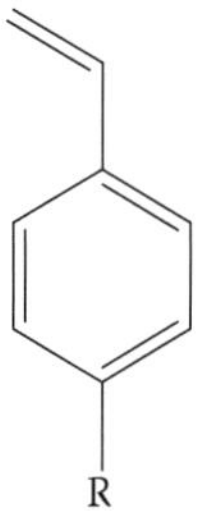	
Styrenes (St)	
R = –H	2–4, 20, 22, 24, 129–142
$–C(CH_3)_3$	143
$–CH_3$	143
–Br	143
–Cl	143
–F	143
$–CF_3$	143
–OAc	139, 143, 144
$–SiMe_3$	145
$–CH_2OCOCH_3$	139
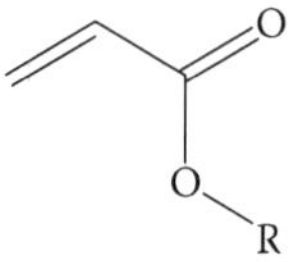	
Acrylates	
R= $–CH_3$	2, 18, 23, 131, 135, 146–148
$–CH_2CH_3$	149
$–CH_2CH_2)_2CH_3$	2, 131, 148–150
$–C(CH_3)_3$	151, 152
$–CH_2CH_2OH$	153, 154
–*iso*bornyl	155
–glycidyl	156
$–CH_2(CH_2)_nC_mF_{2m+1}$	157
$–CH(CH_3)OCH_2CH_3$	158

(*Continued*)

TABLE 10.2 (CONTINUED)
Monomers Polymerized by ATRP

Monomer	References
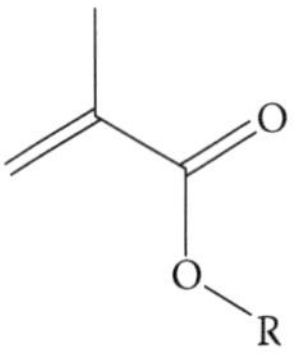	
Methacrylates	
R = $-CH_3$	1, 2, 18–21, 25, 26, 131, 137, 139, 159–163
$-CH_2CH_3$	164, 165
$-CH(CH_3)_2$	166
$-CH_2(CH_2)_2CH_3$	131, 160, 164, 165, 167
$-C(CH_3)_3$	23, 164
$-CH_2H_2OH$	23, 168
$-CH_2CH_2OSiMe_3$	169
–*iso*bornyl	23
–4-nitrophenyl	170
$-CH_2CH_2N(CH_3)_2$	171
$-(CH_2CH_2O)_nCH_3$	172
$-CH_2(CH_2)_nC_mF_{2m+1}$	157, 173
$-CH_2(CH_2)_2SiOEt_3$	174, 175
$-CH_2CH_2SiOMe_3$	176
–Lauryl	177, 178
$-CH_2CHCH_3OH$	179
$-CH_2CHOHCH_2OH$	179
–Glycidyl	180, 181
–Cyclohexyl	182
$-CH(CH_3)OCH_2CH_3$	158
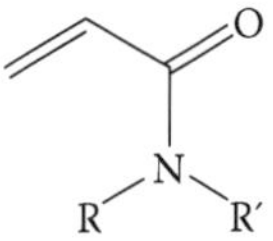	
Acrylamides	
R = $-CH_3$ R′ = $-CH_3$	183–186
–H $-C(CH_3)_3$	187
–H $-CH_2CCH_3OH$	187
–H –H	184, 188, 189

TABLE 10.2 (CONTINUED)
Monomers Polymerized by ATRP

Monomer	References
Miscellaneous Monomers	
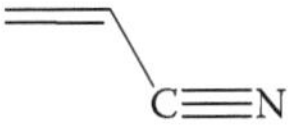 C≡N Acrylonitrile	190–193
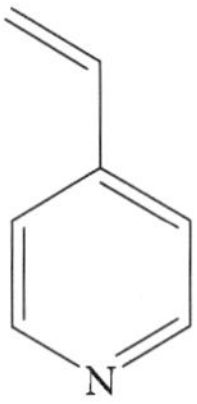 N Vinyl pyridine	194

10.2.3 Ligands

Ligands are one of the most important components of catalyst systems used in ATRP. It helps to solubilize the transition metal salts in organic media and alter the redox potential of the metal center for the appropriate dynamics of exchange between the dormant and active species with atom transfer reaction.[17] For the successful ATRP, the catalyst should meet several prerequisites. First, the catalyst should react with initiator fast and quantitatively to ensure that all the polymer chains start to add monomer at the same time. Second, the catalyst must have moderate redox potential to ensure an appropriate equilibrium between dormant and active species. In general, a low redox potential of the catalyst leads to formation of the high Cu(II) concentration (equilibrium is shifted toward transient radicals). Consequently, a fast and uncontrolled polymerization is observed. In contrast, high redox potential strongly suppresses Cu(II) formation (equilibrium is shifted toward dormant species) via a halogen atom abstraction process leading to very slow polymerization. Third, the catalyst should be less sterically hindered, because excessive steric hindrance around the metal center of catalyst results in a reduction of the catalyst activity. Fourth, a good catalyst should not afford side reactions such as Hoffman elimination, β-H abstraction, and oxidation/reduction of radicals.[17] Nitrogen-based ligands are used extensively in copper-mediated ATRP, whereas phosphorus-based ligands, such as triphenylphosphine (PPh_3), are used with nickel[18, 19], iron[20, 21], rhenium[22], ruthenium[1, 23], rhodium[24, 25], and palladium.[26] Organic acids such as isophtalic acid, iminodiacetic acid, acetic acid, and succinic acid are employed as a cheap and nontoxic ligand in iron-mediated ATRP.[27] For the ruthenium-based ATRP, cyclopentadienyl,[28, 99] indenyl[29, 30], and 4-isopropyl toluene[23] are also used as ligands. Much effort has been devoted to clarifying the relationship between structure and activity of nitrogen-based ligands used in ATRP[17, 31, 32]. It was found that tetradentate ligands,

(N4), form the most active catalysts used in ATRP.[32] In general, activity of nitrogen-based ligands decreases with the number of coordinating sites (N4 > N3 > N2 > N1).[6] Furthermore, ethylene linkage has been found to be the best spacer between two nitrogen atoms of ligand to stabilize the Cu(II) complex.[32] Exceptionally high activity was observed for the complexes with branched tetradentate ligands as compared with their linear analogues.[6] A list containing different types of nitrogen-based ligands used in ATRP has been presented in Table 10.3.

10.2.4 Reverse ATRP

The initiation system for a conventional ATRP process has two potential problems: the organic halides are usually toxic and the oxidation of the catalyst (lower oxidation-state transition-metal compounds) easily occurs. Therefore, an alternative method called reverse ATRP (RATRP), including transition-metal compounds at their higher oxidation states (e.g., as catalysts) and a conventional initiator, such as azobisisobutyronitrile (AIBN), have been developed for copper-based heterogeneous[33, 34] and homogenous[35, 36] polymerization reactions. The generalized reaction mechanism for RATRP is shown in Scheme 10.2.

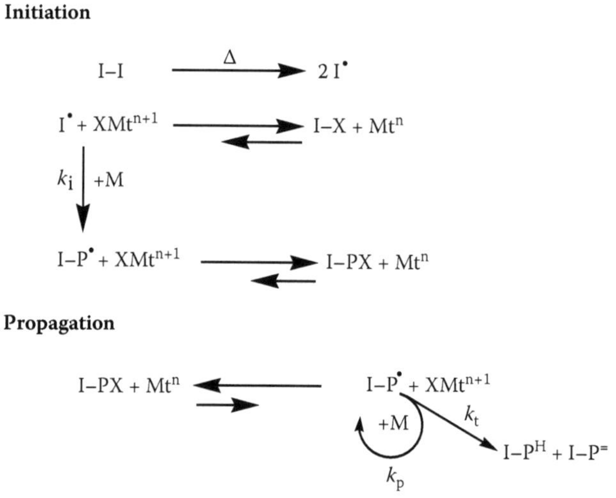

Initiation

$$\text{I–I} \xrightarrow{\Delta} 2\,\text{I}^\bullet$$

$$\text{I}^\bullet + \text{XMt}^{n+1} \rightleftharpoons \text{I–X} + \text{Mt}^{n}$$

$$\downarrow k_i,\ +\text{M}$$

$$\text{I–P}^\bullet + \text{XMt}^{n+1} \rightleftharpoons \text{I–PX} + \text{Mt}^{n}$$

Propagation

$$\text{I–PX} + \text{Mt}^{n} \rightleftharpoons \text{I–P}^\bullet + \text{XMt}^{n+1}$$

$$\text{I–P}^\bullet \ (+\text{M},\ k_p) ;\quad \text{I–P}^\bullet \xrightarrow{k_t} \text{I–P}^{H} + \text{I–P}^{=}$$

SCHEME 10.2

RATRP differs from normal ATRP in its initiation step. The initiating radicals originated from the decomposition of the conventional radical sources can abstract the halogen atom X from the oxidized transition-metal species, XMt^{n+1}, to form the reduced transition-metal species, Mt^n, and the dormant species, I-X, can react with the monomer to create a growing chain, I-P·. The situation then becomes exactly the same as in a classical ATRP. Moineau et al.[37] demonstrated the successful use of an AIBN/Fe(III)/PPh_3 initiating system to form poly(methyl methacrylate) (PMMA). The reverse ATRP initiated by benzoyl peroxide (BPO) is somewhat different from

TABLE 10.3
Nitrogen-Based Ligands Used in ATRP

Ligand	References
R = –H	3, 4
$-CH[(CH_2)_3CH_3]_2$	24, 130
$-CH_2(CH_2)_5CH_3$	24, 130
$-C(CH_3)_3$	24, 130
$-CH_2(CH_2)_2C_4F_9$	157
$-C_{13}H_{27}$	195
R = $-CH_2CH_3$	164, 196
$-C_3H_7$	159, 197
$-C_4H_9$	159
$-C_5H_{11}$	164, 197
$-C_6H_{13}$	164
$-C_7H_{15}$	164
$-C_8H_{17}$	164, 197
$-C_9H_{19}$	164
$-C_{18}H_{37}$	164
$-CH_2CH_2N(CH_2CH_2)_2O$	198
R = –Phenyl R′ = –Hexyl	164
$-CH_3$ –Propyl	159, 164, 196
(N-mesityl pyridine-2-carbaldimine structure)	199

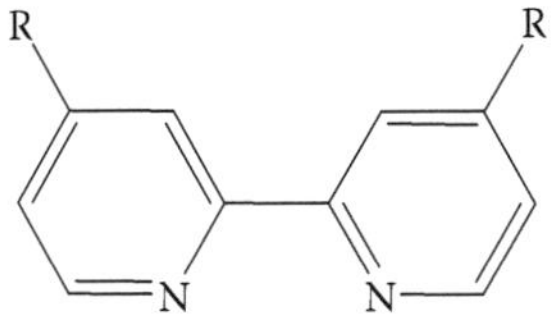

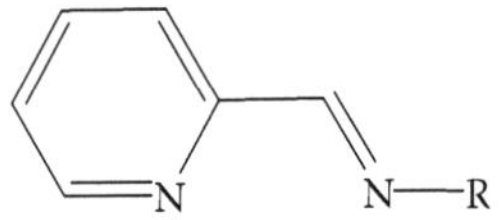

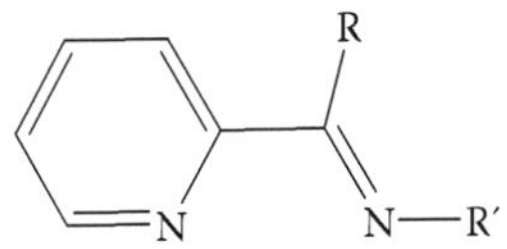

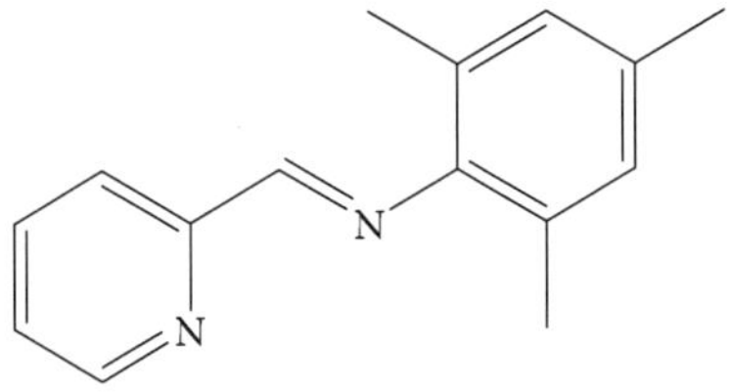

(*Continued*)

TABLE 10.3 (CONTINUED)
Nitrogen-Based Ligands Used in ATRP

Ligand	References
(2-pyridyl)oxazoline structure	47
$R_2N-CH_2CH_2-N(R)-CH_2CH_2-NR_2$	
$R = -CH_3$	200
$-C_2H_5$	201
$-C_4H_9$	201
$-C_6H_{13}$	201
$-C_3H_6C_8F_{17}$	79
$-CH_2CH_2CO_2CH_3$	202
$R_2N-CH_2CH_2-N(R)-CH_2CH_2-N(R)-CH_2CH_2-NR_2$	
$R = -CH_3$	200
$R = -C_2H_5$	201
$-C_3H_7$	201
$-C_4H_9$	201
$-C_5H_{11}$	201
$-C_6H_{13}$	203
$(CH_3)_2N-CH_2CH_2-N(CH_3)_2$	200
bis(2-pyridylmethyl)-N-R amine	
$R = -C_8H_{17}$	204
-(2-pyridyl)methyl	204

TABLE 10.3 (CONTINUED)
Nitrogen-Based Ligands Used in ATRP

Ligand	References
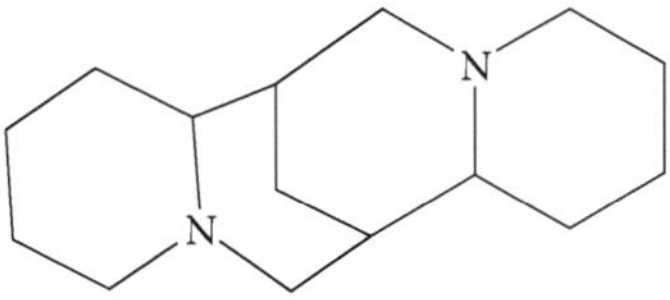	205
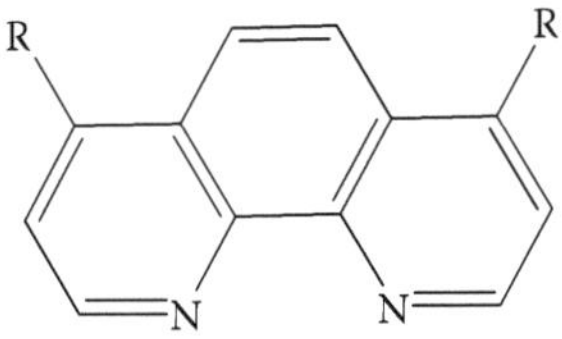 R = –H	206–208
–phenyl	206–208
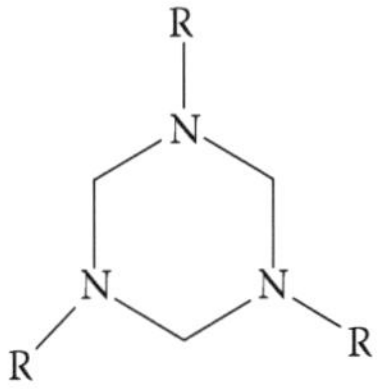 R = $-CH_2CH_2CH_2N(CH_3)_2$	209
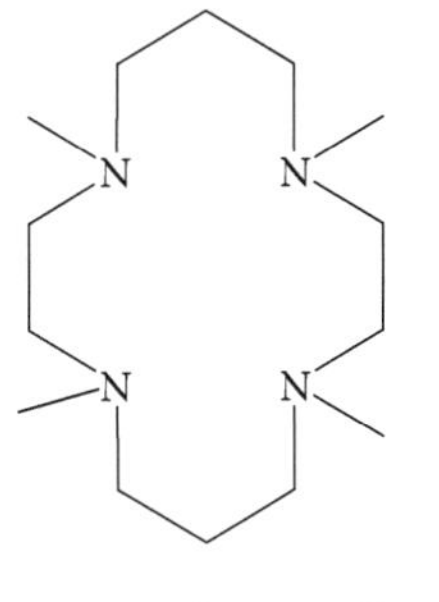	184
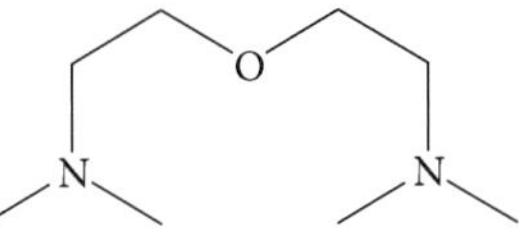	210

(Continued)

TABLE 10.3 (CONTINUED)
Nitrogen-Based Ligands Used in ATRP

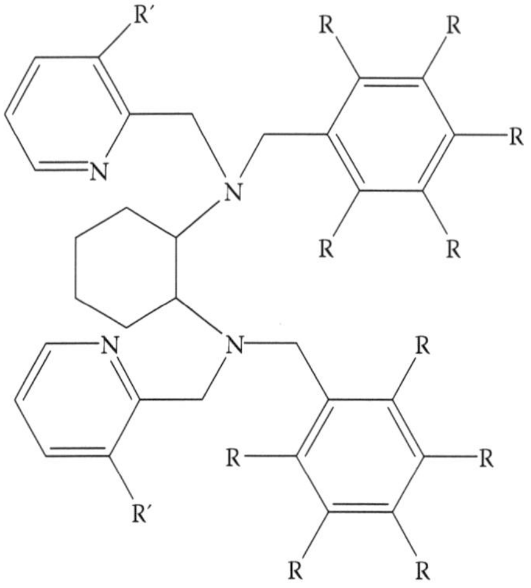

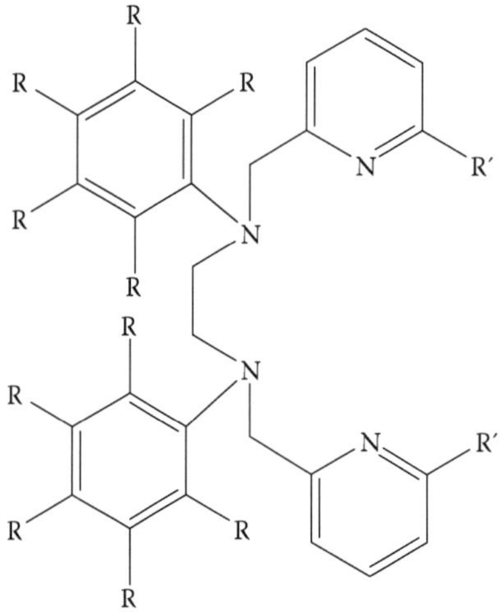

Ligand		References
R = –H	R′ = –H	211
–F	–H	211
–H	$-CH_3$	211
R = –H	R′ = –H	211
R = –F	R′ = –H	211

TABLE 10.3 (CONTINUED)
Nitrogen-Based Ligands Used in ATRP

Ligand	References
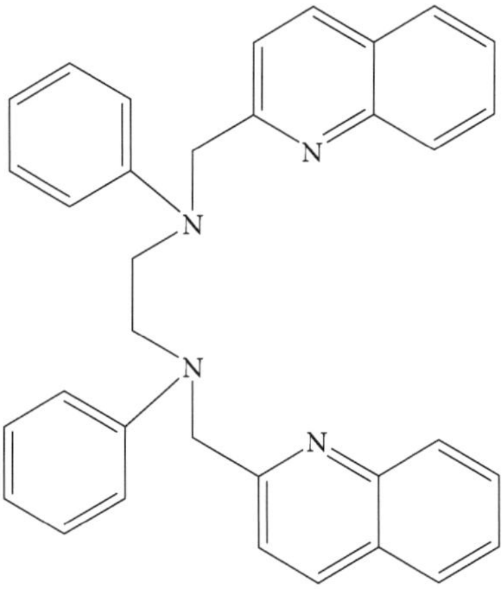 [structure]	211
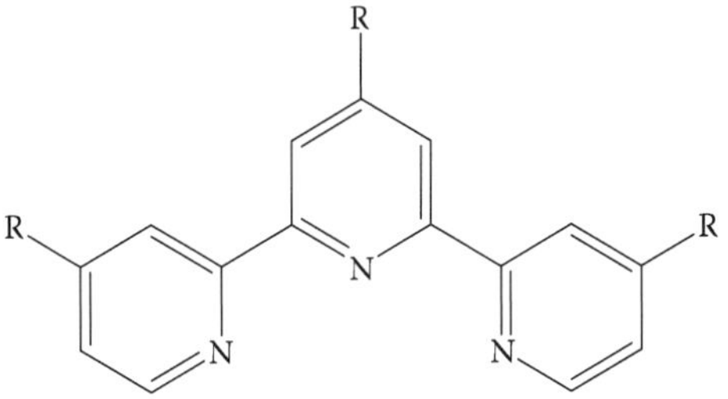 [structure]	
R = –H	212
$-CH(C_4H_9)_2$	212
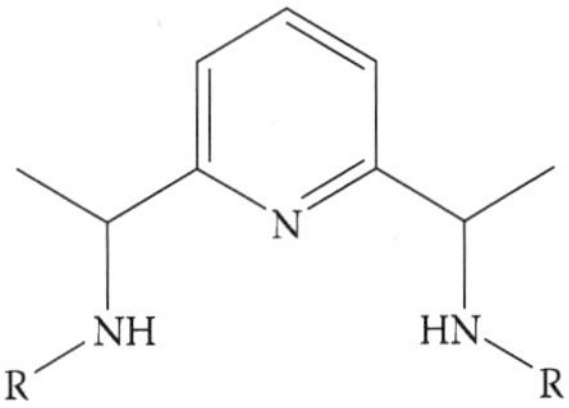 [structure]	
R = –Phenyl	31
$-C_8H_{17}$	31

(*Continued*)

TABLE 10.3 (CONTINUED)
Nitrogen-Based Ligands Used in ATRP

Ligand		References
R = –Phenyl		31
$-C_8H_{17}$		31
R = $-C_4H_9$		213
$-CH_3$		213
$-CH_2CH_2CH_2Si(OMe)_3$		214
–Benzyl		214
–Allyl		214
–2-ethylhexyl		214
R = $-CH_3$	R′ = $-CH_3$	215
$-C_2H_5$	$-C_2H_5$	216
$-C_{18}H_{37}$	$-C_{18}H_{17}$	83
$-CH_3$	$-CH_2CH_2CO_2CH_2CH{=}CH_2$	214
$-CH_3$	$-CH_2CH_2CO_2CH_2Ph$	214
$-CH_3$	$-CH_2CH_2CO_2CH_2CH_2CH_2Si(CH_3)_3$	214
$-CH_3$	$-CH_2CH_2CO_2CH_2CH(CH_2CH_3)(CH_2)_3CH_3$	214
–H	$-CH_2CH_2CO_2CH_2CH{=}CH_2$	214
–H	$-CH_2CH_2CO_2CH_2Ph$	214

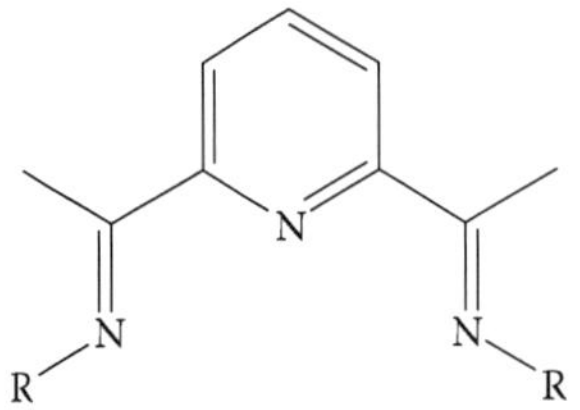

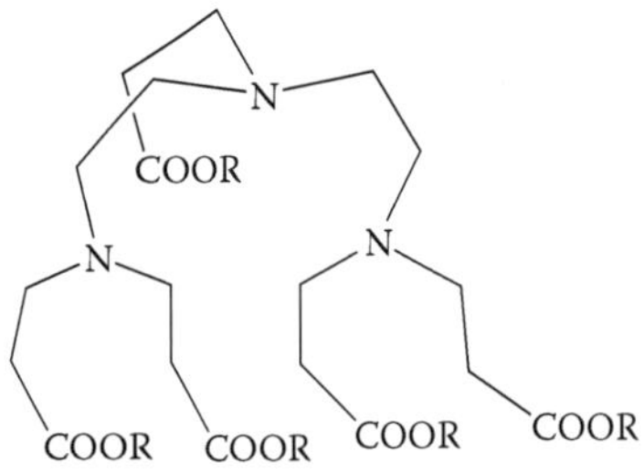

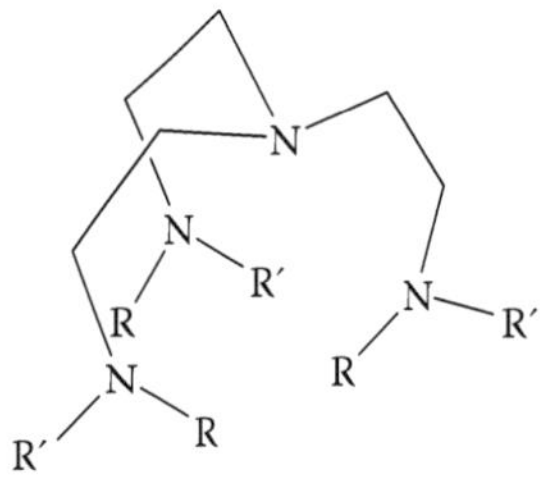

TABLE 10.3 (CONTINUED)
Nitrogen-Based Ligands Used in ATRP

Ligand		References
$-H$	$-CH_2CH_2CO_2CH_2CH_2CH_2Si(CH_3)_3$	214
$-H$	$-CH_2CH_2CO_2CH_2CH(CH_2CH_3)(CH_2)_3CH_3$	214
$-CH_2CH_2CN$	$-CH_2CH_2CN$	
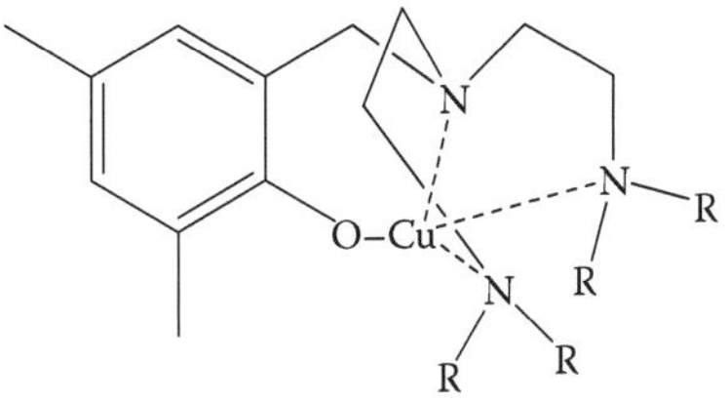 $R = -CH_3$		216
$-CH_2CH_3$		216
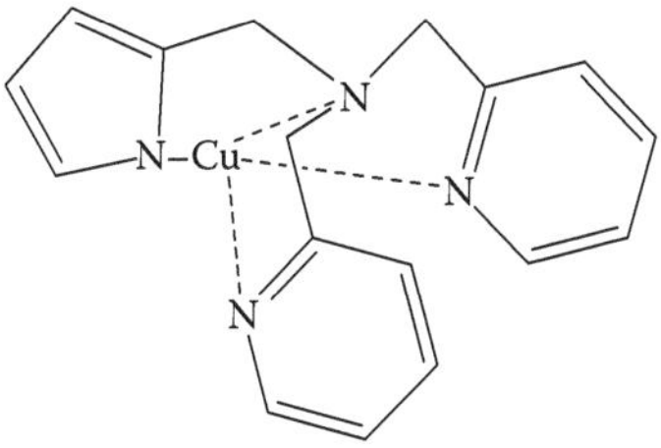		216
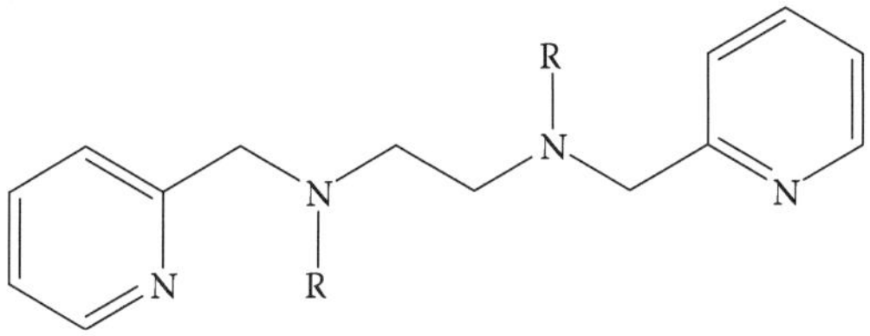 $R = -CH_2CH_2CO_2C_6H_{13}$		217
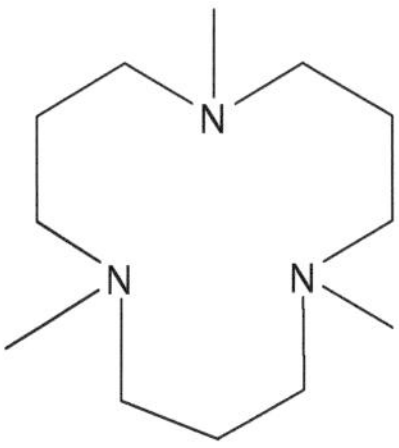		17
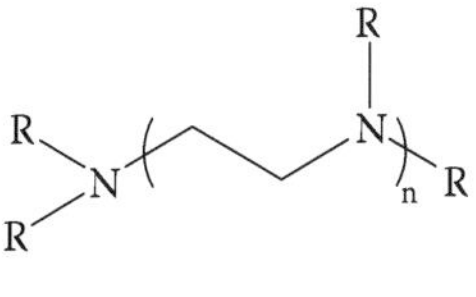		201
$R = -C_2H_5$	$n = 2, 3, 9$	

(*Continued*)

TABLE 10.3 (CONTINUED)
Nitrogen-Based Ligands Used in ATRP

Ligand	References
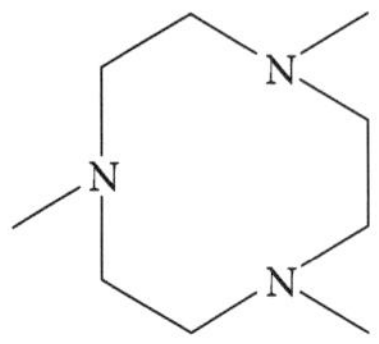 N, N, N (cyclic triamine)	17
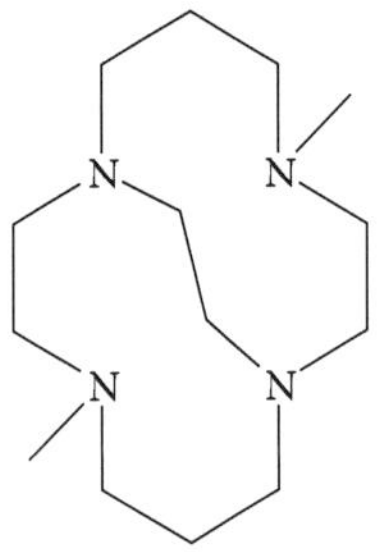 N, N, N, N (bicyclic tetraamine)	32
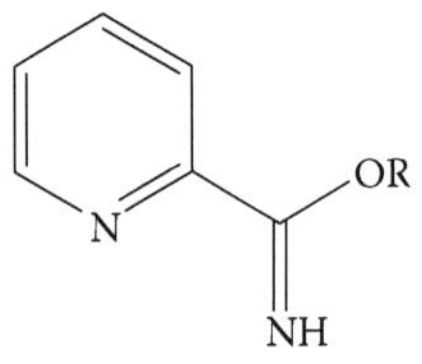 N, OR, NH (pyridine carboximidate)	
R = $-CH_3$	218
$-C_2H_5$	218
$-CH(CH_3)_2$	218
–cyclopentyl	218
–cyclohexyl	218
$-C_8H_{17}$	218
$-CH_2Ph$	218
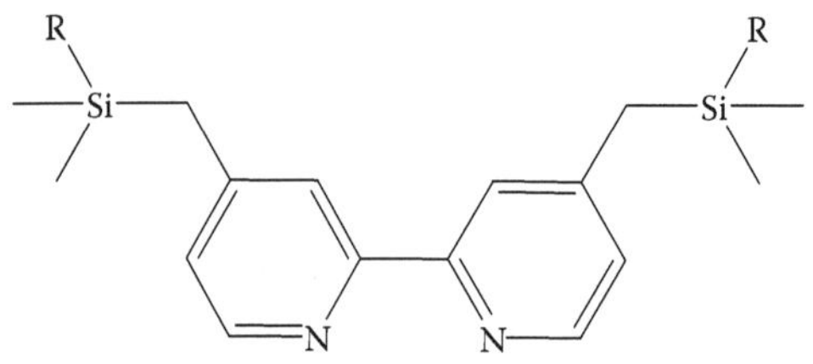 R, Si, R, Si, N, N (silylmethyl bipyridine)	
R = $-C(CH_3)_3$	219
$-C_6H_{13}$	219
$-CH_2C(CH_3)_2Ph$	219

that of AIBN regarding the initiation mechanism. The differences between the BPO and AIBN systems was ascribed to an electron transfer and the formation of a copper benzoate species.[36] For example, the polymerization initiated by the homogeneous BPO/$CuBr_2$ $(dNbpy)_2$ system did not proceed in a controlled manner. In contrast, controlled/"living" polymerization was observed when BPO was used together with CuBr $(dNbpy)_2$. In a heterogeneous system using bpy as the ligand, both CuX and CuX_2 (X = Br or Cl) yielded a controlled polymerization of styrene (St).[38, 39] Wang et al.[40] demonstrated that the BPO/CuBr/bpy and BPO/CuCl/bpy systems can initiate the well-controlled polymerization of MMA at room temperature. The development of new type initiators for RATRP received more interest. In this context, the carbon-carbon bond thermal iniferters, such as 1,1,2,2-tetraphenyl-1,2-ethanediol (TPED) and diethyl 2,3-dicyano-2,3-diphenylsuccinate (DCDPS), have been used successfully in the presence of $FeCl_3(PPh_3)_3$ for the reverse ATRP of methyl methacrylate (MMA) and styrene.[41–43]. Chen and Qiu[44, 45] reported a third ATRP process, namely, *in situ* ATRP of methyl methacrylate (MMA) polymerization with tetraethylthiuram disulfide (TD) or iron(III) tri(diethyldithiocarbamate) [$Fe(dtc)_3$] catalyzed by a $FeCl_3(PPh_3)_3$ complex. No alkyl halide, azo, or peroxide initiators are included in these initiating systems. The initiator (diethylthiocarbamoyl) sulfur chloride (dtc-Cl) and catalyst $FeCl_2$ are both produced *in situ* from the components of the initiating system.

More recently, an aryl diazonium salt compound was also used as an initiator in RATRP. Ucan et al.[46] reported that the controlled polymerization of methyl methacrylate (MMA) was achieved with *p*-chlorobenzenediazonium tetrafluoroborate and Cu(II) or Cu(I)/Cu(II)/*N,N,N′,N″,N″*-pentamethyldietylenetriamine (PMDETA) complex system at various temperatures.

10.2.5 Additives

The main purpose of additives used in ATRP is the acceleration and/or better control of the polymerization. In this regard, when metal alkoxides such as Al(O-*i*-Pr)$_3$,[1] benzoic acid salts or sodium acetate,[47] zero-valent metals,[48] phenols,[49, 50], octanethiol,[51] or reducing monosaccharides[52] were used as additives, a remarkable effect on rate enhancement of polymerization was observed. The effect of metal alkoxides and organic acid salts on the polymerization was explained by the possibility that added compounds interact with the metal catalyst in its higher or lower oxidation state to increase the activity of catalyst.[1, 47] In contrast, the other additives mentioned previously act as reducing agent for the higher oxidation state of metal salts. Thus, the concentration of higher oxidation state metal salts is lowered during the polymerization. This process leads to an increase in the transient radical concentration and therefore, significant rate enhancement is observed. Common solvents, including nonpolar (toluene, xylene, benzene), polar aprotic (diphenyl ether, dimetoxy benzene, anisole, *N,N*-dimethylformamide ethylene carbonate, acetonitrile), and protic polar (alcohols, water), are employed for not only solubilizing the monomers, the produced polymers, and the catalyst, but also to achieve the controlled polymerization condition.

10.2.6 *In Situ* Generation of Cu (I) Species via Electron Transfer Reaction in ATRP

One of the major disadvantages of ATRP is high cost and the easy oxidation of a metal salt in its lower oxidation state. These can be overcome by the recent approaches based on the *in situ* Cu(I) generation via an electron transfer from a reducing agent to a more stable higher oxidation state metal salt. In this regard, Gnanou and Hizal reported that phenols in conjunction with Cu(II)/*N,N,N′,N″,N″*-pentamethyldiethylenetriamine (PMDETA) complex are successfully used to perform copper-mediated ATRP of MMA, styrene, and MA in the presence of a limited amount of air.[53] This approach is different from the lower oxidation state metal-initiated ATRP in two respects:

1. In this system, reactive Cu(I) species are formed in situ via redox process between Cu(II) and phenols. Therefore, the small concentration of Cu(I) promotes the low concentration of transient radicals, and in addition, the initial presence of Cu(II) (persistent radical) facilitates the deactivation process and thus, suppresses undesirable radical–radical coupling reaction.
2. The unavoidable oxidation of catalyst or the oxygen-induced polymerization by the diffused oxygen, as it was reported in lower oxidation state copper-catalyzed ATRP[54, 55] does not play a detrimental role in this system.

A schematic representation of generation and regeneration of the Cu(I) species in the $CuCl_2$/PMDETA-mediated ATRP together with oxygen and phenol consumption

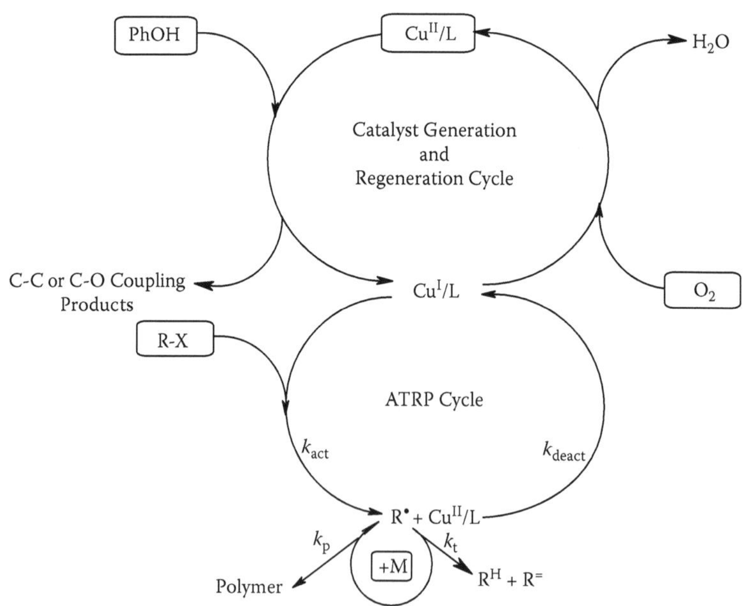

SCHEME 10.3

processes is given in Scheme 10.3. In this figure, the compounds given in rectangles represent the starting materials for the polymerization.

Percec et al.[5, 56, 57] reported for the first time the self-regulated catalytic system using a reaction between aryl sulfonyl chloride and Cu_2O to give Cu(I). Later, Matyjaszewski et al.[58, 59] developed a method based on *in situ* formation of Cu(I) from its higher oxidation state counterpart via combination of ATRP and reverse ATRP processes. In this process, a mixture of an excess amount of alkyl halide and thermal radical initiator together with Cu(II) catalyst complex was used as a dual initiator (simultaneous reverse and normal initiation ATRP (SR&NI ATRP)). However, the major drawback of this method is the dual nature of this initiation system. The free radicals produced by the thermal decomposition of the free-radical initiator and from the alkyl halide both initiate the polymerization; thus, it is not suitable to produce polymers having different functionality and topology. Alternatively, the Matyjaszewski group[60, 61] more recently reported the use of reducing agents, such as ascorbic acid and tin (II) 2-ethylhexanoate, instead of a conventional radical initiator to reduce Cu(II) complex to Cu(I) species as an activator in ATRP (process termed as an activator generated by electron transfer [AGET] ATRP).

10.2.7 Catalyst Recovery in ATRP

As taken into consideration, one of the major challenges faced by the ATRP is the removal and recycling of the catalyst.[62] Many examples of ATRP catalysts immobilized on silica,[63–70] cross-linked polystyrene (PS) beads,[63, 64], or Janda Jel resin can be cited.[71] Moreover, physically adsorbed multidante ligand such as hexamethyltriethylenetetramine (HMTETA) on silica gel has been utilized as supported catalyst for copper-catalyzed ATRP.[72–74] Thereby, removal of catalyst can be done by simple filtration. However, the heterogeneous nature of these processes leads to some limitations over the control of polymerization because of the insufficient deactivation of growing radicals. In an attempt to circumvent this problem, the hybrid catalyst system composed of an immobilized catalyst acting in conjunction with a small amount of soluble catalyst ($Me_6TREN/CuBr_2$) that accelerates the rate of deactivation of the growing radical was used.[75, 76] To demonstrate the utility of this hybrid catalyst system, polymers exhibiting various polymeric architectures, including block and gradient copolymers, were synthesized by copper-catalyzed ATRP.[77] Furthermore, this catalyst system was used for the preparation of polymers with more complex architectures, like star, graft, and bottle-brush copolymers.[78] Other facile approaches to homogenous recoverable catalyst systems, including a fluorous biphasic system,[79] low molecular weight polyethylene,[80] polyethylene with a polyethylene glycol spacer,[81, 82] and a molecular nonfluorous thermomorphic system,[83] were reported. More recently, a reversible catalyst immobilization system via self-assembly of hydrogen bonding was developed for copper-catalyzed LRP of methyl methacrylate (MMA).[84, 85] However, several problems are associated with these aforementioned techniques, the most important of which is that the catalyst has to be protected by inert atmosphere during separation process. Because these catalysts frequently lose their activity as a result of inevitable accumulation of Cu(II) (persistent radical effect)[12] during the first polymerization, a regeneration process of the catalyst using a zero-valent metal was needed.[76] All these make the recycling procedure tedious, time-consuming, and expensive. Therefore, the applicability of these methods at the industrial scale is limited. Recently, Hizal et al.[86] reported that because ATRP of St was successfully performed with an oxidatively stable and easily recoverable catalyst consisting of a silica gel-$CuCl_2$/PMDETA complex, the required Cu(I) species for ATRP was formed by an electron transfer from phenol to Cu(II). Although the use of silica gel-$CuCl_2$/PMDETA catalyst complex with phenol derivatives will have the greatest benefit for the catalyst separation and recycling process in copper catalyzed LRP, a cheap and effective catalytic system for ATRP that can meet the industrial demands for a large-scale production still needs to be developed.

10.3 NITROXIDE-MEDIATED RADICAL POLYMERIZATION (NMP)

Nitroxide-mediated radical polymerization (NMP), also referred as stable free-radical polymerization (SFRP), is one of the versatile methods for the controlled/"living" polymerizations. NMP is generally based on the use of nitroxyl radicals (nitroxides) or alkoxyamines. Solomon and Rizzardo,[87, 88] and Moad and Rizzardo[89] first used alkoxyamines for the polymerization of various vinyl monomers. However, only low molecular weight polymers were obtained. Later Georges et al.[90] showed that high molecular weight polystyrene ($M_n > 50,000$) with low polydispersity (<1.5) could be prepared using benzoyl peroxide (BPO) and a nitroxyl radical such as 2,2,6,6-tetramethyl-1-piperidinyloxy (TEMPO).

Two initiation systems have been generally employed in the NMP. First is a bimolecular system consisting of conventional radical initiator such as BPO or azoisobutyronitrile (AIBN), and a nitroxide (i.e., TEMPO). The conventional radical initiator is decomposed at an appropriate temperature to initiate free-radical polymerization. The initiator-monomer adduct is trapped by the nitroxide leading to formation of the alkoxyamine *in situ*. Second is the unimolecular system using the alkoxyamine that is decomposed into a nitroxide and an initiating radical. This radical subsequently initiates the free-radical polymerization (Scheme 10.4). By using the unimolecular initiator, the molecular weight can be properly controlled, because the number of initiating sites per polymeric chain is defined. In addition, functionalized unimolecular initiators can afford the fully functional groups at the ends of the polymer chain.

The key kinetic aspect of the living radical polymerization systems, including NMP is described through the persistent radical effect (PRE).[11] In the initial stages of the polymerization, a small fraction of the initiating radicals formed from decomposition of the initiator (alkoxyamine) undergo biradical termination (Scheme 10.4). This results in the removal of two initiating radicals from the system and necessitates a small increase in the concentration of mediating radical (persistent radical) relative to the initiating radical/propagating radical.

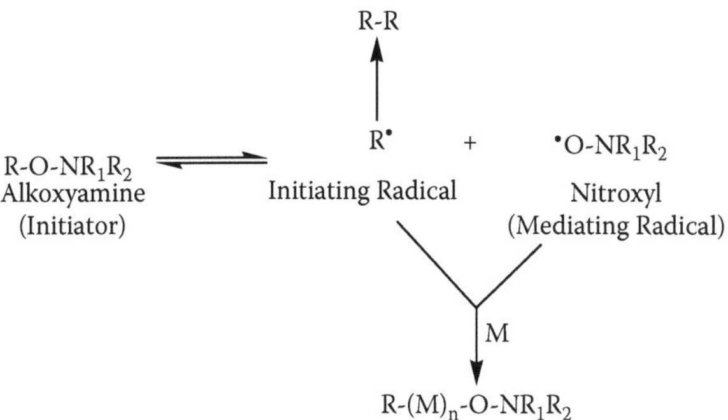

SCHEME 10.4

However, the mediating radical does not undergo coupling or initiate polymerization. This increased concentration of mediating radical is self-limiting because a higher concentration leads to more efficient formation of a polymer with a dormant chain end and a decrease in the amount of radical coupling reaction, leading to the PRE and to the ultimate control over the polymerization process.[91]

Styrenic-type monomers have been the most widely studied using NMP in the presence of TEMPO or its derivatives as mediating radical. This is mainly because the controlled polymerization of acrylates by SFRP is restricted. It was apparent that the structural changes in the nitroxide as a mediating radical were required. On this basis, alicyclic nitroxides were synthesized. These compounds have a hydrogen atom on one of the α-carbons adjacent to nitrogen atom. Alkyl methacraylates can also be polymerized via NMP strategy using NO/NO_2 mixtures.[92]

Numerous studies were performed to elucidate the mechanism and the polymerization kinetics of NMP. The compilation of the existing literature data on all nitroxide radicals as mediating radicals and their use in NMP are presented in Table 10.4.

Advantages of SFRP over other controlled/"living" radical polymerization systems are that the metal catalyst is not necessary and no further purification after polymerization is required. However, major drawbacks of SFRP are the long polymerization time, the limit of the range of the monomers that can be polymerized, and the relatively high polymerization temperature (>125°C). The synthesis of the new nitroxides leads to a wider range of monomers to be polymerized by SFRP at lower temperatures in a shorter polymerization time.

10.4 REVERSIBLE ADDITION-FRAGMENTATION CHAIN TRANSFER POLYMERIZATION (RAFT)

Reversible addition-fragmentation chain transfer (RAFT) polymerization is one of the most recent entrants and one of the most efficient methods in controlled/living radical polymerization. An important advantage of this method over ATRP and NMP

TABLE 10.4
The Structure of Nitroxides and Monomers Used in NMP and Molecular Weight Characteristics of the Resulting Polymers

Nitroxide Structure	Monomer	M_n, M_w, or Molecular Weight Range	M_w/M_n	References
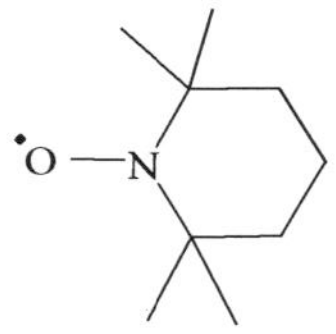	4-*tert*-butyl-4′-(4-vinyl styryl)-trans-stilbene	$M_W = 3700$	1.50	220
	Styrene	$8800 < M_n < 24{,}800$	1.13–1.51	221
	Styrene	$40{,}000 < M_n < 75{,}000$	1.30–1.60	222
	Styrene (dispersion polymerization)	$1000 < M_n < 23{,}100$	1.06–1.30	223
	2-vinylpyridine	$4900 < M_n < 49{,}800$	1.24–2.01	224
	3-vinylpyridine	$32700 < M_n < 226{,}500$	1.08–1.63	225
	4-vinylpyridine	$7000 < M_n < 21{,}200$	1.16–1.40	226
	4-acetoxystyrene	$10{,}100 < M_n < 13{,}620$	1.31–1.63	227
	Vinylferrocene	$M_n < 4800$	1.24–1.80	228
	N,N-dimethylacrylamide	$6540 < M_n < 27{,}300$	1.55–2.68	186
	2-methylene-1,3-dioxepane	(Not living process)		
	Methylmethacrylate	$2900 < M_n < 15{,}300$	1.20–6.50	229
	3 or 4-chloromethylstyrene	$M_n = 22{,}100$	3.11	230
	Styrenesulfonic acid sodium salt	$M_n < 20{,}200$	1.31–1.38	231
	Vinyl acetate	$7200 < M_n < 763{,}000$	1.12–1.33	232
	1,3-butadiene	$M_n < 10{,}000$	<1.25	233
	4-diphenyl-aminostyrene	$M_n = 3200$	3.00	234
	2,5-bis[(4-butyl-benzoyl)oxy] styrene	$196{,}400 < M_n < 13{,}200$	1.20–2.00	235
	p-*tert*-butoxystyrene	$4800 < M_n < 51{,}000$	1.27–1.41	236
		$6900 < M_n < 46{,}100$	1.14–1.17	237
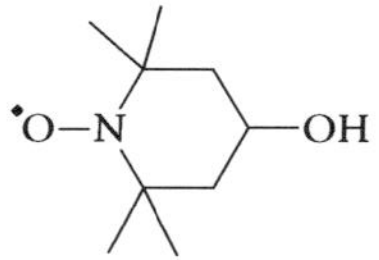	Styrene	$M_n = 2900$	1.09	238
	Styrene	$M_n < 10{,}000$	—	239
	Styrene (mini emulsion)	$M_n < 60{,}000$	—	240

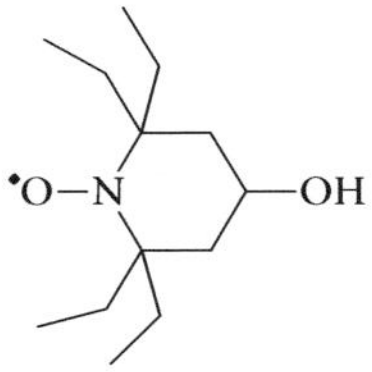	Styrene	$M_n = 11{,}700$	1.36	241
	Styrene	$3300 < M_n < 17{,}000$	1.23–1.36	242
	n-butylacrylate	$M_n = 3400$	1.38	242
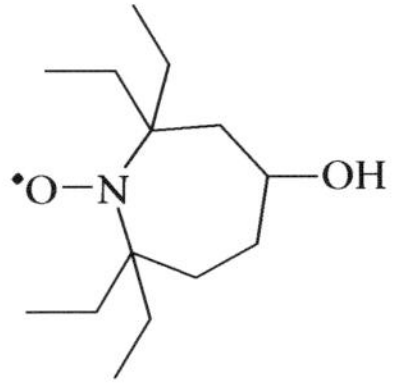	Styrene	$M_n = 10{,}600$	1.64	241
	Styrene	$6400 < M_n < 31{,}000$	1.41–1.62	242
	n-butylacrylate	$M_n = 5300$	1.56	242
	Styrene	$3600 < M_n < 19{,}000$	1.23–1.32	242
	n-butylacrylate	$M_n = 3500$	1.42	242

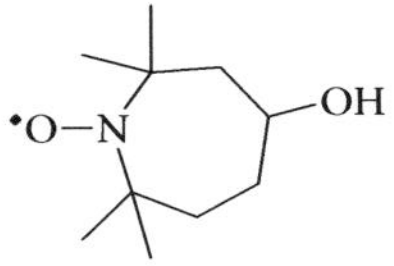

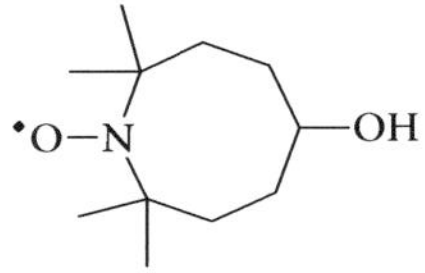

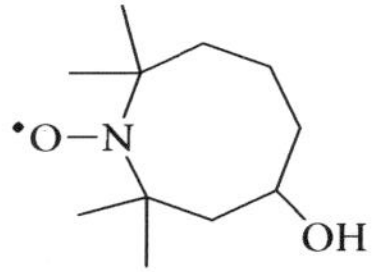

(*Continued*)

TABLE 10.4 (CONTINUED)
The Structure of Nitroxides and Monomers Used in NMP and Molecular Weight Characteristics of the Resulting Polymers

Nitroxide Structure	Monomer	M_n, M_w, or Molecular Weight Range	M_w/M_n	References
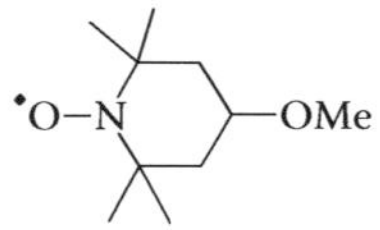 •O–N, OMe	Styrene	$5550 < M_n < 12{,}300$	1.15–2.74	243
	4-bromostyrene	$2930 < M_n < 13{,}000$	1.15–1.23	243
	Chlorostyrenes (2-,3-,4-)	$690 < M_n < 7700$	1.01– 1.34	244
	Methylstyrenes (2-,3-,4-)	$630 < M_n < 6900$	1.01–1.36	245
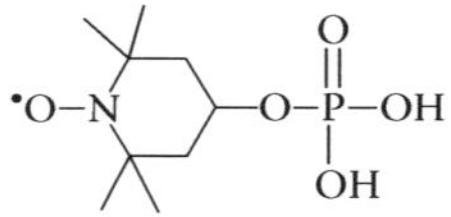 •O–N, O–P(=O)(OH)–OH	Styrene	$M_n < 25{,}000$	—	239
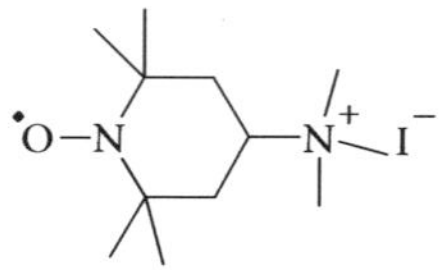 •O–N, N^+ I^-	Sodium 4-styrene sulfonate	$4200 < M_n < 8400$	1.10–1.40	246
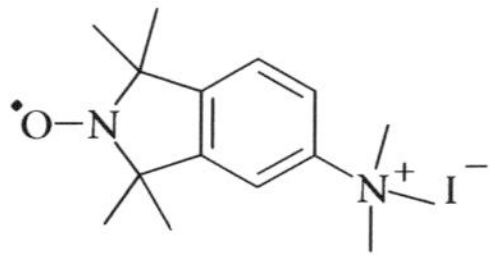 •O–N, N^+ I^-	Sodium 4-styrene sulfonate	$4400 < M_n < 7200$	1.10	246
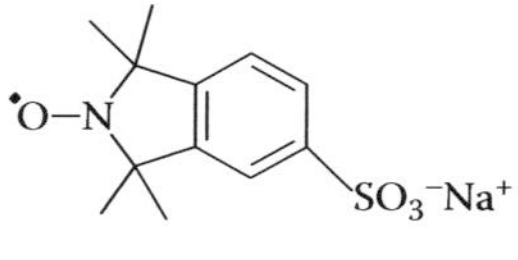 •O–N, $SO_3^-Na^+$	Sodium 4-styrene sulfonate	$4110 < M_n < 8300$	1.03–1.20	246

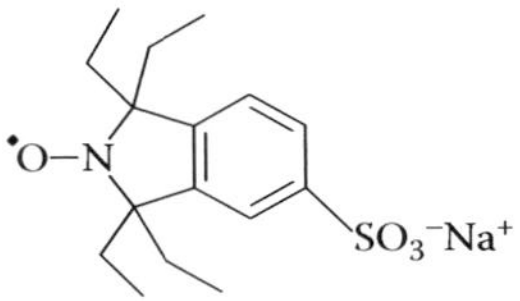	Sodium 4-styrene sulfonate	5800< M_n <7700	1.10	246
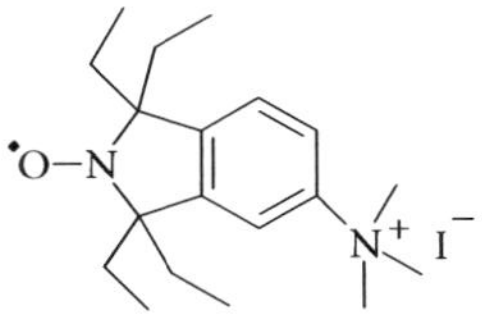	Sodium 4-styrene sulfonate	4400< M_n <7200	1.10	246
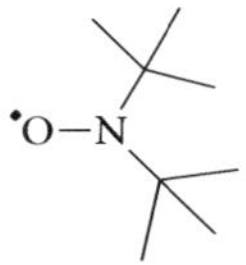	*p-tert*-butylstyrene			
	Styrene	M_n <100,000	1.20–1.30	247
	N-(p-vinylbenzyl)-[*O*-β-D-galactopyranosyl-(1-4)]-D-gluconamide	M_n <87,200	1.10–1.45	248
		2000< M_n <40,000	1.10	249
	3-*O*-acryloyl-1,2:5,6-di-*O*-isopropylidene-α- D -glucofuranoside	3100< M_n <9700	1.20–1.60	250
	Vinyl chloride (suspension)	18,000< M_n <29,000	1.90–2.10	251
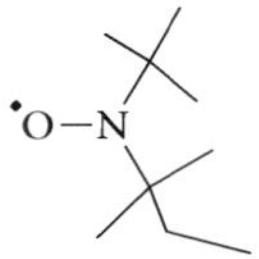	Methylmethacrylate	M_n = 19,200	4.10	230
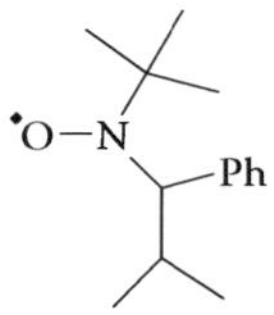	Styrene	M_n <70000	1.08–1.13	252
	Acrylates	M_n = 50,000	1.01–1.10	252
	Acrylonitrile	4500< M_n <55,000	1.12–1.16	252
	N,N-dimetylacrylamide	4000< M_n <48,000	1.12–1.21	252
	Isoprene	M_n <20,000	1.07–1.14	253
	2-vinyl-4,4-dimethyl-5-oxazolone	4200< M_n <33,900	1.08–1.19	254

(*Continued*)

TABLE 10.4 (CONTINUED)

The Structure of Nitroxides and Monomers Used in NMP and Molecular Weight Characteristics of the Resulting Polymers

Nitroxide Structure	Monomer	M_n, M_w, or Molecular Weight Range	M_w/M_n	References
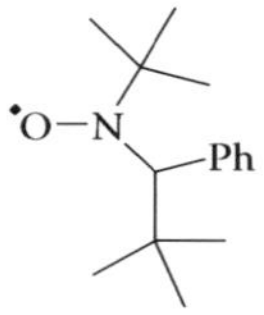 •O–N, Ph	Styrene	$M_n = 30{,}500$	1.39	252
	n-butylacrylate	$M_n = 12{,}000$	1.75	252
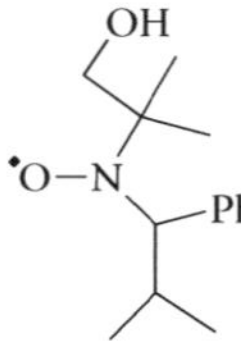 OH, •O–N, Ph	Styrene	$M_n = 80{,}000$	>1.20	252
	n-butylacrylate	$M_n = 27{,}000$	1.40	252
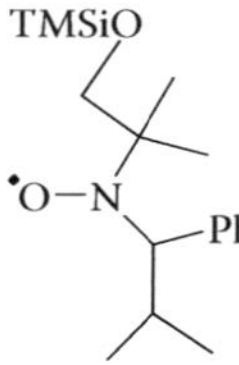 TMSiO, •O–N, Ph	Styrene	$M_n = 22{,}500$	1.15	252
	n-butylacrylate	$M_n = 26{,}500$	1.45	252
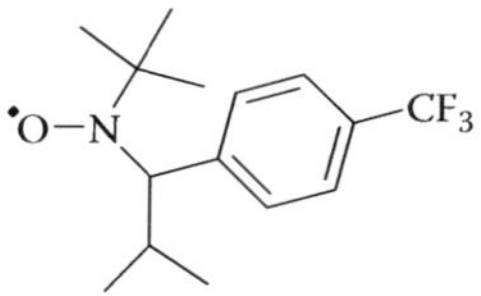 •O–N, CF_3	Styrene	$M_n = 22{,}000$	1.16	252
	n-butylacrylate	$M_n = 29{,}000$	1.50	252

	Styrene	$M_n = 24{,}000$	1.19	252
	n-butylacrylate	$M_n = 32{,}500$	1.55	252
	Styrene	$M_n = 25{,}000$	1.15	252
	n-butylacrylate	$M_n = 27{,}500$	1.55	252
	Styrene	$M_n = 38{,}000$	1.68	252
	n-butylacrylate	$M_n = 95{,}000$	2.10	252
	Styrene	$M_n = 25{,}000$	1.30	252
	n-butylacrylate	$M_n = 38{,}000$	1.70	252
	Styrene	—	1.15–1.20	255
	Isoprene	$M_n = 14{,}000$	1.18	255
	N,N-dimethylacrylamide	$M_n = 15{,}000$	1.12	255
	n-butylacrylate	—	1.10	255

(*Continued*)

TABLE 10.4 (CONTINUED)
The Structure of Nitroxides and Monomers Used in NMP and Molecular Weight Characteristics of the Resulting Polymers

Nitroxide Structure	Monomer	M_n, M_w, or Molecular Weight Range	M_w/M_n	References
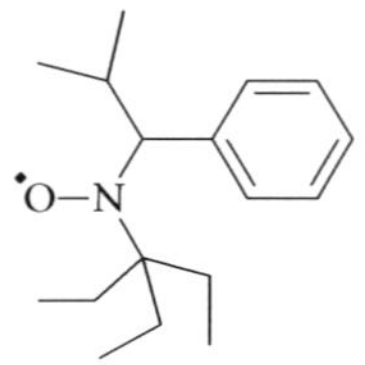	Styrene	9800< M_n <68,200	1.09–1.28	256
	n-butylacrylate	10,200< M_n <43,400	1.15–1.45	256
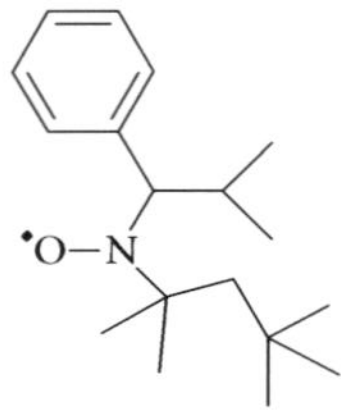	Styrene	8800< M_n <11,700	1.16–1.28	256
	n-butylacrylate	M_n = 13,500	1.18	256
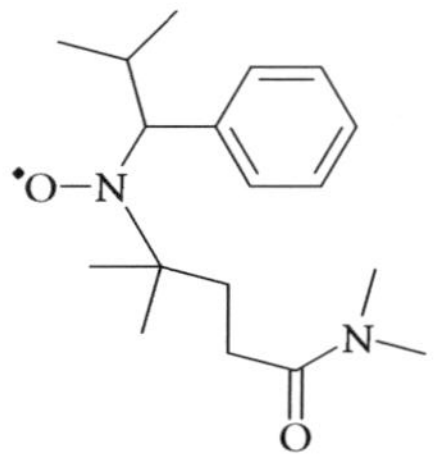	Styrene	3600–6600	1.13–1.15	256
	n-butylacrylate	M_n = 8000	1.20	256

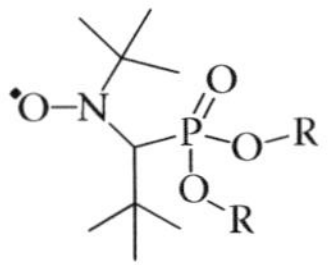	$R = CH_2CH_3$			
	Styrene	M_n <85,000	<1.40	257
	Styrene	M_n = 28,000	1.21	252
	Styrene (mini emulsion)	3280< M_n <63,000	1.47–2.47	258
	n-butyl acrylate	M_n <50,000	<1.50	257
	n-butyl acrylate	M_n = 22,000	1.90	252
	Acrylic acid	M_n <11,000	1.30–1.50	259
	4-vinylpyridine	10,000< M_n <60,000	1.10–1.21	260
	N,N-dimethylacrylamide	4,000< M_n <15,000	1.26–1.54	260
	N,N-dimethylacrylamide	3,400< M_n <37,400	1.07–1.20	261
	$R = CH_2Ph$			
	Styrene	5,800< M_n <11,000	1.07–1.08	262
	$R= CH_2CF_3$			
	Styrene	21,000< M_n <31,000	1.40–1.72	262
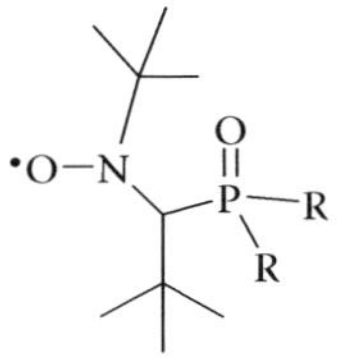	$R = CH_3$			
	Styrene	M_n <60,000	—	263
	$R = CH_2CH_3$			
	Styrene	M_n <80,000	—	263
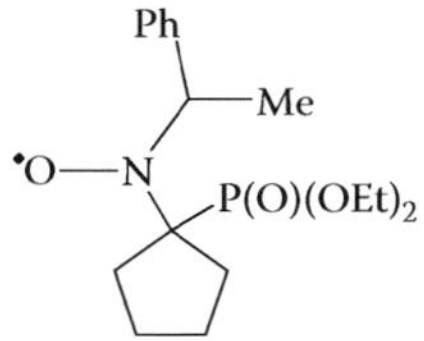	Styrene	18,600< M_n <29,000	1.39–1.42	262

(Continued)

TABLE 10.4 (CONTINUED)
The Structure of Nitroxides and Monomers Used in NMP and Molecular Weight Characteristics of the Resulting Polymers

Nitroxide Structure	Monomer	M_n, M_w, or Molecular Weight Range	M_w/M_n	References
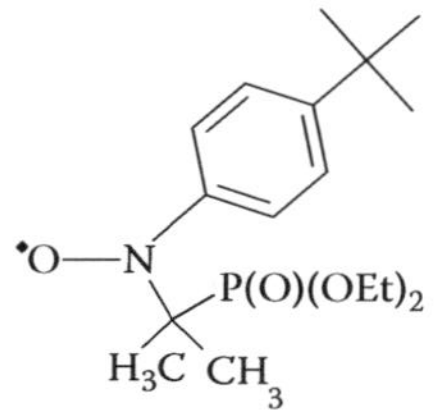 •O—N, $P(O)(OEt)_2$, H_3C CH_3	Styrene	$M_n = 26,000$	1.27–1.70	262
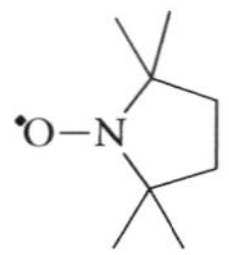 •O–N	Styrene	$M_n < 20,000$	1.20–2.00	264
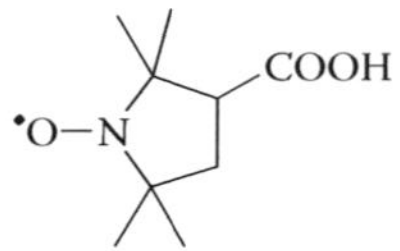 •O–N, COOH	Styrene	$15,000 < M_n < 30,000$	1.20	265
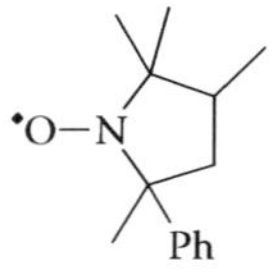 •O–N, Ph	Styrene	$M_n < 60,000$	1.40–1.70	264

	Styrene	40,000< M_n <120,000	1.85–1.40	264
	Styrene	M_n <120,000	1.40–1.60	264
	R = CH_3			
	Styrene	M_n = 23,900	1.34	266
	R = Ph			
	Styrene	M_n = 26,600	1.47	264
	Styrene	M_n <60,000	1.40–1.80	264
	R = p-CF_3-Ph			
	Styrene	M_n = 22,600	1.64	266
	R = p-$(CH_3)_2$N-Ph			
	Styrene	M_n = 24,000	1.16	266
	Styrene	M_n = 1900—16,500	1.17–2.17	267
	Styrene	M_n = 26,500	1.17	252
	n-butylacrylate	M_n = 32,000	2.05	252

(Continued)

TABLE 10.4 (CONTINUED)
The Structure of Nitroxides and Monomers Used in NMP and Molecular Weight Characteristics of the Resulting Polymers

Nitroxide Structure	Monomer	M_n, M_w, or Molecular Weight Range	M_w/M_n	References
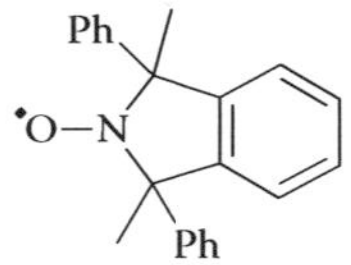	Styrene	$M_n = 33{,}000$	1.71	
	Styrene	$M_n = 22{,}000$	1.49	252
	n-butylacrylate	$M_n = 41{,}500$	2.52	252
	n-butylacrylate	$M_n = 26{,}000$	2.25	
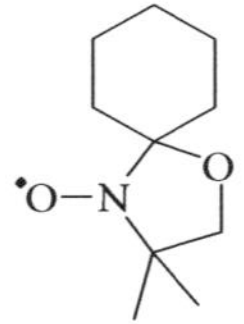	Styrene	$M_n = 50{,}000$	1.72	252
	n-butylacrylate	$M_n = 57{,}000$	3.81	252
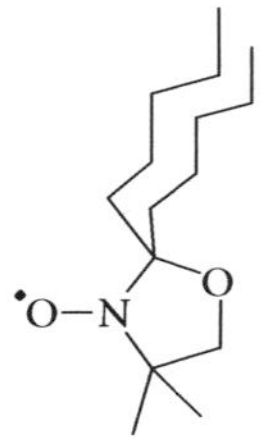	Styrene	$M_n = 42{,}000$	1.65	252
	n-butylacrylate	$M_n = 61{,}000$	2.85	252
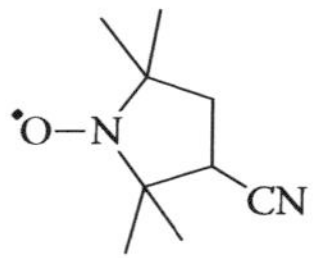	Styrene	$M_n = 20{,}000$	1.25	252
	n-butylacrylate	$M_n = 45{,}000$	1.95	252

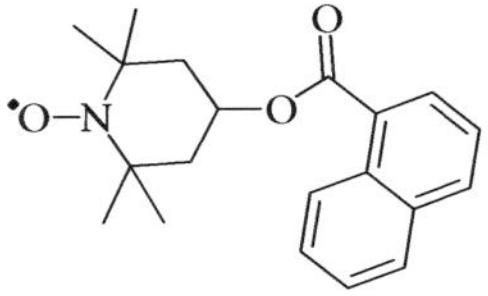	Styrene	$M_n = 7500$	1.11	268
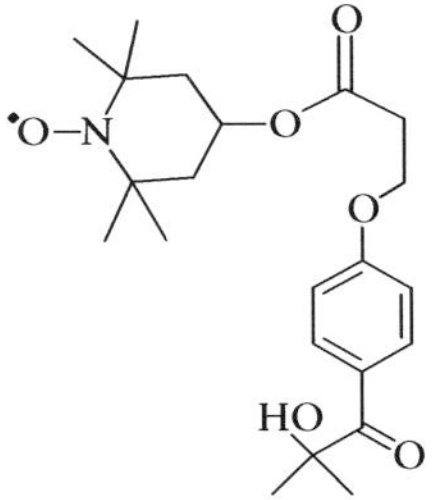	Styrene	$M_n = 7500$	1.11	268
Pyrene	Styrene	$6100 < M_n < 93{,}000$	—	269
	tert-butylacrylate	$4500 < M_n < 141{,}000$	—	269
	Styrene	$800 < M_n < 17{,}500$	1.08–1.09	230
	Methylmethacrylate	$5600 < M_n < 18{,}300$	1.68–1.71	230

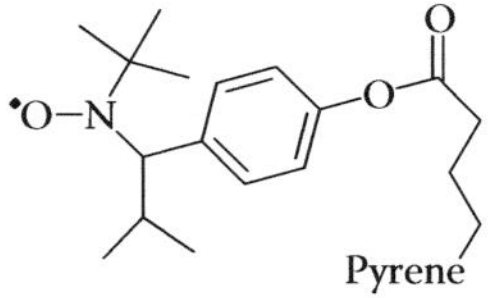

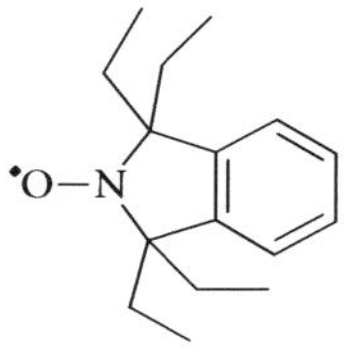

(Continued)

TABLE 10.4 (CONTINUED)
The Structure of Nitroxides and Monomers Used in NMP and Molecular Weight Characteristics of the Resulting Polymers

Nitroxide Structure	Monomer	M_n, M_w, or Molecular Weight Range	M_w/M_n	References
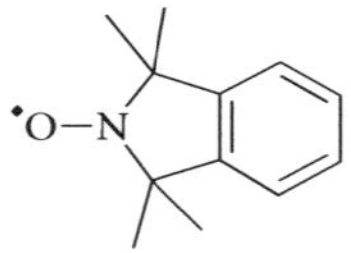	Methylmethacrylate	$M_n = 10{,}500$	2.24	230
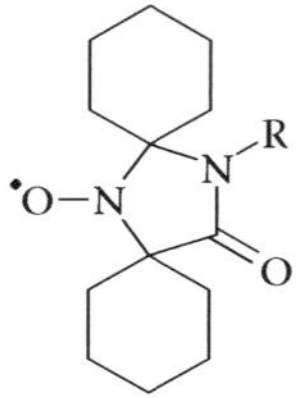	R = H Styrene	$510 < M_n < 16{,}000$	1.23–1.31	230
	R = CH_3 Styrene	$800 < M_n < 16{,}300$	1.09–1.18	230
	R = H Methylmethacrylate	$M_n = 6500$	1.44	230
	R = CH_3 Methylmethacrylate	$M_n = 28{,}800$	1.89	230
	R = CH_3 *tert*-butylacrylate	$1550 < M_n < 9000$	1.32–1.41	230
	R = CH_2Ph Styrene	$M_n < 35{,}000$	1.25–2.30	270

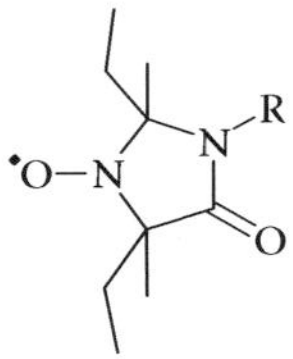	R = H			
	Styrene	$M_n = 16,000$	1.49	230
	R = CH_3			
	Styrene	$500 < M_n < 8100$	1.24–1.36	230
	R = n-C_4H_9			
	Styrene	$1400 < M_n < 6000$	1.22–1.25	230
	R = CH_2Ph			
	Styrene	$1400 < M_n < 5100$	1.23–1.28	230
	R = H			
	Methylmethacrylate	$20400 < M_n < 35,700$	1.57–1.70	230
	R =CH_3			
	Methylmethacrylate	$M_n = 31,000$	1.63	230
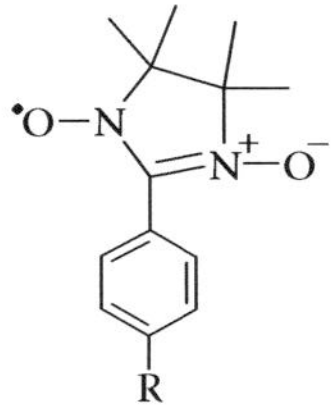	R = CO_2CH_3			
	Styrene	$M_n = 15,000$	1.80–3.50	271
	R = H			
	Styrene	$M_n = 16,000$	2.50–3.40	271
	R = OCH_3			
	Styrene	$M_n = 29,000$	1.60–2.30	271
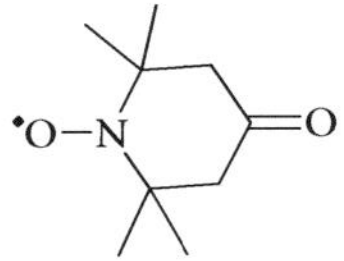	Styrene	$M_n < 17,500$	>1.20 <1.40	272
	Styrene	$M_n = 17,000$	3.30	230
	Methyl acrylate (in ionic liquid)	$M_n < 40,000$	<1.75	273

(*Continued*)

TABLE 10.4 (CONTINUED)
The Structure of Nitroxides and Monomers Used in NMP and Molecular Weight Characteristics of the Resulting Polymers

Nitroxide Structure	Monomer	M_n, M_w, or Molecular Weight Range	M_w/M_n	References
•O–N, =O	Styrene	$9400 < M_n < 62{,}400$	1.09–1.32	241
	n-butyl acrylate	$12300 < M_n < 91{,}300$	1.12–1.34	241
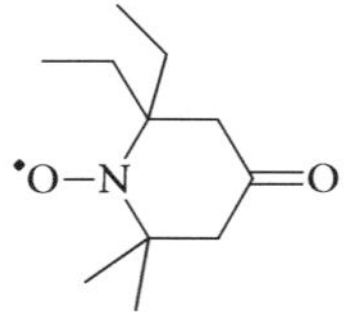 •O–N, =O	Styrene	$M_n = 7400$	1.14	241
•O–N, =O	Styrene	$9000 < M_n < 52{,}000$	1.10–1.53	274
	n-butylacrylate	$11200 < M_n < 15{,}600$	1.44–1.56	274
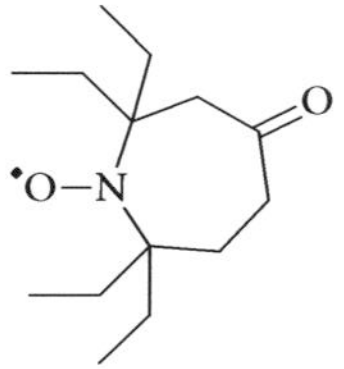 •O–N, =O	Styrene	$M_n = 7700$	1.13	241
•O–N, =O	Styrene	$5000 < M_n < 31{,}500$	1.17–1.34	274
	n-butylacrylate	$M_n = 13{,}800$	1.44	274

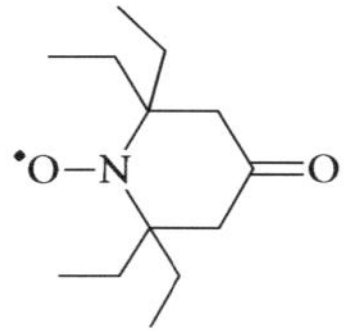

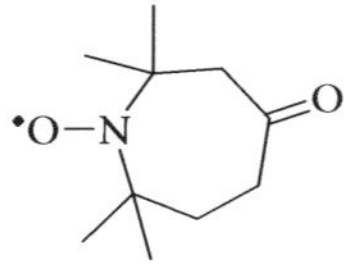

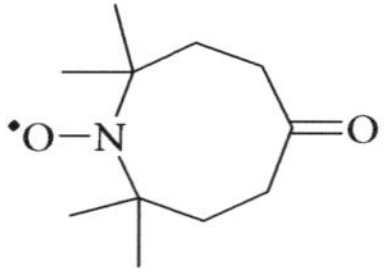

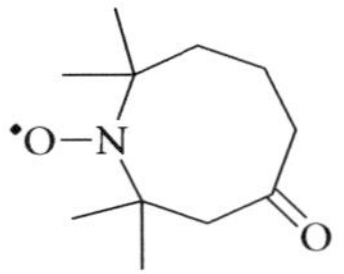	Styrene n-butylacrylate	$5600 < M_n < 23{,}700$ $M_n = 10{,}600$	1.29–1.33 1.80	274 274
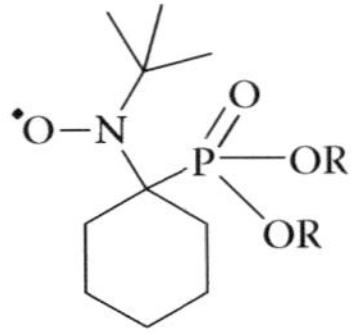	$R = CH_2CH_3$ Styrene	$10{,}000 < M_n < 50{,}000$	1.60	275
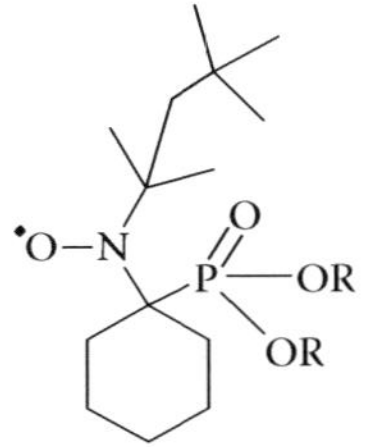	$R = CH_2CH_3$ Styrene R = isopropyl Styrene R = cyclohexyl Styrene	$10{,}000 < M_n < 160{,}000$	1.60	275
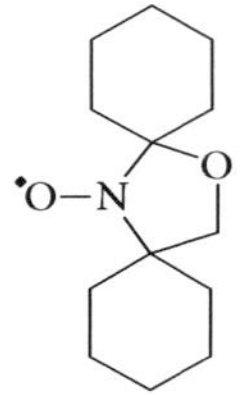	Styrene	$M_n < 35{,}000$	1.50–2.00	276

(Continued)

TABLE 10.4 (CONTINUED)

The Structure of Nitroxides and Monomers Used in NMP and Molecular Weight Characteristics of the Resulting Polymers

Nitroxide Structure	Monomer	M_n, M_w, or Molecular Weight Range	M_w/M_n	References
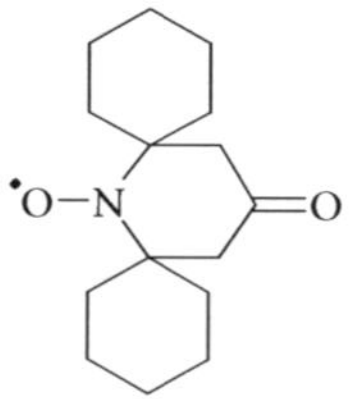 •O–N =O	Styrene	Mn <120,000	<1.36	222
•O–N –OH	Styrene	M_n <125,000	<1.30	222
	n-butyl acrylate	M_n = 24,000	<1.52	277
•O–N –OH	Styrene	M_n <130,000	<1.30	222
	n-butyl acrylate	6130 < M_n <37,400	1.26–1.48	277
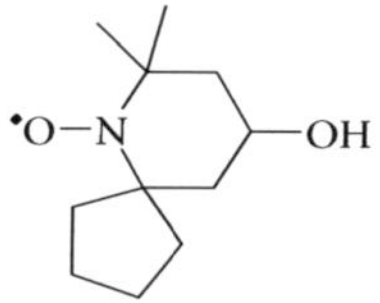 •O–N –OH	Styrene	10,000< M_n <40,000	1.40 –1.60	278

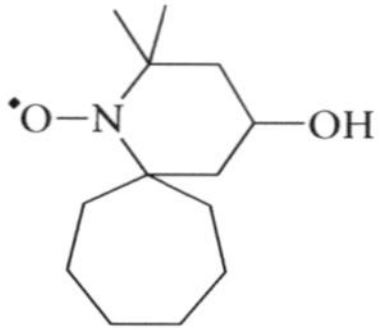

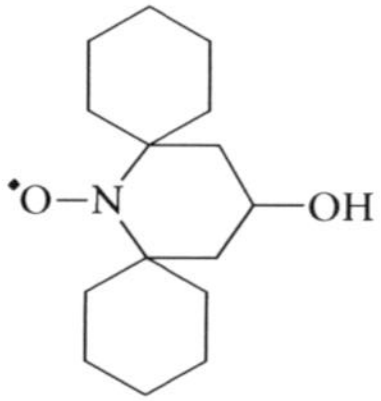

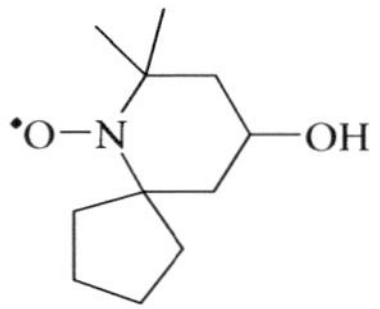	Styrene	10,000< M_n <45,000	1.20–1.40	278
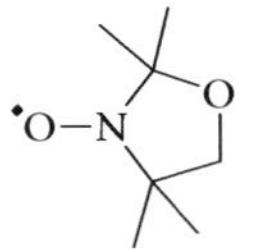	Styrene	4820< M_n <62,900 Not living process	>1.50	276
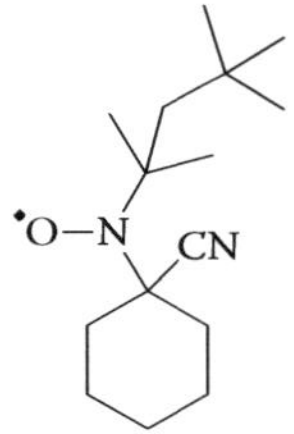	Styrene n-butyl acrylate	M_n <3500 M_n <7000	1.15–1.35 2.05–3.23	279 279
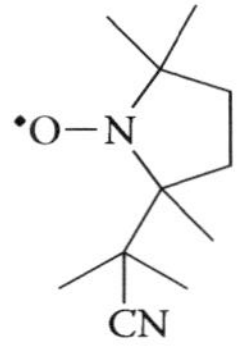	Styrene n-butylacrylate	M_n <4500 M_n <9500	1.15–1.35 1.59–2.14	279 279
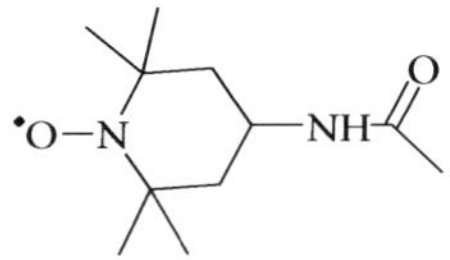	Styrene	M_n <30,000	>1.20 <1.30	272

(*Continued*)

TABLE 10.4 (CONTINUED)

The Structure of Nitroxides and Monomers Used in NMP and Molecular Weight Characteristics of the Resulting Polymers

Nitroxide Structure	Monomer	M_n, M_w, or Molecular Weight Range	M_w/M_n	References
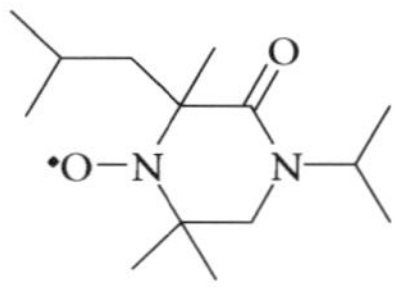	n-butyl acrylate	M_n <6800	1.30	280
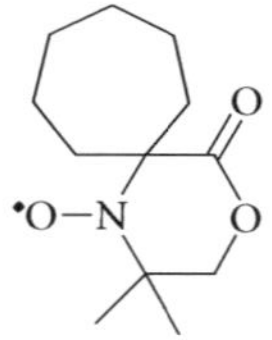	n-butyl acrylate	M_n <1100	2.10	280
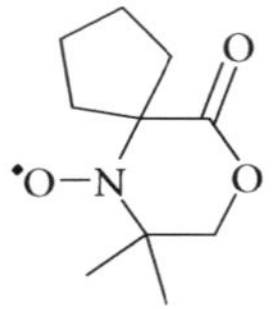	Styrene	M_n <50,000	1.50–2.10	281
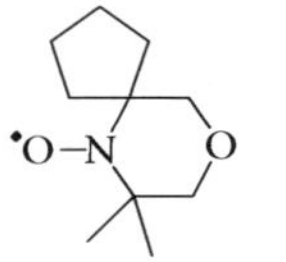	Styrene	M_n <50,000	1.50–2.25	281
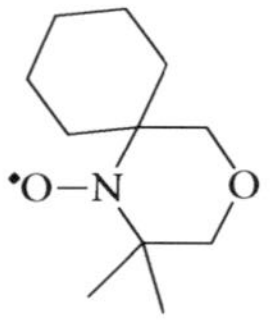	Styrene	M_n <50,000	1.50–1.60	281

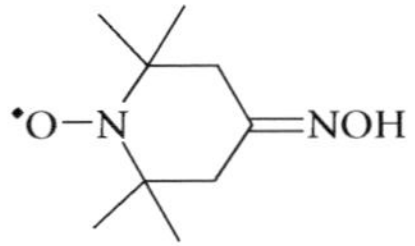

	Styrene	$M_n = 1554$	1.33	282
	n-butyl acrylate	$M_n = 991$	1.46	282
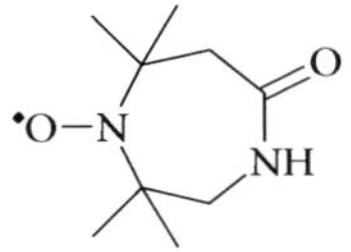	Styrene	$M_n = 2065$	1.34	282
	Styrene	$M_n = 4328$	1.18	282
	n-butyl acrylate	$M_n = 4900$	1.33	282
	Styrene	$M_n = 4180$	1.20	282
	n-butyl acrylate	$M_n = 5507$	1.38	282
	Styrene	$6524 < M_n < 25,475$	1.23–1.64	282
	n-butyl acrylate	$898 < M_n < 36,852$	1.27–2.03	282
	n-butyl methacrylate	$6000 < M_n < 7000$	2.26–2.48	282

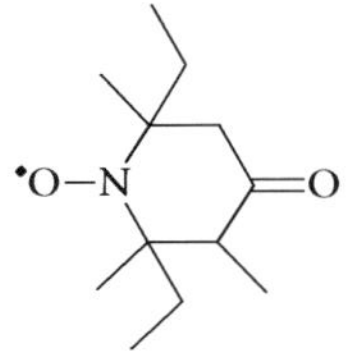

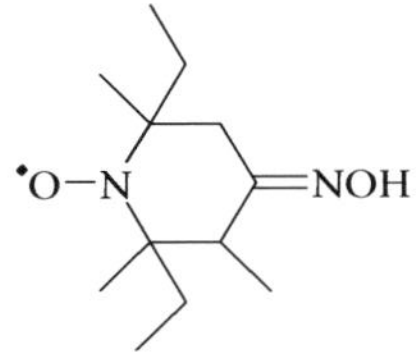

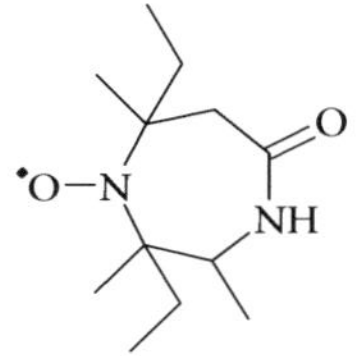

(Continued)

TABLE 10.4 (CONTINUED)
The Structure of Nitroxides and Monomers Used in NMP and Molecular Weight Characteristics of the Resulting Polymers

Nitroxide Structure	Monomer	M_n, M_w, or Molecular Weight Range	M_w/M_n	References
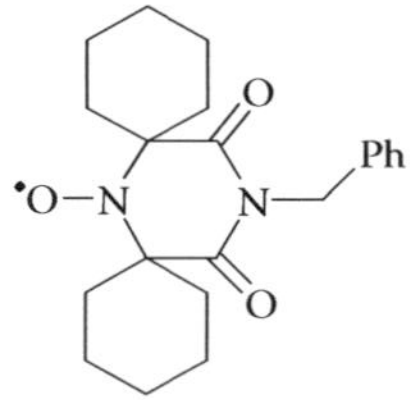	Styrene	M_n <25,000	2.20	283
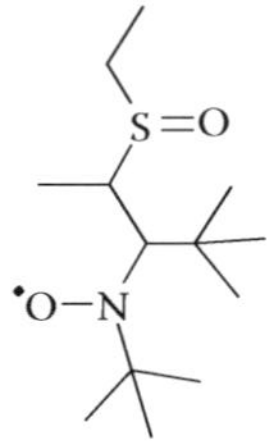	(S_S,R_β,R_α/R_S,S_β,S_α) racemic mixture:			
	R = H			
	ethyl acrylate	25,000< M_n <90,000	1.80–1.60	284
	R = $COOCH_3$			
	ethyl acrylate	M_n <50,000	2.00–2.20	284
	(R_S,R_β,R_α/S_S,S_β,S_α) racemic mixture:			
	R = H			
	ethyl acrylate	25,000< M_n <75,000	~ 1.30	284
	n-butyl acrylate	25,000< M_n <75 000	1.30–1.60	284
	R = $COOCH_3$			
	ethyl acrylate	25,000< M_n <110,000	~1.20	284
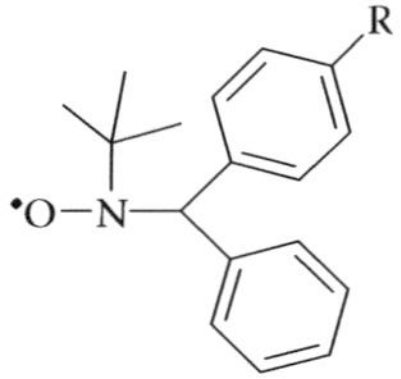	R = H, Cl, OCH_3			
	Styrene	M_n <25,000	1.20–1.40	285

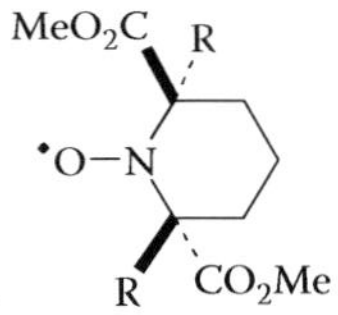 MeO_2C, R, •O–N, R, CO_2Me	racemic, R = Me			
	Styrene	$M_n = 4600$	1.64	242
	racemic, R = Et			
	Styrene	$M_n = 6300$	1.32	242
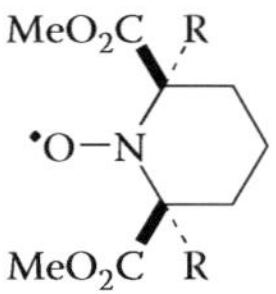 MeO_2C, R, •O–N, MeO_2C, R	R = Me			
	Styrene	$M_n = 7100$	1.57	242
	R = Et			
	Styrene	$M_n = 6100$	1.22	242
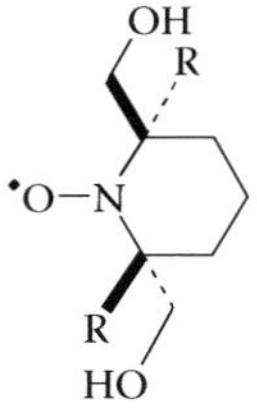 OH, R, •O–N, R, HO	racemic, R = Me			
	Styrene	$M_n = 10{,}300$	1.25	242
	racemic, R = Et			
	Styrene	$10{,}100 < M_n < 46{,}000$	1.24–1.36	242
	racemic, R = Et			
	n-butyl acrylate	$43{,}900 < M_n < 71{,}700$	1.27–1.79	242

(*Continued*)

TABLE 10.4 (CONTINUED)
The Structure of Nitroxides and Monomers Used in NMP and Molecular Weight Characteristics of the Resulting Polymers

Nitroxide Structure	Monomer	M_n, M_w, or Molecular Weight Range	M_w/M_n	References
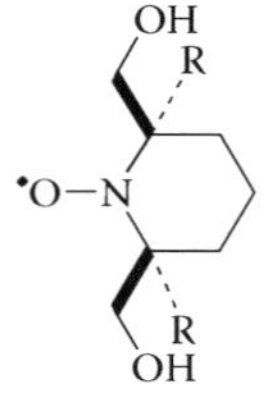	R = Me			
	Styrene	$M_n = 6500$	1.14	242
	R = Et			
	Styrene	$5900 < M_n < 57,000$	1.13–1.32	242
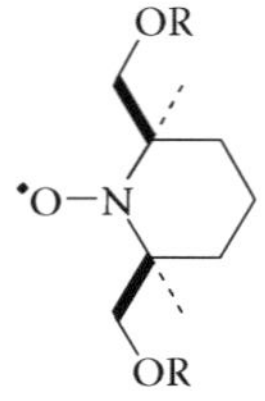	R = Me			
	Styrene	$M_n = 4800$	1.19	242
	R = TBDMS			
	Styrene	$M_n = 7100$	1.19	242
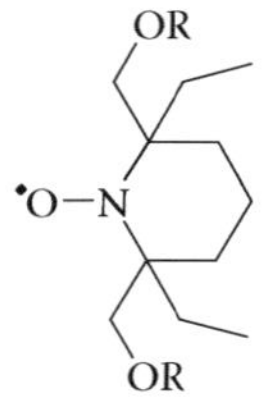	R = TBDMS, trans			
	Styrene	$7300 < M_n < 78,200$	1.11–1.20	242
	R = TBDMS, cis			
	Styrene	$9200 < M_n < 67,400$	1.12–1.22	242
	R = Bn, trans			
	Styrene	$8100 < M_n < 22,100$	1.11–1.17	242
	R = TBDMS, trans			
	n-butyl acrylate	$6000 < M_n < 64,700$	1.12–1.55	242

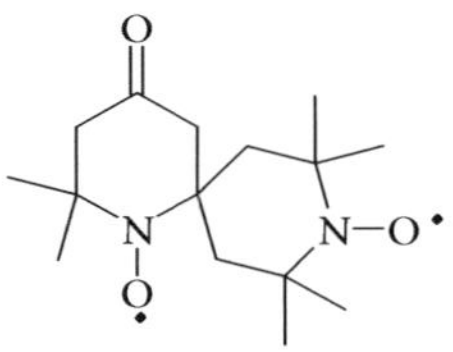	Styrene	—	—	286
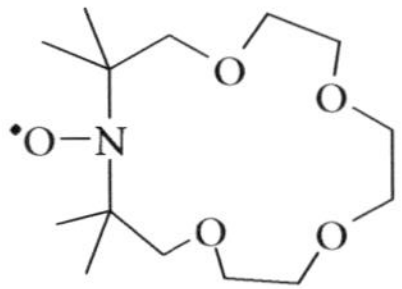	Styrene Styrene (mini emulsion)	$1000 < M_n < 25{,}390$ $3200 < M_n < 29{,}600$	1.18–1.86 2.18–2.86	287
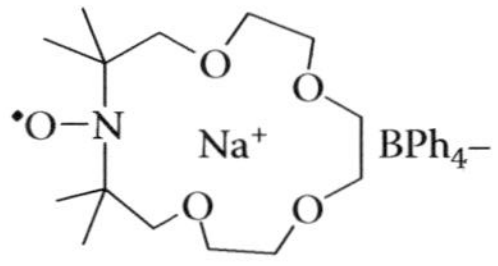 Na^+ BPh_4^-	Styrene	$9100 < Mn < 16{,}550$	1.18–1.42	287

is its tolerance to a wide range of functionalities, namely –OH, –COOH, $CONR_2$, NR_2, SO_3Na, etc., in monomer and solvent. This provides the possibility of performing the polymerization under a wide range of reaction conditions and polymerizing or copolymerizing a wide range of monomers in a controlled manner. The basic concept is based on the earlier studies by the Australian scientists[93] who reported the use of poly(methyl methacrylate) (PMMA) macromonomers as chain transfer agents (CTAs) in radical polymerization. In this process, known as addition-fragmentation chain transfer (AFCT), the propagating radical adds onto a PMMA macromonomer, and a new propagating chain and a new alkene-terminated macromonomer were formed by a chain-transfer reaction (Scheme 10.5).

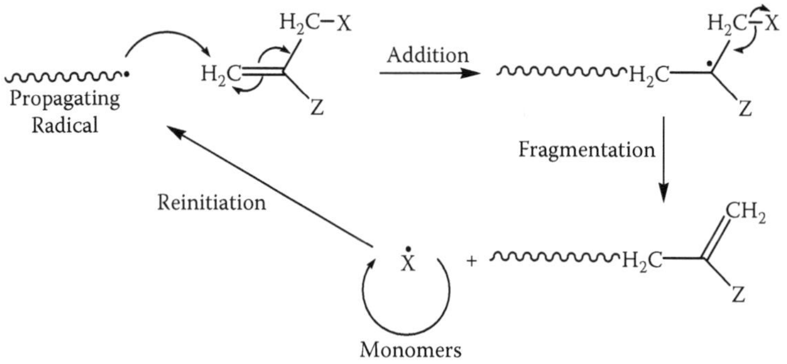

Propagating Radical — $H_2C{=}C(CH_2{-}X)(Z)$ —Addition→ $\sim\sim H_2C{-}\dot{C}(CH_2{-}X)(Z)$ —Fragmentation→ $\dot{X}$ (Monomers) + $\sim\sim H_2C{-}C(=CH_2)(Z)$; Reinitiation

SCHEME 10.5

By using various AFCT agents of the following structures, this process led to the polymers with high polydispersity with the exception of compound 3, which possesses a thiocarbonyl thio group in the structure.

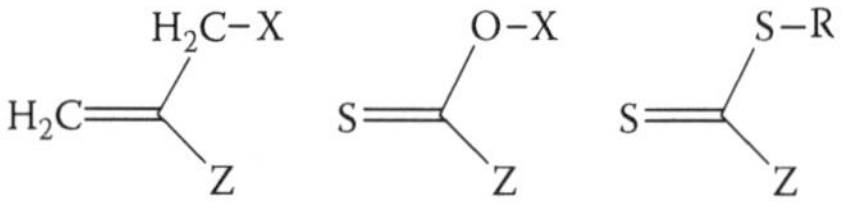

1: $H_2C{=}C(CH_2{-}X)(Z)$ 2: $S{=}C(O{-}X)(Z)$ 3: $S{=}C(S{-}R)(Z)$

It was found that the thiocarbonyl thio compound[93] was the most efficient chain transfer agent and produced a wide range of polymers with predictable molecular weights and very narrow polydispersities. Moreover, at the end of the reaction, the chain ends were still active, reflecting the essential feature of a living polymerization. The process was called RAFT polymerization. A conceptually similar approach was introduced almost at the same time by Zard's group.[94] The use of xanthates as chain transfer agents facilitated the addition of the radical to double bond and the degenerative transfer of radical species to xanthate limited the side product formation. Polymers with narrow molecular weight distribution, >1.05 were obtained. This process was called macromolecular design via interchange of xanthates (MADIX)[95] or degenerative chain transfer.

In RAFT processes, regardless of the name of the method, fragmentation should occur fast enough that the concentration of the intermediate radical remains at a low level, and that the intermediate radical should not act as an initiator or a radical trap.[96] Thus, most of the chains are dormant species that participate in transfer reactions.

TABLE 10.5
Type of RAFT/MADIX Agents Depending on Z Group

Dithioesters	Dithiocarbamate	Trithiocarbonate	Xanthate
Z: C_6H_5	Z: NR′R″	Z: SR′	Z: OR′

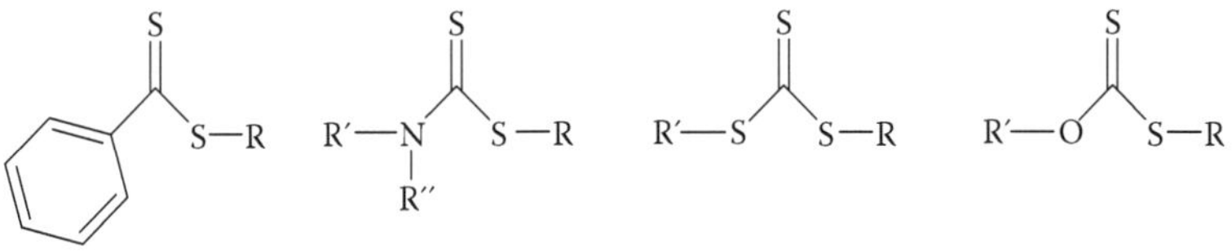

10.4.1 RAFT Agents

General structure of RAFT/MADIX agents has been shown in Scheme 10.6. Depending on the Z group, they were classified into four subgroups such as dithioesters, dithiocarbamates, trithiocarbonates, and xanthates (Table 10.5). The crucial requirement for obtaining polymers with low molecular weight distribution (i.e., M_w/M_n <1.5) is that the termination constant (C_{tr}) of RAFT agent should be larger than 2.[97, 98] However, low M_w/M_n values were also obtained by using monomer feed polymerization technique with a less active RAFT agent.[99]

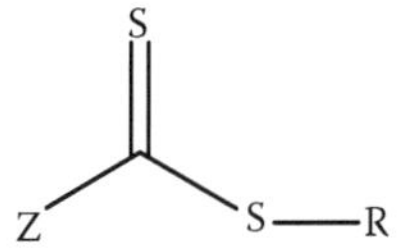

SCHEME 10.6

10.4.2 Synthesis of RAFT Agents

For the preparation of RAFT agents, various synthetic methodologies were used and recently reviewed.[100] These include thionation by cyclic tetrathiophosphate, alkylation, Markovnikoff and Michael addition reactions, via a bis(thiocarbonyl) disulfide, reaction with 1,1′-thiocarbonyl diimidazole (TCDI), and via ATRP. Although dithioesters were readily obtained by almost all synthetic methods, the other RAFT agents were prepared by using selective methods.

10.4.3 Polymerization

The overall RAFT polymerization for vinyl monomers can be represented as in Scheme 10.7.

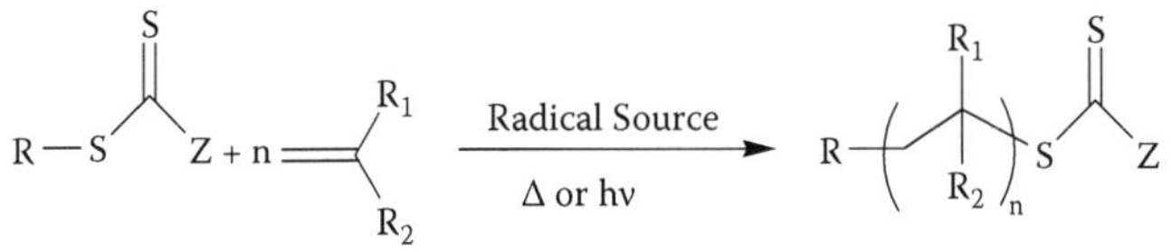

Radical Source

Δ or hν

SCHEME 10.7

The detailed mechanism originally proposed for RAFT polymerization is shown in Scheme 10.8.

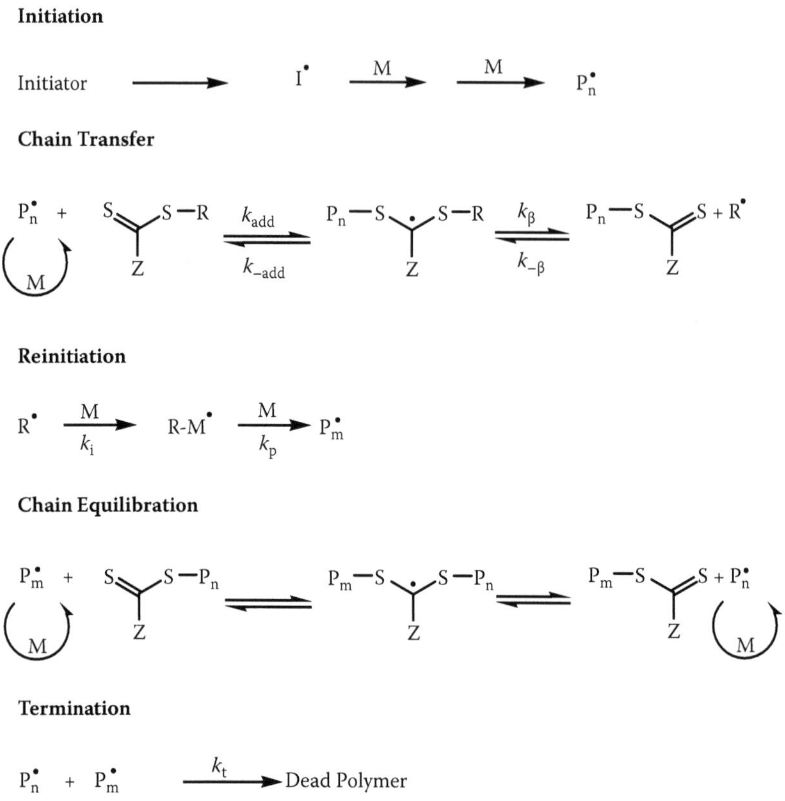

Initiation

$$\text{Initiator} \longrightarrow I^{\bullet} \xrightarrow{M} \xrightarrow{M} P_n^{\bullet}$$

Chain Transfer

$$P_n^{\bullet} \text{ (M) } + S{=}C(Z){-}S{-}R \underset{k_{-add}}{\overset{k_{add}}{\rightleftharpoons}} P_n{-}S{-}\dot{C}(Z){-}S{-}R \underset{k_{-\beta}}{\overset{k_{\beta}}{\rightleftharpoons}} P_n{-}S{-}C(Z){=}S + R^{\bullet}$$

Reinitiation

$$R^{\bullet} \xrightarrow[k_i]{M} R\text{-}M^{\bullet} \xrightarrow[k_p]{M} P_m^{\bullet}$$

Chain Equilibration

$$P_m^{\bullet} \text{ (M) } + S{=}C(Z){-}S{-}P_n \rightleftharpoons P_m{-}S{-}\dot{C}(Z){-}S{-}P_n \rightleftharpoons P_m{-}S{-}C(Z){=}S + P_n^{\bullet} \text{ (M)}$$

Termination

$$P_n^{\bullet} + P_m^{\bullet} \xrightarrow{k_t} \text{Dead Polymer}$$

SCHEME 10.8

In this mechanism, the chain equilibration process is a chain transfer reaction in which radicals are neither formed nor destroyed. The reaction kinetics deviating from those of the conventional free-radical polymerization indicate that RAFT agents do not behave as an ideal chain transfer agent. Reaction conditions should be adjusted so that the fraction of initiator derived chains is negligible. The theoretical degree of the polymerization (DP) can be estimated according to the following equation:

$$\text{DP} = \Delta\text{M}/[\text{RAFT agent}]$$

where ΔM indicates monomer consumption. Positive deviations from the experimental values indicate the partial usage of the RAFT agent. Negative deviations indicate that other polymer chains are also formed to a significant extent by other means (i.e., from initiator).

The RAFT system basically consists of a small amount of RAFT agent and monomer and a free-radical initiator. Radicals stemming from the initiator are used at the very beginning of the polymerization to trigger the degenerative chain transfer reactions that dominate the polymerization. Free radicals affect both the molecular weight distribution of the polymer as the dead polymer chains of uncontrolled molecular weight are formed and the rate of polymerization. Therefore, the concentration of free radicals introduced in the system needs to be carefully balanced. For example, radical concentration was constant between 10^{-6}–10^{-7} (mol l^{-1}) depending on the conditions in the RAFT polymerization of styrene with RAFT agents such as benzyl (diethoxyphosphoryl) dithioformate or benzyl dithiobenzoate.[101] In RAFT polymerization radicals may be generated in three different ways: (1) by decomposition of organic initiators, (2) by the use of an external source (UV–vis or γ-ray), and (3) by thermal initiation. Polymerization temperature is usually in the range of 60–80°C, which corresponds to the optimum decomposition temperature interval of the well-known initiator AIBN. However, even room temperature and high-temperature conditions[102, 103] can also be applied. Generally, a RAFT agent/free-radical ratio of 1:1 to 10:1 yields polymers with narrow molecular weight distributions. Tables 10.6 and 10.7 compile thermal RAFT polymerization of a wide range of monomers in organic and aqueous media using various free-radical initiators.

Photo- and γ-ray-induced reactions, which use light energy to generate radicals in RAFT polymerization, offer a number of advantages compared with thermally initiated ones.

The major advantage is to allow the polymerization to be conducted at room temperature with relatively shorter reaction times. In photoinduced reactions, however, the RAFT agent should carefully be selected, as in some cases control over the molecular weight cannot be attained, particularly at high conversions because it may also decompose under UV light.[104] γ-Ray-induced RAFT polymerization appeared to be more penetrating compared with the corresponding UV-induced processes.[105,106] Typical examples of photo- and γ-ray-induced RAFT polymerizations are collected in Table 10.8.

Radical generation by plasma treatment can also be used for the RAFT polymerization yielding polymers with narrow molecular weight distribution.[107]

10.4.4 Effect of the Structure on the RAFT Agents

For efficient RAFT polymerization, the structure of the RAFT agent should carefully be designed. As is evident from Schemes 10.7 and 10.8, the Z group strongly affects the stability of the thiocarbonyl-thio radical intermediate. The effect of the Z group on the polymerization of a variety of monomers has been investigated by numerous groups[103, 108–114] and recently reviewed.[260] The reactivity of the S=C bond toward radical addition is favored by the strong stabilizing groups. However, stabilization should be finely adjusted so as not to prevent its fragmentation. It was found that the phenyl group appeared to be the most suitable group for most monomers as it balances two criteria. With benzyl group, however, no retardation[110] occurs due to the less stable intermediate and much faster polymerizations were attained, particularly with more reactive monomers.[103, 115, 116] RAFT agents with alkyl groups

TABLE 10.6
Thermal Polymerization of Vinylic Monomers Using RAFT Agents in Organic Solvents

RAFT Agents Z	Z–C(=S)–S–R R	Monomer	References
CH_3	$CH_2C_6H_5$	St	109
		BA	93, 118
	$C(CH_3)_2CN$	St	109
CH_2CH_3	$CH_2C_6H_5$	MA	288
C_6H_5	$C(CH_3)_3$	MMA	289
	$CH(CH_3)CN$	AN	290, 291
	$C(CH_3)_2CN$	AN	292, 291
		MA	293, 289
		MMA	289, 294, 295, 296
		St	109,118,297, 296
		TBDMSMA[a]	298
		Vinyl benzoate	93
	$C(CH_3)_2CH_2C(CH_3)_3$	St	289
		MMA	289
	$CH(CH_3)COOH$	NAM[b]	299
	$C(CH_3)(CN)CH_2CH(CH_3)_2$	MMA	295
	$C(CH_3)(CN)CH_2CH_2CH_2OH$	MMA	93,289
	$C(CH_3)(CN)\ CH_2COOH$	DMAEMA	93
		MMA	300
	$C(CH_3)(CN)CH_2CH_2COOH$	St	300
	$C(CH_3)(CN)CH_2CO_2Na$	$StSO_3Na$	93
	$CH(CH_3)CON(CH_3)_2$	DMA	301
	$C(CH_3)_2CONHCH_2CH_2OH$	MMA	289
	$CH(CH_3)COOCH_3$	MA	293
	$C(CH_3)_2COOC_2H_5$	MMA	289
		St	289,302
	$CH_2C_6H_5$	Ac-Ph-Ome[c]	303
		BA	93, 118, 289
		DMA	301
		MA	289
		MMA	289
		St	304, 109, 289, 101
	$CH(CH_3)C_6H_5$	AA	93
		AN	291
		BA	304, 289
		MA	304, 293
		MMA	289

TABLE 10.6 (CONTINUED)
Thermal Polymerization of Vinylic Monomers Using RAFT Agents in Organic Solvents

RAFT Agents Z	Z–C(=S)–S–R R	Monomer	References
	$C(CH_3)_2C_6H_5$	AN	291
		BA	118
		BMA	93
		BzMA	304
		DMA	301
		MA	305, 306, 293
		MMA	289, 93, 304
		NVP	307
		St	289, 93, 118, 308
		S[d]	306
	$C(CH_3)_2C_6H_4Cl$	MMA	289
	$SC(CH_3)_3$	BA	289
$C_6H_5OCH_3$	$CH_2C_6H_5$	MA	288
C_6H_5CN			
C_6H_3	$CH_2C_6H_5$	St	309
2,6-dimethyl-C_6H_3	$C(CH_3)_2CN$	MMA	295
2,4-dimethyl-C_6H_3			
4-MeO-C_6H_4			
2,4-diMeO-C_6H_3			
3,5-diCF$_3$-C_6H_3	$C(CH_3)(CN)CH_2CH(CH_3)_2$	MMA	295
4-MeO-C_6H_4			
4-F-C_6H_4	$C(CH_3)_2CN$	MMA	295
2,4-diF-C_6H_3			
4-CN-C_6H_4			
3,5-diCN-C_6H_3			
3,5-diCF$_3$-C_6H_3			
4-C_6H_5-C_6H_4			
4-CN-C_6H_4	$C(CH_3)(CN)CH_2CH(CH_3)_2$	MMA	295
$CH_2C_6H_5$	$CH_2C_6H_5$	MA	288
	$CH(CH_3)C_6H_5$	St	297
		MA	103
		NIPAM[e]	103
	$C(CH_3)_2C_6H_5$	AN	291
	$CH(CH_3)COOH$	NAM[b]	299
	$SCH_2C_6H_5$	AN	291

(Continued)

TABLE 10.6 (CONTINUED)
Thermal Polymerization of Vinylic Monomers Using RAFT Agents in Organic Solvents

RAFT Agents $Z-C(=S)-S-R$		Monomer	References
Z	**R**		
$C_{10}H_6$	$CH_2C_6H_5$	St	309
OCH_3	$CH(CH_3)COOCH_3$	VA	310
OCH_3	CH_2COOCH_3	VA	311
OCH_2CH_3	CH_2CN	VA	312
OCH_2CH_3	$CH(CH_3)COO\ CH_2CH_3$	VA	95
OCH_2CH_3	CH_2COOCH_3	VA	311
$OCH(CH_3)_2$			
$OC_6H_4OCH_3$			
$OCH(CH_3)_2$	$CH_2C_6H_5$	VA	313
OC_6F_5		St	109
OC_6H_5			109
OC_2H_5			109
$N(CH_3)_2$	$C(CH_3)_2CN$	MMA	112
$N(C_2H_5)_2$	$CH_2C_6H_5$	St	109
	$CH_2C_6H_5$	St[f]	112
	$CH(CO_2CH_2H_3)_2$	EA	111
		VA	111
$N(C_6H_5)_2$	$CH(COOCH_2CH_3)_2$	NVP	314
$N(CH_3)C_6H_5$	$CH(CO_2CH_2H_3)_2$	EA	111
$N(C_6H_5)_2$		EA	111
		VA	111
$N(C_6H4CH_3)CO_2C(CH_3)_3$	$C(CH_3)CO_2CH_2H_3$	EA	111
$N(C_6H_{11})CO_2CH_3$	$CH(CO_2CH_2H_3)_2$	EA	111
		St	111
		VA	111
$N(C_6H_5)CH_3$	CH_2CN	VA	312
SCH_2CH_3	$CH(CH_3)COOCH_2CH_3$	MA	313
$SCH_2C_6H_5$	$CH_2C_6H_5$	MA	313
		St	109
		St[f]	315
	$CH(CH_3)COOCH_2CH_3$	MA	313
SCH_3	$(CH_2)_4C_6H_2$	St	315
		MA	315
	$C(CH_3)_2CN$	MA	316
		MMA	316
		St	109
		St[f]	316

TABLE 10.6 (CONTINUED)
Thermal Polymerization of Vinylic Monomers Using RAFT Agents in Organic Solvents

RAFT Agents Z	R	Monomer	References
	$CH(C_6H_5)_2COOH$	St[f]	316
$SCH(CH_3)C_6H_5$	$CH(CH_3)C_6H_5$	St[f]	316
$SCH(CH_3)COOCH_3$	$CH(CH_3)COOCH_3$	BA	317
$SCH_2C_6H_5$	$(CH_2CH_2COOCH_2)4C$	MA	315
		St	315
	$CH_2C_6H_5$	MA	316
$SC(CH_3)_2COOH$	$C(CH_3)_2COOH$	NVP	307
$SC_{12}H_{25}$	$C(CH_3)_2COOH$	NVP	307
$SC_{12}H_{25}$	$C(CH_3)_2COOH$	NIPAM	102
Benzyl-9H-carbazole	$CH_2C_6H_5$	MA	318
Benzodioxole	Benzodioxole	St	319
4,4-Dimethyloxazolidin-2-one	$C(CH_3)_2CN$	St	111
		VA	111
	$CH(CO_2CH_2H_3)_2$	St	111
Imidazole	$CH_2C_6H_5$	MA	112
		St[f]	112
2-methyl-benzoimidazole	$CH_2C_6H_5$	St	320
1-methyl-indoline	$CH(CO_2CH_2H_3)_2$	EA	111
Methyl-2-oxooxazolidine-4-carboxylate		St	111
Oxazolidin-2-one		St	111
N-oxide-2-Pyridinyl	$CH_2C_6H_5$	St	321
N-oxide-3-Pyridinyl			
N-oxide-4-Pyridinyl			
5-Oxopyrrolidin-2-yl	$CH(CO_2CH_2H_3)_2$	EA	111
2-phenyl-benzoimidazole	$CH_2C_6H_5$	St	320
4-phenyl-oxazolidin-2-one	$CH(CO_2CH_2H_3)_2$	St	111
1-naphthyl	$C(CH_3)_2CN$	MMA	295
2-naphthyl	$C(CH_3)_2CN$	MMA	295
Pyrrole	$C(CH_3)_2CN$	St	109
		St[f]	112
		MMA	112
	$CH_2C_6H_5$	Ac-Ph-Ome[c]	303
		MA	112
		NIPAAm	322
		St	109
		St[f]	112
	$C(CH_3)_2C_6H_5$	NIPAAm	322

(Continued)

TABLE 10.6 (CONTINUED)
Thermal Polymerization of Vinylic Monomers Using RAFT Agents in Organic Solvents

RAFT Agents Z	Z–C(=S)–S–R R	Monomer	References
2-Pyrrolidinone	$CH_2C_6H_5$	St	109
2-Pyridinyl	$CH_2C_6H_5$	St	321
3-Pyridinyl			
4-Pyridinyl			
4-Pyridinyl	$C(CH_3)_2CN$	MMA	295
4-Pyridinyl –H+ $pTolSO_3$–	$C(CH_3)_2CN$	MMA	295
$PO(OC_2H_5)_2$	$CH_2C_6H_5$	St	101

[a] tert-butyldimethylsilylmethacrylate.
[b] N-Acryloylmorpholine.
[c] Ac-Ph-Ome[c]: N-Acryloyl-L-phenylalanine methyl ester.
[d] 1000 bar.
[e] Additional Lewis acid was used such as $Y(OTf)_3$, $Yb(OTf)_3$, and $Sc(OTf)_3$.
[f] Without initiator.

were also used for styrene,[109, 117] butyl acrylate,[93, 118] and methyl methacrylate,[119] and provided reasonable control over polymerization. The Z group may also be a heteroatom, such as O and N possessing nonbonding electrons. However, in such cases the control of the polymerization is reduced because of the lower reactivity of the double bond toward radical addition. This behavior may be advantageous for the polymerization of fast propagating monomers such as vinyl acetate. Z group of the RAFT agent may be selected according to the following reactivity order as obtained from experimental and theoretical studies.[120]

Phenyl > $S{-}CH_3$ ~ CH_2-phenyl ~ CH_3 ~ N-pyrrolyl > $O{-}CH_2CF_3$ > $O{-}CH_2CH_3$ ~ $N(CH_2CH_3)_2$

Two parameters should be considered for the selection of the R group. First, the R group should be a good leaving group compared with the growing polymeric chain. Second, it should produce re-initiating species exhibiting high reactivity toward the monomer used. The R group also affects the stabilization of the radical intermediate. However, the effect is less pronounced than the Z group. The cumyl and cyanoisopropyl groups appear to be the most efficient for the re-initiation step.[100] Several groups have investigated the use of R groups that mimic the propagating polymeric radical.

TABLE 10.7

Thermal Polymerization of Vinylic Monomers Using RAFT Agents in Water

RAFT Agents Z–C(=S)–S–R

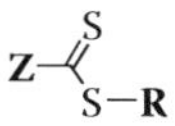

Z	R	Monomer	Temperature (°C)	M_n	*PDI*	References
CH_3	$CH_2C_6H_5$	S	80	35,580	1.38	118
C_6H_5	$CH_2C_6H_5$	S	80	53,210	1.37	118
	$C(CH_3)_2C_6H_5$	S	80	39,850	7.09	118
	$C(CH_3)(CN)CH_2CO_2Na$	$StSO_3Na$	70	<8000	>1.13	93
	$C(CH_3)(CN)CH_2CH_2COOH$	AMPS[a]	70	33,900	1.14	323
		AMBA[b]	70	31,300	1.14	323
	$CH(CH_3)CONHCH_2CH_2SO_3Na$	AAm[c]	70	<38,000	1.04	324
	$CH_2C_6H_4COOH$	MMA	60, 70, 80	<88,000	>1.2	325
		S	80	18,800	1.24	325
	$CH(CH_3)CON(CH_3)_2$	DMA	60, 70, 80	<39,300	>1.14	326
	$C(CH_3)(CN)CH_2CH_2CO_2Na$	DMA	60, 70, 80	<45,570	>1.11	326
OC_2H_5	$CH_2C_6H_5$	S	80	32,370	1.98	118
	$CH(CH_3)C_6H_5$			31,300	2.04	118
S-$C(CH_3)_2COOH$	$C(CH_3)_2COOH$	AM	25–70	<72,000	>1.03	327
		DMA	25	<78,000	>1.05	327

[a] Acrylamidopropylsulfonate.

[b] Acrylamidopropylcarboxylate.

[c] In acetic acid/sodium acetate buffer solution (pH:5).

TABLE 10.8
Photo- and γ-ray-Induced Vinylic Monomers Using RAFT Agents

RAFT Agents: Z–C(=S)–S–R

Z	R	Monomer	λnm	Solvent	Temperature (°C)	References
CH_2CH_3	$CH_2C_6H_5$	MA	$\gamma(^{60}Co)$	—	RT	288
C_6H_5	$CH_2C_6H_5$	MA				
		S				
	$C(CH_3)_2C_6H_5$	S			60	328
$CH_2C_6H_5$	$CH_2C_6H_5$	MA			RT	288
$CH_2C_6H_5$	$CH(CH_3)COOCH_3$	BA	$\lambda = 351$[b]	—	60, 80	329
C_6H_5CN	$CH_2C_6H_5$	MA	$\gamma(^{60}Co)$	—	RT	288
C_6H_5OCH						
$S\text{-}C1_2H_{25}$	$C(CH_3)_2COOH$	MA	$\lambda = 365–405$[a]	Benzene	30	330
$SCH(CH_3)COOCH_3$	$CH(CH_3)COOCH_3$	BA	$\lambda = 351$[b]	—	60, 80	329

[a] (2,4,6-trimethylbenzoyl) dipheny phospine oxide was used as a photoinitiator.
[b] XeF excimer laser pulse.

On the basis of experimental and theoretical studies,[120] the following general principle for the selection of the R group may be proposed.[260]

$$-C(CH_3)_2CN \sim -C(CH_3)_2C_6H_5 > -CH(CO_2Me)C_6H_5 \sim -CH(CN)C_6H_5 > -C(CH_3)_2CO_2Et$$

$$\sim -C(CH_3)_2CONEt_2 > -CH(CH_3)C_6H_5 > -CH(CH_3)CO_2Et > -C(CH_3)_3 > -CH_2C_6H_5 > -CH_2CO_2Et$$

10.4.5 Macromolecular Architectures by RAFT

RAFT polymerization provides the possibility to design specific macromolecular architectures by the variation of the monomer composition and topology. These architectures include end-functional polymers, statistical, gradient, block, comb/brush, star, hyperbranched, and network (co)polymers. Extensive studies have been recently reviewed by Diaz et al.[260] and discussed in the other chapters of this book.

10.5 CONCLUSION

It is clear that, in the past decade, tremendous development has taken place in the field of controlled/living radical polymerization. The fundamental aspects of the three main procedures of controlled/living radical polymerization have been discussed in this chapter. Significant advances in these systems provide the possibility to prepare well-defined polymers with good control over polymer architecture, such as block and graft copolymers, star polymers, macrocycles, and functional polymers. Describing all the studies in the field that have been reported would be impossible. The readers are therefore referred to the books and review articles solely devoted to the individual methods.[121–124]

REFERENCES

1. Kato, M., Kamigaito, M., Sawamoto, M., Higashimura, T. *Macromolecules* 1995, 28, 1721–1723.
2. Wang, J. S., Matyjaszewski, K. *Macromolecules* 1995, 28, 7901–7910.
3. Wang, J. S., Matyjaszewski, K. *Journal of the American Chemical Society* 1995, 117, 5614–5615.
4. Percec, V., Barboiu, B. *Macromolecules* 1995, 28, 7970–7972.
5. Percec, V., Barboiu, B., Bera, T. K., van der Sluis, M., Grubbs, R. B., Frechet, J. M. J. *Journal of Polymer Science Part A-Polymer Chemistry* 2000, 38, 4776–4791.
6. Matyjaszewski, K., Xia, J. H. *Chemical Reviews* 2001, 101, 2921–2990.
7. Kamigaito, M., Ando, T., Sawamoto, M. *Chemical Reviews* 2001, 101, 3689–3745.
8. Percec, V., Barboiu, B., Grigoras, C., Bera, T. K. *Journal of the American Chemical Society* 2003, 125, 6503–6516.

9. Matyjaszewski, K. *Journal of Macromolecular Science-Pure and Applied Chemistry* 1997, A34, 1785–1801.
10. Goto, A., Fukuda, T. *Macromolecular Rapid Communications* 1999, 20, 633–636.
11. Fischer, H. *Macromolecules* 1997, 30, 5666–5672.
12. Fischer, H. *Journal of Polymer Science Part A-Polymer Chemistry* 1999, 37, 1885–1901.
13. Fukuda, T., Goto, A., Ohno, K. *Macromolecular Rapid Communications* 2000, 21, 151–165.
14. Matyjaszewski, K. *Macromolecular Symposia* 1996, 111, 47–61.
15. Matyjaszewski, K. in *Controlled Radical Polymerization,* K. Matyjaszewski, ed., *ACS Symposium Series 685;* American Chemical Society: Washington, DC, 1997, 111, p 258.
16. Davis, K., O'Malley, J., Paik, H. J., Matyjaszewski, K. *Abstracts of Papers of the American Chemical Society* 1997, 213, 320-POLY.
17. Xia, J. H., Zhang, X., Matyjaszewski, K. in *Transition Metal Catalysis in Macromolecular Design;* L. S. Boffa, B. Novak, M., Eds., *ACS Symposium Series 760;* American Chemical Society: Washington, DC, 2000, 768, Chapter 13, pp 207–223.
18. Uegaki, H., Kotani, Y., Kamigaito, M., Sawamoto, M. *Macromolecules* 1998, 31, 6756–6761.
19. Uegaki, H., Kotani, Y., Kamigaito, M., Sawamoto, M. *Macromolecules* 1997, 30, 2249–2253.
20. Matyjaszewski, K., Wei, M. L., Xia, J. H., McDermott, N. E. *Macromolecules* 1997, 30, 8161–8164.
21. Ando, T., Kamigaito, M., Sawamoto, M. *Macromolecules* 1997, 30, 4507–4510.
22. Kotani, Y., Kamigaito, M., Sawamoto, M. *Macromolecules* 1999, 32, 2420–2424.
23. Simal, F., Demonceau, A., Noels, A. F. *Angewandte Chemie-International Edition* 1999, 38, 538–540.
24. Percec, V., Barboiu, B., Neumann, A., Ronda, J. C., Zhao, M. Y. *Macromolecules* 1996, 29, 3665–3668.
25. Moineau, G., Granel, C., Dubois, P., Jerome, R., Teyssie, P. *Macromolecules* 1998, 31, 542–544.
26. Lecomte, P., Drapier, I., Dubois, P., Teyssie, P., Jerome, R. *Macromolecules* 1997, 30, 7631–7633.
27. Zhu, S. M., Yan, D. Y., Van Beylen, M. *Advances in Controlled/Living Radical Polymerization* 2003, 854, 221–235.
28. Neumann, A., Keul, H., Hocker, H. *Macromolecular Chemistry and Physics* 2000, 201, 980–984.
29. Ando, T., Kamigaito, M., Sawamoto, M. *Macromolecules* 2000, 33, 5825–5829.
30. Kamigaito, M., Watanabe, Y., Ando, T., Sawamoto, M. *Journal of the American Chemical Society* 2002, 124, 9994–9995.
31. Matyjaszewski, K., Gobelt, B., Paik, H. J., Horwitz, C. P. *Macromolecules* 2001, 34, 430–440.
32. Tang, W., Matyjaszewski, K. *Macromolecules* 2006, in press ASAP.
33. Wang, J. S., Matyjaszewski, K. *Macromolecules* 1995, *28*, 7572–7573.
34. Wang, W. X., Dong, Z. H., Xia, P., Yan, D. Y., Zhang, Q. *Macromolecular Rapid Communications* 1998, 19, 647–649.
35. Xia, J. H., Matyjaszewski, K. *Macromolecules* 1997, 30, 7692–7696.
36. Xia, J. H., Matyjaszewski, K. *Macromolecules* 1999, 32, 5199–5202.
37. Moineau, G., Dubois, P., Jerome, R., Senninger, T., Teyssie, P. *Macromolecules* 1998, 31, 545–547.

38. Wang, W., Yan, D. in *Controlled/Living Radical Polymerization; Progress in ATRP, NMP, and RAFT. ACS Symposium Series 768;* American Chemical Society, Washington, DC, 2000, Chapter 19, pp 263–275. .
39. Zhu, S. M., Wang, W. X., Tu, W. P., Yan, D. Y. Acta Polymerica 1999, 50, 267–269.
40. Wang, W. X., Yan, D. Y., Jiang, X. L., Detrembleur, C., Lecomte, P., Jerome, R. *Macromolecular Rapid Communications* 2001, 22, 439–443.
41. Qin, D. Q., Qin, S. H., Qiu, K. Y. *Journal of Polymer Science Part A-Polymer Chemistry* 2000, 38, 101–107.
42. Qin, D. Q., Qin, S. H., Chen, X. P., Qiu, K. Y. *Polymer* 2000, 41, 7347–7353.
43. Chen, X. P., Qiu, K. Y. *Macromolecules* 1999, 32, 8711–8715.
44. Chen, X. P., Qiu, K. Y. *Chemical Communications* 2000, 1403–1404.
45. Chen, X. P., Qiu, K. Y. *Chemical Communications* 2000, 233–234.
46. Ucan, R., Tunca, U., Hizal, G. *Journal of Polymer Science Part A-Polymer Chemis. try* 2003, 41, 2019–2025.
47. van der Sluis, M., Barboiu, B., Pesa, N., Percec, V. *Macromolecules* 1998, 31, 9409–9412.
48. Matyjaszewski, K., Coca, S., Gaynor, S. G., Wei, M. L., Woodworth, B. E. *Macromolecules* 1997, 30, 7348–7350.
49. Haddleton, D. M., Clark, A. J., Crossman, M. C., Duncalf, D. J., Heming, A. M., Morsley, S. R., Shooter, A. J. *Chemical Communications* 1997, 1173–1174.
50. Haddleton, D. M., Kukulj, D., Duncalf, D. J., Heming, A. M., Shooter, A. J. *Macromolecules* 1998, 31, 5201–5205.
51. Heuts, J. P. A., Mallesch, R., Davis, T. P. *Macromolecular Chemistry and Physics* 1999, 200, 1380–1385.
52. de Vries, A., Klumperman, B., de Wet-Roos, D., Sanderson, R. D. *Macromolecular Chemistry and Physics* 2001, 202, 1645–1648.
53. Gnanou, Y., Hizal, G. *Journal of Polymer Science Part A-Polymer Chemistry* 2004, 42, 351–359.
54. Acar, A. E., Yagci, M. B., Mathias, L. J. *Macromolecules* 2000, 33, 7700-7706.
55. Nanda, A. K., Hong, S. C., Matyjaszewski, K. *Macromolecular Chemistry and Physics* 2003, 204, 1151–1159.
56. Percec, V., Asandei, A. D., Asgarzadeh, F., Bera, T. K., Barboiu, B. *Journal of Polymer Science Part A-Polymer Chemistry* 2000, 38, 3839–3843.
57. Percec, V., Barboiu, B., van der Sluis, M. *Macromolecules* 1998, 31, 4053–4056.
58. Gromada, J., Matyjaszewski, K. *Macromolecules* 2001, 34, 7664–7671.
59. Li, M., Matyjaszewski, K. *Journal of Polymer Science Part A-Polymer Chemistry* 2003, 41, 3606–3614.
60. Min, K., Gao, H. F., Matyjaszewski, K. *Journal of the American Chemical Society* 2005, 127, 3825–3830.
61. Jakubowski, W., Matyjaszewski, K. *Macromolecules* 2005, 38, 4139–4146.
62. Shen, Y. Q., Tang, H. D., Ding, S. J. *Progress in Polymer Science* 2004, 29, 1053–1078.
63. Haddleton, D. M., Kukulj, D., Radigue, A. P. *Chemical Communications* 1999, 99–100.
64. Kickelbick, G., Paik, H. J., Matyjaszewski, K. *Macromolecules* 1999, 32, 2941–2947.
65. Shen, Y. Q., Zhu, S. P., Pelton, R. *Macromolecules* 2001, 34, 5812–5818.
66. Shen, Y. Q., Zhu, S. P., Zeng, F. Q., Pelton, R. *Journal of Polymer Science Part A-Polymer Chemistry* 2001, 39, 1051–1059.
67. Duquesne, E., Degee, P., Habimana, J., Dubois, P. *Chemical Communications* 2004, 640–641.

68. Nguyen, J. V., Jones, C. W. *Macromolecules* 2004, 37, 1190–1203.
69. Nguyen, J. V., Jones, C. W. *Journal of Polymer Science Part A-Polymer Chemistry* 2004, 42, 1367–1383.
70. Nguyen, J. V., Jones, C. W. *Journal of Polymer Science Part A-Polymer Chemistry* 2004, 42, 1384–1399.
71. Honigfort, M. E., Brittain, W. J. *Macromolecules* 2003, 36, 3111–3114.
72. Shen, Y. Q., Zhu, S. P., Zeng, F. Q., Pelton, R. H. *Macromolecules* 2000, 33, 5427–5431.
73. Shen, Y. Q., Zhu, S. P., Zeng, F. Q., Pelton, R. *Macromolecular Chemistry and Physics* 2000, 201, 1387–1394.
74. Shen, Y. Q., Zhu, S. P., Pelton, R. *Macromolecular Rapid Communications* 2000, 21, 956–959.
75. Hong, S. C., Paik, H. J., Matyjaszewski, K. *Macromolecules* 2001, 34, 5099–5102.
76. Hong, S. C., Matyjaszewski, K. *Macromolecules* 2002, 35, 7592-7605.
77. Hong, S. C., Neugebauer, D., Inoue, Y., Lutz, J. F., Matyjaszewski, K. *Macromolecules* 2003, 36, 27–35.
78. Hong, S. C., Lutz, J. F., Inoue, Y., Strissel, C., Nuyken, O., Matyjaszewski, K. *Macromolecules* 2003, 36, 1075–1082.
79. Haddleton, D. M., Jackson, S. G., Bon, S. A. F. *Journal of the American Chemical Society* 2000, 122, 1542–1543.
80. Liou, S., Rademacher, J. T., Malaba, D., Pallack, M. E., Brittain, W. J. *Macromolecules* 2000, 33, 4295–4296.
81. Shen, Y. Q., Zhu, S. P., Pelton, R. *Macromolecules* 2001, 34, 3182–3185.
82. Shen, Y. Q., Zhu, S. P. *Macromolecules* 2001, 34, 8603–8609.
83. Barre, G., Taton, D., Lastcoueres, D., Vincent, J. M. *Journal of the American Chemical Society* 2004, 126, 7764–7765.
84. Yang, J., Ding, S. J., Radosz, M., Shen, Y. Q. *Macromolecules* 2004, 37, 1728–1734.
85. Ding, S. J., Yang, J., Radosz, M., Shen, Y. Q. *Journal of Polymer Science Part A-Polymer Chemistry* 2004, 42, 22–30.
86. Hizal, G., Tunca, U., Aras, S., Mert, H. *Journal of Polymer Science Part A-Polymer Chemistry* 2006, 44, 77–87.
87. Solomon, D. H., Rizzardo, E., Cacioli, P. *U.S. Pat.* 1986, 4, 581, 429.
88. Solomon, D. H., Rizzardo, E., Cacioli, P. *Eur. Patent Appl.* 1985, 135, 280.
89. Moad, G., Rizzardo, E. *Macromolecules* 1995, 28, 8722–8728.
90. Georges, M. K., Veregin, R. P. N., Kazmaier, P. M., Hamer, G. K. *Macromolecules* 1993, 26, 2987–2988.
91. Hawker, C. J., Bosman, A. W., Harth, E. *Chemical Reviews* 2001, 101, 3661–3688.
92. Detrembleur, C., Claes, M., Jerome, R. *Advances in Controlled/Living Radical Polymerization* 2003, 854, 496–518.
93. Chiefari, J., Chong, Y. K., Ercole, F., Krstina, J., Jeffery, J., Le, T. P. T., Mayadunne, R. T. A., Meijs, G. F., Moad, C. L., Moad, G., Rizzardo, E., Thang, S. H. *Macromolecules* 1998, 31, 5559–5562.
94. Delduc, P., Tailhan, C., Zard, S. Z. *Journal of the Chemical Society-Chemical Communications* 1988, 308–310.
95. Charmot, D., Corpart, P., Adam, H., Zard, S. Z., Biadatti, T., Bouhadir, G. *Macromolecular Symposia* 2000, 150, 23–32.
96. Goto, A., Fukuda, T. *Progress in Polymer Science* 2004, 29, 329–385.
97. Muller, A. H. E., Zhuang, R. G., Yan, D. Y., Litvinenko, G. *Macromolecules* 1995, 28, 4326–4333.

98. Litvinenko, G., Muller, A. H. E. *Macromolecules* 1997, 30, 1253–1266.
99. Krstina, J., Moad, C. L., Moad, G., Rizzardo, E., Berge, C. T. *Macromolecular Symposia* 1996, 111, 13–23.
100. Perrier, S., Takolpuckdee, P. *Journal of Polymer Science Part A-Polymer Chemistry* 2005, 43, 5347–5393.
101. Alberti, A., Benaglia, M., Laus, M., Macciantelli, D., Sparnacci, K. *Macromolecules* 2003, 36, 736–740.
102. Convertine, A. J., Ayres, N., Scales, C. W., Lowe, A. B., McCormick, C. L. *Biomacromolecules* 2004, 5, 1177–1180.
103. Quinn, J. F., Rizzardo, E., Davis, T. P. *Chemical Communications* 2001, 1044–1045.
104. Quinn, J. F., Barner, L., Barner-Kowollik, C., Rizzardo, E., Davis, T. P. *Macromolecules* 2002, *35*, 7620–7627.
105. Bai, R. K., You, Y. Z., Pan, C. Y. *Macromolecular Rapid Communications* 2001, 22, 315–319.
106. Barner, L., Quinn, J. F., Barner-Kowollik, C., Vana, P., Davis, T. P. *European Polymer Journal* 2003, 39, 449–459.
107. Chen, G. J., Zhu, X. L., Zhu, J., Cheng, Z. P. *Macromolecular Rapid Communications* 2004, 25, 818–824.
108. Davis, T. P., Barner-Kowollik, C., Nguyen, T. L. U., Stenzel, M. H., Quinn, J. F., Vana, P. *Advances in Controlled/Living Radical Polymerization* 2003, 854, 551–569.
109. Chiefari, J., Mayadunne, R. T. A., Moad, C. L., Moad, G., Rizzardo, E., Postma, A., Skidmore, M. A., Thang, S. H. *Macromolecules* 2003, 36, 2273–2283.
110. Barner-Kowollik, C., Quinn, J. F., Nguyen, T. L. U., Heuts, J. P. A., Davis, T. P. *Macromolecules* 2001, 34, 7849–7857.
111. Destarac, M., Charmot, D., Franck, X., Zard, S. Z. *Macromolecular Rapid Communications* 2000, 21, 1035–1039.
112. Mayadunne, R. T. A., Rizzardo, E., Chiefari, J., Chong, Y. K., Moad, G., Thang, S. H. *Macromolecules* 1999, 32, 6977–6980.
113. Destarac, M., Bzducha, W., Taton, D., Gauthier-Gillaizeau, I., Zard, S. Z. *Macromolecular Rapid Communications* 2002, 23, 1049–1054.
114. Destarac, M., Brochon, C., Catala, J. M., Wilczewska, A., Zard, S. Z. *Macromolecular Chemistry and Physics* 2002, 203, 2281–2289.
115. Ray, B., Isobe, Y., Matsumoto, K., Habaue, S., Okamoto, Y., Kamigaito, M., Sawamoto, M. *Macromolecules* 2004, 37, 1702–1710.
116. Thomas, D. B., Convertine, A. J., Myrick, L. J., Scales, C. W., Smith, A. E., Lowe, A. B., Vasilieva, Y. A., Ayres, N., McCormick, C. L. *Macromolecules* 2004, 37, 8941–8950.
117. Goto, A., Sato, K., Tsujii, Y., Fukuda, T., Moad, G., Rizzardo, E., Thang, S. H. *Macromolecules* 2001, 34, 402–408.
118. Moad, G., Chiefari, J., Chong, Y. K., Krstina, J., Mayadunne, R. T. A., Postma, A., Rizzardo, E., Thang, S. H. *Polymer International* 2000, 49, 993–1001.
119. Rizzardo, E., Chiefari, J., Mayadunne, R. T. A., Moad, G., Thang, S. H. *ACS Symp. Ser.* 2000, 768, 278.
120. Coote, M. L., Henry, D. J. *Macromolecules* 2005, 38, 1415–1433.
121. Matyjaszewski, K., Davis, T. P. *Handbook of Radical Polymerization,* John Wiley & Sons, Inc, 2002.
122. Hadjichristidis, N., Pispas, S., Floudas, G. *Block Copolymers: Synthetic Strategies, Physical Properties, and Applications,* John Wiley & Sons, Inc, 2003.
123. Matyjaszewski, K., Spanswick, J. *Handbook of Polymer Synthesis Plastics Engineering Series, No 24,* Ed. Kricheldorf, H.R. Nuyken, O. Swift, G. CRC Press 2004–2005, Chapter 17, 895–943.

124. Hadjichristidis, N., Pitsikalis, M., Iatrou, H. *Block Copolymers I* 2005, 189, 1–124.
125. Kim, C. S., Oh, S. M., Kim, S., Cho, C. G. *Macromolecular Rapid Communications* 1998, 19, 191–196.
126. Grigoras, C., Percec, V. *Journal of Polymer Science Part A-Polymer Chemistry* 2005, 43, 319–330.
127. Percec, V., Grigoras, C. *Journal of Polymer Science Part A-Polymer Chemistry* 2005, 43, 5283–5299.
128. Senkal, B. F., Hizal, G., Bicak, N. *Journal of Polymer Science Part A-Polymer Chemistry* 2001, 39, 2691–2695.
129. Patten, T. E., Xia, J. H., Abernathy, T., Matyjaszewski, K. *Science* 1996, 272, 866–868.
130. Matyjaszewski, K., Patten, T. E., Xia, J. H. *Journal of the American Chemical Society* 1997, 119, 674–680.
131. Percec, V., Barboiu, B., Kim, H. J. *Journal of the American Chemical Society* 1998, 120, 305–316.
132. Kotani, Y., Kamigaito, M., Sawamoto, M. *Macromolecules* 1999, 32, 6877–6880.
133. Kotani, Y., Kamigaito, M., Sawamoto, M. *Macromolecules* 2000, 33, 3543–3549.
134. Louie, J., Grubbs, R. H. *Chemical Communications* 2000, 1479–1480.
135. Teodorescu, M., Gaynor, S. G., Matyjaszewski, K. *Macromolecules* 2000, 33, 2335–2339.
136. Kotani, Y., Kamigaito, M., Sawamoto, M. *Macromolecules* 1998, 31, 5582–5587.
137. Takahashi, H., Ando, T., Kamigaito, M., Sawamoto, M. *Macromolecules* 1999, 32, 3820–3823.
138. Simal, F., Demonceau, A., Noels, A. F. *Tetrahedron Letters* 1999, 40, 5689–5693.
139. Kotani, Y., Kamigaito, M., Sawamoto, M. *Macromolecules* 2000, 33, 6746–6751.
140. Watanabe, Y., Ando, T., Kamigaito, M., Sawamoto, M. *Macromolecules* 2001, 34, 4370–4374.
141. Petrucci, M. G. L., Lebuis, A. M., Kakkar, A. K. *Organometallics* 1998, 17, 4966–4975.
142. Brandts, J. A. M., van de Geijn, P., van Faassen, E. E., Boersma, J., van Koten, G. *Journal of Organometallic Chemistry* 1999, 584, 246–253.
143. Qiu, J., Matyjaszewski, K. *Macromolecules* 1997, 30, 5643–5648.
144. Gao, B., Chen, X. Y., Ivan, B., Kops, J., Batsberg, W. *Macromolecular Rapid Communications* 1997, 18, 1095–1100.
145. Chen, X. Y., Jankova, K., Kops, J., Batsberg, W. *Journal of Polymer Science Part A-Polymer Chemistry* 1999, 37, 627–633.
146. McQuillan, B. W., Paguio, S. *Fusion Technology* 2000, 38, 108–109.
147. Davis, K. A., Paik, H. J., Matyjaszewski, K. *Macromolecules* 1999, 32, 1767–1776.
148. Uegaki, H., Kotani, Y., Kamigaito, M., Sawamoto, M. in *Transition Metal Catalysis in Macromolecular Design,* L. S. Boffa, B. Novak, M., Eds., *ACS Symposium Series 760;* American Chemical Society: Washington, DC, 2000, 768, Chapter 12, p 196.
149. Schubert, U. S., Hochwimmer, G., Spindler, C. E., Nuyken, O. *Polymer Bulletin* 1999, 43, 319–326.
150. Moineau, G., Minet, M., Dubois, P., Teyssie, P., Senninger, T., Jerome, R. *Macromolecules* 1999, 32, 27–35.
151. Davis, K. A., Matyjaszewski, K. *Macromolecules* 2000, 33, 4039–4047.
152. Ma, Q. G., Wooley, K. L. *Journal of Polymer Science Part A-Polymer Chemistry* 2000, 38, 4805–4820.
153. Muhlebach, A., Gaynor, S. G., Matyjaszewski, K. *Macromolecules* 1998, 31, 6046–6052.
154. Coca, S., Jasieczek, C. B., Beers, K. L., Matyjaszewski, K. *Journal of Polymer Science Part A-Polymer Chemistry* 1998, 36, 1417–1424.

155. Coca, S., Matyjaszewski, K. *Journal of Polymer Science Part A-Polymer Chemistry* 1997, 35, 3595–3601.
156. Matyjaszewski, K., Coca, S., Jasieczek, C. B. *Macromolecular Chemistry and Physics* 1997, 198, 4011–4017.
157. Xia, J. H., Johnson, T., Gaynor, S. G., Matyjaszewski, K., DeSimone, J. *Macromolecules* 1999, 32, 4802–4805.
158. Van Camp, W., Du Prez, F. E., Bon, S. A. F. *Macromolecules* 2004, 37, 6673–6675.
159. Haddleton, D. M., Jasieczek, C. B., Hannon, M. J., Shooter, A. J. *Macromolecules* 1997, 30, 2190–2193.
160. Granel, C., Dubois, P., Jerome, R., Teyssie, P. *Macromolecules* 1996, 29, 8576–8582.
161. Ando, T., Kato, M., Kamigaito, M., Sawamoto, M. *Macromolecules* 1996, 29, 1070–1072.
162. Matyjaszewski, K., Shipp, D. A., Wang, J. L., Grimaud, T., Patten, T. E. *Macromolecules* 1998, 31, 6836–6840.
163. Nonaka, H., Ouchi, M., Kamigaito, M., Sawamoto, M. *Macromolecules* 2001, 34, 2083–2088.
164. Haddleton, D. M., Crossman, M. C., Dana, B. H., Duncalf, D. J., Heming, A. M., Kukulj, D., Shooter, A. J. *Macromolecules* 1999, 32, 2110–2119.
165. Kotani, Y., Kato, M., Kamigaito, M., Sawamoto, M. *Macromolecules* 1996, 29, 6979–6982.
166. Haddleton, D. M., Waterson, C. *Macromolecules* 1999, 32, 8732–8739.
167. Gaynor, S. G., Qiu, J., Matyjaszewski, K. *Macromolecules* 1998, 31, 5951–5954.
168. Beers, K. L., Boo, S., Gaynor, S. G., Matyjaszewski, K. *Macromolecules* 1999, 32, 5772–5776.
169. Beers, K. L., Gaynor, S. G., Matyjaszewski, K., Sheiko, S. S., Moller, M. *Macromolecules* 1998, 31, 9413–9415.
170. Liu, Y., Wang, L. X., Pan, C. Y. *Macromolecules* 1999, 32, 8301–8305.
171. Zhang, X., Xia, J. H., Matyjaszewski, K. *Macromolecules* 1998, 31, 5167–5169.
172. Wang, X. S., Armes, S. P. *Macromolecules* 2000, 33, 6640–6647.
173. Zhang, Z. B., Ying, S. K., Shi, Z. Q. *Polymer* 1999, 40, 5439–5444.
174. Du, J. Z., Chen, Y. M. *Macromolecules* 2004, 37, 6322–6328.
175. Koh, K., Ohno, K., Tsujii, Y., Fukuda, T. *Angewandte Chemie-International Edition* 2003, 42, 4194–4197.
176. Matyjaszewski, K., Qin, S. H., Boyce, J. R., Shirvanyants, D., Sheiko, S. S. *Macromolecules* 2003, 36, 1843–1849.
177. Raghunadh, V., Baskaran, D., Sivaram, S. *Polymer* 2004, 45, 3149–3155.
178. Xu, W. J., Zhu, X. L., Cheng, Z. P., Chen, J. Y. *Journal of Applied Polymer Science* 2003, 90, 1117–1125.
179. Save, M., Weaver, J. V. M., Armes, S. P., McKenna, P. *Macromolecules* 2002, 35, 1152-1159.
180. Krishnan, R., Srinivasan, K. S. V. *Macromolecules* 2003, 36, 1769–1771.
181. Krishnan, R., Srinivasan, K. S. V. *Macromolecules* 2004, 37, 3614–3622.
182. Munoz-Bonilla, A., Madruga, E. L., Fernandez-Garcia, M. *Journal of Polymer Science Part A-Polymer Chemistry* 2005, 43, 71–77.
183. Senoo, M., Kotani, Y., Kamigaito, M., Sawamoto, M. *Macromolecules* 1999, 32, 8005-8009.
184. Teodorescu, M., Matyjaszewski, K. *Macromolecules* 1999, 32, 4826–4831.
185. Huang, X., Wirth, M. J. *Macromolecules* 1999, 32, 1694–1696.
186. Li, D. W., Brittain, W. J. *Macromolecules* 1998, 31, 3852–3855.
187. Rademacher, J. T., Baum, R., Pallack, M. E., Brittain, W. J., Simonsick, W. J. *Macromolecules* 2000, 33, 284–288.

188. Teodorescu, M., Matyjaszewski, K. *Macromolecular Rapid Communications* 2000, 21, 190–194.
189. Huang, X. Y., Wirth, M. J. *Analytical Chemistry* 1997, 69, 4577–4580.
190. Huang, X. Y., Doneski, L. J., Wirth, M. J. *Analytical Chemistry* 1998, 70, 4023–4029.
191. Matyjaszewski, K., Jo, S. M., Paik, H. J., Shipp, D. A. *Macromolecules* 1999, 32, 6431–6438.
192. Matyjaszewski, K., Jo, S. M., Paik, H. J., Gaynor, S. G. *Macromolecules* 1997, 30, 6398–6400.
193. Barboiu, B., Percec, V. *Macromolecules* 2001, 34, 8626–8636.
194. Xia, J. H., Zhang, X., Matyjaszewski, K. *Macromolecules* 1999, 32, 3531–3533.
195. Collins, J. E., Fraser, C. L. *Macromolecules* 1998, 31, 6715–6717.
196. Haddleton, D. M., Duncalf, D. J., Kukulj, D., Crossman, M. C., Jackson, S. G., Bon, S. A. F., Clark, A. J., Shooter, A. J. *European Journal of Inorganic Chemistry* 1998, 1799–1806.
197. Perrier, S., Berthier, D., Willoughby, I., Batt-Coutrot, D., Haddleton, D. M. *Macromolecules* 2002, 35, 2941–2948.
198. Wang, X. S., Malet, F. L. G., Armes, S. P., Haddleton, D. M., Perrier, S. *Macromolecules* 2001, 34, 162–164.
199. DiRenzo, G. M., Messerschmidt, M., Mulhaupt, R. *Macromolecular Rapid Communications* 1998, 19, 381–384.
200. Xia, J. H., Matyjaszewski, K. *Macromolecules* 1997, 30, 7697–7700.
201. Acar, M. H., Becer, C. R., Ondur, H. A., Inceoglu, S. *Abstracts of Papers of the American Chemical Society* 2005, 230, U4109–U4110.
202. Chu, J., Chen, J., Zhang, K. D. *Journal of Polymer Science Part A-Polymer Chemistry* 2004, 42, 1963–1969.
203. Acar, M. H., Bicak, N. *Journal of Polymer Science Part A-Polymer Chemistry* 2003, 41, 1677–1680.
204. Xia, J. H., Matyjaszewski, K. *Macromolecules* 1999, 32, 2434–2437.
205. Yu, B., Ruckenstein, E. *Journal of Polymer Science Part A-Polymer Chemistry* 1999, 37, 4191–4197.
206. Cheng, G. L., Hu, C. P., Ying, S. K. *Macromolecular Rapid Communications* 1999, 20, 303–307.
207. Cheng, G. L., Hu, C. P., Ying, S. K. *Polymer* 1999, 40, 2167–2169.
208. Destarac, M., Bessiere, J. M., Boutevin, B. *Macromolecular Rapid Communications* 1997, 18, 967–974.
209. Shen, Y. Q., Zhu, S. P., Zeng, F. Q., Pelton, R. H. *Macromolecular Chemistry and Physics* 2000, 201, 1169–1175.
210. Wan, X. L., Ying, S. K. *Journal of Applied Polymer Science* 2000, *75*, 802–807.
211. Johnson, R. M., Ng, C., Samson, C. C. M., Fraser, C. L. *Macromolecules* 2000, *33*, 8618–8628.
212. Kickelbick, G., Matyjaszewski, K. *Macromolecular Rapid Communications* 1999, *20*, 341–346.
213. Zeng, F. Q., Shen, Y. Q., Zhu, S. P., Pelton, R. *Macromolecules* 2000, *33*, 1628-1635.
214. Gromada, J., Spanswick, J., Matyjaszewski, K. *Macromolecular Chemistry and Physics* 2004, *205*, 551–566.
215. Xia, J. H., Gaynor, S. G., Matyjaszewski, K. *Macromolecules* 1998, 31, 5958–5959.
216. Inoue, Y., Matyjaszewski, K. *Macromolecules* 2004, 37, 4014–4021.
217. Ding, S. J., Shen, Y. Q., Radosz, M. *Journal of Polymer Science Part A-Polymer Chemistry* 2004, 42, 3553–3562.

218. Lee, D. W., Seo, E. Y., Cho, S. I., Yi, C. S. *Journal of Polymer Science Part A-Polymer Chemistry* 2004, 42, 2747–2755.
219. Levy, A. T., Olmstead, M. N., Patten, T. E. *Inorganic Chemistry* 2000, 39, 1628–1634.
220. Moroni, M., Hilberer, A., Hadziioannou, G. *Macromolecular Rapid Communications* 1996, 17, 693–702.
221. Georges, M. K., Veregin, R. P. N., Kazmaier, P. M., Hamer, G. K., Saban, M. *Macromolecules* 1994, 27, 7228–7229.
222. Miura, Y., Nakamura, N., Taniguchi, I. *Macromolecules* 2001, 34, 447–455.
223. Gabaston, L. I., Jackson, R. A., Armes, S. P. *Macromolecules* 1998, 31, 2883–2888.
224. Chalari, I., Pispas, S., Hadjichristidis, N. *Journal of Polymer Science Part A-Polymer Chemistry* 2001, 39, 2889–2895.
225. Ding, X. Z., Fischer, A., Brembilla, A., Lochon, P. *Journal of Polymer Science Part A-Polymer Chemistry* 2000, 38, 3067–3073.
226. Bohrisch, J., Wendler, U., Jaeger, W. *Macromolecular Rapid Communications* 1997, 18, 975–982.
227. Barclay, G. G., Hawker, C. J., Ito, H., Orellana, A., Malenfant, P. R. L., Sinta, R. F. *Macromolecules* 1998, 31, 1024–1031.
228. Baumert, M., Frohlich, J., Stieger, M., Frey, H., Mulhaupt, R., Plenio, H. *Macromolecular Rapid Communications* 1999, 20, 203–209.
229. Wei, Y., Connors, E. J., Jia, X. R., Wang, C. *Journal of Polymer Science Part A-Polymer Chemistry* 1998, 36, 761–771.
230. Chong, Y. K., Ercole, F., Moad, G., Rizzardo, E., Thang, S. H., Anderson, A. G. *Macromolecules* 1999, 32, 6895–6903.
231. Kazmaier, P. M., Daimon, K., Georges, M. K., Hamer, G. K., Veregin, R. P. N. *Macromolecules* 1997, 30, 2228–2231.
232. Keoshkerian, B., Georges, M. K., Boilsboissier, D. *Macromolecules* 1995, 28, 6381–6382.
233. Mardare, D., Matyjaszewski, K. *Macromolecules* 1994, 27, 645–649.
234. Pradel, J. L., Boutevin, B., Ameduri, B. *Journal of Polymer Science Part A-Polymer Chemistry* 2000, 38, 3293–3302.
235. Hattemer, E., Brehmer, M., Zentel, R., Mecher, E., Mueller, D., Meerholz, K. *Abstracts of Papers of the American Chemical Society* 2000, 219, U411–U411.
236. Gopalan, P., Pragliola, S., Ober, C., Mather, P., Jeon, H. *Abstracts of Papers of the American Chemical Society* 1999, 218, U499–U500.
237. Ohno, K., Ejaz, M., Fukuda, T., Miyamoto, T., Shimizu, Y. *Macromolecular Chemistry and Physics* 1998, 199, 291–297.
238. Hawker, C. J., Barclay, G. G., Dao, J. L. *Journal of the American Chemical Society* 1996, 118, 11467–11471.
239. Matyjaszewski, K., Gaynor, S. G., Greszta, D., Mardare, D., Shigemoto, T., Wang, J. S. *Macromolecular Symposia* 1995, 95, 217–231.
240. Cunningham, M. F., Tortosa, K., Ma, J. W., McAuley, K. B., Keoshkerian, B., Georges, M. K. *Macromolecular Symposia* 2002, 182, 273–282.
241. Wetter, C., Gierlich, J., Knoop, C. A., Muller, C., Schulte, T., Studer, A. *Chemistry–A European Journal* 2004, 10, 1156–1166.
242. Knoop, C. A., Studer, A. *Journal of the American Chemical Society* 2003, 125, 16327–16333.
243. Yoshida, E., Okada, Y. *Journal of Polymer Science Part A-Polymer Chemistry* 1996, 34, 3631–3635.
244. Yoshida, E., Fujii, T. *Journal of Polymer Science Part A-Polymer Chemistry* 1997, 35, 2371–2378.

245. Yoshida, E., Fujii, T. *Journal of Polymer Science Part A-Polymer Chemistry* 1998, 36, 269–276.
246. Huang, W. L., Charleux, B., Chiarelli, R., Marx, L., Rassat, A., Vairon, J. P. *Macromolecular Chemistry and Physics* 2002, 203, 1715–1723.
247. Jousset, S., Hammouch, S. O., Catala, J. M. *Macromolecules* 1997, 30, 6685–6687.
248. Hammouch, S. O., Catala, J. M. *Macromolecular Rapid Communications* 1996, 17, 683–691.
249. Ohno, K., Tsujii, Y., Miyamoto, T., Fukuda, T., Goto, M., Kobayashi, K., Akaike, T. *Macromolecules* 1998, 31, 1064–1069.
250. Ohno, K., Izu, Y., Yamamoto, S., Miyamoto, T., Fukuda, T. *Macromolecular Chemistry and Physics* 1999, 200, 1619–1625.
251. Wannemacher, T., Braun, D., Pfaendner, R. *Macromolecular Symposia* 2003, 202, 11–23.
252. Benoit, D., Chaplinski, V., Braslau, R., Hawker, C. J. *Journal of the American Chemical Society* 1999, 121, 3904–3920.
253. Benoit, D., Harth, E., Fox, P., Waymouth, R. M., Hawker, C. J. *Macromolecules* 2000, 33, 363–370.
254. Tully, D. C., Roberts, M. J., Geierstanger, B. H., Grubbs, R. B. *Macromolecules* 2003, 36, 4302–4308.
255. Harth, E., Van Horn, B., Hawker, C. J. *Chemical Communications* 2001, 823–824.
256. Studer, A., Harms, K., Knoop, C., Muller, C., Schulte, T. Macromolecules 2004, 37, 27–34.
257. Benoit, D., Grimaldi, S., Robin, S., Finet, J. P., Tordo, P., Gnanou, Y. *Journal of the American Chemical Society* 2000, 122, 5929–5939.
258. Farcet, C., Lansalot, M., Charleux, B., Pirri, R., Vairon, J. P. *Macromolecules* 2000, 33, 8559–8570.
259. Couvreur, L., Lefay, C., Belleney, J., Charleux, B., Guerret, O., Magnet, S. *Macromolecules* 2003, 36, 8260–8267.
260. Diaz, T., Fischer, A., Jonquieres, A., Brembilla, A., Lochon, P. *Macromolecules* 2003, 36, 2235–2241.
261. Schierholz, K., Givehchi, M., Fabre, P., Nallet, F., Papon, E., Guerret, O., Gnanou, Y. *Macromolecules* 2003, 36, 5995–5999.
262. Grimaldi, S., Finet, J. P., Le Moigne, F., Zeghdaoui, A., Tordo, P., Benoit, D., Fontanille, M., Gnanou, Y. *Macromolecules* 2000, 33, 1141–1147.
263. Le Mercier, C., Acerbis, S. B., Bertin, D., Chauvin, F., Gigmes, D., Guerret, O., Lansalot, M., Marque, S., Le Moigne, F., Fischer, H., Tordo, P. *Macromolecular Symposia* 2002, 182, 225–247.
264. Yamada, B., Miura, Y., Nobukane, Y., Aota, M. in *Controlled Radical Polymerization,* K. Matyjaszewski, ed., *ACS Symposium Series 685;* American Chemical Society: Washington, DC, 1998, Chapter 12, p 200.
265. Veregin, R. P. N., Georges, M. K., Hamer, G. K., Kazmaier, P. M. *Macromolecules* 1995, 28, 4391–4398.
266. Cameron, N. R., Reid, A. J., Span, P., Bon, S. A. F., van Es, J. J. G. S., German, A. L. *Macromolecular Chemistry and Physics* 2000, 201, 2510–2518.
267. Puts, R. D., Sogah, D. Y. *Macromolecules* 1996, 29, 3323–3325.
268. Turro, N. J., Lem, G., Zavarine, I. S. *Macromolecules* 2000, 33, 9782–9785.
269. Rodlert, M., Harth, E., Rees, I., Hawker, C. J. *Journal of Polymer Science Part A-Polymer Chemistry* 2000, 38, 4749–4763.
270. Dervan, P., Aldabbagh, F., Zetterlund, P. B., Yamada, B. *Journal of Polymer Science Part A-Polymer Chemistry* 2003, 41, 327–334.
271. Shigemoto, T., Matyjaszewski, K. *Macromolecular Rapid Communications* 1996, 17, 347–351.

272. Han, C. H., Drache, M., Baethge, H., Schmidt-Naake, G. *Macromolecular Chemistry and Physics* 1999, 200, 1779–1783.
273. Ryan, J., Aldabbagh, F., Zetterlund, P. B., Yamada, B. *Macromolecular Rapid Communications* 2004, 25, 930–934.
274. Schulte, T., Studer, A. *Macromolecules* 2003, 36, 3078–3084.
275. Jousset, S., Catala, J. M. *Macromolecules* 2000, 33, 4705–4710.
276. Miura, Y., Mibae, S., Moto, H., Nakamura, N., Yamada, B. *Polymer Bulletin* 1999, 42, 17–24.
277. Miura, Y., Nakamura, N., Taniguchi, I., Ichikawa, A. *Polymer* 2003, 44, 3461–3467.
278. Miura, Y., Ichikawa, A., Taniguchi, I. *Polymer* 2003, 44, 5187–5194.
279. Zink, M. O., Kramer, A., Nesvadba, P. *Macromolecules* 2000, 33, 8106–8108.
280. Puts, R., Lai, J., Nicholas, P., Milam, J., Tahilliani, S., Masler, W., Pourahmady, N. *Abstracts of Papers of the American Chemical Society* 1999, 218, U413–U413.
281. Brinkmann-Rengel, S., Sutoris, H. F., Weiss, H. *Macromolecular Symposia* 2001, 163, 145–156.
282. Nesvadba, P., Bugnon, L., Sift, R. *Journal of Polymer Science Part A-Polymer Chemistry* 2004, 42, 3332–3341.
283. Aldabbagh, F., Dervan, P., Phelan, M., Gilligan, K., Cunningham, D., McArdle, P., Zetterlund, P. B., Yamada, B. *Journal of Polymer Science Part A-Polymer Chemistry* 2003, 41, 3892–3900.
284. Drockenmuller, E., Lamps, J. P., Catala, J. M. *Macromolecules* 2004, 37, 2076–2083.
285. Cuatepotzo-Diaz, R., Albores-Velasco, M., Saldivar-Guerra, E., Jimenez, F. B. *Polymer* 2004, 45, 815–824.
286. Chachaty, C., Huang, W. L., Marx, L., Charleux, B., Rassat, A. *Polymer* 2003, 44, 397–406.
287. Olive, G., Rozanska, X., Smulders, W., Jacques, A., German, A. *Macromolecular Chemistry and Physics* 2002, 203, 1790–1796.
288. Hua, D. B., Ge, X. P., Bai, R. K., Lu, W. Q., Pan, C. Y. *Polymer* 2005, 46, 12696–12702.
289. Chong, Y. K., Krstina, J., Le, T. P. T., Moad, G., Postma, A., Rizzardo, E., Thang, S. H. *Macromolecules* 2003, 36, 2256–2272.
290. Tang, C. B., Kowalewski, T., Matyjaszewski, K. *Macromolecules* 2003, 36, 8587–8589.
291. Liu, X. H., Zhang, G. B., Lu, X. F., Liu, J. Y., Pan, D., Li, Y. S. *Journal of Polymer Science Part A-Polymer Chemistry* 2006, 44, 490–498.
292. An, Q. F., Qian, J. W., Yu, L. Y., Luo, Y. W., Liu, X. Z. *Journal of Polymer Science Part A-Polymer Chemistry* 2005, 43, 1973–1977.
293. Perrier, S., Barner-Kowollik, C., Quinn, J. F., Vana, P., Davis, T. P. *Macromolecules* 2002, 35, 8300–8306.
294. Biasutti, J. D., Davis, T. P., Lucien, F. P., Heuts, J. P. A. *Journal of Polymer Science Part A-Polymer Chemistry* 2005, 43, 2001–2012.
295. Benaglia, M., Rizzardo, E., Alberti, A., Guerra, M. *Macromolecules* 2005, 38, 3129–3140.
296. Dureault, A., Gnanou, Y., Taton, D., Destarac, M., Leising, F. *Angewandte Chemie-International Edition* 2003, 42, 2869–2872.
297. Yang, L., Luo, Y. W., Li, B. G. *Polymer* 2006, 47, 751–762.
298. Nguyen, M. N., Bressy, C., Margaillan, A. *Journal of Polymer Science Part A-Polymer Chemistry* 2005, 43, 5680–5689.
299. D'Agosto, F., Hughes, R., Charreyre, M. T., Pichot, C., Gilbert, R. G. *Macromolecules* 2003, 36, 621–629.

300. Matahwa, H., McLeary, J. B., Sanderson, R. D. *Journal of Polymer Science Part A-Polymer Chemistry* 2006, 44, 427–442.
301. Donovan, M. S., Lowe, A. B., Sumerlin, B. S., McCormick, C. L. *Macromolecules* 2002, 35, 4123–4132.
302. Zheng, G. H., Zheng, Q., Pan, C. Y. *Macromolecular Chemistry and Physics* 2006, 207, 216–223.
303. Mori, H., Sutoh, K., Endo, T. *Macromolecules* 2005, 38, 9055–9065.
304. Chong, Y. K., Le, T. P. T., Moad, G., Rizzardo, E., Thang, S. H. *Macromolecules* 1999, 32, 2071–2074.
305. Arita, T., Beuermann, S., Buback, M., Vana, P. *Macromolecular Materials and Engineering* 2005, 290, 283-293.
306. Arita, T., Buback, M., Vana, P. *Macromolecules* 2005, 38, 7935–7943.
307. Bilalis, P., Pitsikalis, M., Hadjichristidis, N. *Journal of Polymer Science Part A-Polymer Chemistry* 2006, 44, 659–665.
308. Barner-Kowollik, C., Quinn, J. F., Morsley, D. R., Davis, T. P. *Journal of Polymer Science Part A-Polymer Chemistry* 2001, 39, 1353–1365.
309. Dureault, A., Taton, D., Destarac, M., Leising, F., Gnanou, Y. *Macromolecules* 2004, 37, 5513–5519.
310. Theis, A., Davis, T. P., Stenzel, M. H., Barner-Kowollik, C. *Polymer* 2006, 47, 999-1010.
311. Stenzel, M. H., Cummins, L., Roberts, G. E., Davis, T. R., Vana, P., Barner-Kowollik, C. *Macromolecular Chemistry and Physics* 2003, 204, 1160–1168.
312. Rizzardo, E., Chiefari, J., Mayadunne, R. T. A., Moad, G., Thang, S. H. *ACS Symp. Ser.* 2000, 768, 278.
313. Wood, M. R., Duncalf, D. J., Rannard, S. P., Perrier, S. *Organic Letters* 2006, 8, 553–556.
314. Devasia, R., Bindu, R. L., Borsali, R., Mougin, N., Gnanou, Y. *Macromolecular Symposia* 2005, 229, 8–17.
315. Mayadunne, R. T. A., Rizzardo, E., Chiefari, J., Krstina, J., Moad, G., Postma, A., Thang, S. H. *Macromolecules* 2000, 33, 243–245.
316. Mayadunne, R. T. A., Jeffery, J., Moad, G., Rizzardo, E. *Macromolecules* 2003, 36, 1505–1513.
317. Buback, M., Hesse, P., Junkers, T., Vana, P. *Macromolecular Rapid Communications* 2006, 27, 182–187.
318. Hua, D. B., Sun, W., Bai, R. K., Lu, W. Q., Pan, C. Y. *European Polymer Journal* 2005, 41, 1674–1680.
319. Zhou, N. C., Lu, L. D., Zhu, X. L., Yang, X. J., Wang, X., Zhu, J., Cheng, Z. P. *Journal of Applied Polymer Science* 2006, 99, 3535–3539.
320. Yin, H. S., Zhu, X. L., Zhou, D., Zhu, J. *Journal of Applied Polymer Science* 2006, 100, 560–564.
321. Alberti, A., Benaglia, M., Guerra, M., Gulea, M., Hapiot, P., Laus, M., Macciantelli, D., Masson, S., Postma, A., Sparnacci, K. *Macromolecules* 2005, 38, 7610-7618.
322. Schilli, C., Lanzendorfer, M. G., Muller, A. H. E. *Macromolecules* 2002, 35, 6819–6827.
323. McCormack, C. L., Lowe, A. B. *Accounts of Chemical Research* 2004, 37, 312-325.
324. Thomas, D. B., Sumerlin, B. S., Lowe, A. B., McCormick, C. L. *Macromolecules* 2003, 36, 1436–1439.
325. Shim, S. E., Lee, H., Choe, S. *Macromolecules* 2004, 37, 5565–5571.
326. Donovan, M. S., Sanford, T. A., Lowe, A. B., Sumerlin, B. S., Mitsukami, Y., McCormick, C. L. *Macromolecules* 2002, 35, 4570–4572.

327. Convertine, A. J., Lokitz, B. S., Lowe, A. B., Scales, C. W., Myrick, L. J., McCormick, C. L. *Macromolecular Rapid Communications* 2005, 26, 791–795.
328. Barner-Kowollik, C., Vana, P., Quinn, J. F., Davis, T. P. *Journal of Polymer Science Part A-Polymer Chemistry* 2002, 40, 1058–1063.
329. Junkers, T., Theis, A., Buback, M., Davis, T. P., Stenzel, M. H., Vana, P., Barner-Kowollik, C. *Macromolecules* 2005, 38, 9497–9508.
330. Lu, L. C., Yang, N. F., Cai, Y. L. *Chemical Communications* 2005, 5287–5288.

11 Block and Graft Copolymers

Ali E. Muftuoglu, M. Atilla Tasdelen, Yusuf Yagci, and Munmaya K. Mishra

CONTENTS

11.1 INTRODUCTION

Synthesis of materials with novel properties has attracted growing scientific interest in recent years. A great deal of research activity has been devoted to this fundamental issue due to endless demands of the industry for high-tech applications. A number of strategies have been sought by polymer researchers as well as many others, in need of preparing new materials displaying improved physical and chemical properties. Such improved properties might be achieved by combining a number of physical and chemical properties and can usually be adopted by either blending homopolymers or by chemically linking different polymer chains. When two homopolymers are mechanically mixed, formation of an immiscible blend is often the case encountered. Microphase-separated heterogeneous mixtures arising from the incompatibility of the blended polymers can successfully be avoided by the latter strategy, which involves preparation of block and graft copolymers. Thus, engineering macromolecules of various block and graft structures appears to be an elegant approach in achieving polymers with improved physical and chemical properties.

Preparation of block copolymers using living polymerization techniques has been known for quite a long time. Anionic polymerization [1], which provides end-group control and permits the synthesis of polymers with a narrow polydispersity index, is a noteworthy example of these techniques. Yet, it does not offer full control over the molecular weight of the synthesized macromolecules. The limited choice of monomers and the extremely severe reaction conditions lower its applicability.

Recently, advances in controlled polymerization techniques [2–5], such as radical, cationic, metathesis, and group transfer, have allowed for the synthesis of polymers with predetermined molecular weights and low polydispersities. A variety of block copolymers has been prepared by cationic, group transfer, metallocene, and metathesis routes. Among others, controlled radical polymerization routes have been studied extensively, because they can be employed for the polymerization of numerous vinyl monomers under mild reaction conditions. These are atom transfer radical polymerization (ATRP) [6–8], nitroxide-mediated polymerization (NMP) [9–11], and reversible addition-fragmentation chain transfer polymerization (RAFT) [12, 13]. The significant advance in the block and graft copolymer synthesis has come with the advent of controlled radical polymerization (CRP) techniques [14, 15].

One way to prepare block and graft copolymers is by using main chain and side chain macroinitiators, respectively. In this method, reactive sites are produced at the chain ends or side chains, which serve as initiating moieties in the polymerization of a second monomer. In another approach, separately synthesized macromolecular chains having antagonist groups or latent active sites are made to react with each other. If these groups are located at the chain ends, the resulting polymer is a block copolymer. In a similar way, the chains bearing the reactive end-functions can be interacted with pendant moieties of a backbone yielding a graft copolymer.

This chapter covers the synthetic strategies to prepare block and graft copolymers via radical polymerization routes, focusing on the latest, state-of-the-art studies.

11.2 SYNTHESIS OF BLOCK COPOLYMERS

Normally, block copolymers cannot be synthesized by classical free radical copolymerization technique. Spontaneous block copolymer formation upon free radical copolymerization of A and B monomers would only occur when both reactivity ratios r_a and r_b are far larger than unity. Such a system has never been found. Generally, one can get access to block copolymers by the following strategies. According to the most common methodology, monomer A is polymerized completely, after which, monomer B is introduced in the mixture and its polymerization proceeds upon initiation by the active site of the first block. This approach is referred to as sequential monomer addition in controlled/living polymerization methods. For obtaining a well-defined block copolymer, the living site of the first monomer must be effective in initiating the polymerization of the second monomer. That is, simultaneous initiation of all growing B chains must be ensured and the rate of the crossover reaction must be higher than the rate of propagation of monomer B.

Another route involves the use of a difunctional initiator and has been employed in the preparation of ABA symmetric triblock copolymers. In this methodology, a compound possessing two initiating sites is utilized in the formation of the middle block first, followed by the polymerization of the second monomer to synthesize the first and the third blocks. This allows the preparation of the ABA blocks in two, instead of three steps without fractionation or other purification steps. The difunctional initiator must be chosen to initiate the polymerization with the same rate from either direction.

Moreover, AB diblock or ABA triblock copolymers can be prepared by coupling two appropriately end-functionalized chains of A and B homopolymers or AB blocks, respectively. In the latter case, block B should initially contain half the molecular weight of the final desired block B. The efficiency of coupling will be high if the reaction is rapid and the living diblock copolymer is used in excess.

11.2.1 Block Copolymers via Conventional Radical Polymerization

The most widely employed method for the preparation of block copolymers by using conventional radical polymerization routes is the macroinitiator technique. This technique has several advantages. First, macroinitiators can be fully characterized before their use in the free radical step to obtain block copolymers. Moreover, the macroinitiators can be prepared practically by all polymerization methods. In living polymerizations, it is possible to introduce the free radical initiating functionalities into polymers at both initiation and termination steps as presented in Scheme 11.1.

Azo or peroxy groups are incorporated into the free-radical initiating sites of the polymers. Typical examples of such methodology by using living anionic and cationic polymerizations are given in Schemes 11.2 [16] and 11.3 [17], respectively. Tables 11.1 and 11.2 give the compilation of block copolymers prepared by using macroinitiators possessing thermolabile azo and peroxy groups, respectively.

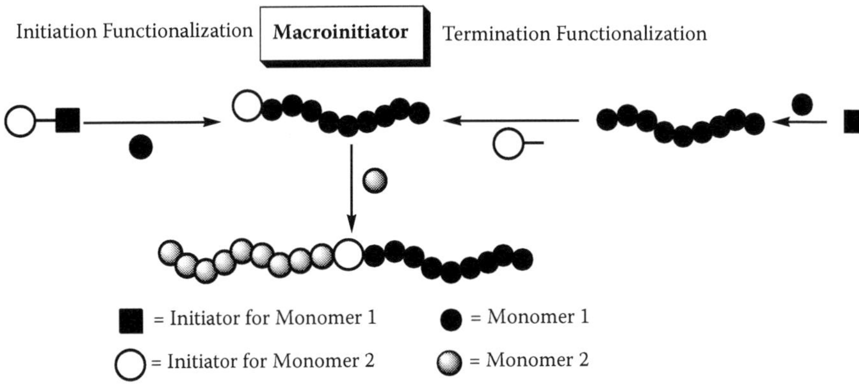

SCHEME 11.1

TABLE 11.1
Block Copolymers Prepared from Macroperoxy-Initiators

1st Segment	2nd Segment	References
PAm	Polyacrylate copolymer	[18, 19]
Polyacrylate copolymer		[18–23]
PBA-co-PAAc		[23]
PEHA		[24]
PEHMA		[22, 24]
PEHA-co-PHEMA		[24]
PEHMA-co-PBMA		[24]
PPFOEA-co-PTFEMA		[25]
PHEA-co-PNMeAm		[21]
PHEA-co-PHEMA		[21]
PHEA-co-PDEGMA		[21, 26]
PLMA		[27]
PLMA-co-PAAc		[22, 24]
PSMA		[18, 22, 24]
PSMA-co-PNMeAm		[28]
PSt-co-PBA		[23, 25]
PSt-co-PMMA		[18]
PVAc		[26]
PVC-co-PVAc		[18, 22]
PEA	PHEMA-co-PEHMA	[20]
PEMA	PEHMA-co-PBMA	[20]
PDFCE	PEMA	[28]
PPFOEA	PBMA	[25]
	PMMA-co-PBA	[25]
	PMMA-co-PHEMA	[25]
PHEA	Polyacrylate copolymer	[21, 26]
PHEMA		[21, 26, 27]
PMMA	PVAc	[29, 30]
	PHEMA	[31]
	PPFOEA	[32]
	PFHEA	[25, 28]
PMMA-co-PSMA	PSMA-co-PEFS	[28]
PMMA-co-PBA	PBMA-co-PPFOEA	[25]
PMMA-co-PEA	PFHEA-co-PPFOEA	[28]
PFHVE	PEGMA-co-PPMA	[28]
PSt	PAN	[33]
	PBVE	[34]
	PBA-co-PHE	[35]
	PBMA-co-PHE	[36]
	PHPMQA	[31]
	PHEMA	[31]
	PMMA	[30, 32, 36–40]
	PNaAMPS	[31]

(Continued)

TABLE 11.1 (CONTINUED)
Block Copolymers Prepared from Macroperoxy-Initiators

1st Segment	2nd Segment	References
	PVAc	[21, 29, 32, 36, 37, 41, 42]
	PVAc-co-PAC	[35]
	PVAC-co-PMAc	[43]
	PVAc-co-PBA	[28]
PENS	PVAc-co-PBA	[28]
PVAc	PBA	[27, 44]
	PBA-co-PEHA	[44]
	PBMA	[18, 22]
	PSt-co-PMAc	[42]
	PSt-co-PMMA	[43]
PVC	PMMA-co-PMFS	[25]
PDMS	PSt, PMMA, PVAc	[45]
PEO	PSt	[46]

PAm: polyacrylamide; PBA: poly(butyl acrylate); PAAc: poly(acrylic acid); PEA: poly(ethyl acrylate); PEHA: poly(ethyl hydroxyacrylate); PHEMA: poly(hydroxylethyl methacrylate); PEHMA: poly(ethylhexyl methacrylate); PBMA: poly(butyl methacrylate); PPFOEA: poly(perfluorodecyl acrylate); PTFEMA: poly(trifluoroethyl methacrylate); PNMeAm: poly(*N*-methylol acrylamide) PDEGMA: poly(diethyleneglycol monomethacrylate); PLMA: poly(lauryl methacrylate); PSMA: poly(stearyl methacrylate); PSt: polystyrene; PVAc: poly(vinyl acetate); PVC: polyvinylchloride; PEMA: poly(ethyl methacrylate); PDFCE: poly(difluorochloro ethylene); PMMA: poly(methyl methacrylate); PPFOA: poly(perfluorooctyl acrylate); PFHEA: poly(perfluorooctyl acrylate); PFHVE: poly(perfluoroheptylvinyl ether); PEGMA: poly(ethyleneglycol monomethacrylate); PPMA: poly(propyl methacrylate); PAN: polyacrylonitrile; PBVE: poly(butylvinyl ether); PHE: poly(methacryloyloxyethyl-t-butyl-peroxy carbonate); PHPMQA: poly(2-hydroxyl-3-methacyloxypropyltrimethlammonium); PNaAMPS: poly(2-acrylamide-2-methylpropane sodium sulfonate); PAC: poly(allyl-t-butylperoxy carbonate); PMAc: poly(methacrylic acid); PMFS: poly(*N*-methyl-*N*-(B-methacryloxyethyl)-perfluorooctyl sulfonamide); PDMS: polydimethylsiloxane; PEO: poly(ethylene oxide).

$$\left[NH_2{-}NH{-}\overset{O}{\overset{\|}{C}}{-}CH_2{-}CH_2{-}\underset{CN}{\overset{CH_3}{C}}{-}N{=}\right]_2$$

$$\downarrow \; 2n \text{ (R-substituted N-carboxyanhydride: ring with R, N–H, O, two C=O)}$$

$$\left[NH{-}\underset{R}{CH}{-}\overset{O}{\overset{\|}{C}}\right]_n{-}NH{-}NH{-}\overset{O}{\overset{\|}{C}}{-}CH_2{-}CH_2{-}\underset{CN}{\overset{CH_3}{C}}{-}N{=}N{-}\underset{CN}{\overset{CH_3}{C}}{-}CH_2{-}CH_2{-}\overset{O}{\overset{\|}{C}}{-}NH{-}NH{-}\left[\overset{O}{\overset{\|}{C}}{-}\underset{R}{CH}{-}NH\right]_n$$

SCHEME 11.2

11.2.2 Block Copolymers via Controlled Radical Polymerization Routes

Well-defined block copolymers can be synthesized by using well-known anionic polymerization techniques, and the first study dates back to as early as 1956, in which styrene and isoprene were successfully blocked anionically [95]. In tailoring well-defined blocks, it is essential that the polymerization be performed in the absence of undesired transfer and termination, which can be provided quite successfully with anionic polymerization. However, it has several drawbacks, such as limited choice of monomers and the extremely severe reaction conditions.

Radical polymerization is the most widely used method in the polymerization of numerous vinyl monomers without the need to provide special reaction conditions required for ionic polymerization systems. It can be employed with ease in industrial applications. However, owing to its inadequacy in the control of chain lengths and in the preparation of block copolymers, it has been a major goal for researchers to establish a controlled mechanism in conjunction with a free radical route. Only after the early works of Otsu [96], Solomon [97], and Georges [98], it was realized that radical polymerizations could be conducted in a controlled manner. This breakthrough opened new pathways in the preparation of tailor-made polymers with desired properties.

Among the controlled radical routes, nitroxide mediated polymerization (NMP), atom transfer radical polymerization (ATRP), and reversible addition fragmentation chain transfer (RAFT) have been studied extensively. In all these methods, the controlled/living character is maintained due to the existence of a dynamic equilibrium between a dormant species present in great amount and a low concentration of active radical sites. Consequently, transfer and termination are minimized. In the absence of chain breaking reactions, the system possesses a living nature, as a result of which, a variety of macromolecular architectures, such as block, graft, star, and hyperbranched (co)polymers with low polydispersities can be achieved. Decreasing the number of active species present in the medium, polymers with predetermined molecular weights can be obtained. Additionally, a successful controlled polymerization system requires that all chains be initiated early and simultaneously, and the exchange between species should be rapid enough to allow only a certain number of monomers to be added each time the chain is active.

Block copolymers can be prepared by controlled polymerization routes as described previously. In the sequential monomer addition strategy, certain experimental conditions governing the growth of the first block should be handled carefully to increase the blocking efficiency. It is essentially important to stop the polymerization of the first monomer before it is used up completely because end functionalities might be lost, resulting in the formation of dead blocks. Another feature to be considered, as in all sequential monomer additions, is the order of introducing

$$\left[Cl-\overset{O}{\overset{\|}{C}}-CH_2-CH_2-\underset{CN}{\overset{CH_3}{C}}-N= \right]_2 \xrightarrow[THF]{2AgBF_4} \underset{(THF^{\oplus})}{O}\sim\sim\sim\overset{O}{\overset{\|}{C}}-CH_2-CH_2-\underset{CN}{\overset{CH_3}{C}}-N=N-\underset{CN}{\overset{CH_3}{C}}-CH_2-CH_2-\overset{O}{\overset{\|}{C}}\sim\sim\sim\underset{(THF^{\oplus})}{O} \quad 2\,BF_4^{\ominus}$$

SCHEME 11.3

TABLE 11.2
Block Copolymers Prepared from Macroazo-Initiators

1st Segment	2nd Segment	References
PAN	PB, PVP	[47]
Polyacrylate copolymer	polyacrylate copolymer	[48]
	polyfluoroacrylate copolymer	[28,49]
PB	PDMI	[50]
PMMA	PB, PBA	[47]
PSt	PAN	[47]
	PB	[47]
	PBA	[51,52]
	PMAc	[53]
	PMA	[54]
	PMMA	[34,47,55]
	PMMA-co-PBMA	[56]
	PNMeVAc	[54]
	polyolefin	[57]
	PSt-co-PSSSt, VPy	[47]
PSt-co-PAN	PSt-co-PB	[58]
PVAc	PAm	[59]
	PB	[47]
Poly(dibenzazepine)	PSt, PMMA, PMA	[60]
Polyamide	PAN	[61]
	PMMA	[34,47,62–65]
	PMMA-co-PVAc	[61]
	PSt	[34,47,51,63–68]
	PSt-co-PAN	[61]
	PSt-co-PIB	[61]
	PVAc-co-PAAc	[61]
	PVIBE-co-PVPp	[61]
	PBMA	[65]
	PHEMA	[65]
	PVAc	[65]
	PEO	[62]
	PPO	[62]
Polyester	polyacrylate copolymer	[48,69–72]
	PMMA	[34,73–83]
	PSt	[45,63,67,76–81,83–85,87–92]
	PB	[47,84]
	PVP	[47]
	PAm	[80,85]
	PAN	[85]
Polyurethane	polyacrylate copolymer	[48,69,70,72]
	PMMA	[48,61,86]
	PSt	[48,86]

(Continued)

TABLE 11.2 (CONTINUED)
Block Copolymers Prepared from Macroazo-Initiators

1st Segment	2nd Segment	References
polysiloxane	polyacrylate copolymer	[87,88]
	PMMA	[83,89]
	PSt	[89]
	PVAc	[89]
	PVC	[90]
polyimide	PSt	[91]
polycarbonate	PSt	[92–94]

PAN: polyacrylonitrile; PB: polybutadiene; PVP: poly(1-vinyl-2-pyrrolidone); PDMI: poly(dimethylitaconate); PBA: poly(butyl acrylate); PMAc: poly(methacrylic acid); PMA: poly(methyl acrylate); PSt: polystyrene; PMMA: poly(methyl methacrylate); PBMA: poly(butyl methacrylate); PNMeVAc: poly(*N*-methyl-*N*-vinylacetate); PSSSt: poly(styrene sodium sulfonate); PAm: polyacrylamide; PVAc: poly(vinyl acetate); PIB: polyisobutylene; PAAc: poly(acrylic acid); PVIBE: poly(vinyl isobutyl ether); PVPp: poly(vinyl propionate); PEO: poly(ethylene oxide); PPO: poly(propylene oxide); PVC: poly(vinyl chloride).

each monomer. Along with the rate constant of cross-addition k_a and the rate constant of propagation k_p of the second monomer, the equilibrium constants K_a and K_b between active and dormant species for the two kinds of monomer units should be taken into consideration. Apparently, the very active monomer should be polymerized first, which is valid for either of the controlled polymerization mechanisms. Different from sequential monomer addition technique utilized in anionic living polymerization, controlled routes allow for the polymerization of monomers A and B to be performed separately in two distinct steps, allowing chain functionalities to be preserved. The first block functions as a macromolecular initiator, a so-called macroinitiator, in the preparation of the second block. The first block can alternatively be synthesized via any of the conventional chain-growth and step-growth polymerization methods. When the monomers to be assembled in a targeted diblock structure do not polymerize by the same mechanism, a need for changing the active site arises and this synthetic approach is referred to as "site transformation." In cases where a bifunctional macroinitiator is employed, the resulting copolymer will be an ABA triblock copolymer.

11.2.3 Block Copolymers via Nitroxide Mediated Polymerization (NMP)

In NMP systems, stable free radicals, such as nitroxides, are used as reversible terminating agents to control the polymerization process. Dormant chains are generated by reversible deactivation of the growing chains through covalent bond formation. At high temperatures, the bond undergoes a homolytic cleavage to produce the active growing chain and the nitroxide radical. Activation is followed by a rapid deactivation, whereby a few monomer units are incorporated to the propagating chain (Scheme 11.4). The counter radicals should be quite stable (i.e., they should be inert

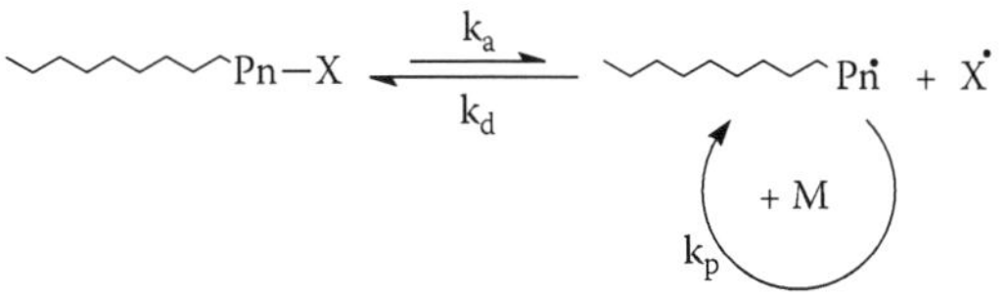

SCHEME 11.4

towards each other, monomers, or side reactions [e.g., H-abstraction]). Bimolecular initiating systems comprising a thermal initiator, such as an azo compound or a peroxide, and a stable free radical, TEMPO, have been utilized in the polymerization of styrene and its derivatives. However, several drawbacks associated with the use of TEMPO limited its utility. These involve its unsuitability for the polymerization of acrylates and its high temperature requirement. These drawbacks could be overcome by the use of unimolecular nitroxide initiators containing both the initiating radical and the alkoxyamine counter radical [10, 99].

Diaz et al. [100] reported the synthesis of poly(*N,N*-dimethylacrylamide-*b*-4-vinylpyridine) [poly(DMAA-*b*-4VP)] by using a β-phosphonylated nitroxide (*N*-*tert*-butyl-*N*-(1-diethylphosphono-2,2-dimethylpropyl) nitroxide, DEPN) as a control agent. The hydrophilic poly(DMAA) block was prepared by polymerizing the monomer along with the initiator (AIBN) and the nitroxide at 110°C, after which the reaction was stopped by cooling the flask and excess monomer was evaporated under reduced pressure (0.5 mmHg). The polymer, recovered as a powder, was used as a macroinitiator in the polymerization of 4VP (Scheme 11.5). The controlled nature of the polymerization was confirmed by the linear plots of both in ($[M]_0/[M]$) versus time and M_n versus conversion, where $[M]_0$, $[M]$, and M_n respectively stand for

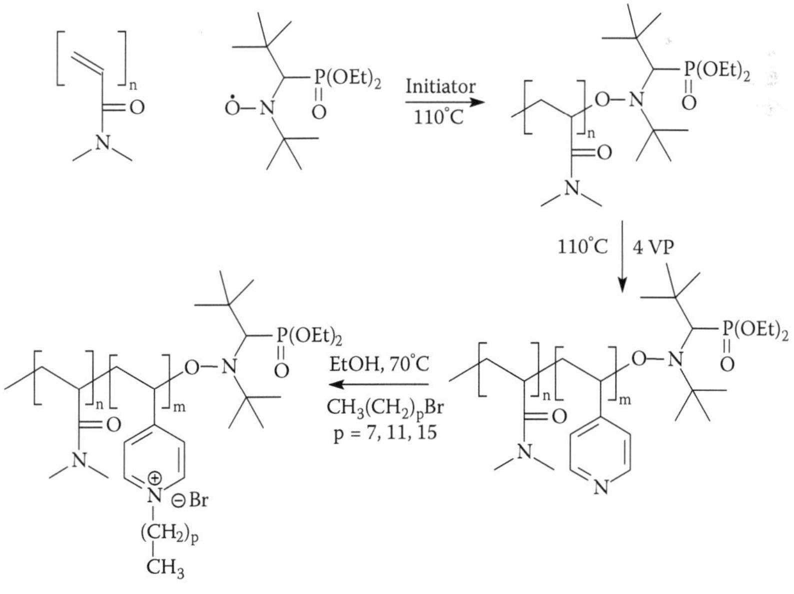

SCHEME 11.5

monomer original and current concentration in the bulk and polymer number-average molar weight. It was found that DEPN provided good control in the homopolymerization of 4VP and DMAA with polymerization rates and conversions (~ 60 %) that were much higher than those observed with TEMPO. Consequently, poly(DMAA-*b*-4VP) with desired lengths of each block was obtained quantitatively.

A variety of block copolymers derived using NMP is given in Table 11.3.

TABLE 11.3
Sequential Addition Synthesis of Block Copolymers by NMP

1st Segment	2nd Segment	References
CMS	St	[101–103]
(St-co-CMI), (St-co-NVC)		[104,105]
tBOSt		[106]
4VP, CMS		[107]
EBPBB		[108]
(St-alt-MAh)		[109]
BrSt and St	St and BrSt	[110]
St	tBuSt	[111]
	PIMS	[112]
	MPCS	[113]
	(St-co-AN)	[114]
	MA	[115]
	DMA	[116]
	BD, IP	[117,118]
	PES	[119]
	multifunctional acryl- and methacryl derivatives	[120]
	PFDS, PFDA	
	StSA	[121]
	BA	[122]
		[123]
AcOSt	MPVB	[124]
SSt	DMAM, SSC	[125]
	4VN	[126]
	Po	[127]
	NVC	[128]
nBA	St	[129]
	HEA	[130]
	(BA-co-St)	[131]
St, nBA	BA, St	[10]
IP	St, MA, MMA	[132]
PPV	(St-co-CMS)	[133]
DMAEA	BA	[134]
St, AcOSt, OSi_2St	Si_2St, Si_2CSt, OSi_2St, St, AcOSt	[47]
AA, St	MA	[135]

TABLE 11.3 (CONTINUED)
Sequential Addition Synthesis of Block Copolymers by NMP

1st Segment	2nd Segment	References
DMA	nBA	[136]
St, CMS	CMS, St	[137]

St: styrene; CMS: chloromethyl styrene; CMI: *N*-cyclohexylmaleimide; NVC: *N*-vinylcarbazole; tBOSt: p-*tert*-butoxystyrene; 4VP: 4-vinylpyridine; EBPBB: 4′-ethylbiphenyl-4-(4-propenoyloxybutyl-oxy)benzoate; MAh: maleic anhydride; BrSt: p-bromostyrene; PIMS: phthalimide methylstyrene; MPCS: 2,5-bis[(4-methoxyphenyl)oxycarbonyl]styrene; AN: acrylonitrile; MA: methyl acrylate; DMA: *N,N*-dimethylacrylamide; BD: 1,3-butadiene; IP: isoprene; PES: 4-(phenylethynyl)styrene; PFDS: (perfluorooctyl-ethylenoxymethylstyrene); PFDA: (1,1,2,2-tetrahydroperfluorodecyl acrylate); StSA: styrene sulfonic acid; AcOSt: 4-acetoxystyrene; MPVB: [(49-Methoxyphenyl) 4-oxybenzoate]-6-hexyl (4-vinylbenzoate); tBuSt: p-tert-butylstyrene; SSt: sodium 4-styrene-sulfonate; DMAM: 4-(di methylamino)methylstyrene; AAc: acrylic acid; SSC: sodium 4-styrenecarboxylate; VN: vinylnaphthalene; Po: (5-(4-acryloyloxyphenyl)-10,15,20-tritolylporphyrin); BA: n-butyl acrylate; HEA: 2-hydroxyethyl acrylate; PPV: 2,5-dioctyloxy-1,4-phenylenevinylene; DMAEA:2-(dimethylamino)ethyl acrylate; OSi_2St: 4-(pentamethyldisiloxymethyl) styrene; Si_2St: 4-(pentamethyldisilyl) styrene; Si_2CSt: 4-(bis(tri methylsilyl)methyl) styrene.

11.2.4 Block Copolymers via Atom Transfer Radical Polymerization (ATRP)

Atom transfer radical polymerization (ATRP), pioneered by Wang and Matyjaszewski [138] and Kato et al. [139], is based on a continuous and reversible halogen transfer (pseudo halogen) between a dormant propagating species, P_n, and a transition-metal (e.g., Cu), complexed by a ligand (e.g., bipyridine), in its lower oxidation state (Scheme 11.6). Halogen transfer is accompanied by a one-electron oxidation of the transition metal, whereby propagation takes place by the addition of monomers to the activated chains. Homogeneity of the polymeric chains is controlled by the fast and simultaneous initiation as well as rapid deactivation of the growing chains. Here, the rate of deactivation should be faster than the propagation rate for an effective control. Low concentration of the active centers, maintained by deactivation, suppresses termination and chain transfer reactions at later stages.

Transition metals involving copper [138], ruthenium [139], nickel [140], iron [141–143], rhodium [144], rhenium [145], and palladium [146] have been employed in conjunction with nitrogen- and phosphorus-based ligands. It was also found that

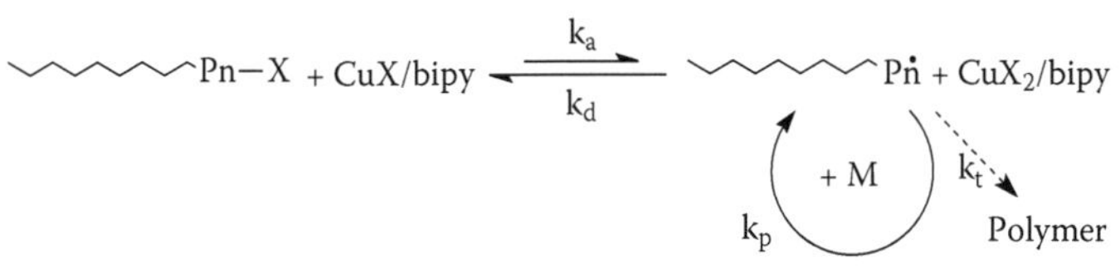

$$\sim\sim Pn{-}X + CuX/bipy \underset{k_d}{\overset{k_a}{\rightleftharpoons}} \sim\sim P\dot{n} + CuX_2/bipy$$

(+ M, k_p; k_t → Polymer)

SCHEME 11.6

nitrogen-based ligands were effective in copper mediated systems, whereas phosphorus-based ligands, particularly PPh_3, were successfully applied with the previously mentioned transition metals. Halogenated compounds possessing activating β-carbonyl, vinyl, phenyl, and cyano groups have been shown to be good ATRP initiators [147–149].

ATRP allows for the facile preparation of di-, tri-, or multi-block copolymers as well as star polymers using bi- or multi-functional initiators carrying the active halogen atoms. Some examples of recent works are presented in Table 11.4. Pyun et al. reported the synthesis of ABA triblock copolymers containing polyhedral oligomeric silsesquioxane (POSS) groups via ATRP [150]. The middle (B) and outer segments (A) were poly(n-butyl acrylate) (pBA) and poly(3-(3,5,7,9,11,13,15-heptaisobutyl-pentacyclo [9.5.1.$1^{3,9}$.$1^{5,1}$5.$1^{7,13}$]-octasiloxane-1-yl)propyl methacrylate (p(MA-POSS)), respectively. Bromo-difunctional poly(n-butyl acrylate), with a number-average molecular weight of 61,700 g/mol, was prepared by the ATRP of BA from a dimethyl 2,6-dibromoheptanedioate initiator and used in the next step as a macroinitiator in the polymerization of MA-POSS to achieve the desired ABA triblock copolymer as depicted in Scheme 11.7.

The macroinitiators prepared by various other routes such as ring-opening, anionic, cationic, and conventional radical processes can also be used in the chain-extensions through ATRP to get diblock or triblock copolymers. For this purpose, a modification reaction from one or both ends of the precursor polymer is usually necessary to incorporate the desired halogen functionality to the chain. A diblock copolymer is produced if an α-functional telechelic is used as the macroinitiator, whereas an ABA triblock copolymer can be obtained from an α,ω-functional telechelic. For example, a poly(ε-caprolactone) macroinitiator was prepared via ring-opening polymerization of ε-caprolactone, followed by a modification reaction on the hydroxyl end group, and used subsequently for the ATRP of 2,5-bis[(4-metho xyphenyl)oxycarbonyl]styrene (MPCS). Finally, diblock copolymers consisting of crystallizable poly(ε-caprolactone) and PMPCS [151] were achieved with relatively narrow polydispersity indices (PDI <1.11) (Scheme 11.8). Polarized optical microscopy (POM) and DSC results confirmed that the diblock copolymer showed liquid

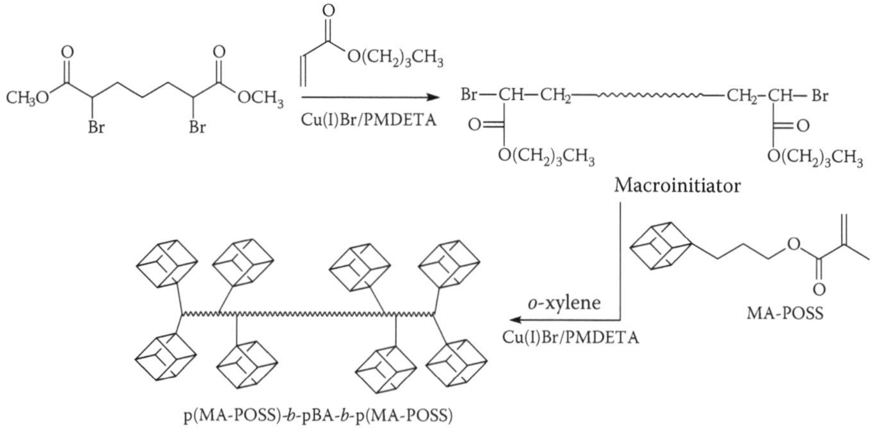

O
O(CH$_2$)$_3$CH$_3$
O O
CH$_3$O OCH$_3$
Br Br
Cu(I)Br/PMDETA
Br—CH—CH$_2$—CH$_2$—CH—Br
O= =O
O(CH$_2$)$_3$CH$_3$ O(CH$_2$)$_3$CH$_3$
Macroinitiator
O
O
MA-POSS
o-xylene
Cu(I)Br/PMDETA
p(MA-POSS)-*b*-pBA-*b*-p(MA-POSS)

SCHEME 11.7

TABLE 11.4
Summary of Block Copolymers Prepared Using ATRP

1st Block	2nd Block	References
BMA	MMA	[161,162]
	St	[163,164]
	St-BMA	[165]
MMA	BMA	[166]
	MA, BA	[167]
	BA	[168,169]
	St	[170,171]
	DMAEMA	[172]
	HEMA	[170,173]
	VP	[174]
	BzMA	[175]
	AN	[176]
	BEMA	[177]
	ArIt	[54]
	tBMA, MMA, DMAEMA	[172,178]
	PMPCS	[179]
MA	MMA	[168]
	DMAEMA	[172]
	DMA	[180]
	BMDO	[181]
nBA	MMA	[140,168,182]
	HPMA	[180]
	St	[183,184]
	TMS-HEA	[185]
	MA	[186]
	MA-POSS	[187]
St	HEMA	[170]
	NPMA	[188]
	MMA	[67]
	BA	[170,189–191]
	tBA	[192–196]
	MA	[138,197]
	MAA, tBMA	[198,199]
	MAIpGlc	[200]
	AcGEA	[201]
	HEMA-co-DMAEMA	[202]
	4VP	[203]
	DMAA	[204]
	PFA	[205]
	DHPA	[206]
tBA	St, MA	[194,207]
	(MA-b-tBA)	[207]
(tBA-b-St)	MA	[193]

(Continued)

TABLE 11.4 (CONTINUED)
Summary of Block Copolymers Prepared Using ATRP

1st Block	2nd Block	References
TMS-HEA	nBA	[185]
LC2		[208]
OEGMA	NaVB	[209]
	MMA	[210]
FOMA	MMA and DMAEMA	[211]
MMA, BA	BA, BMA ,St	[184]
AmS	St	[212]
iBMA		[213,214]
4VP		[215]
LC11		[216]
BA, LA	MMA, iBOA, St	[217]
DMEMA	MMA	[218]
SPMA		[219]
LMA		[220]
AN	St	[221]
	BA	[222]
PEO-b-PMMA	PMPCS	[223]
	St	[224]
PDECVP	St, PDECVP, MMA	[225]
PFS	St	[226]
	MMA	[227]
MPC	DMA, DPA	[228]
	DMA,DEA, DPA, HEMA, HPMA, GMM	[229]
DEA	HEMA	[230]
styrene-based	semifluorinated	[231]
DEAEMA	tBMA	[232]
POEM	BMA	[233]
tBMA, nBMA, OMA,	OMA, tBA	[234]

BMA: butyl methacrylate; MMA: methyl methacrylate; St: styrene; MA: methyl acrylate; BA: butyl acrylate; DMAEMA: (*N,N*-dimethylamino)ethyl methacrylate; DEAEMA: 2-(diethylamino)ethyl methacrylate; HEMA: hydroxylethyl methacrylate; 2VP: 2-vinylpyridine; 4VP: 4-vinylpyridine; BzMA: benzyl methacrylate; AN: acrylonitrile; BEMA: 1-(butoxy)ethyl methacrylate; ArIt: *N*-aryl itaconimide; tBMA: *tert*-butyl methacrylate; PMPCS: 2,5-bis[(4-methoxyphenyl)oxycarbonyl]styrene; DMA: 2-(dimethylamino)ethyl methacrylate; BMDO: (5,6-benzo-2-methylene-1,3-dioxepane); HPMA: hydroxypropyl methacrylate; TMS-HEA: 2-trimethylsilyloxyethyl acrylate; MA-POSS: 3-(3,5,7,9,11,13,-15-heptacyclopentyl-pentacyclo[9.5.1.1.3,91.5,1517,13]-octasiloxane-1-yl)propyl methacrylate; NPMA: p-nitrophenyl methacrylate; MAAc: methacrylic acid; MAIpGlc: 3-O-methacryloyl-1,2 : 5,6-di-O-isopropylidene-D-glucofuranose; AcGEA: 2-(2′,3′,4′,6′- tetra-O-acetyl-beta-D-glucopyranosyloxy) ethyl acrylate; DMAA: N,N-dimethylacrylamide; PFA: perfluorooctyl ethyl acrylate; DHPA: 2,3-dihydroxypropyl acrylate; PFA: 2-perfluorooctyl ethyl acrylate; LC2: 2- [2-(4-cyano-azobenzene-4′-oxy)ethyleneoxy] ethyl methacrylate; NaVB: sodium 4-vinylbenzoate; OEGMA: oligo(ethylene glycol) methacrylate;

FOMA: fluorinated (meth)acrylates; AmS: p(4-amino styrene); iBOA: isobornyl acrylate; iBMA: i-butyl methacrylate; LC11:11-[4-(3-ethoxycarbonyl-coumarin-7-oxy)-carbonylphenyloxy]-undecyl methacrylate; LMA: lauryl methacrylate; LA: lauryl acrylate; SPMA: potassium 3-sulfopropyl methacrylate; PEO: poly(ethylene oxide); PDECVP: poly(dimethyl(1-ethoxycarbonyl)vinyl phosphate); PFS: pentafluorostyrene; DEA: 2-(diethylamino)ethyl methacrylate; DPA: 2-(diisopropylamino)ethyl methacrylate; GMM: glycerol monomethacrylate; MPC: (2-methacryloyloxyethyl phosphorylcholine; POEM: (oxyethylene)-9-methacrylate); OMA: octadecyl methacrylate; PMPCS: 2,5-bis[(4-methoxyphenyl)oxycarbonyl] styrene.

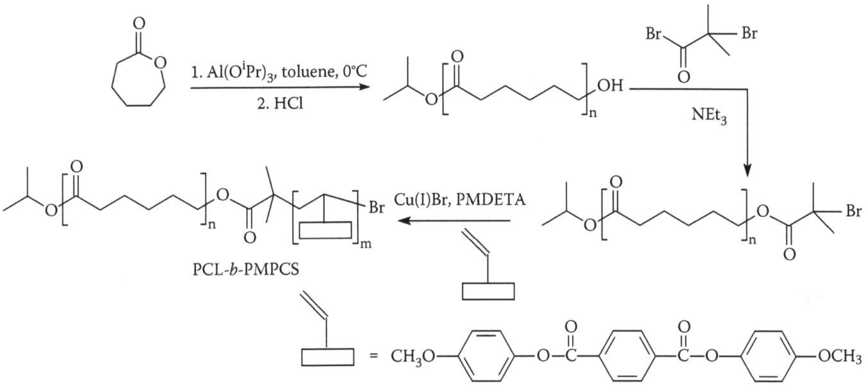

1. $Al(O^iPr)_3$, toluene, 0°C
2. HCl
NEt_3
Cu(I)Br, PMDETA
PCL-*b*-PMPCS
= CH_3O–O–C(=O)–C(=O)–O–OCH_3

SCHEME 11.8

crystalline behavior when the degree of polymerization (DP) of PMPCS block was higher than 44.

Other work, by Kurjata, et al. [152] involves the synthesis of a polydimethylsiloxane (PDMS)-macroinitiator via anionic polymerization of hexamethylcyclotrisiloxane (D3) followed by reaction with 3-(chlorodimethylsilyl) propyl 2-bromo-2-methyl-propanoate propyldimethylchlorosilane. Subsequent use of this macromolecular initiator for the ATRP of oligo[ethylene glycol] methyl ether methacrylate afforded AB type amphiphilic comb-like block copolymers of poly(oligo[ethylene glycol] methyl ether methacrylate) and polydimethylsiloxane (Scheme 11.9).

Another strategy combining two distinct polymerization mechanisms for the synthesis of diblock copolymers is the use of dual initiators. In this approach, a

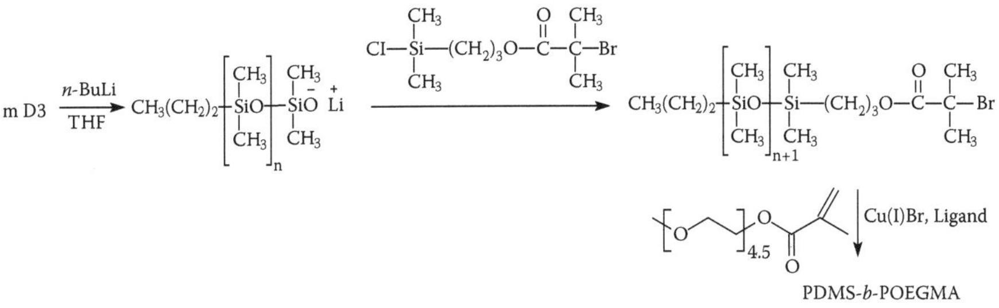

m D3 $\xrightarrow[\text{THF}]{n\text{-BuLi}}$ $CH_3(CH_2)_2$–$[Si(CH_3)_2O]_n$–$Si(CH_3)_2O^-\ Li^+$

$Cl–Si(CH_3)_2–(CH_2)_3O–C(=O)–C(CH_3)_2–Br$

$CH_3(CH_2)_2$–$[Si(CH_3)_2O]_{n+1}$–$Si(CH_3)_2$–$(CH_2)_3O–C(=O)–C(CH_3)_2–Br$

Cu(I)Br, Ligand

4.5

PDMS-*b*-POEGMA

SCHEME 11.9

compound possessing two functional groups is employed to initiate two polymerizations based on different mechanisms. This strategy has the advantage of eliminating the need for transformation of the active chain-end.

In a recent work [153], we reported the use of *N,N*-dimethylaniline end-functional polystyrenes in the preparation of poly(St-*b*-MMA) via photoinduced radical polymerization of MMA and poly(St-*b*-CHO) via radical promoted cationic polymerization of cyclohexene oxide (CHO). For this purpose, 4-(dimethylamino)-benzyl 4-(bromomethyl) benzoate, a dual initiator, was synthesized first by the esterification of 4-(bromomethyl)-benzoyl chloride and [4-(dimethylamino)phenyl]methanol in diethyl ether. In the second step, ATRP of styrene was performed using this initiator in the presence of CuBr/bipyridine complex. The polymer bearing end-chain *N,N*-dimethyl amino moiety was finally used as a macromolecular coinitiator in the photopolymerization of MMA and CHO via radical and radical promoted cationic polymerization routes, respectively, as depicted in Schemes 11.10 and 11.11.

In the former case, macroradical generation was achieved with benzophenone sensitizer by photoexcitation followed by hydrogen abstraction from amino end groups. A visible light initiating system has been utilized in the radical promoted cationic polymerization of CHO. The system involves a xanthene dye (Erythrosin B) as the sensitizer, an aromatic *N,N*-dimethyl amino group and diphenyl iodonium hexafluorophosphate as the radical source and radical oxidizer, respectively.

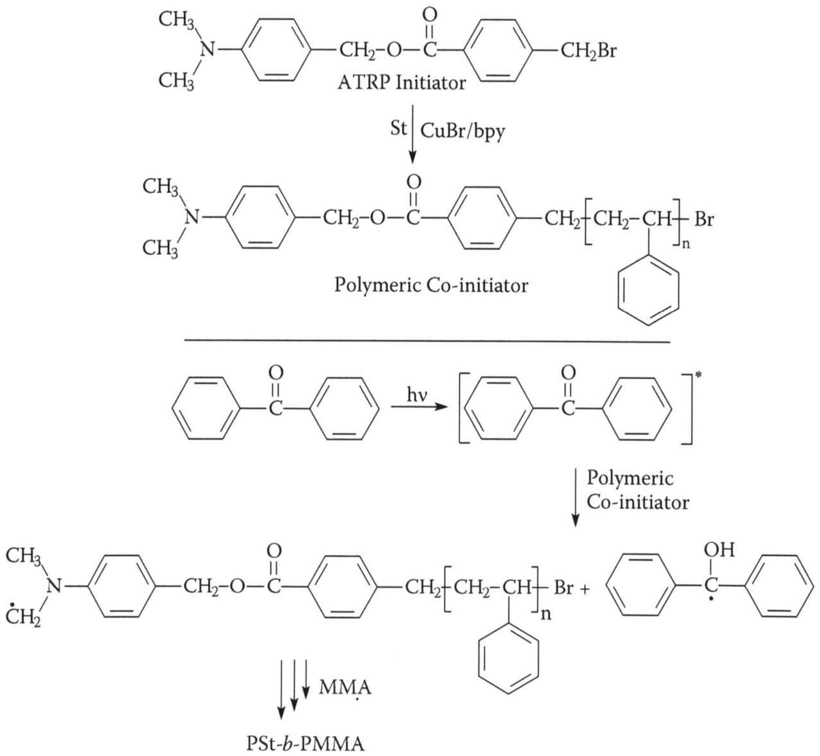

ATRP Initiator

St | CuBr/bpy

Polymeric Co-initiator

hν

Polymeric Co-initiator

MMA

PSt-*b*-PMMA

SCHEME 11.10

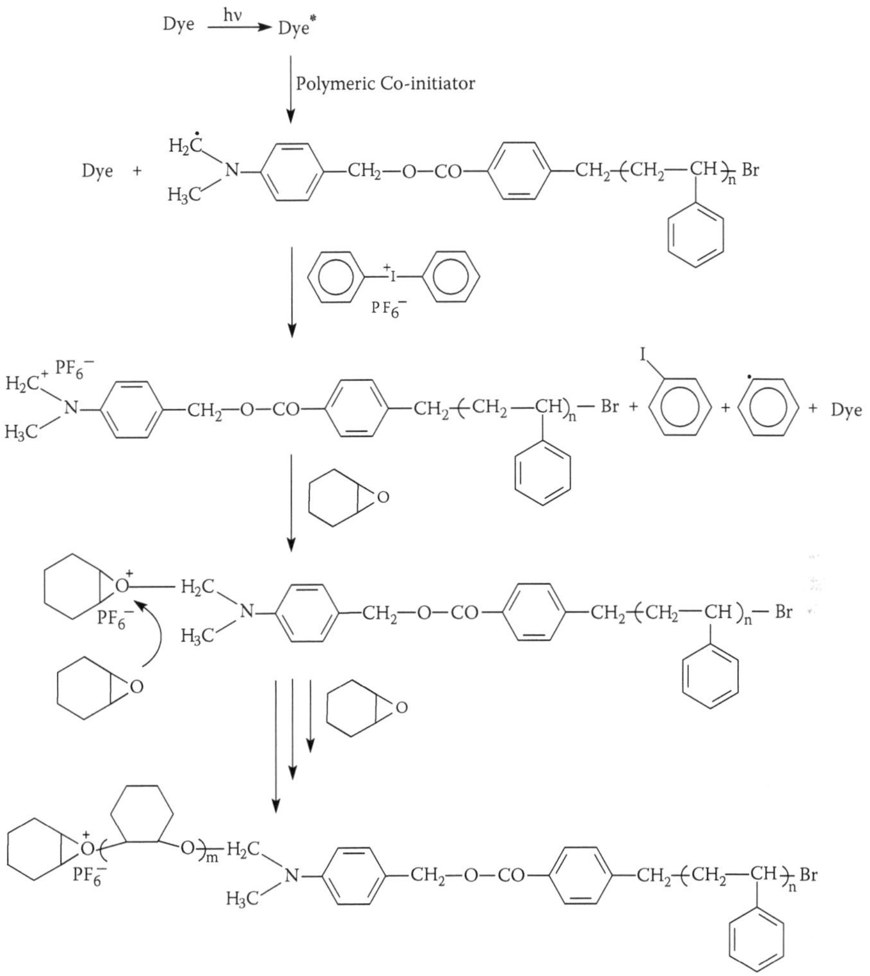

SCHEME 11.11

Although pure block copolymers were obtained via the free radical route, the free radical promoted cationic polymerization, yielding both block and homopolymers.

Du Prez and co-workers reported the sequential two-step synthesis of well-defined block copolymers of polystyrene (PSt) and poly(tetrahydrofuran) (PTHF) by ATRP and cationic ring-opening polymerization (CROP) using 4-hydroxy-butyl-2-bromoisobutyrate as a dual initiator [154]. The bromoisobutyrate group contained in the initiator effectively initiates the ATRP of styrene in conjunction with the Cu(0)/Cu(II)/*N,N,N′,N‴,N″*-pentamethyldiethylenetriamine catalyst, whereas the triflate ester group obtained from the *in situ* reaction of the hydroxyl groups of the initiator with trifluoromethane sulfonic anhydride initiates the CROP of tetrahydrofuran (THF). In this manner, both the PSt-OH and PTHF-Br homopolymers were prepared and used as macroinitiators for the CROP of THF and the ATRP of styrene, respectively, to afford PTHF-*b*-PS block copolymers, as depicted in Scheme 11.12. CROP

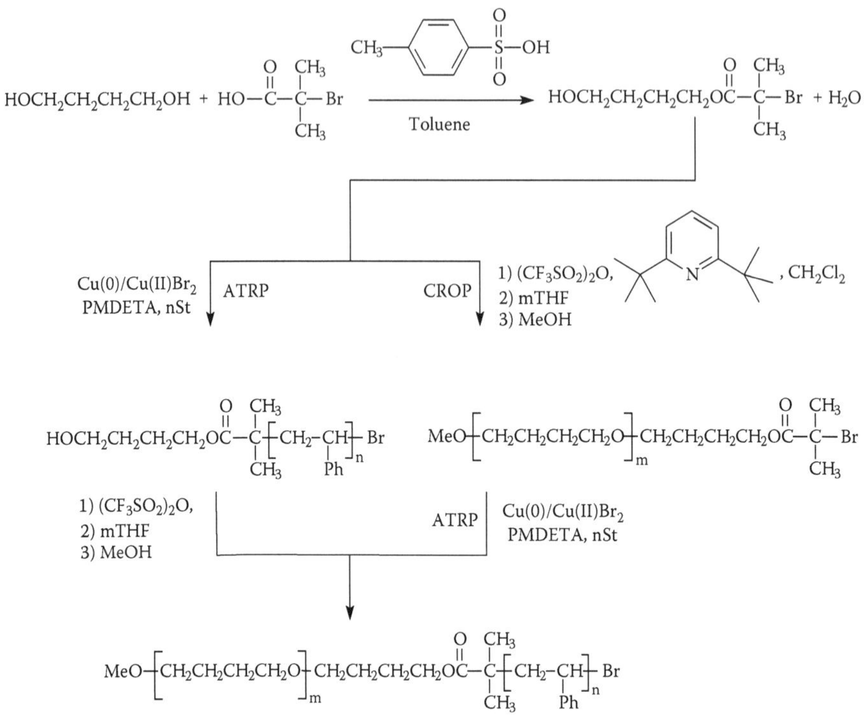

$$HOCH_2CH_2CH_2CH_2OH + HO{-}C(=O){-}C(CH_3)_2{-}Br \xrightarrow[\text{Toluene}]{CH_3{-}C_6H_4{-}SO_2{-}OH} HOCH_2CH_2CH_2CH_2OC(=O){-}C(CH_3)_2{-}Br + H_2O$$

ATRP: $Cu(0)/Cu(II)Br_2$, PMDETA, nSt

CROP: 1) $(CF_3SO_2)_2O$, 2,6-di-tert-butylpyridine, CH_2Cl_2; 2) mTHF; 3) MeOH

$$HOCH_2CH_2CH_2CH_2OC(=O){-}C(CH_3)_2{-}[CH_2{-}CH(Ph)]_n{-}Br$$

$$MeO{-}[CH_2CH_2CH_2CH_2O]_m{-}CH_2CH_2CH_2CH_2OC(=O){-}C(CH_3)_2{-}Br$$

1) $(CF_3SO_2)_2O$, 2) mTHF, 3) MeOH

ATRP: $Cu(0)/Cu(II)Br_2$, PMDETA, nSt

$$MeO{-}[CH_2CH_2CH_2CH_2O]_m{-}CH_2CH_2CH_2CH_2OC(=O){-}C(CH_3)_2{-}[CH_2{-}CH(Ph)]_n{-}Br$$

SCHEME 11.12

of THF using PSt-OH as a macroinitiator produced a mixture of the PSt-*b*-PTHF block copolymer and the remaining PSt macroinitiator due to nonquantitative initiation. On the other hand, PTHF-Br was found to initiate the ATRP of styrene quantitatively, eventually yielding well-defined PTHF-*b*-PSt block copolymers.

Preparation of block copolymers comprising monomers of distinct reactivities requires that the monomers be introduced in a specific sequence if one desires to obtain good control over the polydispersities. The addition should be handled in a manner favoring the cross-propagation (i.e., the initiation of the second block by the first one) over propagation. For instance, a PMMA block should precede the growth of a PSt block, and not the reverse. In this case, the cross-propagation is rather fast, because the equilibrium between active and dormant chains lies more to the active side. On the other hand, the rate of propagation of the St monomer is comparatively slow. As a result, nearly all the chains of the second PSt block are initiated before the propagation takes place. Evidently, this results in a low polydispersity.

Alternatively, a technique called the "halogen exchange" [140, 155–159] can be employed to change this sequence of blocking. In this methodology, the first block (e.g., PSt) is formed using a bromine-functional initiator in combination with a copper (I) bromide catalyst. In the next step, the second monomer (e.g., MMA) is added together with copper (I) chloride. Here, the activation of bromine-functional chain is followed by deactivation with $CuCl_2$ (or CuClBr). Because the carbon-chlorine bond is stronger than that of carbon-bromine, formation of a carbon-chlorine bond predominates in this mixed halogen system, and thereby retards the propagation in comparison to the cross-propagation.

An ABCBA-type pentablock terpolymer was prepared using a bromo-terminated P*t*BuA-*b*-PSt-*b*-P*t*BuA triblock as a macroinitiator in the chain-extension with MMA [160]. The triblock copolymer with M_n = 13,600 and PDI = 1.23 was synthesized beforehand from a bifunctional PSt using the CuBr/PMDETA catalyst system. Subsequent chain-extension with MMA, using the system CuCl/HMTETA afforded the desired PMMA-*b*-P*t*BuA-*b*-PSt-*b*-P*t*BuA-*b*-PMMA terpolymer with M_n = 48,500 and PDI = 1.21. An increase in the PDI was avoided due to the halogen exchange mechanism, which enhanced the rate of cross-propagation relative to the rate of propagation.

11.2.5 Block Copolymers via Reversible Addition Fragmentation Chain Transfer Polymerization (RAFT)

RAFT, developed by Rizzardo and co-workers in the late 1990s, utilizes a chain-transfer-active thiocarbonylthio moiety for the exchange between active and dormant chains [12]. The mechanism is illustrated in Scheme 11.13.

The active species, such as the radicals stemming from the decomposition of the initiator and propagating radicals (P_n^-), are transferred to the RAFT-agents (e.g., the thiocarbonylthio moiety). Meanwhile, an intermediate radical is formed and undergoes a fast fragmentation reaction, giving a polymeric RAFT agent and a new radical. The radical reinitiates the polymerization. The equilibrium is established by successive chain transfer-fragmentation stages. Z and R groups in the thiocarbonylthio compound represent the activating group and homolytically leaving group, respectively. These, in turn, determine the rates of addition and fragmentation. In fact, the choice of RAFT agent for a specified monomer is rather significant and affects the degree of control.

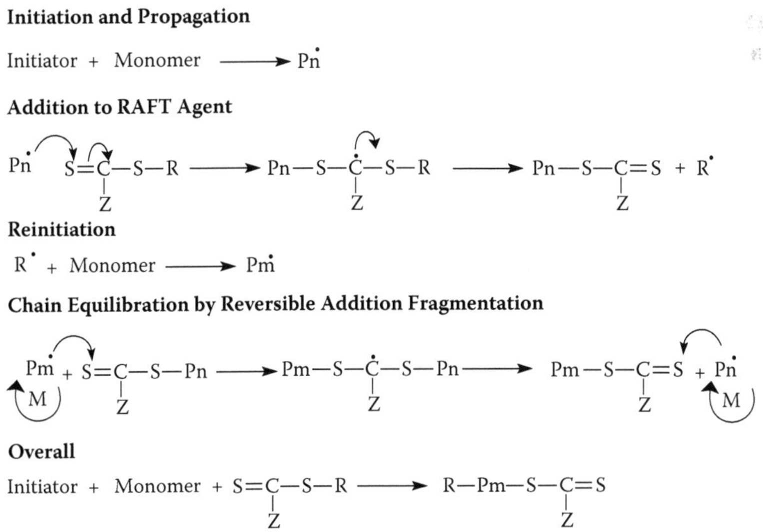

Initiation and Propagation

Initiator + Monomer ⟶ Pn•

Addition to RAFT Agent

Pn• S=C(Z)—S—R ⟶ Pn—S—C•(Z)—S—R ⟶ Pn—S—C(Z)=S + R•

Reinitiation

R• + Monomer ⟶ Pm•

Chain Equilibration by Reversible Addition Fragmentation

Pm• (M) + S=C(Z)—S—Pn ⟶ Pm—S—C•(Z)—S—Pn ⟶ Pm—S—C(Z)=S + Pn• (M)

Overall

Initiator + Monomer + S=C(Z)—S—R ⟶ R—Pm—S—C(Z)=S

SCHEME 11.13

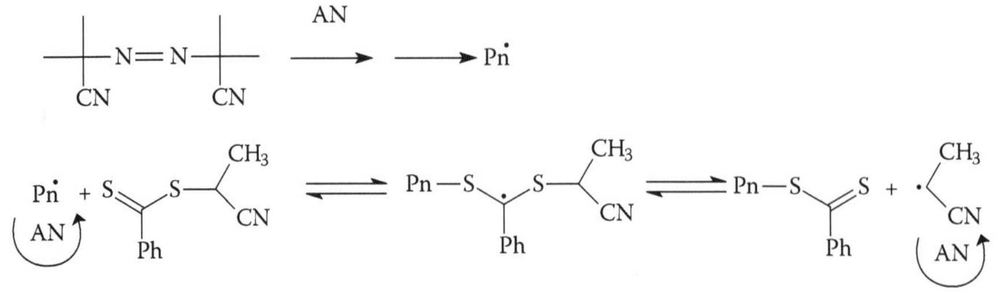

SCHEME 11.14

Matyjaszewski and co-workers reported the preparation of block copolymers of acrylonitrile (AN) and n-butyl acrylate (n-BA) with low polydispersity and controlled molecular weight using a novel RAFT agent, 2-cyanoethyl dithiobenzoate (CED) (Scheme 11.14) [235]. Polymerizations of AN were performed with two different ratios of initiator and RAFT agent at 60°C. Best results were achieved, that is, a 40.2% conversion in 7 h with $Mw/Mn = 1.05$, upon selecting a monomer/initiator/RAFT agent mol ratio of 500:1:3. Attempts to control the polymerization of AN failed when cumyl dithiobenzoate (CDB) was used as the RAFT agent. Moreover, chain extension from poly(*n*-butyl acrylate) (PBA) to AN yielded multimodal GPC traces with high polydispersity signifying a low blocking efficiency. Consequently, PAN macroinitiators were shown to crosspropagate more easily, allowing the preparation of PAN-*b*-PBA block copolymers by chain extension from PAN to *n*-BA.

Convertine et al. [236] were first to polymerize water-soluble vinylpyridine (VP) monomers, specifically 2-vinylpyridine (2VP) and 4-vinylpyridine (4VP), via RAFT, which might be helpful in preparing novel block copolymers with monomers not suitable with other controlled routes. They also showed the ability to synthesize 2VP-4VP and 4VP-2VP block copolymers in a controlled manner by using RAFT (Scheme 11.15). Several other examples of block copolymers prepared by RAFT using sequential monomer addition strategy are presented in Table 11.5.

In another work, PEO containing a xanthate end group was used as a macro RAFT agent in the polymerization of *N*-vinylformamide (NVF) to yield polyethylene-*b*-poly(*N*-vinyl formamide) (PEO-*b*-PNVF) (Scheme 11.16) [237].

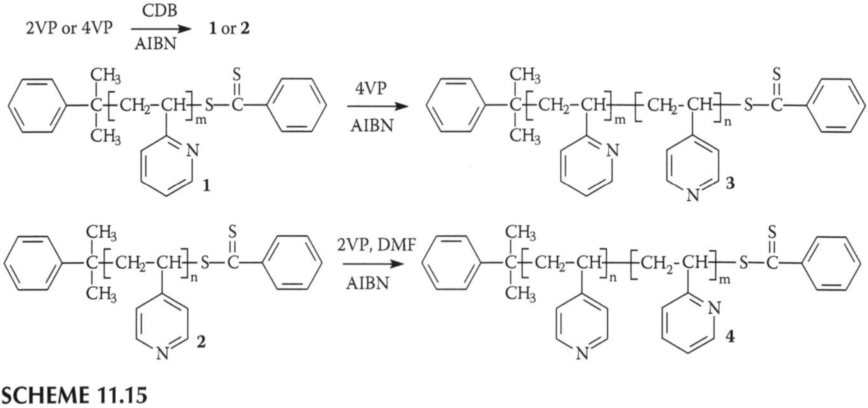

SCHEME 11.15

TABLE 11.5
Synthesis of Block Copolymers by RAFT Using Sequential Addition Strategy

1st Segment	2nd Segment	References
StCo	MA	[240]
AGA	NIPAM, HEA	[241]
St	AA	[242]
	VBCl	[243]
	E3VC	[244]
	MMA	[245]
	BA	[246–248]
	Mah	[249]
	MA	[250]
DMA	SA	[251]
nBA, EOA, EOMA, DMA, NAPy, NIPAM, SSt	NAPy, SB, nBA, VB, DMA	[252]
GMPS	MMA	[253]
MA, nBA, MMA, DMA	EEA	[254]
MMA	DMAEMA	[255]
	St, (St-co-CMS)	[256]
VAc	St	[257]
BA	AOEPCl, HEA	[258]
NIPAM	AA	[259]
DMA	NAS	[260]
4VP	tBA	[261]
MMAGl	HEMA	[262]
AN	PBA	[235]
AEMA, PMMA	MMA, AEMA, BMA, BA, NiPAM	[263]
2VP	4VP	[236]
MMA, BMA, MA, EA, BA, St	PEA	[264]
SSt, VBTMACl	SBz, DMVBA	[265]
St, BA, MA, MMA, BzMA, EO	DMA, MeSt, AA, EA, MAA, St, DMAMEA, BzMA	[266]

StCo: Styrene-Coumarin; MA: methyl acrylate; AGA: acryloyl glucosamine; NIPAM: *N*-isopropylacrylamide; St: styrene; AA: acrylic acid; HEA: hydroxyethyl acrylate; VBCl: 4-vinylbenzyl chloride; Mah: maleic anhydride; E3VC: N-ethyl-3-vinylcarbazole; MMA: methyl methacrylate; BA: n-butyl acrylate; DMA: *N,N*-dimethylacrylamide; SA: sodium acrylate; EOA: poly(ethylene glycol) acrylate; EOMA: poly(ethylene glycol) methacrylate; NAPy: *N*-acryloylpyrrolidine; SSt: sodium 4-styrene-sulfonate; SB: sulfobetaine; VB: vinylbenzoic acid; GMPS: gamma-methacryloxypropyltrimethoxysilane; EEA: 1-ethoxyethyl acrylate; DMAEMA: 2-(dimethylamino)ethyl methacrylate; CMS: chloromethyl styrene; VAc: vinyl acetate; AOEPCl: (2-acryloyloxyethyl phosphorylcholine); NAS: *N*-acryloxysuccinimide; MMAGl: (methyl 6-O-methacryloyl-alpha-D-glucoside); AN: acrylonitrile; HEMA: 2-hydroxyethyl methacrylate; AEMA: 2-(acetoacetoxy)ethyl methacrylate; BMA: butyl methacrylate; 4VP: 4-vinylpyridine; 2VP: 2-vinylpyridine; EA: ethyl acrylate; VBTMACl: (ar-vinylbenzyl)tri methylammonium chloride; BzMA: benzyl methacrylate; SBz: sodium 4-vinylbenzoate; DMVBA:N,N-dimethylvinylbenzylamine; EO: ethylene oxide; MeSt: *para*-methylstyrene; MAA: methacrylic acid.

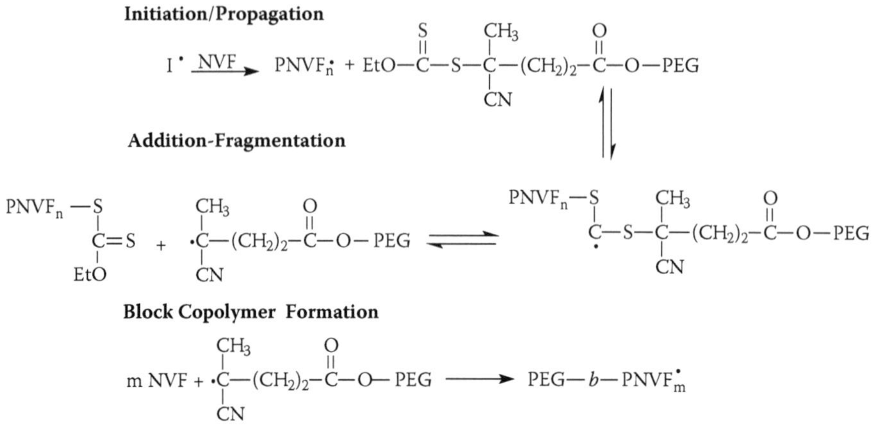

Initiation/Propagation

Addition-Fragmentation

Block Copolymer Formation

SCHEME 11.16

More recently, hydroxy functionalities of PEOs were converted to dithiobenzoyl groups and used as macro RAFT agent in RAFT polymerization of *N*-isopropyl acrylamide (NIPAa) (Scheme 11.17). Depending on the functionality of the initial polymers, AB and ABA type block copolymers with well-defined structures were prepared [238, 239].

In some cases, different controlled polymerization systems are combined to produce block copolymers (see Table 11.6). For example, mechanistic transformations involving ATRP and NMP were performed.

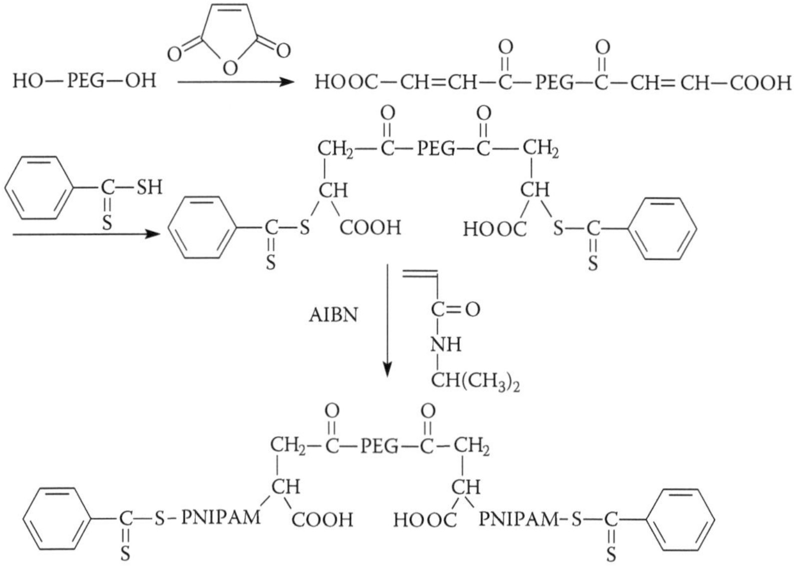

SCHEME 11.17

TABLE 11.6
Some Examples of Polymers Obtained by Combination of Different Controlled Radical Polymerization Mechanisms

Combined Controlled Radical Polymerizations	References
NMP-ATRP	[66,267–270]
Cobalt Mediated - ATRP	[271]
DT-ATRP	[272]
NMP-RAFT	[273]
FRP-ATRP	[274,275]
ATRP-Telomer	[276–278]

NMP: nitroxide-mediated radical polymerization; ATRP: atom transfer radical polymerization; DT: degenerative transfer polymerization; RAFT: reversible addition fragmentation chain transfer polymerization.

11.3 SYNTHESIS OF GRAFT COPOLYMERS

11.3.1 Synthesis of Graft Copolymers via Conventional Radical Polymerization Methods

Graft copolymers are another class of segmented copolymers. As stated previously, no major difference exists between block copolymer synthesis and graft copolymer synthesis. The location of the sites and functions are at the chain ends or on the chains, respectively. Graft copolymers can be obtained with three general methods: (1) grafting-onto, in which side chains are preformed, and then attached to the backbone; (2) grafting-from, in which the monomer is grafted from the backbone; and (3) grafting-through, in which the macromonomers are copolymerized.

11.3.1.1 Graft Copolymers by "Grafting onto" Method

Grafting onto methods involve reaction of functional groups (Y) located at the chain ends of one kind of polymer with other functional groups (X), which are distributed randomly on the main chain of the other polymer (Scheme 11.18).

The method is most suited for the reaction of "living" anionic and cationic polymers with electrophilic and nucleophilic functions carried by a polymer backbone,

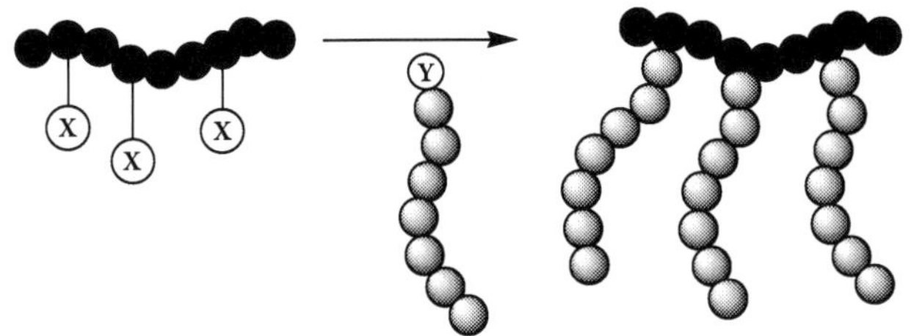

SCHEME 11.18

respectively. Corresponding radical grafting method has been used. In reported studies, usually functional backbone polymers prepared by radical mechanism are reacted with terminally functionalized polymers. The functionalization of polymers by free radical vinyl polymerization has been discussed in detail in another chapter of the book.

11.3.1.2 Graft Copolymers by "Grafting from" Method

Initiating radicals can be formed on a polymer chain upon irradiation with γ-rays, electron beams [279, 280], and UV light [76, 281, 282]. Grafting is usually performed by irradiation of a polymer swollen by a monomer. The method has found a number of applications to modify the properties of the polymers for special uses. For example, surface modification, which is often required to broaden the applicability of plastics, can be achieved by photografting. Surface modification by photografting can bring about improvements in dyeability, adhesiveness, printability, paintability, biocompatibility, antifogging, antistatic, and antistainability properties. Regarding textiles, an increase in wrinkle and flame resistance of fibers may be of commercial importance. Typical examples of photografting are given in Table 11.7.

In this connection, the readers' attention is also directed towards previously published review articles solely devoted to photografting [287, 297–304].

Chemical initiation has also been used in the grafting-from method. In this case, a polymer backbone contains some thermally cleavable bonds such as azo and peroxide linkages (Scheme 11.19). If the polymers are heated in the presence of a second monomer, initiation takes place. In such systems, homopolymer formation is unavoidable because of the concomitant formation of low molar mass radicals.

A wide range of side chain macroinitiators by which graft copolymers can be synthesized via grafting from method are available and their chemistry has been studied in detail [305].

Side chain polymeric photoinitiators are also suitable precursors for this type of graft copolymerization. Both cleavage [306] (Scheme 11.20) and hydrogen abstraction [307] (Scheme 11.21) type photoinitiation have been successfully employed.

In this connection, it should be emphasized that the hydrogen abstraction type photoinitiation has the advantage of producing graft copolymers free from the homopolymer contamination as the process yields only macroradicals.

It is well known that certain cerric salts such as nitrate and sulfate form very effective redox systems in the presence of organic reducing agents such as alcohols, aldehydes, and amines. Application of this technique to polymers with pendant reducing groups provides a versatile method for the grafting-from strategy [308] (Scheme 11.22).

11.3.1.3 Graft Copolymer Synthesis via Macromonomers

Macromonomers are short polymer chains possessing a polymerizable group at one terminus. A great variety of methods involving living polymerization techniques, chain transfer reactions, and end chain modifications have been developed to synthesize such species. Details of these methods have been discussed in earlier reviews and books [167, 290, 305, 309, 310], and are beyond the scope of this book.

TABLE 11.7
Typical Examples of Photografting

Polymer Backbone	Grafted Chain	Reference
PMVK	AN, MMA, VAc	[283]
PMIK, PBVK	St, MMA, AN, VAc	[48]
PSt-co-PVBP, PSt-DEASt	EHMA, DEA	[284]
ABME, PDAA	AA, MA	[285]
BMEA	St, MMA	[286]
PSt-co-PVBB	MMA	[287]
Pst	MMA	[246]
PAm, PAN	AN, Am	[7]
PAN	MEPM	[288]
PVC	MEPM, DAEM	[289]
Cl-MeSt	MEPM, DAEM	[290]
PDMS	HEMA, Am, MAEMA, MAc	[89,291]
PPVS	MMA	[292]
PVA	Am, AAc, AN, St	[38]
PVTClAc, PSt, PC, PA, PPp	St, MMA, CP	[293–295]
D-OH	HEMA, Am, AAc	[296]
PTFE	St, MMA	[60]

PMVK: poly(methyl vinyl ketone); PMIK: poly(3-methyl-3-buten-2-one); PBVK: pol(4,4-dimethyl-1-penten-3-one; AN: acrylonitrile; MMA: methyl methacrylate; VAc: vinyl acetate; St: styrene; EHMA: 2-ethylhexyl methacrylate; DEA: diethylaniline; PSt-co-PVBP: poly(styrene-co-vinylbenzophenone); PSt-co-DEASt: poly(styrene-co-p-N,N-diethylaminostyrene); ABME: allylbenzoin methyl ether; PDAA: propyldiallylamine hydrochloride, diethylaniline; BMEA: poly(α-methylol benzoin methyl ether acrylate); MA: methyl acrylate; AA: acrylic acid; PVBB: poly(styrene-co-p-vinylbenzophenone-p-t-butyl perbenzoate); Am: acrylamide; Cl-MeSt: chloro methylated styrene; MEPM: methoxy poly-ethylene-glycol monomethacrylate; DAEM: 2-(N,N-dimethylamino)ethyl methacrylate; HEMA: 2-hydroxyethyl methacrylate; MAEMA: 2-methylaminoethyl methacrylate; PPVS: poly(phenyl vinyl sulfide); PDMS: polydimethylsiloxane; PVA: polyvinylamine; PVTClAc: poly(vinyl trichloroacetate); PC: polycarbonate; PA: polyamide; PPp: polypeptide; CP: chloroprene; D-OH: dextran; PTFE: polytetrafluoroethylene.

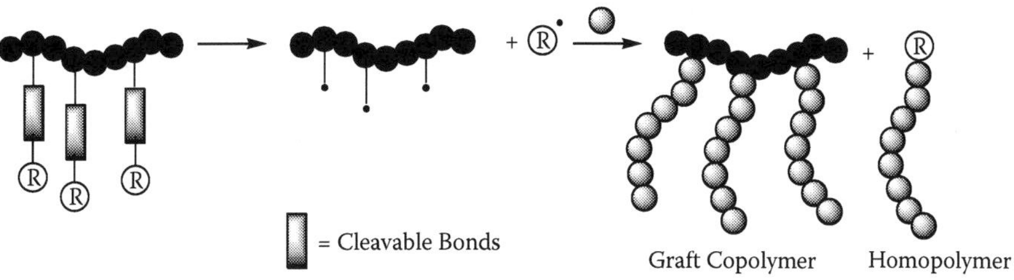

SCHEME 11.19

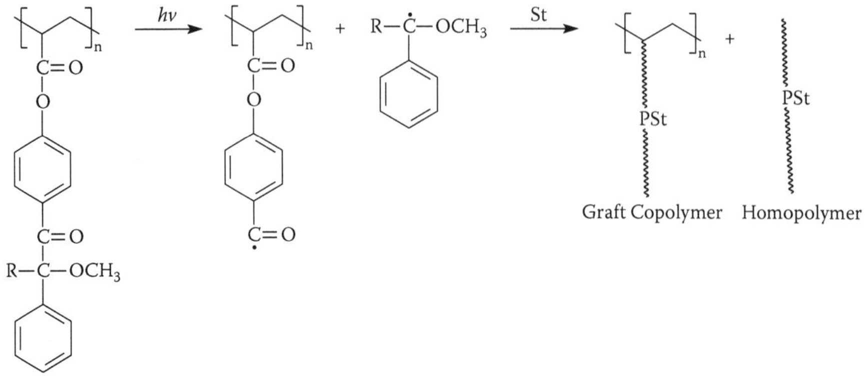

SCHEME 11.20

Macromonomers can be copolymerized with low molecular weight monomers in homogeneous and heterogeneous systems to yield graft copolymers (Scheme 11.23).

In some cases, the copolymerization reactivity of a macromonomer is nearly identical to that of low molecular weight analogue [12, 273]. Their reactivity can be predicted by intrinsic reactivity of the polymerizable end group, which is related to

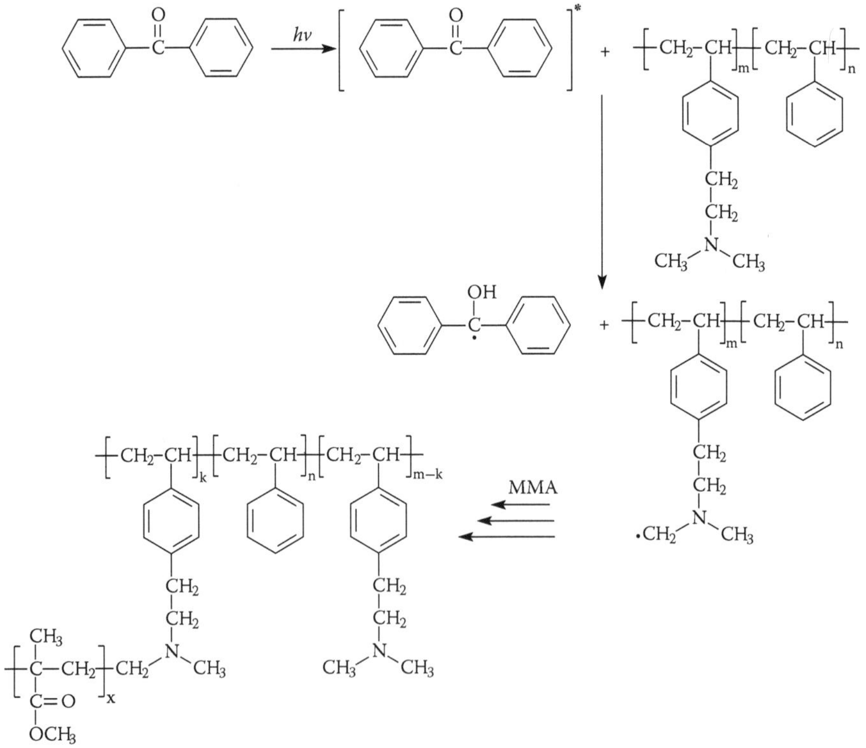

SCHEME 11.21

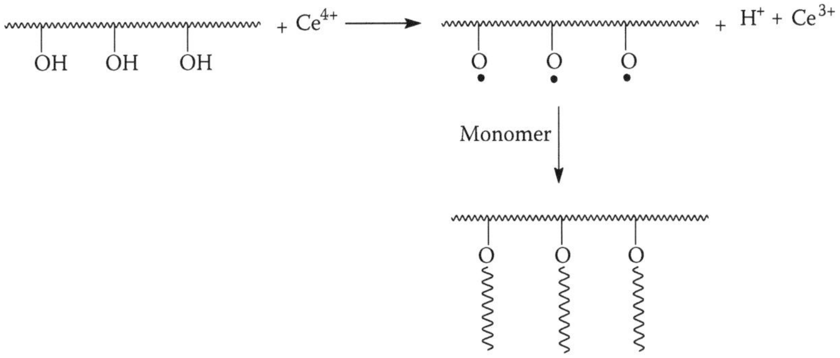

OH OH OH $+ Ce^{4+} \longrightarrow$ O• O• O• $+ H^+ + Ce^{3+}$

Monomer

O O O

SCHEME 11.22

the electronic and steric factors. On the other hand, in some cases macromonomer reactivites are lower due to the thermodynamic repulsion or incompatability of the macromonomer and the trunk polymer from the comonomer. By using macromonomer method, various kinds of graft copolymers can be prepared. At least in principle, no limitation exists in the selection of both macromonomers and comonomers. Moreover, the molecular structure of the produced graft copolymer is well-defined compared with those obtained with other methods. For example, the grafted chain length is controlled by the molecular weight of the macromonomer, and the number of grafted chains and the copolymer composition are controlled by the feed composition and the other copolymerization compositions. Therefore, many applications are promising and literature contains a huge number of reports.

11.3.2 Synthesis of Graft Copolymers via Controlled/Living Radical Polymerization Methods

Preparation of graft copolymers by using controlled/living radical polymerization methods also follows the same synthetic strategy that was described previously for the conventional radical polymerization. Obviously, graft copolymers obtained by using controlled/living radical polymerizations exhibit high structural control. Although earlier studies have only been concerned with ATRP and NMP, today it is possible to synthesize graft copolymers with all the controlled methods. Typical recent examples of graft copolymers prepared by using these methods are presented

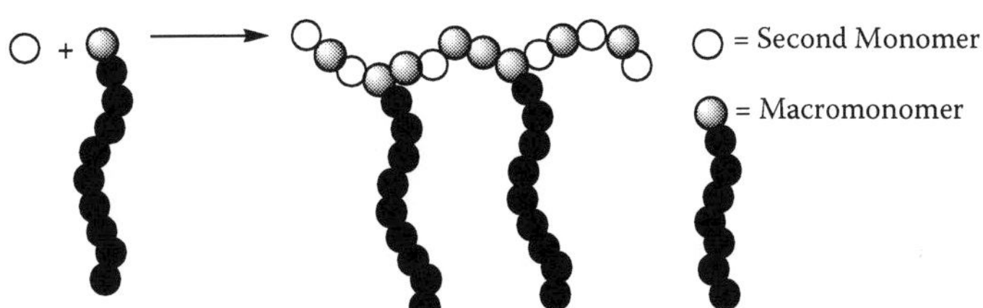

SCHEME 11.23

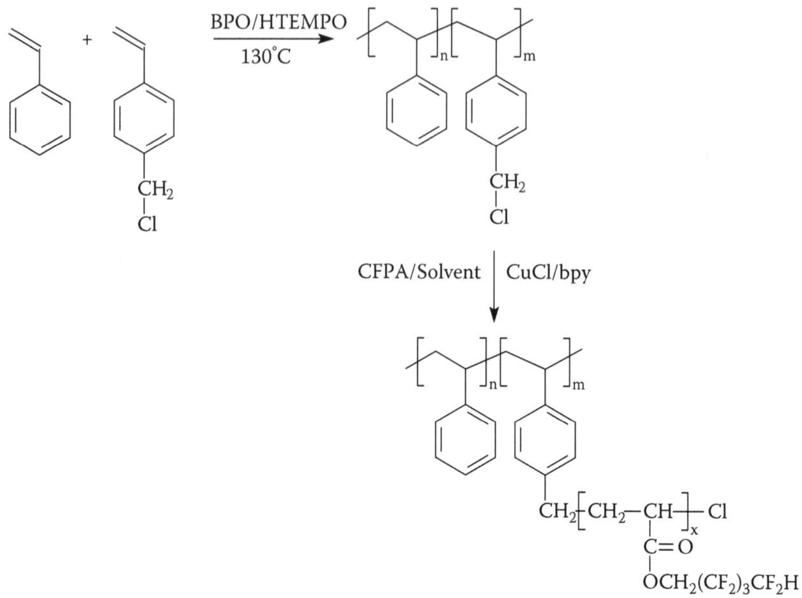

SCHEME 11.24

in Schemes 11.24 and 11.25. A compilation of literature data on these systems is provided in Tables 11.8 and 11.9.

An interesting example which combines two different controlled radical polymerization methods, namely NMP and ATRP, via "Click Chemistry" strategy between anthracene and maleimide [346] has been given at the end of the chapter. Using these Diels-Alder (DA) functional groups, well-defined polystyrene-*g*-poly(methyl methacrylate) (PS-*g*-PMMA) copolymers were successfully prepared. The whole process was divided into two stages; (1) preparation of anthracene and maleimide functional polymers and (2) the use of Diels-Alder reaction of these groups. First, random copolymers of styrene (S) and chloromethyl styrene (CMS) with various CMS contents were prepared by nitroxide mediated radical polymerization (NMP) process. Then,

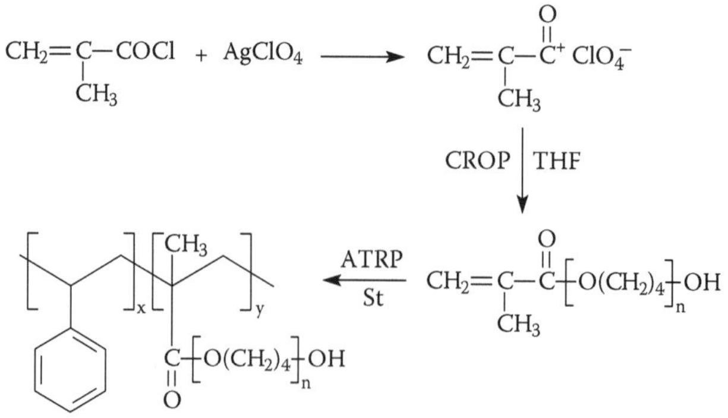

SCHEME 11.25

TABLE 11.8
Graft Copolymers Prepared by Grafting from Technique

Backbone	Grafts	Methods	References
St	St, St-*co*-EOEA, St	FRP/NMP	[311, 312]
DAMA-co-MMA, PMI, MVISC	St	FRP/NMP	[34, 313]
St-CMS	St, EMA, MMA, AMO	FRP/ATRP	[314, 315]
HEA, MAOETMACl	St	FRP/ATRP	[316]
PFS-b-PGMA	PtBA	FRP/ATRP	[317]
PP	St, BMA	Commercial/NMP	[116, 249, 318, 319]
PVC	St, MA, BA, MMA	Commercial/ATRP	[8]
p(IB-co-St)	St	Commercial/ATRP	[320, 321]
PSEP	EMA	Commercial/ATRP	[188]
Polyethylene	St, MMA	Commercial/ATRP	[157, 188, 322]
St	St, BA, MMA	NMP/ATRP	[323]
St-CMS	OFPA	NMP/ATRP	[324]
St, MMA	St	ATRP/NMP	[325]
St-CMS	MMA, BMA, DDMA	DPE/ATRP	[326]
HEMA	BA	ATRP/ATRP	[327]

St: styrene; EOEA: 2-ethoxyethyl acrylate; DAMA: 2-(dimethylamino)ethyl methacrylate; PMI: phenylmaleimide; MVISC: methylvinylisocyanate; CMS: p-chloromethylstyrene; MMA: methyl methacrylate; EMA: ethyl methacrylate; AMO: acryloylmorpholine; HEA: hydroxyethyl acrylate; MAOETMACl: 2-(methyacryloyloxy) ethyl trimethylammonium chloride; PFS: pentafluoro styrene; PGMA: glycidyl methyacrylate; tBA: t-butyl acrylate; BA: n-butyl acrylate; BMA: n-butyl methacrylate; PP: polypropylene; PVC: polyvinyl chloride; IB: isobutylene; PSEP: poly(styrene-b-ethylene-co-propylene); OFPA: 2,2,3,3,4,4,5,5-octafluoropentyl acrylate; DDMA: n-dodecyl methacrylate; HEMA: 2-hydroxyethyl methacrylate.

TABLE 11.9
Graft Copolymers Prepared by Macromonomer Technique

Macromonomer	Comonomer	Mechanism	References
PDMS	MMA	AROP/ATRP	[328–330]/
PCL			[331]
PEO	St	AROP/ATRP	[332,333]/
	MMA		[334]/
	ODMA, ODA		[334]/
	BA		[335]
St, tBA	MMA, St	ATRP/FRP	[106,336]
PAAc, PIBVE, PTHF	MMA	CROP/ATRP	[106,337]
PE	BA	ROMP/ATRP	[338,339]
St, MMA	MMA, St	RAFT/ATRP	[175]
St	NVP	ATRP/FRP	[340]
PMMA	BA	GTP/ATRP-FRP	[341]

(*Continued*)

TABLE 11.9 (CONTINUED)
Graft Copolymers Prepared by Macromonomer Technique

Macromonomer	Comonomer	Mechanism	References
PDMS, PEO	MMA	AROP/RAFT	[328, 330, 342, 343]
PLA	St	AROP/NMP	[344]
PEO	St	AROP/NMP	[113]
2VP	NIPAM	NMP/FRP	[345]

PDMS: polydimethylsiloxane; PCL: poly(ε-caprolactone); MMA: methyl methacrylate; PEO: poly(ethylene oxide); St: styrene; ODA: poly(octadecyl methacrylate); ODMA: poly(octadecyl methacrylate); BA: butyl acrylate; tBA: t-butyl acrylate; PAAc: poly(acrylic acid); PTHF: polytetrahydrofurane; PE: polyethylene; NVP: N-vinyl vinylpyrrolidinone; PLA: polylactide; 2VP: 2-vinylpyridine; NIPAM: N-ispopropylacrylamide.

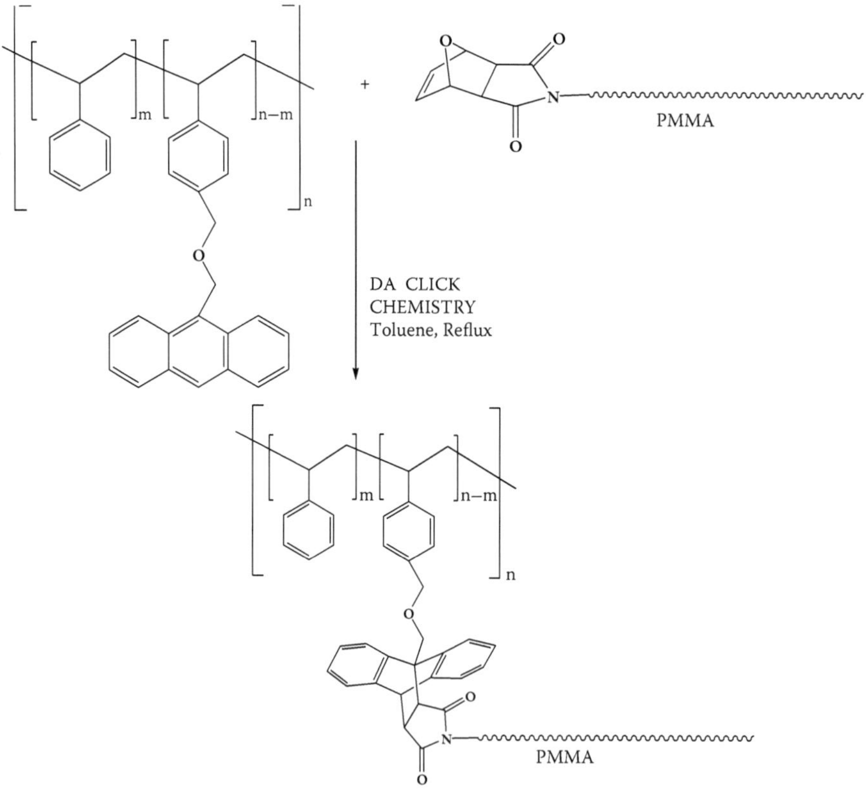

DA CLICK CHEMISTRY
Toluene, Reflux

PMMA

SCHEME 11.26

the chloromethyl groups were converted to anthryl groups via etherification with 9-anthracene methanol. The other component of the click reaction, namely protected maleimide functional poly(methyl methacrylate) (PMMA), obtained by atom transfer radical polymerization (ATRP) using the corresponding functional initiator. Then, in the final stage, PMMA prepolymer was deprotected by *retro* Diels-Alder *in situ* reaction by heating at 110°C in toluene. The recovered maleimide groups and added anthryl functional polystyrene underwent Diels-Alder reaction to form respective (PS-*g*-PMMA) copolymers (Scheme 11.26).

11.4 CONCLUSION

Free-radical polymerization is still a very important process in view of its wide utility in terms of designing various block and graft copolymers as described in this chapter. The design of block and graft copolymers can be achieved via different pathways. Very rapid developments on the area of controlled radical polymerizations in recent years certainly helped to explore the new ways of making block and graft copolymers.

REFERENCES

1. Swarc, M., *Adv. Polym. Sci.*, 2, 275, 1960.
2. Kennedy, J.P. and Ivan, B., *Designed Polymers by Carbocationic Macromolecular Engineering: Theory and Practice,* Hanser, Munich, 96, 1992.
3. Matyjaszewski, K. and Davis, T.P., Eds., in *Handbook of Radical Polymerization,* John Wiley & Sons, Inc, New York 895, Ch. 16, 2002.
4. Webster, O.W., *Makromol. Chem.-Macromol. Symp.*, 70-1, 75, 1993.
5. Novak, B.M., et al., *Adv. Polym. Sci.*, , 102, 47, 1992.
6. Muftuoglu, A.E., et al., *Designed Monomers Polymers*, 7, 563, 2004.
7. Matyjaszewski, K., et al., *Abstr. Pap. Am. Chem. Soc.*, 218, U410, 1999.
8. Paik, H.J., et al., *Macromol. Rapid Commun.*, 19, 47, 1998.
9. Hawker, C.J., et al., *Chem. Rev.*, 101, 3661, 2001.
10. Benoit, D., et al., *J. Am. Chem. Soc.*, 121, 3904, 1999.
11. Fukuda, T., et al., *Macromolecules*, 29, 3050, 1996.
12. Chiefari, J., et al., *Macromolecules*, 31, 5559, 1998.
13. Ganachaud, F., et al., *Macromolecules*, 33, 6738, 2000.
14. Patten, T.E. and Matyjaszewski, K., *Adv. Mater*, 10, 901, 1998.
15. Malmstrom, E.E. and Hawker, C.J. *Macromol. Chem. Phys.*, 199, 923, 1998.
16. Hizal, G. and Yagci, Y. *Polymer*, 30, 722, 1989.
17. Vlasov, G.P., et al., *Makromol. Chem.*, 183, 2635, 1982.
18. Ohmura, H. and Nakayama, M. *Jpn. Kokai*, S56, 1981.
19. Ohmura, H. and Nakayama, M. *Jpn. Kokai*, S56, 1981.
20. Ohmura, H. and Nakayama, M. *Jpn. Kokai*, S55, 1980.
21. Ohmura, H. and Nakayama, M. *Jpn. Kokai*, S56, 1981.
22. Ohmura, H. and Nakayama, M. *Jpn. Kokai*, S56, 1981.
23. Ohmura, H. and Nakayama, M. *Jpn. Kokai*, S56, 1981.
24. Ohmura, H. and Nakayama, M. *Jpn. Kokai*, S55, 1981.
25. Hiramatsu, Y., et al., *Jpn. Kokai*, S60, 1985.
26. Ohmura, H. and Nakayama, M. *Jpn. Kokai*, S55, 1981.
27. Sanui, T. and Matsushima, M. *Jpn. Kokai*, S55, 1980.
28. Oshibe, Y., et al., *Jpn. Kokai*, S60, 1985.
29. Nakayama, M., et al., *Jpn. Kokai*, S56, 1981.

30. Matsushima, M. and Nakayama, M. *Jpn. Kokai*, S55, 1980.
31. Ohmura, H., et al., *Kobunshi Ronbunshu*, 45, 857, 1988.
32. Yamamoto, T., et al., *Polymer*, 32, 19, 1991.
33. Tsvetkov, M.S. and Markovskaya, M., *Izv Vyssh Uchebn Zaved Ser Khim Khim Tekhnol*, 11, 936, 1968.
34. Serhatli, I.E., et al., *Polym. Bull.*, 44, 261, 2000.
35. Suzuki, N., et al., *Kobunshi Ronbunshu*, 44, 89, 1987.
36. Suzuki, N., et al., *Kobunshi Ronbunshu*, 44, 81, 1987.
37. Woodward, A.E. and Smets, G. *J. Polym. Sci., Part A: Polym. Chem.*, 17, 51, 1955.
38. Smets, G., et al., *J. Polym. Sci., Part A: Poly. Chem.*, 55, 767, 1961.
39. Kawai, W., *J. Polym. Sci., Part A: Polym. Chem.*, 15, 1479, 1977.
40. Sugimura, T., et al., *Kogyo Kagaku Zasshi* 69, 718, 1966.
41. Fukushi, K. *Jpn. Kokai*, S60, 1985.
42. Fukushi, K., et al., *Kobunshi Ronbunshu*, 44, 97, 1987.
43. Fukushi, K. and Hatachi, A. *Jpn. Kokai*, S58, 1983.
44. Ohmura, H. *Jpn. Kokai*, S58, 1983.
45. Saigou, T. and Nakayama, M. *Jpn. Kokai*, S63, 1988.
46. Yilgor, I. and Baysal, B.M. *Makromol. Chem. Macromol. Chem. Phys.*, 186, 463, 1985.
47. Fukukawa, K., et al., *Macromolecules*, 38, 263, 2005.
48. Naito, I., et al., *J. Polym. Sci., Part A: Polym. Chem.*, 24, 875, 1986.
49. Fuyuki, T., et al., *Jpn. Kokai*, S61, 1986.
50. Cowie, J.M.G. and Yazdanipedram, M. *Br. Polym. J.*, 16, 127, 1984.
51. Ohta, T. and Sano, S. *Jpn. Kokai*, Hl, 1989.
52. Ohta, T. and Sano, S. *Jpn. Kokai*, Hl, 1989.
53. Fang, T.R., et al., *Polym. Bull.*, 22, 317, 1989.
54. Anand, V., et al., *Polym Int*, 54, 823, 2005.
55. Oppenheimer, C. and Heitz, W., *Angew. Makromol. Chem.*, 98, 167, 1981.
56. Hazer, B., *Angew. Makromol. Chem.*, 129, 31, 1985.
57. Ohta, T., *Jpn. Kokai*, Hl, 1989.
58. Qiu, X.Y., et al., *Angew. Makromol. Chem.*, 125, 69, 1984.
59. Dicke, H.R. and Heitz, W., *Makromol. Chem. Rapid Commun.*, 2, 83, 1981.
60. Bamford, C.H. and Mullik, S.U., *Polymer*, 19, 948, 1978.
61. Craubner, H., *J. Polym. Sci., Part A: Polym. Chem.*, 18, 2011, 1980.
62. Yuruk, H. and Ulupinar, S., *Angew. Makromol. Chem.*, 213, 197, 1993.
63. Denizligil, S., et al., *J. Macromol. Sci. — Pure Appl. Chem.*, 29, 293, 1992.
64. Denizligil, S. and Yagci, Y., *Polym. Bull.*, 22, 547, 1989.
65. Hirano, T., et al., *Polym. J.*, 31, 864, 1999.
66. Inceoglu, S., et al., *Designed Monomers Polymers*, 7, 203, 2004.
67. Schubert, U.S., et al., *Polym. Bull.*, 43, 319, 1999.
68. Onen, A., et al., *Macromolecules*, 28, 5375, 1995.
69. Kumada, H., et al., *Jpn. Kokai*, H2, 1990.
70. Iwamura, G., et al., *Jpn. Kokai*, Hl, 1989.
71. Iwamura, G., et al., *Jpn. Kokai*, Hl, 1989.
72. Iwamura, G., et al., *Jpn. Kokai*, Hl, 1989.
73. Simionescu, C.I., et al., *Eur. Polym. J.*, 23, 921, 1987.
74. Takahashi, H., et al., *J. Polym. Sci., Part A: Polym. Chem.*, 35, 69, 1997.
75. Terada, H., et al., *J. Macromol. Sci. — Pure Appl. Chem.*, A31, 173, 1994.
76. Yagci, Y. and Schnabel, W., *Prog. Polym. Sci.*, 15, 551, 1990.
77. Onen, A. and Yagci, Y., *Angew. Makromol. Chem.*, 181, 191, 1990.
78. Onen, A. and Yagci, Y., *J. Macromol. Sci. – Chem.*, A27, 743, 1990.
79. Hepuzer, Y., et al., *J. Macromol. Sci. — Pure Appl. Chem.*, A30, 111, 1993.
80. Ikeda, I., et al., *Sen-I Gakkaishi*, 53, 111, 1997.

81. Haneda, Y., et al., *J. Polym. Sci., Part A: Polym. Chem.*, 32, 2641, 1994.
82. Eroglu, M.S., et al., *J. Appl. Polym. Sci.*, 68, 1149, 1998.
83. Chang, T.C., et al., *J. Polym. Sci., Part A: Polym. Chem.*, 34, 2613, 1996.
84. Simon, J. and Bajpai, A., *J. Appl. Polym. Sci.*, 82, 2922, 2001.
85. Cakmak, I., et al., *Eur. Polym. J.*, 27, 101, 1991.
86. Cheikhalard, T., et al., *J. Appl. Polym. Sci.*, 70, 613, 1998.
87. Iwamura, G., et al., *Jpn. Kokai*, H2, 1989.
88. Kumada, H., et al., *Jpn. Kokai*, Hl, 1989.
89. Inoue, H. and Kohama, S., *J. Appl. Polym. Sci.*, 29, 877, 1984.
90. Murakami, H., et al., *Jpn. Kokai*, H2, 1990.
91. Ohta, T. and Sano, S. *Jpn. Kokai*, Hl, 1989.
92. Takahashi, H. and Kawasaki, K., *Jpn. Kokai*, S59, 1984.
93. Ohta, T., et al., *Jpn. Kokai*, S64, 1989.
94. Ohta, T., et al., *Jpn. Kokai*, Hl, 1989.
95. Scwarc, M., et al., *J. Am. Chem. Soc.*, 78, 2656, 1956.
96. Otsu, T., et al., *Makromol. Chem.-Rapid*, 3, 133, 1982.
97. Solomon, D.H., et al., in European Patent Application 135280 1985.
98. Georges, M.K., et al., *Macromolecules*, 26, 2987, 1993.
99. Benoit, D., et al., *J. Am. Chem. Soc.*, 122, 5929, 2000.
100. Diaz, T., et al., *Macromolecules*, 36, 2235, 2003.
101. Bertin, D. and Boutevin, B., *Polym. Bull.*, 37, 337, 1996.
102. Lacroix-Desmazes, P., et al., *J. Polym. Sci., Part A: Polym. Chem.*, 38, 3845, 2000.
103. Yoshida, E. and Fujii, T., *J. Polym. Sci., Part A: Polym. Chem.*, 35, 2371, 1997.
104. Baethge, H., et al., *Angew. Makromol. Chem.*, 267, 52, 1999.
105. Baethge, H., et al., *Macromol. Rapid Commun.*, 18, 911, 1997.
106. Yamada, K., et al., *Macromolecules*, 32, 290, 1999.
107. Bohrisch, J., et al., *Macromol. Rapid Commun.*, 18, 975, 1997.
108. Barbosa, C.A. and Gomes, A.S., *Polym. Bull.*, 41, 15, 1998.
109. Benoit, D., et al., *Macromolecules*, 33, 1505, 2000.
110. Yoshida, E., *J. Polym. Sci., Part A: Polym. Chem.*, 34, 2937, 1996.
111. Jousset, S., et al., *Macromolecules*, 30, 6685, 1997.
112. Mariani, M., et al., *J. Polym. Sci., Part A: Polym. Chem.*, 37, 1237, 1999.
113. Wang, Y.B. and Huang, J.L., *Macromolecules*, 31, 4057, 1998.
114. Baumert, M. and Mulhaupt, R., *Macromol. Rapid Commun.*, 18, 787, 1997.
115. Zaremski, M.Y., et al., *Macromolecules*, 32, 6359, 1999.
116. Stehling, U.M., et al., *Macromolecules*, 31, 4396, 1998.
117. Georges, M.K., et al., *Macromolecules*, 31, 9087, 1998.
118. Kobatake, S., et al., *Macromolecules*, 31, 3735, 1998.
119. Sessions, L.B., et al., *Macromolecules*, 38, 2116, 2005.
120. Yin, M., et al., *J. Polym. Sci., Part A: Polym. Chem.*, 43, 1873, 2005.
121. Lacroix-Desmazes, P., et al., *J. Polym. Sci., Part A: Polym. Chem.*, 42, 3537, 2004.
122. Okamura, H., et al., *Polymer*, 43, 3155, 2002.
123. Tortosa, K., et al., *Macromol. Rapid Commun.*, 22, 957, 2001.
124. Bignozzi, M.C., et al., *Macromol. Rapid Commun.*, 20, 622, 1999.
125. Gabaston, L.I., et al., *Polymer*, 40, 4505, 1999.
126. Nowakowska, M., et al., *Macromolecules*, 33, 7345, 2000.
127. Nowakowska, M., et al., *Macromolecules*, 36, 4134, 2003.
128. Nowakowska, M., et al., *Polymer*, 42, 1817, 2001.
129. Listigovers, N.A., et al., *Macromolecules*, 29, 8992, 1996.
130. Bian, K. and Cunningham, M.F., *Macromolecules*, 38, 695, 2005.
131. Farcet, C., et al., *Macromolecules*, 34, 3823, 2001.
132. Benoit, D., et al., *Macromolecules*, 33, 363, 2000.

133. Stalmach, U., et al., *J. Am. Chem. Soc.*, 122, 5464, 2000.
134. Bian, K.J. and Cunningham, M.F., *J. Polym. Sci., Part A: Polym. Chem.*, 44, 414, 2006.
135. Datsyuk, V., et al., *Carbon*, 43, 873, 2005.
136. Schierholz, K., et al., *Macromolecules*, 36, 5995, 2003.
137. Kazmaier, P.M., et al., *Macromolecules*, 30, 2228, 1997.
138. Wang, J.S. and Matyjaszewski, K., *J. Am. Chem. Soc.*, 117, 5614, 1995.
139. Kato, M., et al., *Macromolecules*, 28, 1721, 1995.
140. Moineau, G., et al., *Macromol. Chem. Physic*, 201, 1108, 2000.
141. Ando, T., et al., *Macromolecules*, 30, 4507, 1997.
142. Kotani, Y., et al., *Macromolecules*, 32, 6877, 1999.
143. Matyjaszewski, K., et al., *Macromolecules*, 30, 8161, 1997.
144. Percec, V., et al., *Macromolecules*, 29, 3665, 1996.
145. Kotani, Y., et al., *Macromolecules*, 32, 2420, 1999.
146. Lecomte, P., et al., *Macromolecules*, 30, 7631, 1997.
147. Wang, J.S. and Matyjaszewski, K., *Macromolecules*, 28, 7572, 1995.
148. Percec, V., et al., *Macromolecules*, 30, 8526, 1997.
149. Destarac, M., et al., *Macromol. Chem. Phys.*, 201, 265, 2000.
150. Pyun, J., et al., *Polymer*, 44, 2739, 2003.
151. Zhao, Y.F., et al., *Polymer*, 46, 5396, 2005.
152. Kurjata, J., et al., *Polymer*, 45, 6111, 2004.
153. Muftuoglu, A.E., et al., *J. Appl. Polym. Sci.*, 93, 387, 2004.
154. Bernaerts, K.V., et al., *J. Polym. Sci., Part A: Polym. Chem.*, 41, 3206, 2003.
155. Matyjaszewski, K., et al., *Macromolecules*, 31, 1527, 1998.
156. Matyjaszewski, K., et al., *Macromolecules*, 31, 6836, 1998.
157. Matyjaszewski, K., et al., *J. Polym. Sci., Part A: Polym. Chem.*, 38, 2440, 2000.
158. Tong, J.D., et al., *Macromolecules*, 33, 470, 2000.
159. Moineau, C., et al., *Macromolecules*, 32, 8277, 1999.
160. Davis, K.A. and Matyjaszewski, K., *Macromolecules*, 34, 2101, 2001.
161. Fernandez-Garcia, M., et al., *J. Appl. Polym. Sci.*, 84, 2683, 2002.
162. Granel, C., et al., *Macromolecules*, 29, 8576, 1996.
163. Zou, Y.S., et al., *Acta Polym. Sin.*, 146, 1999.
164. Zou, Y.S., et al., *Chem. J. Chinese U.*, 19, 1700, 1998.
165. Demirelli, K., et al., *Polym.–Plast. Technol. Eng.*, 43, 1245, 2004.
166. Kotani, Y., et al., *Macromolecules*, 29, 6979, 1996.
167. Uegaki, H., et al., *Macromolecules*, 31, 6756, 1998.
168. Shipp, D.A., et al., 31, 8005, 1998.
169. Garcia, M.F., et al., *Polymer*, 42, 9405, 2001.
170. Wang, X.S., et al., *Polymer*, 40, 4157, 1999.
171. Qin, D.Q., et al., *Macromolecules*, 33, 6987, 2000.
172. Zhang, X. and Matyjaszewski, K. *Macromolecules*, 32, 1763, 1999.
173. Beers, K.L., et al., *Macromolecules*, 32, 5772, 1999.
174. Xia, J.H., et al., *Macromolecules*, 32, 3531, 1999.
175. Haddleton, D.M., et al., *J. Am. Chem. Soc.*, 122, 1542, 2000.
176. Alipour, M., et al., *Iran Polym. J.*, 10, 99, 2001.
177. Yuan, J., et al., *Polym. Int.*, 55, 360, 2006.
178. Chatterjee, U., et al., *Polymer*, 46, 1575, 2005.
179. Wang, X.Z., et al., *J. Polym. Sci., Part A: Polym. Chem.*, 43, 733, 2005.
180. Teodorescu, M. and Matyjaszewski, K., *Macromol. Rapid Commun.*, 21, 190, 2000.
181. Yuan, J.Y. and Pan, C.Y., *Chin. J. Polym Sci.*, 20, 171, 2002.
182. Matyjaszewski, K., et al., *J. Polym. Sci., Part A: Polym. Chem.*, 38, 2023, 2000.
183. Borner, H.G., et al., *Macromolecules*, 34, 4375, 2001.

184. Matyjaszewski, K., et al., *Macromolecules*, 33, 2296, 2000.
185. Muhlebach, A., et al., *Macromolecules*, 31, 6046, 1998.
186. Biedron, T. and Kubisa, P., *J. Polym. Sci., Part A: Polym. Chem.*, 40, 2799, 2002.
187. Pyun, J. and Matyjaszewski, K., *Macromolecules*, 33, 217, 2000.
188. Liu, Y., et al., *Macromolecules*, 32, 8301, 1999.
189. Sedjo, R.A., et al., *Macromolecules*, 33, 1492, 2000.
190. Keary, C., et al., *Polym. Int.*, 51, 647, 2002.
191. Cassebras, M., et al., *Macromol. Rapid Commun.*, 20, 261, 1999.
192. Park, K.R., et al., *Polym.–Korea*, 27, 17, 2003.
193. Davis, K.A. and Matyjaszewski, K., *Macromolecules*, 33, 4039, 2000.
194. Davis, K.A., et al., *J. Polym. Sci., Part A: Polym. Chem.*, 38, 2274, 2000.
195. Burguiere, C., et al., *Macromolecules*, 34, 4439, 2001.
196. Abraham, S., et al., *J. Polym. Sci., Part A: Polym. Chem.*, 43, 6367, 2005.
197. Zhao, Y.L., et al., *Polymer*, 46, 5808, 2005.
198. Wang, G.J. and Yan, D.Y., *J. Appl. Polym. Sci.*, 82, 2381, 2001.
199. Krishnan, R. and Srinivasan, K.S.V., *Eur. Polym. J.*, 40, 2269, 2004.
200. Ohno, K., et al., *J. Polym. Sci., Part A: Polym. Chem.*, 36, 2473, 1998.
201. Li, Z.C., et al., *Macromol. Rapid Commun.*, 21, 375, 2000.
202. Guice, K.B. and Loo, Y.L., *Macromolecules*, 39, 2474, 2006.
203. Huang, C.F., et al., *J. Polym. Res.*, 12, 449, 2005.
204. Hua, M., et al., *Acta Polym. Sin.*, 645, 2004.
205. Hikita, M., et al., *Langmuir*, 20, 5304, 2004.
206. Feng, X.S., et al., *Macromol. Chem. Physic*, 202, 3403, 2001.
207. Ma, Q.G. and Wooley, K.L., *J. Polym. Sci., Part A: Polym. Chem.*, 38, 4805, 2000.
208. Han, Y.K., et al., *Macromolecules*, 37, 9355, 2004.
209. Furlong, S.A. and Armes, S.P., *Abstr. Pap. Am. Chem. Soc.*, 219, U437, 2000.
210. Chang, S.M., et al., *Synth. Met.*, 154, 21, 2005.
211. Xia, J.H., et al., *Macromolecules*, 32, 4802, 1999.
212. Gravano, S.M., et al., *Langmuir*, 18, 1938, 2002.
213. Okubo, M., et al., *Colloid Polym. Sci.*, 282, 747, 2004.
214. Kagawa, Y., et al., *Polymer*, 46, 1045, 2005.
215. Yang, R.M., et al., *Eur. Polym. J.*, 39, 2029, 2003.
216. Tian, Y.Q., et al., *J. Polym. Sci., Part A: Polym. Chem.*, 41, 2197, 2003.
217. Richard, R.E., et al., *Biomacromolecules*, 6, 3410, 2005.
218. Chatterjee, U., et al., *Polymer*, 46, 10699, 2005.
219. Masci, G., et al., *Macromolecules*, 37, 4464, 2004.
220. Raghunadh, V., et al., *Polymer*, 45, 3149, 2004.
221. Lazzari, M., et al., *Macromol. Chem. Physic*, 206, 1382, 2005.
222. Tsarevsky, N.V., et al., *Macromolecules*, 35, 6142, 2002.
223. Sun, X.Y., et al., *Polymer*, 46, 5251, 2005.
224. Zhang, H.L., et al., *Macromol. Rapid Commun.*, 26, 407, 2005.
225. Huang, J.Y. and Matyjaszewski, K., *Macromolecules*, 38, 3577, 2005.
226. Jankova, K. and Hvilsted, S., *J. Fluorine Chem.*, 126, 241, 2005.
227. Fu, G.D., et al., *Adv. Funct Mater.*, 15, 315, 2005.
228. Licciardi, M., et al., *Biomacromolecules*, 6, 1085, 2005.
229. Ma, Y.H., et al., *Macromolecules*, 36, 3475, 2003.
230. Cai, Y.L. and Armes, S.P., *Macromolecules*, 38, 271, 2005.
231. Radhakrishnan, K., et al., *J. Polym. Sci., Part A: Polym. Chem.*, 42, 853, 2004.
232. Gan, L.H., et al., *J. Polym. Sci., Part A: Polym. Chem.*, 41, 2688, 2003.
233. Trapa, P.E., et al., *Electrochem. Solid-State Lett.*, 5, A85, 2002.
234. Qin, S.H., et al., *Macromolecules*, 36, 8969, 2003.
235. Tang, C.B., et al., *Macromolecules*, 36, 8587, 2003.

236. Convertine, A.J., et al., *Macromolecules*, 36, 4679, 2003.
237. Shi, L.J., et al., *Macromolecules*, 36, 2563, 2003.
238. Rzayev, J. and Hillmyer, M.A., *Macromolecules*, 38, 3, 2005.
239. Hong, C.Y., et al., *J. Polym. Sci., Part A: Polym. Chem.*, 42, 4873, 2004.
240. Chen, M., et al., *J. Chin. Chem. Soc.*, 53, 79, 2006.
241. Bernard, J., et al., *Biomacromolecules*, 7, 232, 2006.
242. Beattie, D., et al., *Biomacromolecules*, 7, 1072, 2006.
243. Save, M., et al., *Macromolecules*, 38, 280, 2005.
244. Mori, H., et al., *Macromolecules*, 38, 8192, 2005.
245. Kubo, K., et al., *Polymer*, 46, 9762, 2005.
246. Smulders, W.W., et al., *Macromolecules*, 37, 9345, 2004.
247. Smulders, W. and Monteiro, M.J., *Macromolecules*, 37, 4474, 2004.
248. Hao, X.J., et al., *Aust. J. Chem.*, 58, 483, 2005.
249. De Brouwer, H., et al., *J. Polym. Sci., Part A: Polym. Chem.*, 38, 3596, 2000.
250. Hong, C.Y., et al., *J. Polym. Sci., Part A: Polym. Chem.*, 43, 6379, 2005.
251. Xin, X.Q., et al., *Eur. Polym. J.*, 41, 1539, 2005.
252. Mertoglu, M., et al., *Polymer*, 46, 7726, 2005.
253. Mellon, W., et al., *Macromolecules*, 38, 1591, 2005.
254. Hoogenboom, R., et al., *Macromolecules*, 38, 7653, 2005.
255. Fiten, M.W.M., et al., *J. Polym. Sci., Part A: Polym. Chem.*, 43, 3831, 2005.
256. Pan, J.Y., et al., *Chem. J. Chinese U.*, 25, 1759, 2004.
257. Coote, M.L. and Henry, D.J., *Macromolecules*, 38, 5774, 2005.
258. Stenzel, M.H., et al., *Macromol. Biosci.*, 4, 445, 2004.
259. Schilli, C.M., et al., *Macromolecules*, 37, 7861, 2004.
260. Relogio, P., et al., *Polymer*, 45, 8639, 2004.
261. Lokitz, B.S., et al., *Abstr. Pap. Am. Chem. Soc.*, 228, U352, 2004.
262. Albertin, L., et al., *Macromolecules*, 37, 7530, 2004.
263. Krasia, T., et al., *Chem. Commun.*, 538, 2003.
264. Zhuang, R.C., et al., *Acta Polym. Sin.* 288, 2001.
265. Mitsukami, Y., et al., *Macromolecules*, 34, 2248, 2001.
266. Chong, Y.K., et al., *Macromolecules*, 32, 2071, 1999.
267. Tang, C.B., et al., *Macromolecules*, 36, 1465, 2003.
268. Miura, Y., et al., *J. Polym. Sci., Part A: Polym. Chem.*, 43, 4271, 2005.
269. Celik, C., et al., *J. Polym. Sci., Part A: Polym. Chem.*, 41, 2542, 2003.
270. Durmaz, H., et al., *Designed Monomers Polym.*, 8, 203, 2005.
271. Debuigne, A., et al., *Macromolecules*, 38, 9488, 2005.
272. Kaneyoshi, H., et al., *Macromolecules*, 38, 5425, 2005.
273. Bilalis, P., et al., *J. Polym. Sci., Part A: Polym. Chem.*, 44, 659, 2006.
274. Degirmenci, M., et al., *J. Polym. Sci., Part A: Polym. Chem.*, 42, 534, 2004.
275. Erel, I., et al., *Eur. Polym. J.*, 38, 1409, 2002.
276. Destarac, M., et al., *Macromol. Chem. Phys.*, 201, 1189, 2000.
277. Semsarzadeh, M.A. and Mirzaei, A., *Iran Polym. J.*, 12, 67, 2003.
278. Semsarzadeh, M.A., et al., *Eur. Polym. J.*, 39, 2193, 2003.
279. Kaetsu, I., *Radiat. Phys. Chem.*, 25, 517, 1985.
280. Hoffman, A.S., et al., *Radiat. Phys. Chem.*, 27, 265, 1986.
281. Dyer, D.J. *Adv. Polym. Sci.*, 197, 47–65, 2006.
282. Uyama, Y., et al., *Grafting/Characterization Tech./Kinetic Modeling*, 137, 1, 1998.
283. Alexandru, L. and Guillet, J.E., *J. Polym. Sci., Part A: Polym. Chem.*, 13, 483, 1975.
284. Kinstle, J.F. and Watson, S.L., *Abstr. Pap. Am. Chem. Soc.*, 170, 1, 1975.
285. Jackson, M.B. and Sasse, W.H.F., *J. Macromol. Sci. Chem.*, A11, 1137, 1977.
286. Ahn, K.D., et al., *J. Macromol. Sci. Chem.*, A23, 355, 1986.

287. Mukherjee, A.K., et al., *J. Macromol. Sci. Rev. Macromol. Chem.*, C26, 415, 1986.
288. Miyama, H. and Sato, T., *J. Polym. Sci., Part A: Polym. Chem.* 10, 2469, 1972.
289. Okawara, M., et al., *Kogyo Kagaku Zasshi*, 69, 461, 1966.
290. Okawara, M., et al., *Kogyo Kagaku Zasshi*, 66, 1383, 1963.
291. Miyama, H., et al., *J. Biomed. Mater. Res.*, 11, 251, 1977.
292. Kondo, S., et al., *J. Macromol. Sci. Chem.*, A 11, 719, 1977.
293. Ashworth, J., et al., *Polymer*, 13, 57, 1972.
294. Bamford, C.H., et al., *Polymer*, 12, 247, 1971.
295. Bamford, C.H., et al., *Polymer*, 10, 771, 1969.
296. Bamford, C.H., et al., *Polymer*, 27, 1981, 1986.
297. Tazuke, S., et al., *Abstr. Pap. Am. Chem. Soc.*, 62, 1979.
298. Fouassier, J.P., *ACS Sym. Ser.*, 187, 83, 1982.
299. Arthur, J.C., *Developments in Polymer Photochemistry, 2* (N.C. Allen, London), 39, 1981.
300. Arthur, J.C., *Developments in Polymer Photochemistry, 1* (N.C. Allen, London), 69, 1980.
301. Mukherjee, A.K. and Goel, H.R., *J. Macromol. Sci. Rev. Macromol. Chem.*, C25, 99, 1985.
302. Lenka, S., *J. Macromol. Sci. Rev. Macromol. Chem.*, C22, 303, 1982.
303. Samal, R.K., et al., *J. Macromol. Sci. Rev. Macromol. Chem.*, C26, 81, 1986.
304. Mishra, M.K., *J. Macromol. Sci. Rev. Macromol. Chem.*, C22, 409, 1982.
305. Yagci, Y. and Mishra, M.K., *Macromolecular Design: Concept and Practice*, M. K. Mishra, Ed., Polymer Frontiers International, New York, Ch. 6, 229, 1994.
306. Angiolini, L., et al., *Polymer*, 40, 7197, 1999.
307. Muftuoglu, A.E., et al., *Turk. J. Chem.*, 28, 469, 2004.
308. Mino, G. and Kaizerman, S., *J. Polym. Sci., Part A: Polym. Chem.*, 31, 242, 1958.
309. Rempp, P.F. and Franta, E., *Adv. Polym. Sci.*, 58, 1, 1984.
310. Ito, K. and Kawaguchi, S., *Branched Polymers I*, 142, 129, 1999.
311. Hawker, C.J., *Angew. Chem. Int. Ed.*, 34, 1456, 1995.
312. Appelt, M. and Schmidt-Naake, G., *Macromol. Mater. Eng.*, 289, 245, 2004.
313. Sun, Y., et al., *J. Polym. Sci., Part A: Polym. Chem.*, 39, 604, 2001.
314. Coskun, M. and Temuz, M.M., *J. Polym. Sci., Part A: Polym. Chem.*, 41, 668, 2003.
315. Temuz, M.M. and Coskun, M., *J. Polym. Sci., Part A: Polym. Chem.*, 43, 3771, 2005.
316. Guerrini, M.M., et al., *Macromol. Rapid Commun.*, 21, 669, 2000.
317. Fu, G.D., et al., *Macromolecules*, 38, 2612, 2005.
318. Miwa, Y., et al., *Macromolecules*, 32, 8234, 1999.
319. Miwa, Y., et al., *Macromolecules*, 34, 2089, 2001.
320. Fonagy, T., et al., *Macromol. Rapid Commun.*, 19, 479, 1998.
321. Fonagy, T. and Ivan, B., *Abstr. Pap. Am. Chem. Soc.*, 216, U811, 1998.
322. Liu, S.S. and Sen, A., *Macromolecules*, 34, 1529, 2001.
323. Grubbs, R.B., et al., *Angew. Chem. Int. Ed.*, 36, 270, 1997.
324. Liu, B., et al., *Macromol. Chem. Phys.*, 202, 2504, 2001.
325. Liu, B., et al., *Macromol. Chem. Phys.*, 22, 373, 1999.
326. Stoeckel, N., et al., *Polym. Bull.*, 49, 243, 2002.
327. Beers, K.L., et al., *Macromolecules*, 31, 9413, 1998.
328. Borner, H.G. and Matyjaszewski, K., *Macromol. Symp.*, 177, 1, 2002.
329. Shinoda, H., et al., *Macromolecules*, 34, 3186, 2001.
330. Shinoda, H., et al., *Macromolecules*, 36, 4772, 2003.
331. Shinoda, H. and Matyjaszewski, K., *Macromolecules*, 34, 6243, 2001.
332. Wang, X.S. and Armes, S.P., *Macromolecules*, 33, 6640, 2000.
333. Wang, X.S., et al., *Chem. Commun.*, 1817, 1999.

334. Neugebauer, D., et al., *Macromolecules*, 39, 584, 2006.
335. Neugebauer, D., et al., *Polymer*, 44, 6863, 2003.
336. Deng, G.H. and Chen, Y.M., *J. Polym. Sci., Part A: Polym. Chem.*, 42, 3887, 2004.
337. Guo, Y.M., et al., *Polymer*, 42, 6385, 2001.
338. Hong, S.C., et al., *J. Polym. Sci., Part A: Polym. Chem.*, 40, 2736, 2002.
339. O'Donnell, P.M. and Wagener, K.B., *J. Polym. Sci., Part A: Polym. Chem.*, 41, 2816, 2003.
340. Matyjaszewski, K., et al., *J. Polym. Sci., Part A: Polym. Chem.*, 36, 823, 1998.
341. Roos, S.G., et al., *Macromolecules*, 32, 8331, 1999.
342. Li, Y.G., et al., *Polym Int*, 53, 349, 2004.
343. Shinoda, H. and Matyjaszewski, K., *Macromol. Rapid Commun.*, 22, 1176, 2001.
344. Hawker, C.J., et al., *Macromol. Chem. Phys.*, 198, 155, 1997.
345. Wohlrab, S. and Kuckling, D., *J. Polym. Sci., Part A: Polym. Chem.*, 39, 3797, 2001.
346. Gacal, B., et al., *Macromolecules*, 39, 5330, 2006.

Part III

Technical Processes of Vinyl Polymerization

12 Continuous Processes for Radical Vinyl Polymerization

Kyu Yong Choi

CONTENTS

12.1 INTRODUCTION

Free-radical polymerization of vinyl monomers takes place through intermediates having an unpaired electron known as free radicals. Many vinyl monomers are readily polymerized by free-radical mechanisms because free-radical polymerization is relatively less sensitive to impurities compared with ionic polymerizations. Free radicals can be generated in a number of ways, including organic or inorganic initiators and even without added initiators (e.g., thermal and photoinitiation). Over 50 different organic peroxides and azo initiators in over 100 different formulations are produced commercially. Initiators are selected based on several factors: polymerization rate, reaction temperature, solubility, and polymer properties.

A typical free-radical homopolymerization process consists of the following reactions:

Initiation

$$I \xrightarrow{k_d} 2R$$

$$R + M \xrightarrow{k_1} P_1$$

Propagation

$$P_1 + M \xrightarrow{k_p} P_2$$

$$P_n + M \xrightarrow{k_p} P_{n+1}$$

Chain termination

$$P_n + P_m \xrightarrow{k_{tc}} M_{n+m}$$

$$P_n + P_m \xrightarrow{k_{td}} M_n + M_m$$

Chain transfer

$$P_n + M \xrightarrow{k_{fm}} M_n + P_1$$

$$P_n + X \xrightarrow{k_{fx}} M_n + X\cdot$$

In the preceding, M is the monomer, I is the initiator, R is the primary radical, P_n is the live polymer radical with n monomer repeat units, M_n is the dead polymer with the n monomer repeat units, and X is the solvent, impurity, or chain transfer agent. At high monomer conversion, the polymer's mobility decreases and termination reactions become diffusion controlled ("gel effect"). As a result, the polymerization rate increases rapidly and the polymer's molecular-weight distribution becomes broad.

Commercial free-radical polymerization processes are subdivided into bulk (mass), solution, suspension, emulsion, dispersion, and precipitation polymerization. In bulk polymerization, where no solvent is present, polymers may be soluble in their own monomers (e.g., polystyrene, poly(methyl methacrylate), poly(vinyl acetate)) or insoluble in their monomers (e.g., poly(vinyl chloride), polyacrylonitrile). Although pure polymers are obtainable, high viscosity with an increasing monomer conversion limits the maximum solid content in the reactor. In solution polymerization, solvent miscible with a monomer dissolves the polymer (e.g., styrene in ethyl benzene), the viscosity of a polymerizing solution is relatively low, and the polymerization takes place homogeneously. In suspension polymerization, the organic monomer phase is dispersed as small droplets by mechanical agitation and polymerized to hard solid polymer particles with monomer-soluble initiators. Each monomer droplet acts like a single microbatch polymerization reactor. The polymer particle size or its distribution is governed by mechanical agitation and surface stabilizer. Emulsion polymerization differs from suspension polymerization in two important respects: (1) the initiator (water soluble) is located in the aqueous phase, and (2) the polymer particles produced are typically of the order of 0.1μ in diameter, which is about 10 times smaller than the smallest encountered in suspension or dispersion polymerization. The use of water in both suspension and emulsion polymerization reactors facilitates the removal of polymerization heat.

A variety of polymerization reactors are used in industrial polymerization processes. They are, for example, continuous reactors, semibatch reactors, and batch reactors. The choice of reactor type or configuration for a given polymerization reaction depends on many factors, such as polymerization mechanism, thermodynamic properties of monomers and polymerizing fluid, production rate, reaction conditions (e.g., temperature, pressure, viscosity, phases, etc.), heat removal capacity, product properties, investment and operating cost, operability, and controllability. Whereas batch reactors are useful for small-to-intermediate volume polymers or specialty polymers, continuous reactors are suitable for a large-scale production of commodity vinyl polymers such as polystyrene, poly(methyl methacrylate), poly(vinyl acetate), and polyethylene. Compared with batch reactors, in which batch-to-batch variations in product quality can be a problem, continuous reactors have advantages in that polymer quality control through process automation can be achieved more effectively. For example, in a batch copolymerization reactor, a composition drift occurs because monomers have different reactivities. In a continuous reactor operating at steady state, all polymer molecules are made under the same reaction condition, and, thus, compositional heterogeneity can be prevented.

In continuous industrial free-radical polymerization processes, many different types of reactors are used [1]. They are continuous-flow stirred tank reactors, tower reactors, horizontal linear flow reactors, tubular reactors, and screw reactors. In some processes, different types of reactors are used together in a reactor train. In stirred tank reactors, no spatial concentration and temperature gradients exist, whereas in linear flow or tubular reactors, concentration and temperature vary in the direction of flow of the reacting fluid. Specially designed reactors such as screw reactors or extruder reactors are also used to produce specialty vinyl polymers. In this chapter, some important characteristics of continuous reactors used in industrial free-radical polymerization processes are discussed.

12.2 CONTINUOUS POLYMERIZATION REACTORS FOR FREE-RADICAL POLYMERIZATION OF VINYL MONOMERS

12.2.1 Continuous Stirred Tank Reactors

Continuous stirred tank reactors (CSTRs) are perhaps the most widely used in industrial continuous free-radical polymerization processes. Monomers, solvents, initiators, and additives (e.g., chain transfer agents) are continuously fed to a mechanically agitated reactor and the product solution is removed continuously from the reactor. In a CSTR, the reaction mixture is backmixed by a mechanical stirrer and its effluent temperature and composition are the same as the reactor contents if perfect backmixing is achieved in the reactor. The liquid level is normally held constant by controlling the product withdrawal rate. At steady state, the polymerization rate and polymer properties are time-invariant; thus, uniform quality product is obtainable. However, some variations in the reaction conditions should also be expected (e.g., feed purity, feed temperature, cooling water source temperature [seasonal variations], etc.); thus, an efficient closed-loop control system is required to regulate such variations and to keep the reactor at its target operating conditions. In a large, continuous stirred tank reactor, varying degrees of imperfect mixing (e.g., segregation, short-circuiting, and stagnation) or temperature nonuniformity may exist in the reactor. In such a case, some inconsistency in the product properties may result. Because the mixing is provided by mechanical agitation, stirred tank reactors are suitable for relatively low-viscosity fluids. For high-viscosity fluids, specially designed reactors and agitators (e.g., helical ribbons, anchors, scroll agitators) are required for efficient mixing and heat transfer.

In industrial polymerization processes, multiple CSTRs are also commonly used. Each reactor is operated at different reaction conditions to achieve desired final polymer properties. In a typical CSTR polymerization system, the reaction heat is removed through a jacket in which cooling fluid is circulated. To provide an additional heat removal capacity, internal cooling coils, external heat exchanger, and reflux condenser may be installed.

Continuous stirred tank reactors are used commercially for solution, bulk (mass), and emulsion polymerization of vinyl monomers. In bulk homogeneous polymerization processes (e.g., polystyrene), the reactor system usually consists of a single CSTR or multiple CSTRs and an extruder-type devolatilizer to remove unreacted monomer, which is then recycled to the reactor. As monomer conversion increases, the viscosity of the polymerizing fluid increases and the overall heat removal efficiency decreases. When styrene is polymerized in bulk in a stirred tank reactor, monomer conversion is limited to about 30–40% due to an increasing viscosity of the polymerizing fluid above this conversion level. However, the overall monomer conversion is limited to about 30–40% due to an increasing viscosity of the polymerizing fluid above this conversion level. However, the overall monomer conversion can be very high because unreacted monomer is constantly recycled to the reactor.

Figure 12.1 presents some examples of continuous stirred tank reactor systems for free-radical vinyl polymerization processes. In the bulk styrene polymerization process depicted in Figure 12.1(a) [2], styrene monomer, stripped of inhibitor added for transportation, is supplied to a prepolymerization reactor with an organic initiator. The monomer-polymer mixture is then fed to a series of stirred tank reactors operating at higher temperatures than in the prepolymerization reactor. At low temperatures, the molecular weight of the polymer decreases. The unreacted monomer and low-molecular-weight polymers or oligomers are removed in a devolatilizer. The volatile monomer is then distilled off and recycled to the polymerization reactor. The purified polymer is palletized in an extruder. Although only two CSTRs are depicted in Figure 12.1a, more than two CSTRs can be used. Instead of stirred tank reactors, tower reactors can also be combined in a reactor train with CSTRs. When a series of CSTRs or tower reactors are used, the polymerization temperature is progressively raised through the reaction zone to deal with a polymeric fluid of increasing viscosity.

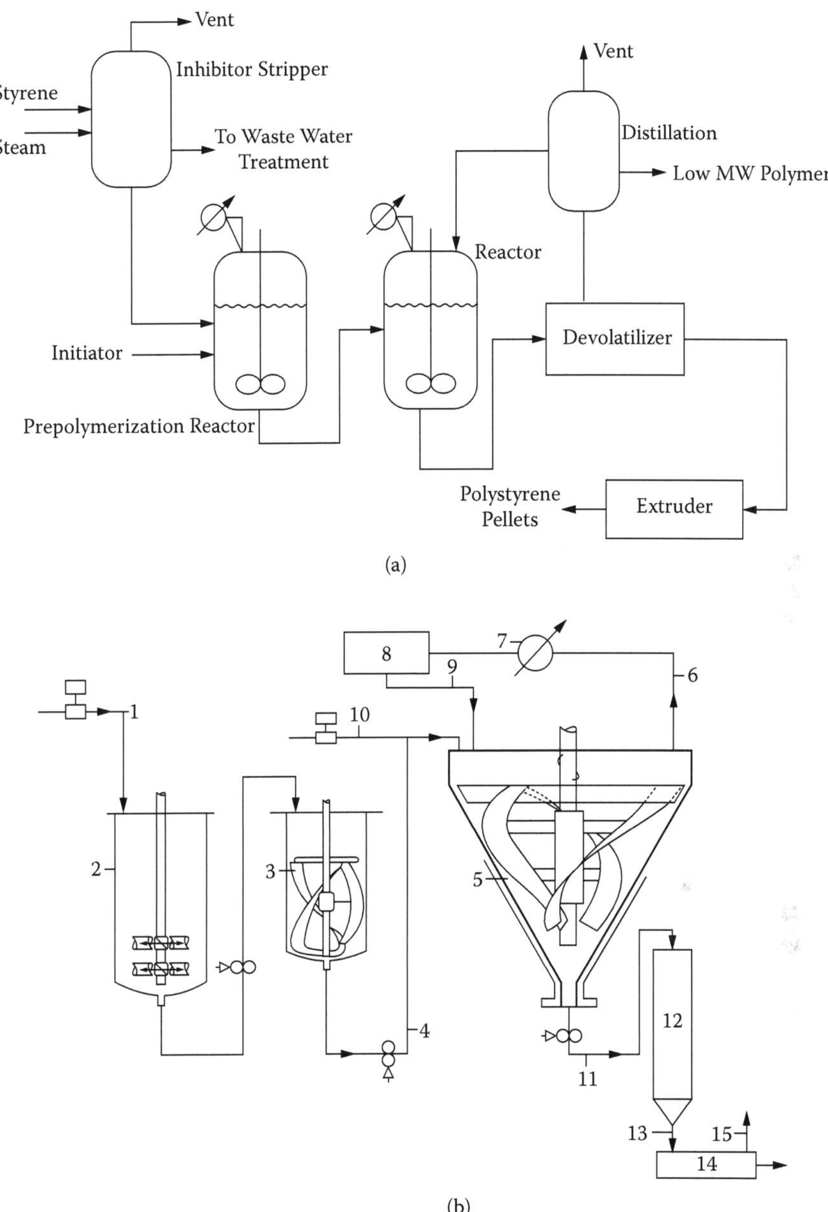

FIGURE 12.1 Examples of continuous stirred tank reactors for free radical polymerization.

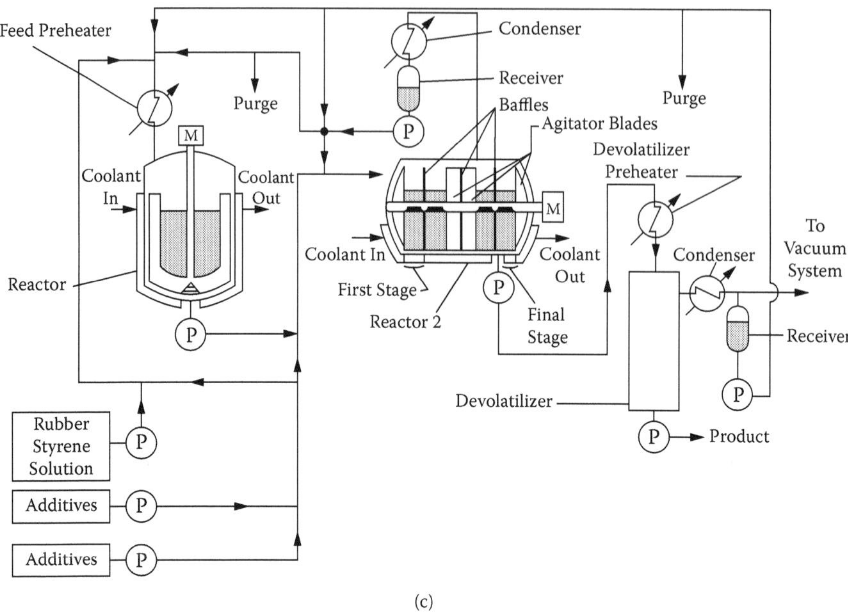

FIGURE 12.1 (Continued).

A cascade of polymerization reactors is used for the production of high-impact polystyrene (HIPS). Impact polystyrene is a polymer toughened by a rubber within the polystyrene matrix. A continuous reactor system for mass polymerization of styrene with polybutadiene rubber is illustrated in Figure 12.1b [1]. In the first reactor agitated by a turbine impeller, the conversion of styrene is maintained at about 8%, which is slightly ahead of the phase-inversion point. The second reactor is fitted with a scroll agitator and conversion is increased to about 23%. The first two reactors are used to avoid problems with rubber-phase particle size and gel formation due to excessive backmixing of the reaction mixture near the phase-inversion point. The polymerization is continued in the third reactor operating at 173°C and 86% conversion with a 2.4-h residence time. This reactor is cooled by a refluxing styrene monomer. Finally, the reaction is completed in an unagitated tower reactor to about 97% conversion at 207°C.

In the polymerization process depicted in Figure 12.1c [3], a fresh feed of 8% polybutadiene rubber in styrene is added with antioxidant and recycled monomer to the first reactor operating at 124°C and about 18% conversion at about 40% fillage. The agitator is a horizontal shaft on which a set of paddles is mounted. Because the temperature in each compartment can be varied, it is claimed that the linear flow behavior provided by the reactor staging results in more favorable rubber-phase morphology than would be the case if the second reactor were operated as a single continuous stirred tank reactor.

In operating a continuous stirred tank reactor, maintaining a desired polymerization temperature is often the most important objective. It is because the polymerization rate and many of the polymer properties are strongly dependent on temperature. For example, polymer molecular weight decreases as the reaction temperature

TABLE 12.1
Jacket Surface Areas of Agitated Polymerization Reactors

Reactor Volume (U.S. gallons)	Cooling Area (ft^2)
2,000	180
3,000	245
3,700	292
6,300	392
16,500	744

Source: N. Platzer, *Ind. Eng. Chem.*, 62(1), 6 (1970).

is increased. If the reaction heat is not properly removed, excessive pressure buildup and/or thermal runaway may occur. In a jacketed reactor, the removal of reaction heat becomes increasingly difficult as the reactor volume increases because of the reduced heat transfer area/reactor volume ratio. Table 12.1 illustrates the jacket surface areas of agitated polymerization reactors. Large polymerization reactors require, in addition to a cooling jacket, the installation of internal cooling baffles or internal cooling coils and, frequently, of reflux condensers.

When continuous stirred tank reactors are used for polymerization, less than 100% conversion is usually obtained. Thus, a large quantity of un-reacted monomers and solvent are separated from the polymerizing mixture, purified, and recycled to the reactor. Continuous reactors are also useful in manufacturing copolymers. Figure 12.2 illustrates a schematic of a continuous copolymerization reactor with recycle [5]. Here, two monomers (methyl methacrylate and vinyl acetate) and an initiator are supplied to the reactor with solvent and chain transfer agent. In this process where both polymer molecular weight and copolymer composition are controlled, the process disturbances (e.g., impurities, compositional variations) caused by the recycle stream must be properly regulated to ensure consistent copolymer quality. To do so, an advanced multivariable control technique can be applied to minimize process interactions and disturbances [5].

12.2.2 Multizone Stirred Reactors

Multizone stirred reactors have been used for the polymerization of ethylene to low-density polyethylene (LDPE) and for the polymerization of styrene and its comonomers. In general, the reactor is a vertical or a horizontal vessel of a large length/diameter ratio. The reactor is divided into several equal or nonequal volume internal compartments, and reactor internals are designed to provide good radial mixing but also some axial segregation. Thus, the flow and mixing characteristics of the multizone reactors are somewhere between those of a single stirred tank reactor and a plug flow reactor (e.g., tubular reactor) and they are sometimes called linear flow reactors (LFRs). The main motivation for using the multizone reactors is that the polymerization conditions (e.g., temperature and composition) can be widely varied along the length of the reactor to obtain desired polymer yield and polymer properties. For example, comonomers or other reactive additives can be

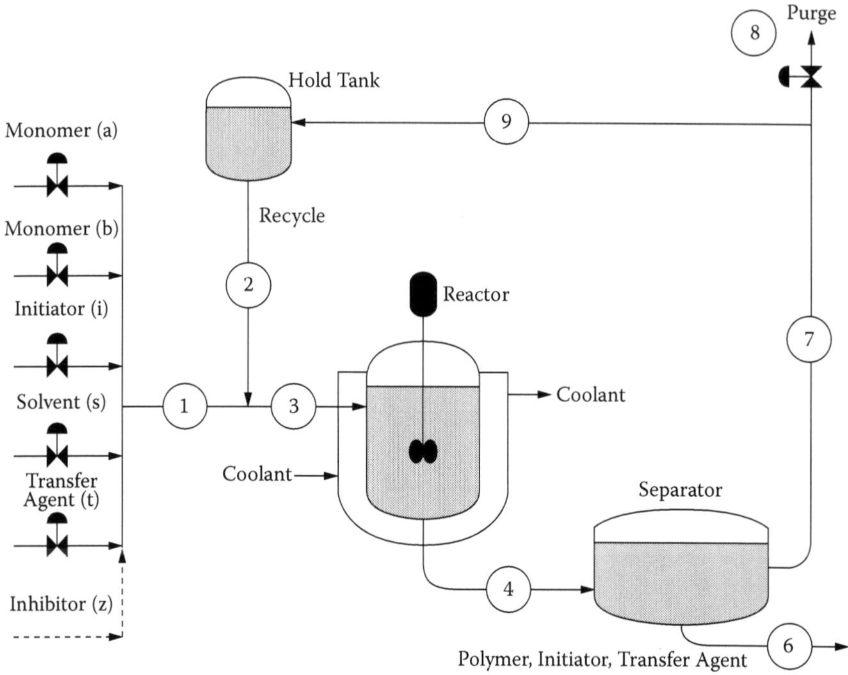

FIGURE 12.2 Continuous copolymerization reactor with recycle.

injected in different concentrations to each compartment. In designing and operating tower reactors or linear flow reactors, it is important to ensure uniform radial mixing in the reactor vessel. If any radial velocity gradients are present, a buildup of a high-viscosity polymer layer may occur, lowering the heat transfer efficiency. Because any changes in the upstream reactor or reaction zones affect the performance of the downstream reactors, it is also important to design reactor control systems that will offset any process upsets. Thus, the design and operation of the reactor becomes more complicated than using a single-zone stirred tank reactor.

12.2.2.1 Polyethylene Reactors

Low-density polyethylene (density = 0.915–0.935 g cm^{-3}) has long been manufactured by free-radical polymerization using continuous autoclave reactors. The autoclave reactor shown schematically in Figure 12.3(a) is a typical multizone ethylene polymerization reactor. The reactor is typically a vertical cylindrical vessel with a large *L/D* ratio. The reacting fluid is intensely mixed by an agitator that consists of a vertical shaft to which impeller blades are attached to ensure efficient mixing of monomers, polymers, and initiators. High-pressure ethylene is supplied to each reaction compartment with a peroxide initiator. Depending on the polymer properties desired, the feed rates of ethylene and initiator to each reaction zone are varied. Because the polymerization occurs at high pressure (1500–2000 atm), the reactor wall thickness is large;

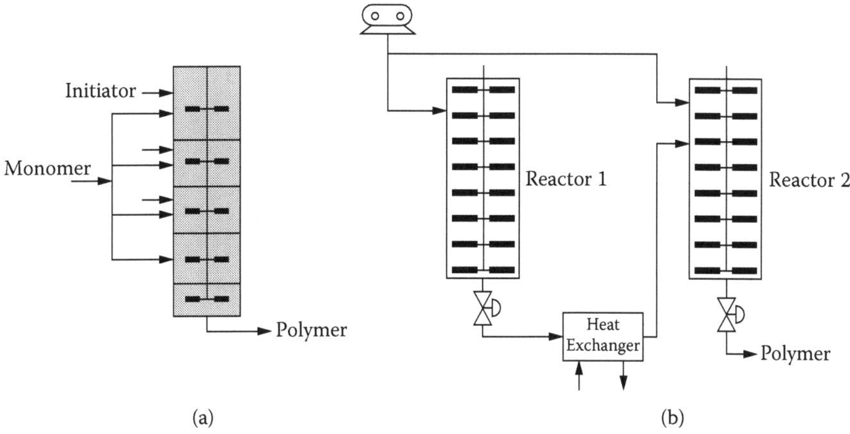

FIGURE 12.3 Multizone reactors for high-pressure ethylene polymerization.

thus, the reactor is operated adiabatically. The heat of polymerization (21.2 kcal mol^{-1}) is removed by the bulk flow of the polymerizing fluid. In general, the feed ethylene temperature is substantially lower than the reactor temperature. The polymerization temperature in each zone is controlled by regulating the initiator injection rate. Cold monomer feeds may be added to a few selected reaction compartments to remove additional reaction heat. As the polymerization rate is very high, the reactor residence time is very short (e.g., 1–3 min). In high-pressure ethylene polymerization, polymer properties (e.g., molecular weight, molecular-weight distribution, short-chain and long-chain branching frequencies, etc.) are strongly affected by polymerization temperature, pressure, and initiator type. More than one autoclave reactor can also be used for LDPE production. In the polymerization process shown in Figure 12.3(b) [6], the reaction mixture of about 15% ethylene conversion is cooled by a heat exchanger placed between the two autoclave reactors. Because additional reaction heat is removed by the external heat exchanger, a higher ethylene conversion (25%) has been claimed to be obtainable. Similar multiple reactor configurations are also reported in patent literature [7]. In ethylene polymerization processes, a loss of uniform mixing or the presence of impurities in the reacting fluid may cause a rapid decomposition of ethylene and polyethylene. Therefore, it is crucial to maintain perfect backmixing in each reaction zone. Strong nonlinear behavior of high-pressure autoclave reactors and the effect of micromixing have been subjects of research by some workers [8–11].

12.2.2.2 Tower Reactors for Styrene Polymerization

The tower reactors similar to the ethylene polymerization reactors are used in other free-radical vinyl polymerization processes. Figure 12.4(a) is a schematic of the tower reactor for bulk styrene polymerization developed by Farben in the 1930s [4]. The prepolymers prepared in batch prepolymerization reactors to about 33–35% conversion are transferred to a tower reactor where the temperature profile is controlled from 100°C to 200°C by jackets and internal cooling coils. No agitation

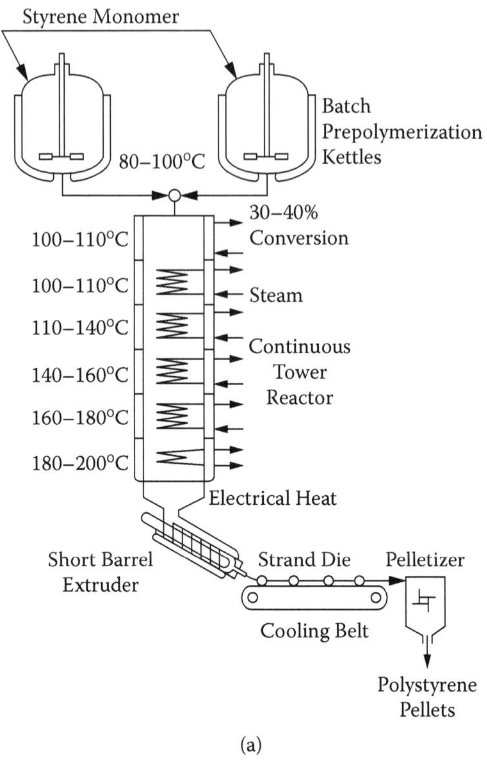

(a)

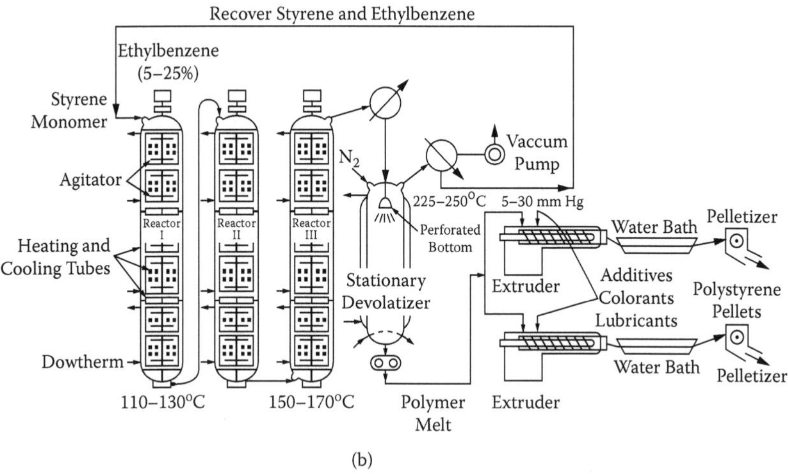

(b)

FIGURE 12.4 Tower reactors for styrene bulk polymerization.

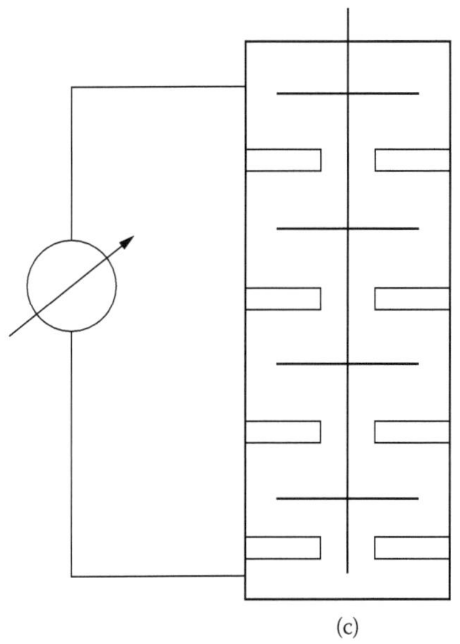

(c)

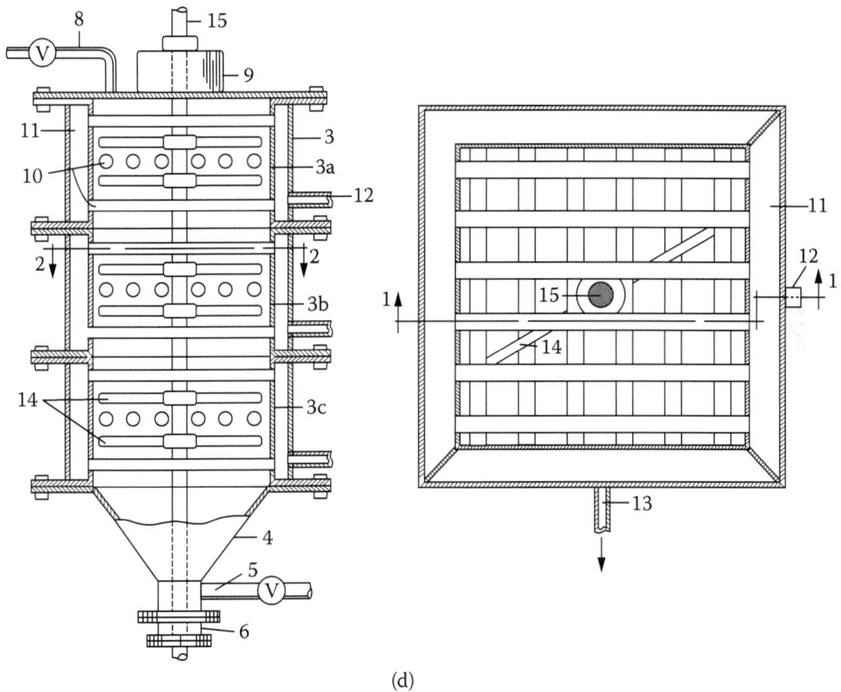

(d)

FIGURE 12.4 (Continued).

device is in the tower reactor. The product is then discharged from the bottom of the tower by an extruder, cooled, and pelletized.

Figure 12.4(b) also presents a continuous bulk styrene polymerization reactor system that consists of a series of towers using slow agitation and grids of pipes through which a mixture of diphenyl oxide is circulated for temperature control [4]. In this reactor system, each reactor is mildly stirred and operating at different temperatures. Ethyl benzene (5–25%) may be added to a styrene feed stream to reduce the viscosity of a polymer solution and to ease heat transfer. A vacuum degasser removes the residual styrene and ethylbenzene, which are recycled to the first reactor.

In a multizone reactor for the manufacture of impact polystyrene as illustrated in Figure 12.4(c) [12], a portion of product stream is cooled in an external heat exchanger and recycled to the top of the reactor. At a high recirculation rate, mixing is between backmixed and plug flow. In this reactor, horizontal rod agitators or layers of tubes are installed at each level to create shearing action throughout the mass of rubber-styrene solution undergoing polymerization [13]. This reactor is called the recirculated stratified agitated tower. Figure 12.4(d) is a stratifying polymerization reactor patented by Dow Chemical Company [14]. In this reactor, the revolving rods prevent channeling and promote plug flow.

12.2.2.3 Horizontal Linear Flow Reactors

Another type of linear flow reactor system for the synthesis of high-impact polystyrene is depicted in Figure 12.5 [1]. Here, the first-stage backmixed reactor (CSTR) is maintained just beyond the phase-inversion point (98°C, 14% solids) and the dissolved styrene reacts to form either a graft copolymer with the rubber or a homopolymer in the linear flow reactor train. Note that a portion of the effluent (130°C, 35% solids) from the second reactor is recycled to the first reactor. The temperature of the polymerizing mixture is gradually increased as it travels through the linear flow reactors and the final conversion of about 72% is achieved.

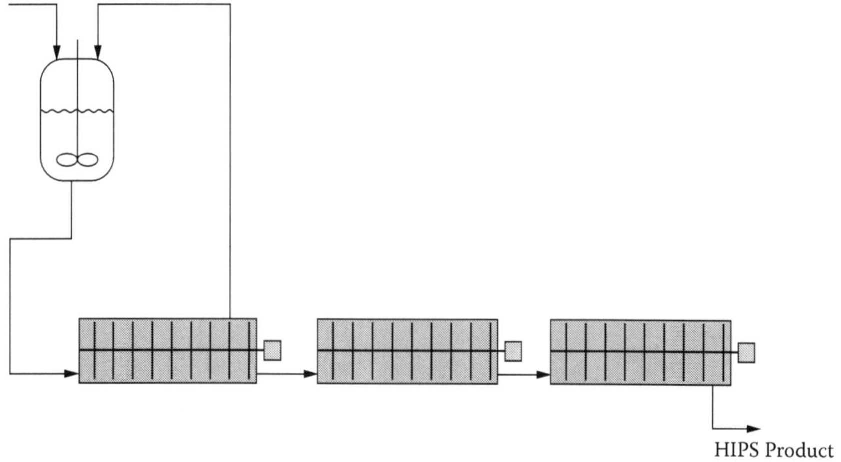

FIGURE 12.5 Continuous reactor system for mass polymerization of high-impact polystyrene.

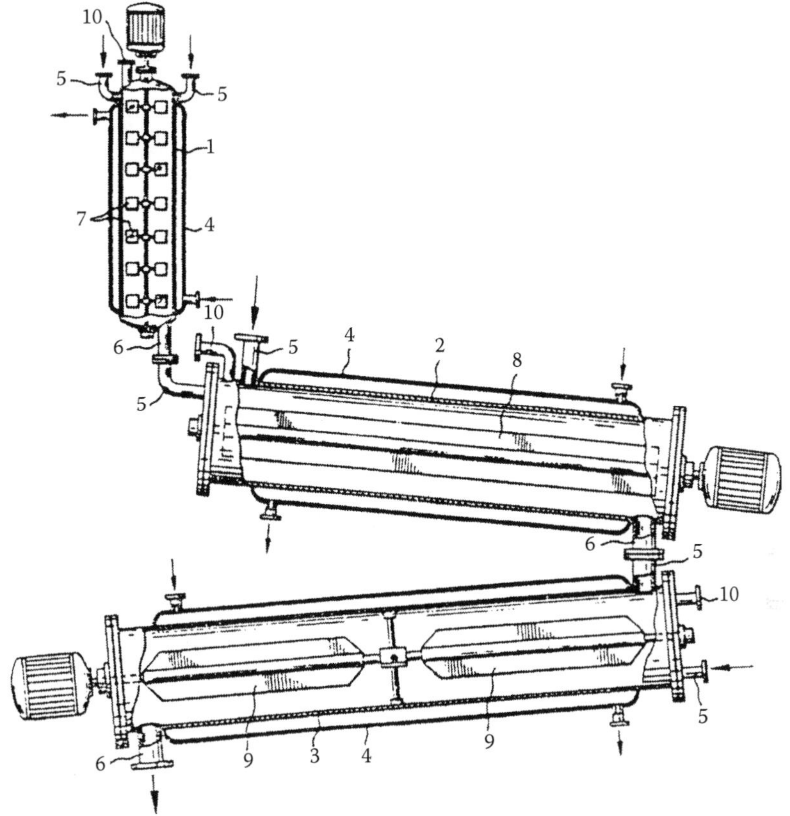

FIGURE 12.6 Continuous suspension polymerization reactor for vinyl chloride.

12.2.2.4 Continuous Suspension Polymerization Reactors

Suspension polymerization of vinyl monomers is usually performed in batch reactors. However, the feasibility of continuous suspension polymerization has been reported in some literature. A multiple-reactor system for continuous suspension polymerization of vinyl chloride is illustrated in Figure 12.6 [15]. Monomer, water, initiator, and suspending agents are fed to a vertical tower reactor equipped with a multistage stirrer. The reaction mixture of about 10% conversion is then transferred to the second and third reactors, which contain blade stirrers. Each reactor is jacketed for heat removal. Plug flow of the polymerization mixture is maintained in the reaction zones.

12.2.2.5 Continuous Emulsion Polymerization Reactors

Emulsion polymers (latex) have long been produced by continuous processes due to the low viscosity of latex polymers. Certain emulsion polymerization systems (e.g., polyvinyl chloride, polyvinylacetate) often exhibit large and sustained oscillations in conversion, polymer, and latex properties such as number of particles, particle size, and molecular weight. Figure 12.7(a) illustrates the conversion and number of

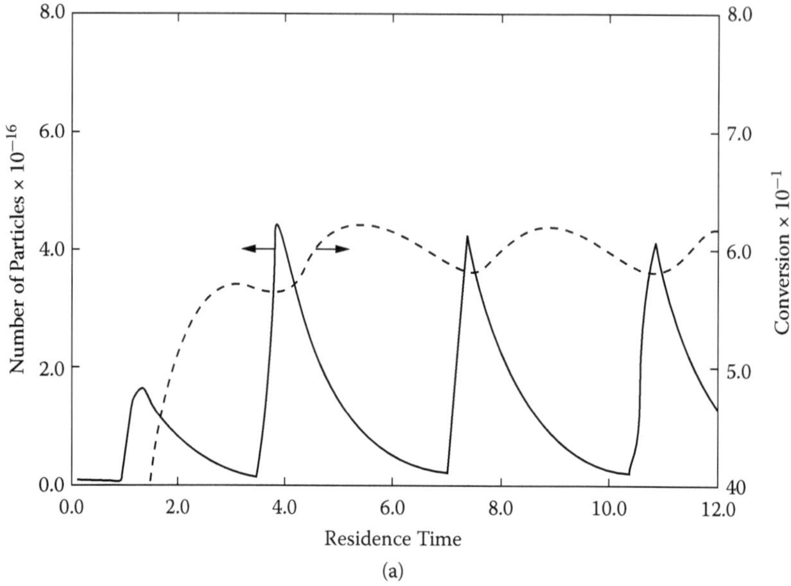

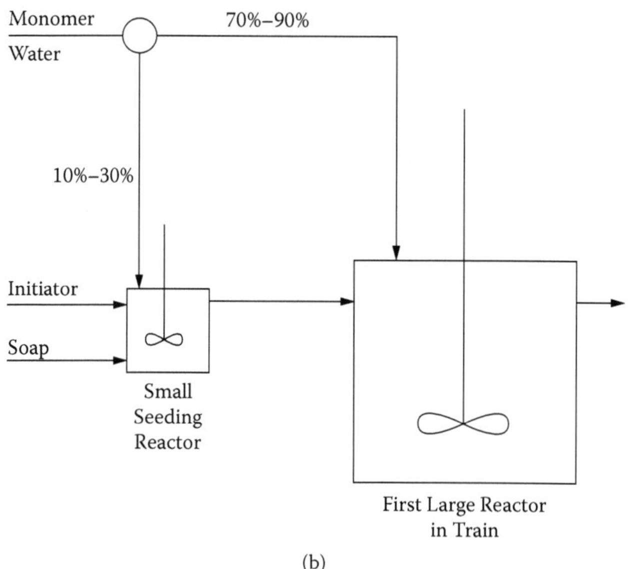

FIGURE 12.7 Continuous emulsion polymerization (a) conversion and number of particles versus residence time for a single CSTR; (b) reactor system with a small seeding reactor; (c) conversion and number of particles versus time with a seeding reactor.

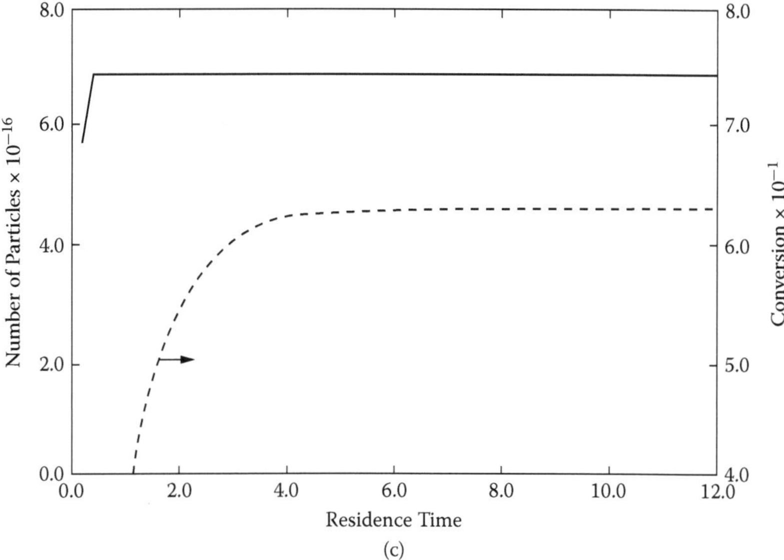

FIGURE 12.7 (Continued).

polymer particles versus reactor residence time for a single CSTR. Such oscillatory behaviors are due to the periodic formation and depletion of soap micelles, which lead to short periods of rapid particle generation followed by long periods in which no nucleation occurs. Because the polymer particles are not covered adequately by soap, particle agglomeration occurs during the periods of rapid particle nucleation. This problem can be solved by reconfiguring the process system. For example, a very small seeding reactor precedes the main polymerization reactor, as presented in Figure 12.7(b). Only a portion of the monomer and water are fed to the seeding reactor, and the remainder is fed to the main reactor. Then, particle generation is completed in the seeding reactor by using high soap and initiator concentrations. Only particle growth occurs in the main reactor. Figure 12.7(c) is the resulting monomer conversion and the number of polymer particles versus reactor residence time [16]. This example illustrates that the understanding of polymerization mechanism and kinetics is crucial in designing efficient continuous polymerization reactor systems.

12.2.3 Tubular Reactors

Vinyl monomers can be polymerized in tubular reactors. The main difference between tubular reactors and stirred tank reactors is that backmixing of the reactor's content is minimal in the tubular reactor. Tubular reactors have certain advantages over stirred tank reactors: design simplicity, good heat transfer capability, and narrow molecular-weight distribution of the product polymer due to minimal backmixing. The large surface area/volume ratio is particularly advantageous for the dissipation of heat generated by exothermic polymerization. However, empty tubular reactors are not widely used for commercial production of vinyl polymers except for high-pressure ethylene polymerization. One of the major problems in using the empty tubular reactors for the polymerization of vinyl monomers is that the viscosity of the polymerizing fluid increases significantly as monomer conversion increases. Consequently, large radial and axial temperature gradients may exist and the velocity profile is significantly distorted. Then, a buildup of a slowly moving liquid layer occurs at the reactor walls, causing a large variation in residence time distribution, plugging problems, poor heat transfer, and poor product quality [17]. These phenomena have been the subject of extensive theoretical modeling and experimental studies [18–20]. To minimize the radial velocity gradient, internal mixing elements such as paddles or static mixers can be installed in the tubes [21, 22]. Recently, a 30,000-ton/year polystyrene plant has been constructed using static mixer reactors (Figure 12.8) [23]. In the first polymerization stage, monomer and solvent are fed to the recirculating loop reactor, consisting of SMR static mixer-reactors. This reactor consists of bundles of intersecting tubes through which the heat transfer medium flows. In addition to the radial mixing of the product stream at low shear rate, high heat transfer coefficients are obtained, and a large internal heat transfer surface is formed by the tube. Therefore, relatively small temperature gradients exist in the reactor. The reaction temperature and polymerization rates are controlled via the temperature of the heat transfer fluid flowing through the tube bundles of the static mixer-reactors. In the second polymerization stage, the monomer/polymer solution flows to the plug flow reactors filled with static mixers. In these reactors, the monomer/polymer solution is radially mixed to such an extent that the temperature, concentration, and velocity over the reactor cross section are kept nearly constant. The polymerization temperature is increased gradually, resulting in increased conversion. It has been claimed that the uniform molecular weight of the final polymer is the result of a well-controlled time-temperature history.

12.2.3.1 High-Pressure Ethylene Polymerization Reactors

The most well-known tubular polymerization reactor system is the one used in high-pressure ethylene polymerization processes for the production of low-density polyethylene. Figure 12.9 depicts a classical high-pressure, high-temperature tubular ethylene polymerization reactor system. The reactor can be a long single tube, or a tube with multiple feed streams along its length, or a bundle of tubes, mounted vertically or horizontally. To provide plug flow, a high length-to-diameter ratio (250:1 to 12,000:1) is employed. Purified ethylene is compressed to 1000–3500 atm and mixed with a free-radical initiator (oxygen) and the residence time is as low as 20 sec. The first section of the tubular reactor is heated because the heat generated by polymerization is insufficient. The second section of the reactor is cooled to provide the desired temperature profile. Because the fluid velocity is very high, a pressure drop along the tube length is significant and it affects the propagation rate. Consequently, the molecular-weight distribution of the polymer is broader than if a stirred autoclave reactor is used [4]. Chain transfer agents such as ketones, aldehydes, alcohols, hydrogen, or chlorinated compounds are added to narrow molecular-weight distribution. The reaction product is discharged into high and low-pressure separators.

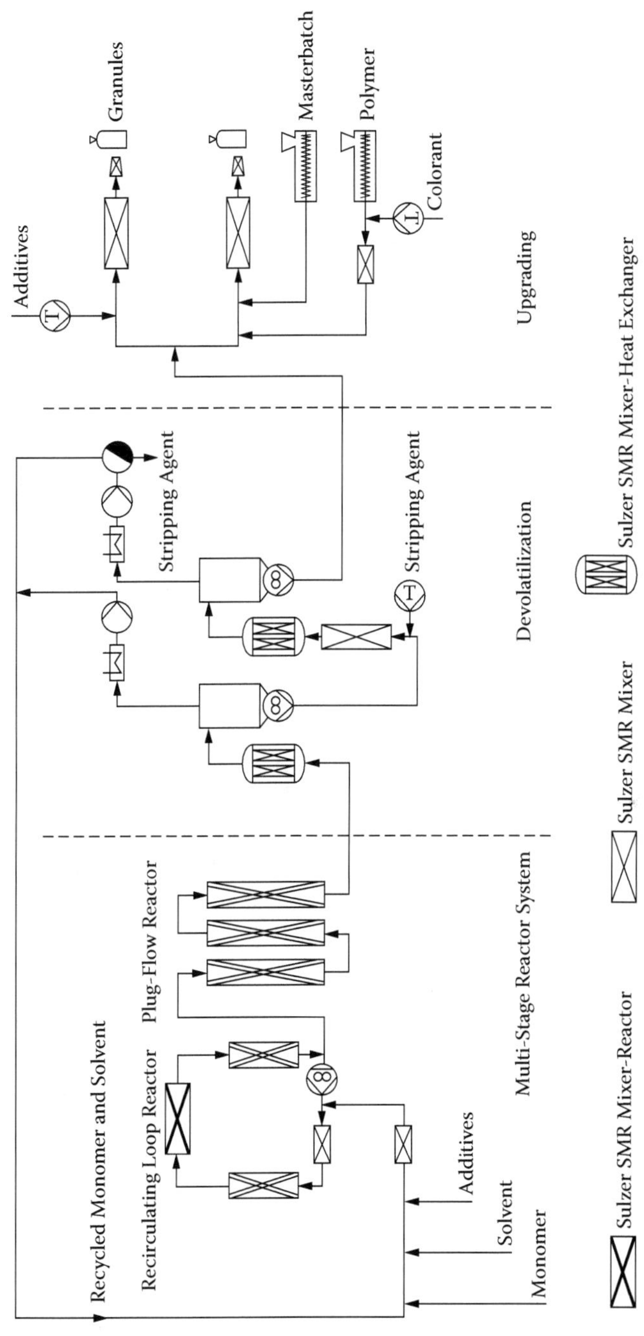

FIGURE 12.8 Schematic flow sheet of the bulk polystyrene process with static mixers.

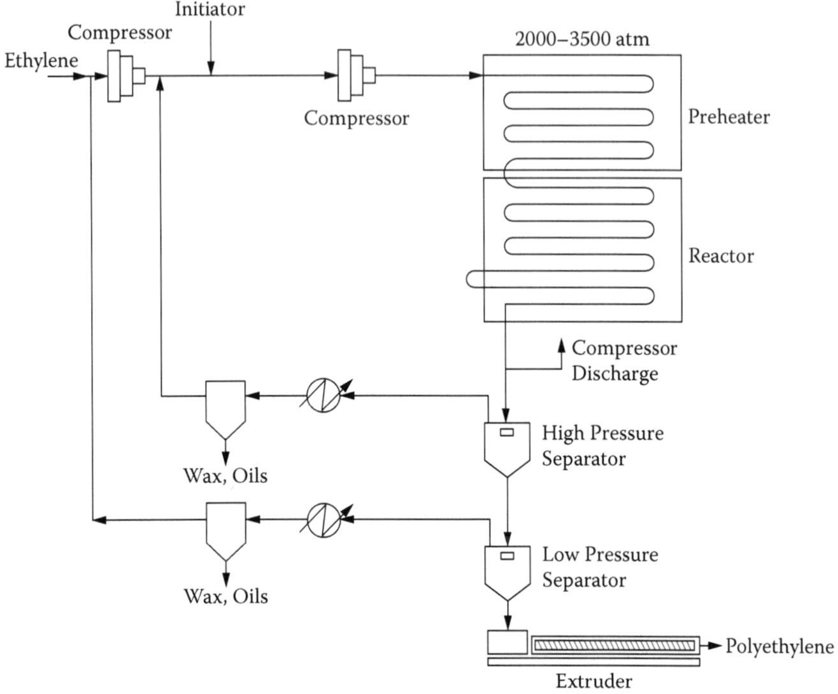

FIGURE 12.9 A tubular reactor system for free-radical ethylene polymerization.

12.2.3.2 Continuous Tubular Suspension Polymerization Reactors

Suspension polymerization is mostly performed in batch reactors. However, some reports have been published on the continuous suspension polymerization in different types of reactors. For continuous suspension polymerization, certain requirements apply: (1) narrow residence time distribution to achieve high conversion, (2) good mixing of the two phases to obtain polymers with proper particle size distribution, (3) no dead space and gas phase within the reactor to avoid reactor fouling, and (4) large heat transfer surface area for heat removal. Figure 12.10 is a schematic of a pilot plant-scale continuous suspension polymerization reactor system [24]. The reactor consists of a tube with a blade stirrer. Vinyl acetate containing organic initiator, and a water-containing dispersion agent are pumped in parallel flow through the tube reactor from top to bottom. The conversion was obtained above 90%, and good particle size distribution was obtained.

12.2.3.3 Tubular Emulsion Polymerization Reactors

Because a tubular reactor offers minimal backmixing, it can produce latex polymers with a narrow particle size distribution. A large amount of literature has been published about emulsion polymerization in tubular reactors (nonagitated and

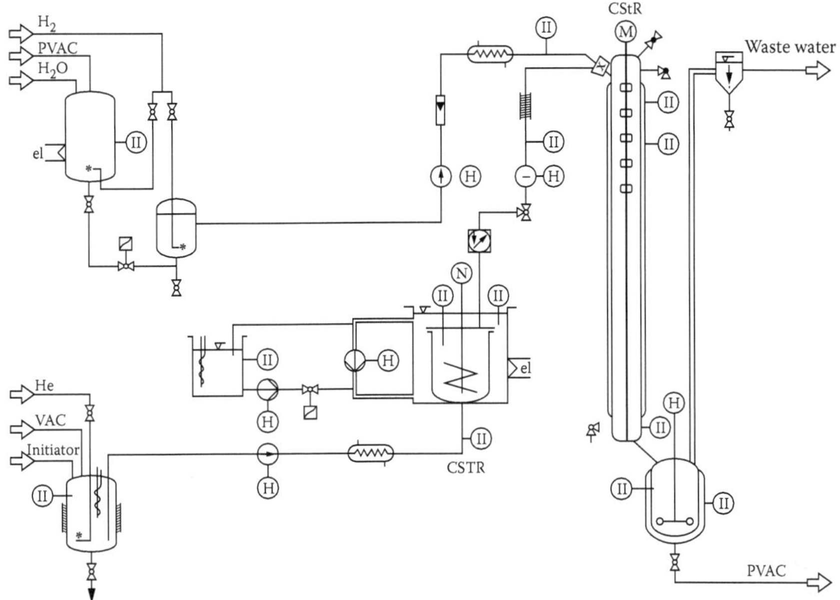

FIGURE 12.10 Flow diagram of a continuous suspension polymerization reactor system.

agitated) [25–27]. One of the major problems in operating a tubular reactor for high conversion emulsion polymerization is the occurrence of fouling and plugging. In a recent study, it was reported that the use of a pulsation source can eliminate the reactor fouling and plugging problems in a laboratory-scale tubular emulsion polymerization reactor to obtain a narrow particle size distribution at high monomer conversion [28, 29].

12.2.4 Continuous Screw Reactors

Twin-screw extruders have long been used by the plastic compounding industry for mixing and dispersing of one or more components into a polymer matrix. Because twin-screw extruders provide excellent control of mixing characteristics and gas, liquid, or solid feed can be added at any point along the extruder length, they are also used to produce high-performance specialty polymers. The batch size versatility of extruders compared with a full-scale polymerization reactor is particularly attractive for the manufacture of specialty polymeric materials. In the reactive extrusion process, the appropriate monomer(s), prepolymers, and initiator(s) are fed to an extruder where the polymerization takes places, and the resulting polymer is forced into a mold or a die to give a finished article. Figure 12.11 illustrates a two-stage extruder reactor where the grafting of maleic anhydride is vented, and in the second stage, the reaction mixture is pressurized to the die. The feeds to the reactor consist of a peroxide master batch and the feedstock to be grafted. As the polymer feeds are being melted and mixed, the temperature rises with distance down the reactor.

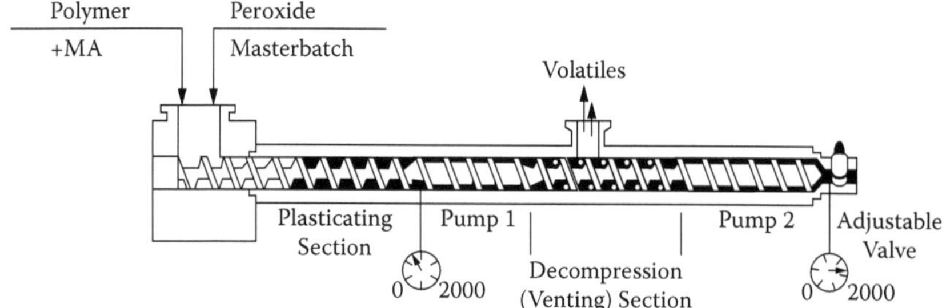

FIGURE 12.11 Extruder reactor for the grafting of maleic anhydride.

12.3 DESIGN AND OPERATION OF CONTINUOUS FREE-RADICAL POLYMERIZATION REACTORS

At steady state, continuous polymerization reactors produce the polymers of constant properties. In reality, however, the reactor is subject to some external disturbances such as variations in feed composition, feed temperature, and cooling water source temperature. Therefore, accurate feedback control of the reactor is required to maintain consistent product quality and reactor stability. When several different-grade products are manufactured, the reactor conditions must be changed for each grade.

It is also known that continuous polymerization reactors may exhibit highly nonlinear dynamics such as multiple steady states, autonomous oscillations, and strong parametric sensitivity. Thus, it is important to understand the steady state and dynamic behaviors of a reactor to maintain reactor stability and to achieve smooth transitions for grade changes.

As described earlier, the control of polymerization temperature is often the most important objective in operating a polymerization reactor. In Figure 12.12, three commonly used configurations for heat removal in continuous stirred tank reactors are illustrated [31]. In the jacketed reactor (Figure 12.12(a)), temperature is controlled by both internal cooling coils and a jacket. The jacket temperature is regulated in response to the reactor temperature. For a large reactor vessel, some time delays may be present between the jacket temperature and the resulting reactor temperature, causing a potential temperature instability. Figure 12.12(b) shows the autorefrigerated reactor in which controlled vaporization of the monomer and solvent serves to remove the heat of polymerization. The reactor temperature and pressure are maintained very close to the bubble point. Figure 12.12(c) shows a temperature control scheme for the reactor with an external heat exchanger. A buildup of high-viscosity polymer layers on the low-temperature heat exchanger surfaces can lower the heat transfer efficiency.

The design of the reactor control system is an integral part of the polymerization process design. Some important problems and needs in designing effective controls for industrial polymerization processes are: (1) on-line measurements, (2) severely nonlinear processes, (3) modeling and identification for control system design, (4) modeling for simulation and operator training, and (5) process monitoring and

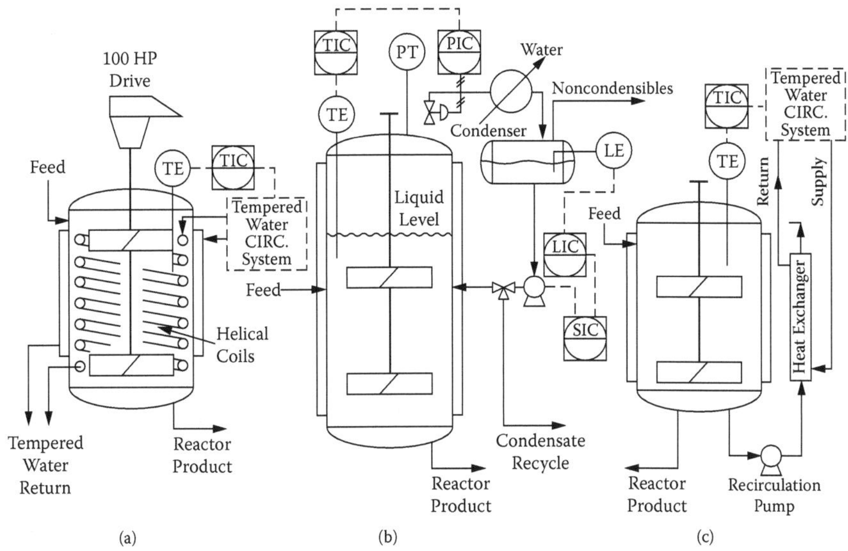

FIGURE 12.12 Polymerization reactor temperature control schemes. (a) temperature control via cooling jacket and coils; (b) temperature control via vaporization of solvent/monomer; (c) temperature control via forced circulation of syrup through external heat exchanger.

diagnosis [32]. Some commercial software packages are available for the modeling of a variety of continuous free-radical polymerization processes.

In industrial continuous polymerization processes, it is often required to produce the polymers of different properties. Obviously, certain reaction parameters must be changed. If the reactor residence time is large (e.g., several hours), it becomes crucial to bring the reactor from one steady-state operating condition to a new steady-state condition as rapidly as possible to minimize the production of transition products. In general, process variables such as temperature, pressure, and bulk phase composition change much faster than those related to polymer properties (e.g., polymer molecular weight, molecular-weight distribution, copolymer composition, morphology, etc.). Unfortunately, these polymer properties are difficult to monitor on-line and they are usually measured infrequently by off-line laboratory analysis. Because continuous polymerization reactors produce a large quantity of polymer products, such time delays in property measurement may cause a significant loss of productivity. To solve such problems, advanced state estimation techniques have recently been developed [33–36]. For example, a dynamic process model (first-principles model based on mass and energy balances) is used on-line in conjunction with a state estimator such as an extended Kalman filter to calculate the polymer properties (e.g., molecular-weight averages, copolymer composition, particle size distribution, etc.) with infrequent process measurements with some time delays. Then, the predicted or estimated polymer properties are used to regulate reactor control variables and to keep the reactor at its target operating conditions.

REFERENCES

1. R. H. M. Simon and D. C. Chappelear, *Am. Chem. Soc. Symp. Ser.*, 104, 71 (1979).
2. Radian Corporation, *Polymer Manufacturing*, Noyes Data Corp., Park Ridge, NJ, 1986.
3. D. E. Carter and R. H. M. Simon, U.S. Patent 3,903,202 (1975).
4. N. Platzer, *Ind. Eng. Chem.*, 62(1), 6 (1970).
5. J. P. Congalidis, J. R. Richards, and W. H. Ray, *AIChE J.*, 35(6), 891 (1989).
6. I. Suzuki, T. Kamei, and R. Sonoda, U.S. Patent 3,875,128 (1975).
7. H. Sutter, K. U. Haas, and W. P. Ledet, U.S. Patent 4,607,086 (1986).
8. C. Georgakis and L. Marini, *Am. Chem. Soc. Symp. Ser.*, 196, 591 (1982).
9. L. Marini and C. Georgakis, *AIChE J.*, 30, 401 (1984).
10. B. G. Kwag and K. Y Choi, *Ind. Eng. Chem. Res.*, 33(2), 211 (1994).
11. B. G. Kwag and K. Y. Choi, *Chem. Eng. Sci.*, 49(24B), 4959 (1994).
12. D. B. Priddy, *Adv. Polym. Sci.*, 111, 67 (1994).
13. J. L. Amos, *Polym. Eng. Sci.*, 14(1), 1 (1974).
14. D. L. McDonald et al., U.S. Patent 2,727,884 (1955).
15. H. Klippert, E. Tzschoppe, S. Paschalis, J. Weinlich, and M. Englemann, U.S. Patent 4, 424,301 (1984).
16. A. Penlidis, J. F. MacGregor, and A. E. Hamielec, *AIChE J.*, 31(6), 881 (1985).
17. S. Lynn, *AIChE J.*, 23, 387 (1977).
18. J. W. Hamer and W. H. Ray, *Chem. Eng. Sci.*, 41, 3083 (1986).
19. J. W. Hamer and W. H. Ray, *Chem. Eng. Sci.*, 41, 3095 (1986).
20. C. J. Stevens and W. H. Ray, *Am. Chem. Soc. Symp. Ser.*, 404, 337 (1989).
21. K. T. Nguyen, E. Flaschel, and A. Renken, in *Polymer Reaction Engineering*, K. H. Reichert and W. W. Geiseler, Eds., Hanser, New York, 1983, p. 175.
22. W. J. Yoon and K. Y. Choi, *Polym. Eng. Sci.*, 36(1), 65 (1996).
23. W. Tauscher, *Sulzer Tech. Rev.*, 2 (1991).
24. K. H. Reichert, H. U. Moritz, C. Gabel, and G. Deiringer, in *Polymer Reaction Engineering*, K. H. Reichert and W. W. Geiseler, Eds., Hanser, New York, 1983, p. 154.
25. A. L. Rollin, I. Patterson, R. Huneault, and P. Bataille, *Can. J. Chem. Eng.*, 55, 565 (1979).
26. R. Lanthier, U.S. Patent 3,551,396 (1970).
27. M. Ghosh and T. H. Forsyth, *Am. Chem. Soc. Symp. Ser.*, 24 367 (1976).
28. D. A. Paquet Jr. and W. H. Ray, *AIChE J.*, 40(1), 73 (1994).
29. D. A. Paquet Jr. and W. H. Ray, *AIChE J.*, 40(1), 88 (1994).
30. X. Xanthos, *Reactive Extrusion*, Hanser, New York, 1992.
31. L. S. Henderson III and R. A. Cornejo, *Ind. Eng. Chem. Res.*, 28, 1644 (1989).
32. B. A. Ogunnaike, *IEEE Control System*, 41 (April 1995).
33. K. J. Kim and K. Y. Choi, *J. Process Control*, 1, 96 (1991).
34. H. Schuler and Z. Suzhen, *Chem. Eng. Sci.*, 41 2681 (1985).
35. W. H. Ray, *IEEE Control Syst. Mag.*, 3 (August 1986).
36. D. K. Adebekun and F. J. Schork, *Ind. Eng. Chem. Res.*, 28 1846 (1989).

13 Technical Processes for Industrial Production

Kyu Yong Choi, Byung-Gu Kwag, Seung Young Park, and Cheol Hoon Cheong

CONTENTS

13.1 INTRODUCTION

Polymerization of vinyl monomers is of enormous industrial importance. These vinyl polymers are mostly thermoplastics and they are used in a wide variety of end-use applications. Many vinyl monomers are polymerized by free-radical, ionic, and coordination polymerization mechanisms. Among these, free-radical polymerization is the most widely used in industrial production of vinyl polymers. Ionic polymerization is generally used to manufacture specialty polymers. Free-radical polymerization is advantageous over other processes in that it is less sensitive to impurities in the raw materials, and the rate of polymerization as well as polymer properties can be controlled by the choice of initiator and polymerization conditions.

In the homopolymers of vinyl or olefinic monomers, polymer architecture represented by molecular-weight distribution, molecular-weight averages, and long-chain and short-chain branching has a significant impact on the physical, mechanical, and rheological properties of polymers. These properties are strongly influenced by specific polymerization process conditions. The properties of vinyl polymers are also varied widely by co-polymerizing two or more vinyl monomers or diens.

One common factor in most of the free-radical polymerization processes is that polymerization reactions are highly exothermic and the viscosity of the reacting mass increases significantly with conversion. Thus, mixing and heat removal are the key process or reactor design factors. Also, the polymerization kinetics and mechanism are quite complex and often poorly understood despite many years of commercial production of vinyl polymers. This implies that depending on the particular chemical and physical characteristics of the polymer system, reactor types and process operating conditions must be properly designed and controlled.

Both batch and continuous reactors are used in industrial vinyl polymerization processes. Agitated kettles, tower reactors, and linear flow reactors are just a few examples of industrially used polymerization reactors. The choice of reactor type depends on the nature of polymerization systems (homogeneous versus heterogeneous), the quality of product, and the amount of polymer to be produced. Sometimes, multiple reactors are used and operated at different reaction conditions. Whichever reactor system is used, it is always necessary to maximize the productivity of the process by reducing the reaction time (batch time or residence time) while obtaining desired polymer properties consistently.

For the industrial production of vinyl polymers, *mass (bulk), solution, suspension, and emulsion* processes are commonly used. In the following some important characteristics of these processes are briefly described.

Mass (bulk) Polymerization: If the polymer is soluble in its own monomer or solvent in all proportions (e.g., polystyrene, poly(methyl methacrylate), poly(vinyl acetate)), a single homogeneous phase is present in the reacting medium (homogeneous mass and solution processes). Some polymers, such as poly(vinyl chloride), poly(acrylonitrile), and poly(vinylidene chloride), do not dissolve in their own monomers, and they precipitate from the liquid almost immediately after the polymerization is started (heterogeneous mass process). Because no additives such as suspension stabilizer or emulsifier are necessary, the polymer prepared by mass process is in its purest form. However, the viscosity of the polymerizing mass increases rapidly and significantly with an increase in conversion in homogeneous bulk processes, making the mixing and the reactor temperature control difficult. Therefore, specially designed agitators are required to handle high-viscosity polymer fluid. In many industrial processes, mass polymerization is performed in continuous reactors, monomer conversion is low, and unreacted monomer is recovered and recycled. In heterogeneous mass processes (e.g., poly(vinyl chloride)), the final polymer product is recovered as a powder.

Solution Polymerization: Solution polymerization is usually performed in a continuous-flow reactor system because of low-solution viscosity. Monomer is mixed with organic solvent and the polymer produced dissolved completely in the medium. The product from the solution process is used as a solution itself or as an intermediate for other applications. Because the solution viscosity is low, heat transfer through the reactor jacket is efficient. The overall polymerization kinetics of solution polymerization is quite similar to those of bulk polymerization, except for the effect of solvent chain transfer. If chain transfer to solvent occurs significantly, polymer molecular weight decreases.

Suspension polymerization: Most of the vinyl polymerizations are highly exothermic and the removal of reaction heat is an important reactor and process design consideration. In a batch suspension polymerization, which is often called *pearl or bead* polymerization, organic monomer is dispersed in water as discrete droplets (10 μm–5 mm) by mechanical agitation. Each droplet contains monomer(s), an organic-soluble initiator, a chain transfer agent, or other additives, and when heated to a desired reaction temperature, polymerization occurs to high conversion. The interfacial tension, the intensity of agitation, and the design of the stirrer and reactor system dictate the dispersion of monomer droplets. The aqueous phase serves as a suspension medium and also as heat sink. The reaction heat released from the polymer particles is effectively removed by the water surrounding the polymer particles and, thus, isothermal reaction can be achieved. Therefore, the reactor temperature control in the suspension process is relatively easier than in the bulk process.

The liquid droplets dispersed in water undergo constant collision and some of the collisions result in coalescence. If the coalescence is not controlled or excessive, suspension stability is lost and undesirable particle agglomeration occurs. To prevent the coalescence or agglomeration of monomer droplets and polymer particles and to obtain uniform-sized particles, a suspension stabilizer (protective colloid or dispersant) is added to the aqueous phase. The most important issues in suspension polymerization are the control of polymer particle size distribution and the resulting polymer particle morphology. If the polymer is insoluble in its monomer (e.g., poly(vinyl chloride)), the heterogeneous polymerization takes place in the droplets and the development of particle morphology becomes quite complex. The kinetics of suspension polymerization is quite similar to those of mass polymerization. In suspension polymerization, mixing is critical to preventing the sedimentation of droplets due to the increase in density as polymerization proceeds. Any accumulation of stagnant mass at the bottom zone of the reactor may cause the increases in polymerization rate and heat generation, leading to the rapid monomer vaporization. Some critical issues on suspension polymerization have been reviewed recently [1, 2].

Emulsion polymerization: High polymerization rate and high polymer molecular weight are simultaneously obtainable in emulsion polymerization. Due to the heterogeneous nature of emulsion polymerization, chemical and physical phenomena in emulsion polymerization are far more complex than in other polymerization processes. As in suspension polymerization, the monomer is dispersed in water by mechanical agitation as small droplets (0.05–2.0 mm) in emulsion polymerization. However, several fundamental differences can be found between the emulsion and suspension processes. In suspension process, polymerization occurs in each monomer droplet containing an organic-soluble initiator; in emulsion polymerization, negligible or no reaction takes place in the monomer droplets. It is required that monomer is only slightly soluble in water.

Particle nucleation and particle growth are important steps in emulsion polymerization because they affect the overall polymerization rate and polymer properties. Thus, initiator concentration and surfactant type/concentration have significant effects on the polymerization kinetics. In a batch emulsion process, particle nucleation and growth steps can be separated to some extent by employing a multistage reaction process. In a continuous process, both particle nucleation and growth steps occur simultaneously unless a seed reactor is provided to separate these two effects. In general, latex particle size distributions obtained by batch and continuous processes are quite different.

One of the key ingredients in emulsion polymerization is surfactant or emulsifier that has both hydrophilic and hydrophobic ends. The monomer droplets are stabilized by a monolayer of surfactant at the monomer–water interface. If the surfactant concentration exceeds a critical level (critical micelle concentration), the aggregates of surfactant molecules, called micelles, are formed. The size of micelles is about 100 Å, which consists of approximately 5–100 surfactant molecules. The emulsifier molecules in the micelles are oriented so that the hydrophobic ends of the surfactant molecules are oriented toward the center of micelle, whereas the hydrophilic ends extend out into the aqueous phase. Because the micelles are much smaller than the monomer droplets, the total surface area of the micelles is generally one to three orders of magnitude larger than that of the monomer droplets. As a result, the monomer droplets hardly absorb radicals from the aqueous phase and little reaction occurs in the dispersed monomer droplets. In emulsion polymerization, a water-soluble initiator such as potassium persulfate is used. Small amounts of monomer dissolved in the aqueous phase diffuse into the center of micelles. As the whole mixture in the reactor is heated, initiator decomposes in the aqueous phase to primary radicals.

When a free radical generated in the aqueous phase is captured by the micelle-containing monomer, polymerization starts. As polymerization continues, more and more emulsifier is required to stabilize the growing monomer-swollen polymer particle until the micelles disappear. The monomer dissolved in water may polymerize to low-molecular-weight oligomers in the aqueous phase. These oligomers with chain lengths less than 5 may diffuse to growing latex particles, or they may be emulsified by surfactant molecules and form a new polymer particle. With an increase in monomer conversion, dispersed monomer droplets become smaller and the polymer latex particles become larger. The monomer concentration in monomer-swollen polymer latex particles remains constant when the separate monomer droplets are present in the aqueous phase. Eventually, the monomer droplets disappear and the polymerization rate decreases because no additional supply of monomer to polymer particle is available. Figure 13.1 illustrates the particle growth mechanism [3].

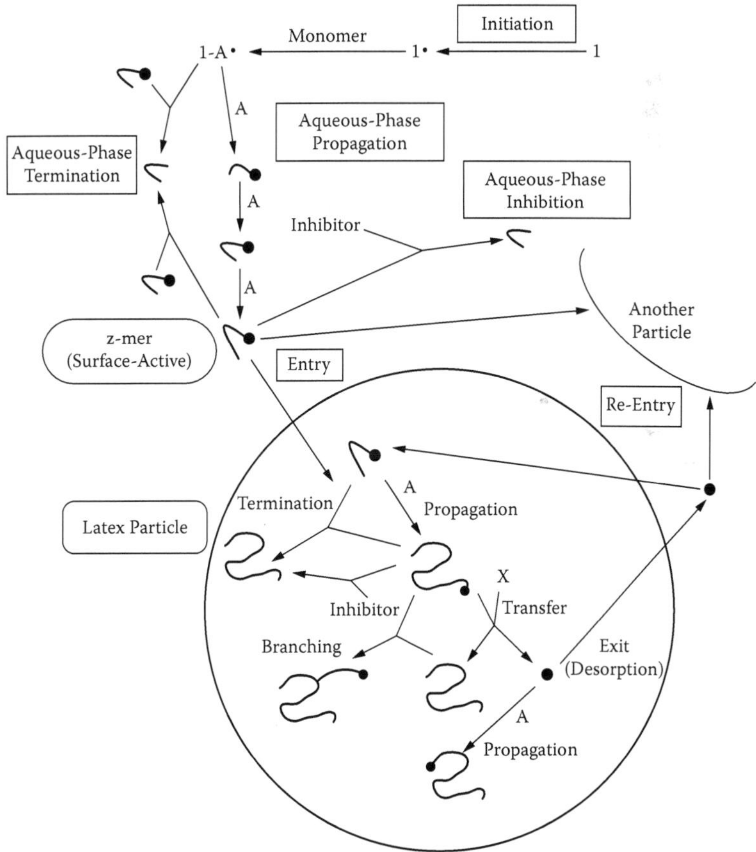

FIGURE 13.1 Particle growth in emulsion polymerization. From J. P. Congalidis and J. R. Richards, paper presented at the Polymer Reaction Engineering III Conference, March 1997. With permission.

TABLE 13.1
Types of Commercial Polymerization Processes

Polymer	Homogeneous Processes		Heterogeneous Processes		
	Mass (Bulk)	Solution	Suspension	Emulsion	Mass (Precipitation)
Polyethylene	■				
Polystyrene	■	■	■	■	
Poly(methyl methacrylate)	■	■	■	■	
Polyacrylonitrile					■
Poly(vinyl chloride)			■	■	■
Poly(vinyl acetate)	■	■	■	■	
Poly(tetrafluoroethylene)			■	■	
Poly(vinyl/vinylidene chloride)		■	■	■	■

Emulsion polymerization kinetics has been the subject of extensive research since the 1940s, but many aspects of emulsion polymerization are still not completely understood. The discussion of emulsion polymerization kinetics is beyond the scope of this chapter, and many excellent references on emulsion polymerization kinetics are available [4, 5].

In this chapter, various industrially important free-radical polymerization processes are reviewed. Table 13.1 presents the types of commercial polymerization processes for the polymers discussed in this chapter.

The manufacturing of the polymers to be discussed in this chapter are commercially well established. It is not the objective of this chapter to provide an extensive review of polymerization kinetics and mechanisms that are already well addressed or reviewed elsewhere. Instead, the discussion will be focused on the technical aspects of polymerization processes reported in open literature.

Although the polymerization processes for many vinyl polymers are well established, the polymer industry is continuously pursuing improved process technology. For example, batch reaction time or residence time (in a continuous reactor process) needs to be minimized while desired polymer properties are obtained. Consistency in the polymer product quality is another important process control objective in operating polymerization reactors. It is also desired that a product-grade slate should be diversified. However, achieving such goals is not always straightforward or easy. In many polymerization processes, polymer properties or quality control parameters are difficult to monitor on-line and making appropriate process adjustments is not a trivial matter when some deviations in the product properties from their target values are detected. High exothermicity and process nonlinearity also cause the design and operation of polymerization reactors difficulty. The kinetics of polymerization, particularly those of heterogeneous polymerization, are often not completely understood. Recent academic and industrial research activities indicate that some of these problems can be solved by using process models in conjunction with advanced computer control techniques.

13.2 HIGH-PRESSURE ETHYLENE POLYMERIZATION PROCESSES

Low-density polyethylene (LDPE, 0.915–0.935 g cm^{-3}) is produced industrially by either high-pressure free-radical polymerization or transition-metal-catalyzed low-pressure processes (e.g., gas-phase and slurry-phase Ziegler–Natta processes). Although the latter is gaining increasing popularity in recent years with the development of high-activity catalysts, high-pressure free-radical processes are still industrially very important and widely used. High-pressure polyethylene processes are characterized by high reaction pressure (1000–3000 atm) and high reaction temperature (150–300°C). Both stirred autoclave reactors and tubular reactors are commonly used in the industry. Commercial high-pressure ethylene polymerization reactor systems are discussed in Chapter 11. In general, tubular reactors give a more stable operation, whereas autoclave reactors often tend to be quite unstable with more frequent ethylene decomposition reactions. It has been known that polyethylenes made by different processes exhibit quite different molecular architecture and, thus, final end-use properties.

LDPE manufactured by high-pressure free-radical polymerization technology is characterized by the presence of long-chain and short-chain branches. Polymer density, molecular-weight distribution, and branching frequencies (short chain and long chain) are strongly influenced by reaction pressure. The short branches are primarily ethyl and *n*-butyl groups. Up to 15 to 30 such side groups/1000 carbon atoms in the chain occur in LDPE. The number of short branches has a major effect on the density of polymer. Long-chain branches are much less frequent than short branches. The long-chain branches have important effects on polymer processability, clarity of polyethylene film, drawdown of coating resins, and service strength [6]. For typical polyethylenes, M_n (number-average molecular weight) is in the range 5000–40,000 and M_w (weight-average molecular weight) is in the range 50,000–800,000. With increasing pressure, propagation rate increases more rapidly than the termination rates and backbiting reactions, leading to higher density, less branching, higher molecular weight, and fewer vinyl end groups.

Polyethylene made by high-pressure technology is often copolymerized with small amounts of comonomers (e.g., propylene, butane-1, hexane 1, octane-1, vinyl acetate, and acrylic acid). The ethylene-vinyl acetate copolymers are used in film, wire or cable coating, and molding applications. Copolymers of ethylene and acrylic acid are treated with compounds of sodium, potassium, zinc, and so forth to form salts attached to the copolymer chain. Such copolymers are often called ionomers.

Some technological problems involved in building large LDPE production units are process operation, size or compressors, reactor structure, high-pressure valves, and safety problems [7]. Due to high exothermicity of the polymerization reaction, the removal of reaction heat is a critical design problem. Factors that affect the heat removal include: reactor surface/volume ratio, reaction mixture and feed ethylene temperature difference, thickness of the polyethylene layer at the inner wall of the reactor, reaction mixture flow rate, and reactor material heat conductance. It should be noted that the thickness of the laminar layer at the reactor wall is affected by the reaction mixture flow rate.

When ethylene is polymerized by free-radical mechanism, high pressure and high temperature are required. Organic initiators and oxygen are used as free-radical generators. The general kinetic scheme for free-radical ethylene copolymerization is represented as follows:

Initiation

$$I \rightarrow 2R$$

$$O_2 + M \rightarrow 2R$$

$$R + O_2 \rightarrow RO_2 \rightarrow 2R$$

$$3M_i \rightarrow R_i$$

Propagation

$$R_j(n) + M_i \rightarrow R_i(n + 1)$$

$$R_i(n) + R_j \rightarrow P(m + n)$$

Chain termination

$$R_i(n) + R_j(m) \rightarrow P^{=}(n) + P(m)$$

$$R_i(n) + R_j(m) \rightarrow P^{=}(m) + P(n)$$

Chain transfer to monomer

$$R_j(n) + M_i \rightarrow R_i^{=}(1) + P(n)$$

Chain transfer to chain transfer agent

$$R_j(n) + X \rightarrow X^* + P(n)$$

β-Scission of terminal radicals

$$R_j(n) \rightarrow P_{i(j)}^{=}\ (n - 1 + R_j(1)$$

Chain transfer to polymer

$$R_j(n) + P_i(m) \rightarrow R_i'(m) + P(n)$$

$$R_i'(m) + M_j \rightarrow R_i'(m + 1)$$

$$R_j(n) + P_i(m) \rightarrow R_j'(m) + P(n)$$

Backbiting (short-chain branching)

$$P_j(n) \rightarrow R_i'(n)$$

$$R_i'(n) + M_j \rightarrow R_j'(n + 1)$$

$$(R(n) \rightarrow R'(n))$$

β-Scission of internal radical centers

$$R'_j(n) \rightarrow P^{=}(m) + R_{j(i)}(n-m)$$

$$R'_j(n) \rightarrow P^{=}(n-m) + R_{j(i)}(m)$$

Explosive decomposition

$$M_j \rightarrow \text{carbon} + \text{hydrogen} + \cdots$$

$$P(n) \rightarrow \text{carbon} + \text{hydrogen} + \cdots$$

Where I is the initiator, R is the primary radical, M is the monomer (ethylene), $R_n(n)$ is the growing polymer radical with chain length n ending with monomer j, P(n) is the dead polymer with chain length n, and P-(n) is the dead polymer with chain length n and a terminal double bond.

As discussed previously, chain transfer to polymer, backbiting, and β-scission reactions lead to long-chain and short-chain branches in the polymer. The intramolecular chain transfer reaction ("backbiting") occurs when the end of the polymer chain coils backward to abstract a hydrogen radical from the fifth carbon atom back in the polymer chain, and the chain growth starts there. The intermolecular chain transfer to polymer leads to long-chain branching. The kinetic scheme depicted previously can be used to develop a comprehensive kinetic model to predict not only the polymerization rate or conversion but also the resulting polymer properties.

Sometimes, polymerization occurs in two phases (ethylene phase and polymer phase). The polyethylene produced in the polymer base is thought to be highly branched and of higher molecular weight. The degree of miscibility decreases as temperature is decreased.

In high-pressure polyethylene (PE) processes, a variety of free-radical generating initiators are used. Depending on the polymerization, temperature costs account for a significant fraction of total process operating costs. The need to minimize the specific initiator consumption rate (i.e., grams of initiator injected per kilogram of PE produced) by employing optimal process operating conditions and initiator types always exists. For some free-radical initiators used in high-pressure polyethylene processes (e.g., *tert*-butyl peroctoate, *tert*-butyl 3, 5, 5-trmethylperhexanoate, di-*tert*-butyl peroxide), it has been reported that the specific initiator consumption rate is a nonlinear function of reaction temperature. For example, as shown in Figure 13.2, at low reaction temperatures, the specific initiator consumption rate decreases with increasing reaction temperature; however, as the reaction temperature is further increased, the specific initiator consumption rate increases. It is generally believed that such a nonlinear dependence of the specific initiator consumption rate on reaction temperature in a stirred autoclave reactor is due to imperfect mixing of the reacting fluid in the reactor (Figure 13.2).

13.2.1 Continuous Autoclave Reactors

In high-pressure autoclave processes, either single-stage or multistage reactor systems are used. Multistage reactors are advantageous in that the polymerization conditions

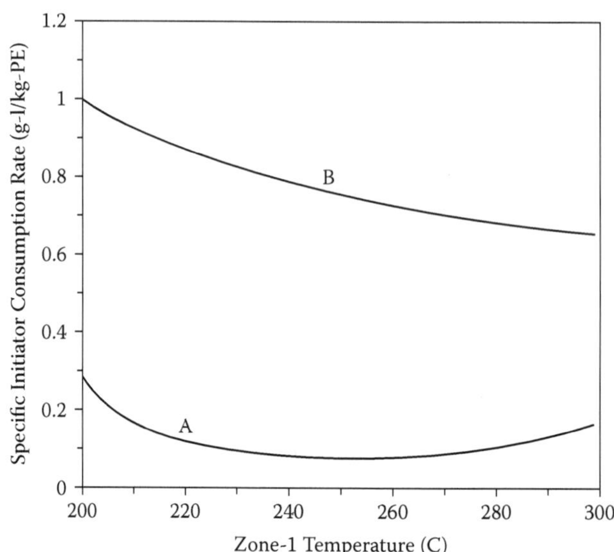

FIGURE 13.2 Specific initiator consumption rate for two common free-radical initiators. From B. G. Kwag and K. Y. Choi, *Ind. Eng. Chem. Res.*, 33, 211 (1994). With permission.

in each compartment or reaction zone can be varied to broaden the product-grade slate. When a multistage or multicompartmented stirred autoclave reactor system is used, the reactor is typically a vertical cylindrical vessel with a large L/D radio. A single initiator or a mixture of initiators is used. In a typical high-pressure process, fresh ethylence, after primary compression, is combined with recycled ethylence and with a comonomer. The mixture is then pressurized to the desired reactor pressure in the second compression state. The reacting fluid (ethylene and polyethylene mixture) is intensely mixed by an agitator that consists of a vertical shaft to which impeller blades are attached to ensure efficient mixing of monomer, polymers, and initiators.

In general, the overall residence time is very short (1–3 min) and the reactor operates adiabatically. About 18–20% of monomer conversion is obtained. The heat of polymerization (21.2 kcal mol^{-1}) is removed by injecting cold monomer feed and by the bulk flow of ethylene and polyethylene mixture in the reactor. The reactor temperature is controlled by regulating the flow rate of initiator injected into the reactor. When a multistage reactor system is used, keeping the temperature in each zone at its desired level is important. This is because the resulting polymer properties (e.g., molecular weight, molecular-weight distribution, and short-chain and long-chain branching frequencies) are strongly dependent on reaction temperature as well as pressure and initiator type. Cold monomer feeds may be added to a few selected reaction compartments to remove additional reaction heat.

For a single continuous autoclave reactor of volume V to which ethylene and initiator are supplied, the following simple modeling equations can be derived:

$$\frac{dM}{dt} = \frac{1}{\theta}(M_f - M) - k_p MP \tag{13.1}$$

$$\frac{dI}{dt} = \frac{q_i}{V}(I_r - I) - k_d I \tag{13.2}$$

$$\frac{dT}{dt} = \frac{1}{\theta}(T_f - T) + \frac{(-\Delta H_r)}{\rho C_p} k_p MP \tag{13.3}$$

where θ is the residence time, T is the reactor temperature, T_f is the feed ethylene temperature, M_f is the feed ethylene concentration, M is the monomer concentration in the reactor, I_f is the feed initiator concentration, I is the initiator concentration in the reactor, k_p is the propagation rate constant, k_d is the initiator decomposition rate constant, q_i is the initiator feed rate, ΔH_r is the heat of polymerization, ρ is the fluid density, C_p is the fluid heat capacity, and P is the concentration. At steady state, we can show from the preceding equations that the monomer conversion (x) is directly dependent on the reactor and ethylene feed temperatures as follows:

$$x = \frac{\rho C_p}{(-\Delta H_r) M_r}(T - T_r) \tag{13.4}$$

Notice that the monomer conversion is determined by the temperature difference between the reactor content and the monomer feed. It is also easy to show that the corresponding initiator feed rate is given by

$$q_i = \left(\frac{x}{\theta k_p (1-x)}\right)^2 \frac{Vk_t}{2\theta} \left(\frac{f_i k_d I_f}{1 + \theta k_d}\right)^{-1} \tag{13.5}$$

where k_t is the termination rate constant and f_i is the initiator efficiency factor (fraction of radicals available for chain initiation). Similar equations can be derived for a multizone reactor system.

In high-pressure ethylene polymerization, the kinetic rate constants are also dependent on pressure. For example, the propagation rate constant is given by

$$k_p = 4.8 \times 10^5 \exp\left(\frac{-4450 + 0.31P}{T}\right) \tag{13.6}$$

where k_p is in $m^3\ mol^{-1}\ sec^{-1}$, P is in bar, T is in K. It should be emphasized that a great deal of inconsistency exists in the reported values of the kinetic rate constants for high-pressure ethylene polymerization.

In autoclave processes, a mixture of fast initiator (low-temperature initiator) and slow initiator (high-temperature initiator) is used. Some examples of the commercial initiators used in high-pressure polyethylene processes are shown in Table 13.2.

When the reactor consists of more than one compartment, the behavior of downstream compartments are strongly dependent on those of upstream compartments. For the two-compartment continuous autoclave reactor, Figure 13.3 illustrates the temperature of the second compartment as a function of the first compartment for several volume ratios of the two compartments [18]. Here, the volume of zone 1

TABLE 13.2
Examples of Commercial Initiators Used in High-Pressure PE Processes

Peroxide	Mol. Wt. (g mol^{-1})	Active Oxygen (wt%)	1-Hour Half-Life Temperature (°C)
Fast Initiators			
t-Butyl perneodecanoate	244.4	6.55	64
t-Amyl perneodecanoate	258.4	6.19	62
Di-(2-ethyl-hexyl)peroxy dicarbonate	346.5	4.62	47
Slow Initiators			
Di-*tert*-butyl peroxide	146.2	10.7	147
Di-*tert*-amyl peroxide	174.2	8.7	142
t-Butyl hydroperoxide	90.1	14.1	200
3,4-Dimethyl-3,4-diphenyl hexane	266.2	—	210

(first compartment) is fixed and only the zone 2 (second compartment) volume is varied. Notice that as the volume ratio is increased, the zone 2 temperature increases for a fixed zone 1 temperature, particularly in the zone 1 temperature range 200–250°C. For large zone 2 volumes, the fluid residence time increased, and as a result, ore initiators decompose to radicals, which in turn accelerates the chain propagation reaction. Figure 13.3 also illustrates that the temperature difference between zone 1 and zone 2 is quite large for low zone 1 temperatures.

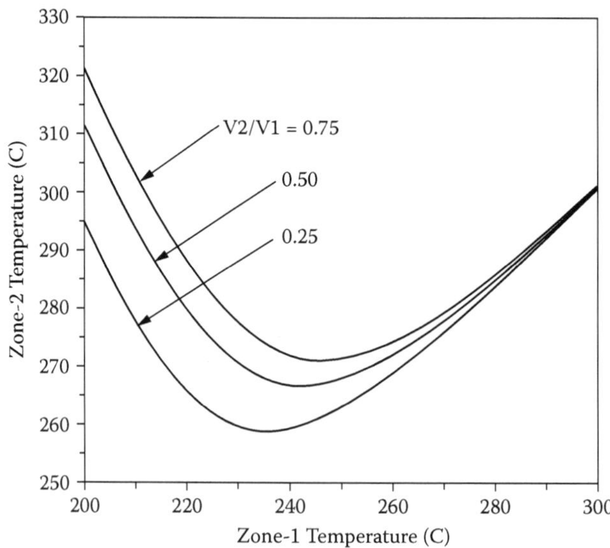

FIGURE 13.3 Effect of zone 2–zone 1 volume ratio on zone 2 temperature. From B. G. Kwag and K. Y. Choi, *Ind. Eng. Chem. Res.*, 33, 211 (1994). With permission.

Although the high-pressure polyethylene process has been used in the polymer industry for years, many aspects of polymerization are still not completely understood. With a growing importance for tightly controlling the polymer properties or developing new polymer grades by changing the process operating conditions, the need for a thorough understanding of the process is becoming more important than ever.

One of the well-known phenomena in high-pressure polyethylene processes is a rapid ethylene decomposition reaction or thermal runaway, known as “decomp.” At 300°C, ethylene, and even polyethylene decompose to carbon, methane, hydrogen, and other hydrocarbon by-products. When the decomposition reaction takes place, the reactor pressure builds up quickly and the reactor must be vented, shut down, and flushed for a long period of time before a new startup is initiated. The resulting economic loss will be quite significant. The following reactions are believed to occur when ethylene is thermally decomposed [9]:

$$2C_2H_2 \rightarrow C_2H_3\cdot + C_2H_5\cdot$$

$$C_2H_5 \Leftrightarrow C_2H_2 + H\cdot$$

$$C_2H_5\cdot + C_2H_5 \rightarrow C_2H_6 + C_2H_5\cdot$$

$$H\cdot + C_2H_4 \rightarrow H_2 + C_2H_3\cdot$$

$$C_2H_3\cdot + C + CH_3\cdot$$

$$CH_3\cdot + C_2H_4 \rightarrow CH_4 + C_2H_3\cdot$$

$$2CH_3\cdot \rightarrow C_2H_6$$

$$CH_3\cdot + C_2H_3\cdot \rightarrow C_2H_2 + CH_4$$

$$2C_2H_{3\cdot}\cdot \rightarrow C_2H_2 + C_2H_4$$

The thermal runaway can be caused by the formation of local hot spots. The causes for local hot spots are, for example, feed impurity, excess initiator in feed, poor feed distribution, inadequate mixing, mechanical friction, poor reactor temperature control system, feed temperature disturbance, and so forth [9]. Imperfect mixing in the polyethylene reactor has been considered as a primary cause for the runaway reaction phenomena. Short reactor residence time and comparable macromixing times may lead to imperfect micromixing. Then there can be different polymerization rates, and different concentration and temperature gradients in the mixing zones [10]. Recently, a three-dimensional computational fluid dynamics (CFD) approach has been used by Read and co-workers [10] to analyze the micromixing phenomena in a single high-pressure LDPE reactor. According to their computer simulation study, characteristics of the autoclave reactor system include a steep concentration of the initiator profile close to the inlet, a temperature that increases going down the reactor, a maximum in the radical concentration, a conversion that increases down the reactor, and great sensitivity to the composition of the initiator.

In a stirred autoclave process, the polymerization is usually performed in a single-phase region to facilitate the heat removal and to avoid fouling and forming cross-linked polymers. The presence of a viscous polymer-rich phase can also be the cause for the thermal runaway reaction via the autoacceleration effect [11]. When phase separation occurs in the reactor, chain termination in highly viscous polymer-rich phase becomes severely diffusion controlled, resulting in a rapid increase in the propagation rate.

Buildup of polymer deposits on the reactor surfaces often occurs if dead spots are present in the reactor. In the polymer deposits, polymerization and long-chain-branching reactions continue to occur, contributing to gel formation of fish eyes that decrease the polymer quality.

In some instances, polymerization is performed in a two-phase region to produce LDPE exhibiting superior film properties because of narrower molecular-weight distribution and less long-chain branches. In an autoclave reactor, phase separation can be achieved by lowering the pressure or by adding an inert antisolvent such as nitrogen to the reaction mixture [12].

It should also be noted that recently a new high-pressure autoclave process has been developed by Exxon Chemical Company to produce linear low-density polyethylene using metallocene catalyst technology. Because the metallocene catalyst is a single-site catalyst, the molecular-weight distribution of the resulting polyethylene is very low ($M_w/M_n \approx 2.0$). The polymerization is performed in a staged autoclave reactor at 1000–2000 atm and 150–250°C with 30–120 sec of reactor residence time [13].

13.2.2 Tubular Reactors

The tubular high-pressure ethylene polymerization reactor system consists of a long (more than a mile) narrow jacketed spiral tube with multiple feed streams along its length, ethylene compressors, and flash separators. To provide a plug flow profile in the reactor, a high length/diameter ratio (250–12,000) is used. For approximately 120,000 tons h^{-1} plant, the tube diameter of around 50 mm is used. The unreacted ethylene in the polymerizing mixture is separated and recycled to the reactor. The polymerization pressure is typically of about 2500–3000 atm and temperatures in the range 150–330°C, which is close to its safety limit (345°C). Compressed pure ethylene is mixed with a free-radical initiator (oxygen) and the reactor residence time is as low as 20 sec. The first section of the tubular reactor is heated because the heat generated by polymerization is insufficient. Because the fluid velocity is very high, a pressure drop along the tube length is significant and it affects the propagation rate. As a result, polymer molecular-weight distribution is broader than by using a stirred autoclave reactor [14].

For certain polymer grades, the inner reactor tube is fouled due to polymer deposition. The polymer deposition occurs due to the cooling of the reaction mixture through wall heat transfer. The fouling is faced mainly by increasing the coolant inlet temperature in the corresponding reactor jacket to melt the polymer wall deposits [15]. The time scale of the fouling–defouling cycles is about 2–12 h. A common method for minimizing the fouling is the practice of pressure-pulsing the reactor. About once a minute, the pressure at the end of the reactor is suddenly dropped for several seconds by partially opening a special valve. The pressure pulse then transmits itself through the long tubular reactor, causing sudden increases in the flow velocities that tear the polyethylene deposits from the tube walls [6].

Because untreated ethylene is recycled, impurities may accumulate in the system that may affect the overall production of the primary radicals and the molecular-weight developments of the polyethylene product. Chain transfer agents such as ketones, aldehydes, alcohols, hydrogen, or chlorinated components may be added to narrow molecular-weight distribution.

In a tubular polyethylene reactor, the high-pressure valves are an important part of the process technology. For example, the difference in pressure at the release valve at the end of the tubular reactor can amount to 3000 atm. The temperature difference across the valve can be up to 60°C. When the release valve is opened, the flow increases severalfold and the regulating power of the valve must be very high and can be 100 mega-pounds. The valve adjusting must take place in milliseconds and must be exact to 10^{-3} mm [7].

After polymerization, the reacting mass is transferred to a high-pressure separator (150–200 atm). The polymer-rich stream withdrawn from the bottom of the high-pressure separator undergoes a second separation step at near-atmospheric pressures in a low-pressure separator. The phase separation in the high-pressure separator and in separators of the recycle line makes it possible to remove the polymer or wax from the mixture. To improve the separation efficiency, the pressure or the temperature in the separators must be decreased. Then, the recycle gas contains only traces of wax, and a lower amount of ethylene remains dissolved in the polymer melt from the high-pressure separator.

Figure 13.4 illustrates the phase-equilibrium curves for mixtures of ethylene and two LDPEs of different molecular weights [16]. The cloud curves give the pressure at the cloud point when the mixture starts to separate. The coexistence curves give the composition of the phases. The left branch of a coexistence curve shows the composition of the ethylene-rich light phase and the right branch gives the composition of the polymer-rich dense phase. Outside the cloud curve, the mixture is homogenous (single phase). Because polymers have a molecular-weight distribution, a phase separation in mixtures of polymers is always accompanied by a fractionation. As a result, the polymer in the polymer-rich phase has a higher polydispersity than the polymer in the ethylene-rich phase.

13.3 POLYSTYRENE

Polystyrene is one of the most important commodity polymers and perhaps the most well-known and most extensively studied polymers. Styrene can be polymerized by free-radical polymerization, ionic (cationic and anionic) polymerization, and coordination polymerization. Free-radical polymerization is most frequently used to produce atactic polystyrene. Ionic polymerization is used to prepare polystyrene of narrow molecular-weight distribution. Because styrene reacts readily with many other vinyl monomers and rubbers, a wide variety of styrene copolymers are commercially available. Many styrene copolymers are commercially available, but the following four styrene polymers are of particular importance:

- Polystyrene homopolymer: a clear, colorless, and brittle amorphous polymer (often called "crystal polystyrene" or "general-purpose polystyrene").
- Styrene–acrylonitrile copolymer (SAN): a random copolymer having strong chemical resistance, heat resistance, and mechanical properties.

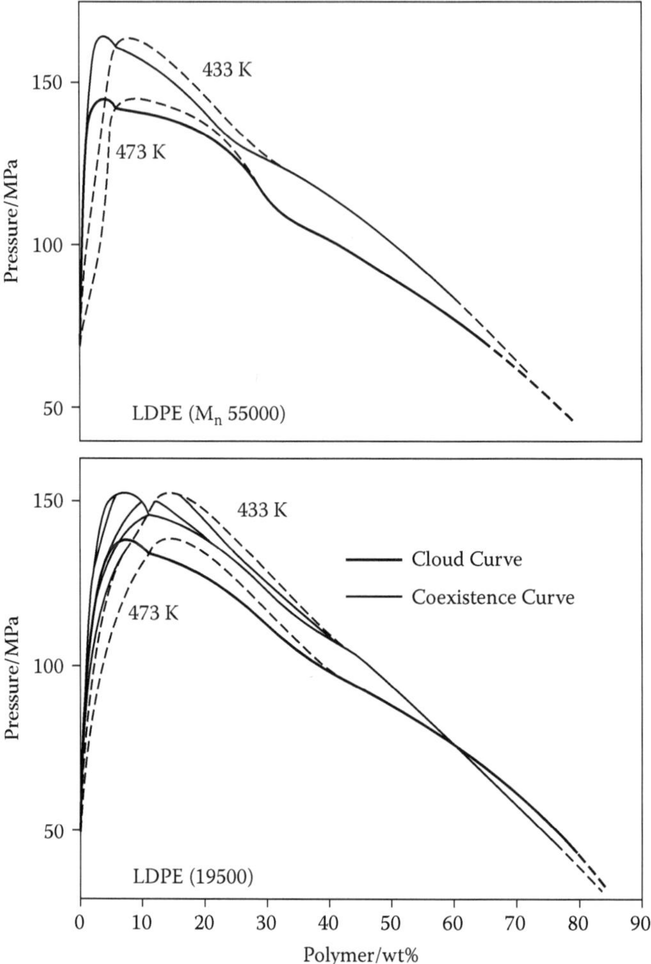

FIGURE 13.4 Phase equilibrium curves of ethylene and LDPE mixtures. From G. Luft, paper presented at Polyethylene World Congress, 1992. With permission.

- High-impact polystyrene (HIPS): an amorphous two-phase polymer with rubber particles in a polystyrene matrix to impart added impact strength.
- Acrylonitrile–butadiene–styrene copolymer (ABS): a rubber-modified SAN copolymer; SAN forms the matrix phase and is grafted to portion of rubber.

The SAN and ABS copolymers contain approximately 25 wt% of acrylonitrile and polybutadiene rubber in amounts up to 20 wt%. Other styrene copolymers of industrial importance include styrene–maleic anhydride copolymer (SMA), styrene–divinylbenzene copolymer, acrylic–styrene–acrylonitrile terpolymer, and styrene–butadiene copolymer. Recently, metallocene catalysts have been developed to synthesize syndiotactic polystyrene (sPS). The polymerization process and process conditions have major effects on polymer properties and process economy. For styrene homopolymerization and copolymerization, various types of polymerization reactors are used commercially.

13.3.1 Styrene Homopolymerization

Polystyrene has a glass transition temperature of 100°C and is stable to thermal decomposition to 250°C. Styrene homopolymers are manufactured by suspension, mass (bulk), and solution polymerization processes. The polymerization can be initiated thermally at high temperatures or by a free-radical initiator. Typical molecular weights for polystyrene range from 200,000 to 300,000. When heated above 100°C, styrene generates free radicals and polymerizes to amorphous polystyrene. Thermal polymerization is advantageous in that because no additives are needed, the resulting polymers are pure. Thermal polymerization of styrene has been studied by many workers, and several reaction mechanisms have been proposed. Among them, the mechanism proposed by Mayo [17] is the most widely accepted. This mechanism involves the formation of the Diels–Alder adduct, followed by a molecular-assisted homolysis between the adduct and another styrene

$$M + M \rightarrow AH$$

$$AH \rightarrow 2M$$

$$AH + M \rightarrow R\cdot + R\cdot'$$

where M is the monomer and AH is 1-phenly-1, 2, 3, 9-tetrahydronaphathalene (Dies-Alder adduct). The overall initiation rate is expressed as

$$R_i = 2k_iM^3$$

The kinetic scheme for free-radical styrene homopolymerization initiated by a chemical initiator is represented as follows:

Initiation

$$I \xrightarrow{k_d} 2R$$

$$R + M \xrightarrow{k_1} P_1$$

Propagation

$$P_1 + M \xrightarrow{k_p} P_2$$

$$P_n + M \xrightarrow{k_p} P_{n+1}$$

Termination

$$P_n + P_m \xrightarrow{K_{tc}} M_{n+m}$$

$$P_n + P_m \xrightarrow{k_{td}} M_n + M_m$$

TABLE 13.3
Typical Operating Conditions for Styrene Homopolymerization

Process	Temperature (°C)	Pressure	Reaction Time (h)
Suspension	110–170	Reduced	5–9
Mass (bulk)	80–200	Slightly reduced to 10–20mmHg	12–18 (batch)2–8 (continuous)
Solution	90–130	Atmospheric to 10–20 mmHg	6–8

Chain transfer

$$P_n + M \xrightarrow{k_{fm}} M_n + P_1$$
$$P_n + S \xrightarrow{k_{fs}} M_n + S\cdot$$

where I is the initiator, R is the primary radical, M is the monomer, P_n is the polymer radical with n monomer units, M_n is the dead polymer with n monomer units, S is the solvent.

In styrene polymerization, termination is mostly by combination (coupling). Quite often, a mixture of initiators or multifunctional initiators is used to achieve a high polymerization rate and high polymer molecular weight [18–20]. Typical operating conditions for styrene homopolymerization processes are presented in Table 13.3 [21].

13.3.1.1 Suspension Polymerization

Suspension polymerization of styrene is mostly performed in a batch reactor. A styrene monomer containing initiators(s) and chain transfer agents is dispersed in water as fine droplets by mechanical agitation. A blowing agent (e.g., *n*-pentane) can be added to the monomer if expandable polystyrene beads are desired. Each monomer droplet acts like a microbulk polymerization reactor. As polymerization proceeds, the monomer droplets become hard polymer particles or beads. To prevent the agglomeration of suspended particles, a water-soluble suspending agent is added. Because the heat generated in the suspended polymerizing particles can be dissipated effectively through, the polymer particle size and its distribution are dictated by agitation, type of suspending agent, and its concentration.

A typical monomer/water ratio in suspension processes ranges from 1:4 to 1:1 and the particle size ranges from 0.1 to 1.0 mm. After polymerization, the polymer beads are washed with acid to remove initiator residue and suspension stabilizers and palletized using an extruder where any remaining monomer is further removed.

13.3.1.2 Mass (Bulk) Polymerization

Polystyrene manufactured by a mass process has high clarity and good electrical insulation properties. The main difficulties in polymerizing styrene by a mass process include heat removal and handling of a highly viscous polymerizing mass. If an initiator is used, the polymerization temperature is below 100°C, but a higher temperature is employed when the polymerization is thermally initiated. When a stirred tank reactor is used, monomer conversion per pass is limited to 30–50%. The polymerizing mass is transferred to an extruder/devolatilizer where unreacted monomer is recovered and recycled to the reactor. The stirred reactor is equipped with a reflux condenser operating under reduced pressure.

Both circulated and noncirculated tower reactors are used for bulk styrene polymerization. In a noncirculated tower reactor with no agitation device, styrene is polymerized in bulk to about 33–35% conversion. The reactor temperature profile is controlled from 100°C to 200°C by jackets and internal cooling coils. From the inlet to the reactor outlet, the temperature is progressively increased to reduce the viscosity of the polymerizing mass and to keep the polymerization rate high. In a noncirculated tower reactor system, more than one tower reactors are often used and the flowing reacting mass is slowly agitated. The product is discharged from the bottom of the tower by an extruder, cooled, and palletized. A small amount of ethyl benzene (5–25%) may be added to a styrene feed stream to reduce the viscosity of a polymer solution and to ease heat transfer.

13.3.1.3 Solution Polymerization

In solution polymerization of styrene, the viscosity of the polymerizing solution is much lower than in the mass process; thus, temperature control is easier. Ethyl benzene is the most commonly used solvent and its concentration in the fed stream is about 5–25%. After polymerization, unreacted styrene and solvent are removed from the polymer and recycled. Three reactor types are used for solution polymerization of styrene as depicted in Figure 13.5 [22].

Continuous plug flow reactors such as recirculated stratified agitated tower reactors are multistaged, having a temperature profile of 100–170°C. Recirculated coil and ebullient reactors are single staged and operated isothermally.

13.3.2 Styrene Copolymerization

Styrene homopolymer is a brittle polymer. Styrene copolymers are industrially of significant importance because a wide variety of polymer properties can be obtained

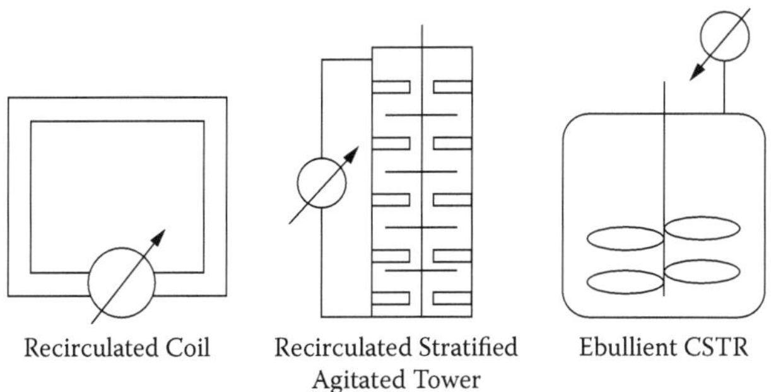

FIGURE 13.5 Continuous solution polymerization reactors. From D. B. Friddy, *Adv. Polym. Sci.*, 111, 67 (1994). With permission.

by copolymerizing styrene with rubbers (diens) and other vinyl monomers. When two vinyl monomers are copolymerized, the copolymer composition is determined by the following four propagation reactions:

$$M_1^* + M_1 \xrightarrow{K_{11}} M_1^*$$

$$M_1^* + M_2 \xrightarrow{K_{12}} M_2^*$$

$$M_2^* + M_1 \xrightarrow{K_{21}} M_1^*$$

$$M_2^* + M_2 \xrightarrow{K_{22}} M_2^*$$

where M_1 and M_2 are the comonomers, M_1^* and M_2^* are the polymer chain ending in a radical derived from an M_1 and M_2 monomer, respectively. The copolymer composition is determined by the reactivity ratios defined as

$$r_1 \equiv \frac{k_{11}}{k_{12}}$$

$$r_2 \equiv \frac{k_{22}}{k_{21}}$$

The mole fraction (F_1) of monomer M_1 in the polymer phase can then be expressed as follows:

$$F_1 = \frac{r_1 f_1^{\,2} + f_1 f_2}{r_1 f_1^{\,2} + 2 f_1 f_2 + r_2 f_2^2} \tag{13.7}$$

where f_1 and f_2 are the mole fractions of monomer 1 and monomer 2 in the bulk phase, respectively. The reactivity ratios for various copolymerization systems are listed in the *Polymer Handbook* [23]. Quite often, different values of reactivity ratios are reported by different workers for the same copolymerization system. Equation (13.7) is valid for other binary copolymerization processes.

Figure 13.6 illustrates the copolymer composition curves for several styrene–comonomer systems. This figure shows that a high degree of composition drift and composition nonhomogeneity may occur in a styrene–acrylonitrile (SAN) copolymerization system for certain copolymer compositions.

When styrene is copolymerized with rubber to impact polymers (HIPS and ABS), rubber particles are imbedded into a polystyrene matrix. These soft rubber particles grafted onto a rigid polystyrene body are not compact but contain occluded matrix material. The grafting reaction can be schematically represented as

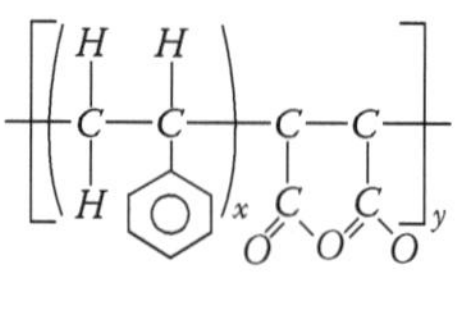

$$\left[\left(\mathrm{CH_2{-}CH(C_6H_5)}\right)_x - \mathrm{CH{-}CH}(\mathrm{C{(=O)}{-}O{-}C{(=O)}})\right]_y$$

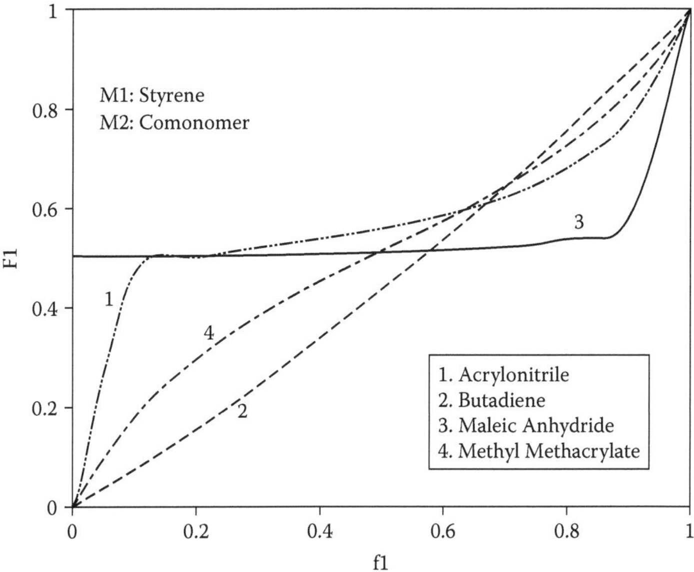

FIGURE 13.6 Copolymer composition curves.

In graft copolymerization using an initiator, grafting occurs, preferentially via the primary radicals, whereas in thermal graft copolymerization grafting is initiated by polymer radicals [24, 25].

13.3.3 Styrene–Acrylonitrile Copolymer

The styrene–acrylonitrile (SAN) is one of the largest volume copolymers of styrene. SAN copolymers are amorphous, transparent, and glossy random copolymers produced by batch suspension, continuous mass (or solution), and emulsion polymerization processes. The molecular weight and the acrylonitrile content of the copolymer are the key factors in determining polymer properties. SAN has improved tensile yield, heat distortion, and solvent residence than polystyrene because of the incorporation of acrylonitrile. For example, SAN is resistant to aliphatic hydrocarbons, alkalines, battery acids, vegetable oils, foods, and detergents. SAN is attacked by some aromatic hydrocarbons, ketones, esters, and chlorinated hydrocarbons [26]. Emulsion processes are used to produce SAN diluent for ABS resins. SAN produced by mass and suspension processes are primarily used for molding applications. A variety of copolymer properties or grades is available, depending on the molecular weight and the copolymer composition (styrene/acrylonitrile ratio).

The major problems in SAN processes are related to reactor temperature control, mixing, and copolymer composition control. The viscosities of concentrated SAN solutions are higher than polystyrene of the same molecular weight; thus, heat removal becomes more difficult. In a continuous reactor process, the mixing of low-viscosity reacting fluid can be difficult. The polymerization rate is higher than that of styrene homopolymerization. If mixing is not homogeneous. SAN degrades, resulting in coloring and contamination. To avoid composition drift in a batch copolymerization process, SAN copolymers are often manufactured at the azeotropic point (see Figure 13.6), where monomer and polymer have the same composition. However, if the desired copolymer composition is not the azeotropic composition, the copolymer composition varies with conversion (or composition in the bulk phase). Composition drift in SAN copolymers is undesirable because SAN copolymers of different compositions are incompatible and cause phase separation. Therefore, it is crucial to monitor the bulk-phase composition and to make some corrective actions to prevent the copolymer composition drift. For example, more reactive monomer or comonomer can be added to the reactor during polymerization to keep the monomer/comonomer ratio constant. The polymerization reactors that can be used for continuous mass processes are loop reactors and continuous stirred tank reactors (CSTR) with an anchor agitator.

13.3.3.1 Emulsion Process

Both batch (or semicontinuous) and continuous emulsion process are used to manufacture SAN latex. In a batch process, a styrene, acrylonitrile, and aqueous solution of a water-soluble initiator, e.g., potassium persulfate), emulsifier, and chain transfer agent (molecular-weight regulator, e.g., dodecyl mercaptan) are charged into a stirred tank reactor. The weight ratio of styrene/acrylonitrile is generally kept between 70:30 and 85:15. If the desired copolymer composition corresponds to the azeotropic composition, copolymer composition drift will be minimal. Otherwise, a monomer mixture having a different styrene/acrylonitrile ratio from the initial charge must be added continuously during polymerization to keep the ratio constant because any wall deviation in the bulk monomer phase composition can easily result in a significant composition heterogeneity. For example, two SAN copolymers differing more than 4% in acrylonitrile content are incompatible, resulting in poor physical and mechanical properties [27]. To calculate the monomer feeding policy, a dynamic optimization technique can be used with a detailed process model [28]. In such a process, the reactor temperature can also be varied to minimize the batch reaction time while maintaining the copolymer composition and molecular weight and/or molecular-weight distribution at their target values.

In a continuous emulsion process, two or more stirred tank reactors in series are used. Separate feed streams are continuously added into each reactor. The reactors are operated at about 68°C. The latex is transferred to a holding tank (residence time of about 4 h) before being steam-stripped to remove unreacted monomers. In a continuous process, the residence time distribution is generally broad. A large holding tank placed downstream of the reactors provides extra time to the reaction mixture and reduces the molecular-weight distribution.

13.3.3.2 Suspension Process

Suspension polymerization is performed by using a single reactor or two parallel reactors. A mixture of monomers, monomer-soluble initiator (peroxides and azo compounds), and any additives (e.g., chain transfer agents) is dispersed in water by mechanical agitation in the presence of a suspension stabilizer. The suspension polymerization temperatures range from 70°C to 125°C. The reactor temperature is increased gradually during the batch. SAN copolymer particles of 10–3000 μm are obtained. To keep the copolymer composition constant, a mixture of monomers is added into the reactor as in emulsion processes.

13.3.3.3 Mass Process

The mass process has some advantages over emulsion and suspension processes in that it is free of emulsifiers and suspending agents and thus produces SAN copolymers having higher clarity and good color retention. As no solvent is used, the viscosity of the reacting mass increases with conversion, and the removal of polymerization heat becomes a problem when the conversion of monomers exceeds ≤ 60–70%. The mass polymerization is performed in a jacketed reactor equipped with a reflux condenser at a temperature ranging from 110°C to 210°C. In a continuous process, 50–65% conversion obtained in a single reactor is further increased to 65–90% in the second reactor, which is typically a horizontal linear flow reactor. To handle the highly viscous reaction mass, the reaction temperature in the linear flow reactor is increased along the direction of flow. Mass polymerization can be initiated either thermally or chemically by organic initiators. The polymer product is fed to a film evaporator, where unreacted monomers are recovered.

13.3.4 High-Impact Polystyrene

Styrene homopolymer is a rigid but brittle polymer with poor impact strength. Rubber is grafted to polystyrene to improve the impact strength of polystyrene. Fine rubber particles (0.5–10 μm) are dispersed in the polymer matrix. The incorporated rubber particles are cross-linked and contain grafted polystyrene. The inner structure of the polymer is determined by the polymerization process. Because the refractive indices of the rubber and the polystyrene phase are different, HIPS polymer is a translucent-to-opaque white polymer that exhibits high-impact strength and is resistant to wear. *Cis*-polybutadiene is the most common rubber used in the manufacture of HIPS. The properties of HIPS depend on the amount and type of rubber as well as many other reaction variables. Rubber particles that are too small or too large may cause a loss of impact strength. A glossy HIPS polymer needs to balance the small particle size against impact strength. High rubber content, large particle size, high matrix molecular weight, and the choice of plasticizer improves the HIPS polymer's resistance to environmental stress crack agents [29].

Impact polystyrene is manufactured commercially by suspension, mass, and solution processes. In a mass process, polybutadiene rubber (2–5 μm) is dissolved in styrene. As styrene is polymerized, phase separation occurs, and a rubber-rich phase and a polystyrene-rich phase are formed. The reaction mixture becomes opaque because of the difference in refractive indices between the two phases. Initially, polybutadiene in styrene is the continuous phase, and polystyrene in styrene is the discontinuous phase. When the phase volumes reach approximately equal volumes and sufficient shearing agitation exists, phase inversion occurs. Then, polystyrene in styrene becomes the continuous phase and polybutadiene in styrene becomes the discontinuous phase. At the phase-inversion point, a change in viscosity is observed [30]. The cohesion barrier attributable to the solution viscosity is overcome by the shearing agitation [31]. If shearing agitation is not adequate, phase inversion does not occur and a cross-linked continuous phase that produces gel is formed. Figure 13.7 is a phase

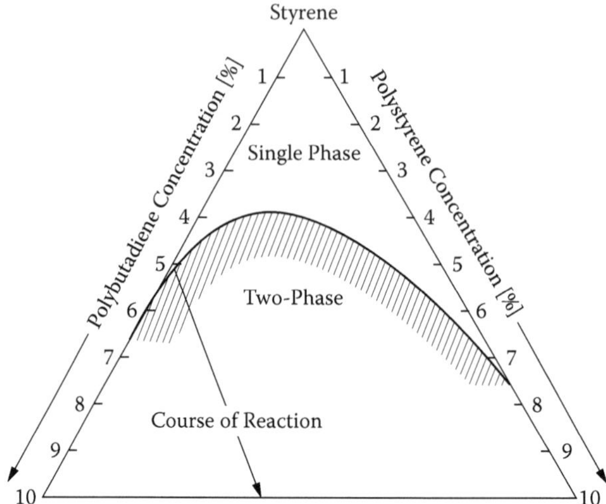

FIGURE 13.7 Phase diagram of styrene–polystyrene–polybutadiene system. From W. A. Ludwico and S. L. Rosen, *J. Polym. Sci. Polym. Chem. Ed.*, 14, 2121 (1976). With permission.

diagram of the styrene–polystyrene–polybutadiene system [32]. The following equation has also been proposed to calculate the maximum degree of grafting (f) [33]:

$$f = \frac{V_R}{V_S} \ln\left(1 + \frac{V_R}{V_S} x\right) \tag{13.8}$$

where V_R and V_S are the phase volume of rubber and of polystyrene in styrene, respectively, and x is the conversion.

During polymerization, some of the free radicals react with the rubber, which is then grafted to the polystyrene chain. The grafting at the interface strongly affects particle size, morphology, and toughness of the polymer. The grafted rubber with its side chains accumulates at the interface between the two phases and functions as an oil-in-oil emulsion stabilizer. Figure 13.8 illustrates various rubber particle structures in impact polystyrene [30].

The kinetics of graft copolymerization substantially correspond to those of styrene homopolymerization except at low rubber concentrations and at high conversions due to cross-linking reactions [34]. Figure 13.9 illustrates a schematic of a network of polybutadiene and polystyrene [35].

At high conversions, the salvation of the rubber and the gel effect (Trommsdorf effect) cause an increase in the molecular weight of the grafted polystyrene. The viscosity change with styrene conversion in the manufacture of HIPS is shown in Figure 13.10 with typical rubber particle morphology [34].

Like a styrene homopolymer, HIPS are manufactured commercially by using agitated tower reactors, a stratifier, and back-mixed reactors [36–40]. HIPS polymers are often manufactured by a multistage process. Figure 13.11 illustrates a continuous mass process for the manufacture of impact polystyrene [40]. In the first stage (prepolymerizer), operating at 125–140°C, a solution of rubber and styrene is charged into the reactor with initiator, antioxidants, and other additives. Dissolving rubber

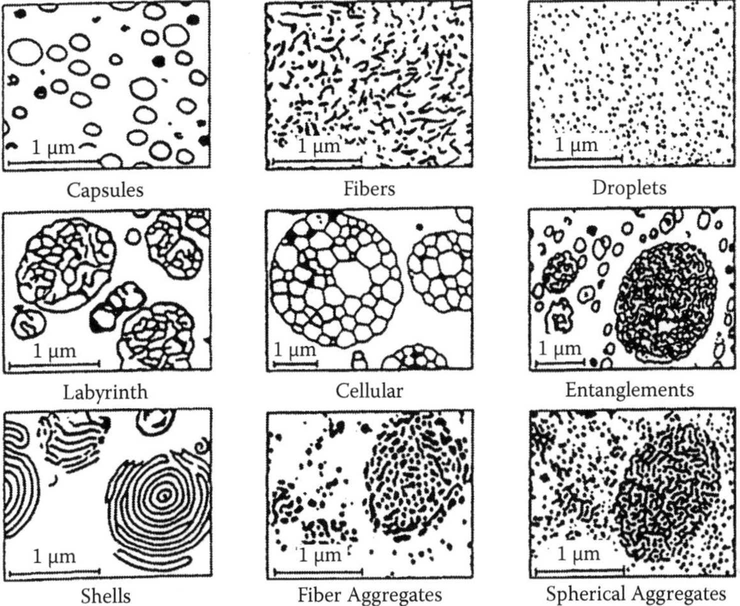

FIGURE 13.8 Various rubber particle structures in impact polystyrene. From W. D. Watson and T. C. Wallace, *ACS Symp. Ser.*, 285, 363 (1985). With permission.

in styrene is a slow process and must be done with enough shearing agitation to effect phase inversion and to adequately size the rubber particles. The phase inversion occurs in the first stage. The polymerization is performed to a conversion that is close to the point where increasing viscosity seriously limits mixing and temperature control. The agitation in the prepolymerization reactor must be sufficient to shear rubber particles. The speed of the agitator necessary to develop the shear is determined by the viscosity of the reaction medium which is a function of the temperature and the conversion level. The total monomer conversion in this stage is from about 25% to 35%. The product from the prepolymerization stage is then transferred to the second-stage reactor (intermediate zone), where the maximum monomer conversion reaches about 65–85% at 140–160°C in the two second-stage reactors shown in Figure 13.11 (marked by 4 and 11), and styrene vapor is condensed and recycled. The third–stage rector (final polymerizer) is a vertical or horizontal cylinder or tower

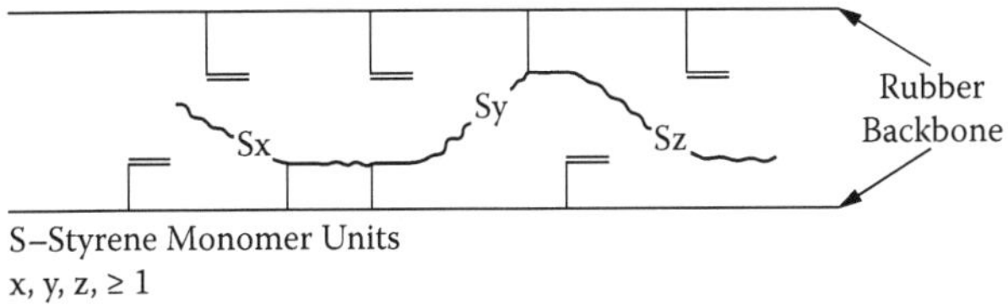

FIGURE 13.9 A network of polybutadiene and polystyrene. From D. J. Stein, G. Fahrbach, and H. Adler, *Adv. Chem. Ser.*, 142, 148 (1975). With permission.

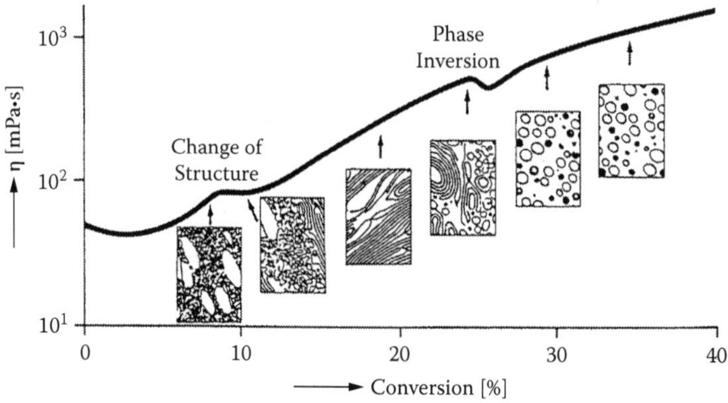

FIGURE 13.10 Viscosity–conversion curve for HIPS process. From A. Echte, F. Half, and J. Hambrecht, *Angew. Chem. Int. Ed. Engl.*, 20, 344 (1981). With permission.

reactor. It is an adiabatic plug flow reactor in which 85–90% conversion is reached. To ensure plug flow, only very slow stirring is allowed. In a vertical reactor, the polymerizing mass from the second stage is fed to the top of the reactor and flows downward. The temperature of the reacting mass increases gradually from 175°C to 215°C.

Figure 13.12 illustrates another example of a continuous bulk HIPS process where the polymerization is carried out in multiple stages [41]. Rubber is dissolved in styrene in the first stage operating under atmospheric pressure at temperatures

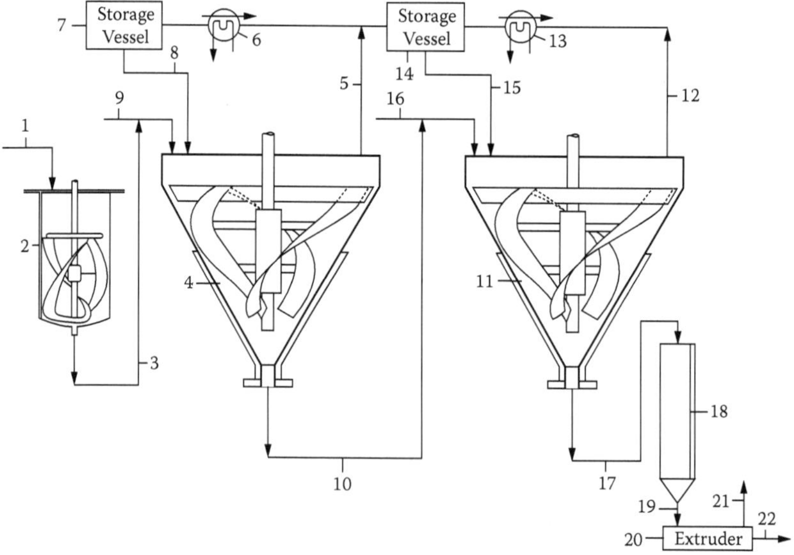

FIGURE 13.11 Continuous mass impact polystyrene process. From G. Gawne and C. Ouwerkerk, U.S. Patent 4,011,284 (1977). With permission.

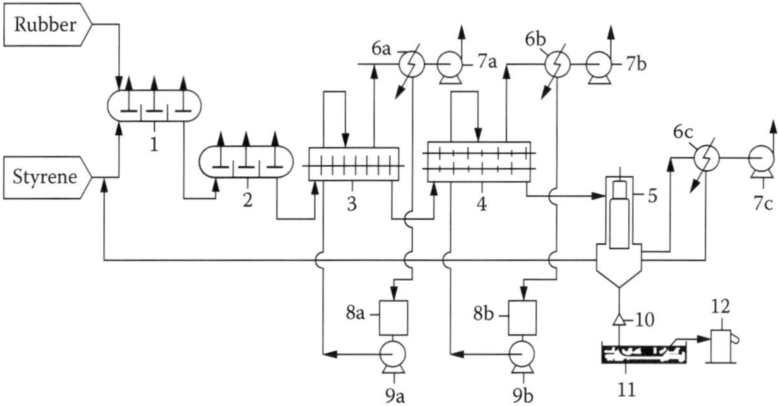

FIGURE 13.12

raised stepwise from 20°C to 110°C. Then, the rubber solution is polymerized in the second stage (prepolymerization stage) at 100–130°C under atmospheric pressure to effect the phase inversion of rubber. The conversion in the prepolymerization stage is controlled in the range 25–40% and the polymerization heat is removed by use of an external cooling jacket. The prepolymer is further polymerized in the third stage at 100–200°C under reduced pressure, and the conversion is controlled in the range 45–60%. In this reactor, the reaction heat is removed by spraying the monomer onto the solution. The evaporated monomer is condensed and recycled. In the final-stage polymerization reactor, polymerization is performed at 100–230°C at 1 atm to obtain a 70–85% conversion.

The following correlation has been proposed for the calculation of the apparent viscosity of a HIPS prepolymer system [42]:

$$\frac{1}{\eta} = \frac{0.696(1-\Phi_{ps})}{\eta_R + 0.311(\Phi_{ps}/\eta_{ps})} \tag{13.9}$$

where ϕ_{ps} is the polystyrene/styrene phase volume, η_R is the rubber phase viscosity, and η_{ps} is the polystyrene phase viscosity.

Impact polystyrene can be prepared by suspension polymerization. However, no shearing agitation occurs within the individual polymer particles. Thus, a prepolymerization with shearing agitation needs to be performed before suspension polymerization to obtain good polymer properties.

13.3.5 ABS Processes

ABS originally stood for an acrylonitrile–butadiene–styrene terpolymer. However, ABS is now used as a general term for a class of multicomponent polymers containing elastomeric rubber particles dispersed in a matrix of rigid copolymer. Because rubber particles and rigid copolymer matrix are incompatible, mechanical blending of rubbers with rigid vinyl copolymers is not effective. Instead, elastomeric rubber particles (polybutadiene rubber, styrene–butadiene rubber, acrylonitrile–butadiene rubber) are grafted by styrene and acrylonitrile copolymer. Typically, a grafting efficiency of 30–40% provides optimal polymer properties. Grafting occurs by direct attack of initiator radicals through hydrogen abstraction, and chain transfer onto the rubber polymer through double bonds. The degree of grafting is proportional to the surface area of the dispersed rubber particles. Because the surface/volume ratio increases with a decrease in particle size, a higher degree of grafting is achieved with smaller rubber particles. Process parameters that influence the degree of grafting are the concentrations of monomers, chain transfer agent, surfactant, polymerization temperature, and monomer/rubber ratio. The lower the monomer concentration and monomer/rubber ratio, the higher the grafting efficiency becomes. The degree of grafting is increased with an increase in reaction temperature, whereas it decreased with an increase in surfactant concentration. Increased rubber content and larger rubber particle size give rise to higher impact strength. ABS having a higher impact strength generally lowers tensile strength and modulus, and heat resistance. Other than impact strength, the following properties are important in manufacturing ABS: surface gloss, heat distortion temperature, flame retardancy, and weatherability.

The ABS graft polymers are commercially produced by emulsion polymerization and mass suspension polymerization processes although solution polymerization is used for some special types of polymer. A continuous emulsion–mass process has also been developed recently. In general, ABS manufactured by the emulsion process exhibits higher impact strength than the polymer made by the mass suspension process.

13.3.5.1 Emulsion Polymerization

The emulsion polymerization process (batch or continuous) is widely used for producing ABS because various grades of ABS resins can be manufactured. In a continuous process, two to six reactors are used in series with rubber latex feed added to either the first reactor or the first two reactors. In a typical ABS emulsion process, the styrene/acrylonitrile copolymer (SAN) and polybutadiene (PBL) are separately prepared by emulsion processes. More than one reactor can be used to produce PBLs of different rubber particle sizes (e.g., 0.1 μm and 0.4 μm). The two PBL lattices are separately graft polymerized and the resulting lattices are steam-stripped. Then, they are blended with SAN latex, coagulated, dewatered, and dried.

The monomer feeds containing monomers, a water-soluble initiator (e.g., potassium persulfate and redox systems), chain transfer agents (e.g., *tert*-dodecyl mercaptan) and an emulsifier (e.g., disproportionated potassium and sodium rosinates, tridecycloxy(polyoxyethylene)phosphate) are separately added to each reactor. Rubber latex may also be imbibed with styrene or styrene/acrylonitrile mixture before the mixture is charged to the first reactor. In a batch process, monomers with initiator and chain transfer agents are often added continuously during the batch operation. The rubber lattices used in the emulsion polymerization process contain a high concentration of butadiene (60–90%) and have a solid content of 20–50% with an average latex particle size of 0.2–0.4 μm. To obtain a bimodal rubber particle size distribution in the resulting rubber latex, a mixture of two lattices, one having a large and the other a small particle size, may be used. In general, styrene/acrylonitrile weight ratio ranges from 25:75 to 40:60, but for the production of ABS with high styrene content, a styrene/acrylonitrile ratio ranging from 70:30 to 75:25 is employed and the ratio is kept constant during polymerization. Methyl acrylate and methyl methacrylate are also used to make a highly transparent graft polymer. To prepare the polymer with a high acrylonitrile content, water-insoluble azo compounds are used to minimize the polymerization of acrylonitrile in aqueous phase due to the high solubility of acrylonitrile in water.

13.3.5.2 Mass Suspension Polymerization

The mass suspension polymerization process requires a lower investment than the emulsion process. The mass suspension ABS polymers have better melt flow characteristics and light stability. In general, ABS manufactured by the mass suspension process exhibit low impact strength, poor surface gloss, and poor weatherability. To improve impact strength and gloss, they are often bended with emulsion ABS resin in a mixture of styrene and acrylonitrile. The resulting dispersion is then blended with a second dispersion prepared by a organic extraction of the ABS latex with a mixture of styrene and acrylonitrile. The blended dispersion is then devolatilized to remove monomers and extruded into polyblend resin pellets. High shearing forces may also be applied during the mass polymerization stage to reduce the rubber particle size and, thus, to improve surface gloss.

The mass suspension ABS polymerization process consists of two stages. In the first-stage mass polymerization, 5–15% rubber (polybutadiene or SBR containing about 75% butadiene) dissolved in styrene is graft polymerized with styrene and acrylonitrile at 98–115°C without an initiator or at 70–95°C with an organic-soluble initiator for 3–6 h to about 20–35% conversion. The reaction mass is then transferred to a second-stage suspension polymerization reactor. In the suspension reactor, polymerization is performed at higher temperatures (80–150°C) in the presence of organic initiators(s). Polyvinyl alcohol (PVA) is commonly used as suspension stabilizer. In batch processes, the polymerization temperature is gradually increased during the batch to achieve a high conversion of monomers.

13.3.5.3 Mass Polymerization

Mass polymerization can be thermally initiated or initiated by organic peroxides. The polymerization temperature, generally higher than that of emulsion polymerization, is increased in stages from 100°C to as high as 240°C. In a continuous mass process, grafted rubber dispersion and monomers are fed to a jacketed reactor equipped with a helical ribbon-type agitator and a reflux condenser. The prepolymer is then transferred to the second-stage polymerization reactor, gradually increased as the polymer mass flows. The reaction product is continuously charged to a vented extruder and pelletized.

13.3.6 Styrene–Maleic Anhydride Copolymers

When styrene is copolymerized with maleic anhydride, which does not homopolymerize, a completely alternating copolymer is obtained. Maleic anhydride, randomly incorporated into the polystyrene backbone, increases the glass transition temperature and heat distortion temperatures (>260°F). The copolymers are stable during injection molding to temperatures above 550°F [26].

$$\sim CH_2{-}CH{=}CH{-}X + \cdot R\sim \longrightarrow \sim CH_2{-}\underset{R}{\underset{|}{CH}}{-}\dot{C}H{-}X$$

$$\searrow \sim \dot{C}H{-}CH{=}CH{-}X + RH$$

13.3.7 Syndiotactic Polystyrene by Metallocenes

With recent development of metallocenes for α-olefin polymerization, the synthesis of syndiotactic polystyrene (sPS) with metallocene catalysts has attracted strong research interest. Although isotactic polystyrene was synthesized in the 1950s by Natta et al. [43] with a $TiCl_4/Al_4(C_2H_5)_3$ catalyst [42], the polymer is highly crystalline with a melting temperature of 240°C and its crystallization temperature was too slow for any commercial applications. The synthesis of sPS was first reported by Ishihara and co-workers [44]. Syndiotactic polystyrene is highly crystalline polymer with a melting temperature of about 270°C. Its glass transition temperature is similar to atactic polystyrene. However, unlike isotactic polystyrene, syndiotactic polystyrene exhibits a relatively fast crystallization rate, low specific gravity, low dielectric constant, high elastic modulus, and excellent resistance to chemicals. The sPS may find some applications in the automotive, electronic, and packaging industries. Because sPS is brittle when used alone, sPS needs to be reinforced with fiberglass, or elastomers need to be used as structural materials.

Syndiotactic polystyrene can be prepared using group IV metal (Ti, Zr) compounds such as methylaluminoxane (MAO). It has been suggested that Ti^{3+} species are active sites for producing sPS. Aluminum alkyls are ineffective for preparing syndiotactic polystyrene. The syndiotactic configuration of polystyrene arises from phenyl–phenyl repulsive interactions between the last inserted monomer unit of the growing chain and the incoming monomer. However, it is generally known that zirconium compounds are less active and efficient than titanium compounds. A large number of catalytic compounds or compositions have been investigated by many researchers, and different catalysts exhibit different polymerization kinetic behaviors and polymer properties. In addition, it is still an active research area. Therefore, it is not easy to generalize the syndiotactic styrene polymerization kinetics. An excellent review of recent literature on the synthesis of syndiotactic polystyrene has been published by Pó and Cardi [45].

Syndiotactic polystyrene is prepared in an organic diluent such as toluene. As the reaction progresses, gel-like, or solid precipitates are formed, and the reaction proceeds in a heterogeneous phase. Due to the formation of solid polymers insoluble in diluent liquid, it is thus important to control fouling formation in a polymerization reactor. The polymer recovered is treated with acidified alcohol to deactivate the catalyst. The product polymer contains some atactic polystyrene, which is extracted out in boiling acetone or methylethylketone (MEK). It has been reported that the fraction of syndiotactic sites is far smaller than atactic sites (1.7 % versus 20% of total titanium), and the syndiospecific propagation rate constant is far larger than the propagation rate constant for atactic sites [46, 47]. Some authors suggest that syndiospecific active sites change into aspecific sites at high reaction temperatures. The syndiotactic fraction is often measured by a MEK-insoluble fraction. Nuclear magnetic resonance is used to determine syndiotacticity, which represents the percentage of syndiotactic sequences (diads, triads, tetrads, or pentads) relative to the total number of sequences along the polymer chain.

13.4 POLY(METHYL METHACRYLATE) AND COPOLYMERS

Acrylic homopolymers [(—CH_2 CH(COOR)—$)_n$] and copolymers are synthesized from acrylates and methacrylates. Through copolymerization, the polymer properties are widely varied from soft, flexible elastomers to hard, stiff thermoplastics and thermosets. Acrylic polymers are produced in many different forms including sheet, rod, tube, pellets, beads, film, solutions, lattices, and reactive syrups.

The poly(methyl methacrylate) (PMMA) homopolymer is completely amorphous and has high strength and excellent dimensional stability due to the rigid polymer chains. PMMA has exceptional optical clarity, very good weatherability and impact resistance, and is resistant to many chemicals. PMMA is manufactured industrially by bulk, solution, suspension, and emulsion processes. The bulk polymerization is used to manufacture PMMA sheets, rods, tubes, and molding powders. The solution polymerization is used to prepare polymers for used as coatings, adhesives, impregnates, and laminates. Small polymer beads (0.1–5 mm) made by suspension polymerization are used as molding powders and ion-exchange resins. MMA is often copolymerized to reduce brittleness and improve processability. Emulsion polymerization is used to produce aqueous dispersion of polymers for paint, paper, textile, floor polish, and leather industries. Typical operating conditions for various MMA polymerization processes are presented in Table 13.4.

13.4.1 Bulk Polymerization

When pure MMA is polymerized, viscosity increases significantly and the Trommsdorf effect (diffusion-controlled termination, autoacceleration) becomes pronounced. Thus, conventional stirred reactors are inadequate for the bulk polymerization. The PMMA sheet is made by bulk (or cast) polymerization of MMA with an organic initiator (peroxide or azo compounds) in a mold consisting of two glass plates separated by a flexible gasket that is also the confining wall of the mold. The entire mold assembly is placed in an air oven and heated. In this process, clear liquid monomers of partially polymerized homogeneous syrups are polymerized directly into shapes

TABLE 13.4
Typical Operating Conditions for some MMA Polymerization Processes

Process	Temperature (°C)	Reaction Time (h)	Conversion(1%)
Mass	30–90	8–24	20–50
Solution	136–143	6–8	>95
Suspension	30–75	3–6	>95
Emulsion	25–90	>2.5	>95

such as sheets, rods, tubes, or blocks. Polymers manufactured by bulk process are normally better in clarity, homogeneity, and color than those produced by other processes such as suspension, solution, and emulsion. Due to the high reaction exotherm, the boiling monomer may leave bubbles in the sheet and some shrinkage occurs on polymerization. Thus, the polymerization is performed below the boiling point of MMA (100.5°C).

In a continuous sheet-casting process, the casting syrup is confined between stainless-steel belts by using a flexible gasket and the belts run progressively through polymerization and annealing zones. The belt moves at about 1 m min^{-1}, which results in about 45 min of residence time in the curing zone (70°C) and about 10 min in the annealing zone (110°C) [48]. Pressure is maintained to prevent the boiling of MMA monomer.

13.4.2 Suspension Polymerization

Methyl methacrylate mixed with a small amount of initiator is dispersed in water by agitation as 0.1–5-mm droplets, stabilized by organic or inorganic protective colloids or surface stabilizer. The monomer/water ratio generally ranges from 50:50 to 25:75. Lower ratios are not practical for economical production. The reactor content is headed under a nitrogen atmosphere to the desired polymerization temperature. The heat of polymerization is effectively dissipated from the polymerizing droplet to the aqueous phase. Each droplet can be viewed as a microbatch bulk polymerization reactor, and the bulk MMA polymerization kinetics can be applied. As monomer conversion increases, the polymer particles become sticky and they tend to agglomerate.

The chain termination in methyl methacrylate polymerization is almost exclusively via a disproportionation mechanism:

$$P_n + P_m \xrightarrow{k_{td}} M_n + M_M^{=}$$

where P_n is the growing polymer radical, M_n is the dead polymer, and $M_M^{=}$ is the dead polymer with a terminal double bond. It is important to achieve a high monomer conversion because the removal of unreacted monomer from polymer is difficult.

In batch suspension polymerization, controlling the polymer particle (bead) size and its distribution is important. In the early stage of suspension polymerization, the monomer droplet size is determined as a result of the dynamic equilibrium between breakage by shear or turbulence forces and coalescence by interfacial tension or adhesion forces [49]. As the shearing action in an agitator reactor can be nonuniform, an equilibrium condition is only possible if all the particles move through a zone of maximum shear. The elastic deformation of colliding particles may produce a greater surface disturbance, leading to a fusion of droplets or particles. Several empirical correlations are available for the estimation of particle size in suspension polymerization [2]. The polymer particle size and distribution are strongly affected by geometric factors (e.g., reactor type, reactor height/diameter ratio, stirrer type and geometry, baffle, etc.), rector operating parameters (e.g., reaction time, stirrer speed, monomer/water ratio, temperature, stabilizer type and concentration, additives, etc.), and physical properties of the aqueous phase (e.g., interfacial tension, density, viscosity, pH, etc.) [2]. For example, particle size distribution becomes narrower and shifts toward smaller sizes with increases in stirring speed and suspending agent concentration. At lower temperatures, the particle size distribution is controlled by the viscosities of the dispersed phase and the aqueous phase at a fixed agitator speed; thus, lower temperatures favor larger particles. At higher temperatures, the polymerization rate is so high the viscosity of the dispersed phase (droplets) is increased before the droplets reach their equilibrium size and the particle size increases with temperature. The average polymer particle size decreases as stabilizer concentration or agitation speed is increased [49].

In suspension polymerization operation, the detection of suspension failure is important. In general, the first indications of suspension failure are abnormal agitator drive power readings; however, this does not occur until polymer buildup on and/or near the agitator is sufficient to disturb the mixing behavior [50].

Suspension polymerization of MMA is industrially performed using batch reactors. Although some literature on continuous suspension polymerization processes has been published, no large-scale continuous suspension process is currently used in industry. Some problems of continuous suspension polymerization are achieving high monomer conversion, avoiding fouling of the reactor wall and pipes, and achieving polymers with the desired particle size distribution and molecular-weight distribution [51]. Technically, narrow residence time distribution is necessary to obtain high conversions, and good mixing of the two phases is important to obtain polymers with the proper particle size distribution. To avoid fouling, dead space should be avoided [52].

13.4.3 Solution Polymerization

Solution polymerization of acrylic esters is usually performed in large agitated reactors in an organic solvent. Both propagation and termination rates are affected by the nature of the solvent, but the rate of initiator decomposition is almost independent of the solvent. It has been reported in literature that solvents such as benzene, toluene, and xylene enhance the rate of MMA polymerization. A typical recipe for the copolymerization of MMA is presented in Table 13.5 [48]. The solution polymerization

TABLE 13.5
Typical Recipe for the Copolymerization of MMA

Materials	Parts
Xylene (solvent)	28.4
Ethoxyethanol (solvent)	14.1
Methyl methacrylate (monomer)	23.3
2-Ethylhexyl methacrylate (comonomer)	15.5
Hydroxyethyl methacrylate (comonomer)	17.6
tert-Butyl perbenzoate (initiator)	1.1

is conducted at 140°C by adding the monomer and initiator mixture uniformly over 3 h. After the addition of reactants is complete, the reaction is continued for 2 more hours at the same temperature.

The autoacceleration effect (Trommsdorf effect) is less pronounced in solution polymerization than in bulk or suspension polymerization due to lower viscosity of the polymerizing solution. To prevent a thermal runaway reaction, the reactants are often added gradually to the reactor. The polymer molecular weight is controlled through the use of a chain transfer agent and by initiator concentration and type. Monomer concentration, solvent type, and reaction temperature also affect the molecular weight.

13.4.4 Emulsion Polymerization

Emulsion homopolymerization or copolymerization of MMA is usually performed in a pressurized batch reactor with a water-soluble initiator and surfactant. The polymerization temperature may be varied from 85°C to 95°C to achieve high conversion. Bacterial attack, common in acrylic polymer latex, can be avoided by pH adjustment and the addition of bactericidal agents.

13.5 POLYACRYLONITRILE

Acrylonitrile is readily polymerized to polyacrylonitrile (–[$CH_2CH(CN)$]–) by free-radical polymerization with organic initiators (peroxides and azo compounds). Rigidity, low permeability, high tack as adhesives, strong resistance to chemicals and solvents, heat resistance, and so forth are mostly due to the polar nature of polyacrylonitrile. Polyacrylonitrile is often copolymerized with halogen-containing monomers (e.g., vinyl chloride, vinylidene chloride) to impart flame retardancy. The nitrile group may also be reactive, leading to a colored naphtyridine group [53].

Like poly(vinyl chloride), polyacrylonitrile does not dissolve in its own monomer; thus, when bulk polymerized, the polymer precipitates from the monomer solution. The precipitated polymer particles, if not stabilized, tend to agglomerate to form a polymer paste or slurry. Two polymerization reaction loci are used: a polymer-free monomer phase and a polymer phase containing dissolved monomer. The polymer phase may be saturated with monomer. The physical properties of the polymer and monomer phase have a significant effect on the polymerization rate. The polymer particles formed in bulk polymerization of acrylonitrile consist of polymeric phase made up of 94% polyacrylonitrile and 6% acrylonitrile [54]. Due to its extremely low solubility in acrylonitrile, the polyacrylonitrile polymer is believed to grow on the polymer particle surface from the very low conversion. If the coagulation of initial small precipitated particles occurs, the reaction site will be either the outer or inner surface of a coagulated particle.

The polymerization of acrylonitrile also exhibits strong autocatalytic behavior (Trommsdorf effect), making the heat removal difficult as conversion increases. The autocatalytic effect becomes pronounced at high initiator concentrations and low reaction temperatures. It is generally believed that the rapid rate increase is due to the formation of precipitated polymeric phase where radical occlusion occurs. The reduced radical mobility in the interior of the particles reduces chain termination. The degree of radical occlusion is largest for glassy polymers like polyacrylonitrile which are not highly swollen by the monomer. The degree of occlusion decreases with chain transfer to the monomer or transfer agent and if the reaction temperature is higher than the glass transition temperature of the monomer–polymer mixture [55]. Figure 13.13 illustrates the typical polymerization rate curve [56].

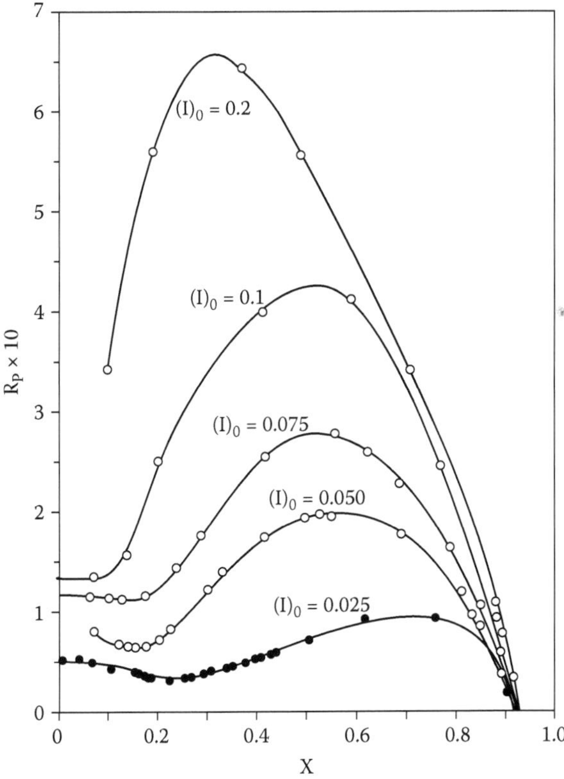

FIGURE 13.13 Polymerization rate curves for bulk polymerization of acrylonitrile (60°C with AIBN initiator). From L. H. Garcia-Rubio and A. E. Hamielec, *J. Appl. Polym. Sci.*, 23, 1397 (1979). With permission.

Notice that the initial rate acceleration period is followed by an almost constant rate period. Then, the rate increases rapidly to a maximum before falling to a zero reaction later. The polymer molecular-weight averages increase with conversion. The removal of polymerization heat becomes a critical factor in operating a polymerization reactor. Extensive kinetic study and process modeling have been reported in the literature [56, 57].

Polyacrylonitrile can be produced by continuous bulk, continuous slurry, and emulsion polymerization processes. In a continuous slurry process, small monomer droplets are suspended in an aqueous medium. In this process, heat removal is more efficient than in the bulk process. In one example [58], a 0.3% H_2SO_4 aqueous solution, a catalyst solution (15% Na_2SO_3 and 4.22% $NaClO_3$ in water), and a monomer solution are continuously supplied into an agitated reactor. At 35°C, 90% monomer conversion is achieved at 1.7 h of residence time. A portion of the polymerizing mixture is circulated from the bottom of the reactor through an external heat exchanger to remove the heat of polymerization.

In the emulsion polymerization process, redox catalyst is commonly used to achieve a rapid polymerization at low temperatures (20–60°C). The polymer is recovered by coagulation with a salt. Acrylonitrile can also be polymerized by solution process with dimethylformamide as a solvent.

13.6 POLY(VINYL CHLORIDE)

Poly(vinyl chloride) (PVC) is one of the oldest yet most important thermoplastic polymers. Commercially, PVC is manufactured by three main processes: suspension, mass, and emulsion polymerization. The polymers produced by mass and suspension processes are similar and are used in similar applications. The suspension process is currently the dominating industrial PVC process (about 75% of the world's PVC capacity). The emulsion polymerization process is used to produce specialty resins for paste applications (e.g., coated fabric, roto-molding, slush molding). PVC manufactured by the mass process has properties similar to those of the suspension process.

As PVC is insoluble in vinyl chloride monomer (VCM), the polymerization is a heterogeneous process. The polymer phase separates from the monomer phase at a conversion of 0.1% and the polymerization occurs in both the monomer phase and the polymer phase. It has been suggested that the radicals and polymer formed in each phase grow and terminate without any transfer of active radicals between the phase [59]. Also suggested is that radical transfer occurs between the phases by sorption and desorption phenomena [60]. It is believed that the monomer phase contains only trace amounts of polymer due to its insolubility in the monomer. The polymerrich phase is at equilibrium with the monomer as long as a free-monomer phase exists. Thus, for the monomer conversion up to about 77%, the polymer phase separates from the monomer and forms a series of agglomerated polymer particles. The volume of the monomer phase decreases as the polymer phase grows and absorbs the monomer. As the free-monomer phase disappears, the reactor pressure starts to drop. At a higher conversion, the monomer in the polymer-rich phase continues to polymerize.

The most prominent feature of the PVC process is the development of a complex particle morphology. Thus, it is necessary to understand the mechanism of particle nucleation, growth, and aggregation. The mechanism of PVC grains can be summarized as follows [61]:

- Stage 1: Primary radicals are formed by the decomposition of initiator. Macroradicals with a chain length of more than 10–30 monomer units precipitate from the monomer phase (at ≈ 0.001% conversion). The reaction mixture consists mainly of a pure monomer.
- Stage 2: Microdomains (0.10–0.02 μm, <0.01% conversion) are produced by the aggregation of precipitated macroradicals and macromolecules.
- Stage 3: Aggregation of microdomains produces domains (0.1–0.3 μm, <1% conversion; primary particle nuclei) stabilized by negative charge. The limiting size of a domain depends on the agitation speed and additives used. The new domains continue to be formed. This stage is completed at 5–10% conversion.
- Stage 4: Primary particles formed by the aggregation and growth of domains. Primary particles grow with conversion at almost the same rate. The diameter of the spherical particles is about 0.8–1.0 μm. The growth process continues until the formation of a continuous network in droplets (about 1530% conversion).
- Stage 5: Primary particles grow and aggregate until the free-monomer phase disappears. The diameter of the final primary particles is about 1.2–1.5 μm (50–70% conversion). As the monomer phase disappears, the reactor pressure starts to decrease.
- Stage 6: Primary particles fuse together as agglomerates (5–10 μm) until final conversion is reached.

No clear boundary line exists between stages, and two neighboring stages may occur simultaneously during polymerization. The mechanism of morphology development is illustrated in Figure 13.14 [62].

The kinetics of vinyl chloride polymerization are very complex, as illustrated in the following table [61]:

Initiation	$1 \rightarrow 2R^{\cdot}$
Propagation:	
Head-to-tail propagation	$P_n^{\cdot} + M \rightarrow P_{n+1}^{\cdot}$
Head-to-head propagation	$P_n^{\cdot} + M \rightarrow P_{n+1}^{*}$
Chlorine shift reaction	$P_n^{*} \rightarrow (P_n^{\cdot})'$
Tail-to-tail propagation	$P_n^{*} + M \rightarrow P_{n+1}^{\cdot}$
Formation of chloromethylbranches	$(P_n^{*})' + M \rightarrow P_{n+1}^{\cdot}$
Splitting off chlorine radical	$(P_n^{*})' + M_n + Cl\cdot$
Initiation of polymer radicals bychlorine radical	$Cl\cdot + M \rightarrow P_1^{\cdot}$
Chlorine radical transfer to polymer	$P_n^{\cdot} + Cl\cdot \rightarrow P_n' + HCL$
Propagation toward formation of a chain branch	$P_n^{\cdot} + M \rightarrow P_{n+1}^{\cdot}$
Formation of an internal double bond	$P_n^{\cdot} \rightarrow M_n + Cl\cdot$
Chain transfer to polymer	$P_n^{\cdot} + M_m \rightarrow M_n^{\cdot} + P_m^{\cdot}$
Formation of long-chain branch	$P_n^{\cdot} + M \rightarrow P_{n+1}^{\cdot}$
Termination	
Combination	$P_n^{\cdot} + P_m^{\cdot} \rightarrow P_{n+m}^{\cdot}$

Disproportionation	$P_n^{\cdot} + P_m^{\cdot} \rightarrow M_n + M_m$
Primary radical termination	$P_n^{\cdot} + R^{\cdot} \rightarrow M_n$
Termination with Cl·	$P_n^{\cdot} + Cl\cdot + \rightarrow M_n$
Other reactions	
Backbiting reaction	$P_n^{\cdot} \rightarrow P_{n,b}$
Propagation toward formation of a short branch	$P_{n,b}^{\cdot} + M \rightarrow P_{n+1,b}^{\cdot}$

As illustrated in this table, some side reactions occur in vinyl chloride polymerization due to rearrangement effects in the polymer chains, unsaturated structures in the polymer chains (1.5–3.0 double bonds per 1000 monomer units), and the tertiary chlorine structures formed by chain transfer to polymer. The chlorobutyl groups

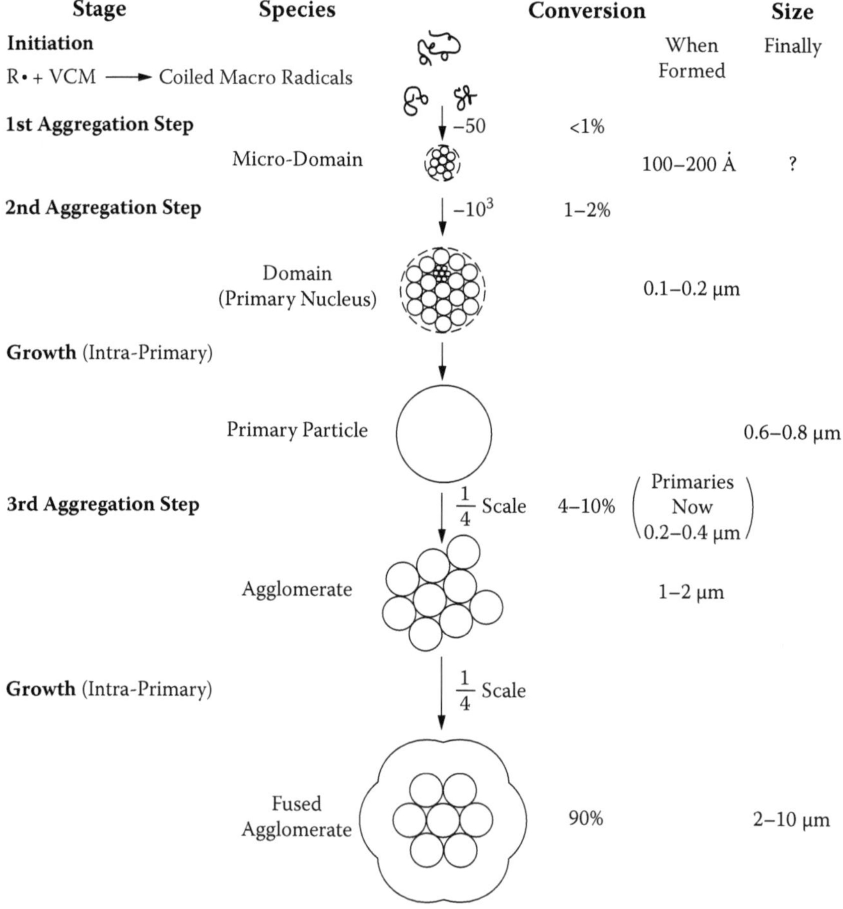

FIGURE 13.14 Development of PVC particle morphology. From M. W. Allsopp, *Pure Appl. Chem.*, 53, 449 (1981). With permission.

(2–3 per 1000 monomer units) formed by backbiting reactions contribute to the thermal instability of the polymer through the tertiary chlorine on the backbone [63].

The polymer properties to be controlled in commercial PVC manufacturing processes are molecular weight, grain size and its distribution, grain porosity, density, purity, color and thermal stability, and electrical resistivity. Common PVC grades contain 2–3 pendant chloromethyl groups and about 0.5 long branches, formed via hydrogen abstraction from the polymer, per 1000 carbon atoms [64]. Due to the heterophasic nature of PVC polymerization processes, the development of grain morphology is very complex. The kinetics of vinyl chloride polymerization have been reviewed by several authors [61, 65].

13.6.1 Suspension Polymerization

Suspension polymerization of vinyl chloride (VCM) is performed in a stirred batch reactor using one or several monomer-soluble organic initiators (e.g., peresters, peroxydicarbonates, peroxides, azo compounds). Initiators are chosen to have decomposition rates that produce polymerization rates to match reactor heat removal capabilities. The monomer is dispersed by agitation in droplets of 50–20 μm and each monomer droplet acts like a microbulk polymerization reactor. One or more protective colloids (or suspension stabilizers) such as a modified cellulose or a partially hydrolyzed polyvinyl acetate (PVA) are added to prevent particle agglomeration and to control the size and morphology of polymer particles. A typical recipe for suspension polymerization is presented in Table 13.6 [66]. Buffers, chain transfer agents, or commoners can be added if required. Final PVC grains are about 100–180 μm in diameter. Each PVC grain consists of an agglomerate of a number of smaller grains. Each agglomerate consists of a large number of smaller PVC particles (3–5 μm) that are aggregated into a porous network. The PVC grain is surrounded by a wall or skin approximately 2 μm thick at low conversion. This well-defined membrane of skin has been shown to be a PVA/PVC graft copolymer [67]. At about 30% conversion, the skin is quite strong and stable. During this time, the polymer density changes.

In a typical commercial PVC suspension process, polymerization begins with the addition of water, suspending agent, and initiator to an air-free reactor. Then, the reactor is evacuated to remove oxygen and then VCM is added to the agitated mixture. Agitation is the primary reactor design variable affecting product quality. The mixing blades are usually either retreat curve or turbine types that pump the slurry from the bottom of the reactor upward around the sides of the reactor. Reactors are

TABLE 13.6
Typical Recipe for Suspension Polymerization

Ingredient	VCM (wt%)
Water	100–130
VCM	100
Suspension stabilizer	0.05–0.15
Initiator	0.03–0.07

usually baffled. It is generally known that the product particle size and distribution is established early in the reaction.

Initially, hot water or stream is circulated through the reactor jacket to bring the reactor contents to the desired temperature rapidly. As the reaction heat is released, cooling water is circulated through the jacket to control the reactor temperature. A reflux condenser may be used to improve the reactor's heat removal capability. It is desirable to have a constant rate of polymerization and, thus, a constant rate of heat generation. A mixture of initiators having different decomposition characteristics may be used to keep constant rate of heat release. The effectiveness of the initiator depends on its ability to generate active radicals in the monomer-swollen polymer phase. Any side reaction that removes the initiator from the reaction site will reduce the initiator's effectiveness and may lead to increased fouling [63]. Constant temperature is maintained until 85–90% conversion is obtained. Around 65–70% conversion, most of the unreacted monomer is imbibed in the polymer, and as polymerization proceeds, the pressure within the grain falls and the grain collapses with folding and rupturing of the surface, accompanied by intrusion of water.

Because small polymer particles are produced, folding of the reactor surface must be minimized. When the fouling occurs, the heat transfer coefficient for the cooling surfaces decreases and the reactor temperature control becomes difficult. Furthermore, the deposits are peeled off from the inner surface of the reactor and mixed into the polymer. As a result, the quality of the product deteriorates. Figure 13.15 illustrates the variations in the jacket heat transfer coefficient with conversion [68]. Notice that the jacket heat transfer coefficient decreases rapidly as the monomer conversion approaches 60%. The decrease of the heat transfer coefficient depends on the slurry concentration.

To prevent wall fouling, the following approaches can be taken [63]: prevent the deformation of the monomer droplet; prevent the adhesion of the polymer particles to a reactor surface; prevent the adsorption of monomer to a reactor surface; and the polymerization either at the wall or in the aqueous solution. For example, the fouling at the reactor walls can be reduced by using glass-coated reactors, reactors' internals with smooth surfaces, fouling suppressants, correct choice of initiator, buffer, and ph. A reflux condenser may be installed to aid the reactor's heat removal capacity. After polymerization, the polymer is steam-stripped to remove residual monomer and then dried. Continuous stripping can be done in a trayed stripping tower using steam to heat and strip the slurry. For the drying of PVC particles, fluidized-bed dryers, rotary and flash-fluid-bed dryers are used. In a rotary dryer, concurrent flow of drying air and wet cake is used. A flash-fluid-bed dryer is a two-stage dryer that uses a high-temperature air stream to entrain the wet cake in a duct and dry it through the constant rate section [68]. In a PVC process, high monomer conversion is required to achieve high productivity, and polymer morphology must be properly controlled (e.g., particle shape, size and its distribution, surface characteristics, internal structure, porosity, and bulk density).

In suspension polymerization, the size and size distribution of polymer particles depend on the agitation speed and surface stabilizer. As the stirring speed is increased, the monomer droplets become smaller and the droplet size distribution becomes narrower. If the stirring speed is too high, the droplet size increases.

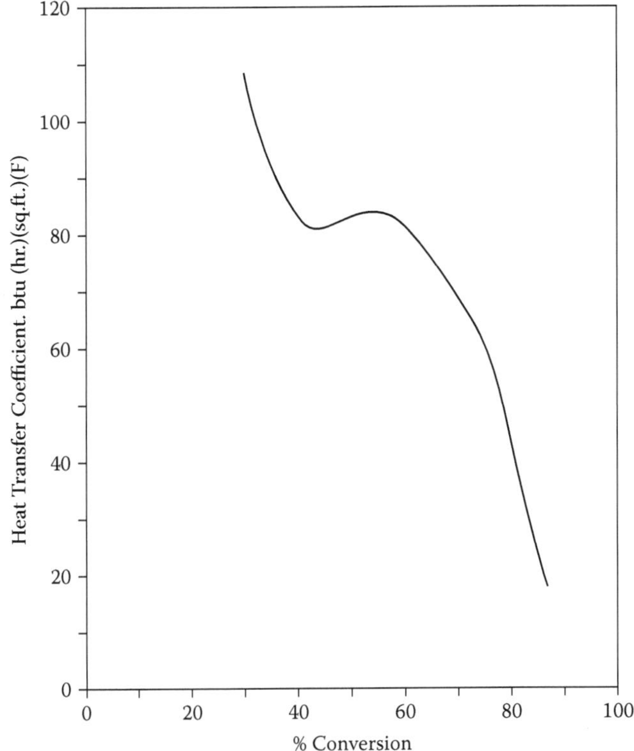

FIGURE 13.15 Reactor heat transfer coefficient versus conversion. From J. B. Cameron, A. J. Lundeen, and J. H. Mcculley Jr., *Hydrocarbon Process.*, 39 (March 1980). With permission.

The use of stabilizer yields smaller particles and narrower particle size distribution. The colloidal stability of domains is reduced and earlier aggregation occurs as the polymerization temperature is increased. The size of the primary particles increases and the number decreases, less able to resist droplet contraction, and, consequently, grains with lower porosity are produced [2, 69]. The high porosity of the PVC particles leads to the rapid absorption of the additives (e.g., plasticizers). In a typical PVC sample, 1 particle in about 10^5 particles will have a low porosity and a slow rate of plasticizer adsorption. These particles of low porosity may cause the formation of "fish eyes" in a flexible film. A uniformly porous polymer is also desired for ease of VCM removal and processing. Large gels, if present, make the VCM removal difficult and may cause nonuniformity in the final product. Polymer grains with a higher packing density and lower porosity are used in rigid applications to maximize extruder outputs.

Unreacted monomer is removed by evacuation and the polymer recovered from the water slurry. The polymer slurry is centrifuged to produce a wet cake having a water content of 18–25%. A two-stage fluidized bed may be used to dry the polymer to below 0.3 wt% water content [68] by centrifuging and drying. The residual monomer concentration is reduced to less than 10 ppm. The total cycle takes approximately 7–10 h. The polymerization time is limited by the reactor's heat removal capabilities and by the rate of reaction heat release.

13.6.2 Mass Polymerization

In the mass polymerization process, no water is used as in suspension processes. Therefore, in general, the mass polymerization reactor is smaller than the suspension polymerization reactor. As PVC is insoluble in its own monomer, polymer starts to precipitate from the liquid and the medium becomes opaque as soon as the reaction commences. As soon as the polymer chains start growing, they make up small primary particles about 0.1 μm in diameter. These particles then coagulate to form larger polymer grains that continue to grow during polymerization. Stirring has a strong effect on the formation of polymer grains. At a high stirring speed, no new polymer particle is formed as the conversion reaches about 2%. As the polymerization proceeds, the liquid monomer is absorbed by PVC particles, and at about 30% conversion, the medium becomes powdery.

The PVC grains produced by suspension and mass processes are generally different in shape. In the suspension process, the polymer particle morphology is determined primarily by the interaction between the primary and secondary suspending agents. In the mass process, the particle morphology is controlled by the temperature and agitation in the reactor and by the effects of additives in the first-stage reactor [63]. Figures 13.16 and 13.17 illustrate the PVC particles made by two different processes. For a given resin density, PVC made by the mass process exhibits

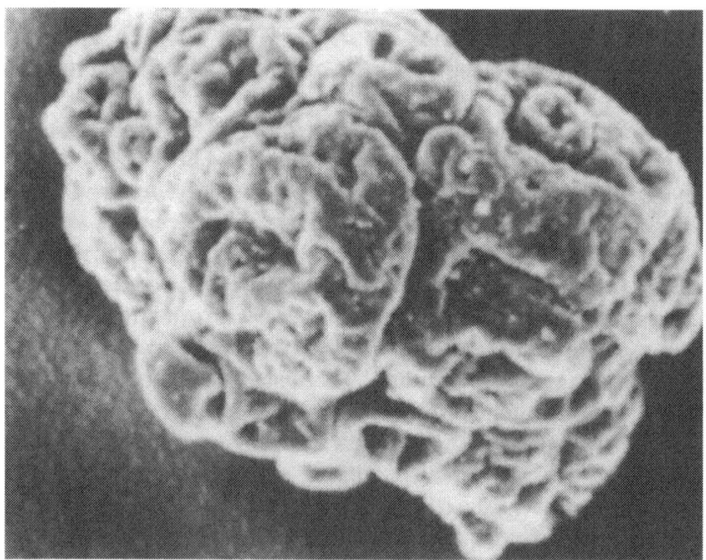

FIGURE 13.16 PVC grain by mass process. From N. Fischer and L. Goiran, *Hydrocarbon Process.*, 143 (May 1981). With permission.

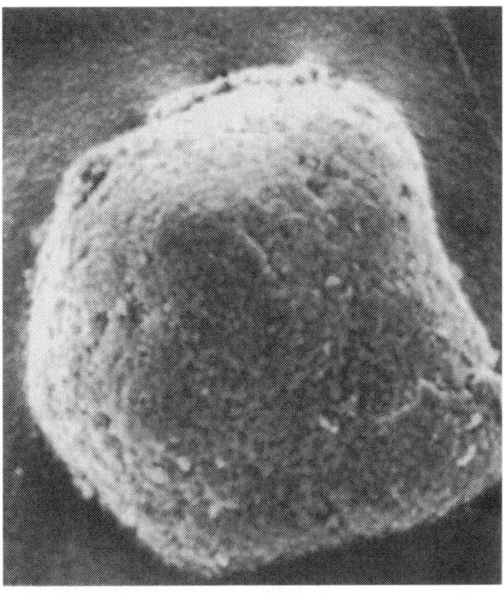

FIGURE 13.17 PVC grain by suspension process. From N. Fischer and L. Goiran, *Hydrocarbon Process.*, 143 (May 1981). With permission.

larger porosity. Figure 13.18 compares the porosity and density of mass and suspension PVC products [69].

Industrial mass PVC processes were developed by many companies. Because mass polymerization begins in a liquid phase and progresses rapidly to a final

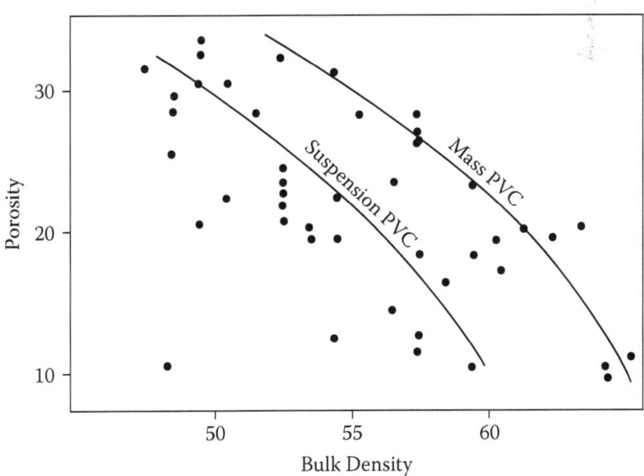

FIGURE 13.18 Relationship between porosity and density for mass and suspension PVC. From N. Fischer and L. Goiran, *Hydrocarbon Process.*, 143 (May 1981). With permission.

powder phase, stirring devices must be specially designed to fit such conditions. In a two-step process, each phase of the process is dealt with in a separate reactor. In the first step, prepolymerization is performed in a stirred tank reactor fitted with a turbine agitator with flat vertical blades and baffles. The prepolymerization reactor is loaded with a mixture of fresh and recycled monomer (about 50% of the total monomer to be polymerized), initiator, and any additives required. The reactor is heated to the prespecified reaction temperature. Particle size is homogeneous and becomes smaller with more vigorous agitation. When the conversion reaches about 7–8%, particle nucleation is completed. These polymer particles are cohesive enough to allow the transfer of the entire contents of the rector to a second-stage reactor, where polymerization is finished. In the second-stage reactor, as the conversion exceeds 25% the medium becomes powdery. The second reactor is typically a vertical or a horizontal rector specially designed to stir the powder phase at low speed. Vertical reactors have several advantages: The reactor content can be emptied more rapidly; the reactor can be cleaned more quickly; the reactor temperature can be measured more readily. In practice, only half the total monomer is added in the prepolymerizer and the rest is added in the second reactor.

The mass polymerization is performed at 40–70°C and 70–170 psi, and the monomer is in equilibrium with the vapor phase. The reaction temperature is kept constant by maintaining constant reactor pressure. It is necessary to have enough monomer to permit the heat transfer by vaporization and a free surface (reactor walls, agitators, and condensers) for the recondensation of the monomer. The condensed monomer is readsorbed immediately by the PVC particles because, unlike the suspension process, no colloidal membrane ("skin") surrounds the particle. The residual monomer is degassed directly in the polymerization reactor until equilibrium between the reactor pressure and the pressure of the recovery condenser is achieved. Compressor degassing is then followed until a high vacuum is reached in the reactor (around 100 mm Hg). Finally, the vacuum is broken with nitrogen or water vapor.

13.6.3 Emulsion Polymerization

Emulsion polymerization of vinyl chloride is initiated by a water-soluble initiator such as potassium persulfate. Initially in the reactor, monomer droplets are dispersed in the aqueous phase (continuous phase) containing initiator and surfactant (emulsifier). As the reactor content is heated, the initiator decomposes into free radicals. When the surfactant concentration exceeds the critical micelle concentration (CMC), micelles are formed. Free radicals or oligomers formed in the aqueous phase are then captured by these micelles. Vinyl chloride monomer is slightly soluble in water. As the monomer dissolved in water diffuses into micelles containing radicals, polymerization occurs. With an increase in monomer conversion in the polymer particles, separate monomer droplets become smaller and eventually they disappear. The monomer concentration in polymer particles is constant as long as liquid monomer droplets exist. The rate of emulsion polymerization is represented by

$$R_p = \frac{k_p[M]_p \bar{n} N_p}{N_A} \tag{13.10}$$

where $[M]_p$ is the monomer concentration in a polymer particle, $\bar{n}$ is the average number of radicals per particle, N_p is the number of polymer particles, and N_A is Avogadro's number.

The formation of primary radicals governs the rate of initiation and particle population. Because radical generation occurs in the aqueous phase, whereas radical termination occurs in the polymer particles, the polymerization rate and molecular weight can be increased at the same time. In vinyl chloride emulsion polymerization, the emulsifier greatly affects the polymerization kinetics and the physicochemical and colloidal properties of the polymer. The average polymer particle size is of the order 0.1–0.3 μm, which is the size of primary particle nuclei in bulk and suspension polymerizations. The following is a summary of the typical kinetic features of batch vinyl chloride emulsion polymerization [61].

1. The number of the latex particles is independent of the initiator concentration.
2. The number of latex particles varies strongly with the emulsifier concentration.
3. The number of latex particles becomes constant after 5–10% conversion.
4. The rate of polymerization increases with increasing initiator concentration and the reaction order of initiator varies between 0.5 and 0.8.
5. The average number of radicals per particle is less than 0.5 and usually of the order of 0.01–0.001 (Smith–Ewart Cast I kinetics), indicating that radical desorption from polymer particles into the water phase is significant.
6. At high conversion, autoacceleration (Trommsdorf effect) occurs.
7. The polymer molecular weight is independent of particle number and size and initiator concentration.

The emulsion latex, at about 45% solids from the batch stripper, is processed through a thin-film evaporator to increase the latex solids to about 60–65%. Then the latex is spray-dried and aggregated to about 20–50 μm dry particles.

13.7 POLYVINYL ACETATE

The poly(vinyl acetate) (PVAc) homopolymer ({[$CH_2CH(C(-O)CH_3)$]}) is manufactured by polymerizing vinyl acetate (VA) using organic initiators in an alcohol, ester, or aromatic solvents. Redox polymerization of vinyl acetate is industrially used for emulsion polymerization. With a redox system, the activation energy of polymerization is greatly reduced and, thus, low-temperature polymerization becomes possible. Examples of redox initiator systems are hydrogen (activated) with peroxide, hydrogen, and palladiumsol with peroxide, sodium perchlorate (sodium sulfite), peroxides and organic-metal salts, peroxide and titanium sulfate, metal salt-sulfuric acid-benzoyl peroxide, *p*-chlorosulfuric acid-aminebenzoyl peroxide, azobisisobutyronitrile-*p*-chlorobenzene sulfuric acid, and so on [70]. By hydrolysis or alcoholysis, PVAc is converted to poly(vinyl alcohol) (PVA). Vinyl acetate is also used as comonomer for ethylene and vinyl chloride polymerizations.

Industrially, vinyl acetate is polymerized by emulsion, suspension, and solution polymerizations. Due to high exothermicity and the occurrence of branching at a high polymer/monomer ratio, bulk polymerization is not important industrially. Because the rate constant of chain transfer to polymer in vinyl acetate polymerization is larger than that in other vinyl polymerization, branching of PVAc is of great practical importance. The chain transfer to polymer occurs mainly on hydrogen atoms of the acetyl group, because the radical so formed is stabilized by the neighboring carbonyl group. The properties of poly(vinyl alcohol) derived from PVAc are also strongly affected by branching. In vinyl acetate polymerization, branch points are introduced in polymer molecules by reaction of polymer radicals with dead polymers and by terminal double bond polymerization. The chain branching exerts a strong influence on the molecular-weight distribution of polymer. The high reactivity of PVAc polymer radical is attributed to its low degree of resonance stabilization. It has been reported that branching reactions are highly sensitive to reactor residence time distribution and mixing effects [71]. The polymer molecular weight increases with conversion in bulk vinyl acetate polymerization, whereas the molecular weight of the corresponding poly(vinyl alcohol) remains unchanged with conversion. This indicates that the branching occurs exclusively at the acetoxy group of poly(vinyl acetate) [72]. It was also reported that the hydrogen atoms in the a-position of the main polymer chain are more reactive than those on the acetoxymethyl group.

A kinetic scheme for vinyl acetate polymerization can be described as follows [71]:

Initiation

$$I \xrightarrow{k_d} 2R$$

$$R + M \xrightarrow{k_1} P_{1,0}$$

Propagation

$$P_{n,b} + M \xrightarrow{k_p} P_{n+1,b}$$

$$P^{=}_{n,b} + M \xrightarrow{k_p} P^{=}_{n+1,b}$$

Chain transfer to monomer

$$P_{n,b} + M \xrightarrow{k_{tr,m}} M_{n,b} + P^{=}_{1,0}$$

$$P^{=}_{n,b} + M \xrightarrow{k_{tr,m}} M^{=}_{n,b} + P^{=}_{1,0}$$

Chain transfer to solvent

$$P_{n,b} + S \xrightarrow{k_{tr,s}} M_{n,b} + P_{1,0}$$

$$P^{=}_{n,b} + S \xrightarrow{k_{tr,s}} M^{=}_{n,b} + P_{1,0}$$

Chain transfer to polymer

$$P_{n,b} + M_{m,c} \xrightarrow{k_{tr,p}} M_{n,b} + P_{m,c+1}$$

$$P^{=}_{n,b} + M_{m,c} \xrightarrow{k_{tr,p}} M^{=}_{n,b} + P_{m,c+1}$$

$$P_{n,b} + M^{=}_{m,c} \xrightarrow{k_{tr,p}} M^{=}_{n,b} + P^{=}_{m,c+1}$$

$$P^{=}_{n,b} + M^{=}_{m,c} \xrightarrow{k_{tr,p}} M^{=}_{n,b} + P^{=}_{m,c+1}$$

Terminal double bond polymerization

$$P_{n,b} + M^{=}_{m,c} \xrightarrow{k_{pdb}} M_{n+m,b+c+1}$$

$$P^{=}_{n,b} + M^{=}_{m,c} \xrightarrow{k_{pdb}} P^{=}_{n+m,b+c+1}$$

Termination by disproportionation

$$P_{n,b} + P_{m,c} \xrightarrow{k} \tfrac{1}{2} M^{=}_{n,b} + \tfrac{1}{2} M_{n,b} + \tfrac{1}{2} M^{=}_{m,c} + \tfrac{1}{2} M_{m,c}$$

$$P_{n,b} + P^{=}_{m,c} \xrightarrow{k} \tfrac{1}{2} M^{=}_{n,b} + \tfrac{1}{2} M_{n,b} + M^{=}_{m,c}$$

$$P^{=}_{n,b} + P_{m,c} \xrightarrow{k} M^{=}_{n,b} + \tfrac{1}{2} M^{=}_{m,c} + \tfrac{1}{2} M_{m,c}$$

$$P^{=}_{n,b} + P^{=}_{m,c} \xrightarrow{k} M^{=}_{n,b} + M^{=}_{m,c}$$

In the preceding scheme, I is the initiator; R is the primary radical; $P_{n,b}$ is the live polymer radical with n monomer units and b branches; $M_{n,b}$ is the dead polymer with n monomer units and b branches; $P^{=}_{n,b}$ is the live polymer radical with n monomer units, b branches, and a terminal double bond; $M^{=}_{n,b}$ is the dead polymer with n monomer units, b branches, and a terminal double bond; M is the monomer; and S is the solvent.

The previous kinetic model allows the calculation of monomer conversion, polymer molecular weight, and branching frequency. The effects of polymerization conditions and the reactor types (e.g., batch, continuous segregated, and continuous micromixed reactors) have been investigated using the previous kinetic model [71].

The overall density of the branches (number of branches per monomer molecule polymerized) can be calculated using the following method. Let N_0 be the total number of monomer molecules (both polymerized and unpolymerized), and x is the fraction of monomer molecules polymerized:

$$x \equiv \frac{N_0 - N}{N_0} \tag{13.11}$$

where N is the number of monomer molecules when the fractional monomer conversion is x. In addition, define b as the total number of branches. Then,

$$\frac{db}{dt} = k_{tr,p} P N_0 x \tag{13.12}$$

$$\frac{dx}{dt} = k_p P (1 - x) \tag{13.13}$$

From these equations, we obtain

$$\frac{db}{dx} = \left(\frac{k_{tr,p}}{k_p}\right) N_0 \frac{X}{1 - X} \equiv C_p N_0 \frac{X}{1 - X} \tag{13.14}$$

Upon integration,

$$\rho = \frac{b}{N_0 x} = -C_p\left(1 + \frac{1}{x}\ \text{In}\ (1-x)\right) \tag{13.15}$$

This equation indicates that branching is negligible at low monomer conversion. For vinyl acetate polymerization, the following equation has also been proposed [73]:

$$n_B = C_p \overline{DP_B x} \tag{13.16}$$

where n_B is the average number of branches per molecule grafted onto the polymer backbone, C_p is the polymer chain transfer constant, $\overline{DP}_B$ is the average degree of polymerization of the polymer backbone, and x is the degree of conversion of the monomer. Then the fraction (F) of the new polymer formed that is grafted to the polymer backbone is given approximately by

$$F = \frac{C_p[B/M]}{C_s([S]/[M]) + C_m} \tag{13.17}$$

where [B/M] is the ratio of the weight of the polymer to the weight of the monomer, [S]/[M] is the ratio of the moles of solvent to the moles of monomer, C_s is the solvent chain transfer constant, and C_m is the monomer chain transfer constant. Figure 13.19 depicts the variation of number of branch points with number-average molecular weight of PVAc [74] and the number of branches per molecule and degree of conversion [73]. When branching occurs, the number-average polymer molecular weight is little affected, but the weight-average molecular weight increases. Therefore, the polydispersity (M_w/M_n) increases (i.e., broadening molecular-weight distribution).

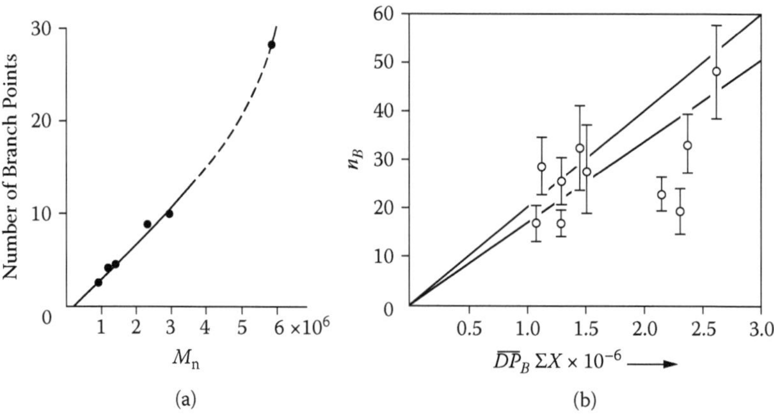

FIGURE 13.19 Effect of molecular weight and conversion on branching. From G. C. Berry and R. G. Craig, *Polymer*, 5, 19 (1964); H. W. Melville and P. R. Sewell, *Makromol. Chem.*, 32, 139 (1959). With permission.

13.7.1 Emulsion Polymerization

The emulsion polymerization of vinyl acetate (to homopolymers and copolymers) is industrially most important for the production of latex paints, adhesives, paper coatings, and textile finishes. It has been known that the emulsion polymerization kinetics of vinyl acetate differs from those of styrene or other less water-soluble monomers largely due to the greater water solubility of vinyl acetate (2.85% at 60°C versus 0.054% for styrene). For example, the emulsion polymerization of vinyl acetate does not follow the well-known Smith–Ewart kinetics and the polymerization exhibits a constant reaction rate even after the separate monomer phase disappears. The following observations have been reported for vinyl acetate emulsion polymerization [78]:

1. The polymerization rate is approximately zero order with respect to monomer concentration at least from 20% to 85% conversion.
2. The polymerization rate depends on the particle concentration to about 0.2 power.
3. The polymerization rate depends on the emulsifier concentration with a maximum of 0.25 power.
4. The molecular weights are independent of all variables and mainly depend on the chain transfer to the monomer.
5. In unseeded polymerization, the number of polymer particles is roughly independent of conversion after 30% conversion.

Poly(vinyl acetate) latex can be formed with anionic, cationic, and nonionic surfactants, or protective colloids, or even without added surfactant. The ionic strength of the aqueous phase affects the stability of the polymer particles and the polymerization rate. As the ionic strength is increased, the electrostatic repulsive energy barrier is reduced so that latex coagulates. The average latex particle sizes of commercially produced PVAc are 0.2–10 μm and the viscosity is in the range 400–5000 mPa·s [48]. The commonly used water-soluble initiators are ammonium persulfate, potassium persulfate, hydrogen peroxide, and water-soluble azo compounds. The aqueous phase is usually buffered to pH 4–5 with phosphate or acetate to stabilize the decomposition of initiators and to minimize monomer hydrolysis. During the emulsion polymerization, particle formation continues until about 80% conversion is reached. The latex properties (e.g., viscosity, rheology, and solubility) are strongly affected by the degree of grafting. Continuous reactors are commonly used for the emulsion polymerization of vinyl acetate.

13.7.2 Solution Polymerization

Bulk polymerization of vinyl acetate is difficult because removing the heat of polymerization is difficult and the occurrence of branching at high polymer/monomer ratios leads to insolubilization. Thus, solution polymerization is preferred industrially in the manufacture of PVAc for adhesive applications or as an intermediate product for poly(vinyl alcohol) production. In solution polymerization of vinyl acetate, it has been known that the solvent has a strong effect on the final polymer molecular weight and the nature of the polymer end groups. The organic solvents that can be employed in solution polymerization of vinyl acetate include benzene, methanol, ethanol, methyl acetate, ethyl acetate, ketones, *tert*-butanol, and water. Among these, methanol is the preferred solvent because poly(vinyl acetate) prepared in methanol solution is used as intermediate for the production of poly(vinyl alcohol).

13.7.3 Suspension Polymerization

The suspension polymerization of vinyl acetate is an important industrial process. In suspension polymerization, organic monomer phase is dispersed as fine droplets in an aqueous media. Each monomer droplet, containing organic initiator, acts like a micropolymerization reactor. Homopolymers of vinyl acetate are readily obtained by suspension polymerization. However, vinyl acetate is moderately soluble in water and, thus, coalescence of the suspended polymer particles may occur unless proper reaction conditions (e.g., initiator concentration, suspending agent, stirring rate, pH) are employed. The optimal pH range of 4–5 should be maintained to prevent the hydrolysis of VA during the polymerization. (The hydrolysis leads to the formation of acetaldehyde — a strong chain transfer agent.) The polymer particle size distribution is influenced by the concentration of protective colloid and the agitation and agitator geometry.

13.8 POLY(TETRAFLUOROETHYLENE)

Poly(tetrafluoroethylene) (PTFE) is a straight linear polymer with the formula—(CF_2CF_2), having strong chemical resistance, heat resistance, low friction coefficient (antistick property), and excellent electrical insulation properties. Teflon is a trade name for homopolymers and copolymers of PTFE and is the largest volume fluoropolymer. The polymer has an extremely high molecular weight (10^6–10^7) and its high thermal stability is due to the strong carbon–fluorine bond. The close packing of the fluorine atoms around the carbon backbone provides a protective shield, making the polymer resistant to corrosion. PTFE is produced in three different forms (granular, fine powder, and aqueous dispersion) by suspension and emulsion polymerization processes [48]. Emulsion polymerization produces the PTFE polymer either as an aqueous dispersion or a fine powder. Although granular PTFE can be molded in various forms, the polymer produced by aqueous dispersion is fabricated by dispersion coating or conversion to powder for paste extrusion. The tetrafluoroethylene (TFE) monomer is obtained by pyrolysis of chlorodifluoromethane and is a colorless, tasteless, odorless, and nontoxic gas with heat of polymerization of –172 kJ mol^{-1}. Homopolymers of TFE and its copolymers can also be prepared in the solid state using actinic radiation as initiator.

In suspension polymerization of TFE, an unstable dispersion is formed in the early stages of polymerization. Without a dispersing agent and vigorous agitation, the polymer coagulates partially. As a result, the polymer is stringy, irregular, and variable in shape. The solid polymer recovered is then ground to appropriate particle sizes.

When TFE is polymerized in an aqueous medium with an initiator and emulsifier, PTFE is obtained in a fine-powder form. TFE monomer gas is supplied to a mechanically agitated reactor containing water, an initiator, an emulsifier, and any

TABLE 13.7
Typical Polymerization Conditions

Product Form	Pressure (psi)	Temperature (°C)
Fine powder	1000	55–240
Aqueous dispersion	15–500	0–95

comonomer. PTFE is polymerized using a free-radical initiator. Termination is predominantly by coupling (radical combination), and chain transfer is negligible. As a result, PTFE has a very high molecular weight. It is important to prevent premature coagulation, although the final polymer is recovered as fine-powder resin. The thin dispersion rapidly thickens into a gelled matrix and coagulates into a water-repellent agglomeration. Typical polymerization conditions are presented in Table 13.7.

When TFE is polymerized by suspension polymerization, PTFE granules are obtained. Suspension polymerization is performed in a batch-agitated reactor with or without a dispersing agent. Vigorous agitation is required to keep the polymer in a partially coagulated state. After polymerization, the polymer is separated from the aqueous medium, dried, and ground to the desired size. Unfortunately, very little has been published in open literature concerning the kinetic aspects of TFE free-radical polymerization.

Finely divided powders of PTFE can also be prepared through gamma irradiation of a gaseous TFE monomer.

13.9 POLY(VINYL/VINYLIDENE CHLORIDE)

Vinylidene homopolymers and copolymers are known as Saran. The vinylidene chloride (VDC) monomer (CH_2-CCl_2) is a colorless liquid with a characteristic sweet odor and soluble in most polar and nonpolar organic solvents. Poly(vinylidene chloride) (PVDC) is a chemical-resistant polymer having high impermeability to gases and vapors. PVDC is highly crystalline in the normal-use temperature range (0–100°C) because it has a linear symmetrical chain structure which is independent of the polymerization temperature. Thus, PVDC can be oriented into high-strength fibers and films [75]. Due to its extreme thermal instability, PVDC cannot be heated to the molten state for longer than a few seconds without degradation.

Poly(vinylidene chloride) can be polymerized by solution, slurry, suspension, and emulsion processes. Like PVC, PVDC is not soluble in its own monomer. Therefore, when VDC is bulk homopolymerized, the rapid development of turbidity occurs in the reaction medium due to the presence of PVDC crystals at the beginning of polymerization. The heterogeneity of the reaction process exerts a marked influence on the polymerization kinetics. With the progress of reaction to about 20% conversion, it becomes a thick paste. As the conversion increases further, the crystalline phase grows and the liquid slurry becomes a hard, porous solid mass [76]. During the final stage, hot spots can develop in the mass due to the high reaction rate and poor heat transfer. Unlike in the PVC and polyacrylonitrile systems where spherical aggregates are formed, anisotropic growth of polymer particles takes place in the PVDC system.

The following heterogeneous polymerization scheme has been proposed [77]:

Initiation in the liquid phase	$I \xrightarrow{k_d} 2R^{\cdot}$
Propagation in the liquid phase	$R^{\cdot} + M \xrightarrow{k_p} P$
Termination in the liquid phase	$P \xrightarrow{k_c}]$
Radical precipitation	$P \xrightarrow{k_c} P_s$
Propagation at solid-liquid interface	$P_s + M \xrightarrow{k_{ps}} P_s$
Termination at the solid-liquid interface	$2P_s + \xrightarrow{k_{ts}} D$

where D is the dead polymer.

If the solid polymer phase contains no monomer at high conversion except that adsorbed on its surface, the interior of the polymer crystals should remain inaccessible unless the polymer is heated to a high enough temperature for chain motion in the crystalline phase to occur. In addition, if the polymerization occurs only on the solid surface and the radicals precipitate before terminating in the liquid phase, the polymerization rate will increase with conversion because of the increase in surface area. Then, the number of polymer particles, particle shape, and morphology will influence the polymerization kinetics. Assuming that a rectangular lamellar particle grows on the edges only, Wessling proposed the following rate equation [76]:

$$R_p = \left(\frac{f_i k_i k_{ps}^2}{k_{ts}}\right) C_1 [I]_0^{1/2} \left(\frac{N}{M_0}\right)^{1/4} \left(\frac{m}{M_0}\right)^{1/4} \tag{13.18}$$

where $[1]_0$ is the initiator concentration, M_0 is the number of moles of monomer present initially, m is the number of moles of monomer converted to polymer, N is the number of particles, and C_1 is a morphology factor defined as

$$C_1 = \left[\frac{2d(q+1)}{V_M}\left(\frac{hV_p}{q}\right)^{1/2}\right]^{1/2} \tag{13.19}$$

where q is the ratio of large lamellar dimensions, h is the fold length, d is the thickness of the reaction zone, and V_M is the molar volume.

The PVDC homopolymer is difficult to process. Thus, copolymers of vinylidene chloride–vinyl chloride, vinylidene chloride-alkyl acrylate, and vinylidene chloride–acrylonitrile, which are easier to process than PVDC homopolymer, are widely used in industrial processes. The choice of comonomer significantly affects the properties of the copolymer. Table 13.8 illustrates the reactivity ratios of some important monomers (monomer 1 = VDC) [76].

The introduction of a comonomer such as vinyl chloride into the polymer chains reduces the crystallinity of the polymer to some extent. Practically, amorphous

TABLE 13.8
Reactivity Ratios of Some Important Monomers

Monomer	r_1	r_2
Acrylonitrile	0.49	1.20
Butadiene	<0.45	1.9
Butylacrylate	0.88	0.83
Ethyl vinyl ether	3.2	0.0
Maleic anhydride	9.0	0.0
Methylacrylate	1.0	1.0
Methyl methacrylate	0.24	2.53
Styrene	0.14	2.0
Vinyl acetate	5.0	0.05
Vinyl chloride	3.25	0.3

copolymer is obtained with a 3:1 ratio of vinylidene chloride/vinyl chloride mixture [75]. Both emulsion and suspension free-radical polymerization processes are used for commercial production of PVDC copolymer. Because the vinylidene chloride is easily oxidized, the polymerization is usually performed at less than 50°C.

When vinyl chloride and vinylidene chloride are copolymerized, the resulting polymer is a heterogeneous mixture of copolymers of different composition due to the large difference between the reactivity ratios of the two monomers ($r_{vcd} = 0.3$, $r_{vdc} = 3.2$). To obtain a homogeneous product, the faster-polymerizing monomer should be added during the polymerization to maintain the composition of the monomer mixture constant.

13.9.1 Emulsion Polymerization

Emulsion polymerization itself is a heterogenous system. However, in the emulsion of polymerization of VDC, polymer precipitates in the latex particles and thus the reaction is also heterogeneous. If the polymer is isolated and used as a dry powder, low soap recipes of marginal colloidal stability are used, whereas if the polymer is to be used as a latex, a higher surfactant concentration is required. PVDC copolymer latex made by the emulsion process is used as a coating compound applied to various substrates. Redox initiator systems are normally used for VDC emulsion polymerization at low temperatures. The emulsion polymerization should be performed at pH <9 because the polymer is attacked by an aqueous base [76]. The polymer may be recovered in dry-powder form by coagulating the latex with an electrolyte, followed by washing and drying. A typical recipe for emulsion polymerization is presented in Table 13.9 [21].

The role of activator is to promote the initiator decomposition so that a lower reaction temperature can be used to obtain a high-molecular-weight polymer within reasonable reaction times. At 30°C, 95–98% conversion is achieved in a batch reactor after 7–8 h of reaction time and an average latex particle diameter is 100–150 nm. After polymerization, the emulsion is coagulated, washed, and dried. During the copolymerization in a batch reactor, copolymer composition drift may occur.

TABLE 13.9
Typical Recipe for Emulsion Polymerization

Material	Parts (in wt.)
Vinylidene chloride	78
Vinyl chloride (comonomer)	22
Water	180
Potassium persulfate (initiator)	0.22
Sodium bisulfite (activator)	0.11
Dihexyl sodium sulfosuccinate (emulsifier)	3.58
Nitric acid	0.07

To prevent it, more reactive monomer or monomer/comonomer mixture may be added into the reactor during the course of polymerization.

To describe the kinetics of VDC emulsion polymerization, the classical Smith–Ewart model and the surface growth model have been used. In the surface growth model, the polymerization is assumed to occur in a restricted zone at the particle surface, not in the core of the polymer particle. The reaction zone can be an adsorbed monomer layer or a highly swollen surface. Then, the particle would grow from the surface outward. Figure 13.20 illustrates a conversion–time curve for the batch emulsion polymerization of VDC using a Redox initiator system [$(NH_4)S_2O_8/Na_2S_2O_5$] with sodium lauryl sulfate as the emulsifier. Notice that three distinct stages take place [76].

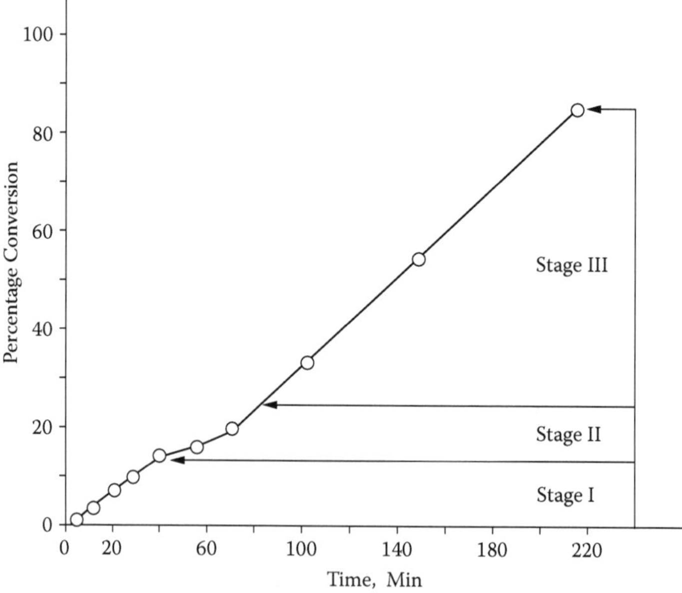

FIGURE 13.20 Emulsion polymerization of VDC. From R. A. Wessling, *Polyvinylidence Chloride,* Gordon and Breach Science Publishers, New York, 1977. With permission.

13.9.2 Suspension Polymerization

Extrusion- and molding-grade resins of PVDC are manufactured by suspension polymerization at about 60°C to 85–90% conversion for 30–60 h. Suspension polymers are purer than emulsion polymers; however, polymerization time is significantly longer and a high-molecular-weight copolymer is difficult to obtain. The average size of polymer particles is between 150 and 600 μm. The initiator should be uniformly dissolved in the monomer phase before droplets are formed by mechanical agitation. If initiator distribution is nonuniform, some monomer droplets polymerize faster than others, leading to monomer diffusion from slow-polymerizing droplets to fast-polymerizing droplets [76]. The fast-polymerizing droplets form dense, hard, glassy polymers that are extremely difficult to fabricate because adding stabilizers or plasticizers is very difficult.

REFERENCES

1. E. Vivaldo-Lima, P. E. Wood, A. E. Hamielec, and A. Penlidis, *Ind. Eng. Chem. Res.*, 36, 939 (1997).
2. H. G. Yuan, G. Kalfas, and W. H. Ray, *J. Macromol. Sci. — Rev. Macromol. Chem.*, C31 (2/3), 215 (1991).
3. J. P. Congalidis and J. R. Richards, paper presented at the Polymer Reaction Engineering III Conference, March 1997.
4. R. G. Gilbert, *Emulsion Polymerization*, Academic Press, New York, 1995.
5. D. C. Blackley, *Emulsion Polymerization*, Wiley, New York, 1975.
6. L. F. Albright, in *Processes for Major Addition-Type and Their Monomers*, 2nd ed., Robert E. Krieger Publ. Co., Malabar, FL, 1985.
7. K. H. Imhausen, F. Schöffel, J. Zink, J. Falbe, W. Paper, and W. Zoller, *Hydrocarbon Process.*, 155 (November 1976).
8. B. G. Kwag and K. Y. Choi, *Ind. Eng. Chem. Res.*, 33, 211 (1994).
9. S. X. Zhang, N. K. Read, and W. H. Ray, *AIChE J.*, 42(10), 2911 (1996).
10. N. K. Read, S. X. Zhang, and W. H. Ray, *AIChE J.*, 43(1), 104 (1997).
11. B. Folie and M. Radosz, *Ind. Eng. Chem. Res.*, 34(5), 1501 (1995).
12. D. Constantin and J. P. Machon, *Eur. Polym. J.*, 14, 703 (1978).
13. B. Folie, paper presented at the Polymer Reaction Engineering III Conference, March 1997.
14. N. Platzer, *Ind. Eng. Chem.*, 62(1), 6 (1970).
15. C. Kiparissides, G. Verros, and A. Pertsinidis, *Chem. Eng. Sci.*, 49(24B), 5011 (1994).
16. G. Luft, paper presented at Polyethylene World Congress, 1992.
17. F. R. Mayo, *J. Am. Chem. Soc.*, 90, 1289 (1969).
18. K. Y. Choi, W. R. Liang, and G. D. Lei, *J. Appl. Polym. Sci.*, 35, 1562 (1988).
19. W. J. Yoon and K. Y. Choi, *J. Appl. Polym. Sci.*, 46, 1353 (1992).
20. W. J. Yoon and K. Y. Choi, *Polymer*, 33 (21), 4582 (1992).
21. *Polymer Manufacturing*, Noyes Data Corp., Park Ridge, NJ, 1986.
22. D. B. Friddy, *Adv. Polym. Sci.*, 111, 67 (1994).
23. J. Brandrup and E. H. Immergut, *Polymer Handbook*, 3rd ed., Wiley, New York, 1989.
24. J. P. Fischer, *Angew. Makromol. Chem.*, 33, 36 (1973).
25. J. L. Locatelli and G. Riess, *Angew. Makromol. Chem.*, 35, 57 (1974).
26. *Modern Plastics Encyclopedia*, McGraw-Hill, New York, 1990.
27. G. E. Molau, *Polym. Lett.*, 3, 1007 (1965).
28. D. Butala, K. Y. Choi, and M. K. H. Fan, *Comput. Chem. Eng.*, 12(11), 1115 (1988).

29. R. A. Bubeck, C. B. Arends, E. L. Hall, and J. B. Vander Sande, *Polym. Eng. Sci.*, 21(10), 624 (1981).
30. W. D. Watson and T. C. Wallace, *ACS Symp. Ser.*, 285, 363 (1985).
31. G. F. Freeguard and M. Karmarkar, *J. Appl. Polym. Sci.*, 15, 1657 (1971).
32. W. A. Ludwico and S. L. Rosen, *J. Polym. Sci. Polym. Chem. Ed.*, 14, 2121 (1976).
33. S. L. Rosen, *J. Appl. Polym. Sci.*, 17, 1805 (1973).
34. A. Echte, F. Half, and J. Hambrecht, *Angew. Chem. Int. Ed. Engl.*, 20, 344 (1981).
35. D. J. Stein, G. Fahrbach, and H. Adler, *Adv. Chem. Ser.*, 142, 148 (1975).
36. D. E. Carter and R. H. M. Simon, U.S. Patent 3,903,202 (1975).
37. *Hydrocarbon Process.*, 163 (November 1985).
38. C. L. Mott and B. A. Kozakiewicz, U.S. Patent 4,221,883 (1980).
39. K. Brenstart, K. Buchholz, and A. Echte, U.S. Patent 3,568,946 (1972).
40. G. Gawne and C. Ouwerkerk, U.S. Patent 4,011,284 (1977).
41. C. Fukumoto, T. Furukawa, and C. Oda, U.S. Patent 4,419,488 (1933).
42. S. Y. Zhiqiang and P. Zuren, *J. Appl. Polym. Sci.*, 32, 3349 (1986).
43. G. Natta, F. Danusso, and D. Sianesi, *Makromol. Chem.*, 28, 253 (1958).
44. N. Ishihara, T. Seimiya, M. Kuramoto, and M. Uoi, *Macromolecules*, 19, 2464 (1986).
45. R. Pó and N. Cardi, *Prog. Polym. Sci.*, 21, 47 (1996).
46. J. C. W. Chien, Z. Salajka, and S. Dong, *J. Polym. Sci. Polym. Chem. Ed.*, 29, 1243 (1991).
47. J. C. W. Chien, Z. Salajka, and S. Dong, *J. Polym. Sci., Polym. Chem. Ed.*, 29, 1253 (1991).
48. *Encyclopedia of Polymer Science and Engineering*, Vol. 1, Wiley, New York, 1985.
49. G. Kalfas, H. G. Yuan, and W. H. Ray, *Ind. Eng. Chem. Res.*, 32, 1831 (1953).
50. R. H. M. Simon and G. H. Alford, *Appl. Polym.*, 26, 31 (1975).
51. K. H. Reichert, U.-H. Moritz, Ch. Gabel, and G. Deiringer, in *Polymer Reaction Engineering*, K. H. Reichert and W. Geiseler, Eds., Hanser, Munich, 1983, p. 153.
52. K. H. Reichert and H.-U. Moritz, *J. Appl. Polym. Sci.: Appl. Polym. Symp.*, 36, 151 (1981).
53. L. Patron, C. Mazzolini, and A. Moretti, *J. Polym. Sci.*, C42, 405 (1973).
54. D. N. Bort, G. F. Zvereva, and S. I. Kuchanov, *Polym. Sci. USSR*, 20, 2050 (1979).
55. M. R. Juba, *ACS Symp. Ser.*, 104, 267 (1979).
56. L. H. Garcia-Rubio and A. E. Hamielec, *J. Appl. Polym. Sci.*, 23, 1397 (1979).
57. L. H. Garcia-Rubio and A. E. Hamielec, *J. Appl. Polym. Sci.*, 23, 1413 (1979).
58. W. C. Mallison, U.S. Patent 2,847,405 (1958).
59. G. Talamini and P. Gasparini, *Makromol. Chem.*, 117, 140 (1968).
60. J. Ugelstad, H. Flogstad, T. Hertzberg, and E. Sund, *Makromol. Chem.*, 164, 171 (1973).
61. T. Y. Xie, A. E. Hamielec, P. E. Wood, and D. R. Woods, *Polymer*, 32(3), 537 (1991).
62. M. W. Allsopp, *Pure Appl. Chem.*, 53, 449 (1981).
63. M. Langsam, in *Encyclopedia of PVC*, Vol. 1, L. I. Nass and C. A. Heiberger, Eds., Marcel Dekker, New York, 1985.
64. K. B. Abbås, *Pure Appl. Chem.*, 53, 411 (1981).
65. J. Ugelstad, P. C. Mørk, and F. K. Hansen, *Pure Appl. Chem.*, 53, 323 (1981).
66. P. V. Smallwood, in *Concise Encylopedia of Polymer Science and Engineering*, Wiley, New York, 1990.
67. P. V. Smallwood, *Polymer*, 27, 1609 (1986).
68. J. B. Cameron, A. J. Lundeen, and J. H. Mcculley Jr., *Hydrocarbon Process.*, 39 (March 1980).
69. N. Fischer and L. Goiran, *Hydrocarbon Process.*, 143 (May 1981).
70. M. K. Lindermann, in *Vinyl Polymerization, Part I*, G. E. Ham, Ed., Marcel Dekker, New York, 1967.
71. T. W. Taylor and K. H. Reichert, *J. Appl. Polym. Sci.*, 30, 227 (1985).
72. S. I. Nozakura, Y. Morishima, and S. Murahashi, *J. Polym. Sci.*, A-1, 10, 2781 (1972).
73. G. C. Berry and R. G. Craig, *Polymer*, 5, 19 (1964).
74. H. W. Melville and P. R. Sewell, *Makromol. Chem.*, 32, 139 (1959).

75. C. B. Havens, *Ind. Eng. Chem.*, 42, 315 (1950).
76. R. A. Wessling, *Polyvinylidene Chloride,* Gordon and Breach Science Publishers, New York, 1977.
77. R. A. Wessling and I. R. Harrison, *J. Polym. Sci.*, A-1, 9, 3471 (1971).
78. K. H. S. Chang, M. H. Litt, and M. Nomura, in *Emulsion Polymerization of Vinyl Acetate*, M. S. El-Aasser and J. W. Vanderhoff, Eds., Applied Science Publishing, Englewood, NJ, 1981, p. 89.

Part IV

Vinyl Polymer Technology

14 Vinyl Polymer Degradation

Chapal K. Das, Rathanasamy Rajasekar, and Chaganti S. Reddy

CONTENTS

14.1 INTRODUCTION

In recent decades, synthetic polymeric materials, because of their unique physical properties, have rapidly replaced the more traditional materials such as steel and non-ferrous metals, as well as natural polymeric materials such as wood, cotton, and natural rubber. One of the limiting factors in the application of polymers at high temperatures is their tendency to not only become softer but also to thermally degrade. However, one weak aspect of synthetic polymeric materials compared with steel and other metals is that these materials are combustible under certain conditions. Most of the synthetic polymers are highly thermally sensitive due to the limited strength of the covalent bonds that make up their structures. The breaking of chemical bonds under the influence of heat is the result of overcoming bond dissociation energies. Thermal degradation can present an upper limit to the service temperature of polymers much as the possibility of mechanical property loss. Indeed unless correctly prevented, significant thermal degradation can occur at temperatures much lower than those at which mechanical failure is likely to occur.

Degradation of polymeric materials [1–5] is an important issue from both the academic and the industrial viewpoints. Understanding the degradation of polymers is of paramount importance for developing a rational technology of polymer processing and higher-temperature applications. Controlling degradation requires understanding of many different phenomenological factors like synthetic conditions, processing conditions, including chemical mechanisms, the influence of polymer morphology, the complexities of oxidation chemistry, also the effects of stabilizers, fillers and other additives. Most commercial polymers are manufactured by processes involving chain polymerization, poly-addition, or poly-condensation reactions. These processes are generally controlled to produce individual polymer molecules with defined

- Molecular weight (or molecular weight distribution)
- Degree of branching
- Composition

Once the initial product of these processes is exposed to further shear stress, heat, light, air, water, radiation, or mechanical loading, the chemical reactions started in the polymers will have the net result of changing the chemical composition and the molecular weight of the polymer. These reactions in turn lead to a change in the physical and optical properties of the polymer. In practice, any change of the polymer properties relative to the initial desirable properties is called degradation. In this sense, "degradation" is a generic term for any number of reactions that are possible in a polymer. Degradation of polymers can be brought about by either physical factors, such as heat, light, and mechanical stress, or by chemical agents such as oxygen, ozone, acids, and alkalis. In the following sections of this chapter, emphasis will be made to discuss elaborately the different types of degradations encountered in polymeric materials and their mechanistic aspects.

14.2 THERMAL DEGRADATION

Thermal degradation of polymers *is molecular deterioration as a result of overheating*. At high temperatures, the components of the long chain backbone of the polymer can begin to separate (molecular scission) and react with one another, to change the properties of the polymer. It is part of a larger group of degradation mechanisms for polymers that can occur from a variety of causes such as:

- Heat (thermal degradation and thermal oxidative degradation in the presence of oxygen)
- Light (photo degradation)
- Oxygen (oxidative degradation)
- Weathering (generally UV degradation)

All polymers will experience some type of degradation during service and this will result in a steady decline in their properties. In fact, degradation is inevitable and the resulting chain reaction will accelerate, unless the cycle is interrupted in some manner. The only real variable is how long it is going to take for thermal degradation to become evident and result in a loss in properties, which is significant enough for the end-user to notice.

14.2.1 Mechanistic Aspects of Thermal Degradation of Polymers

Many initial stages of degradation exist as subsets of thermal degradation. Following is a list, with brief descriptions of each.

- Random Initiation — This occurs in the middle of a polymer chain, at an unspecified point.
- Terminal (end) Initiation — This occurs at the end of a polymer chain, when a monomer is volatized out of the reaction medium (i.e., evaporates).
- De-propagation — This occurs very similarly to terminal initiation, but the process continues and monomers keep volatizing out of the medium.
- Uni-molecular Termination — A short polymer chain breaks up into products, but this rarely is accounted for in the bulk phase.
- Termination by Recombination — Two polymer radicals join together to form non-radical products.
- Termination by Disproportionation — Two polymer radicals share radicals and yield two non-radical polymers.
- Intermolecular Transfer — A polymer and a polymer radical yield two polymers and a polymer radical.

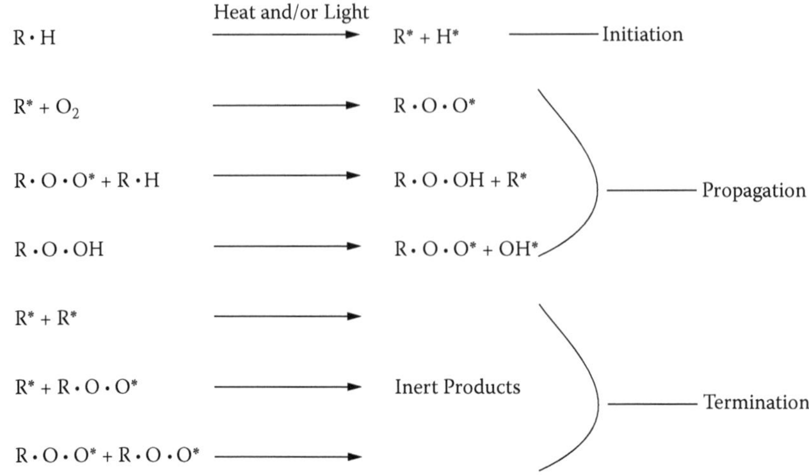

SCHEME 14.1 Consequences of thermal degradation.

Most of these reactions yield radicals, which are the basis of all chemical degradation. A general mechanism for thermal degradation is shown in Scheme 14.1.

The chemical reactions involved in thermal degradation lead to physical and optical property changes relative to the initially specified properties. Thermal degradation generally involves changes in the molecular weight (and molecular weight distribution) of the polymer and typical property changes include:

- Reduced ductility and embrittlement
- Chalking
- Color changes
- Cracking
- General reduction in most other desirable physical properties

The dominant mechanism of degradation and the degree of resistance to degradation depends on the application and the polymer concerned. The results are the same for most polymer families and significant property degradation can occur, when thermal degradation does occur.

14.3 MECHANICAL DEGRADATION

A polymer can be degraded mechanically by two processes, mostly by machining processes, and rarely by ultrasonic processes. Machining processes take advantage of the visco-elastic behavior of most polymers. The range includes five regions. Familiarity with the regions is the best way to prevent entering deformation ranges.

- Glassy region — Chain segments are frozen in place and movement is minimal.
- Transition region — Chain segments undergo short-range diffusional motion, and alterations occur as a function of applied forces.

R• + Monomer (M)	⟶	RM•	———	Addition
R• + M →	⟶	Polymer + M•	———	Monomer Transfer
R• + Solvent (S)	⟶	Polymer + S•	———	Solvent Transfer
R• + R′XH	⟶	Polymer + R′X•	———	Scavenger Transfer
R• + O_2	⟶	RO_2•		
RO_2 + RH	⟶	RO_2H + R•		
RO_2H	⟶	RO• + OH		
RO_2• + ~~~HC = CH~~~	⟶	~~~HC(—RO_2•)—CH~~~		

SCHEME 14.2 Radical reactions of mechanical degradation.

- Rubbery plateau — Short-range diffusional motion is rapid, but entanglements act as cross-links, inhibiting motion.
- Rubbery flow — Molecules as a whole are now moving, long-range entanglements are slipping, and applied stress leads to permanent deformation.
- Liquid flow — Long-range changes in the chain occur instantaneously, and deformation is completely irreversible.

Two elementary chemical reactions occur during mechanical degradation. Primarily, the whole polymer is divided into two radicals, purely by virtue of the mechanical forces placed upon it. Subsequently, the radicals that are formed can undergo any of a multitude of reactions, as outlined in Scheme 14.2.

Ultrasonic processes induce main-chain rupture by cavitations. Cavitations occur under the influence of ultrasound, in which bubbles are formed and collapse in the materials. The rapid collapse sends out shock waves into the material and the main-chain rupture is believed to stem from this rapid motion of solvent molecules, to which the macromolecules cannot adjust.

14.4 CHEMICAL DEGRADATION

The chemical degradation of polymeric materials is really an encompassing topic that can refer to thermal degradation or oxidative degradation as well. The basic issue in reference to chemical degradation is solvolysis. Solvolysis is concerned with the breaking of C-X bonds where X is a non-carbon atom, mostly O, N, P, S, Si, or a halogen. Solvolysis agents, such as water, alcohols, ammonia, etc., rupture the main chain of the polymer and break it down. A common solvolysis reaction is hydrolysis, in which one of the degradation products is water. Water soluble polymers succumb easily to hydrolysis, whereas water insoluble polymers are degraded only on the surface of the specimen.

Different kinds of chemical agents can attack polymers. Those with the most practical application to study from a corrosive point of view are atmospheric pollutants. Other agents that can degrade polymers include ozone, acids, alkalis, and halogens.

Reactants		Products		Reaction
$PH + NO_2$	$\longrightarrow$	$P\cdot + HNO_2$	—	Hydrogen Abstraction
$P\cdot + NO_2$	$\longrightarrow$	PNO_2	—	Addition of Nitro Groups
$P\cdot + NO_2$	$\longrightarrow$	$P{-}O{-}N=O$	—	Formation of Nitrite Ester Groups
$PONO$	$\longrightarrow$	$PO\cdot + NO$	—	Rupture of O-N Bond
$PO\cdot + NO_2$	$\longrightarrow$	$PONO_2$	—	Formation of Nitrate Ester Groups
$PO\cdot$	$\longrightarrow$	$F_1 + F_2$	—	Mainchain Cleavage (Beta-Scission)
$PO\cdot$	$\longrightarrow$	$\sim\overset{H}{\overset{\mid}{C}}=O + \cdot CH_2\sim$		
$PO\cdot + PH$	$\longrightarrow$	$POH + P\cdot$	—	Formation of Hydroxyl Groups

SCHEME 14.3 Elementary reactions of chemical degradation.

Atmospheric pollutants, such as nitrogen dioxide (NO_2) and sulfur dioxide (SO_2), nearly always react with polymers in the presence of air in this practical application of outdoor exposure. Simultaneous scission and cross linking occurs when unsaturated polymers with accessible linear sections, where the rate of reaction is dependent upon the temperature and partial pressure of oxygen. A mechanism is set forth for the interaction between polyethylene and NO_2 by Schnabel [6], with a similar proposal for SO_2, which is depicted in Scheme 14.3).

Ozone is another atmospheric pollutant that can degrade polymers, notably natural rubber, by the oxidation of its double bonds. Car tires usually have anti-ozonants that saturate the rubber to inhibit premature breakdown.

14.5 OXIDATIVE DEGRADATION

Two types of oxidation occur in the case of polymeric materials: direct oxidation and auto-oxidation.

Direct oxidation refers to reactions that occur spontaneously under standard temperature and pressure, such as the oxidation of polymers with metal ions as functional groups. The exposure of these polymers to oxidizing agents such as potassium permanganate, sulfuric acid, or nitric acid leads directly to oxidation. The oxidation of these metal ions may deteriorate the mechanical properties of the polymers, because direct oxidation leads to preferential oxidation of crystallization sites in polymers that are only partially amorphous.

Auto-oxidation occurs with polymers with molecular oxygen in their structure. This leads to the presence of free radicals. In commercial plastics, the main concern is that of thermal degradation of hydro-peroxides (R–OOH) that tend to be present in many polymers in small quantities. Because R–OOH groups decompose at relatively low temperatures, the presence of extremely reactive hydroxide ions (OH–) is common.

Hydroxide radicals are the initiation step for a multitude of hydrolysis reactions or addition reactions, such as in aromatic compounds. Auto-oxidation of many polymers is often modeled with a scheme originally developed for natural rubber as

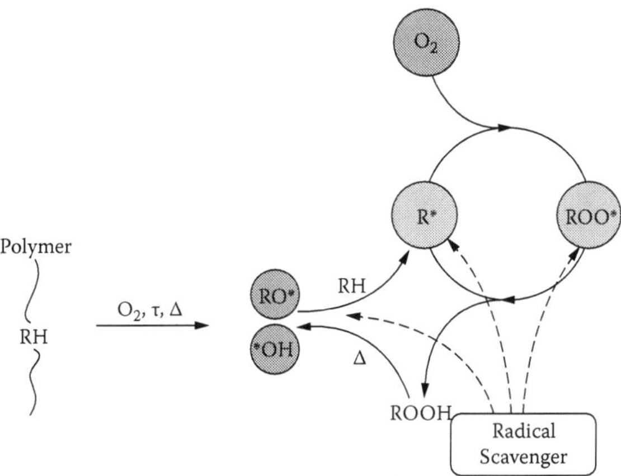

SCHEME 14.4 Auto-oxidation of polymers.

illustrated in Scheme 14.4. The important aspect of this figure is that once oxidation starts, which it always will, it sets off a chain reaction that accelerates degradation unless stabilizers are used to interrupt the oxidation cycle.

Trace quantities of metals in polymers, whether impurities or minor parts or the polymer chain, can also catalyze the oxidative process. Metals are often present in the polymeric matrix as ions, which can pull a hydrogen molecule off the matrix, leaving a free radical. Free radicals start the chain scission, which leads to degradation.

14.6 PHOTO-DEGRADATION

Exposure to sunlight and some artificial lights can have adverse effects on the useful life of plastic products. UV radiation can break down the chemical bonds in a polymer. This process is called photo-degradation and ultimately causes cracking, chalking, color changes, and the loss of physical properties. Photo-degradation, once started, essentially follows the same scheme as presented previously. Because photo-degradation generally involves sunlight, thermal oxidation takes place in parallel with photo-oxidation.

Photo-degradation differs from thermal oxidation, in that it can be started by absorption of UV light. Most pure polymers are theoretically incapable of absorbing UV light directly. Trace amounts of other compounds within the polymer, such as degradation products or catalyst residues, can however absorb UV. For this reason, effective thermal and processing stabilization is a prerequisite for effective long-term light stabilization.

14.7 BIOLOGICAL DEGRADATION

The bio-degradation of polymers on a natural level is familiar to everyone in the form of incomplete degradations such as fossil fuels, and complete degradations such as those of proteins and lipids, for nutritional purposes.

Unfortunately for environmentalists, man-made polymers are theoretically biodegradable but essentially inert. Microbial degradation has been found to be effective for low molecular weight polymers, whereas macromolecular matrices are scarcely affected. It has been shown that contrary to naturally occurring polymers, synthetic polymers are attacked only from the ends of the chain, which would account for the much slower rate of degradation.

Natural products that are particularly susceptible to biological attack include:

- Industrial gums — hydrolyzable by bacteria and fungi
- Natural rubber — almost completely consumable by soil microorganisms
- Starch — degraded easily by bacteria and fungi
- Cellulose — attack by biological agents, through enzymatic hydrolysis

14.7.1 Enzymatic Degradation

Enzymes are mostly involved in the chemical mode of polymer degradation, pertaining to the decomposition of polymers that are a part of organized living species. Organized species have evolved to the point where they have enzymes that can break down certain polymers (such as proteins) in their digestive systems, which have become highly specific to their biological process.

Enzymatic degradation occurs by a catalytic process. Molecular conformation is very important to the specificity of the enzyme, and the enzyme can be rendered inactive (denatured) very quickly by varying the pH, temperature, or solvent. Some enzymes require the presence of other enzymes (co-enzymes) to be effective, in some cases forming association complexes in which the coenzyme acts as a donor or acceptor for a specific group.

In synthetic polymers, chain ends tend to be deep in the polymer matrix, and because enzymes tend to attack only at the chain ends, the process is quite slow.

14.7.2 Microbial Degradation

Microorganisms, such as fungi and bacteria, degrade polymers by ingesting the carbon molecules that are the backbone of the polymer. Given either an aerobic (fungi) or an anaerobic (both fungi and bacteria) environment, microbial degradation is theoretically possible. Due to the hydrophobicity of many plastics, however, optimal growth conditions (water, light) are rarely found and the microorganisms usually do not penetrate deeper than the surface of a part.

14.8 POLYMER DEGRADATION BY HIGH-ENERGY RADIATION

Radiative degradation can affect polymers in two ways, chain scission and cross linking. Photochemical radiation, such as light at 2800 angstroms, has the ability to break C-C bonds and instigate both scission and cross-linking. Stronger radiation, such as high-energy electrons, X-rays, and gamma rays, can initiate the formation of radicals, which can un-saturate the chemical bonding and lead to both chain scission and cross-linking. The predominance of a hugely cross-linked mass or a jumble of

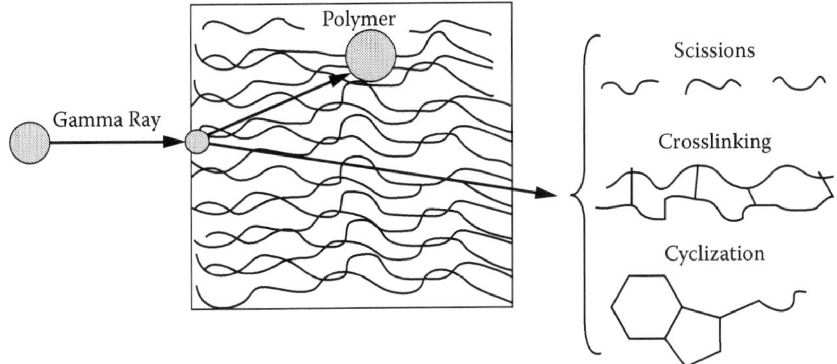

SCHEME 14.5 Radiation effect on polymers.

low-molecular-weight fragments depends mostly upon the rate of cross linking, a function of the polymer.

Two types of radiative degradation can occur, namely photolysis and radiolysis. Photolysis occurs under the influence of photochemical radiation, light in the wavelength of 100 to 10,000 angstroms which imparts energy in the range of 100–1000 kcal/mole. Degradation of the polymer chain by this method is due to the absorption of energy in discrete energy units by specific functional groups that are on the chain.

Radiolysis occurs under the influence of ionizing radiation, i.e., X-rays, electron beams, and gamma rays. Radicals form by a random ray striking a particular atom and initiating an energy transfer. The presence of a radical initiates un-saturation in chemical bonding, chain scission, and cross linking, as well as volatile fragments that initiate the formation of more radicals, perpetrating the reactions. When a gamma ray hits a polymer electron, it is deviated and loses a part of its energy, which is transmitted to the polymer electron, which initiates some chemical reactions such as scissions, cross-linking, and oxidation in presence of air or oxygen, cyclization, isomerization, amorphization, etc., as presented in Scheme 14.5.

14.9 DEGRADATION OF INDIVIDUAL VINYL POLYMERS

14.9.1 Poly (Vinyl Chloride)

14.9.1.1 Historical Aspects of Poly (Vinyl Chloride)

Poly (vinyl chloride), PVC, is a synthetic polymer prepared from vinyl chloride monomer (VCM),

$$\mathrm{CH_2{=}CHCl} \longrightarrow \left[\!-\mathrm{CH_2}-\mathrm{CHCl}-\!\right]_n$$

where n = 700–1500 which has a unique position amongst all the polymers produced today. The polymerization of vinyl chloride monomer (VCM) has been well known since 1872. Baumann [7] was the first who produced poly (vinyl chloride) (PVC) by accident. He exposed VCM to sunlight and obtained a white solid material that could be heated up to 130°C without decomposition. However, when exposed to higher temperatures the material started to melt and simultaneously produced a considerable amount of acid vapor, finally resulting in a black-brownish colored material.

In the late 1920s [8,9], vinyl chloride copolymers were introduced in the market. In addition, the possibility of plasticizing PVC, by using esters such as tritolyl phosphate and dibutyl phthalate, resulting in more flexible material was discovered at that time. The introduction of emulsion and suspension polymerization systems for PVC resins, both around 1933, have been substantial advancements, just like the development of the so-called "easy processing" suspension resins, which were capable of absorbing plasticizers without gelation. In the early 1930s, PVC had already been introduced in small quantities in different types of products both in the United States and Germany. However, large-scale production began in Germany in 1937. After 1939, especially during the World War II, the production started to attain commercial significance on a worldwide basis. Due to deficiency of essential conventional materials and the demand for certain properties, which were not possessed by any of the available materials during that period of time, this war promoted the development of a lot of plastics. Growth of the production of PVC has been so rapid after 1939 that most countries with any degree of industrialization produce some vinyl resin. The production and use of rigid PVC increased dramatically in the early 1960s, due to large improvements in both heat-stabilizing agents and processing equipment [10]. The great versatility of VCM monomers and the low production costs of vinyl polymers are the two major reasons for their large share of the plastic market. The vinyl polymers, especially PVC, can be converted into many different products exhibiting an extremely wide range of properties both physical and chemical by using modifying agents, such as plasticizers, fillers, and stabilizers [11]. The products can range from a flexible garden hose to a rigid drainpipe, from flexible sheets for raincoats to rigid sheets for packing, from soft toys to upholstery. Eventually, PVC compositions have succeeded in displacing materials such as rubber, metals, wood, leather, textiles, conventional paints and coatings, ceramics, glass, etc.

14.9.1.2 Current Position

Among the most important commercial thermoplastics, such as polyethylene (PE), polypropylene (PP), and polystyrene (PS), PVC occupies almost the third position in the world in terms of commercial as well as application point of view [10,12]. Today, PVC is manufactured mainly by three different polymerization processes: suspension, bulk or mass, and emulsion polymerization. However, the suspension process embraces almost 80% of all commercial production of PVC [13]. Polymerization of VCM occurs according to a free-radical addition mechanism, which includes initiation, propagation, chain transfer to monomer, and bimolecular termination steps as presented in Scheme 14.6 [14].

PVC possesses many desirable characteristics that have allowed it to achieve its present status. Despite its enormous technical and economical importance, PVC also possesses many problems [15–17]. Among the commercially available thermoplastics,

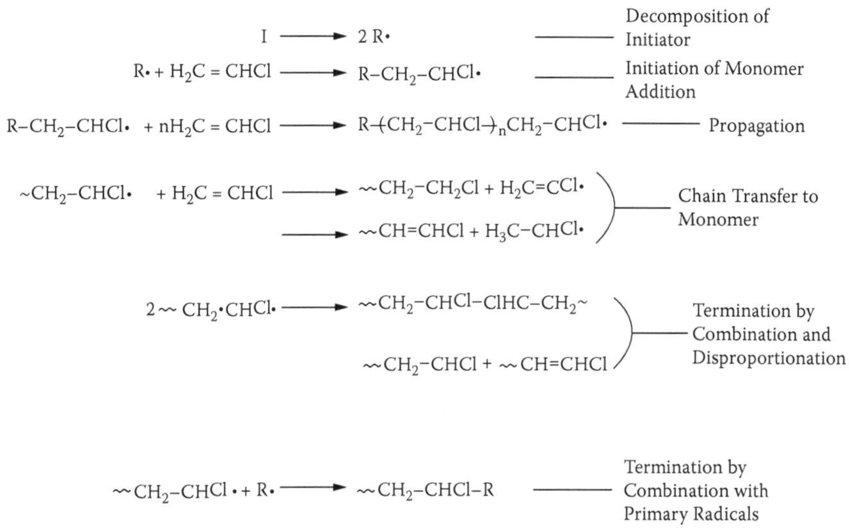

$$I \longrightarrow 2\,R\bullet \quad \text{——— Decomposition of Initiator}$$

$$R\bullet + H_2C=CHCl \longrightarrow R{-}CH_2{-}CHCl\bullet \quad \text{——— Initiation of Monomer Addition}$$

$$R{-}CH_2{-}CHCl\bullet + nH_2C=CHCl \longrightarrow R{-}(CH_2{-}CHCl)_n CH_2{-}CHCl\bullet \quad \text{——— Propagation}$$

$$\sim CH_2{-}CHCl\bullet + H_2C=CHCl \longrightarrow \sim CH_2{-}CH_2Cl + H_2C{=}CCl\bullet$$
$$\longrightarrow \sim CH{=}CHCl + H_3C{-}CHCl\bullet \quad \text{——— Chain Transfer to Monomer}$$

$$2 \sim CH_2\bullet CHCl\bullet \longrightarrow \sim CH_2{-}CHCl{-}ClHC{-}CH_2\sim$$
$$\sim CH_2{-}CHCl + \sim CH{=}CHCl \quad \text{——— Termination by Combination and Disproportionation}$$

$$\sim CH_2{-}CHCl\bullet + R\bullet \longrightarrow \sim CH_2{-}CHCl{-}R \quad \text{——— Termination by Combination with Primary Radicals}$$

SCHEME 14.6 VCM polymerization: Free radical addition mechanism.

PVC is almost certainly the least naturally stable polymer in commercial use. During processing, storage, and utilization, PVC degrades as it is exposed to high temperatures, high mechanical stresses, or ultraviolet light, all in the presence of oxygen. Degradation of the polymer occurs by successive elimination of hydrogen chloride (HCl), which is called dehydrochlorination, yielding long-chain polyenes as depicted in Scheme 14.7. This consequently causes discoloration, deterioration of the mechanical properties, and a lowering of the chemical resistance.

Therefore, PVC requires stabilization for practically any technical applications. Stabilization mainly proceeds by the addition of compounds, which contain transition metals like lead, tin, and zinc.

14.9.1.3 The Reasons for Low Thermal Stability of PVC

PVC is a linear polymer formed by head-to-tail addition of VCM molecules to the growing polymer chain. Thermogravimetric analysis on low molecular model compounds, such as 2,4,6-trichloroheptane, 2-chloropropane, and 2,4-dichloropentane, corresponding to the regular head-to-tail structure of PVC containing secondary chlorines only, demonstrates that these model compounds are stable up to at least 200–300°C. On the other hand, commercially available PVC would start to degrade around 120°C, if it is not stabilized before processing [15,18–21]. It is well known that the quality or thermal stability of PVC decreases when monomer conversion increases.

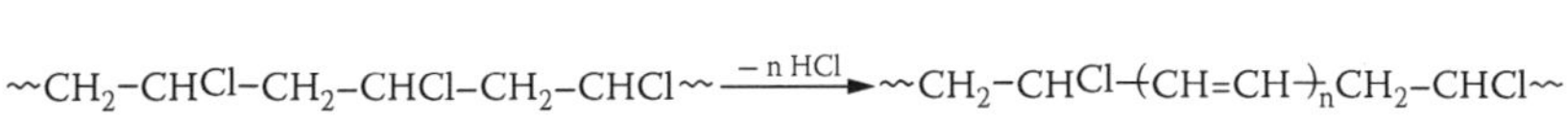

$$\sim CH_2{-}CHCl{-}CH_2{-}CHCl{-}CH_2{-}CHCl\sim \xrightarrow{-\,n\,HCl} \sim CH_2{-}CHCl{-}(CH{=}CH)_n CH_2{-}CHCl\sim$$

SCHEME 14.7 Dehydrochlorination of PVC.

VCM is polymerized in a batch-wise process, which means that the monomer supply gets more and more consumed with increasing monomer conversion. Consequently, side-reactions by the macro-radicals will increasingly occur, resulting in the formation of many different types of structural irregularities. Some of these defects have been demonstrated to have a dramatic influence on the thermal stability [11,16,22]. The frequently occurring structural defects in PVC are a wide range of branches, which are formed by various routes. Some of them appear to affect the thermal stability, whereas others are completely harmless. The frequently occurring branches and the most important types of branches concerning the lower thermal stability of PVC are outlined next.

- Chloromethyl (MB) [22–32]
- 1,2-dichloroethyl branches (EB) [22, 23, 33–35]
- 2,4-dichloro-n-butyl branch (BB) [27, 36–38]
- Long chain branching (LCB) [36, 37, 39]
- Diethyl branches (DEB) [34, 40–42]
- Internal allylic chlorines [37, 39, 43, 44]
- Tacticity 45–48]

14.9.1.4 Degradation of PVC

A considerable number of studies have been performed on the mechanism of degradation of PVC. However, it is still yet to be completely understood. Thermal decomposition [49–53] of PVC results in an intense discoloration of the polymer, which is a result of the formation of long conjugated polyene sequences that absorb radiation in the visible region [54,55]. It is generally accepted that during degradation of PVC, HCl molecules are eliminated in succession along the polymer chain yielding these conjugated polyenes. The dehydrochlorination process involves three successive steps. It starts with a relatively slow initiation of HCl removal, which is followed by a rapid zipper-like elimination of HCl and thus the formation of polyenes, which is finally terminated. Dehydrochlorination is initiated mainly by structural defects such as allylic and tertiary chlorine atoms present in the backbone of PVC. After the first elimination of HCl, allylic chlorine has been formed, which is a very active moiety, supporting the fast zipper-like elimination of HCl. The main issue is the type of mechanism by which the overall dehydrochlorination process takes place. To elucidate the mechanism of dehydrochlorination of PVC, considerable studies have been performed on some small aliphatic model compounds for PVC. Based on these model studies various schemes have been proposed, which can be classified as involving unimolecular eliminations, ionic, free-radical chain, or polaron mechanisms.

Many authors [53,56–59] have proposed the unimolecular mechanism for dehydrochlorination of PVC. Especially Fisch and Bacaloglu [60] who suggested the six-centered concerted mechanism of dehydrochlorination, catalyzed by HCl as depicted in Scheme 14.8 [61].

Ionic and quasi-ionic mechanisms for dehydrochlorination of PVC have been described elaborately in the earlier studies [18,20,62–65]. A few proposed mechanisms are shown in Scheme 14.9, such as the mechanism for dehydrochlorination of

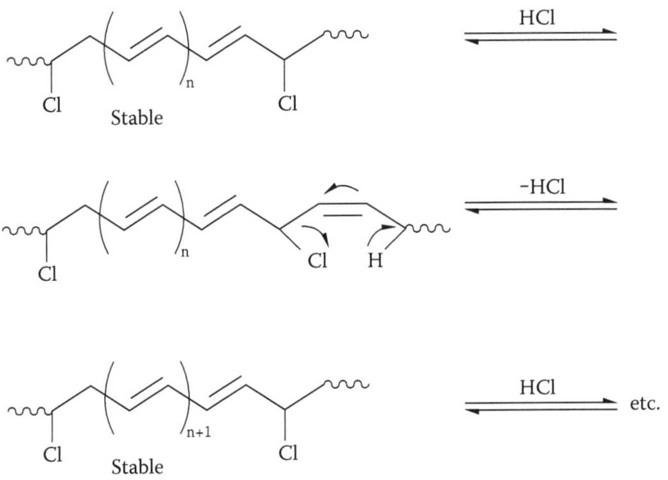

SCHEME 14.8 Dehydrochlorination of PVC: Six-centered concerted mechanism.

PVC involving an ion-pair; a quasi-ionic route via a four-center transition state are shown along with the well-established catalysis of dehydrochlorination by HCl [22].

Both the unimolecular and the ionic routes for dehydrochlorination of PVC are mainly driven by the presence of structural defects, like allylic chloride and tertiary chloride in the backbone of PVC macromolecular chains as shown in Scheme 14.10.

In addition, the radical mechanism for dehydrochlorination of PVC has been discussed in the earlier studies [55,58,66–69]. As the attack of a chlorine radical on the PVC chain is supported by many of them, this mechanism is depicted in Scheme 14.11 [22]. The chlorine radical abstracts a methylene hydrogen atom, forming HCl. The newly formed macroradical will dissociate chlorine radical adjacent to this radical, which results in the formation of a new double bond and new chlorine radical. The newly released chlorine radical will attack the neighboring methylene group immediately.

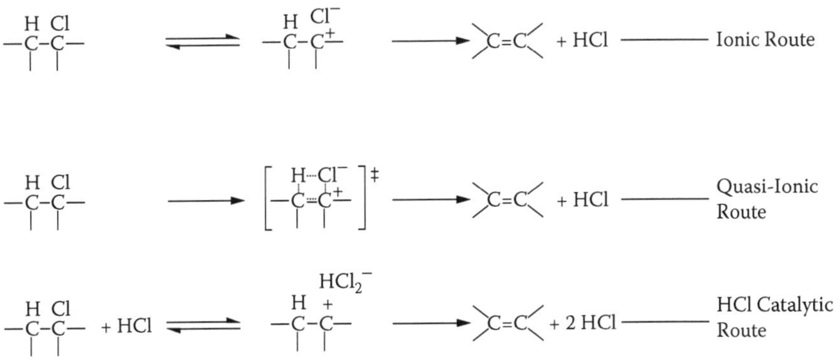

SCHEME 14.9 Dehydrochlorination of PVC: Ionic and quasi-ionic mechanisms.

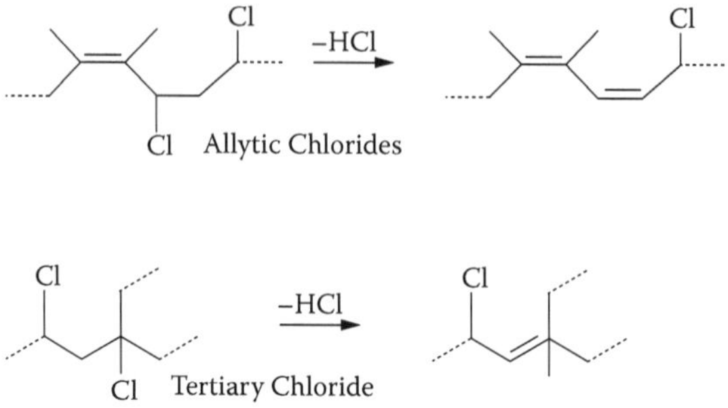

Cl $\xrightarrow{-HCl}$ Cl

Cl Allytic Chlorides

Cl $\xrightarrow{-HCl}$ Cl

Cl Tertiary Chloride

SCHEME 14.10 Dehydrochlorination of PVC: Unimolecular and ionic routes.

This cycle will repeat many times, which results in formation of a polyene sequence in the chain.

Ion-radical (polaron) mechanisms for dehydrochlorination of PVC have also been proposed in the earlier works as presented in Scheme 14.12 [70–74]. Tran [74], who reviewed the ion–radical mechanism of thermal dehydrochlorination of PVC, mentioned that not only polarons can propagate dehydrochlorination, but also solitons and bi-polarons.

Photodegradation of PVC [55,58,67,75–77] is considered to occur according to a radical mechanism. Initiation of degradation occurs by excitation of the polymer molecular chains by irradiation of the material with UV light. This results in elimination of a chlorine atom from PVC. The eliminated chlorine can promote dehydrochlorination of PVC at another polymer chain. The extent of degradation depends primarily on the presence of photosensitive chromophores in the polymer chain, because irregular structures and impurities (e.g., carbonyl groups, hydroperoxides, metal salts, and unsaturation) are often present in processed polymeric materials [78–80]. Whatever the nature of the chromophores initially, absorbing UV light in the original material led to a development of polyene structures in PVC. These structures rapidly accumulate in photolyzed PVC and act as predominant absorbing chromophores due to their large extinction coefficients.

PVC as obtained from the manufacturer is converted into finished products by high shear mixing processes such as milling, intensive mixing, and extrusion. These processes will always occur in the presence of a certain amount of oxygen. PVC suffers from chain scission during these processing conditions of high mechanical

$$Cl\bullet + \sim CH{=}CH{-}CH_2{-}CHCl\sim \xrightarrow{-HCl} \sim CH{=}CH{-}\dot{C}H{-}CHCl\sim \longrightarrow$$

$$\sim(CH{=}CH)_2CH_2{-}CHCl\sim + Cl\bullet \xrightarrow{-HCl} \sim(CH{=}CH)_2\dot{C}H{-}CHCl\sim \longrightarrow$$

$$\sim(CH{=}CH)_3CH_2{-}CHCl\sim + Cl\bullet \xrightarrow{-HCl} \text{etc.}$$

SCHEME 14.11 Dehydrochlorination of PVC: The radical mechanism.

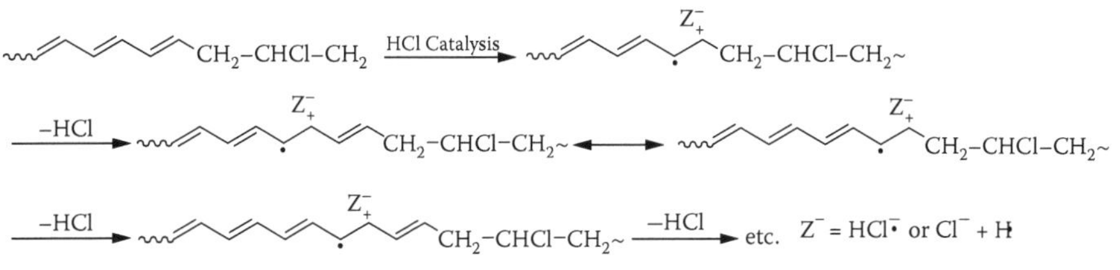

Z^{-}_{+}

~~CH$_2$–CHCl–CH$_2$ —HCl Catalysis→ ~~CH$_2$–CHCl–CH$_2$~

—–HCl→ ~~CH$_2$–CHCl–CH$_2$~ ←→ ~~CH$_2$–CHCl–CH$_2$~

—–HCl→ ~~CH$_2$–CHCl–CH$_2$~ —–HCl→ etc. Z^{-} = HCl$^{\bar{\cdot}}$ or Cl^{-} + H

SCHEME 14.12 Ion–radical mechanism for dehydrochlorination of PVC.

shear, which are high enough to set up critical stresses in the polymer [81,82]. It has been shown that radical acceptors added during processing act as stabilizers against mechanical degradation [83]. Thus, mechano-chemical degradation of PVC [84–89] comprises not only chain scission and cross-linking, but also initiation of radical chain dehydrochlorination reactions, and thus high-energy deformations during processing constitute another important consideration in the prevention of degradation of PVC.

14.9.2 Poly (Vinyl Alcohol)

14.9.2.1 Introduction

Herrmann and Haehnel were the first to prepare poly (vinyl alcohol) in 1924 by saponifying poly (vinyl esters) with stoichiometric amounts of caustic soda solution [90]. PVA is one of the few linear, non-halogenated, aliphatic polar polymers. In 1937, Herrmann, Haehnel, and Berg also reported that poly (vinyl alcohol can be prepared from poly (vinyl esters) by transesterification with absolute alcohols in the presence of catalytic amounts of alkali [91]. The transesterification principle is still used by all poly (vinyl alcohol) producers.

14.9.2.2 Structure and Crystallinity

The backbone of the poly (vinyl alcohol) macromolecular chain possesses mainly 1, 3-diol units. The content of 1,2-diol units in poly (vinyl alcohols) obtained by hydrolysis of poly (vinyl acetates) is about less than 1–2%. The content of 1,2-diol sequences can be minimized by lowering the polymerization temperature of vinyl acetate [92] or using other vinyl esters such as vinyl benzoate [93] or vinyl formate [94]. The content of 1,2-diol units influences the degree of swelling of poly (vinyl alcohol) in water [95].

Poly (vinyl alcohol) is reported to possess carboxyl and carbonyl end groups, which are more abundant than initiator end groups [96]. It is also reported to possess a slight number of branches in the backbone structure. Poly (vinyl alcohol) is a crystalline polymer [97]. The degree of crystallinity depends strongly on the structure and the previous history of the poly (vinyl alcohol). Acetyl groups and other incorporated groups of any kind can lower the crystallinity [98]. The degree of crystallinity of fully saponified poly (vinyl alcohols) is increased by heat treatment, which also lowers the solubility in water [99]. This effect is less marked for poly (vinyl alcohol) that possesses acetyl groups.

14.9.2.3 Degradation of Poly (Vinyl Alcohol)

A great deal of research has been performed in the past few years on the degradation of vinyl polymers because of their commercial importance. It is well understood that polymers like poly (vinyl chloride), poly (vinyl acetate) and poly vinyl alcohol) degrade by zip elimination of low molecular weight compounds like HCl, H_2O, CH_3COOH, etc., at relatively high rates even at processing temperatures. In all cases, degradation led to the formation of polyene sequences along the polymer backbone.

In earlier works, Meissner and Heublein [100] reported that PVA starts degradation at about 100°C, forming conjugated ketoallyl structures of the type —$(C{=}C)_n$—CO—. Finch [101] reported the formation of keto functional groups in the poly (vinyl alcohol) backbone by tertiary hydrogen transfer and abstraction followed by hydrolysis reaction during degradation process. Upon heating poly (vinyl alcohol) above the decomposition temperature, the polymer begins a rapid zip elimination of H_2O [102,103]. This process coupled with melting causes the material to foam or intumesce as it decomposes. This and other decomposition reactions cause color changes and cross linking to yield insoluble yellow to black rigid foam-like residues. The sequence of reactions that occurs during degradation of poly (vinyl alcohol) is shown in Scheme 14.13 [104].

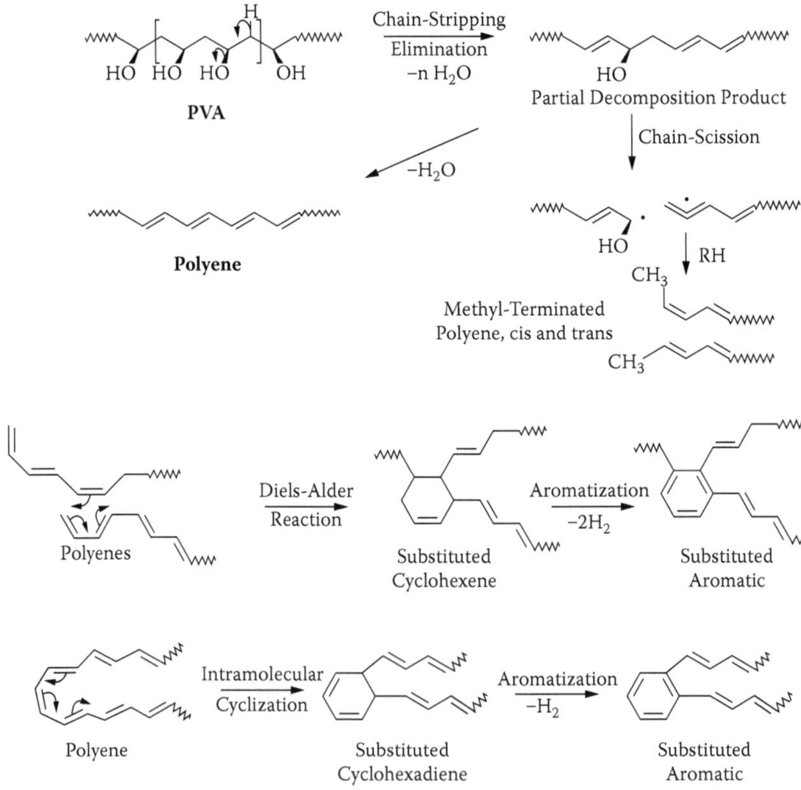

SCHEME 14.13 Poly (vinyl alcohol) degradation: The sequence of reactions.

14.9.3 Poly (Vinyl Acetate)

14.9.3.1 Introduction

Poly (vinyl acetate) is amorphous, odorless, and tasteless and possesses highlight fastness and weather resistance. It is also known to be one of the most important commercial polymers because of its low cost and high performance of the products, combined with a wide range of properties and applications. The glass transition temperature (T_g) of poly (vinyl acetate) depends to a certain extent on the molecular mass. The polymers are soluble in many esters, ketones, cyclic ethers, phenols, halogenated aliphatic hydrocarbons, and methanol. The most important uses of poly (vinyl acetate) are in emulsion paints and adhesives. Other uses include as antinoise compounds, chewing gum bases, concrete adhesives, binder for nonwoven fabrics, fibrous leather substitutes, and coatings for paper.

14.9.3.2 Degradation of Poly (Vinyl Acetate)

Despite its enormous technical and economic importance, its degradation at high temperatures is studied extensively by many authors [102,105]. In the present scenario, one reason for studying the thermal degradation of poly (vinyl acetate) is the indispensability of polymer processing and recycling using enhanced temperatures. Poly (vinyl acetate) normally degrades under vacuum at an appreciable rate at temperatures about 227°C. The thermal degradation of poly (vinyl acetate) goes through a very complex process which includes the following reactions: chain scission, radical recombination, C-H bond cleavage, hydrogen abstraction, radical addition, conjugated double bond formation, cyclization, aromatization, and graphitization [106].

Grassie et al. [107] have reported that the rate constant for the elimination of CH_3COOH from poly (vinyl acetate) is much lower than the rate constant for the elimination of HCl from poly (vinyl chloride). Detailed information about kinetic mechanism of polyene formation along the backbone of poly (vinyl acetate) during degradation was deduced from the time dependence of the polyene distribution. The method for the determination of single polyene sequences and double bonds was described in the earlier works [108].

14.9.3.3 Degradation of Poly (Methyl Methacrylate)

Sun et al. [109] performed a series of studies on the degradation mechanisms and pyrolysis behavior of acrylic polymers (i.e., poly (methyl methacrylate) and poly (methacrylic acid)). Degradation experiments were performed for polymers alone and for polymer-ceramic mixtures in nitrogen or oxygen atmospheres. They found the mechanism of thermal degradation of poly (methyl methacrylate) to be depolymerization, and furthermore they observed that the reaction was accelerated by the presence of oxygen (thermal-oxidative degradation). In the case of poly (methyl methacrylate) and alumina mixture, a surface reaction between the polymer and the ceramic was indicated by Fourier transform infrared spectroscopy (FTIR) and gas chromatography (GC) results, but not in poly (methyl methacrylate) and silica mixture. The thermal degradation mechanism of poly (methacrylic acid) found by Sun et al. is very different compared with that of poly (methyl methacrylate). However, similar surface reactions were observed in polymer/ceramic mixtures. In general, the results agree with the mechanism proposed by Bakht [110]. Similar work performed in an oxygen atmosphere showed different results due to oxidative degradation.

14.9.4 Poly (Vinyl Butyral)

14.9.4.1 Introduction

Poly (vinyl butyrals) are the most commercially important poly (vinyl acetals). They are available in numerous types. Two important processes are mainly prepared for the synthesis of poly (vinyl butyrals). One begins with aqueous poly (vinyl alcohol) solutions, which are reacted with n-butyraldehyde in the presence of mineral acid. The difficulty with this method is to achieve a sufficiently homogeneous polymer. In the second process, an alcoholic poly (vinyl alcohol) suspension is acetalied. In this method, a homogeneous poly (vinyl butyral) solution is obtained and the solution can be freed from insoluble by filtration.

The interest in poly (vinyl butyrals) is based on their structure, which is composed of reactive, markedly hydrophobic and hydrophilic polymer units, the ratio between the two groups being variable within a wide range. The hydrophobic vinyl butyral units led to the polymer's good thermoplastic processability, solubility in numerous solvents, elasticity and toughness, as well as compatibility with many other resins and plasticizers. The hydrophilic vinyl alcohol groups are responsible for the high adhesion to inorganic materials such as glass and metals, high strength, cross-linking ability, and anti-corrosive action.

Poly (vinyl butyrals) are amorphous and transparent. At low degrees of acetalization, they are water soluble. The most important uses of poly (vinyl butyral) are in the manufacture of laminated glass sheets and paints. Because of their elasticity and good adhesion, poly (vinyl butyrals) are important raw materials for paints, particularly for metal coatings. They are frequently blended with phenolic, alkyd, epoxy, urea, or melamine resins. Poly (vinyl butyrals) are outstandingly effective as primers with high corrosion protection. In wash primers, poly (vinyl butyrals) are used in the form of dispersants.

14.9.4.2 Degradation of Poly (Vinyl Butyral)

Bakht [110] studied the thermal degradation of PVB copolymers having different degrees of vinyl butyral substitution in the temperature range of 473–723 K. Thermal volatilization analysis (TVA), thermogravimetric analysis (TG), and infrared spectroscopy (IR) were used to investigate the composition of the volatile products formed due to the decomposition of PVB. He observed mainly water and butyraldehyde as byproducts during thermal degradation of PVB. He proposed both a free-radical mechanism and a molecular elimination mechanism to interpret the formation of butyraldehyde.

Sun et al. [109] performed a series of studies on the mechanism of degradation and pyrolysis behavior of PVB. Masia et al. [111] investigated the effect of various oxides on PVB degradation behavior in air and in argon atmospheres. They found that the oxides used have significant catalytic effects on both thermal and thermal-oxidative degradation of PVB. They concluded that besides surface chemistry, the surface structure, number of hydroxyl groups, percentage of free water, and amount of oxygen adsorbed on the oxide surface and surface impurities would influence degradation. Cima et al. [112] conducted studies on the binder distribution and the effect of different atmospheres on binder degradation. Some evidence [113] indicates that bound water on ceramic surfaces promotes thermal degradation. This effect is significant if the filler surface is hydrophobic (e.g., silica, clay, etc.), and the polymer has groups that are sensitive to hydrolysis. Water that is tightly bound [113] to the filler surface exhibits acidic properties, which leads both to the acceleration of hydrolysis of the macromolecules and to their acidolysis, especially at elevated temperatures. The recent work of Howard et al. [114] sheds some light on the interaction of PVB on alumina surface. They found that low molecular weight PVBs burned cleanly and the hydroxyl and acetate functional groups affected the adsorption of PVB on alumina. They examined the adsorption characteristics using variable-temperature FTIR. Graehev et al. [115] performed spectroscopic studies on the thermal degradation of PVB and found that the degree of substitution of alcohol groups by butyral groups greatly influenced degradation. Yang et al. [116] showed that the thermal properties of composites at elevated temperatures can be predicted more accurately if degradation is taken into consideration. In their work, they also proposed a model for predicting the density of laminates as a function of the process variables, lamination temperature, pressure, and time. An earlier article [117] discussed the steam oxidation of carbon left behind due to thermal degradation of polymers in ceramics. The binder thermal decomposition process involves chemical reactions as well as heat and mass transfer. It is also affected by the substrate's geometry, apparent density, void, volume, and particle size as well as distribution. A literature review indicated a lack of quantitative information on the kinetics of the thermal degradation of polymers in ceramic surfaces. The earlier article [117] also provides quantitative information on the kinetics of thermal degradation of PVB and PVB in alumina, mullite, and silica. In addition, this chapter also discusses the effect of heating rates and pellet compression pressure on the thermal degradation of PVB in alumina, mullite, and silica composites.

14.10 CHARACTERIZATION OF POLYMER DEGRADATION — ANALYTICAL TECHNIQUES

Apart from method for molecular weight and molecular weight distribution determinations, which are powerful in detecting degradation in linear polymers, a wealth of conventional analytical methods are used, which are frequently applied to demonstrate the chemical changes that occur in polymers. As far as chemical changes in bulk polymers are concerned, spectroscopic methods, such as infrared (IR) and ultraviolet (UV) absorption spectroscopy, are used to identify the formation and disappearance of chromophoric groups. Nuclear magnetic resonance (NMR) is one of the powerful spectroscopic methods for the analysis of structural changes that occur in polymers during degradation and chemical reactions. Electron spin resonance (ESR) spectroscopy is mostly preferred for the detection of intermediates and free radicals.

In the field of polymer degradation, various other analytical techniques are used, such as thermal gravimetric analysis (TGA), differential thermal analysis (DTA), and differential calorimetry (DSC). Mass spectrometry (MS) and electron spectroscopy for chemical analysis (ESCA) were useful in the analysis of polymer degradation. Thermal analysis combined with evolved gas analysis (EGA) has been used for some time. Thermogravimetry (TG) coupled with Fourier transform infrared (FTIR) spectroscopy (TG/FTIR), thermogravimetry (TG) coupled with mass spectrometry (TG/MS), and thermogravimetry (TG) coupled with GC/MS offers structural identification of compounds evolving during thermal degradation of polymers. These evolved gas analysis (EGA) techniques allow identifying the chemical pathway of the degradation reaction by determining the decomposition products. A brief description of the previously mentioned analytical techniques will be covered in the following sections.

14.11 THERMAL GRAVIMETRIC ANALYSIS

Thermal gravimetric analysis (TGA) enables the investigation of thermal decomposition of an analyte. In TGA, the temperature of a sample is controlled. The change in sample mass is measured as a function of temperature. Typically, temperature is ramped linearly. It can provide information both on the temperatures over which the material being investigated decomposes, and also the relative mass fractions of the various constituents within a sample. Furthermore, the rate of mass loss as a function of temperature can be employed to study reaction kinetics and reaction mechanisms. Thus, TGA is an important tool used to determine thermal stability, reaction kinetics, and stoichiometry and degradation of polymers.

The technique is of particular interest in the investigation of polymers because they often contain several distinct phases in an unknown proportion. Furthermore, in creating products from polymers, it is essential to determine their working temperature ranges. Determining decomposition temperature puts an upper bound on the working temperature range. TGA is often used in conjunction with mass spectrometry or Fourier transform infrared adsorption spectroscopy (FTIR). These methods allow the identification of volatile materials being released.

In summary, TGA involves the recording of mass change as a function of temperature. Studying mass change as a function of temperature in turn provides information on thermal decomposition, degradation, reaction kinetics, and composition. Attention is now turned to methods that provide more quantitative analyses of thermodynamics and kinetics.

14.12 DIFFERENTIAL SCANNING CALORIMETRY AND DIFFERENTIAL THERMAL ANALYSIS

Differential scanning calorimetry (DSC) and differential thermal analysis (DTA) are employed to identify the kinetics, enthalpies, and temperatures of onset of reactions and phase transitions. In polymer samples, DSC and DTA are useful in identifying glass transition temperatures, melt temperatures, and solvent evaporation. When used in conjunction with thermal gravimetric analysis (TGA), DSC and DTA provide additional information in identifying mass loss steps.

Differential scanning calorimetry was first developed in the early 1960s at Perkin Elmer. Two identical furnaces are employed in DSC. One contains the sample and the other the reference. The temperature difference between the sample and the reference are once again determined. The temperature difference is driven to zero via a feedback loop that proportions power between the two furnaces. A second feedback loop is used to drive the average temperature of the two furnaces along a temperature profile. In DSC, the difference in power provided to each furnace is plotted as a function of temperature. Because sample and reference are both maintained at the same temperature, DSC is appropriate for quantitative analysis. This is because both sample and reference are maintained at the same temperature.

14.13 MASS SPECTROMETRY

Mass spectrometry (MS) is employed to perform elemental and molecular identification based on charge-to-mass ratio on molecules of up to approximately 100,000 Daltons. Polymers generally have higher molecular weights, so MS is used for the identification of decomposition products, and solvent and ionic content.

14.14 COUPLED INSTRUMENTAL TECHNIQUES FOR THE CHARACTERIZATION OF POLYMER DEGRADATION

The nature and amount of a volatile product or products formed during the thermal degradation of materials can be assessed by evolved gas analysis technique (EGA) [118]. EGA technique allows us to identify and analyze the gaseous species evolved in the process of combustion and/or pyrolysis, in which a series of chemical reactions occur as a function of temperature and are analyzed using thermal analytical methods and/or multiple techniques. EGA is normally used to assign the chemical pathway of degradation reactions by determining the composition of the decomposition products from various materials. Two approaches are generally preferred for EGA, simultaneous analysis and combined analysis. In the simultaneous analysis approach, two methods are employed to examine the materials at the same time, such as TG-FTIR and TG-MS on-line analysis, in which decomposition products that evolve from pyrolyzed materials can be monitored simultaneously. The combined analysis technique, on the other hand, employs more than one sample for each instrument, and real time analysis is not possible. In the simultaneous and combined analysis of EGA, two major forms of sampling procedures can be used to semi-quantitatively and qualitatively determine the decomposition products, which are continuous mode and intermittent mode. Continuous mode involves introducing the gaseous products directly into the detector system using an interface coupling [119,120], such as TG-MS and TG/FTIR. In this case, the products can be continuously and simultaneously examined by the detector. Intermittent mode (or batch mode) involves trapping evolved gaseous species at a low temperature or in an absorption chamber [121,122] and the trapped products are then introduced into a detector for identification, such as a TG-GC-MS or Pyrolysis-GC-MS system. The intermittent mode allows the investigator to optimize the detector parameters to make the best choice for different samples. The detector systems need not to modify for the intermittent mode. The main advantage of the continuous mode is the ability to conduct simultaneous and continuous real time analysis. The ability to measure the simultaneous and continuous mass loss of materials and monitor the gaseous products responsible for the mass loss allows TG-FTIR and TG-MS techniques to be widely employed in the field of material science, including the study of coal pyrolysis and combustion, polymer degradation, evaluation of hazardous materials, thermal stability of materials, and formation of pharmaceuticals [123]. Remarkable work [124,125] on characterizing materials, determining kinetic parameters, and quantitative analysis of degradation products has been performed with a TG-FTIR coupled system. TG-MS coupled systems had also been widely employed in the characterization of materials. The development of this instrumentation and its application has been discussed in an excellent review [126]. However, the fact that one cannot distinguish between all the evolved gases due to overlapping in either FTIR spectra (similar wavelengths) or MS spectrum (identical masses) is an obvious disadvantage of these techniques. Diatomic molecules that do not possess a permanent dipole moment are not infrared active, and therefore cannot be identified using infrared absorption. In addition, overlapping peaks, such as the O-H stretching vibration in water and carboxylic acids, presents a problem in the analysis of mixtures. On the other hand, the mass spectrometer is very sensitive and can identify the mass range of the species in a sample. In addition, it has a faster response time than an FTIR system. The detection limit of the MS system is usually about less than 1 ppm, which is more sensitive than an FTIR system having a general detection limit of approximately 10 ppm. However, the MS system has a relatively high cost and because it measures only mass-to-charge ratios, it does not have the ability to identify isomers. Another major disadvantage of TG-FTIR and TG-MS techniques is their inability to detect high molecular mass compounds. TG-(GC-MS) or Pyrolysis-(GC-MS) techniques, however, are capable of identifying high molecular mass compounds released from materials during the combustion process, although they cannot provide the time- and temperature-dependent gas evolution profiles that can be obtained from TG-FTIR and TG-MS techniques. Therefore, the combination of TG-FTIR-MS-(GC-MS) techniques makes it possible to determine the composition and distribution of almost all the gaseous products evolved from materials, and then evaluate the decomposition mechanisms of the polymeric materials.

REFERENCES

1. Grassie, N., *Chemistry of High Polymer Degradation Process*, Butterworths, London, 1956.
2. Madorsky, S.L., *Thermal Degradation of Organic Polymers*, Interscience, New York, 1964.
3. Geuskens, G., Ed., *Degradation and Stabilization of Polymers*, Halsted, New York, 1975.
4. Jellinek, H. H. G., Ed., *Aspects of Degradation and Stabilization of Polymers*, Amsterdam, 1978.
5. White, J. R. and Trunbull, A., *J. Mater. Sci.*, 1994, 29, 584–613.
6. Schnabel, W., *Polymer Degradation: Principles and Applications*, Hanser, New York, 1981.
7. Baumann, E., *Ann. Chem. Pharm.*, 1872, *163*, 308–322.
8. Matthews, G., *Vinyl and Allied Polymers, Volume 2: Vinyl Chloride and Vinyl Acetate Polymers*, Iliffe Books, London, 1972.

9. Nass, L. I., *Encyclopedia of PVC*, Marcel Dekker Inc.: New York, 1976.
10. Cameron, J. B., Ludeen, A. J., McCully, J. H., Jr., and Schwab, P. A., *J. Appl. Polym. Sci.: Appl. Polym. Symp.*, 1981, *36*, 133–150.
11. Naqvi, M. K., *J. Macromol. Sci.-Rev. Macromol. Chem. Phys.*, 1985, *C25*, 119–155.
12. Summers, J. W., *J. Vinyl. Add. Tech.*, 1997, *3*, 130–139.
13. Zimmermann, H., *J. Vinyl. Add. Tech.*, 1996, *2*, 287–294.
14. Xie, T. Y., Hamielec, A. E., Wood, P. E., and Woods, D. R., *J. Vinyl. Tech.*, 1991, *13*, 2–25.
15. Braun, D., *Pure Appl. Chem.*, 1971, *26*, 173–192.
16. Braun, D., *Dev. Pol. Degr.*, 1981, *3*, 101–133.
17. Mukherjee, A. K. and Gupta, A., *J. Macromol. Sci.-Rev. Macromol. Chem. Phys.*, 1981, *C20*, 309–331.
18. Asahina, M. and Onozuka, M., *J. Polym. Sci., Part A*, 1964, *2*, 3505–3513.
19. Asahina, M. and Onozuka, M., *J. Polym. Sci., Part A*, 1964, *2*, 3515–3522.
20. Macoll, A., *Chem. Rev.*, 1969, *69*, 33–60.
21. Mayer, Z. *J. Macromol. Sci.-Rev. Macromol. Chem. Phys.*, 1974, *11*, 263–292.
22. Starnes, W. H., Jr. and Girois, S., *Polymer Yearbook*, 1995, *12*, 105–131.
23. Rigo, A., Palma, G., and Talamini, G., *Makromol. Chem.*, 1972, *153*, 219–228.
24. Baker, C., Maddams, W. F., Park, G. S., and Robertson, B., *Makromol. Chem.*, 1973, *165*, 321–323.
25. Abbas, K. B., Bovey, F. A., and Schilling, F. C., *Makromol. Chem., Suppl.*, 1975, *1*, 227–234.
26. Bovey, F. A., Abbas, K. B., Schilling, F. C., and Starnes, W. H., Jr., *Macromolecules*, 1975, *8*, 437–439.
27. Abbas, K. B., *J. Macromol. Sci.-Phys.*, 1977, *B14*, 159–166.
28. Starnes, W. H., Jr., Schilling, F. C., Abbas, K. B., Plitz, I. M., Hartless, R. I., and Bovey, F. A., *Macromolecules*, 1979, *12*, 13–19.
29. Starnes, W. H., Jr., Schilling, F. C., Abbas, K. B., Cais, R. E., and Bovey, F. A., *Macromolecules*, 1979, *12*, 556–562.
30. Bovey, F. A., Schilling, F. C., and Starnes, W. H., Jr., *Polymer Preprints*, 1979, *20*, 160–163.
31. Hjertberg, T. and Wendel, A., *Polymer*, 1982, *23*, 1641–1645.
32. Hjertberg, T., Sorvik, E. M., and Wendel, A., *Makromol. Chem., Rapid Commun.*, 1983, *4*, 175–180.
33. Starnes, W. H., Jr. *Pure Appl.Chem.*, 1985, *57*, 1001–1008.
34. Starnes, W. H., Jr., Schilling, F. C., Plitz, I. M., Cais, R. E., Freed, D. J., Hartless, R. I., and Bovey, F. A., *Macromolecules*, 1983, *16*, 790–807.
35. Starnes, W. H., Jr., Wojciechowski, B. J., Velaquez, A., and Benedikt, G. M., *Macromolecules*, 1992, *25*, 3638–3642.
36. Bovey, F. A., Schilling, F. C., McCrackin, F.L., and Wagner, H.L., *Macromolecules*, 1976, *9*, 76–80.
37. Starnes, W. H., Jr., Schilling, F. C., Plitz, I. M., Cais, R. E., and Bovey, F. A., *Polymer. Bull.*, 1981, *4*, 555–562.
38. Hjertberg, T. and Sorvik, E. M., Formation of anomalous structures in poly (vinyl chloride) and their influence on the thermal stability: effect of polymerization temperature and pressure, in *Polymer Stabilization and Degradation*, Klemchuk, P. P., Ed., American Chemical Society: Washington, D.C., 1985, pp. 259–284.
39. Hjertberg, T. and Sorvik, E. M., *Polymer*, 1983, *24*, 673–684.
40. Starnes, W. H., Jr., Chung, H., Wojciechowski, B. J., Skillicorn, D. E., and Benedikt, G. M., *Polym. Prepr.*, 1993, *34*, 114–115.
41. Starnes, W. H., Jr., Zaikov, V. G., Chung, H. T., Wojciechowski, B. J., Tran, V. H., Saylor, K., and Benedikt, G. M., *Macromolecules*, 1998, *31*, 1508–1517.

42. Starnes, W. H., Jr., *Prog. Polym. Sci.*, 2002, *27*, 2133–2170.
43. Abbas, K. B. and Sorvik, E. M., *J. Appl. Polym. Sci.*, 1976, *20*, 2395–2406.
44. Hjertberg, T. and Sorvik, E. M., *Polymer*, 1983, *24*, 685–692.
45. Guyot, A., Roux, P., and Tho, P. Q., *J. Appl. Polym. Sci.*, 1965, *9*, 1823–1840.
46. Guyot, A., Roux, P., and Tho, P. Q., *J. Appl. Polym. Sci.*, 1968, *12*, 639–653.
47. Millan, J., Carranza, M., and Guzman, J., *J. Polym. Sci. Symp.*, 1973, *42*, 1411–1418.
48. Millan, J., Madruga, E. L., and Martínez, G., *Angew. Chem.*, 1975, *45*, 177–184.
49. Marks, G. C., Benton, J. L., and Thomas, C. M., *Soc. Chem. Ind. (London) Monograph*, 1967, *26*, 204–235.
50. Nolan, K. P. and Shapiro, J. S., *J. Polym. Sci. Symp.*, 1976, *55*, 201–209.
51. Cooray, B. B. and Scott, G., *Eur. Polym. J.*, 1980, *16*, 169–177.
52. Nagy, T. T., Kelen, T., Turcsanyi, B., and Tudos, F., *Polymer. Bull.*, 1980, *2*, 77–82.
53. Bacaloglu, R. and Fisch, M. H., *Polym. Degrad. Stab.*, 1995, *47*, 33–57.
54. Marval, C. S., Sample, J. H., and Roy, M. F., *J. Am. Chem. Soc.*, 1939, *61*, 3241–3244.
55. Winkler, D. E., *J. Polym. Sci.*, 1959, *35*, 3–16.
56. Svetly, J., Lukas, R., and Kolinsky, M., *Makromol. Chem.*, 1979, *180*, 1363–1366.
57. Baum, B. and Wartman, L. H., *J. Polym. Sci.*, 1958, *28*, 537–546.
58. Bengough, W. I. and Sharpe, H. M., *Makromol. Chem.*, 1963, *66*, 45–55.
59. Amer, A. R. and Shapiro, J. S., *J. Macromol. Sci.-Chem.*, 1980, *A14*, 185–200.
60. Fisch, M. H. and Bacaloglu, R., *J. Vinyl. Add. Tech.*, 1999, *5*, 205–217.
61. Starnes, W. H., Jr., Wallach, J. A., and Yao, H., *Macromolecules*, 1996, *29*, 7631–7633.
62. Roth, J. P., Rempp, P., and Parrod, J., *J.Polym.Sci.*, 1963, *C4*, 1347–1366.
63. Haddon, R. C. and Starnes, W. H., Jr., *Polym. Prepr.*, 1977, *18*, 505–509.
64. Starnes, W. H., Jr., Haddon, R. C., Hische, D. C., Plitz, I. M., Schosser, C. L., Schilling, F. C., and Freed, D., *J. Polym. Prepr.*, 1980, *21*, 138–139.
65. Raghavachari, K., Haddon, R. C., and Starnes, W. H., Jr., *J. Am. Chem. Soc.*, 1982, *104*, 5054–5056.
66. Arlman, E. J., *J. Polym. Sci.*, 1954, *12*, 547–558.
67. Bengough, W. I. and Sharpe, H. M., *Makromol. Chem.*, 1963, *66*, 31–44.
68. Stromberg, R. R., Straus, S., and Achhammer, B. G., *J. Polym. Sci.*, 1959, *35*, 355–368.
69. Tudos, F., Kelen, T., Nagy, T. T., and Turcsanyi, B., *Pure Appl. Chem.*, 1974, *38*, 201–226.
70. Tran, V. H. and Guyot, A., *Polym. Degradability Stability*, 1991, *32*, 93–103.
71. Tran, V. H., Guyot, A., Nguyen, T. P., and Molinié, P., *Polym. Degradability Stability*, 1992, *37*, 209–216.
72. Tran, V. H., Garrigues, C., Nguyen, T. P., and Molinié, P., *Polym. Degradation Stability.*, 1993, *42*, 189–203.
73. Owen, E. D., and Al-Awar, M. M., *Polymer*, 1994, *35*, 2840–2843.
74. Tran, V. H., *J. Macromol. Sci.*, 1998, *C38*, 1–52.
75. Campbell, J. E. and Rauscher, W. H., *J. Polym. Sci.*, 1955, *18*, 461–478.
76. Stapfer, C. and Granick, J. D., *J. Polym. Sci., Part A*, 1971, *9*, 2625–2636.
77. Abbas, K. B. and Sorvik, E. M., *J. Appl. Polym. Sci.*, 1973, *17*, 3577–3594.
78. Starnes, W. H., Jr., *ACS Symp. Ser.*, 1981, *151*, 197–215.
79. Adeniyi, J. B. and Scott, G., *Polym. Degrad. Stab.*, 1987, *17*, 117–129.
80. Xu, P., Zhou, D., and Zhao, D., *Eur. Polym. J.*, 1989, *25*, 575–579.
81. Geddes, W. C., *Rubber Chem. Technol.*, 1967, *40*, 177–216.
82. Nagy, T. T., Ivan, B., Turcsanyi, B., Jenckel, E., and Tudos, F., *Polymer. Bull.*, 1980, *3*, 620.
83. Ouchi, I., *J. Polym. Sci., Part A*, 1965, *3*, 2685–2692.
84. Decker, C., *J. Appl. Polym. Sci.*, 1976, *20*, 3321–3336.
85. Decker, C., *Eur. Polym. J.*, 1984, *20*, 149–155.
86. Daniels, V. D. and Rees, H. H., *J. Polym. Sci., Polym. Chem. Ed.*, 1974, *12*, 2115–2122.
87. Balandier, M. and Decker, C., *Eur. Polym. J.*, 1978, *14*, 995–1000.

88. Decker, C. and Balandier, M., *Eur. Polym. J.*, 1982, *18*, 1085–1091.
89. Decker, C. and Balandier, M., *Makromol. Chem.*, 1982, *183*, 1263–1278.
90. Herrmann, W.O. and Haehnel, W., *Chem. Forschungsgemeinschaft*, 1924, DE-OS 450 286.
91. Herrmann, W.O., Haehnel, W., and Berg, H., *Chem. Forschungsgemeinschaft*, DE 642 531, 1937; *Chem. Abstr.*, 31, 59059.
92. Noro, K. and Takida, H., *Kobunshi Kagaku*, 1962, *19*, 261.
93. Ito, T. and Nomo, K., *Kobunshi Kagaku*, 1968, *15*, 310; *Chem. Abstr.*, 1960, *54*, 8140d.
94. Fujii, K., Imoto, S., Ukida, J., and Matsumoto, M., *J. Polym. Sci., Part B*, 1963, *1*, 497.
95. Ukida, J. and Naito, R., *Kogyo Kagaku Zasshi*, 1955, *58*, 717, *Chem. Abstr.*, 1956, *50*, 8245h.
96. Shiraishi, M., *Kobunshi Kagaku*, 1962, *19*, 676.
97. Bunn, C.W., *Nature (London)*, 1966, *161*, 929.
98. Hayashi, S., Nakano, C., and Motoyama, T., *Kobunshi Kagaku*, 1963, *20*, 303, *Chem. Abstr.*, 1964, *61*, 5802b.
99. Sakurada, I., Nukushima, Y., and Sone, Y., *Kobunshi Kagaku*, 1955, *12*, 506.
100. Meissner, H. and Heublein, G., *Acta Polymerica*, 1986, *37*, 323.
101. Finch, C.A., *Polyvinylalcohol, Properties and Uses*, New York, John Wiley & Sons, 1973, 495.
102. Anders, H. and Zimmerman, H., *Polym. Degradability Stability*, 1987, *18*, 111–122.
103. Cullis, C.F. and Hirschler, M.M., *The Combustion of Organic Polymers*, Clarendon Press, Oxford, 1981, pp. 117–119.
104. Gilman, J.W., David, L. V, and Kashiwagi, T., in *Fire and Polymers II: Materials and Test for Hazard Prevention, Am. Chem. Soc., ACS Symp. Ser. 599*, August 21–26, 1994, Washington, D.C.
105. Costa, L., Avatanco, M., Bracco, P., and Brunella, V., *Polym. Degradation Stability*, 2002, *77*, 503–510.
106. McNeill, I.C. and McGuiness, R.C., *Polym. Degradation Stability*, 1983, *5*, 303–316.
107. Grassie, N., McLaraen, I.F., and McNeill, I.C., *Eur. Polym. J.*, 1970, *6*, 679, 865.
108. Anders, H., Zimmermann, H., and Behnisch, J., *Proc. VII Eur. Symp. Polym. Spectrosc.*, Dresden, 1985.
109. Sun, Y. N. et al., Pyrolysis behavior of acrylic polymers and acrylic polymer-ceramic mixtures, in *Ceramic Transactions, Vol I, Ceramic Powder Science, IIA*, Messing, G. L. et al., Eds., American Ceramic Society, Westerville, OH, 1988, p. 538.
110. Bakht, M. F., *Pak. J. Sci. Ind. Res.*, 1983, *26*, 35.
111. Masia, S. et al., *J. Mater. Sci.*, 1989, *24*, 1907.
112. Cima, M. J. et al., Firing-atmosphere effects on char content from alumina-polyvinyl butyral films, in *Ceramic Transactions, Vol. I., Ceramic Powder Science, IIA*, Messing, G. L. et al., Eds., American Ceramic Society, Westerville, OH, 1988, p. 567.
113. Bryk, M. T., *Degradation of Filled Polymers: High temperature and Thermal-Oxidative Processes*, Ellis Horwood, New York, 1991.
114. Howard, K. E. et al., *J. Am. Ceram. Soc.*, 1990, *73*, 2543.
115. Graehev, V.I. et al., A spectroscopic study of the kinetics of thermal oxidative degradation of poly (vinyl butyral), *Vysokomol. Soyed.*, 1974, *A16*, 2, 317.
116. Yang, T. C. et al., Heat capacity of composites of alumina, mullite and silica. Presented at the Polym. Process. Soc. Meet., West Virginia University, Morgantown, WV, August 1993.
117. Boddu, M. V. et al., *J. Am. Ceram. Soc.*, 1990, *73*, 1701.

118. Wendlandt, W. W., *Thermal Analysis*, 3rd ed., John Wiley & Sons, New York, 1985, pp. 461, 588.
119. Chiu, J., *Anal. Chem.*, 1968, *40*, 1516.
120. Zitomer, F., *Anal. Chem.*, 1968, *40*, 1091.
121. Flath, R. A., *Guide to Modern Methods of Instrumental Analysis*, Gouw, T. H., Ed., Wiley Interscience, New York, 1972, Ch. 9.
122. Freeman, S. K., *Ancillary Techniques of Gas Chromatography*, Ettre, L. S. and McFadden, W. H, Eds., Wiley Interscience, New York, 1969, Ch. 6.
123. Prins, R. B. New techniques, news and applications, *North Am. Thermal Anal. Soc. Notes*, Winter Issue, 1987/1988, *19(4)*, 48, 51.
124. Putzing, C. L., Leugers, M. A., Mckelvy, M. L., Mitchell, G. E., Nyquist, R. A., Papenfuss, R. R., and Yurga, L., *Anal. Chem.*, 1994, *66(12)*, 51.
125. Carangelo, R. M., Solomon, P. R., and Gerson, D., *Fuel*, 1987, *66*, 960.
126. Friedman, H. L., *Thermochem. Acta*, 1970, *1*,199.

15 Fiber-Filled Vinyl Polymer Composites

Chapal K. Das, Madhumita Mukherjee, and Tanya Das

CONTENTS

15.1 FIBER-FILLED VINYL POLYMERS

15.1.1 Introduction

15.1.1.1 Vinyl Polymer

From 1930 to 1940, the initial industrial development of four of today's major thermoplastics took place: poly (vinyl chloride) (PVC), the poly olefins, poly (methyl methacrylate), and polystyrene. In the past, all these materials were known as ethenoic polymers because they can be considered formally as the derivative of ethylene. However, they were often termed incorrectly as vinyl polymers [1].

In current years, the concept of vinyl polymers has changed a lot. According to the recent concepts vinyl polymers are polymers made from vinyl monomers; that is, small molecules containing carbon–carbon double bonds (i.e., vinyl groups $-CH{=}CH_2$). Where one hydrogen atom is substituted with other groups, depending on the substituent, vinyl polymers can be categorized into different categories. If the R substituent is an olefin monomer ($CH_2 = CHR$), H, alkyl, aryl, or halogen, the corresponding polymer is a poly olefin. If the R is a cyanide group or carboxylic group or its ester or amide, the substance is called an acrylic polymer. Thus, the vinyl group may form part of the allyl group and can be contained in all acrylates. Broadly, vinyl polymers are considered as those poly olefins in which the R substituent in the olefin is bonded to the unsaturated carbon skeleton through an oxygen atom (vinyl alcohol, vinyl esters, vinyl acetals, vinyl ethers, and vinyl cinnamate) or a nitrogen atom (vinyl pyrrolidone, vinyl carbazole) [2]. Because of containing the double bonds, vinyl monomers can be polymerized, forming polymers that are vinyl polymers. In these polymers formed by addition polymerization of the vinyl monomers, the double bonds are turned into single bonds, which is an instance of addition polymerization. No vinyl groups are included in the resulting polymer.

In recent years, polystyrene, ethylene propylene co-polymers, poly tetra fluoro ethylene and poly methyl methacrylate have been considered vinyl polymers.

Vinyl polymers constitute an important part of the plastics industry. On the basis of the physical and chemical properties, vinyl polymers are used in many fields, such as the paper and textile industries in the treatments for paper and textiles, as adhesives, and in many special applications. Among various vinyl polymers, some important polymers and their preparation and uses are discussed next.

15.1.1.1.1 Poly (Vinyl Alcohol)

Poly (vinyl alcohol) (PVA) is prepared by alcoholysis of poly (vinyl acetate), because vinyl alcohol monomer does not exist in the free state. Either an acid or base catalysis is used for alcoholysis. Catalyst concentration regulates the degree of alcoholysis. The presence of hydroxyl groups attached to the main chain renders the polymer hydrophilic. PVA therefore dissolves in water to a varying degree, according to the degree of hydrolysis. As the degree of hydrolysis increases, solubility of PVA decreases appreciably. Fully hydrolyzed PVA is soluble only at higher temperatures (>85°C), possibly due to extensive hydrogen bonding and high degree of crystallinity.

$$H_3C-[CH_2-CH(OH)]_n-CH_3$$

Poly (vinyl alcohol)

PVA is available commercially in different grades depending on the degree of hydrolysis and molecular weight.

It serves as a non-ionic, surface-active agent and is used in suspension polymerization as a protective colloid. It is also used in paper sizing, textile sizing, cosmetics, paper coatings, and adhesives. Moreover, completely hydrolyzed grades are used as water-resistant adhesives. Because PVA film has little tendency to adhere to any other plastics or polymers, it is used as a stick-preventing agent in molds. Partially or incompletely hydrolyzed grades are used as insecticides, disinfectants, and bath salts.

PVA fibers have been generated in Japan, which are also given the name vinyl or vinylon fibers.

15.1.1.1.2 Poly (Vinyl Acetate)

Poly (vinyl acetate) (PVAc) is very soft and possesses excessive cold flow, which prevents its use in molded plastics because the T_g of 280°C of this polymer is either slightly above (at various temperature) or below the ambient temperature.

PVAc is prepared mainly by free radical-initiated chain reaction (i.e., emulsion polymerization).

$$H_3C-[CH_2-CH(O-C(=O)-CH_3)]_n-CH_3$$

Poly (vinyl acetate)

It finds its uses in emulsion paints, adhesives for textiles, paper, and woods. In addition, it is used as a sizing material and as a permanent starch. Depending upon the molecular weight and nature of co-monomers (acrylate, fumerate, and maleate), various commercial grades of PVAc are available. Two of the most widely used vinyl acetate polymers in the plastic industry are ethyl vinyl acetate (EVAc) co-polymers and vinyl chloride–vinyl acetate (VAcVC) co-polymers.

EVAc co-polymers are very flexible and possess excellent toughness. It is advantageous to use EVAc co-polymers commercially over natural rubber for easy processing because no vulcanization is needed for PVAc co-polymers, despite the fact that it is slightly less flexible than natural rubber.

15.1.1.1.3 Poly (Vinyl Acetal)

Poly (vinyl acetals) are manufactured by the reaction of poly (vinyl alcohol) and aldehydes. They may also be produced directly from poly (vinyl acetate) with no separation of the alcohol. The reaction with aldehydes generating poly (vinyl acetal) proceeds randomly involving a pair of neighboring hydroxyl groups on the polymer chains. As a result, some hydroxyl groups remain unreacted and isolated, and contain unreacted hydroxyl groups and acetals groups. Moreover, poly (vinyl acetal) constitutes residual acetate groups as a result of incomplete hydrolysis of poly (vinyl acetates) to the poly (vinyl) alcohol, which is used in the acetalization reaction. The relative proportions of these three groups thus significantly regulates the precise properties of that polymer.

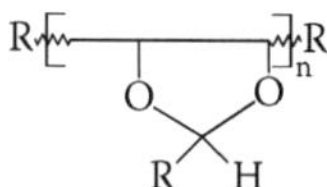

Poly (vinyl acetal)

15.1.1.1.4 Poly (Vinyl Butyrate)

Poly (vinyl butyrate) (when the R group is propyl group, then the resulting polymer is poly (vinyl butyrate)) possesses some interesting characteristic features such as high strength, high stability, clearness, good adhesive properties, and resistance to moisture. Thus, poly (vinyl acetals) are mainly used as an adhesive in the interlayer between glass plates of laminated safety glass and bullet-proof ingredients.

15.1.1.1.5 Poly (Vinyl Formal)

Poly (vinyl formal) (produced by the reaction between the formaldehyde) with low hydroxyl and acetates groups finds uses in wire enamel and in structural adhesives. In both cases, poly (vinyl formal) is used by mixing it with phenolic resin and curing it by heating.

15.1.1.1.6 Poly (Vinyl Cinnamate)

Poly (vinyl cinnamate) is synthesized by the Schotten–Baumann reaction, by the reaction between poly (vinyl alcohol) in sodium or potassium hydroxide solution, and cinnamoyl chloride in methyl ethyl ketone. The resulting material is a co-polymer of vinyl alcohol and vinyl cinnamate. The polymer becomes crosslinked in exposure to light.

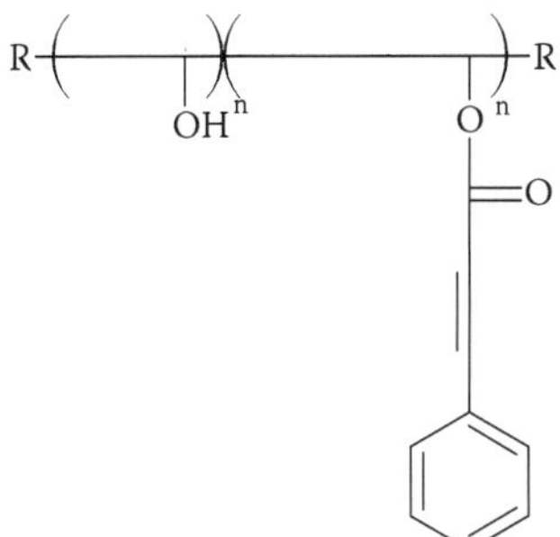

Poly (vinyl cinnamate)

The resulting polymer can be applied in photography, lithography, and related fields as a photo resistant.

15.1.1.1.7 Poly (Vinyl Ethers)

Poly (vinyl ethers) are produced by the polymerization of the vinyl alkyl ether monomers, which are generated from acetylene and the corresponding alcohols. The polymerization takes place by cationic initiation using Friedel–Craft-type catalysts.

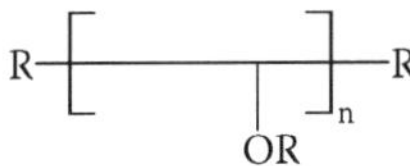

Poly (vinyl ethers)
(Where R = methyl, ethyl)

Depending upon the alkyl substituents, poly (vinyl ethers) can be categorized into different types (where R may be methyl, ethyl, isobutyl), and they find uses in various fields. Poly (vinyl methyl ether) is used as a sensitizer in the manufacture of rubber latex dipped goods.

Ethyl and butyl derivatives find their applications as adhesives. Pressure-sensitive adhesive tapes that are made from poly (vinyl ethyl ether) in conjunction with antioxidants possess a higher shelf life than that of similar tapes made from natural rubber.

15.1.1.1.8 Poly (Vinyl Carbazole)

Poly (vinyl carbazole) is prepared by polymerization of vinyl carbazole by using free radical initiation or Ziegler–Natta catalysis. Poly (vinyl carbazole) is used in xerography as it is an excellent electrical insulator; it has a high softening point and good photoconductivity.

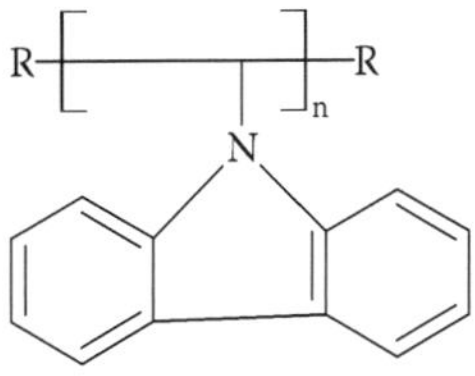

Poly (vinyl carbazole)

15.1.1.1.9 Poly (Vinyl Pyrrolidone)

Poly (vinyl pyrrolidone) is made by free-radical-initiated polymerization of N-vinyl pyrrolidone in aqueous solution to produce a solution containing 30% polymer.

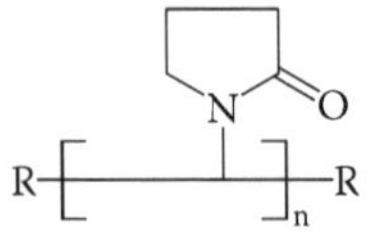

Poly (vinyl pyrrolidine)

Poly (vinyl pyrrolidone) is used in cosmetics due to its ability to form a loose addition compound with many substances. It is used as a blood plasma substitute in an emergency. It also finds its application in textile treatments because of its affinity for dyestuffs.

15.2 TYPES OF FIBERS

Fibers are the strongest polymers among the three different types of polymers (rubbers, plastics, and fibers). Among the natural polymers that have extreme industrial importance, the foremost place is occupied by the fibers, which may be of any type, either plant or animal origin. The natural or synthetic fibers can be used very appropriately as textile materials for their unique properties, which is associated not only with their high mechanical strength (tensile strength in the range of 20,000–150,000 psi), but also with their useful properties such as high thermal insulation, dynamic stability, softness, and flexibility. Natural fibers, such as wool, silk, cellulose, and cotton, as well as synthetic polyamides, polyester fibers, and acrylic fibers constitute this class of polymers. For a fiber possessing high tensile strength and modulus, the polymeric material must constitute combined properties of high molecular regularity and high organized energy density, which can be manifested through polar structures in the repeat units promoting a high order of permanent crystallinity (in the full processing temperature zone) on cold drawing. The crystalline melting point, T_m of this particular polymer must normally be within the range 200–300°C to make it appropriate for hot processing and for its trouble-free spinning into fibers. The polymers for fiber processing must be chemically inert to common solvents to take advantage of dry cleaning of the fibrous material. Moreover, these particular polymers must possess appreciably high molecular weight to ensure high mechanical properties such as tensile strength and other related properties. Extensive intermolecular hydrogen bond formation is the governing factor in artificial fiber technology.

15.2.1 DISCUSSION

Fiber-reinforced composites (FRC) are one of the oldest and most widely used composite materials. FRCs have been studied and developed mainly because of their vast structural potential, which includes:

1. Light weight
2. High strength
3. High stiffness

4. High impact resistance
5. Excellent heat or thermal resistance
6. Low electrical conductivity
7. Good wear resistance
8. Excellent corrosion resistance
9. Low cost

Most of the structural elements or laminates made of fibrous composites consist of several distinct layers of unidirectional lamina. Each lamina is usually made of the same ingredient materials (e.g., resin and glass), but an individual layer may differ from another layer. The properties of the FRC mainly depend on several factors such as: (1) relative volumes of the constituent materials, (2) form of the reinforcement used such as continuous and discontinuous fibers, and woven or non-woven reinforcement, and (3) orientation of fibers with respect to common reference axes. In addition, hybrid laminates can be made, consisting of layers having different fibers and/or matrix materials. Fiber-reinforced polymer (FRP) composites are widely used in various fields of marine, aerospace, transportation, and communications applications, because of their high strength and stiffness. The main applications of FRP composites include the:

1. *Automotive industry*: Piston rod, engine blocks, frames, etc.
2. *Medical field*: wheelchairs, prostheses, orthopedics, etc.
3. *Electrical industry*: cable wires, motor brushes, etc.
4. *Sports industry*: tennis racquets, skies, fishing rods, bicycle frames, motor cycle frames, etc.
5. *Defense industry*: missile nozzles, bulletproof jackets, etc.
6. *Textile industry*: shuttles

The fiber is an important constituent in composites. A great deal of research and development has been performed with fibers, keeping in mind the effects of types, volume fraction, architecture, and orientations. The fibers can be chopped, woven, stitched, and/or braided. They are usually treated with sizing materials to improve the bond as well as binders to improve the handling. Some important sizing materials are frequently starch, gelatin, oil, or wax. Some inorganic materials, such as silane and permanganate, are also used as sizing materials. The most widespread types of fibers used in advanced composites for structural applications are fiberglass, aramid, and carbon. Fiberglass is the least expensive and carbon is the most expensive. The cost of aramid fibers is about the same as the lower grades of the carbon fiber.

15.2.2 Glass Fibers

The glass fibers are divided mainly into three categories: (1) E-glass, (2) S-glass, and (3) C-glass. Glass fibers are mainly made by the following steps (see Scheme 15.1).

1. Silica, limestone, boric acid, and other minor ingredients are mixed first.
2. The mixture is heated to very high temperature until the melting of the actual mixture occurs.

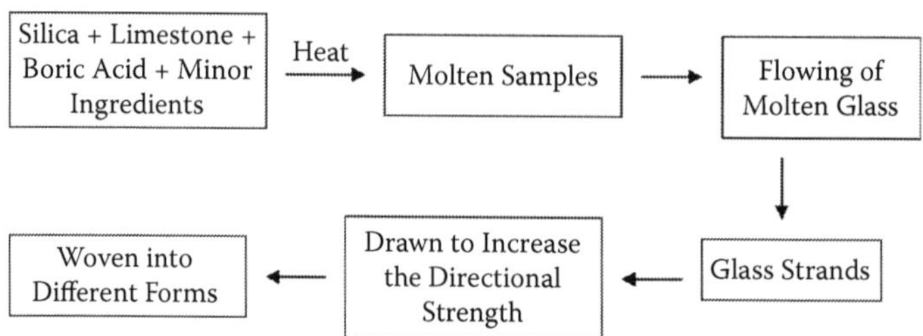

SCHEME 15.1 Flow chart for the preparation of glass fibers.

3. Then, the molten glass is allowed to flow through fine holes in platinum plate.
4. The glass strands are then cooled, gathered, and wound.
5. Next, the fibers are drawn to increase the directional strength.
6. Finally, the fibers are woven into various shapes for use in composites.

The E-glass is mainly used in electrical fields and is less expensive. The S-glass is utilized in high strength FRC. S-glass is approximately 40% stronger than E and possesses more temperature resistance. The C-glass is for high corrosion resistance, and it is mainly used in civil engineering applications. Of the previous three categories, E-glass is the most common reinforcement material used in civil applications. Although the glass material creeps under a sustained load, it can be designed to perform satisfactorily. The fiber itself is regarded as an isotropic material and has a lower thermal expansion coefficient than that of steel (see Table 15.1).

15.2.3 Aramid Fibers

Aramid fibers are synthetic organic fibers consisting of aromatic polyamides. They have excellent fatigue and creep resistance. Several marketable grades of aramid fibers (Kevlar fiber) are available. Among them, the two most commonly used grades in structural applications are Kevlar®(1)* 29 and Kevlar® 49. The Young's modulus curve for Kevlar® 29 is linear to a value of approximately 83 GPa but then becomes slightly curved upward to a value of approximately 100 GPa at rupture. Alternatively, for Kevlar® 49, the curve is linear to a value of approximately 124 GPa at rupture (see Table 15.2). Fibers are anisotropic materials. Being anisotropic in nature, Kevlar fibers possess a lower magnitude of transverse and shear modulus than those in the longitudinal direction. It is very difficult to process the fibers to attain good chemical and mechanical bonding with the resin.

15.2.4 Carbon Fibers

The graphite or carbon fibers are made mainly from three types of polymer precursors: (1) polyacrylonitrile (PAN) fibers, (2) rayon fibers, and (3) pitch.

*Kevlar® is a registered trademark of E.I. du Pont de Nemours & Co.

TABLE 15.1
Mechanical Properties of Glass Fibers

Properties	E-Glass	S-Glass
Density (g/cm^3)	2.60	2.5
Tensile strength (MPa)	1720	2530
Elongation at break (%)	2.4	2.9
Young's modulus (MPa)	72,000	87,000

The tensile stress–strain curves of those fibers are linear to the point of break. Although many carbon fibers are available on the open market, they can be randomly divided into three grades as presented in Table 15.3. They have lower thermal expansion coefficients than both the glass and aramid fibers. The carbon fiber is an anisotropic material, and its transverse modulus is an order of magnitude less than its longitudinal modulus. The material has a very high fatigue and creep resistance.

The elongation at break of the carbon fibers is much lower than the other fibers and tensile strength decreases with increasing the modulus as well. Because of the brittleness of the material for processing at higher modulus, it is very critical in joint and connection details possessing high stress concentration. As a result, carbon fiber composite laminates are more effective with adhesive bonding that eliminates mechanical fasteners.

15.3 RESIN SYSTEMS

The resin is another important ingredient in composites. The two classes of resins are the thermoplastics and thermosets. A thermoplastic resin remains solid at room temperature. It melts when heated and solidifies when cooled. The long-chain polymers do not chemically crosslink. Because they do not cure permanently, they are unwanted for structural application. Conversely, a thermosetting resin will cure permanently by irreversible crosslinking at elevated temperatures. This characteristic makes the thermoset resin composites very desirable for structural applications. The most common resins used in composites are the unsaturated polyesters, epoxies, and vinyl esters; the least common ones are the polyurethanes and phenolics.

TABLE 15.2
Mechanical Properties of Kevlar Fibers of Different Grades

Properties	Kevlar 29	Kevlar 49
Density (g/cm^3)	1.44	1.44
Tensile strength (MPa)	2270	2270
Elongation at break (%)	2.8	1.8
Young's modulus (MPa)	83,000–100,000	124,000

TABLE 15.3
Mechanical Properties of Carbon Fibers

Properties	High Strength	High Modulus	Ultra-High Modulus
Density (g/cm^3)	1.8	1.9	2.0–2.1
Tensile strength (MPa)	2480	1790	1030
Elongation at break (%)	1.1	0.5	0.2
Young's modulus (MPa)	230,000	370,000	520,000

Among the various resin systems, the vinyl ester resins were developed to take advantage of both the workability of the epoxy resins and the fast curing of the polyesters. Vinyl ester has higher physical properties than polyesters but costs less than epoxies. The acrylic esters are dissolved in a styrene monomer to produce vinyl ester resins, which are cured with organic peroxides. A composite product containing a vinyl ester resin can withstand high toughness demand and offers excellent corrosion resistance.

Fiber-reinforced composites are manufactured by using various patented manufacturing techniques.

15.3.1 Pultrusion

Pultrusion is a continuous, preset closed-molding process that is cost effective for high volume production of constant cross section parts. Due to consistency of cross section, resin dispersion, fiber distribution, and alignment, excellent composite structural materials can be fabricated by pultrusion. The basic process usually involves pulling of continuous fibers through a bath of resin, blended with a catalyst, and then into pre-forming fixtures where the section is partially pre-shaped and excess resin is removed. It is then passed through a heated die, which determines the sectional geometry and finish of the final product. The profiles produced with this process can compete with traditional metal profiles made of steel and aluminum for strength and weight.

In pultrusion, process fiber roving and mats are pulled out through a resin bath followed by a passing into a heated die. High temperature is maintained inside the die. At this elevated temperature inside the die, the composite matrix is cured into a definite structural shape of constant cross section. On the other hand, we can say that the pultrusion process involves the continuous-length fiber reinforcements, which are impregnated with an activated resin, and are pulled through a heated die that shapes the material. The heated die activates the catalyst and, as a result, the resin is cured. The solid cured profile is cut to length in-line, as part of the continuous process.

Fiber-reinforced composites containing uniform cross section with outstanding longitudinal strength and stiffness is generally manufactured by the pultrusion technique. In advanced pultrusion, processing and materials technology offers the production of composites possessing excellent complex profiles of considerable dimensions. The fibers that are used for reinforcement are typically "wetted out" using one of two processing methods: (1) using a resin bath, and (2) injecting a resin at the front of the die.

Resin injection is advantageous over a resin bath because of the reduction of styrene emissions. It is also one of the more environmentally friendly techniques available.

15.3.2 Lay-Up Process

The lay-up process mainly includes a hand or machine building-up of mats of fibers. In this process, the resin matrix holds the constituent mats of fibers together tightly and permanently. The technique allows numerous layers of different fibers of different orientations to be built up to a desired sheet thickness and product shape.

The lay-up process includes two types of techniques:

1. *Hand Lay-Up*: The major parts of the fiber reinforced polymer (FRP) composites are produced by using the hand lay-up technique. Some examples of applications where hand lay-up processes are used include: boats, aircraft skins, picnic tables, car bodies, diesel truck cabs, hard shell truck, bed covers, and interior materials. The main features of the hand lay-up process are: (1) it is labor intensive, and (2) the process requires well-ventilated facilities with labor-friendly equipment because the plastic resins produce toxic fumes.

   The hand lay-up process is mainly used to produce parts from an open, glass reinforced mold. The surface of the mold is treated with several layers of wax followed by spraying them with a pigmented polyester resin, which is called a gel coat. The gel coat duplicates the mold surface. The glass fiber is layered over the gel coat of which each layer is saturated by using polyester resin that is particularly formulated to cure at ambient temperature. The polyester resin must be catalyzed by a peroxide catalyst, which is toxic and corrosive. Each glass layer is pressed by hand with rollers that hold the polyester resin tightly into the glass fiber. Approximately three to nine layers are added and permitted to cure together depending on the desired strength of the particular composite part. Once the resin is cured, the substance is removed from the mold for trimming and post-mold surface preparation. The manufacture of a high-performance airplane requires both carbon fiber and epoxy plastic by hand lay-up process.

   The composite part produced by the hand lay-up technique can be constructed into various forms. Moreover, this technique does not require any special apparatus. The products can be effortlessly designed as a weekend backyard project. Molds produced from the hand lay-up process can be easily modified and cut into desired part for prefabrication, and thus can be applied to produce various surface textures. Undercuts and straight-wall elimination is needed for maintaining the inflexible nature of the final product. In post-molding operations, any opening must be machined for that purpose, along with keeping all corners at a large radius. The structure generated from the hand lay-up technique is often stiffened with the additions of honeycombed, rigid foam blocks.
2. *Spray Lay-Up Processes*: Spray lay-up is an open-molding composites fabrication process. It proceeds in a faster rate than that of the hand lay-up process. In this process, resin and reinforcements are sprayed onto a mold.

The matrix and fibers are applied either separately or simultaneously "chopped" in a combined stream from a chopper gun. The fibers fed in this process are randomly oriented and the process is generally automated. Workers roll out the spray-up to compress the laminate. Wood, foam, or other core materials are then added. Thus, a secondary spray-up layer imbeds the core between the laminates (sandwich construction). After the completion of lay-up of the fibers, the plastic must harden or cure in a reasonable time at room temperature.

The spray lay-up process results in structures of low specific strength, which usually do not belong on the end product. This leads to the very few applications of this process in aerospace. Spray lay-up is useful in joining back-up structures to composite face sheets on composite tools. Spray lay-up also finds only limited use for obtaining fiberglass splash from transfer tools.

15.3.3 The Filament Winding Process

The filament winding technique is comprised of the winding of resin-impregnated fiber or tape on a mandrel surface in an accurate geometric shape. This is performed by rotating the mandrel, followed by positioning the fibers precisely on the mandrel surfaces. Winding of continuous fibrils of carbon fiber, fiberglass, or other materials in very precise patterns leads to composites with stronger properties and lighter weights than steel.

The operation of filament winding machines mainly follows the principles of controlling machine motion through various axes of motion, which are the spindle or mandrel rotational axis, the cross or radial carriage motion axis, and the horizontal carriage motion axis. When the composites need more modified and accurate fiber placement, then further additional axes are added, which typically include rotating eye axis or a yaw motion axis.

The filament winding process was designed mainly to produce nose cones, fuselage structures, and missile casings. In recent times, polymer industries other than defense and aerospace have discovered the strength, advantages, and versatility of the filament winding technique.

15.4 TYPES OF VINYL POLYMERS

Vinyl polymers are generally classified into three categories: vinyl plastics, thermosets, and rubbers. The classification of vinyl polymers is represented in Scheme 15.2.

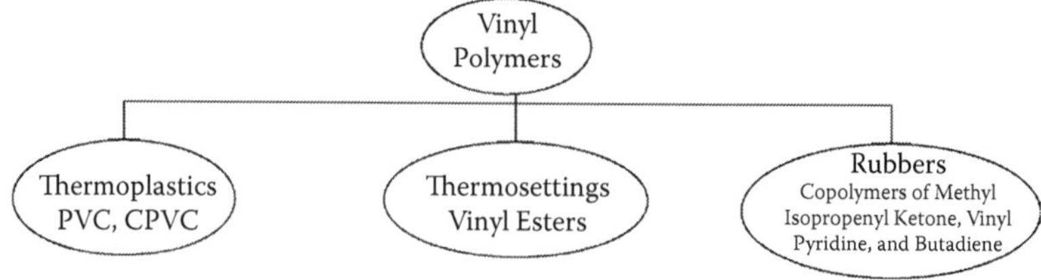

SCHEME 15.2 Classification of vinyl polymers.

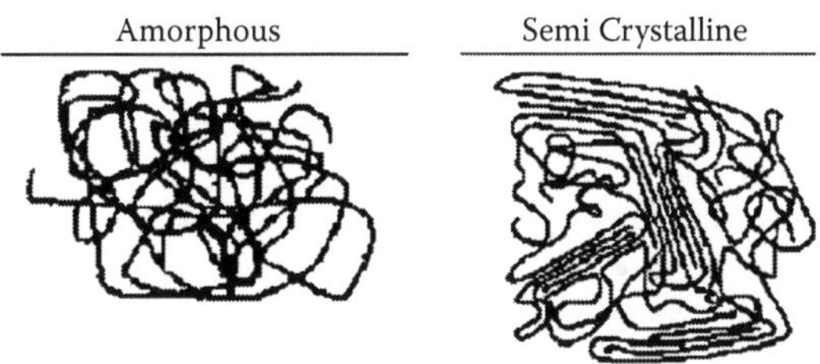

FIGURE 15.1 Thermoplastic vinyl polymers: (1) amorphous and (2) semi-crystalline.

15.4.1 Thermoplastics

Thermoplastics may be defined as the materials that can be softened or plasticized repeatedly on application of thermal energy, without much change in properties if treated with certain precautions (e.g., polystyrene, polyolefins, linear polyesters, nylons, poly (vinyl chloride), polyethers, etc.). Thermoplastics normally remain fusible and insoluble after cycles of heating or cooling. They are capable of being molded and remolded repeatedly. Thermoplastics offer many advantages over traditional materials, including low processing costs, low energy for manufacture, low density, and the ability to make complex shapes relatively easily.

Thermoplastic materials generally fall within two classes of molecular arrangement: amorphous and semi-crystalline (Figure 15.1).

15.4.1.1 Poly (Vinyl Chloride)

Poly (vinyl chloride) (PVC) is an example of an amorphous thermoplastic and polypropylene (PP) is a semi-crystalline polymer.

PVC is one of the world's most widely used plastics because of its versatile nature. PVC can be used either in rigid compounds or blended with plasticizers to produce a flexible derivative. It possesses exceptional flexibility in formulation and processing, which produces flexible to rigid plastics (i.e., PVC can be classified into two categories). Rigid PVC has a wide range of properties, and it can be used in different products. PVC has not been generally used as a potential reinforcing agent, but it has huge applications as a matrix for reinforced plastic products such as reinforced hose, conveyor beltings, and heavy-duty fabrics. For effective processing, some stabilizers and lubricants must be compounded with PVC.

PVC is made mainly by suspension polymerization. However, mass polymerization may be used in a few cases, and plastisols and organosols are produced by emulsion polymerization.

Potential plasticizers are used to soften the thermoplastic resins, and for that purpose, flexible PVC compounds contain plasticizers. PVC plasticizers (additives) must be high-boiling solvents such as dioctyl phthalate (DOP) and didecyl phthalate (DDP). PVC degrades relatively easily compared with other thermoplastics, and the degradation products are corrosive. For this reason, special care should be taken for the processing of PVC using some plasticizers because the plasticizers also act as processing aids.

Rigid PVC compounds normally consist of resin, a stabilizer, and a modifier such as chlorinated polyethylene. Powdered and palletized compounds are made by combining the ingredients in a high-intensity mixer and on twin-screw extrusion lines, respectively.

The main disadvantage of PVC is its low heat resistance. PVC degrades very easily at very low temperatures. The performance of the PVC can be improved by coupling it with any rubber materials (i.e., ABS for high-temperature applications).

Major uses of the rigid PVC lie in building and construction applications. Most of the PVC is extruded into products such as pipe, siding, and window profiles. Rigid PVC is molded into bottles and made into thermoforming boxes and blister packs. Packaging is another use for PVC. Flexible PVC is used as food wrap. Other uses of PVC are cable coating, flooring, toys, and garden hoses. Plastisols and organosols are used in casting, spray and dip coating, and slush molding.

15.4.1.2 Chlorinated Vinyl Polymers

Chlorinated PVC (CPVC) resin possesses high thermal stability under load and is more combustion resistant than PVC. CPVC is used to make dark color window frames, pipe and fittings for portable water and industrial chemicals, and housing for appliances (non-stick utensils) and business equipment.

15.4.1.3 Poly Vinylidene Chloride

Poly vinylidene chloride (PVdC) is the co-polymer of vinylidene chloride with vinyl chloride or other monomers. Poly vinylidene chloride exhibit exceptional barrier resistance to oxygen, carbon dioxide, water, and many organic solvents. Care should be taken in processing PVdC, which includes keeping temperatures below about 204.44°C and using corrosion-resistant metals wherever hot resin contacts the equipment surface. PVdC is used in food wrapping and medical packaging in the form of monolayer films. The co-extruded and sheet forms are used as barrier layers. PVdC coatings are applied to containers to prevent gas transmission, but have to be removed before the packages are recycled.

15.4.2 Thermosetting Vinyl Resins

Among all plastic materials, thermosetting materials are very important because of several advantages: (1) high thermal stability, (2) high creep resistance, (3) excellent resistance to deformation under load, (4) high dimensional stability, and (5) high rigidity and hardness. Along with these advantages, lightweight and excellent insulating properties make the thermoset materials very attractive in polymer industries. The thermosetting materials are formed by the compression and transfer molding methods, together with the more recent evolution of thermoset injection molding techniques. These methods offer low processing cost and mechanized properties of thermosetting polymers.

Similar to all polymer composites, thermosetting composites consist of two major parts: (1) a resin/or matrix system, which generally contains such components as curing agents, hardeners, inhibitors, and plasticizers; and (2) fillers and/or reinforcements, which may consist of minerals or organic fibers, and/or inorganic or organic chopped cloth or paper.

The matrix system usually is the main part in a composite system. It determines to a great extent the cost, heat resistance, electrical qualities, dimensional stability, chemical resistance, decorative possibilities, and flammability. Fillers and reinforcements affect all these properties to changeable degrees, but their most dramatic effects are seen in strength and toughness and, sometimes, electrical qualities. The main types of vinyl thermosetting resins are presented next.

15.4.2.1 Vinyl Esters

Vinyl esters are tough and inflexible unsaturated thermoset plastics. Vinyl esters can be cured by both peroxide catalyzed addition polymerization of vinyl groups and anhydride crosslinking of hydroxyl groups at room temperature or elevated temperatures. The cured bisphenol-A vinyl esters possess high chemical resistance, whereas epoxy novolac vinyl esters are characterized by solvent and heat resistance and flame retardance. All types of vinyl esters in general are tough and flexible in a wide range. Vinyl ester thermoset plastics are mainly used as the matrix in glass or other fiber reinforced plastics. They provide exceptionally high strength properties in highly corrosive or chemical environments when compared with other commercial reinforced plastic matrices such as thermosetting polyesters. Accepted processes used include filament winding, transfer molding, pultrusion coating, and laminating. Uses include structural composites, sheet molding compounds, and chemical apparatus.

Vinyl ester resins bond the best features of both polyester and epoxy resins, with a few concessions. Strength is similar to epoxies, but the resins are less expensive and easier to handle. They are characterized by high reinforced plastics strength and outstanding resistance to forceful media (including both acids and alkalis) at high temperatures. They also possess good impact and fatigue resistance and have low water permeability, thus reducing blistering. They also possess excellent electrical and thermal insulation properties. Compatibility and bond strength to glass graphite and aramid fibers are good (Figure 15.2).

They contain unsaturation only at the end of the chain and not in the repeat units. They also contain less ester linkages, which give improved chemical resistance, whereas terminal double bonds give a tougher and more resilient resin structure. Vinyl ester may also be chemically modified to give specific but more limited properties, yet at a low price.

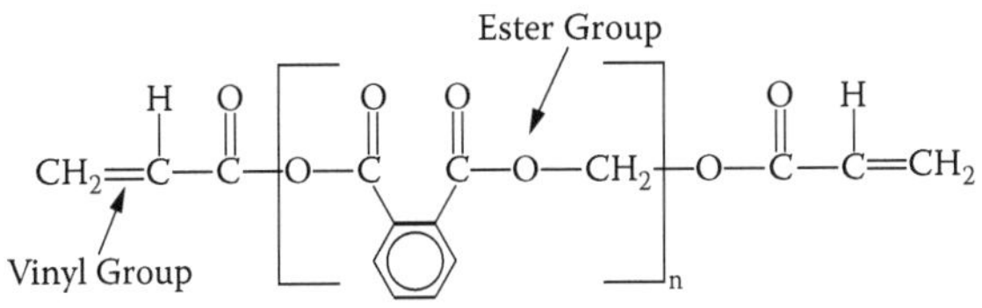

FIGURE 15.2 Poly vinyl ester.

The most frequent applications are chemical-resistant equipment, tanks and pipes, and structural automobile parts. Resins can also be formulated for SMC and BMC systems, with excellent specific properties, such as dimensional stability, heat resistance, and oil resistance.

15.4.2.2 Rubbers

Useful plastic properties may be exhibited by a wide range of molecular weight polymers (low polymer and high polymer). Useful rubbery properties are exhibited by those polymers that are basically high polymeric in nature. A polymer behaving as rubbery material must have the T_g in the range of –40°C to –80°C for reasonable advantages and common uses. The rubbery materials are polymers of low cohesive energy density and poor molecular symmetry, at least in the unstrained state, which is the reason behind the amorphousness of that particular polymer. This phenomenon permits the polymer to attain enough degree of freedom of molecular motion so that deformation of high magnitude takes place rapidly. Although these features and requirements for elastomers entail high local or segmental mobility, the full or gross molecular motion as in the flow region must be low in realistic elastomers. For fulfilling the demand of regaining the original shape and dimension on stress release, restricted chain slippage must be assured. This restricted chain slippage is technically achieved by introducing the widely spaced primary valance crosslinks in the chain molecular system. This is due to the fact that on application of forces, large deformations can take place without the rupture of primary bonds. The difference between rubbers, plastics, and fibers is not really basic or very inherent; it is instead a matter of degree. Very minute or normal variations in chemical structure or physical parameters very often cause a significant variety of properties, which may transform a rubber into a resin or plastics or vice versa, and a plastic to a fiber or vice versa. Thus, (1) reasonable to extensive cyclization, halogenation, or hydro-halogenation of natural rubbers converts them into a plastic variety; (2) controlled chlorosulphonation of polyethylene leading to the introduction of Cl and $–SO_2Cl$ groups in the chain molecule by treating it simultaneously with chlorine and sulphur dioxide gas under pressure breaks its molecular symmetry and converts this general-purpose plastic into a synthetic rubber without disturbing the polarity. For example, PVC is transformed into a leathery or rubbery product by mixing it with appropriate liquid plasticizers [3].

Methyl isopropenyl ketone and certain vinyl pyridine derivatives have been copolymerized with butadiene to give special-purpose rubbers.

15.5 TYPES OF FIBERS

Fibers are generally classified in two categories, namely natural and synthetic fibers, which in turn are known as man-made or artificial fibers. These two categories are further classified into various types as depicted in Scheme 15.3.

15.5.1 Natural Fibers

Fibers, which initiate or originate from natural sources such as plants, animals, and minerals, do not need any fiber configuration or renovation and are classified as natural fibers.

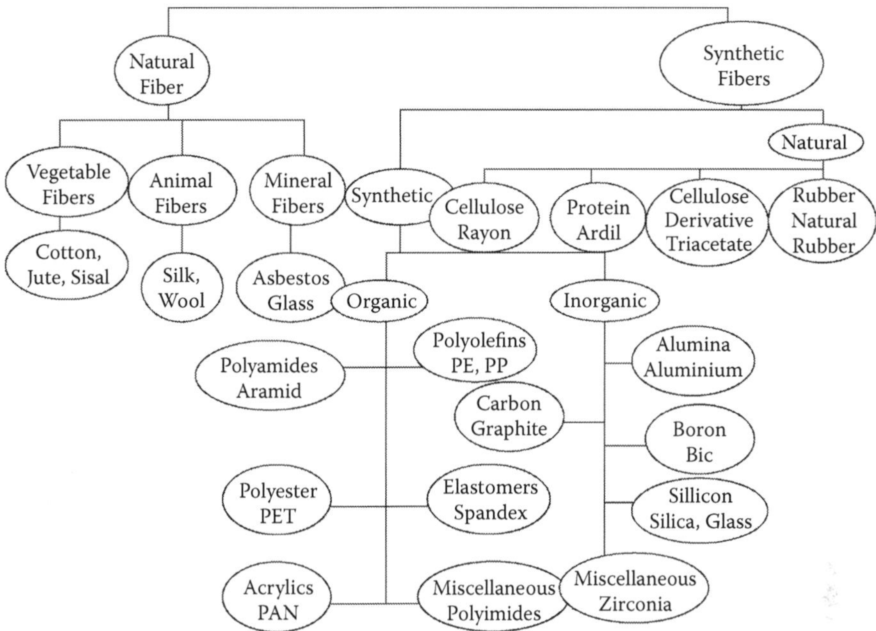

SCHEME 15.3 Classification of fibers.

The growing environmental awareness and new rules and regulations are forcing industries to look for more ecologically friendly materials for their products. Natural fibers have been confirmed as suitable reinforcement materials for composites because they combine good mechanical properties with environmental advantages (i.e., they are plentiful and environmental) [4]. Abundant and renewable fibers are suited for common applications such as storage of consumer goods and low-cost housing. Natural fibers are further classified into categories based on their origin.

Representative properties of some important natural fibers are given in Table 15.4. The tensile strength of a fiber is usually expressed in terms of tenacity, which is grams (force) per denier. A denier is a measure of fiber thickness given as the fiber weight in grams per 9000 m. The corresponding SI unit is the tex (g per 1000 m). Tenacity is then expressed as nM/tex.

15.5.2 Mineral Fibers

Fibers originating from naturally occurring minerals are called mineral fibers [5]. Asbestos is a mineral fiber. Among the various types of asbestos, the most common type is chrysotile, which is a chemically hydrated magnesium silicate with an idealized formula (Mg_6 $(OH)_4$ Si_2 O_5). It has terrific weathering characteristics and is resistant to most of the chemicals, except strong acids and bases. In composites with plastics, it improves creep resistance and heat deflection temperature, and imparts flexural strength and modulus.

TABLE 15.4
Properties of Various Types of Fibers

Fiber	Sp Gr	Length (Cm)	Tensile Strength *10^{-3}, PSI	Elastic Modulus *10^{-6}, PSI	Heat Resistance (°C)
		Natural Organic			
Cotton	1.6	5.08	50–100	—	135v
Sisal	1.3	60.96	120	—	100v
Wool	1.3	38.1	29	—	100v
		Synthetic Organic			
Polyester	1.4	—pq	100	—	248u
Acrylic	1.2	—pq	50	—	232t
Polyamide	1.1	—pq	70–120	—	248t
Cellulose acetate	1.3	—pq	25	—	260t
Regenerated cellulose (rayon)	1.5	—pq	30–105	—	204s
		Natural Organic			
Asbestos	2.5	Up to 10–16	0.02	100–200	1520s
Metal Refractories					
Steel	7.8	—pq	200–400	20–30	1602t
Aluminum	2.8	—pq	60–90	10	655t
Molybdenum	10.2	Up to 1.27	—	42	2590t
Magnesium	1.8	—pq	40	6	648t
Tungsten	19.3	Up to 2.54	200	58	3395t
		Synthetic- Inorganic			
Conventional glass	2.6	—pq	400	10.5	315r
Beryllium glass	2.6	—pq	280	12–20	814s
Quartz	2.2	—pq	100–350	10–25	1924r
Carbon	1.8	—pq	20	1–4	3423t

p = filament, q = staple, r = softens, s = decomposes, t = melts, u = sublimes, v = used up to this temperature.

15.5.3 Animal Fibers

Fibers obtained from living organisms are known as animal fibers. For example, wool is an animal fiber obtained mainly from the fleece of domestic sheep [6]. Wool is warm (i.e., its fibers do not conduct heat), elastic, crease resistant, absorbent, and strong.

Silk is another animal fiber [6]. Silk, unlike other natural fibers such as cotton, flax, wool, etc., does not have a cellular structure. In this respect and in the way it is formed, silk closely resembles synthetic fibers.

15.5.4 Plant Fibers

Fibers originated from various parts of the plants are known as plant fibers [7]. Plant fibers are classified mainly into three types, depending on the part of the plant from which they are extracted.

1. *Blast or stem fibers*: Blast or stem fibers are the fibrous bundles in the inner bark of the plant stem. Common examples are jute, ramie, flax, mesta, etc.
2. *Leaf fibers*: Leaf fibers run along the length of the leaves. Examples are sisal, pineapple, screw pine, abaca, etc.
3. *Fruit fibers*: Examples are coir, cotton, kapok, areca nut, and oil palm.
4. All other plant fibers include roots, leaf segments, flower heads, seed hulls, and short-stem fibers.

15.5.5 Synthetic Fibers (Man-Made Fibers)

Fibers in which the basic chemical units have been shaped by chemical synthesis followed by fiber formation are called synthetic fibers [5]. For example, nylon is a synthetic fiber forming polyamide, in which less than 85% of the amide linkages are directly attached with two aromatic rings. Similarly, Kevlar fiber is an aromatic polyamide (aramid). Kevlar is the Du Pont trade name of poly (p-phenylene diamine terephthalamide), which is very popular for its high strength and modulus.

Glass fibers are the most usually used reinforcing fibers for polymeric matrix composites [8]. The two types of glass fibers commonly used in the fiber reinforced plastics (FRP) industry are E-glass and S-glass. One more type of glass fiber is C-glass, mainly used for its corrosion resistance toward acids and bases.

15.5.5.1 Regenerated Fibers

An additional type of man-made fiber is regenerated fibers. These fibers, which are regenerated either from organic or inorganic systems, are classified as regenerated fibers (see Scheme 15.3). Polymers from natural sources have been dissolved and regenerated after passage through a spinneret to form fibers. For example, rayon is a manufactured fiber composed of regenerated cellulose in which different substituents are replaced with not more than 15% of hydrogen of the hydroxyl groups [5]. These fibers are used for reinforcing plastics and rubbers.

15.6 FIBER REINFORCEMENT ON VINYL POLYMERS

15.6.1 Short-Fiber Reinforcement

Short fibers generally lie in the fiber length range of 3 mm or less. High-performance composites are generally made from continuous fibers, but there are many applications for which the requirements are less demanding, or for which the appropriate manufacturing route cannot handle long fibers. It is then natural to consider using short fibers. Some reinforcing fibers are also accessible in the form of short filaments. Thermoplastics and thermoset molding compounds that contain haphazardly arranged chopped fibers, usually glass, are cheap materials that are finding widespread use in mainly non-load-bearing applications. Although the stiffness of these materials is usually greater than those of the base resins from which they are compounded, their strengths are rarely much higher than those of the plain (filled) matrices. An important virtue of such materials, however, distant from their cheapness and moldability, is that they are substantially tougher than unreinforced polymeric materials. They are approximately isotropic, and the lack of obvious planes of weakness that occur in laminates means that the cracking process resembles that of conventional homogeneous materials more than that of anisotropic laminates.

Short-fiber reinforcement includes both natural and synthetic fiber reinforcements [9–10]. Over the past few years, developments of biofiber-based composites have attracted the attention of scientists and technologists because of environmental and economical advantages with respect to raw material utilization from renewable resources [11]. Advantages of biofibers over traditional reinforcing materials such as glass, carbon, talc, and mica are low density, low cost, high toughness, acceptable specific properties, and biodegradability [12]. Cinelli et al. [13] reported on short corn fiber reinforcements using polyvinyl alcohol as the continuous matrix. PVA, a hydrolysis product of polyvinyl acetate, is well suited for blending with natural polymers because it is a highly polar synthetic polymer and is also considered to be biodegradable [14, 15]. For these reasons, PVA has been widely applied to improve mechanical properties in composites with natural polymers [16–19].

From the study of Cinelli et al., it is clear that corn fiber appreciably affects the mechanical properties as well as thermal properties of the CF/PVA composites. This effect is more pronounced in presence of plasticizers. They have used glycerol, pentaerythritol, and polyethylene glycol 200 as plasticizers. By using plasticizers, compounding could be achieved at lower temperatures, by which fiber decomposition during processing can be avoided. They have also used corn fibers, which had a thermal stability suitable for processing at high temperatures. According to the observation of Cinelli et al., the composites containing these plasticizers had tensile bars that were uniform and flexible. By varying the amount of the plasticizers (polyethylene glycol), the elongation at break (EL) and Young's modulus (YM) values vary appreciably. As the amount of plasticizers increased, the EL of the composite increased appreciably, whereas its YM value decreased appreciably. Moreover, by increasing plasticizer content, ultimate tensile strength (UTS) of the composite decreased remarkably. The difference in the plasticizer ratio affected the mechanical properties of the tensile bars. Glycerol is a more effective plasticizer for PVA than pentaerythritol, as indicated by the decrease in elongation at break and appreciable enhancement of ultimate tensile strength. According to their observations, composites containing pentaerythritol in higher proportions than glycerol was much less flexible, as indicated by the sharp increase in YM.

Compston et al. [20] reported on the effect of matrix toughness and loading rate on the mode-II interlaminar fracture toughness of glass-fiber/vinyl-ester composites. They studied glass fiber/rubber toughened vinyl ester composites for this purpose. They used the Irwin–Kies expression for fracture energy to calculate the expressions for mode-II critical strain energy-release rate [21]. They compared the composite G_{IIC} at the maximum load point with the order of G_{IC} matrix, and the comparison did not show any significant effect of loading rate or matrix toughness. Fracture surface studies revealed the similar deformation features in each composite, which also supported the mechanical test results. They also observed that the absence of a clear loading rate effect is consistent with the bulk of published mode-II work, but the effect of the absence of a clear matrix is not consistent. They suggested from this study that failure in these glass fiber/vinyl ester composites is interface controlled (i.e., as the matrix toughness becomes apparent an unstable failure initiated before an increase in composite G_{IIC}). Moreover, from this study, they reported that rubber-toughened vinyl ester matrices in the glass-fiber composites would not have improved the resistance to mode-II dominated impact induced dilamination. Instead, through-thickness damage in an impacted composite structure is likely to involve mixed-mode loading.

Mukherjee et al. [22] demonstrated the effect of short Kevlar fiber reinforcement on the thermal and mechanical properties of ethylene propylene co-polymer (EP) matrixes with or without surface modification of fibers. They used direct fluorination and oxy-fluorination as surface modification technique. According to their observation, the surface-treated Kevlar fiber reinforced EP demonstrate better thermal stability in comparison with the untreated Kevlar/EP, which is more prominent in the case of an oxy-fluorinated composite. Surprisingly, they have noticed that tensile strength of the untreated Kevlar fiber/EP composite decreases in comparison to the neat polymer. Nevertheless, the enhancement of the tensile strength was observed in the case of surface modified Kevlar fiber/EP composites. This may be due to the better adhesion between the fiber and the matrix due to surface modification of the fiber.

They used the 100% EP matrix and 1.43% Kevlar fiber (original [X], fluorinated [Y], and oxy-fluorinated [Z)]) and mainly concentrated on the thermal, crystalline, mechanical, and dynamic mechanical performance of those composites.

To study the thermal parameters and decomposition pattern, they performed differential scanning calorimetry/thermogravimetric analysis (DSC/TGA). From these studies, it was clearly observed that all short Kevlar fiber reinforced EP composites (X, Y, and Z) show a higher thermal stability than the neat polymer (EP). Moreover, fiber surface modification leads to a much higher thermal stability of the composites. All the composites. along with the neat polymer, show two-step degradation in a TG plot (Figure 15.3). From the figure, it is also clear that oxy-fluorinated Kevlar fiber reinforced EP shows maximum thermal stability among the all EP/Kevlar composites. They offered the following explanation: Due to surface modification of the Kevlar fiber, some functional groups are generated onto the Kevlar surface, which in turn enhances the fiber/matrix adhesion, leading to the higher thermal stability of the surface modified Kevlar fiber/EP composites. In addition to that, the highest thermal stability of the oxy-fluorinated Kevlar fiber/EP composite (Z) can be considered as the generation of the long-living peroxy ($RO_2\cdot$) [23–26] radicals onto the Kevlar surface, leading to the graft co-polymerization and resulting in the enhanced adhesion between the fiber and the matrix in the case of the particular composite.

From the differential scanning calorimetry (DSC) study, some depression of the melting point of the composites was observed. This may be due to the dilution effect that arises because of the incorporation of the Kevlar fiber into the EP matrix. Table 15.5 gives the variation of melting point values of the modified and unmodified Kevlar fiber reinforced composites (X, Y, and Z), along with the neat matrix (EP). From Table 15.5, it is also clearly evident that the heat of fusion (i.e., enthalpy values) of all the composites (X, Y, and Z) is substantially lower than that of the pure matrix (EP). The enthalpy values of modified fiber reinforced composites (Y, Z) further increase in comparison with the unmodified Kevlar fiber/EP composite (EP).

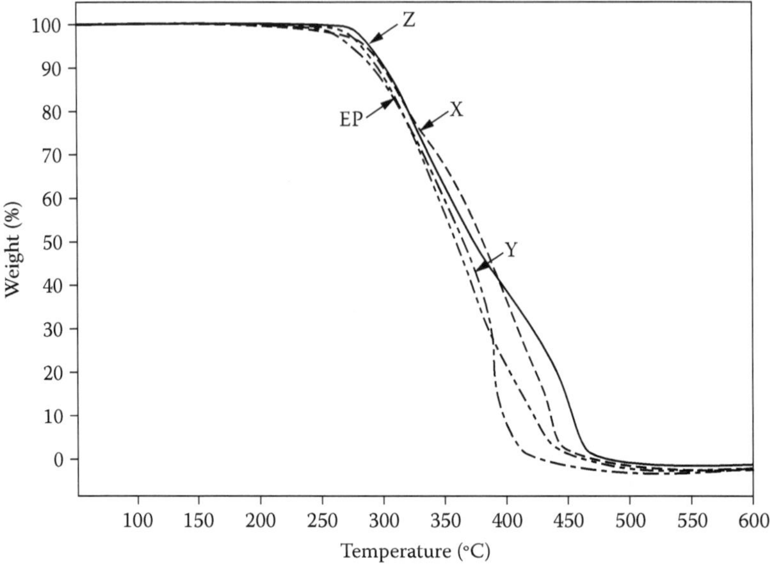

FIGURE 15.3 TG study of pure EP (EP), Kevlar fiber reinforced EP (X), fluorinated Kevlar fiber reinforced EP (Y), and oxy-fluorinated Kevlar fiber reinforced EP (Z).

An X-ray diffraction (XRD) study was performed by the authors to understand the crystalline behavior of the unmodified and modified Kevlar fiber/EP composites. The respective parameters are tabulated in Table 15.6, and the X-ray defractogram is represented in Figure 15.3. From the table, it is clearly observed that the percentage of crystallinity of the fiber reinforced composites (X, Y, and Z) decreases appreciably in comparison with the neat matrix (EP). This may be due to the hindrance of diffusion or migration of EP molecular chain onto the growing matrix surface of polymer crystal by Kevlar fiber. With surface modification of Kevlar fiber, the percent crystallinity of the composites increases appreciably, which is also evident from Table 15.6, because it is known that the percentage of crystallinity is directly proportional to the heat of fusion (enthalpy). The observation of the authors regarding the trend of percentage of crystallinity of the composites in the XRD study is quite in line with that of the results obtained from the DSC study. The authors have offered

TABLE 15.5
Thermal Parameters of EP/Kevlar Composites

Sample	T_m (°C)	H_f (J/g)
Pure EP (EP)	165.0	73.0
Kevlar/ EP (X)	162.3	69.2
Fluorinated Kevlar/EP (Y)	162.4	68.5
Oxy-fluorinated Kevlar/EP (Z)	164.1	70.0

TABLE 15.6
X-ray Parameters of EP/Kevlar Composites

Sample Code	% Crystallinity	Peak Angle (2θ in degree)				
		θ_1	θ_2	θ_3	θ_4	θ_5
EP	74.00	14.20	16.95	18.70	21.30	21.90
X	52.00	14.20	17.0	18.70	21.32	21.88
Y	58.00	14.20	17.05	18.68	21.34	21.95
Z	60.00	14.10	16.90	18.66	21.20	21.88

the explanation of this observation as the better surface adhesion between the fiber and matrix in presence of functional group on the Kevlar surface. This means that surface modification significantly affects the surface characteristics of Kevlar fiber, owing to the generation of an active reaction site onto the fiber surface. This effect is more pronounced in the case of oxy-fluorinated Kevlar fiber/EP composite (Z). From Figure 15.4 and Table 15.6, it is also evident that the peak position of the composites, with or without surface modified Kevlar fiber, does not shift significantly.

The dynamic mechanical thermal analysis was performed by the authors to investigate the dynamic mechanical behavior of the composites (X, Y, and Z), along with the pure polymer (EP). The storage modulus versus the temperature curve of the EP/Kevlar composites is depicted in the Figure 15.5. From this figure, it is clearly seen that storage modulus of every composite, along with the neat polymer, is decreasing with increasing temperature due to the decrease in segmental mobility

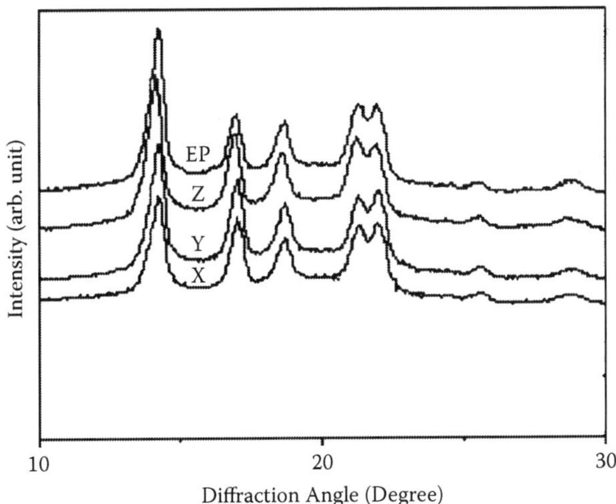

FIGURE 15.4 X-ray diffractograms of EP/Kevlar Kevlar composites: (1) pure EP (EP), (2) unmodified Kevlar/EP (B), (3) fluorinated Kevlar/EP (E)m, and (4) oxy-fluorinated Kevlar/EP (H).

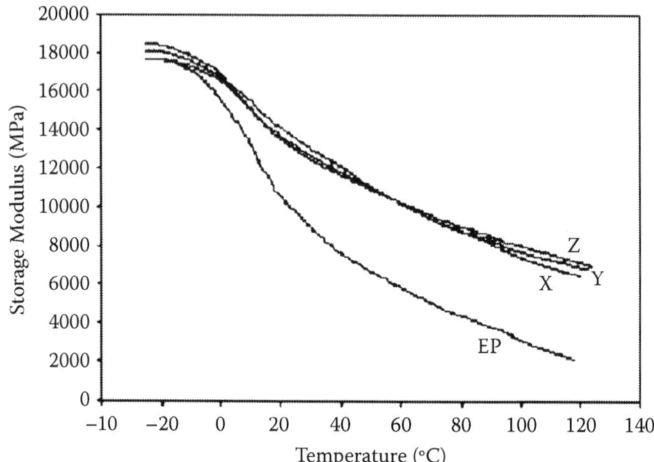

FIGURE 15.5 Storage modulus versus temperature curve of (1) pure EP (EP), (2) unmodified Kevlar fiber reinforced EP (X), (3) fluorinated Kevlar fiber reinforced EP (Y), and (3) oxy-fluorinated Kevlar fiber reinforced EP (Z).

of the matrix phase. All EP/Kevlar composites show greater storage modulus than the pure EP because of the reinforcement imparted by the Kevlar fiber into the EP matrix, which permits stress transfer from matrix to fiber. The storage modulus of every composite (X, Y, and Z) shows a sharp decrease in the region of 0–10°C [27], which is the glass transition temperature of the EP matrix. It is clearly evidenced that the storage modulus, particularly in the case of oxy-fluorinated Kevlar fiber reinforced EP (Z), increases especially in glass region due to the incorporation of more functional group onto the surface of the Kevlar fiber, resulting better fiber/matrix adhesion, which in turn reduces the molecular mobility of EP.

The authors also studied the variation of tan delta with temperature of unmodified and modified (by fluorination and oxy-fluorination) Kevlar fiber reinforced EP composites. Table 15.6 and 15.5 represent the results obtained from the tan delta study. From Figure 15.6, it is very much evidenced that tan delta value of every composite (except X) decreases in comparison to the neat sample (EP). It is well known that loss peaks of the polymer composites indicate the mechanical energy dissipation capacity of the particular material. Thus, from this study, the authors have suggested that in the case of oxy-fluorinated Kevlar fiber/EP composite (Z) showing lowest loss peak among the all composite shows lowest mechanical energy dissipation capacity. Thus, the molecular mobility of that particular composite (Z) decrease and in turn the mechanical loss is reduced after incorporation of the oxy-fluorinated Kevlar fiber into the EP matrix. From Figure 15.5, it is also clear that one relaxation peak at approximately 10°C appears in the case of every composite along with the neat polymer (EP), which correlates with a α-transition temperature of EP. From Table 15.6 and Figure 15.6, it is observed that the α-transition (T_g) value of every composite shifts to a higher temperature side for all composites. The authors have suggested that due to the incorporation of Kevlar fibers (modified and unmodified),

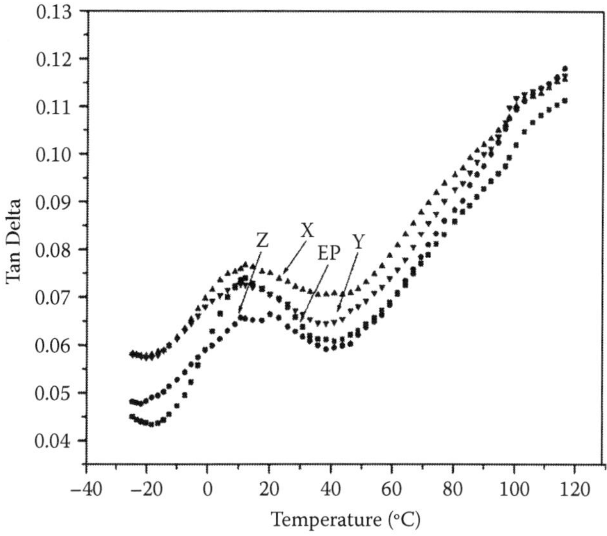

FIGURE 15.6 Tan delta versus temperature curve for (1) pure EP (EP), (2) unmodified Kevlar fiber reinforced EP (X), (3) fluorinated Kevlar fiber reinforced EP (Y), and (4) oxy-fluorinated Kevlar fiber reinforced EP (Z).

the mobility of the matrix chain decreases, and due to surface modification of Kevlar fibers enhancement of fiber matrix adhesion at the fiber interface occurs, which in turn increases the T_g value of modified Kevlar reinforced EP composites (Y, Z). From the aforementioned figure, we can also see that in the case of composites (X, Y, and Z), the transition peaks broaden appreciably in comparison to the pure polymer (EP). The broadening of the loss peak can be attributed to the better fiber/matrix adhesion at the interface because the matrix adjacent to the fibers can be considered to be in different state as compared with the bulk polymer matrix, which may cause the disturbance of the relaxation of the matrix polymer.

The authors also concentrated on the mechanical properties of the modified and unmodified Kevlar fiber/EP composites. The mechanical properties of composites along with the matrix polymer are displayed in Table 15.7. According to their

TABLE 15.7
Variation of Glass Transition Temperature of EP/Kevlar Composites

Sample Code	T_g (°C)
Pure EP (EP)	11.44
Unmodified Kevlar/EP (X)	12.45
Fluorinated Kevlar/EP (Y)	12.78
Oxy-fluorinated Kevlar/EP (Z)	13.12

observations, the matrix possesses higher tensile strength as compared with the EP/unmodified Kevlar composite. The authors offered an explanation for this behavior — that this may arise from the very basic ordered structure of EP, which is also evident from the percentage crystallinity value obtained from XRD and DSC studies mentioned earlier. The fact is that the mechanical properties, such as tensile strength and others, of polymeric materials strongly depend on their microstructure and crystallinity. Thus, crystallinity enhances the tensile properties of pure EP. According to the authors' observations, the tensile strength decreases sharply upon the addition of Kevlar fibers (sample X) into the EP polymer matrix as compared with neat polymer (EP), but it increases gradually in the case of fluorinated Kevlar fiber/EP (sample Y) and oxy-fluorinated Kevlar fiber/EP composites (Z). This depression of tensile strength in the presence of unmodified Kevlar fiber may be due to the result of two opposing effects: the reinforcement of the matrix by fibers and dilution of the matrix.

The authors have mentioned that in the case of unmodified Kevlar fiber/EP composite (sample X), the matrix is not restrained properly by fibers due to poor adhesion between the fiber and matrix, leading to the concentration of the localized strains on the matrix. As a result, the fibers are pulling out from the matrix leaving it diluted by non-reinforcing, debonded fibers. Nevertheless, in the case of modified Kevlar/EP composites (i.e., Y and Z), the matrix is adequately restrained and the stress is more evenly distributed. The authors have ascribed this phenomenon to the fact that the functional groups affect the surface characteristics of the fibers, which results in the enhancement of the adhesion between the fiber and matrix. This is more pronounced in the case of oxy-fluorinated Kevlar fiber reinforced EP (Z). For the modified Kevlar fiber reinforced EP (Y and Z), the reinforcement is predominant over the crystallization effect.

Mukherjee et al. also studied the effect of unmodified and modified (by fluorination and oxy-fluorination) short Kevlar fiber loading on the properties of EP composites [22]. They studied the thermal, crystalline, and dynamic properties of EP/Kevlar composites with respect to fiber loading (see Table 15.8). For their study, they used 35 g EP as bulk matrix and three different proportions (0.25 g, 0.50 g, and 1.0 g) of unmodified and modified (fluorinated and oxy-fluorinated) Kevlar fiber. The mixing is performed in a Brabender mixer at 200°C at 60 r.p.m. for 10 min. After that, the mixers are molded in compression molding at 200°C under 10 MPa pressure for 10 min.

The authors studied the crystalline properties of the composites by using an XRD study. They reported that the percentage of crystallinity increases appreciably in the

TABLE 15.8
Mechanical Properties of Unmodified and Modified Kevlar/EP Composites

Sample	Tensile Strength (MPa)	Elongation at Break (%)
Pure EP (EP)	27	125
X	20	6.85
Y	30	5.13
Z	33	4.02

case of modified fiber reinforced EP composites, in comparison with the unmodified reinforced EP composite, which is quite similar with their previous observation. Nevertheless, with increasing the fiber loading, the percentage of crystallinity increases appreciably, possibly due to the nucleating effect of the fiber at higher fiber content. However, in every case, the percentage of crystallinity shows lower value in comparison to the neat polymer because of the very basic ordered structure of the EP matrix.

To investigate the thermal properties of the composites (i.e., the effect of modified and unmodified Kevlar fiber loading on the EP matrix), the authors performed DSC and TG studies. From the TG analysis, it was observed that the modified Kevlar fiber reinforced EP composites demonstrate much higher thermal compared with the unmodified composite, which may be due to the better adhesion between the fiber and the matrix at the interface. Besides, with fiber loading, the thermal stability of the composites increases appreciably, possibly due to the incorporation of a more thermal, stable Kevlar aromatic moiety.

From a DSC study, the authors noticed that all the modified Kevlar fiber reinforced composites show a higher heat of fusion value (enthalpy) than the unmodified composites, and the gradation of the enthalpy value is also observed with increasing fiber loading in the matrix, which is in agreement with the results obtained by the author in the XRD study (i.e., enhanced trend of percentage of crystallinity).

Dynamic mechanical thermal analysis (DMTA) was performed by the authors, and they reported that storage modulus of the modified Kevlar fiber reinforced EP increases when compared with the unmodified derivative. The reason behind this is the fact that surface modification leads to the generation of the functional groups onto the Kevlar surface, which in turn increase the storage modulus value of the particular composites. Increase in fiber loading also shows enhancement of storage modulus, but in the case of unmodified Kevlar, fiber reinforced EP containing 0.25% of Kevlar fiber shows the lowest storage modulus value among all the composites. This has been explained by the authors as the phase separation resulting from the agglomeration of the fibers.

Mukherjee et al. also studied the effect of fiber surface modification (by fluorination and oxy-fluorination) of the Kevlar fiber on the properties of synditactic polystyrene (s-PS) system. They studied the thermal and dynamic mechanical behavior of the s-PS/Kevlar composite.

The composites were prepared by mixing 3.5 g of the unmodified and modified (fluorination and oxy-fluorinated) Kevlar fiber into 600 g of the s-PS used as a matrix with a twin-screw extruder and molded in the injection molding machine.

Studies of the thermal properties indicate that thermal stability of the modified Kevlar fiber reinforced s-PS composites increases as compared with the unmodified s-PS/Kevlar composite, due to the better surface adhesion between the fiber and the matrix. It can also be explained as the graft co-polymerization occurs at the fiber/matrix interface in the case of oxy-fluorinated/s-PS composite, leading to the most thermally stable composite among all the composites.

From a DSC study, Wang et al. observed that the crystallization peak appears at approximately 140.7°C for pure s-PS (Figure 15.7) [28]. On introduction of unmodified and surface modified Kevlar fibers, significant changes in the crystalline behavior

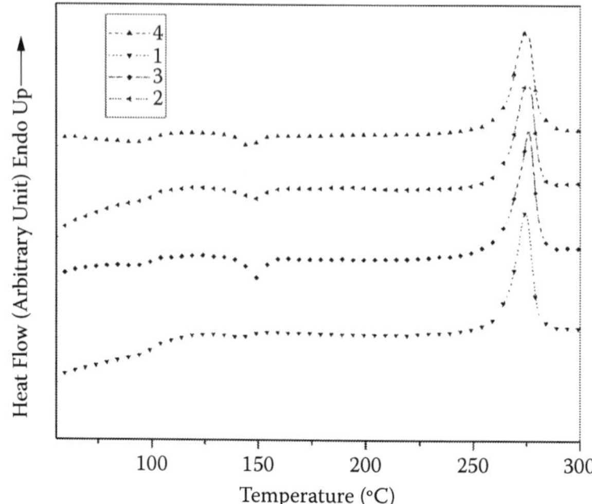

FIGURE 15.7 DSC picture of (1) pure s-PS, (2) s-PS/unmodified Kevlar, (3) s-PS/fluorinated Kevlar, and (4) s-PS/oxy-fluorinated Kevlar fiber.

of the composites were observed by the authors. They emphasized that crystallization temperature is the maximum in the case of fluorinated Kevlar fiber/s-PS composite (approximately 148.9°C) and minimum in the case of oxy-fluorinated Kevlar fiber/s-PS composite. The authors provided an explanation for this observation: oxy-fluorination causes the graft co-polymerization at the interface of fiber/matrix, leading to the branching or co-polymerization or crosslinking and resulting in the lowering of the crystallinity of the particular composite, among others.

The authors also studied the dynamic mechanical behavior of the s-PS/Kevlar composites along with the neat polymer matrix. According to their observations, the storage modulus of the composites increases appreciably in the case of all composites in comparison with the neat matrix polymer. This may be due to the reinforcing effect of the fibers, which results in the effective stress transfer between the fiber and the matrix. Fiber surface modification results in increase in the storage modulus of the composites because of the introduction of the functional group onto the Kevlar surface, resulting in better fiber/matrix adhesion.

15.6.2 Long-Fiber Reinforcement

Long-fiber reinforced thermoplastics (LFT) are one of the fastest growing materials in the polymer composites industry. LFTs possess some attractive characteristics, such as high specific tensile properties and recyclability, which leads LFT to replace the metals and thermoset composites in several areas of modern life. Moreover, due to the advantages of choosing a wide range of fibers as reinforcing agents and the matrix from a vast group of thermoplastic resins, the properties of the LFTs can be altered according to the customers' desires. LFTs find their huge applications in every aspect of life, which includes from deep inside the ocean to outer space. Fibers in the reinforced composites are relentlessly broken up into very short lengths. Because of this, the thermoplastic molding composites contain very short lengths of fiber (typically 0.3 mm). It is well known that the mechanical properties of the fiber reinforced composites are closely related to the length of the reinforcing fibers.

Closed molded, discontinuous long-fiber reinforced composites exhibit some attractive features such as greater strength, stiffness, and impact properties in contrast to short-fiber reinforced thermoplastics, which include processability at high volume, ability to fill complex geometries, inherent recyclability, and the capability of partial integration. Long-fiber reinforced thermoplastic (LFT) composites have one of the highest growth rates in the polymer material areas, behind a projected 30% growth from 2000–2004 [29]. Thermoplastic composites usually encompass a cost effective commodity matrix such as PP, polyethylene (PE), nylon, etc., reinforced with glass, carbon, or aramid fibers. E-glass is the most commonly used reinforcing agent because of the predominance of the cost/performance ratio over the weight/performance ratio in the aerospace industry. Thermoplastic composites used in these fields can be of various types such as (1) short-fiber reinforced thermoplastics (SFRT), (2) long-fiber reinforced thermoplastics (LFT), and (3) glass mat thermoplastics (GMT). Injection molded SFRT composites containing the fiber lengths less than 4 mm (before the fiber breaks during processing) are very common in the case of the aforementioned composites. Today, the full advantage of the reinforcing fiber is still not realized properly because of the low fiber aspect ratio. Injection molded LFTs also suffer from severe fiber length dilapidation in the plastication and injection stages. Moreover, some difficulties arise in processing mechanism with high fiber content and initial high fiber lengths (exceeding of 13 mm) resulting high melt viscosity. LFTs possessing the high-fiber aspect ratio are far more advantageous over the short-fiber reinforced counterpart because of their excellent strength. Injection-compression and extrusion-compression molding techniques are used for processing long fiber reinforced thermoplastic composites. In LFTs, the fiber lengths initially remain in a range greater than 13 mm, depending on the desired properties, fiber concentration, and processing techniques. LFTs present numerous advantages in contrast to GMTs such as: (1) GMTs lead to the possibility to work without semi-finished mats (e.g., in-line extrusion), which are less labor intensive; (2) GMTs possess lower compression forces due to a decrease in melt viscosity, which results in more capital cost in tooling and machinery in comparison to the LFTs. Besides, LFTs propose a higher surface quality, less part rejection due to an increase in the ability to fill complex features, and an incorporated recyclability. Another advantage of LFTs is greater emancipation in choosing fiber and matrix materials. In the compression molding process, fibers are oriented in the in plane orientation during flow, which can plague the consolidated element. Preferential orientation during compression molding results in the reduction of strength and stiffness in critical areas, which prevent warping through anisotropic tightening upon cooling [30]. However, high fiber aspect ratios and more fiber loading cause the depression of fiber mobility in the melt, resulting in a decrease in the degree of preferential orientation. A high degree of preferential orientation may still occur depending on mold geometry, charge location, and pre-orientation of the fibers in the composites. Bartus mentioned this phenomenon in his work using finite element methods to simulate the compression molding process with CAD press-thermoplastic [31]. The growing applications of LFTs in automotive and other industries demands the determination of impact response in these materials, which increases to ensure the safety and stability of designed structures [32]. Sheet molding compounds (SMC) were first used in this application. However, it was not successful because of the fragile nature of the resulting components. As for those applications, the component must be flexible in nature, and it must withstand impact from stones and other objects [33]. Many authors have attempted to characterize the impact performance of discontinuous and randomly oriented LFT composites. The intrinsic fiber architectures in LFTs offer an accurate characterization of the complex failure mechanisms. Most efforts in understanding the impact performance and failure mechanisms of LFTs are mainly based on Charpy and Izod impact testing, which involves low-velocity drop tower impact testing [34]. Very little work has been done on the effect of intermediate velocity blunt object impact (BOI) on LFTs. In this case, intermediate velocities are greater than low velocity drop tower impacts or pendulum type impacts, yet slower than high velocity ballistic type impacts. The velocity range is the key factor for the effect of blunt objects, such as rocks and debris traveling at highway speeds for automotive applications, as well as impact induced by debris from natural calamities, such as hurricanes and tornadoes for storm shelters and military housing applications.

Broyles et al. [35] reported on the static and mechanical performance of vinyl ester and carbon fiber composites processed with different sizing effects. They used pultruded long carbon fibers for easy processing. They used two different sizing agents for this purpose. Their study reported that the thermoplastic sizing materials with high T_g offered many processing-related advantages in comparison with the traditional oligomeric epoxy sizing agent by (1) producing a stiff carbon fiber that is less susceptible to damage during the pre-wet-out portion of the pultrusion process, (2) optimizing the interaction with the vinyl ester matrix, which lowered the contact angle between the fiber and the matrix, thus reducing the wet-out time, in addition to the better fiber/matrix adhesion, and (3) swelling and relaxing with the vinyl ester matrix, thus making it rubbery, which in turn allowed the fibers to be compacted and compressed at the pultrusion die. According to Broyles et al., the previously mentioned processability improvements cause better surface finish and fiber alignment. They also reported that the mechanical performance of the final composite improved with the variation of the composite's sizing agents.

Again, Verghese et al. [36] reported on the environmental durability of pultruded carbon fiber/vinyl ester composites processed with different fiber sizing agents. The sizing agents they used were poly (vinylpyrrolidone) (PVP) and modified polyhydroxyether of bisphenol A. They concluded that the hydrogen bonding between the sizing material (PVP) and the matrix improved the water resistance of the resulting composite in comparison with the other sizing agent. The interphase of the fiber matrix composite has played an important role in regulating the moisture uptake characteristics and the residual strength of the composite. Traditionally used sizing material (G′) led to the larger amount of voids in comparison with the thermoplastic sizing agents, which in turn resulted in the lower drop in tensile strength after saturation.

15.6.3 Woven Cloth Reinforcements

Woven reinforcement is constructed by interlacing fibers, filaments, or yarns to form fabric patterns such as baskets, satin, leno weaves, scrim, etc. These different weaving patterns can be utilized to provide different processing and/or directional properties, such as filling threads, in which the threads lie in the so-called machine direction, and warp threads in which the threads are lying in the transverse direction or at 90° to the filling threads.

A reinforcing woven fabric consists of two types of warps: (1) warps of reinforcing filamentary yarns arranged to form a high-density portion of warps, and (2) low-density portion of warps in the transverse direction and wefts of reinforcing filamentary yarns extending obliquely to the warps. A fiber reinforced composite material is formed using the preformed material, which is made by using multiple numbers of the reinforcing woven fabrics and a beam particularly suitable as the fiber reinforced composite material. The high-density portion of warps in the reinforcing woven fabric satisfies the strength and rigidity against bending or tensile stress required for the flange of the beam. The obliquely extending wefts in the low-density portion of warps can satisfy the strength against shear stress required for the web of the beam, when a plurality of the reinforcing woven fabrics are laminated to form the preformed material for the beam.

Li [37] has shown the processing of sisal fiber reinforced composites by resin transfer method (RTM). In his study, he used glass and carbon fabric as reinforcing agents and vinyl ester as matrix and silane, and a permanganate treatment was performed for the surface modification of the sisal fiber. According to his observations, fiber surface treatments show a positive effect on the permeability of the sisal textile. Though both the coupling agents show similar effects on permeability of the vinyl ester composite, Li has explained the reasons in different ways in two cases. According to Li, by introducing strong polar groups onto the fiber surface, silane treatments increase the permeability of the modified sisal fiber/vinyl ester composite. Permanganate treatment breaks down the fabric into small fibers, thus increasing the flow channel in the composite and, as a result, permeability increases. Li has also reported that fiber volume fraction has a governing effect in increasing the permeability of the resulting composite. Further, RTM is an efficient method for over-compressing molding in the case of sisal fiber/vinyl ester composites, creating fewer voids, thus giving rise to better mechanical properties of the composite.

Again, Akil et al. [38] discussed the effect of strain rate on the compression behavior of a woven fabric S2-glass-fiber-reinforced vinyl ester composite. They reported the quasi-static and high strain rate compression behavior of an S2-glass woven fabric/vinyl ester composite determined in in-plane and through-thickness directions. They concluded that at higher strain rates, an increase in modulus and failure strength was observed in both direction and in in-plain direction, and failure occurred by axial splitting followed by kink banding at quasi-static strain rates, whereas only axial splitting occurred at high strain rates. They also observed that in the through-thickness direction, failure modes at quasi-static and high strain rates appeared to be very similar, which was attributed to the strain rate sensitive matrix (i.e., vinyl ester) properties.

15.6.4 Hybrid Composites

Incorporation of two or more types of fibers within a single matrix or vice versa is known as hybridization, and the composite that results is referred to as hybrid or hybrid composites. The behavior of hybrid composites is guided by the weighted sum of individual components, in which a more favorable balance exists between the inherent advantages and disadvantages [39, 40].

Hybrid composites can be classified based on fabrication methods [41]. The following are the main types of hybrid constructions (see Figure 15.8):

1. *Mixed fiber tows*: In this processing method, two types of fibers are uniformly mixed throughout the resin without any specific concentration of either type of fibers.
2. *Mixed fiber ply (intraply)*: This type of hybrid composite is made by the discrete layers of a mixture of the individual fibers.

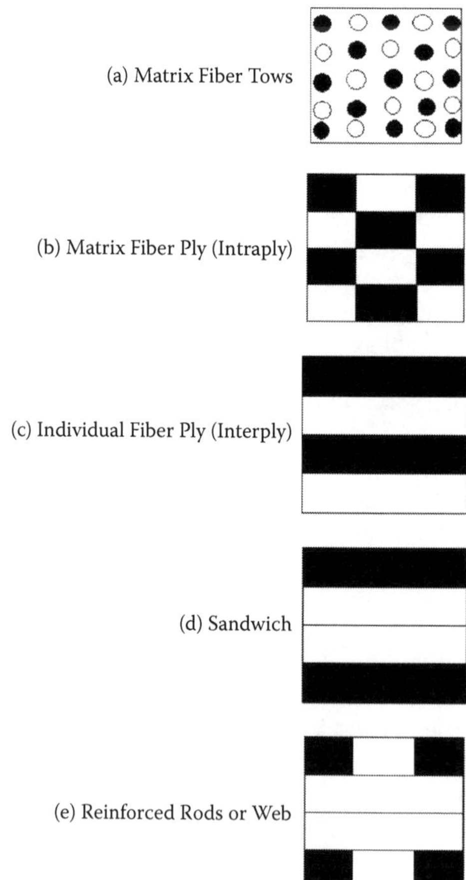

FIGURE 15.8 Different hybrid configuration.

3. *Individual fiber ply (interply)*: In this type of hybrid composite, the fiber layers are arranged alternatively.
4. *Sandwich*: This arrangement has two types of fibers in which one is sandwiched between the other types. Here, one type of fiber at the center acts as a nucleus, and the other at the periphery acts as a shell.
5. *Reinforced rods or webs*: In this type, a constituent fiber web or rod is selectively positioned on the continuous medium of other type of fibers.

The properties of the hybrid composites are solely dependent on the constituent materials and they are directly proportional to their additive properties.

The characteristic property of the hybrid composites can be calculated from the equation given next.

$$E_H = E_1V_1 + E_2V_2 \tag{15.1}$$

Here, E is the characteristic property of hybrid composites, V_1 and V_2 are the hybrid volume fraction of reinforcements of the individual composites, and E_1 and E_2 are the characteristic properties of the individual composites.

Herzog et al. [42] reported about the durability of the fiber reinforced polymer (FRP) composite-wood hybrid products. For this, they used a composites pressure resin infusion system of fabrication to produce E-glass/vinyl ester (fiber reinforced composite)-wood hybrid composite. The authors studied the properties that include shear stress, the percentage of wood failure experienced in shear, and the delamination of the glass fiber reinforced polymer composite-wood interface when subjected to an accelerating aging test. They compared all these properties obtained from the composite pressure resin infusion system of fabrication with those of hybrid composite obtained from a controlled system comprised of an adhesive-bonded procured fiber reinforced composite sheet. This study reported that composite-pressure resin infusion system-fabricated hybrid possesses shear strength equal to or greater than the composite made from a conventional fabrication method. The authors explained it as the graded interface and/or enhanced dimensional stability achieved here by resin impregnation. However, when the vinyl ester bond lines created in a composite pressure resin infusion system are exposed to severe stresses due to repeated water saturation and drying, it is observed that the bonds fail to meet the requirements for structural wood adhesives intended for exterior exposure. In this study, they recommended that the fiber reinforced polymer used should be thick enough to eliminate this type of failure during testing.

As mentioned in Section 15.6.1, Cinelli et al. [13] reported on short corn fiber reinforcements using polyvinyl alcohol as the continuous PVA matrix. In the same study, they used starch in different proportions to reduce the cost as well as the amount of synthetic polymer in the composites and for the ease of preparation and biodegradation. According to the authors, the addition of starch influenced the thermal stability of the composite, making it suitable for processing at high temperature. In addition, starch improved the interfacial adhesion between the fiber and the matrix. The authors explained these phenomena by saying that additional starch reduced the elongation of the composite, thus decreasing the flow of the continuous phase PVA.

15.6.5 *In Situ* Composites

Fine fibrils that reinforce the matrix effectively giving rise to the development of polymer composites are known as in situ composites [43]. *In situ* composites are very similar to the short-fiber reinforced thermoplastic composites. It is known that the stiffness and the mechanical strength of polymers can be improved considerably by short-fiber reinforcement. Among them, short glass fibers are much more important in this regard. Conventional polymer composites reinforced with higher volume content of glass fiber are generally very difficult to process because of their high melt viscosity. Other difficulties regarding the short glass fiber polymer composites include wear of processing facility, which results in the abrasion of fiber reinforcement leading to fiber breakage. Thus, *in situ* composites are considered highly competitive with the glass reinforced polymers, when processability and low density are taken into consideration. Several aspects involved in the processing to develop *in situ* composites with superior properties include rheological behavior crystallization, compatibility and the processing-structure-property relationship of the LCP thermoplastic blends. The development of *in situ* composites is hampered by high melting temperatures of LCP, which are appreciably higher than that of the commodity engineering resins. At high processing temperature the resin becomes unstable and degrades during extrusion or injection molding, whereas LCP with lower melting temperature are beneficial.

The present authors and co-workers are working in the relative field of *in situ* fibrillation with varying degrees of processing conditions. They have chosen s-PS as the continuous phase and LCP as the filler. From their study, we can predict that during the twin-screw extrusion LCP fibrillization occurs at a maximum in the interior portion of the extrudate. They also observed that an optimum extrusion speed should be maintained to achieve the maximum LCP fibrillization for obtaining maximum adhesion between LCP and s-PS matrix. They also emphasized that a high rate of injection molding is accompanied by high inter-planar spacing and a large crystal size in the s-PS crystalline phase. In addition, the authors found that the orientation of fibrils and the rate of fibrillization are different for different injection molding rates.

15.7 APPLICATIONS OF FIBER-REINFORCED VINYL POLYMERS

Fiber-reinforced vinyl polymers have many applications in various fields, due to the possibility of an excellent combination of high thermal and mechanical stability of both the vinyl resin matrices and the fibers. Carbon fiber/vinyl ester composites have traditionally been used in high-end aerospace applications. Industries such as offshore oil exploration, civil infrastructure, etc., have increased their use of composite materials. For composites to find acceptance in civilian markets, such as infrastructure, the cost of the processing steps and raw materials must be reduced while maintaining or improving their performance. For this purpose, vinyl ester/carbon fiber composites are being used for cost effectiveness.

Increased agricultural productivity worldwide has promoted the utilization of agricultural products as eco-friendly, plastic materials. Bio-degradable polymeric materials based on agricultural wastes, crop surpluses, and co-products can be used as economical alternatives for use in the fabrication of consumer articles either as their raw components or after modifications [44–49]. In this regard, natural fiber-reinforced polyvinyl composites are broadly used as biodegradable polymer composites for agricultural and food technology industries. As mentioned earlier, Cinelli et al. reported about the reinforcement of the corn fiber in the PVA matrix. Natural fibers can be added to composites to grant stiffness, increase degradability, and lower the cost of the final product. In particular, the market related to single-use products, such as cups, containers, transplanting pots, and cutlery, appear suitable for applications of these composites [50]. Degradable-plastic composites are emerging materials that offer benefits to the environment by minimizing waste that would otherwise be deposited in landfills.

Lower-cost glass-fiber-reinforced composites are increasingly used for structural applications. As discussed by Davies [51], the marine environment is one of the largest areas of possible application of glass-fiber composites (for examples, see [52, 53]). Applications include civil and military surface vessels, offshore structures, and underwater structures. Aging in corrosive marine environments and delamination as a result of wave impact or shock loading are among the problems that have to be addressed for these composites [54]. In this respect, vinyl ester resin is attractive for use as a composite matrix. It has demonstrated better resistance to hydrolysis [23] and greater retention of mechanical properties after accelerated aging tests compared with the more popular polyester resins [55]. In addition, the toughness of commercial vinyl esters can be considerably enhanced by the addition of relatively small amounts of liquid rubber [56, 57].

15.8 ADVANTAGES

Advantages of bio-fiber-reinforced vinyl polymers over other fiber-reinforced thermoplastics are low density, low cost, high toughness, acceptable specific properties, and biodegradability.

Long-fiber reinforced vinyl polymer composites offer numerous advantages. such as the possibility to work without semi-finished materials, making it less labor intensive, and lower compression forces due to a decrease in melt viscosity, which results in lower cost in tooling and machinery.

Fiber reinforced vinyl polymers coupled with appropriate compatibilizing agents propose high thermal, mechanical, and dimensional stability, and weathering durability of the resulting polymer composites.

Biodegradable fiber reinforced vinyl polymers benefit the environment by degrading the waste materials.

15.9 LIMITATIONS

Fiber reinforced vinyl polymer composites mainly suffer from the limitations arising from the processes of the manufacturing. Indeed, those composites are noted for the variability of the mechanical properties they exhibit unless they have been produced under the most controlled conditions. The variability of those fiber reinforced materials produced by hand lay-up methods is more marked than that of composites made by mechanical processes. Any composites consisting of widely thermal expansion coefficient, which is heated during manufacture, may develop sufficiently high residual stresses to crack a brittle matrix from cooling. The most common defects that occur in processing fiber reinforced vinyl polymer composites include: (1) incorrect state of resin cure, (2) incorrect overall fiber volume fraction, (3) non-uniform fiber distribution, and (4) missaligned or broken fibers.

15.10 FIBER REINFORCEMENT ON POLYOLEFINES

15.10.1 Short-Fiber Reinforcement

Short-fiber reinforced polymeric composites gain importance due to the advantages in outstanding mechanical properties, processing, and low cost. Many researchers have analyzed the short-fiber reinforcement in poly olefin matrixes.

Mohanty et al. studied the viscoelastic behavior of short, natural-fiber thermoplastic composite for interfacial adhesion, mechanical, thermal, and weathering performance. They have used jute and sisal fiber reinforcements in PP and high-density poly ethylene (HDPE) matrixes [58]. They have focused on the mechanical, thermal, and viscoelastic properties of surface modified jute and sisal fiber reinforcement in the PP and HDPE matrixes, respectively, by using MAH-g-PP (MAPP) and MAH-g-PE (MAPE) as compatibilizers. It was observed that a substantial enhancement occurred in the mechanical properties of the MAPP-treated sisal PP composites as compared with the jute PP system. The cooler interaction as well as the covalent linkage between the hydroxyl groups of sisal fibers and MAPP resulted in increases in tensile strength as well as the flexural and impact strength, when compared with the untreated composites. The mechanical properties of the composites further increased with the increase in MAPP concentration. According to their observation, for sisal/PP composites, a relatively higher concentration of MAPP is required to modify the fiber matrix interfacial region as compared with the jute/PP system. This behavior is primarily attributed to high polar characteristics of sisal fibers, owing to the presence of more pre hydroxyl groups [59]. However, with the further increase in the MAPP concentration, marginal deterioration occurred in mechanical strength, with a significant drop in tensile modulus. They have explained it by a similar argument of self-entanglement of the PP chains and slippage of the molecules [60].

15.10.2 Long-Fiber Reinforcements

As mentioned in Section 15.6.2, long-fiber-reinforced thermoplastics offer very attractive features in the recent scenario. Here we will discuss the long-fiber reinforcements in the polyolefin matrices.

Bartus et al. [61] have investigated the performance of long-fiber reinforced thermoplastics subjected to transverse intermediate velocity blunt object impact. They used long glass fibers as reinforcing agents and PP as continuous phase. They reported that the energy dissipation between the fiber and the matrix took place by the mechanism, namely fiber fracture, fiber de-bonding, fiber pullout, and matrix fracture. According to them the damage initiated around the periphery of the impactor as a result of high transverse shear stresses. In this regard they have assumed a uniform stress distribution occurred around the periphery of the projectile and the failure occurred in areas with preferential fiber orientation to the tangent impact area. Away from the impactor, the major failure occurred by simultaneous tearing across planes of preferential fiber orientation.

They have also observed in their study that the increase in contact stresses decreased the critical energy required for perforation. According to the authors, this might be a proposed mechanism for a decrease in energy dissipation for the conically tipped projectiles. They have also mentioned that critical energy dissipation energy for samples subjected to conical projectile impact was appreciably less than samples impacted by flat projectiles. According to the authors, however, samples impacted by conical projectiles showed a higher degree of damage. They have offered an explanation for this type of behavior of the composite material as being the enhancement of the fiber strain energy at the point of impact for impacted samples with the conical projectiles, which caused the fiber fractures by releasing greater strain energy than the energy required to create a single crack plane resulting in crack bifurcation to dissipate the excess energy.

Bartus et al. also pointed out that impactor velocity did not influence the critical energy dissipation in the specimens, which is indicative of the fact that materials were not responsive to the loading rate in the range investigated. According to the authors, critical energy dissipation bears a linear relationship with respect to areal density for each projectile geometry. Besides, fiber orientation may also play a critical role in dissipating the energy. The energy dissipation capacity of the impacted long-fiber reinforced thermoplastics decreases as areal density decreases and as fiber orientation increases.

From their study, they concluded that the materials that exhibit a high degree of anisotropy, or rapid changes in fiber orientation, will be prone to impact damage. A random orientation most likely offers the maximum impact strength by creating a tortuous path for crack propagation. Finally, long-fiber-reinforced polyolefin composites with random fiber orientation possess greater damage tolerance due to crack blunting.

15.10.3 Woven Cloth Reinforcement

Woven fabric plays an important role in the modern art of living. It is widely used as reinforcement in automotive, wind turbine blades, marine building, aircraft, sports equipment, etc. Woven fabric composites offer more balanced properties and high impact resistance [62, 63]. However, woven fabric reinforced composites possess flexure of tow at the crossover point between tows. It is well known that for the fiber reinforced composites, stress concentration occurs at the crossover points. Thus, woven fabric composites show complex mechanical response and complex damage behavior under applied load. The mechanical stability of these composites is less as compared with cross-ply laminates [64–69] as the mechanical properties of woven fabrics are governed by various factors such as: (1) interlace parameters (i.e., architecture pattern, yarn size, yarn spacing length, fiber crimp angle, and volume fraction of fiber bundles); and (2) laminate parameters such as stacking orientation and overall fiber volume fraction. To reduce stress concentration and, as a result, to improve the mechanical properties of woven fabric composites, the appropriate design of the structure of those composites is highly recommended.

Jennifer [70] et al. reported about the glass-PP woven fabric composites. In their study they have discussed a friction model for thermostamping commingled glass-PP woven fabrics. In this research, they used Stribeck theory and the Hersey number calculations to consider the velocity and normal force of the tool as they have the greatest effect on the friction coefficient. They used the initial fabric temperature through the use of appropriate power-law parameters to calculate the viscosity of the composite. Because the tool temperature is lower than the melting point of the PP, it cannot be directly considered with the viscosity calculation. They have presented a fabric–friction model using a specific material and posited whether one can use this model for other fabric too. According to the authors, the final element model should be modified as the material model is refined. The authors mentioned that the results obtained from the parametric study indicated that with the increase of the velocity of the punch, the reaction force increased appreciably due to the increase in the friction force at all metal/fabric interfaces.

15.10.4 Hybrid Composites

A general discussion about hybrid composites was previously presented in Section 5.5.4. Poly olefin hybrid composites are discussed in this section. Hybrid composites may contain both natural and synthetic fiber or either of them. Many researchers have reported the hybrid composites using natural and synthetic fibers [71–73]. Comparatively high hydrophilic nature and low mechanical properties of natural fibers can be partly alleviated by hybridizing them with glass fibers. Several investigations have already been reported in the field of natural/synthetic fiber hybrid composites [74–77]. Studies indicate that an increase in glass fiber context resulted in an increase in improvement in properties of air glass hybrid composites. Thus, for example, 20% air hybridization brought about 130% increase in ultimate tensile strength, 60–90% improvement in flexural strength, 82% in flexural modulus, and 53% increase in impact strength [74–77]. Varma and co-workers [78] evaluated the enhancement in the properties of coir/polyester composites by incorporating glass well mixed with coir. They found that the addition of relatively small volume of glass (0.05) enhances the tensile strength by about 100%. They also observed that glass fiber addition reduces moisture uptake of coir fiber and enhances the resistance to weathering of coir fiber reinforced composites. Kalaprasad et al. [79] investigated the effect of three coupling agents, namely silane, titanate, and toluene di-isocyanate (TDI) on the mechanical properties of jute/glass hybrid composites. It is well understood that the way of hybridization with small quantity of synthetic fibers, high strength, high toughened, and long-term durable products can be made from natural fibers and polymers using advanced technology. Kalaprasad et al. demonstrated the hybrid effect in the mechanical properties of short sisal/glass hybrid fiber reinforced low-density polyethylene composites [79]. They equipped the hybrid composites with the low-density polyethylene (LDPE) reinforced with intimately mixed short sisal and glass fibers by solution mixing technique. The changes in the mechanical properties of LDPE with the incorporation of a semi-synthetic fiber, namely glass fiber, and natural fiber, namely sisal, have been discussed, stressing the influence of fiber orientation and composition by Kalaprasad et al. All the properties except elongation at break were found to increase volume from the action of glass fiber.

The hybrid effect was calculated using the additive rule of hybrid mixtures. A positive hybrid effect was observed in all the properties except for elongation at break, which indicates an indifferent deviation. The positive hybrid effect has been explained in terms of the increased fiber dispersion and orientation with an increase in volume fraction of glass fibers. The modification of sisal fiber by NaOH treatment increases the mechanical properties of longitudinally oriented composites. Again, Kalaprasad et al. [80] show the influence of short glass fiber addition on the mechanical properties of sisal reinforced low-density polyethylene. They reported on the enhancement in the mechanical fiber reinforced polyethylene composites by the incorporation of short glass fibers. For longitudinally oriented composites, the tensile strength has been found to increase by about 80% by the addition of quite a small volume fraction, approximately 0.03, of glass. The tensile strength of alkali treated sisal fiber containing composites shows an improvement of more than 90% by the addition of the same volume fraction of glass. The flexural strength is found to increase by about 60% for the same composition. It is also observed that water absorption tendency of the composite decreases by the process of hybridization.

15.10.5 *In Situ* Fiber Reinforcement

Today, the LCP incorporation into a thermoplastic matrix has been the subject of many research works. The *in situ* or self-reinforcing blends are very attractive due to the advantages over the glass fiber reinforcements because of the lower processing energy requirements and recyclability.

Tjong et al. [81] investigated the isothermal and non-isothermal crystallization of LCP/PP and compatibilized LCP/PP blends. They used maleic anhydride grafted PP (MA-g-PP) as compatibilizer. They observed that LCP phase does not react as a nucleating agent in the case of PP/LCP blend under isothermal crystallization. In the presence of MA-g-PP, the LCP domain acts as a heterogeneous nucleating agent for PP spherulites, followed by three-dimensional growths. They also reported that the PP crystallinity in the compatibilized system is slightly less than that of an uncompatibilized blend. According to the authors, the impact strength of the compatibilized system increases in comparison to the uncompatibilized blend.

Again, Farasoglou et al. [82] reported on the processing conditions and the compatibilizing effects on reinforcement of PP/LCP blends. They also used MA-g-PP as a compatibilizer. According to their observations, an optimum concentration of compatibilizer should be needed to achieve the maximum mechanical stability of the aforementioned blends. They offered an explanation for this kind of behavior from the PP/LCP blends (i.e., a specific range of MA-g-PP content exists, in which maximum fibrillation occurs, resulting in the best compatibility between the fiber and the matrix).

15.11 APPLICATIONS

One of the major fields of application for natural fiber reinforced poly olefin composites can be found in the automotive industry. The aerospace industry pays particular attention to high-temperature/humidity environments due to their deleterious effects on material properties. Thus, fiber reinforced poly olefin composites satisfy this need. Due to the good combination of properties, fiber-reinforced poly olefin composites are used particularly in the automotive and aircraft industries, and in the manufacturing of spaceships and sea vehicles. The aramid fiber reinforced poly olefin composites have huge applications in the boat industry, sports-car bodies, and in many secondary structures in the aircraft industry. Fiber reinforced poly olefin composites are becoming the obvious choice in structural applications such as aerospace, automotive, civil, and defense industries. In addition to their high specific strengths, their greatest advantage is the tailor-made capability, which allows the designer to teach directional strength and stiffness to the structural components of the system. Woven fabric poly olefin composites are widely used in automotive, wind turbine blades, marine building, aircraft, sports equipment, etc.

15.12 ADVANTAGES

Fiber reinforced poly olefin composites have their advantages and importance due to the outstanding mechanical properties, processing, and low cost. They offer tremendous thermal stability and are weather resistant.

Fiber reinforcing *in situ* poly olefin blends are very attractive due to the advantages over the glass fiber reinforcements because of the lower processing energy requirements and recyclability.

Kevlar fiber reinforced poly olefin composites provide high strength and good dimensional stability as well as high impact resistance of the resulting composites.

In situ poly olefin composites offer good rheological behavior, high compatibility, high crystallization, and good processing structure property relationship between the LCP and poly olefin matrix. The incorporation of LCP into the poly olefin matrix results in *in situ* fibrilization, which results in shear thinning viscosity at low shear rate and low melt viscosities for flow-induced orientation of LCP fibrils. Fiber reinforced poly olefin composites provide flexibility in selection and changing of styling and product aesthetic considerations. Natural fiber reinforced tooling is frequently 2–5 times cheaper than that of metals which significantly reduces amortization cost. Similar to metal fiber reinforced poly olefin, composites have indefinite shelf lives. The high torsional stiffness of various vehicles, particularly high-speed aircraft, can be satisfied by using fiber reinforced poly olefins. They possess outstanding corrosion resistance.

15.13 LIMITATIONS

The limitations of the fiber reinforced poly olefin are quite similar to those of the vinyl ester composites, as mentioned earlier. Major defects regarding the fiber reinforced poly olefin composites are associated with the manufacturing process of the composites. The defects can be categorized as

1. Incorrect state of matrix cure, which results most often from variation in local exotherm temperatures in thick or complex sections during autoclaving
2. Misaligned or broken fibers throughout the matrix

3. Non-uniform fiber distribution with resultant matrix rich regions
4. Mechanical damage, gaps or overlaps, or other faults in the fiber arrangement
5. Incorrect overall fiber volume fraction
6. Disbonded inter-laminar regions
7. Poor fiber matrix/interfacial matrix adhesion, which mainly depends on voids present at the fiber matrix interface.

REFERENCES

1. Brydson, J.A., The evolution of vinyl plastics, in *Plastics Materials*, 7th ed., Butterworth-Heinemann, 6, Ch. 1.
2. Chanda, M. and Roy, S.K., Industrial polymers, in *Plastic Technology Handbook,* 2nd ed., Marcel Dekker, Inc., New York, 1993, 427, Ch. 4.
3. Ghosh, P., Basic concepts of high polymer systems, in *Polymer Science and Technology (Plastics, Rubbers, Blends and Composites)*, 2nd ed., Tata McGraw-Hill Publishing Company Limited, New Delhi, India. 23, Ch. 1.
4. Bledzki, A.K. and Gassan, J., Composites reinforced with cellulose fibers, *Progress in Polymer Science*, 24, 2, 1999.
5. Milewski, J.V. and Katz, H. S., *Handbook of Reinforcement for Plastics*, Van Nostrand Reinhold Company, New York, 1987.
6. Sadov, F., Korchagin, M., and Matetsky, A., *Chemical Technology of Fibrous Materials*, Mir Publishers, Moscow, 1978.
7. Satyanarayana, K.G., Pai, B.C., Sukumaran, K., and Pillai, S.G.K., *Handbook of Ceramics and Composites*, Vol. 1, Chermisinoff, N.D., Ed., Marcel Dekker, Inc., New York, 1990, 339.
8. Mallick, P.K., *Fiber Reinforced Composites*, Marcel Dekker, Inc., New York, 1988.
9. Lopez-Anido, R. and Karbhari, V.M., Fiber reinforced composites in civil infrastructure, in *Emerging Materials for Civil Engineering Infrastructure: State of the Art*, Lopez-Anido, R. and Naik, T.R., Eds., Reston VA, ASCE Press, 2000, Ch. 2.
10. Podolny, W. Jr., Winds of change and paradigms of obsolescence, in *Proc. The National Steel Bridge Symposium*, Chicago, 1996.
11. Narayan, R. and Rowell, R.M., in *Emerging Technologies for Materials and Chemicals from Biomass*, Vol. 1, Schultz, T.P. and Narayan, R., Eds., ACS Symp., 476, 1992.
12. Mohanty, A.K., Misra, M., and Hinrichsen, G., *Macromol, Mater. Sci.*, 2000, 1, 276.
13. Cinelli, P., Chiellini, E., Lawton, J.W., and Imam, S.H., Properties of injection molded composites containing corn fiber and poly (vinyl alcohol), *J. Polym. Res.*, 13, 107, 2006.
14. Krupp, L.R. and Jewell W., *J. Environ. Sci. Technol.*, 26, 193, 1992.
15. Matsumura, S., Tomizawa, N., Toki, A., Nishikawa, K., and Toshima, K., *Macromolecules*, 32, 7753 (1999).
16. Otey, F.H., Mark, A.M., Mehltretter, C.L., and Russell, C.R., *Ind. Eng. Chem.*, 13, 90, 1974.
17. Wang, X.J.R., Gross, A., and Mc Carthy, S.P., *Environ. J. Polym. Degradation*, 3, 161, 1995.
18. Coffin, D.R., Fishman, M.L., and Ly, T.V., Thermomechanical properties of blends of pectin and poly (vinyl alcohol), *J. Appl. Polym. Sci.*, 61, 71, 1996.
19. Zhiqiang, L., Yi, F., and Xiao-Su, Y., Thermoplastic starch/PVAl compounds: preparation, processing, and properties, *J. Appl. Polym. Sci.*, 74, 2667, 1999.

20. Compston, P., Jar, P.-Y.B., Burchill, P.J., and Takahashi, K., The effect of matrix toughness and loading rate on the mode-II interlaminar fracture toughness of glass-fibre/vinyl-ester composites, *Composites Sci. Technol.*, 61, 321, 2001.
21. Irwin, G.R. and Kies, J.A., Critical energy rate analysis of fracture strength, *J. Welding*, 19, 193, 1954.
22. Mukherjee, M., Das, C.K., and Kharitonov, A.P., Fluorinated and oxy-fluorinated short Kevlar fiber reinforced ethylene propylene polymer, *Polym. Composites*, 27, 205, 2006.
23. Florin, R.E., Electron spin resonance spectra of polymers during fluorination, *J. Fluorine. Chem.*, 14, 253, 1979.
24. Kolpakov, G.A., Kuzina, S.I., Kharitonov, A.P, Moskvin, Yu, L., and Mikhailov, A.I., *Sov. J. Chem. Phys.*, 9, 2283, 1992.
25. Kuzina, S.I., Kharitonov, A.P., Moskvin, Yu, L., and Mikhailov, A.I., *Russ. Chem. Bull.*, 45, 1996, 1623.
26. Kharitonov, A.P., *Pop. Plast. Packag.*, 42, 75, 1997.
27. Lopez Manchado, M.A., Valentini, L., Biagiott, J., and Kenny, J.M., Thermal and mechanical properties of single-walled carbon nanotubes–polypropylene composites prepared by melt processing, *Carbon*, 43, 1499, 2005.
28. Wang, C., Lin, C.C., and Tseng, L.C., Miscibility, crystallization and morphologies of syndiotactic polystyrene blends with isotactic polystyrene and with atactic polystyrene, *Polymer*, 47, 390, 2006.
29. Marsh, G., Composites on the road to big time, *Reinf Plast.*, 47, 33, 38, 2003.
30. Saito, M., Kukula, S., and Kataoka, Y., Practical use of the statistically modified laminate model for injection moldings. Part 1: method and verification. *Polym. Composites*, 19, 497, 1998.
31. Bartus, S.D., Long-fiber-reinforced thermoplastic: process modeling and resistance to blunt object impact. A thesis, The University of Alabama at Birmingham, Materials Science and Engineering, December 2006.
32. Lee, S.M., Cheon, J.S., Im, Y.T., Experimental and numerical study of the impact behavior of SMC plates, *Composite Struct.*, 47, 551, 1999.
33. Thomason, J.L., and Vlug, M.A., Schipper, G., and Krikor, H.G.L.T., Influence of fibre length and concentration on the properties of glass fibre-reinforced polypropylene. Part 3: strength and strain at failure, *Composites: Part A*, 27, 1075, 1996.
34. Thomason, J.L. and Vlug, M.A, Influence of fibre length and concentration on the properties of glass fibre-reinforced polypropylene: 4. Impact properties. *Composites*, 28, 277, 1997.
35. Broyles, N.S., Verghese, K.N., Davis, E.R.M., Lesko, J.J., and Riffle, J.S., Pultruded carbon fiber/vinyl ester composites processed with different fiber sizing agents. Part I: processing and static mechanical performance, *J. Mater. Civ. Eng.*, 17, 320, 2005.
36. Verghese, K.N.E., Broyles, N.S., Lesko, J.J., Davis, R.M., and Riffle, J.S., Pultruded carbon fiber/vinyl ester composites processed with different fiber sizing agents. Part II: enviro-mechanical durability. *J. Mater. Civ. Eng.*, 17, 334, 2005.
37. Li, Y., *Mater. Manufacturing Process.*, 21, 181, 2006.
38. Akil, O., Yildirim, U., Güden, M., and Hall, I.W., Effect of strain rate on the compression behavior of a woven fabric S2-glass fiber reinforced vinyl ester composite, *Polym. Test.*, 22, 883, 2003.
39. Bunsell, A.R. and Harris, B., Hybrid carbon and glass fibre composites, *Composites*, 5, 157, 1974.
40. Summerscales, J. and Short, D., Carbon fibre and glass fibre hybrid reinforced plastics, *Composites*, 9, 157, 1978.
41. Short, D. and Summerscales, J., Hybrids — a review. Part 1: techniques, design and construction, *Composites*, 10, 215, 1979.

42. Herzog, B., Goodell B., Lopez-Anido, R., and Gardner, D.J., *Forest Prod. J.*, 55, 11, 2005.
43. Kiss, G., *Polym. Eng. Sci.*, 27, 4, 1987.
44. Bozell, J.J., Chemicals and materials from renewable resources, in *ACS Symp. Ser. 784*, American Chemical Society, Washington D.C., 2001.
45. Rowell, R.M., Schultz, T.P., and Narayan, R., Emerging technologies for materials and chemicals from biomass, in *ACS Symp. Ser. 476*, American Chemical Society, Washington, D.C., 1992.
46. Fishman, M.L., Friedman, R.B., and Huang, S.J., Polymers from agricultural co-products, *ACS Symp. Ser. 575*, American Chemical Society, Washington, D.C., 1994.
47. Chiellini, E., Cinelli, P., Antone, S.D., and Ilieva, V. I., *Polymer*, 47, 538, 2002.
48. Chiellini, E., Chiellini, F., and Cinell, P., in *Degradable Polymers—Principle and Applications*, 2nd ed., Scott G., Ed., Kluwer Press (in press).
49. Grass, J.E. and Swift, G., Agricultural and synthetic polymers. biodegradability and utilization, *in ACS Symp. Ser. 433*, American Chemical Society, Washington, D.C., 1990.
50. Doane, W.M., *J. Polym. Matter*, 11, 229, 1994.
51. Davies, P., Application of composites in a marine environment: status and problems, in *Durability Analysis of Structural Composite Systems*, Cardon, A. H., Ed., A.A. Balkema, Rotterdam, 1996.
52. Davies, P. and Lemoine, L., Eds., Nautical construction with composite materials, *Proc. 3rd IFREMER Conf.*, Paris, 1992.
53. Shenoi, R.A. and Wellicome, J.F., Eds. *Composite Materials in Marine Structures*, Cambridge Ocean Technology Press, Cambridge, 1993.
54. Dow Plastics, Derakane epoxy vinyl ester resins, technical product information.
55. Guiterrez, J., Le Ley, F., and Hoarau, P., A study of the aging of glass fiber resin composites in a marine environment: nautical construction with composite materials. *Proc. 3rd IFREMER Conf.*, Paris, 1992, p. 338.
56. Pham, S. and Burchill, P.J., Toughening of vinyl ester resins with modified polybutadienes, *Polymer*, 36, 3279, 1995.
57. Siebert, A.R., Guiley, C.D., Kinloch, A.J., Fernando, M., and Heijnsbrock, E.P.L., Elastomer-modified vinyl ester resins: impact and fatigue resistance, in *Toughened Plastics II: Novel Approaches in Science and Engineering*, Riew, C.K. and Kinloch, A.J., Eds., American Chemical Society, Washington, D.C.,1987.
58. Mohanty, S., Ph.D. Thesis, Central Institute of Plastic Engineering and Technology (CIPET) Bhubaneshwar, India, 2004.
59. Rowell, R.M., Tillman, A.M., Simonson, R.A., and Wood, J., *Chem. Tech.*, 6, 427, 1986.
60. Colum, X., Carrasco, F., Pags, P., and Canavate, J., Effects of different treatments on the interface of HDPE/lignocellulosic fiber composites, *Composites Sci. and Tech.*, 63, 161, 2003.
61. Bartus, S.D. and Vaidya, U.K., Performance of long fiber reinforced thermoplastics subjected to transverse intermediate velocity blunt object impact, *Composite Struct.*, 67, 263, 2005.
62. Mouritz, A.P., Bannister, M.K., Falzon, P.J., and Leong, K.H., Review of applications for advanced three-dimensional fiber textile composites, *Composites: Part A*, 30, 1445, 1999.
63. Rudd, C.D., Turner, M.R., Long, A.C., and Middleton, V., Tow placement studies for liquid composite moulding, *Composites: Part A*, 30, 1105, 1999.
64. Whitcomb, J., Kondagunta, G., and Woo, K., Boundary effects in woven composites, *J. Composite Mater.*, 29, 507, 1995.

65. Kuhn, J.L. and Charalambides, P.G., Elastic response of porous matrix plain weave fabric composites. Part I: modeling, *J. Composite Mater.*, 32, 1426, 1998.
66. Ishikawa, T. and Chou, T.W., Stiffness and strength behaviour of woven fabric composites, *J. Mater. Sci.*, 17, 3211, 1982.
67. Todo, M., Takahashi, K., Beguelin, P., and Kausch, H.H., Strain-rate dependence of the tensile fracture behaviour of woven-cloth reinforced polyamide composites, *Composite Sci. Technol.*, 60, 763, 2000.
68. Gao, F., Boniface, L., Ogin, S.L., Smith, P.A., and Greaves, R.P., Damage accumulation in woven-fabric CFRP laminates under tensile loading. Part 1: observations of damage accumulation, *Compos. Sci. Technol.*, 123, 59, 1999.
69. Fujii, T., Amijima, S., and Okubo, K., Microscopic fatigue processes in a plainweave glass-fibre composite, *Composite Sci. Technol.*, 49, 327, 1993.
70. Gorczyca-Cole, J.L., Sherwood, J.A., and Chen, J., *Composites: Part A*, 38, 393, 2007.
71. Mohan R., and Kishore, C., *Reinforced Plast. Composites*, 4, 186, 1985.
72. Pavithran, C., Mukherjee, P.S., and Brahmakumar, M., *Reinforced Plast. Composites*, 10, 71, 1991.
73. Shah, A.N. and Lakkad S.C., *Fiber Sci. Tecnol.*, 15, 41, 1981.
74. Prasad, S.V., Pavithran, C., and Rohatgi, P.K., *J. Mater. Sci.*, 18, 1443, 1983.
75. Varma, D.S., Varma, M., and Varma, I.K., *38th IUPAC Int. Symp. Macromolecules*, Amsterdam, 1985.
76. Varma, D.S., Varma, M., and Varma, I.K., *J. Reinforced Plast. Composites*, 4, 419, 1985.
77. Varma, M., Ph.D. Thesis, IIT, New Delhi, 1985.
78. Varma, I.K., Ananthakrishnan, S.R., and Krishnamurthy, S., Composites of glass/modified jute fabric and unsaturated polyester resin, *Composites*, 20, 383, 1989.
79. Kalaprasad, G., Thomas, S., Pavithran, C., Neelakantan, N.R, and Balakrishnan, S., hybrid effect in the mechanical properties of short sisal/glass hybrid fiber reinforced low density polyethylene composites, *J. Reinforced Plast. Composites*, 15, 48, 1996.
80. Kalaprasad, G., Joseph, K., and Thomas, S., Influence of short glass fiber addition on the mechanical properties of sisal reinforced polyethylene composites, *J. Composite Mater.*, 31, 5, 1997.
81. Tjong, S.C., Chen, S.X., Li, R.K.Y., Crystallization kinetics of compatibilized blends of a liquid crystalline polymer with polypropylene, *J. Appl. Polym. Sci.*, 64, 707, 1997.
82. Farasoglou, P., Kontou, E., Spathis, G., Gomez Ribelles, J.L., and Gallego Ferrer, G., Processing conditions and compatibilizing effects on reinforcement of polypropylene-liquid crystalline polymer blends, *Polym. Composites*, 21, 84, 2000.

16 Particulate-Filled Vinyl Polymer Composites

Chaganti S. Reddy, Ram N. Mahaling, and Chapal K. Das

CONTENTS

16.1 INTRODUCTION TO VINYL POLYMERS

Vinyl polymers are derived from *vinyl monomers* (i.e, small molecules containing carbon-carbon double bonds). Vinyl polymers are the largest family of polymers. Vinyl polymerization is the linking together of unsaturated compounds (monomers) to make chain polymers as indicated through the scheme:

$$n[CH_2{=}CHX] \rightarrow -[\text{-}CH_2 - CHX-]_n-$$

The α-carbon atom ($CH_2{=}CXY$) substituted vinyl monomers produce high-performance polymers but the substances which are substituted on both carbon atoms of the ethylinic bond ($RCH{=}CHR^1$) show reduced polymerization.

16.2 POLYMERIZATION ASPECTS OF VINYL POLYMER

The rate of polymerization and the molecular weight of the polymer produced are important features of polymerization. These are determined by the initiation reaction, which starts the growing polymer chain, the propagation reaction that leads to high molecular weight polymer, and the deactivation processes, termination, and transfer that produce the final stable polymer molecules. The three kinds of propagation are determined by the nature of initiation employed. Vinyl polymerization depends on the free radical, anionic, and cationic processes, which, in turn, depend on whether the unpaired electron of the propagating chain (terminal atom) carries a negative or positive charge.

In radical polymerization, the initiating species have no role in the propagation of the main chain, whereas in ionic systems, the counterion, which, owing to electrostatic attraction must remain in close proximity to the active center, may have a major influence on the polymerization. Ziegler–Natta catalysis and metallocene catalysis are the two common catalysts used for vinyl polymerization, although there is also a special class of ionic catalyst known as the *coordination catalyst.*

The polymerization is free radical or ionic, depending on the monomer and the nature of the electrolyte. The commodity plastics like polyethylene and polypropylene are classified as vinyl polymers along with polystyrene, polyvinyl chloride, poly(methyl methacrylate), etc. Teflon is also known as a vinyl polymer.

16.2.1 Historical Background of Vinyl Polymer

The world's most versatile plastic had a rather humble beginning: A rubber scientist during the early 1920s stumbled onto a new material with fantastic properties during his search for a synthetic adhesive. Waldo Semon was intrigued with his finding, and experimented by making golf balls and shoe heels out of the versatile material called poly(vinyl chloride) (PVC). Soon after his discovery, PVC-based products, such as insulated wire, raincoats, and shower curtains, hit the market. As more uses for vinyl were discovered, industry developed more ways to produce and process the new plastic [1].

The 1930s: The application of polymers such as PVC was limited to gaskets and tubing.

The 1940s: Joining industries across the world during the 1940s, PVC manufacturers turned their attention to assisting the war effort. Vinyl-coated wire was widely used aboard U.S. military ships, replacing wire insulated with rubber. Vinyl's versatility and flame-resistant properties led to several commercial uses after World War II.

The 1950s and 1960s: Five companies were exclusively involved in PVC at the century's midpoint, and innovative uses for vinyl continued to be found during the 1950s and 1960s. An advanced vinyl-based latex was used on boots, fabric coatings, and inflatable structures and also found applications in the building trades.

The 1970s: Vinyl products quickly became a staple of the construction industry; the plastic's resistance to corrosion, light, and chemicals made it a promising candidate for building applications. PVC piping was soon adapted for transporting water to thousands of homes and industries, aided by improvements in the material's resistance to extreme temperatures. Twenty companies were producing vinyl by 1980.

Today: Vinyl is among the largest-selling plastics in the world, and the industry employs more than 100,000 people in the United States alone. Vinyl's low cost, versatility, and performance make it the material of choice for a number of industries such as health care, communications, aerospace, automotive, retailing, textiles, and construction. Rigid as pipe or pliable as plastic wrap, vinyl is a leading material of the 21st century. Vinyl, or PVC, is the world's most versatile plastic; therefore, it is used to make everything from food wrap to autobody parts. Vinyl is made up of two simple building blocks: chlorine, based on common salt, and ethylene, from crude oil. The resulting compound, ethylene dichloride, is converted at very high temperatures to vinyl chloride monomer (VCM) gas. Through the chemical reaction known as polymerization, VCM becomes a chemically stable powder, PVC resin.

Vinyl is one of the most versatile polymers because of its recycling properties, which are essential for our day-to-day lives in a pollution-free environment. Even after a useful life span of decades, vinyl products can be recycled into new applications lasting decades more. Significant amounts of vinyl scrap (coming from polymer manufacturers and plastic processors) are collected and recycled. Overall, more than 99% of all manufactured vinyl compound ends up in a finished product, due to widespread post-industrial recycling. Post-consumer vinyl recycling continues to grow as an increasing number of recycling programs are equipped to handle vinyl bottles. The Vinyl Institute has supported the development of technology to automatically separate one plastic from another — a key to making plastics recycling successful. Several vinyl recycling systems have been developed around the nation by innovative companies with the encouragement of the Vinyl Institute. Researchers have found that vinyl can easily be separated from other plastics automatically because technology can spot its unique chlorine chemical composition. If you look into the vinyl recycling statistics: According to a 1999 study by Principia Partners, more than 1 billion pounds of vinyl were recovered and recycled into useful products in North America in 1997. About 18 million pounds of that was post-consumer vinyl diverted from landfills and recycled into second-generation products [2].

16.3 INTRODUCTION TO COMPOSITE MATERIALS

Composites are multiphase materials in which the different phases are artificially blended to attain properties that the individual components alone cannot attain. Generally, polymer–matrix composites are used for lightweight materials/structures (aircrafts, sporting goods, etc.) electronic enclosures, vibration damping, etc. [3]. Generally, composite materials are made up of filler (one or more) in a particular matrix. Fillers may be particulate or fibrous in nature. Polymer–matrix nanocomposites are a class of reinforced polymers comprised of low loading of nanometric-sized filler particles, which give them improved technical properties, such as tensile strength modulus, thermal, and barrier properties. Such properties have made them highly valuable in components such as panels and as barrier and coating materials in automobile, civil, and electrical engineering, as well as packaging.

Nanometer scale platy materials were exfoliated into polymers for better performance. The organic modifier compatibilizes the hydroxide surface to polymer matrices and spaces the crystalline layers apart to minimize the energy needed for exfoliation during the compounding process. The storage modulus of the polyethylene elastomer containing 5% organic modified MAH is better than that filled with 20% micron size talc particles. The performance of PP, ABS, Nylon 6,6, and PC nanocomposites will also be discussed [4].

A constant need exists for stronger, lighter, less expensive, and more versatile polymer composites to meet the demands of industrial consumers such as the automobile and aerospace industry. Polymer composites, such as glass or carbon fiber reinforced thermosets and thermoplastics, are very common. In addition to glass fibers, many other inorganic and organic materials, both natural and synthetic, are commonly used. Talc and other platy materials add rigidity, increase the heat deflection temperature, and reduce the thermal expansion of the polymer. Generally, the effectiveness of reinforcing fillers in composites is inversely proportional to the size and directly proportional to the aspect ratio of the filler. Therefore, the geometry of the particle is an important factor. The greater the surface to volume ratio of the filler, the greater will be the effectiveness of the filler. From a theoretical point of view, a small platy material with a high aspect ratio should yield the most effective reinforcing fillers. Typically, the dimension through the plate of a monolayer is measured in nanometers or tens of nanometers. Another factor in the effectiveness of reinforcing fillers is how well the material is dispersed in the polymer matrix. The more closely a monolayer dispersion is approached, the more effective the filler is in its reinforcing role and other benefits, such as barrier properties. A group at Toyota Research Laboratory [5] reported on monolayer dispersion of modified clay particles in nylon, with significant improvements in properties at low loading. One of the many platy particles that can be used as nanofillers is a class of compounds known as layered double hydroxides. The surfaces of layered double hydroxide materials are covered with hydroxyl groups that make the particles hydrophilic and allow the layers to hydrogen bond. Some chloride anions are also exchanged on the surface. When these anions are exchanged with fatty acids, sulfates, phosphates, or their salts, the hydrogen bonding is reduced and the surfaces are made more compatible to non-polar polymers. It is possible to achieve comparable composite properties at much lower loading of filler and/or achieve properties that are not obtainable with conventional reinforcing fillers. This report discusses the modification and dispersion of one type of platy material, the layered double hydroxide, into polyolefin matrixes and into the more polar polymers ABS, nylon, and PC [4].

Much research has been done in the development of the polymer composites and to study the bulk properties of a polymer. There is no significant effect without the additives. The most well-known examples are the rubber or elastomer composites, where variations in the choice of additives can produce such widely differing products as tires, battery boxes, o-rings, profiles, etc.

16.3.1 Additives

Most polymers, including vinyl, require additives during the manufacturing process. Specific additives such as colorants, heat and light stabilizers, impact modifiers, etc., are responsible for specific properties.

Plasticizers are used as softening agents and provide low temperature flexibility and weldability. Phthalate ester plasticizers have been safely used for more than 50 years. Plasticizers have a very wide range of applications starting with wire and cable products, toys, and synthetic leathers, as well as in pharmaceuticals and personal care products such as cosmetics and lotions. Their versatility and excellent performance prolongs the products' service lives, gives them resistance, and helps reduce spoilage and waste [2].

The structure of particulate-filled polymers appears to be simple. The homogeneous distribution of particles is responsible for the end-use properties as well as applications. The most important of these are aggregation and orientation of anisotropic filler particles. The first is related to the interactions acting in a particulate-filled polymer.

Particle–particle interactions induce aggregation, whereas matrix–filler interaction leads to the development of an interphase with properties different from those of both components. Both influence composite properties significantly. Secondarily, Van der Waals forces play a crucial role in the development of these interactions. Their modification is achieved by surface treatment. Occasionally, reactive treatment is also used, although it is less important in thermoplastics than in thermoset matrices. In the following sections of this chapter, attention is focused on interfacial interactions, their modification, and their effect on composite properties.

16.4 TYPES OF PARTICULATE FILLERS USED IN VINYL POLYMERS

The term filler is usually applied to solid additives incorporated into the polymer to modify its physical (usually mechanical) properties [6]. Air and other gases are considered fillers for cellular polymers. There are several types of fillers recognized in polymer technology, summarized in Figure 16.1. The properties of the matrix strongly influence the effect of filler on the composite's properties. The reinforcing effect of the filler increases with decreasing matrix stiffness.

Particulate fillers are generally of two types: "inert" fillers and reinforcing fillers. The term inert filler is something of a misnomer because many properties may be affected by the incorporation of such a filler. For example, in plasticized PVC compound, the addition of an inert filler reduces the die swell on extrusion, increases modulus with hardness, improves electrical insulation properties, and reduces tackiness. Inert fillers will also usually substantially reduce the cost of the compound. Among the fillers used are calcium carbonates, china clay, talc, and barium sulfate. For normal uses, such fillers should be quite insoluble in any liquids with which the polymer compound is liable to come in contact.

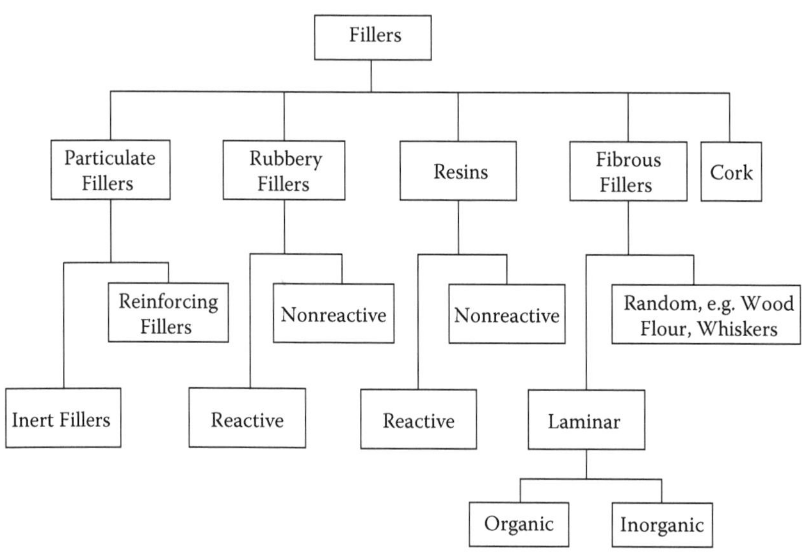

FIGURE 16.1 Fillers classification.

In an elastomer-based composites system, it is observed that the finer the particle size, the higher the values of such properties as tensile strength, modulus. and hardness. Again, this depends upon the optimum loading of filler. The particle shape also has an influence, such as in china clay, which is of a plate-like structure that tends to be oriented during processing to give products that are anisotropic.

Non-black fillers are mainly used in white or light colored compounds, where reinforced mechanical properties are required. The entire range of non-black fillers may be divided into the order of their reinforcing properties: activated silica (pyrogenic or precipitated silicas), activated aluminum silicate, activated calcium silicate, activated calcium carbonate, and magnesium carbonate.

16.5 PREPARATIVE METHODS OF PARTICULATE-FILLED COMPOSITES

In the field of nanofiller-based composites, the control of the morphology is essential to obtain the desired properties. As such, the processing is a key step, which was extensively studied. This determines if the fillers are actually dispersed at the nano level or not, and if the reinforcement can actually benefit from the "*nano-effect*" previously reported.

Four major routes can be followed for the processing of nanocomposites based on preformed inorganic particles: solution method, melt-mixing/melt-blending, sol-gel method, and *in situ* polymerization (see Figure 16.2). Preferential use of one over the other depends on the type and physical state of the polymer, and their possible ways of interactions with the fillers.

Additionally, the inorganic phase can be formed *in situ* in the polymer via the sol-gel process; the structures are then rather different.

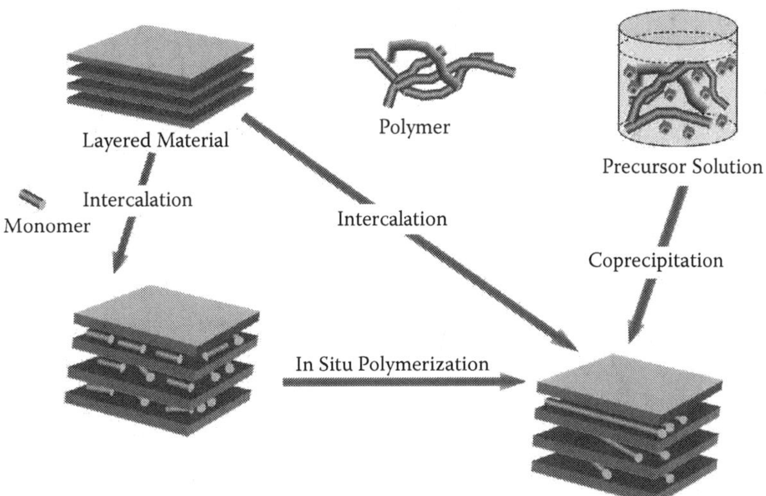

FIGURE 16.2 Different synthetic approaches for the preparation of polymer nanocomposites based on layered silicates.

16.5.1 Solution Method

The preparation in solution consists of the dispersion of the preformed fillers into a solvent (with possible adjunction of a compatibilizing agent), followed by the addition of the polymer soluble in the same solvent (i.e., thermoplastic or pre-polymer precursor of a thermosetting network). The solvent is then removed (critical stage). Its evaporation must be complete, otherwise it may damage the quality of the final material. Examples of the use of this technique can be found in the literature for the preparation of nanocomposites based on polyamide or polyethylene matrices.

16.5.2 Melt-Mixing/Melt-Blending

For thermoplastic polymers, the fillers can be introduced directly in the melt by using high shearing devices, with no significant modification of the traditionally used process. The introduction of layered fillers into polyolefin could be achieved using, for example, a twin-screw extruder with an optimized screw profile and the adjunction of a compatibilizing agent. The use of polypropylene grafted with a maleic anhydride of high molecular weight was reported as efficient for improving the layer dispersion [7–9].

This is an industrially accepted and economically viable method for large-scale production of polymer nanocomposites, especially polyolefin nanocomposites.

16.5.3 Sol-Gel Process

The sol-gel process was extensively studied with respect to the synthesis of epoxy-silica hybrids (i.e., *in situ* formed silica phase) [10,11]. Silane precursors of the silica phase are introduced into the organic medium. The most commonly used precursor in the literature is tetraethoxy silane (TEOS). The sol-gel process consists of successive hydrolysis and condensation leading to the formation of a glassy phase, which reinforces the polymer. The two phases are formed simultaneously, and the microstructure obtained results from the competition between the formation of the silica and the polymerization of the matrix. The size and morphology of the dispersed silica phase obtained depends on the order of incorporation of the components, their concentration, and the type of catalyst used: *cluster-cluster* growth regime under acidic conditions and *cluster-monomer* aggregation regime under basic conditions [12]. The microstructure of nanocomposites obtained by the sol-gel process is observed as significantly different from that of processes such as *melt-blending*, etc. (based on preformed particles).

16.5.4 *In Situ* Polymerization or Polymerization Filling Technique

In situ polymerization consists of mixing the preformed fillers with the monomers that act as reactive solvents, followed by the polymerization. The critical step in this process is the dispersion of the particles; different tools can be used as a function of the organic medium, its physical state, and viscosity. This route is used for the preparation of nanocomposites based on polyamides, polyester, epoxies, etc.

16.6 SURFACE MODIFICATION OF PARTICULATE FILLERS

Surface modifications of particulate fillers: This is an overview of particulate filler production and use. Fillers are used in polymers for a variety of reasons: cost reduction, improved processing, density control, optical effects, thermal conductivity, control of thermal expansion, electrical properties, magnetic properties, flame retardancy, and improved mechanical properties, such as hardness and tear resistance. For example, in cable applications, fillers such as metakaolinite are used to provide better electrical stability, whereas others, such as alumina trihydrate, are used as fire-retardants.

Each filler type has different properties and these in turn are influenced by the particle size, shape, and surface chemistry. Filler characteristics are discussed from costs to particle morphology. Particle specific surface area and packing are important aspects. Filler loading is also critical and this is discussed. The terminology used in this field is explained and, where appropriate, illustrated. Practical aspects of filler grading are described. For example, the use of an average particle size on data sheets can be misleading as it may not accurately reflect particle size distribution. Different measuring conditions can also give rise to variations in apparent particle size. The principal filler types are outlined. These include carbon black, natural mineral fillers, and synthetic mineral fillers. The use of clay in nanocomposites is outlined. Carbon blacks are very important fillers, especially in the rubber industry. Filler surface modification is an important topic. Most particulate fillers are inorganic and polar, which can give rise to poor compatibility with hydrocarbon polymers and processing problems, among other effects. The main types of modifying agent and their uses are described, from fatty acids to functionalized polymers.

Fillers are also discussed in relation to different polymer types. For example, in flexible PVC, because of the plasticizer, the filler has little effect on processing. This allows relatively high filler levels to be incorporated. Illustrations are included to explain concepts from microscopic filler structure to the effects of fillers on polymer properties.

To achieve high efficiency of particulate fillers in the development of high-performance polymer composites/vinyl polymer composites, surface modification is badly needed. This has been widely investigated in recent years. For better adhesion between the polymer and the filler, many factors are taken into account, such as polar-polar attractions.

Particle size or surface area is a factor of the greatest importance in reinforcement because it can vary over such a wide range. The area of interface between solid and elastomer per cubic centimeter of compound depends on the surface area per gram of the filler and also on the amount of filler in the compound. This introduces another significant factor to the already wide range of variation.

The nature of the solid surface may vary in a chemical sense, having different chemical groups (hydroxyl or metalloxide in white fillers), organic, carboxyl, and quinine, or lactone groups in carbon black, etc. In a physical sense, they may be different in adsorptive capacity and in energy of adsorption.

In polymers of a polar nature, such as PMMA, PVC, PVDF, PVF, etc., a stronger interaction will occur with filler surfaces showing dipoles such as OH groups or a chlorine atom. With the commodity plastics, no dramatic influences on reinforcement are noticeable when carbon black surfaces are chemically modified.

Chemical surface groups play an important role in their effective rate of cure with many vulcanizing systems. In specific cases, the properties are strongly affected by the concentration of hydroxyl or other oxygen-containing groups on the surface of carbon black. However, overall, the physical adsorption activity of the filler surface is of much greater importance than its chemical nature for the mechanical properties of the general-purpose polymer.

16.6.1 Factors Controlling the Properties of Particulate-Filled Polymers

The characteristics of particulate-filled polymers are determined by the properties of their components, composition, structure, and interactions [7]. These four factors are equally important and their effects are interconnected [13]. The specific surface area of the filler, for example, determines the size of the contact surface between the filler and the polymer, and thus, the amount of the interphase formed. Surface energetics influence the structures, the effect of composition on properties, as well as the mode of deformation. A relevant discussion on the previously mentioned properties and interactions in particulate-filled polymers cannot be performed without defining the role of all the factors that influence the properties of the composite and the interrelation among them. A relevant discussion on the previously mentioned four properties is elaborately discussed in the following sections.

16.6.2 Component Properties

The properties of the matrix strongly influence the effect of filler on the composite's properties. The reinforcing effect of the filler increases with decreasing matrix stiffness. In elastomers, true reinforcement takes place, and both stiffness and strength increases [14]. This is demonstrated well in Figure 16.3, where tensile yield stress of $CaCO_3$ composites is plotted against composition for two different matrices. LDPE is reinforced by the filler, whereas yield stress of PVC continuously decreases with increasing filler content [15]. For the sake of easier comparison, the data are plotted on a relative scale, related to the yield stress of the matrix. The direction of the change in yield stress or strength is determined by the relative load-bearing capacity of the components [16,17]. In weak matrices, the filler carries a significant part of the load; it reinforces the polymer. The extent of stress transfer depends on the strength of the adhesion between the components. If this is weak, the separation of the interfaces takes place even under the effect of small external loads [18,19]. With increasing matrix stiffness, the effect of interaction becomes dominant. The relative load-bearing capacity of the components is determined by this factor. In a stiffer matrix, larger stresses develop around the inclusions, and the probability of dewetting increases. In such matrices, dewetting is usually the dominating micro-mechanical deformation process. The structure of crystalline polymers may be significantly modified by the introduction of fillers. All the important aspects of the structure (e.g., crystallite and spherulite

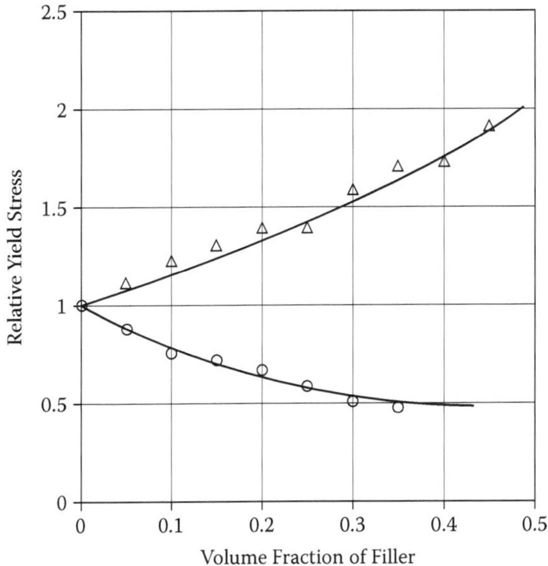

FIGURE 16.3 Effect of matrix properties on the tensile yield stress of particulate-filled composites: (O) PVC, (Δ) LDPE. Filler: $CaCO_3$, r = 1.8 μm.

size, and crystallinity change [on filling]) are altered because of nucleation [20]. A typical example of this is the extremely strong nucleation effect of talc in PP [21,22], which is demonstrated in Figure 16.4. The nucleating effect is characterized by the peak temperature of crystallization, which increases significantly upon the addition of the filler. Elastomer-modified PP blends are presented as a comparison; crystallization temperature decreases in this case. Talc also nucleates polyamides. Increasing the crystallization temperature leads to an increase in lamella thickness and crystallinity, whereas the size of the spherulites decreases on nucleation. Increasing nucleation efficiency results in increased stiffness and decreased impact resistance [23,24].

Numerous filler characteristics influence the properties of composites [25,26]. Chemical composition and, in particular, purity of the filler both have a direct and an indirect effect on its application possibilities and performance. A trace of heavy metal acts as contamination and decreases stability. Insufficient purity leads to discoloration. High purity $CaCO_3$ has the advantage of a white color, whereas the grey shade of talc-filled composites excludes them from some fields of application.

The thermal properties of fillers differ significantly from those of thermoplastics. This has a beneficial effect on productivity and processing. Decreased heat capacity and increased heat conductivity reduce cooling time [27]. Changing thermal properties of the composites results in a modification of the skin-core morphology of crystalline polymers and, thus, in the properties of injection molded parts as well. On the other hand, large differences in the thermal properties of the components leads to the development of thermal stresses, which also influence the performance of the composite under external load.

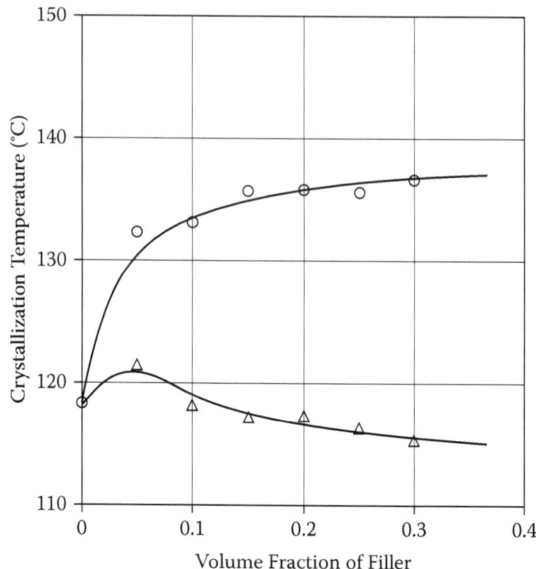

FIGURE 16.4 Nucleation effect of talc in PP: (O) talc, (Δ) EPR.

Although a number of filler characteristics influence composite properties, the particle size, specific surface area, and surface energies must again be mentioned here. All three also influence interfacial interactions. In the case of large particles and weak adhesion, the separation of the matrix-filler interface is easy. Debonding takes place under the effect of a small external load. Small particles form aggregates that cause deterioration in the mechanical properties of the composites. Specific surface area, which depends on the particle size distribution of the filler, determines the size of the contact surface between the polymer and the filler. The size of this surface plays a crucial role in interfacial interactions and the formation of the interphase.

16.6.3 Composition

Composition (i.e., filler content of composites) may change over a wide range. The effect of changing composition on composite properties is clearly seen in Figure 16.3. The interaction of various factors determining composite properties is also demonstrated by this figure. The same property may change in a different direction as a function of matrix characteristics, but interfacial adhesion may also have a similar effect. The goal of filling is either to decrease price or to improve properties (e.g., stiffness, dimensional stability, etc.). These goals require the introduction of the largest possible filler content, which, however, may lead to the deterioration of other properties. An optimization of properties must be performed during the development of composites. Numerous models are available, which describe how various properties of the composites vary with composition [13,26,28]. With the help of these models, composite properties may be predicted and an optimization can be performed.

16.6.4 Structure

The structure of particulate-filled polymers appears to be simple. In most cases, a homogeneous distribution of particles is assumed. This, however, rarely occurs and often special particle related structures develop in the composites. The most important of these are aggregation and orientation of anisotropic filler particles. The first is related to the interactions acting in a particulate-filled polymer.

16.6.5 Interfacial Interactions

Particle–particle interactions induce aggregation, whereas a matrix–filler interaction leads to the development of an interphase with properties different from those of both components. Both influence composite properties significantly. Secondarily, Van der Waals forces play a crucial role in the development of these interactions. Their modification is achieved by surface treatment. Occasionally, reactive treatment is also used, although it is less important in thermoplastics than in thermoset matrices. In the following sections of this chapter, attention is focused on interfacial interactions, their modification, and their effect on composite properties.

16.6.6 Modification of Interfacial Interactions

The easiest way for the modification of interfacial interactions is the surface treatment of fillers. The compound used for the treatment (coupling agent, surfactant, etc.) must be selected according to the characteristics of the components and the goal of the modification. The latter is very important because surface modification is often regarded as a magical tool that can solve all the problems of processing technology and product quality. It must always be kept in mind that in a particulate-filled polymer, two kinds of interactions take place: *particle–particle* and *matrix–filler* interaction. Surface treatment modifies both, and the properties of the composites are determined by the combined effect of the two. The types of material as well as the amount of the material used for the treatment must also always be optimized both from the technical and the economic aspect. In the following sections, surface modifications will be divided into four categories and will be discussed accordingly.

16.6.7 Non-Reactive Treatment — Surfactants

The oldest and most often used modification of fillers is the coverage of their surface with a small molecular weight organic compound [26,29]. Amphoteric surfactants are usually used. These have one or more polar groups and a long aliphatic chain. A typical example (Figure 16.5) is the surface treatment of $CaCO_3$ with stearic acid [26,29–31]. The principle of the treatment is the preferential adsorption of the polar group of the surfactant onto the surface of the filler. The high-energy surfaces of inorganic fillers can often develop special interactions with the polar groups of the surfactants. Preferential adsorption is promoted, to a large extent, by the formation of ionic bonds between stearic acid and the surface of $CaCO_3$. Nevertheless, in other cases, hydrogen or even covalent bonds can also form. Surfactants diffuse to the surface of the filler even from the polymer melt, which is further proof for the preferential adsorption [32]. Because of their polarity, reactive coupling agents also

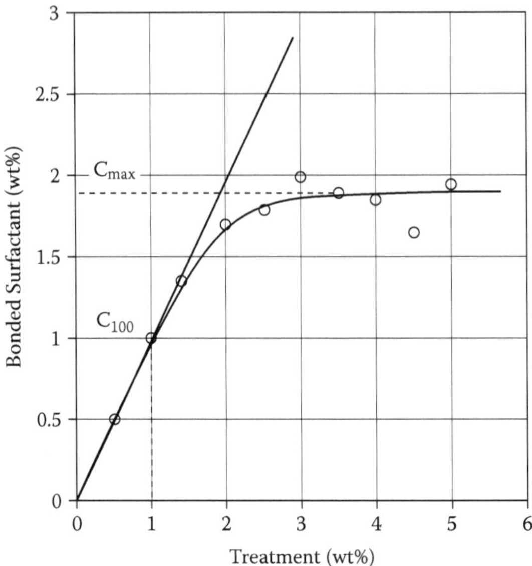

FIGURE 16.5 Typical dissolution curve of stearic acid adsorption on the surface of a $CaCO_3$ filler: r = 1.8 μm.

adsorb onto the surface of the filler. If lack of reactive groups does not make possible chemical coupling with the polymer, these exert the same effect on composite properties as their non-reactive counterparts [33–35].

One of the crucial questions of non-reactive treatment, which is very often neglected, is the amount of surfactant to be used. It depends on the type of the interaction, the size of the treating molecule, its alignment to the surface, and on some other factors. Determination of the optimum amount of surfactant is essential for the efficiency of the treatment. An insufficient amount of surfactant does not bring about the desired effect, whereas excessive amounts lead to processing problems as well as to the deterioration of the mechanical properties and appearance of the product [36,37].

The specific surface area of the filler is an important factor that must be taken into consideration during surface treatment. The proportionally bonded surfactant depends linearly on it [36]. ESCA studies performed on the surface of a $CaCO_3$ filler covered with stearic acid have indicated that ionic bonds form between the surfactant molecules and the filler surface, and that the stearic acid molecules are oriented vertically to the surface [36]. These experiments have demonstrated the importance of both the type of interaction and the alignment of surfactant molecules to the surface. A further proof for the specific character of surface treatment is supported by the fact that talc and silica adsorb significantly smaller amounts of stearic acid per unit surface area than $CaCO_3$. The lack of specific interaction in the form of ionic bond formation results in a significantly lower amount of proportionally bonded molecules [36]. As a result of the treatment at the surface, free energy of the filler decreases drastically [29,38].

As an effect of non-reactive treatment, particle–particle interaction and the matrix–filler interaction decreases. The consequence of this change is a decrease in

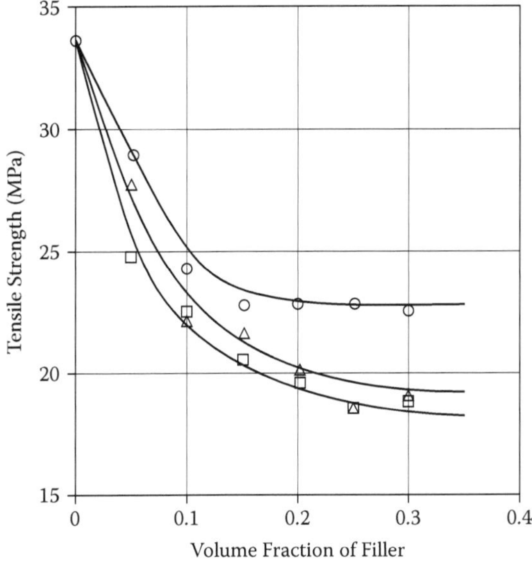

FIGURE 16.6 Effect of non-reactive surface treatment of a $CaCO_3$ filler with stearic acid on its interaction with PP: (O) non-treated, (Δ) 75%, (□) 100% surface coverage, r = 1.8 μm.

the yield stress and strength as well as improved deformability [29,30]. This is demonstrated by Figure 16.6, which illustrates the decrease of tensile strength of PP-$CaCO_3$ composites with increasing surface coverage of the filler. Adhesion and strong interaction, however, are not always necessary or advantageous to prepare composites with the desired properties. Plastic deformation of the matrix is the main energy absorbing process during an impact, which decreases with increasing adhesion [39–41].

16.6.8 Reactive Treatment — Coupling Agents

Reactive surface treatment involves chemical reaction of the coupling agent with both the polymer matrix and filler. The considerable success of silanes in glass reinforced thermosets has led to their application in other fields. They are used, or at least experimented with, in all kinds of composites irrespective of the type, chemical composition, or other characteristics of the components. Reactive treatment, however, is even more complicated than non-reactive treatment. Polymerization of the coupling agent and development of chemically bonded and physisorbed layers render the identification of surface chemistry, characterization of the interlayer, and optimization of the treatment very difficult [42]. Despite these difficulties, significant information has been collected on silanes [43]; much less is known about the mechanism and effect of other coupling agents such as titanates, zirconates, etc. [26].

Silane coupling agents are successfully used with fillers and reinforcements, which have reactive –OH groups on their surface (e.g., glass fibers, glass flakes and beads, mica, and other silicate fillers) [44–46]. Use of silanes with fillers, such as $CaCO_3$, $Mg(OH)_2$, wood flour, etc., were tried, but proved to be unsuccessful in most cases [47,48]. Sometimes, contradictory results were obtained with glass and other siliceous fillers [7]. Acidic groups are preferable for $CaCO_3$, $Mg(OH)_2$, $Al(OH)_3$, and $BaSO_4$. Talc cannot be treated successfully either with reactive or non-reactive agents because of its inactive surface. Only broken surfaces contain a few active –OH groups. The chemistry of silane modification has been extensively studied and described in the literature [42,43]. Model experiments have indicated that a multilayer film forms on the surface of the filler; the first layer is chemically coupled to the surface, this is covered by crosslinked silane polymers, and the outer layer is a physisorbed silane. The matrix polymer may react chemically with the coupling agent, but interdiffusion of the matrix and the polymerized silane-coupling agent (i.e., physical interaction) also takes place [49,50].

Recent studies have indicated that the adsorption of organofunctional silanes is usually accompanied by polycondensation. The high-energy surface of non-treated mineral fillers bonds water from the atmosphere, which initiates the hydrolysis and polymerization of the coupling agent [51,52]. Although the chemistry of silane modification of reactive silica fillers is well documented, much less is known about the interaction of silanes with polymers. It is more difficult to create covalent bonds between a coupling agent and a thermoplastic, because the latter rarely contains reactive groups. The most reactive are the polycondensation polymers (i.e., polyesters or polyamides), whereas transesterification reactions may lead to coupling. Literature references indicate some evidence of reactive coupling. Indeed, the strength of polyamide and polycarbonate composites increases on aminosilane treatment [52–54]. The increased strength is always interpreted as an improvement of properties. However, the change in deformability and impact characteristics are usually not mentioned; they frequently decrease as an effect of reactive treatment, as has also been shown [54].

Although one would expect that coupling occurs through the reaction of free radicals as suggested by Widmann et al. [54], the latest results indicate that PP oxidizes during processing, even in the presence of stabilizers, and the formed acidic groups react with aminosilanes leading to reactive coupling [55]. Figure 16.7 is the linear plot of relative yield stress against the volume fraction of the filler for three different treatments in PP-$CaCO_3$ composites. It is obvious that stearic acid acts as a surfactant, whereas the aminosilane is a reactive coupling agent.

Considering the complexity of the chemistry involved, it is not surprising that the amount of coupling agent and surface coverage also have an optimum here, which is similar to the surfactants in non-reactive surface treatment. This has been proven by Trotignon et al. [45], who found a strong increase and a maximum in the strength of PP-mica composites as a function of the silane used. Optimization of the type and amount of coupling agent is also crucial in reactive treatment, and although “proprietary” treatments may lead to some improvement in properties, they might not be optimal or cost effective.

The improper choice of coupling agent may lead to insufficient or even detrimental effects. In some cases, hardly any change in the properties are observed, or the effect can be clearly attributed to the decrease of surface tension due to the coverage of the filler surface by an organic substance (i.e., non-reactive treatment) [34, 35].

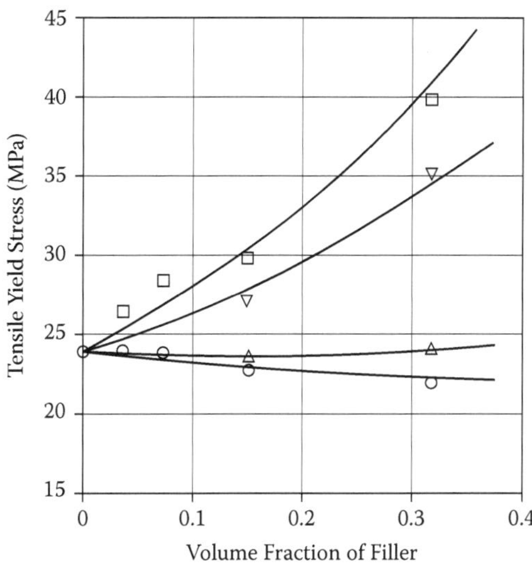

FIGURE 16.7 Effect of interdiffusion of the functionalized polymer with the matrix on the mechanical properties of PP-cellulose composites. Molecular weight of MA-PP: (O) non-treated, (Δ) 350, (∇) 4500, (□) 3.93×10^4.

16.6.9 Polymer Layer — Interdiffusion

Numerous results indicate that physical adsorption of the polymer on the filler surface leads to the development of a rigid interphase. Chemical coupling with a single covalent bond has the same effect. The coverage of filler surface with a polymer layer, which is capable of interdiffusion with the matrix, proved to be very effective both in stress transfer and in forming a thick diffused interphase with acceptable deformability. In this treatment, the filler is usually covered by a functionalized polymer, preferably by the same polymer as the matrix. The functionalized polymer is attached to the surface of the filler by secondary, hydrogen, ionic, and sometimes by covalent bonds. The polymer layer interdiffuses with the matrix. Entanglements are formed and strong adhesion is created. Because of its increased polarity, in some cases, reactivity-, maleic-anhydride, or acrylic-acid-modified polymers are used, which adsorb on the surface of most polar fillers, even from the melt. Most frequently, this treatment is used in polyolefin composites because other treatments often fail, and functionalization of these polymers is relatively easy. Often, a very small amount of modified polymer is sufficient to achieve significant improvement in stress transfer [56,57].

The importance of interphase interdiffusion and entanglement density is clearly demonstrated by the experiments of Felix and Gatenholm [58], who introduced maleic anhydride modified PP (MA-PP) into PP-cellulose composites and achieved an improvement in the yield stress. The increase was proportional to the molecular weight of MA-PP as depicted in Figure 16.7. Increased adhesion is clearly shown by

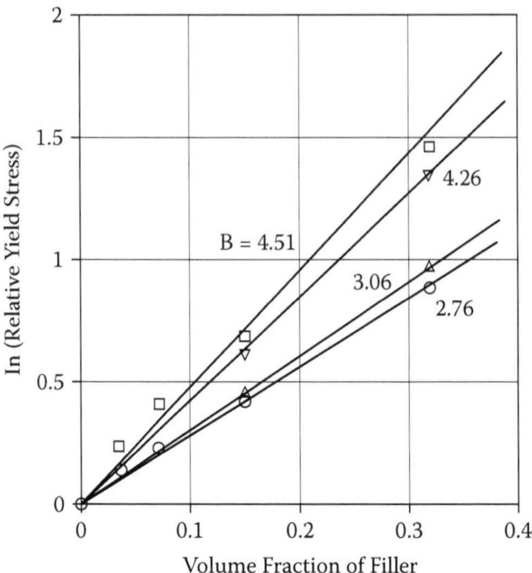

FIGURE 16.8 Increased adhesion due to the coverage of the filler with a functionalized polymer: (O) non-treated, (Δ) 350, (∇) 4500, (□) 3.93 × 10^4.

the increase in parameter B, which reflects the strength of the interaction in this case as presented in Figure 16.8. Model calculations have demonstrated that the improvement of properties has an upper limit and a plateau is reached at a certain, but not too high, molecular weight. The maximum effect of functionalized PP was found with fillers of high-energy surfaces [7–9], with those capable of specific interactions (e.g., ionic bond with $CaCO_3$ [57,59], or chemical reactions with wood flour, kraft lignin, or cellulose [56,58].

16.6.10 Soft Interlayer — Elastomers

Incorporation of hard particles into the polymer matrix creates stress concentration, which induces local micro-mechanical deformation processes. Occasionally, these might be advantageous for increasing plastic deformation and impact resistance, but usually they cause deterioration in the properties of the composite. Encapsulation of the filler particles by an elastomer layer changes the stress distribution around the particles and modifies the local deformation processes. Encapsulation can take place spontaneously. It can be promoted by the use of functionalized elastomers, or the filler can be treated in advance.

The coverage of the filler with an elastomer layer has been studied mostly in PP composites, but occasionally in other polymers as well. PP has a poor low-temperature impact strength, which is frequently improved by the introduction of elastomers [60]. Improvement in impact strength, however, is accompanied by a simultaneous decrease of modulus, which cannot be accepted in certain applications; a filler or reinforcement is added to compensate the effect.

Although most of the literature dealing with these materials agrees that the simultaneous introduction of two different types of materials (elastomer, filler) is beneficial, this is practically the only similarity that can be found. A large number of such systems were prepared and investigated, but the observations concerning their structure, the distribution of the components, and their effect on composite properties are rather controversial. In some cases, separate distribution of the components and independent effects were observed [61,62]. In other cases, the encapsulation of the filler by the elastomer was more evident [63,64]. Despite its importance, the structure of the composites is not always clear [65,66]. These different structures naturally lead to dissimilar properties as well. Composition dependence of shear modulus agrees well with the values predicted by the Lewis–Nielsen model in the case of separate dispersion of the components, whereas large deviations were observed when filler particles are embedded into the elastomer [67]. Embedding of the filler particles was also observed in the dynamic mechanical spectra of the composites. Here, the encapsulated filler particles stretched the elastomer, leading to an increased relaxation peak maximum at approximately –50°C. Model calculations have indicated that a maximum of 70% of the filler particles could be embedded into the elastomer, whereas, even in the cases when separate dispersion of the components dominates, at least 5–10% of the particles were encapsulated. The results indicate that the mere combination and homogenization of the components cannot achieve complete encapsulation or separate dispersion.

Extensive investigations [62,68,69] have indicated that the final structure is determined by the relative magnitude of adhesion and shear forces during the homogenization of the composite. Similar principles can be used as for the estimation of aggregation, because the same forces are effective in the system. If shear forces are larger than adhesion, separate distribution of the components occurs, whereas strong adhesion leads to an embedded structure. The three parameters that influence structure are adhesion (W_{AB}), particle size (r), and shear rate ($.\gamma$). Fillers usually have a wide particle size distribution — very often in the 0.3-10 μm range. Consequently, composites will always contain small particles, which are encapsulated, and large particles, which are separately distributed [68].

All the experiments performed on such systems have demonstrated the primary importance of adhesion in structure formation and in the resulting effect on properties. Numerous attempts have been made to obtain composites with exclusive structures (i.e., *complete coverage* or *separate distribution*). In most cases, PP or elastomers modified with maleic anhydride or acrylic acid were used to enhance adhesion between selected components; separate dispersion in the case of MAPP, embedding with MA-EPDM [8,9,69,70]. Similar contradictions appear in the case of some of the properties, especially with respect to impact resistance. In three-component PE composites, a better impact resistance-modulus ratio could be achieved, but Jancar and Kucera [59] observed lower impact strength with AA-EPDM than with a non-modified elastomer. Although the relationship of structure and impact resistance has not yet been determined unambiguously, experiments performed on a series of PP bumper compounds containing a filler and an elastomer have proven that a simultaneous increase in impact resistance and stiffness can be achieved by the introduction of a separate elastomer phase (see Figure 16.9). Such

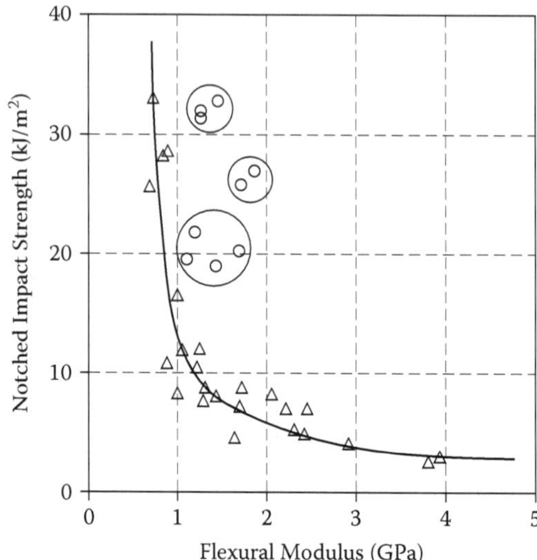

FIGURE 16.9 Improved stiffness and impact resistance in PP composites containing a filler and an elastomer: (Δ) usual behavior, (O) improved properties.

composites show a positive deviation from the usual, inverse correlation of stiffness, and impact resistance.

In today's scenario, most of the research is focused on nanofillers instead of macroscopic fillers. This is to improve the intrinsic characteristics of polymeric resins at a much lower loading level. The incorporation of lighter nanofillers in a polymer matrix is one way to develop lightweight and high-strength materials.

16.7 INFLUENCE OF PARTICLE SIZE FROM MICRO-NANO METER SCALE

Polymer nano-composite is a rapidly developing topic with attractive results for different applications. The weakness of the polymer could be improved through the use of nano-fillers as compared with micro-fillers. Inorganic particles are well known for the enhancement of the mechanical and tribological properties of polymers, and this issue has been widely investigated in the past decades. It has been found that the particle size plays important role in the improvement of the wear resistance, tensile, modulus, thermal, as well as mechanical properties. Reducing the particle size to the nanoscale level is assumed to significantly improve the composite efficiency; nanoparticle-filled polymers, the so-called polymer nanocomposites, are very promising materials for various applications. Polymer nanocomposites are characterized by their huge interfacial surface area, which may result in a peculiar physical network structure of three-dimensional interphase.

Researchers have used several studies to enhance the modulus and hardness of polymers, from micro- to nanoscale. Furthermore, by either diminishing the particle size or enhancing the particle volume fraction, the strength can be improved. In some cases, however, the fracture toughness and modulus remain fairly independent of the particle size [71], even when going down to the nanoscale [72–74]. This applies to the field of vinyl polymer composites in particular, which has a wide range of applications and the maximum amount of plastics consumed worldwide. The filler used in the polymer industry worldwide is 20–30% loading, but the same properties can be achieved by the use of 3–5% loading, if it is in nanoscale.

16.7.1 Nanofiller Reinforcement

The difference between the behaviors of micro- and nano-reinforced polymers can be analyzed by observing the specific changes in properties in nanoscale. In nanoscale, polymer chain lengths approach the filler dimensions so that they display particular interactions influencing the macroscopic behavior of the materials. The large number of parameters involved can have a synergistic action. Indeed, many parameters have been taken into account for the reinforcement efficiency of filler into a given medium [75].

- Chemical nature of the fillers
- Shape and orientation of the fillers
- Average size, size distribution, and specific area of the particles
- Volume fraction
- Dispersion state
- Interfacial interactions
- Respective mechanical properties of each phase

These parameters, detailed in the following paragraphs, act either in a competitive or in a cooperative way, and their effects cannot always be separated. Thus, they can complicate the understanding of the mechanisms involved for predicting the final material behavior.

16.7.2 Specific Effect of Nanofillers

As it is now well known, nanometric fillers are expected to improve the properties of materials significantly more, even at lower loading than conventional micro-fillers. This is due to:

1. The size effect: large number of particles with smaller interparticle distances and high specific surface area results in larger interfacial area with the matrix
2. The interactivity and/or potential reactivity of the nanofillers with the medium
3. The possible structural organization of the nanofillers

Furthermore, the requisite properties are achievable for low filler contents. This allows the conservation of the processing route used for neat systems, which is a substantial advantage in an industrial context.

To highlight the difference between micro- and nanofillers concerning the effect of specific surface area, let us quote the example of a precipitated silica that typically displays a specific surface area lying between 50 and 400 m^2/g, whereas natural silica micro-particles have a specific surface area in the range of 2 m^2/g [76].

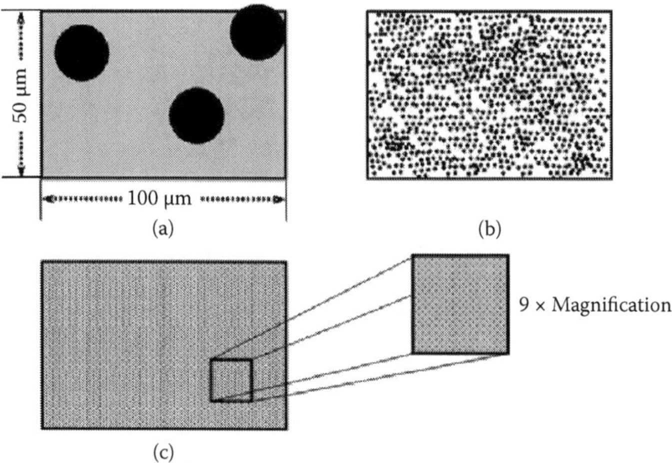

FIGURE 16.10 A schematic representation of dispersion of 10 μm (a), 1 μm (b), and 1 nm (c) particles in a constant volume — observation from SEM.

Figure 16.10 illustrates the effect of miniaturization of the fillers on the surface area, on the number of particles, and on the interparticle distance: As the size decreases, the number of particles per unit surface area and the surface area of the filler increases. These trends are obviously observed only in the case of individually dispersed particles, which is quite challenging to achieve.

The increased interfacial area and the reduction of the interparticle distance leads to a modification of the structure of the polymer in the neighborhood of the filler, which can substantially modify the properties of the polymer even at lower filler content. The reduction of the mobility of the polymer chains leads to the formation of boundary layers, where the properties are really different from that in the bulk [77,78], and an increase of the crystallinity of semi-crystalline polymers, because particles act as nucleating agents [79].

The apparent volume fraction of filler appears to increase because (1) the immobilized stiff interface plays an important role, and (2) because of the occluded polymer chains between the fillers, making it sometimes possible to reach the percolation threshold at very low filler levels.

16.7.3 Ability of Interactions/Reactions of the Nanofillers with the Medium

A direct consequence of the large surface area of nanofiller is the large interfacial zone where physico-chemical interactions can take place between the polymer and filler. Indeed, the silanol content is the same for any given silica specific surface area (1.8 nm^2, for surface area between 50 m^2/g and 400 m^2/g). Thus, with increasing specific surface area, the number of sites for grafting or interaction with the medium increases markedly. This can also result in a modification of the kinetics of the reactions.

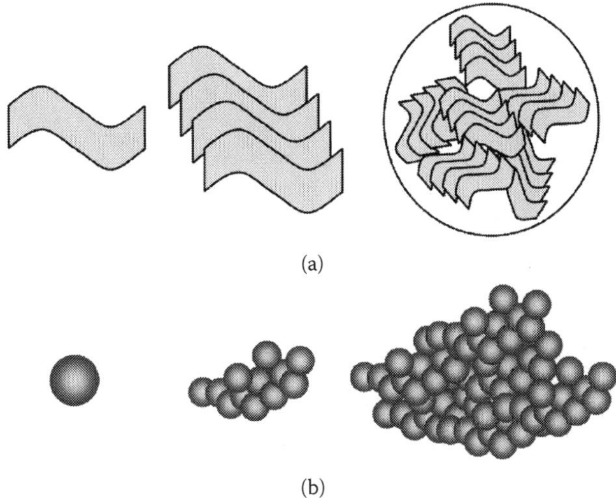

FIGURE 16.11 Similarities in multi scale organization of structural nanofillers: (a) montmorillonite, (b) nanosilica.

It is possible to balance the interparticular versus particle–polymer interactions, or reactions, via surface modification of the fillers. In case the particle is covered by reactive groups, they may act as additional crosslinking sites in thermosets, with a high functionality.

16.7.4 Structural Organization

The structural organization of the fillers is well known for montmorillonites (Figure 16.11). For fumed silica/precipitated silica, the processing induces the three-dimensional structural organization: primary particles gathered in individual aggregates that can form agglomerates.

In montmorillonite-based composites, the large-scale arrangement of the platelets into the polymer matrix is traditionally described as *exfoliation* (or *delamination*), *intercalation*, or *tactoids* (also referred to as conventional composites when the dispersion is not actually achieved at the nanoscale) with decreasing levels of dispersion (Figure 16.12). A complete exfoliation is generally difficult to achieve in thermoplastic formulations, such as polyolefins. The morphology obtained depends on the efficiency of the processing and on the organophillic surface modification of the montmorillonites, which increases the interactions with the polymer [80].

The properties of the final material, including the viscosity of the suspensions, can be closely related to the possibility of structural organization of the fillers.

16.7.5 Geometric Characteristics

Table 16.1 compares typical values of the geometric characteristics of some of the commonly used nanofillers. In all the cases, at least one dimension exists in the scale of a few nanometers, but the number, *d*, of dimensions in this scale varies depending

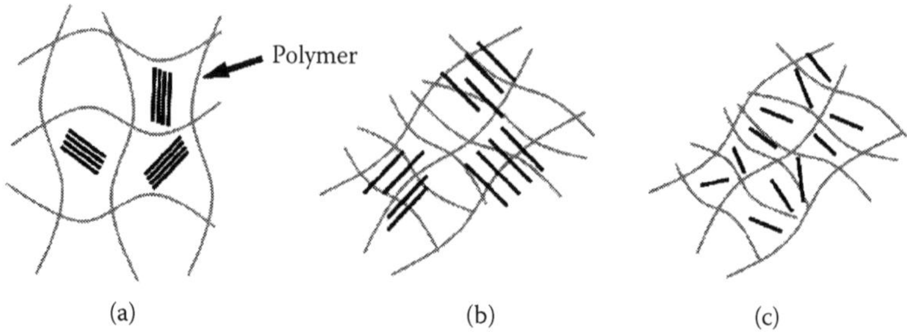

FIGURE 16.12 Traditional morphologies of montmorillonites-based polymer nanocomposites: (a) conventional, (b) intercalated, and (c) exfoliated/delaminated.

on whether the filler displays a spherical ($d = 3$), fibrillar ($d = 2$), or lamellar ($d = 1$) geometry.

However, the values given in Table 16.1 for the shape factor and size of fumed silica are those of the primary particles. In reality, isolated primary particles do not exist, and, therefore, one should consider an aggregate as an elemental object accounting for the effect of fumed silica. Due to the fractal geometry exhibited, the geometric parameters defined using Euclidian geometry is insufficient.

Generally, the higher the shape factor, the lower the percolation threshold for the same geometry of particles. Lamellar fillers, such as montmorillonites, exhibit a shape factor typically in the range 600–1000. It is difficult to compete with their efficiency, especially when properties such as gas barrier are targeted. However, the

TABLE 16.1
Geometric Characteristics of Various Nanofillers

Nanofiller	Geometry of the Nanofiller	Characteristic Dimensions	Shape Factor (L/Φ or L/T)	Specific Surface Area (m²/g)
Carbon black	Spherical	Φ = 250 nm	1	7–12
Carbon nanotubes	Tube	Φ = 1–50 nm L =10–100 μm	>1000	—
Cellulose whiskers	Fibrillar	Φ = 10–20 nm L =1 μm	100–150	150
Graphite	Lamellar	L = 6–100 μm	6–100	6–20
Montmorillonite	Lamellar	L = 0.6–1 μm e =1 nm	600–1000	700–800
Precipitated/ Sol- gel silica	Spherical	Φ = 5–40 nm	1	50–400
Talc	Lamellar	L = 1–20 μm e =0.2–6 μm	5-20	2–35
TiO_2	Spherical	Φ = 8–40 nm	1	7–160

"*tortuosity*" effect involved is strongly affected by the exploitation degree and the orientation of the fillers that can be controlled by the process.

In a similar way, an increase in the length of crack propagation due to the deviation from the fillers induces an increase of the surface developed and of the energy consumed by fracture. This generally enhances the fracture toughness.

Blending organic polymeric materials with inorganic filler on a molecular scale can form nanocomposites. In this process, the components are combined with at least one dimension in the range of 1—100 nm [81–87]. However, when the dimension of the inorganic reinforcement approaches nanoscale [88–90], the difficulty of homogeneously dispersing the inorganic part in the organic matrix dramatically increases, owing to the lack of affinity between the inorganic fillers and the organic polymers. Thus, efforts on compensation for this flaw are urgent and important for the development of new nanocomposites. This is due to an agglomeration of nanoparticles and the high melt viscosity of the matrix. To overcome these limitations, an initial strategy has been proposed by Hausslein and Fallick [91], which is based on the filler encapsulation by a polymer coating. A second approach relies upon the chemical modification of the filler surface by functional silanes and titanate esters, which are able to promote adhesion to the polymer matrix [92–97]. Moreover, the high cost of the functional silane compounds also limits their applicability in mass production.

A research team led by Chapal Kumar Das focused on surface functionalization of nanofillers, especially nanosilica, with low-cost organic functional modifiers for the development of polymer nanocomposites, especially polyethylene, polypropylene, and the copolymers of ethylene and propylene, by adopting the economically and industrially viable *melt blending* method.

16.8 SURFACE FUNCTIONALIZATION OF NANOSILICA WITH DIGLYCIDYL ETHER OF BISPHENOL-A (EPOXY RESIN)

The reaction of silica particles with epoxy resin (DGEBA), an organic functionalizing agent, was performed as follows. Nanosilica (NS) particles were suspended in methyl isobutyl ketone solvent. Epoxy resin was then added to the resulting solution. The weight ratio of nanosilica and epoxy resin was taken as 40:60. After adding 1000 ppm of $SnCl_2$ as a catalyst to this suspension, it was introduced into a three-necked, round-bottomed flask equipped with a mechanical stirrer, a water condenser, and a thermometer. The reaction was performed at 140°C for 2 h. The solvent was then removed with a rotary evaporator, and the product was dried in a vacuum oven for 1 h and used further as a surface functionalized nanofiller for the synthesis of novel nanocomposites based on polyolefins.

It is well known that NS is hydrophilic in nature, and the surface of NS particles possess three types of silanol groups: vicinal, geminal, and isolated silanol groups (Si-OH). The high bond strength of Si-O renders the surface of silica too acidic in nature and, as such, highly reactive toward Lewis bases. Figure 16.13 depicts the IR spectra of NS and epoxy resin (DGEBA) functionalized nanosilica (ENS).

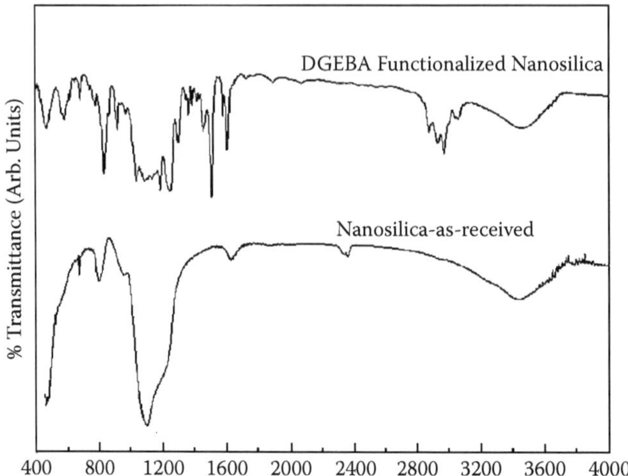

FIGURE 16.13 IR spectra of nanosilica and epoxy resin (DGEBA) functionalized nanosilica.

NS exhibits a strong IR band at 1000-1100 cm^{-1}, corresponding to the Si-O-Si stretching and a broad IR band at about 3436 cm^{-1}, corresponding to the O-H stretching of surface silanol groups. The spectrum of ENS in comparison to the spectrum of NS shows new stretching vibration bands at 1178 cm^{-1}, 1234 cm^{-1} (Si–O–C), 827 cm^{-1}, 2931 cm^{-1} (1,4-substituted benzene ring), and 1097 cm^{-1}, 1138 cm^{-1} (Si-O, Si-O-C). As expected, the epoxy resin is chemically connected to the surface of NS as presented in Scheme 16.1.

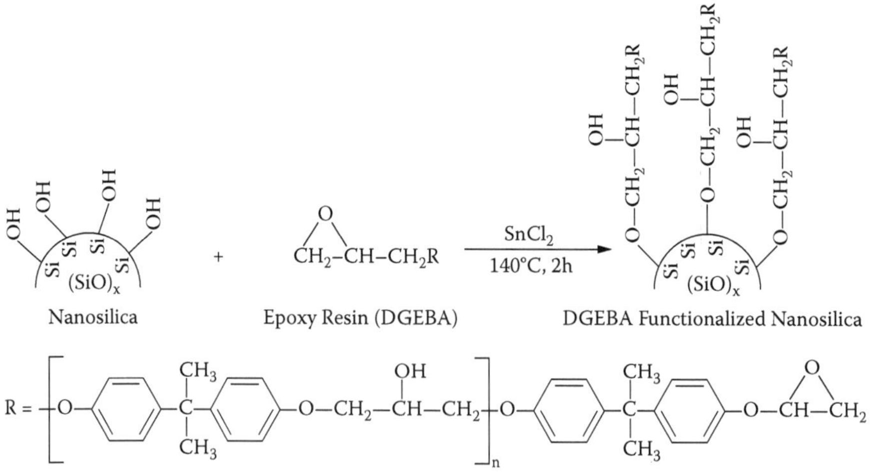

OH OH OH OH Si Si Si Si Si $(SiO)_x$ + CH_2–CH–CH_2R (O epoxide) $\xrightarrow[140°C,\ 2h]{SnCl_2}$ O–CH_2–CH(OH)–CH_2R (×3) Si Si Si Si Si $(SiO)_x$

Nanosilica Epoxy Resin (DGEBA) DGEBA Functionalized Nanosilica

R = [–O–C_6H_4–C(CH_3)$_2$–C_6H_4–O–CH_2–CH(OH)–CH_2–]$_n$ O–C_6H_4–C(CH_3)$_2$–C_6H_4–O–CH–CH_2 (O epoxide)

SCHEME 16.1 Reaction mechanism of epoxy resin functionalization on nanosilica.

16.8.1 Treatment of Zinc-Ion-Coated Nanosilica with Epoxy Resin

To improve the dispersion of the zinc-ion (Zn-ion)-coated nanosilica (ZNS) in the matrix and to increase the hydrophobic nature of the surface, the Zn-ion-coated nanosilica was first modified with epoxy resin. The modification of Zn-ion-coated nanosilica was performed as follows. First, the epoxy resin is dissolved in methyl isobutyl ketone and to this solution Zn-ion-coated nanosilica was added, then the entire suspension was stirred for 2 h at 200°C. After the reaction, the solvent was removed using a rotary evaporator, and the entire mixture was dried under a vacuum; the dried powder was used further for compounding with polyolefins for the synthesis of novel nanocomposites based on polyolefins.

Figure 16.14 is the FTIR spectra of Zn-ion-coated structural silica and DGEBA encapsulated Zn-ion-coated structural silica. In comparison to the spectrum of Zn-ion-coated structural silica (as received), stretching vibration bands can be observed at 1168 cm^{-1}, 1234 cm^{-1} (Si-O-C), 827 cm^{-1}, 2938 cm^{-1} (1,4-substituted Benzene ring), and 1589 cm^{-1} (COO-M+), in the case of DGEBA-encapsulated, Zn-ion-coated structural silica. This proves, as expected, that Zn-ion activity toward the oxirane group of DGEBA forms a chemical network over Zn-ion-coated structural silica. Based on FTIR assessment and SEM morphological study (Section 16.5.3), we proposed a tentative model structure of DGEBA encapsulation over Zn-ion-coated structural silica. This is presented in Scheme 16.2.

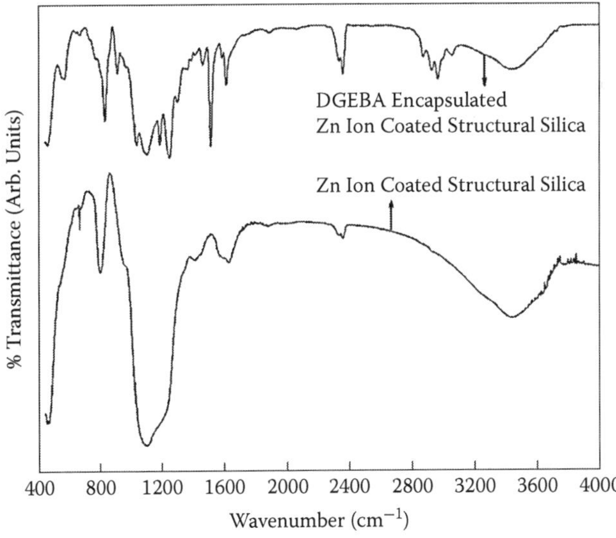

FIGURE 16.14 IR spectra of Zn-ion-coated structural silica and DGEBA encapsulated Zn-ion-coated structural silica.

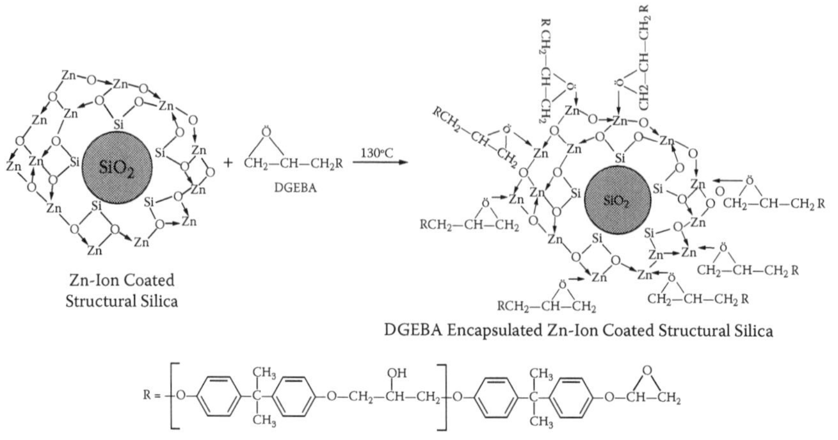

SCHEME 16.2 A tentative model structure of DGEBA encapsulation on Zn-ion-coated structural silica.

16.9 NANOSILICA-FILLED VINYL POLYMER COMPOSITES

16.9.1 Synthesis of Nanocomposites Based on Polyolefins

Nanocomposites based on polyolefins were prepared using the "melt mixing" method in a sigma-internal-mixer. The nanofillers loading level was kept constant (i.e., 2.5 wt%) in all the synthesized polyolefin nanocomposites. In the case of low-density polyethylene (LDPE) and metallocene ethylene-octene copolymer (mPE), the blending of polymers with nanofillers was performed at 150°C for 15 min at a fixed internal rotor speed of 100 r.p.m. The nanocomposites thus obtained were hot pressed using a compression mold at 150°C for 10 min and at a constant pressure of 10 MPa. The composite sheets thus obtained were allowed to cool down to room temperature at a rate of 2°C/min under the applied pressure. This allowed slow crystallization of the samples. In the case of polypropylene (PP) and propylene–ethylene copolymer (EP), the blending of polymers with the nanofillers was performed at 200°C for 15 min. The obtained nanocomposites were hot pressed using a compression mold at 200°C for 15 min and a constant pressure of 10 MPa.

A brief overview of synthesis and characterization of polyolefin–nanosilica nanocomposites [98–105] developed in the Das laboratory is summarized in the form of flowcharts in Figures 16.15 through 16.24.

16.10 MECHANICAL PROPERTIES OF PARTICULATE-FILLED VINYL POLYMER COMPOSITES

The mechanical properties of PP composites containing non-treated fillers are determined mainly by their particle characteristics (i.e., particle size and particle size distribution). Incorporation of small-sized particulate fillers leads to increased strength, and sometimes modulus improvement, whereas deformability and impact strength

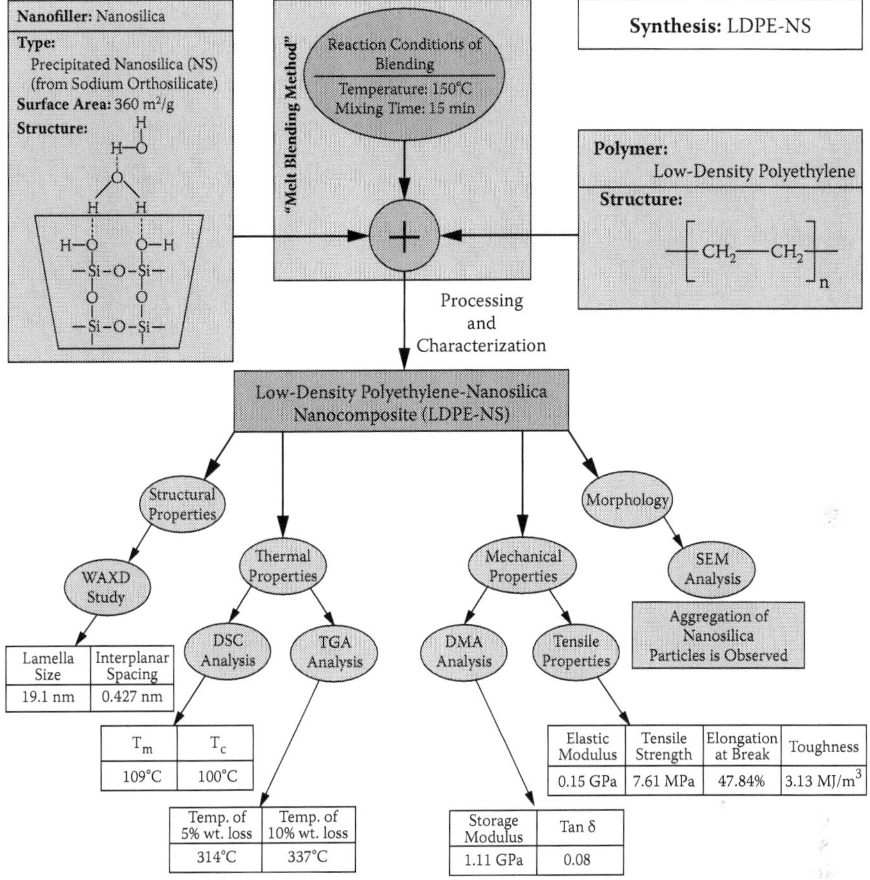

FIGURE 16.15 Low-density polyethylene-nanosilica nanocomposite (LDPE-NS).

decrease with decreasing particle size. Particle size alone, however, is not sufficient for the characterization of any filler; the knowledge of the particle size distribution is equally important [27]. Generally, in the case of large particles, the volume in which stress concentration is effective increases with particle size. In addition, the strength of matrix–filler adhesion also depends on it. The other end of the particle size distribution (i.e., the amount of small particles) is equally important. Aggregation tendency of fillers increases with decreasing particle size. Extensive aggregation, however, leads to insufficient homogeneity, rigidity, and lower impact strength. Aggregated filler particles act as crack initiation sites under dynamic loading conditions. The nanocomposites reportedly exhibit superior mechanical properties over the conventional composites [106–109], and their thermal stability [110,111] also improves with the introduction of nanoscale inorganic reinforcements.

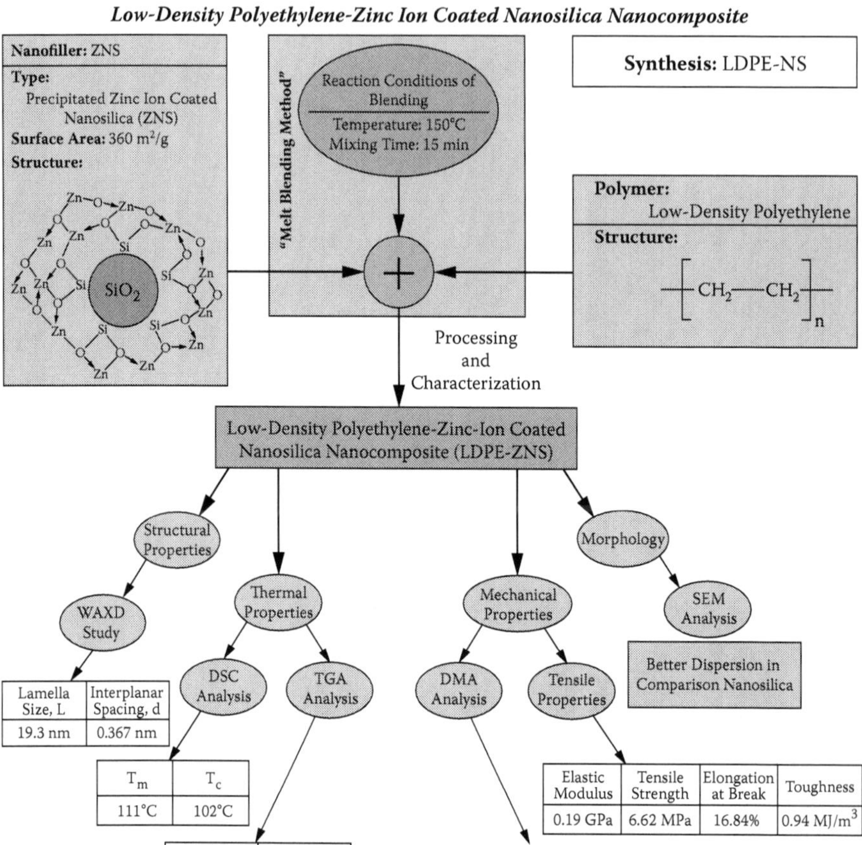

FIGURE 16.16 Low-density polyethylene-zinc-ion-coated nanosilica nanocomposite (LDPE-ZNS).

The specific surface area of fillers is closely related to their particle size distribution. However, it also has a direct impact on composite properties. Adsorption of both small molecular weight additives, as well as that of the polymer, is proportional to the size of the matrix–filler interface [25]. Adsorption of additives may change stability, whereas matrix–filler interaction significantly influences mechanical properties — namely, yield stress, tensile strength, and impact resistance [16,17].

Anisotropic particles reinforce polymers, and the effect increases with the anisotropy of the particle. In fact, fillers and reinforcements are very often differentiated by their degree of anisotropy (i.e., aspect ratio). Plate-like fillers, such as talc and mica, reinforce polymers more than spherical fillers; the influence of glass fibers is even

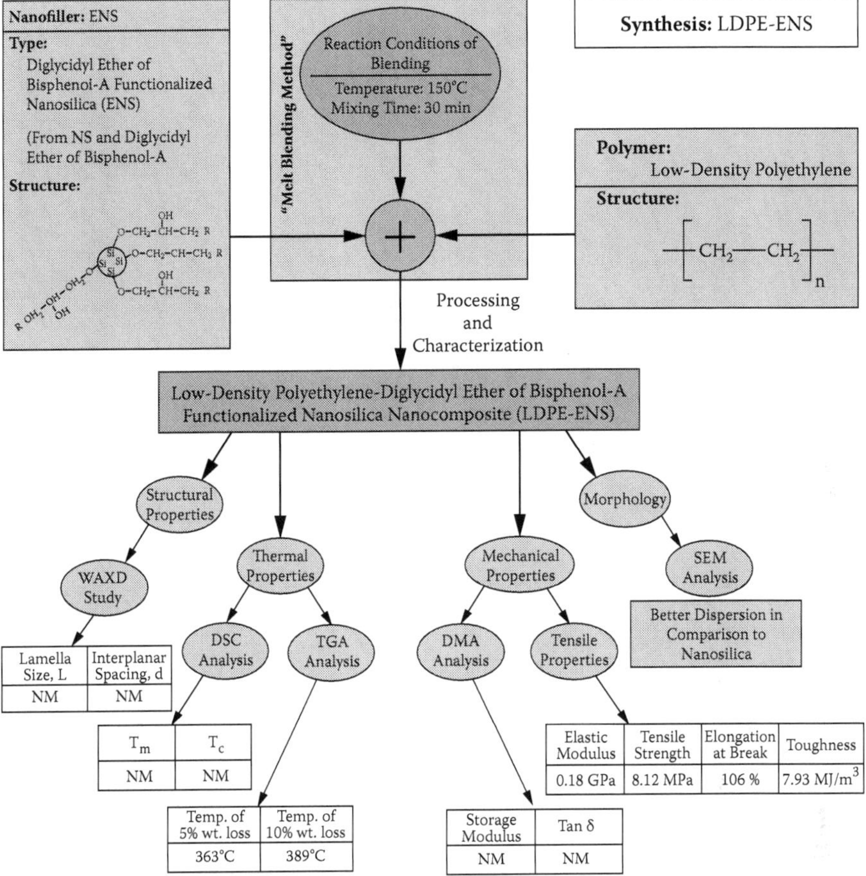

FIGURE 16.17 Low-density polyethylene-diglycidyl ether of bisphenol-A functionalized nanosilica nanocomposite (LDPE-ENS).

stronger [28]. Anisotropic fillers are oriented during processing, thereby enhancing their reinforcing effect. This effect depends very much on the extent of orientation.

The hardness of a filler has a strong effect on the wear of the processing equipment, but this is also influenced by the size and shape of the particles, the composition, viscosity, and speed of processing. [27]. Surface-free energy (i.e., surface tension) of the filler determines both the matrix–filler and particle–particle interactions. Whereas the former has a pronounced effect on the mechanical properties, the latter determines aggregation. Both interactions can be modified by surface treatment.

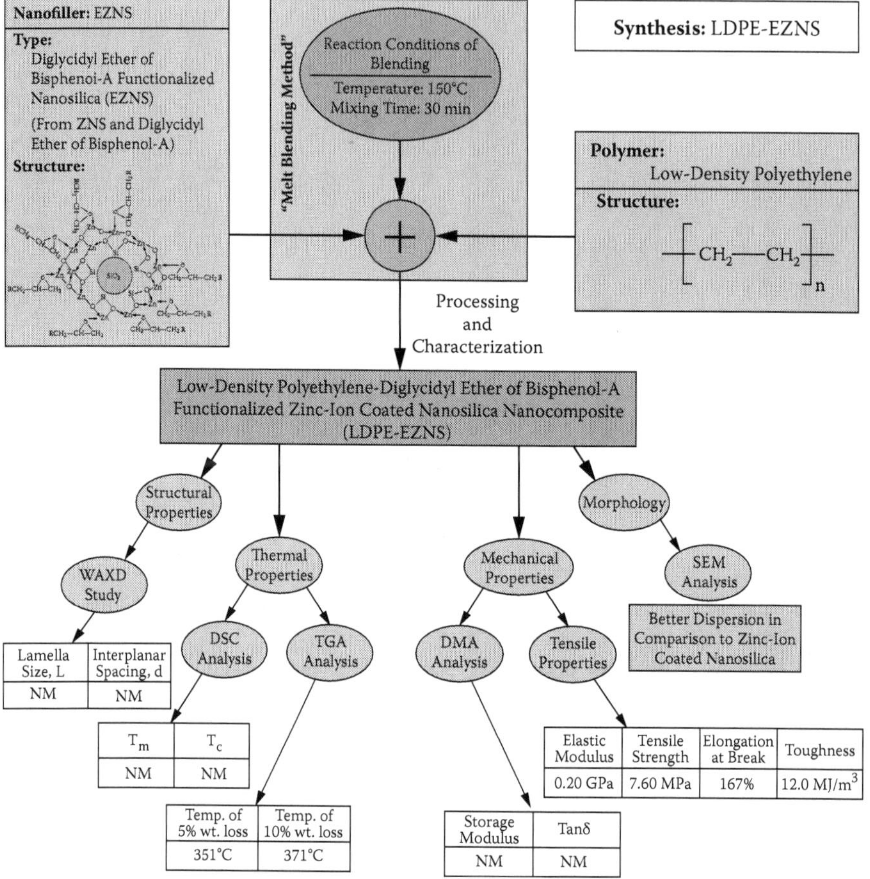

FIGURE 16.18 Low-density polyethylene-diglycidyl ether of bisphenol-A functionalized zinc-ion-coated nanosilica nanocomposite (LDPE-EZNS).

16.11 GAS-PHASE POLYMERIZATION OF PROPYLENE/ETHYLENE IN THE PRESENCE OF SURFACE-TREATED NANOFILLERS

Another way of tailoring the microstructure of polyolefins and mechanical properties is by metallocene polymerization catalysis in presence of nanofillers. The polymerization filling technique (PFT) invented by Howard [112] has drawn much attention from polymer and material scientists working in the field of polyolefin nanocomposites. PFT consists of attaching a Ziegler–Natta type catalyst onto the surface of an inorganic filler so that the olefin monomer can be polymerized from the filler surface. This provides high filler loading capability to the polymer, and, at the same time, maintains appreciable mechanical properties in the final nanocomposite. More recently,

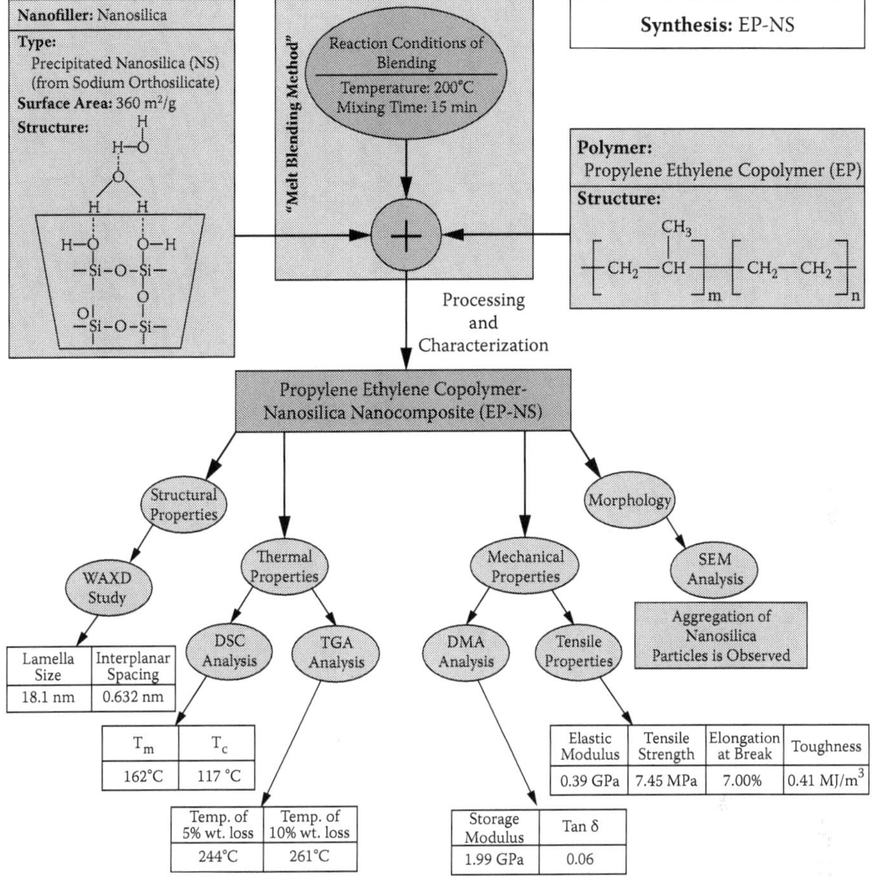

FIGURE 16.19 Propylene ethylene copolymer-nanosilica nanocomposite (EP-NS).

research led by a team headed by Robert Jerome [113–115] has focused on the development of nanocomposites, especially polyethylene, by adopting the aforementioned technique using a single-site metallocene catalyst. Much of the current research related to this area is directed toward the development of "*in situ*" nanocomposites of olefins by polymerizing them with metallocenes in the presence of surface-treated fillers, all in the slurry phase. However, the gas-phase polymerization technique is a highly preferred method for the large-scale industrial production of polyolefins and their copolymers, especially metallocene linear low-density polyethylene.

McKenna and Statton [116] have investigated the gas-phase polymerization of ethylene using a silica-supported rac-$Me_2Si[Ind]_2ZrCl_2$/methylaluminoxane (MAO) at a temperature between 40°C and 80°C, using NaCl as a bed support and triethylaluminum (TEA) as a scavenger for impurities. Recently, Dealy and Wissbrun [117]

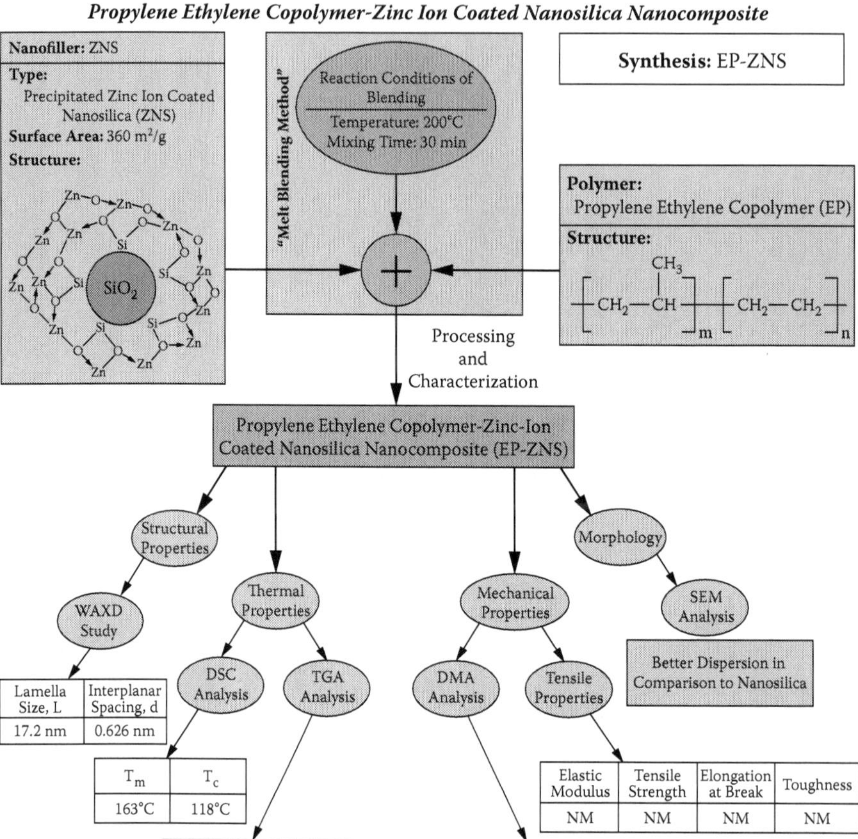

Lamella Size, L	Interplanar Spacing, d
17.2 nm	0.626 nm

T_m	T_c
163°C	118°C

Temp. of 5% wt. loss	Temp. of 10% wt. loss
259°C	286°C

Storage Modulus	Tan δ
1.18 GPa	0.06

Elastic Modulus	Tensile Strength	Elongation at Break	Toughness
NM	NM	NM	NM

FIGURE 16.20 Propylene ethylene copolymer-zinc-ion-coated nanosilica nanocomposite (EP-ZNS).

investigated gas-phase polymerization of ethylene using a zirconocene catalyst supported on mesoporous molecular sieves. Kinloch and Young [118] have also studied the influence of polymerization conditions on the gas-phase polymerization of ethylene with silica supported zirconocene [Cp_2ZrCl_2] catalyst.

Wunderlich [119] investigated the gas-phase polymerization of propylene using a solid catalyst complex obtained from Et[Ind]$_2$ZrCl$_2$ and methylaluminoxane (MAO), with NaCl as the dispersion medium.

To the best of our knowledge, no report is currently available in open literature regarding the development of polyethylene/polypropylene nanocomposites by gas-phase polymerization of olefin monomers, in the presence of a surface-activated nanofiller that acts as bed support using a nanosilica supported bis(cyclopentadienyl)

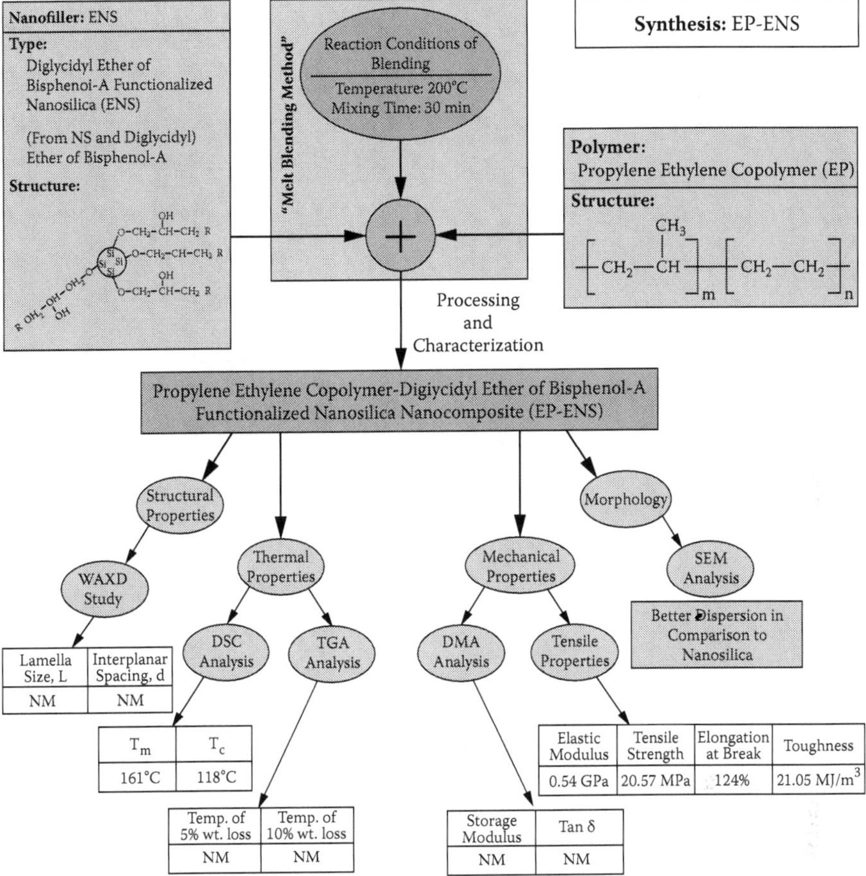

Lamella Size, L	Interplanar Spacing, d
NM	NM

T_m	T_c
161°C	118°C

Temp. of 5% wt. loss	Temp. of 10% wt. loss
NM	NM

Storage Modulus	Tan δ
NM	NM

Elastic Modulus	Tensile Strength	Elongation at Break	Toughness
0.54 GPa	20.57 MPa	124%	21.05 MJ/m^3

FIGURE 16.21 Propylene ethylene copolymer-diglycidyl ether of bisphenol-A functionalized nanosilica nanocomposite (EP-ENS).

zirconium (IV) dichloride catalyst. Our approach for the development of polyethylene and polypropylene nanocomposites differs from others in the following aspects:

1. The gas-phase polymerization methodology does not require solvents, eliminates polymer purification, and utilizes monomer feed stocks efficiently.
2. The novelty of this approach is the use of nanosilica supported zirconocene catalyst in gas-phase polymerization of ethylene in presence of surface activated fillers for the development of *in situ* polyethylene nanocomposites.

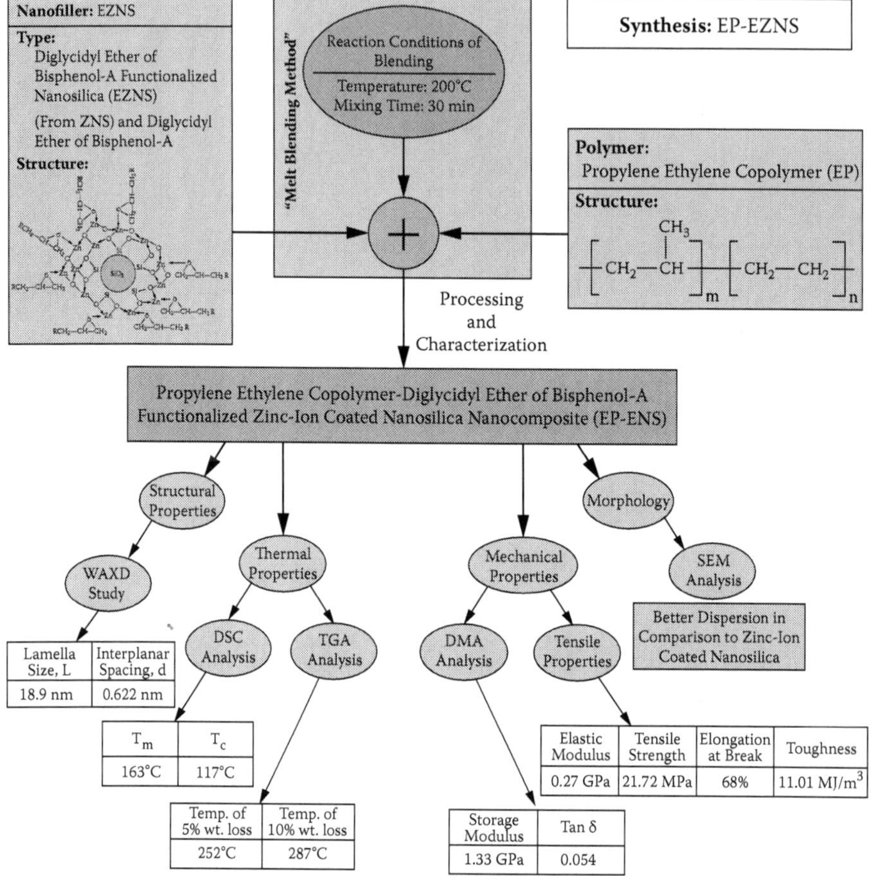

Lamella Size, L	Interplanar Spacing, d
18.9 nm	0.622 nm

T_m	T_c
163°C	117°C

Temp. of 5% wt. loss	Temp. of 10% wt. loss
252°C	287°C

Storage Modulus	Tan δ
1.33 GPa	0.054

Elastic Modulus	Tensile Strength	Elongation at Break	Toughness
0.27 GPa	21.72 MPa	68%	11.01 MJ/m^3

FIGURE 16.22 Propylene ethylene copolymer-diglycidyl ether of bisphenol-A functionalized zinc-ion-coated nanosilica nanocomposite (EP-EZNS).

3. In the case of slurry-phase methods, after the final product has been obtained, purification is always needed because the possibility of traces of solvents remaining in the product exists; this is eliminated in our approach.
4. One noteworthy observation of our approach is that the gas-phase polymerization technique is a highly preferred method for the large-scale industrial production of polyolefins and their copolymers, especially metallocene linear low-density polyethylene.

The authors have successfully prepared *in situ* polypropylene nanocomposites by gas-phase polymerization technique, and reported it in their earlier communication [120,121].

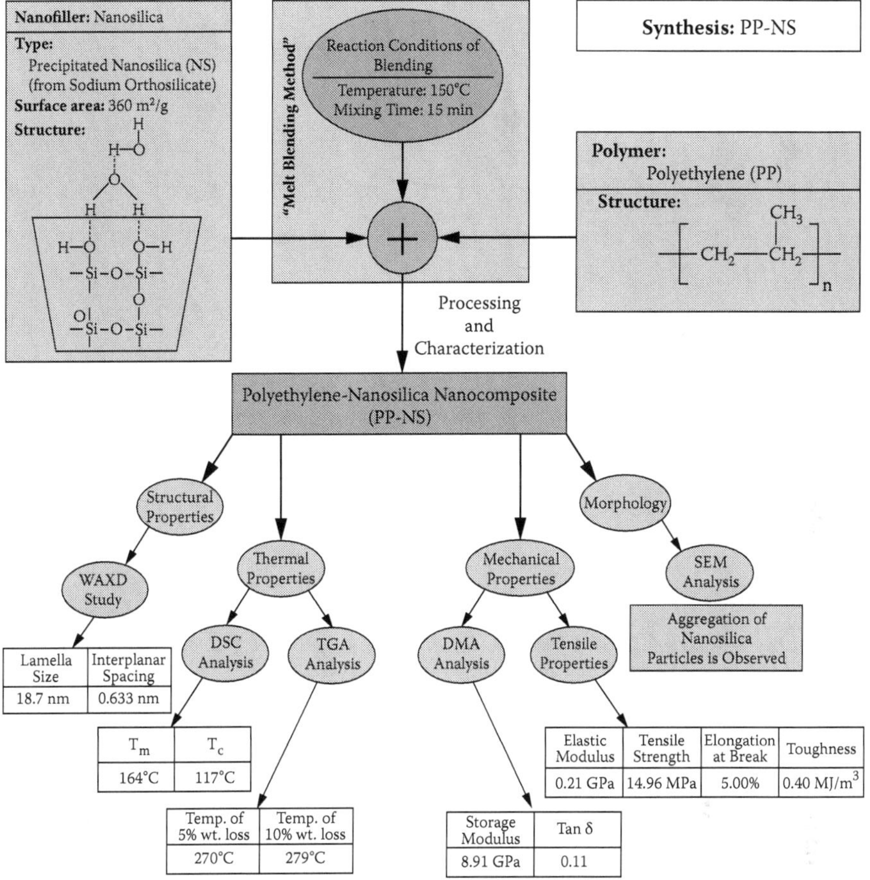

FIGURE 16.23 Polypropylene-nanosilica nanocomposite (PP-NS).

16.12 PROPYLENE POLYMERIZATION

Three different polymerizations were conducted in presence of the MAO-treated fillers. The products obtained were Closite-20A-filled polypropylene (CFPP), kaolin-filled polypropylene (KFPP) and nanosilica-filled polypropylene (SFPP). These were synthesized by gas-phase polymerization of propylene using a nanosilica-supported zirconocene catalyst in the presence of the respective MAO-treated nanofillers. Propylene polymerization was conducted in a 2-liter stainless steel reactor. The reactor was first dried by heating it up to 120°C under vacuum for 1h and then cooled to 70°C under nitrogen flow; 10 g of previously vacuum-dried, MAO-treated nanofiller was added into the reactor, and 0.039 g of nanosilica-supported zirconocene was

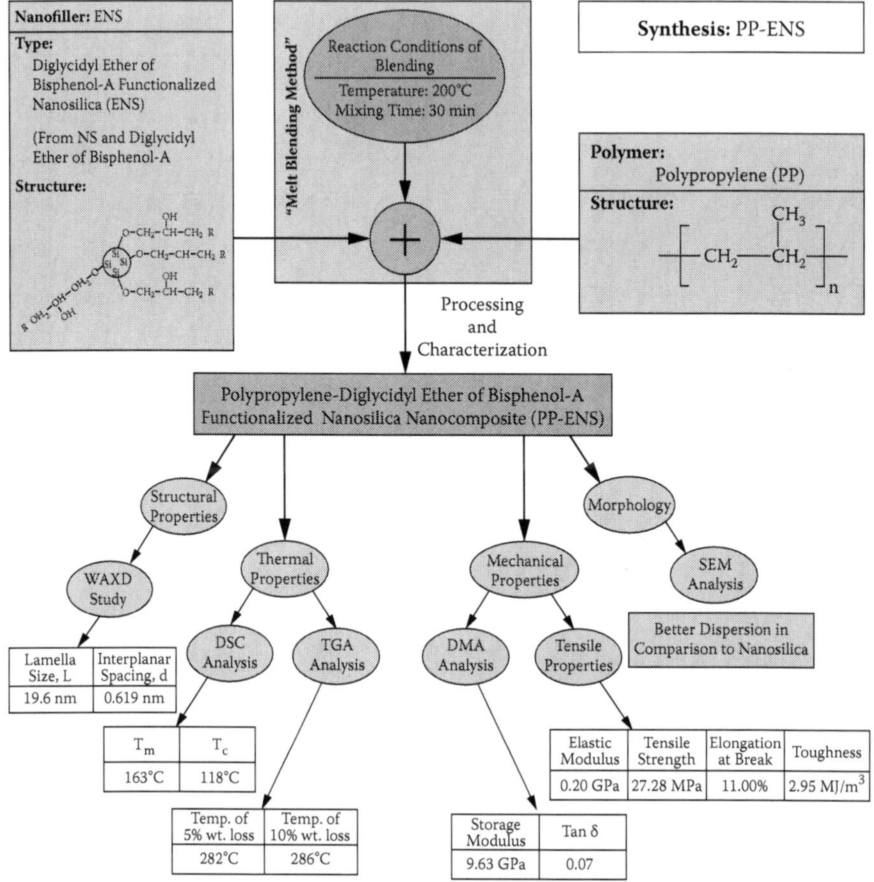

FIGURE 16.24 Polypropylene-diglycidyl ether of bisphenol-A functionalized nanosilica nanocomposite (PP-ENS).

added into the reactor through a Teflon capillary under nitrogen flow. The mass was then stirred for 15 min at 70°C. Thereafter, 5 ml of MAO was injected into the reactor. Then the entire mass was stirred for 5 min at 70°C. For a uniform polymerization of propylene, a continuous flow of propylene gas was maintained through the bottom of the reactor at 8-bar pressure. The polymerization was performed for 5 h at a constant temperature of 70°C. Polymerization was stopped by releasing the pressure and cooling down the reactor to the room temperature. The solid mass of the composite formed in the reactor was then collected and dried at 100°C for 1 h under vacuum. The dried composite samples are in the form of fine powders. The typical results of gas-phase polymerization are listed in Table 16.2.

TABLE 16.2
Propylene Polymerization Results

	In situ Polypropylene Nanocomposites		
	CFPP	KFPP	SFPP
Nanosilica-supported-Zirconocene catalyst (mmol)	0.034	0.034	0.034
MAO treated nanofiller (g)	10	10	10
Total MAO (mmol)	17	17	17
[Al]/[Zr]	500	500	500
Activity (Kg/mol. Zr)	411 ± 3	441 ± 5	470 ± 4
Nanofiller abundance (%)	41	40	38
Viscosity molecular weight, $M_v \times 10^{-4}$ (g/mol)[†]	6.90	3.80	19.40

[a] Viscosity molecular weight of extracted polypropylene measured at 30°C in benzene.

REFERENCES

1. http://www.wimancorp.com.
2. http://www.azom.com.
3. Chung, D. L., *Composite Materials*, Springer-Verlag London, 2004.
4. Chou, C. J., Read, A. E., Garcia-Meitin, E. I., Bosnyak, C. P., *Polymer Nanocomposites*. The Dow Chemical Company. (http://www.4spe.org).
5. Kojima, Y., Usuki, A., Kawasumi, M., Okada, A., Fukushima, Y., Kurauchi, T., and Kamigatio, O., *J. Mater. Res.*, 8, 1185, 1993.
6. Brydson, J. A., *Plastics Materials Technology and Engineering*, 1999.
7. Mader, E. and Freitag K. H., *Composites*, 21, 397, 1990.
8. Kelnar, I., *Angew. Makromol. Chem.*, 189, 207, 1991.
9. Chiang, W. Y. and Yang, W. D., *J. Appl. Polym. Sci.*, 35, 807, 1988.
10. Matejka, L. et al., Formulation and structure of the epoxy-silica hybrids, *Polymer*, 40, 171, 1998.
11. Matejka, L. and Kdusek, J., Structure evaluation in epoxy-silica hybrids: sol-gel process, *J. Non. Cryst. Solids*, 226, 114, 1998.
12. Kang, S. et al. Preparation and characterization of epoxy composites filled with functionalized nanosilica particles obtained via sol-gel process, *Polymer*, 42, 879, 2001.
13. Pukanszky, B., in *Polypropylene: Structure, Blends and Composites*, Vol. 3, In: Karger-Kocsis, J., Ed., Chapman and Hall, London, 1995, p. 1.
14. Krysztafkiewicz, A., *Surface Coatings Technol.*, 35, 151, 1988.
15. Voros, G. Fekete, E., and Pukanszky, B. J., *Adhesion*, 64, 229, 1997.
16. Pukanszky, B. et al., *Interfaces in polymer, ceramic, and metal matrix composites*, Elsevier, New York, 1988, p. 467.
17. Pukanszky, B., *Composites*, 21, 255, 1990.
18. Vollenberg, P., Heikens, D., and Ladan, H. C. B., *Polym. Compos.*, 9, 382, 1988.

19. Pukanszky, B. and Voross, G., *Compos. Interfaces*, 1, 411, 1993.
20. Riley, A.M. et al., *Plast. Rubber. Process. Appl.*, 14, 85, 1990.
21. Menczel, J. and Varga, J., *J. Thermal Anal.*, 28, 161, 1983.
22. Fujiyama, M. and Wakino, T., *J. Appl. Polym. Sci.*, 42, 2739, 1991.
23. Pukanszky, B., Mudra I., and Staniek, P. J., *Vinyl Additive Technol.*, 3, 53, 1997.
24. Pukanszky, B. and Fekete, E., *Adv. Polym. Sci.*, 139, 109, 1999.
25. Schlumpf, H.P., *Chimia*, 44, 359, 1990.
26. Rothon, R., *Particulate-filled polymer composites*, Longman, Harlow, 1995.
27. Schlumpf, H. P., *Kunststoffe*, 73, 511, 1983.
28. Nielsen L. E., *Mechanical properties of polymers and composites*, Marcel Dekker, New York, 1974.
29. Pukanszky, B., Fekete, E., and Tudos, F., *Makromol Chem., Macromol Symp.*, 28, 165, 1989.
30. Jancar, J. and Kucera, J., *Polym. Eng. Sci.*, 30, 707, 1990.
31. Jancar, J., *J. Mater. Sci.*, 24, 3947, 1989.
32. Marosi, G. et al., *Colloids Surf.*, 23, 185, 1986.
33. Maiti, S.N. and Mahapatro, P.K., *J. Appl. Polym. Sci.*, 42, 3101, 1991.
34. Bajaj, P., Jha, N.K. and Jha, R.K., *Polym. Eng. Sci.*, 29, 557, 1989.
35. Bajaj, P., Jha N.K. and Jha, R.K., *Br. Polym. J.*, 21, 345, 1989.
36. Fekete, E. et al., *J. Colloid Interface Sci.*, 135, 200, 1990.
37. Raj, R.G. et al., *J. Appl. Polym. Sci.*, 38, 1987, 1989.
38. Papirer, E., Schultz, J., and Turchi, C., *Eur. Polym. J.*, 12, 1155, 1984.
39. Allard, R.C., Vu-Khanh, T., and Chalifoux, J.P., *Polym. Composites*, 10, 62, 1989.
40. Bramuzzo, M., Savadori, A., and Bacci, D., *Polym. Composites*, 6, 1, 1985.
41. Pukanszky, B. and Maurer, F.H.J., *Polymer*, 36, 1617, 1995.
42. Ishida, H. and Koenig, J.L., *J. Polym. Sci., Polym. Phys.*, 18, 1931, 1980.
43. Plueddemann, E.P., *Silane Coupling Agents*, Plenum, New York, 1982.
44. Yue, C.Y. and Cheung, W.L., *J. Mater. Sci.*, 26, 870, 1991.
45. Trotignon, J.P. et al., *Polymer Composites*, Walter de Gruyter, Berlin, 1986, p. 191.
46. Matienzo, L.J. and Shah, T.K., *Surf. Interface Anal.*, 8, 53, 1986.
47. Vollenberg, P.H.T. and Heikens, D., *J. Mater. Sci.*, 25, 3089, 1990.
48. Raj, R.G., Kokta, B.V., and Daneault, C., *Int. J. Polym. Mater.*, 12, 239, 1989.
49. Ishida, H. and Miller, J.D., *Macromolecules*, 17, 1659, 1984.
50. Ishida, H. and Kumar, G., Eds., *Molecular Characterization of Composite Interfaces*, Plenum, 1985.
51. Demjen, Z., Pukanszky, B., Foldes, E., and Nagy, J., *J. Interface Colloid. Sci.*, 194, 269, 1977.
52. Sadler, E.J. and Vecere, A.C., *Plast. Rubber Process. Appl.*, 24, 271, 1995.
53. Zolotnitsky, M. and Steinmetz, J.R., *J. Vinyl Additive Technol.*, 1, 109, 1995.
54. Widmann, B., Fritz, H.G., and Oggermuller, H., *Kunststoffe*, 82, 1185, 1992.
55. Demjen, Z., Pukanszky, B., and Nagy, J. Jr., Possible coupling reactions of functional silanes and polypropylene, *Polymer*, 40, 1763, 1999.
56. Takase, S. and Shiraishi, N., *J. Appl. Polym. Sci.*, 37, 645, 1989.
57. Jancar, J., Kummer, M., and Kolarik, J., in *Interfaces in Polymer, Ceramic, and Metal Matrix Composites*, Ishida, H., Ed., Elsevier, New York, 1988, p. 705.
58. Felix, J.M. and Gatenholm, P., *J. Appl. Polym. Sci.*, 50, 699, 1991.
59. Jancar, J. and Kucera, J., *Polym. Eng. Sci.*, 30, 714, 1990.
60. Karger-Kocsis, J., Kallo, A., and Kuleznev, V.N., *Polymer*, 25, 279, 1984.
61. Lee, Y.D. and Lu, C.C., *J. Chin. Inst. Chem. Eng.*, 13, 1, 1982.
62. Kolarik, J. and Lednicky, F., in *Polymer Composites*, Sedlacek, B., Ed., Walter de Gruyter, Berlin, 1986, p. 537.
63. Stamhuis, J.E., *Polym, Composites*, 9, 280, 1988.

64. Gupta, A.K., Kumar, P.K., and Ratnam, B.K., *J. Appl. Polym. Sci.*, 42, 2595, 1991.
65. Serafimov, B., *Plaste Kautsch*, 33, 331, 1986.
66. Faulkner, D.L.,, *J. Appl. Polym. Sci.*, 36, 467, 1988.
67. Kolarik, J., Lednicky, F., and Pukanszky, B., in *Proc. 6th ICCM/2nd ECCM*, Vol. 1, Matthews, F.L.,, Buskell, N.C.R., Hodgkinson, J.M., and Morton, J., Eds., Elsevier, London, 1987.
68. Pukanszky, B., Tudos, F., Kolarik, J., and Lednicky, F., *Polym. Composites*, 11, 98, 1990.
69. Chiang, W.Y., Yang, W.D. and Pukanszky, B., *Polym. Eng. Sci.*, 32, 641, 1992.
70. Kolarik, J., Lednicky, F., Jancar, J., and Pukanszky, B., *Polym. Commun.*, 31, 201, 1990.
71. Coloney, A., Kausch, H. H., Kaiser, T., and Beer, H.R., *J. Mater. Sci.*, 22, 381, 1987.
72. Ng, C.B., Schadler, L.S., and Seigel, R.W., *Nanostructured Mater.*, 12, 57, 1999.
73. Ng, C.B. et al., *Adv. Compos. Lett.*, 10,101, 2001.
74. Ou, Y., Yang. F.. and Yu, Z., *J. Polym. Sci. B. Phys.*, 36, 789, 1998.
75. Krishnamoorti, R. and Vaia, R.A., *Polymer Nanocomposites: Synthesis, Characterization, and Modeling*, American Chemical Society, Washington, D.C., 2002.
76. Katz, H.S. and Milewski, J.V., Eds., Handbook of Fillers Plastics,, Van Nostrand Reinhold, New York, 1987.
77. Dutta, N.K., Choudhury, N.R., and Haider, B., "High resolution solid-state NMR investigation of the filler-rubber interactions. Part 1: High-speed 1H magic-angle spinning NMR spectroscopy in carbon black filled styrene-butadiene rubber, *Polymer*, 35, 20 4293, 1994.
78. Ou, Y.C. et al., Efeects of alkylation of silicas on interfacial interaction and molecular motion between silicas and rubber, *J. Appl. Polym. Sci.*, 1321, 1996.
79. Kausch, H.H. and Beguelia, P., Deformation and fracture mechanisms in filled polymers, *Macromol. Mater. Eng.*, 169, 79, 2001.
80. Pinnavaia, T.J. and Beall, G.W., Eds., *Polymer Clay Nanocomposites*, John Wiley & Sons, New York, 2001, p. 4.
81. Komarneni, S., *J. Mater. Chem.*, 2, 1219, 1992.
82. Schmidt, H., *J. Non-Crystl. Solids*, 73, 681, 1985.
83. Novak, B.M., *Adv. Mater.*, 5, 422, 1993.
84. Mark, J.E., *Polym. Eng. Sci.*, 36, 2905, 1996.
85. Wang, Z. and Pinnavaia, T.J., *Chem. Mater.*, 10, 1820, 1998.
86. LeBaron, P.C., Wang, Z., and Pinnavaia, T.J., *Appl. Clay Sci.*, 15, 11, 1999.
87. Zeng, Q.H., Wang, D.Z., Yu, A.B. and Lu, G.Q., *Nanotechnology*, 13, 549, 2002
88. Hackett, E., Manias, E., and Giannelis, E.P., *Chem. Mater.*, 12, 2161, 2000.
89. Ou, Y., Yang, F., and Yu, Z., *J. Polym. Sci., Part B*, 36, 789, 1998.
90. Okada, A., Usuki, A., Kurauchi, T., and Kamigaito, O., in *1995 ACS Symp. Ser.*, Vol. 585, American Chemical Society, Washington, D.C., 1995, p. 55.
91. Hausslein, R.W, and Fallick, G., *J. Appl. Polym. Symp.*, 11, 119, 1969.
92. Gahde, J. et al., *Polym. Sci. (USSR)*, 19, 1446, 1977.
93. Solomon, D.H. and Rosser, M.J., *J. Appl. Polym. Sci.*, 9, 1261, 1965.
94. Velasco-Santos, C., Martinez-Hemandez, A.L., Lozada-Cassou, M., Alvarez-Castillo, A., and Castano, V. M., *Nanotechnology*, 13, 495, 2002.
95. Carrot, G. et al., *Macromolecules*, 35, 8400, 2002.
96. Urzua-Sanchez, O. et al., *Polym. Bull.*, 49, 39, 2002.
97. Lin, J., Siddiqui, J.A. and Ottenbrite, R.M., *Polym. Adv. Technol.*, 12, 285, 2001.
98. Reddy, C. S., Novel nanocomposites based on polyolefins: effect of surface functionalization of nanosilica on the structural, mechanical, dynamic mechanical and thermal properties of polyolefin composities, PhD dissertation, Indian Institute of Technology Kharagpur, India, 2007.

99. Reddy, C. S. and Das, C. K., *Composite Interfaces*, 11 (8–9), 687, 2005.
100. Reddy, C. S., et al., *Macromolecular Research*, 13 (3), 22, 2005.
101. Reddy, C. S., Narkis, M. and Das, C. K., *Polymer Composites*, 26 (6), 806, 2005.
102. Reddy, C. S. and Das, C. K., *Polymers and Polymer Compositie*s, 14 (3), 281, 2006.
103. Reddy, C. S. and Das, C. K., *Polymer Plastic Technology and Engineering*, 45 (7), 815, 2006.
104. Reddy, C. S. and Das, C. K., *Journal of Applied Polymer Science*, 102 (3), 2117, 2006.
105. Reddy, C. S. and Das, C. K., *Polymer International*, 55 (8), 923, 2006.
106. Hsiue, G.H., Chen, J.K., and Liu, Y. L., *J. Appl. Polym. Sci.*, 76, 1609, 2000.
107. Chen, J.P. et al., in *ACS Symp. Ser.*, Vol. 585, American Chemical Society, Washington, D.C., 1995, p. 297.
108. Mirabella, F.M., Bafna, A., Rufener, K., and Mehta, S., *2002 Proc. Nanocomposites*, San Diego, Sept. 2002.
109. Tyan, H.L., Leu, C.M., and Wei, K.H., *Chem. Mater.*, 13, 222, 2001.
110. Lee, A. and Lichtenhan, J.D., *J. Appl. Polym. Sci.*, 73, 1993, 1999.
111. Gilman, J.W., *Appl. Clay Sci.*, 15, 31, 1999.
112. Howard, Jr., U.S. Patent 4,187,210, 1980.
113. Alexander, M. et al., *Macromol. Chem. Phys.*, 202, 2239, 2001.
114. Alexander, M. et al., *Polymer*, 43, 2123, 2002.
115. Dubosis, P., Alexander, M., and Jerome, R., *Macromol. Symp.*, 194, 13, 2003.
116. McKenna, G.B. and Statton, W.O., *J. Polym. Sci. Macromol. Rev.*, 11, 1, 1976.
117. Dealy, J.M. and Wissbrun, K.F., *Melt Rheology and Its Role in Plastics Processing*, Van Nostrand Reinhold, New York, 1989.
118. Kinloch, A.J. and Young, R.J., *Polymer Fracture*, Elsevier, London, 1983, p. 147.
119. Wunderlich, B., *Crystal Nucleation, Growth, Annealing*, Academic Press, New York, 1976.
120. Reddy, C. S. and Das, C. K., *Journal of Macromolecular Science: Part A: Pure and Applied Chemistry*, 43 (9), 1365, 2006.
121. Reddy, C. S. and Das, C. K., *Journal of Polymer Research*, 14, 129, 2007.

17 Vinyl Polymer Applications and Special Uses

Chapal K. Das, Sandeep Kumar, and Tanmoy Rath

CONTENTS

17.1 INTRODUCTION

The term "vinyl" generally refers to the monomers that are the building blocks of this class of polymer, with vinyl monomers being small molecules containing carbon–carbon double bonds. Polyethylene (PE), polystyrene (PS), polypropylene (PP), polymethylmethacrylate (PMMA), polytetrafluoroethylene (PTFE), polyvinyl chloride (PVC), polyvinyl acetate (PVAc), polyvinylidene chloride (PVDC), polyvinyl alcohol (PVA), polyvinylacetals, polyvinyl fluoride (PVF), polyvinyl pyrrolidone (PVP), polyvinylcarbazole (PVK), and polyvinyl ethers are all examples of compounds referred to generically as vinyl polymers. PE, the simplest vinyl polymer, is made from the monomer ethylene. When polymerized, the ethylene molecules are joined along the axes of their double bonds to form a long chain of many thousands of carbon atoms containing only single bonds between atoms. By adding HCl, HF, CH_3COOH, or CH_3OH to acetylene, one can form vinyl chloride, vinyl fluoride, vinyl acetate, and vinyl methyl ether, respectively. These monomers are polymerized through various methods, and the resulting thermoplastics are processed through injection molding and extrusion.

Henry Victor Regnault (1810–1878) first produced PVC in 1835 by combining ethylene dichloride with potassium hydroxide in alcohol. Ostromislen researched it again in 1912. Klatte, a German chemist, patented it in 1913. Klatte also discovered vinyl acetate and vinyl esters. Reppe discovered vinyl ethers in 1927. The first vinyl polymer produced in large amounts was polyvinyl chloroacetate, which was used as a lacquer during World War I. PVAc adhesives were introduced in 1926.

Probably the most familiar of these vinyls, however, is PVC. It was soon discovered that a plasticizer mixed with PVC yields a flexible plastic. By 1931, IG Farbenindustrie had developed an industrial production process using Klatte's methods. Production of PVC, which IG Farbenindustrie called PC, began in Germany in 1932 and moved to the United States shortly after that. Production of PVC began in France in 1941 under the name Rhoval. PVC is weather-resistant, is an excellent electrical insulator, and is non-flammable. These characteristics lent themselves to applications on military hardware during World War II.

Today, both rigid and plasticized PVC are in great demand. Both are often used in co-polymers — most commonly co-polymers of vinyl chloride and vinyl acetate. These have been used in everything from records to vinyl floor tiles. Vinyl chloride is also blended with ethylene, propylene, vinylidene chloride, ethyl, n-butyl, or 2-ethylhexyl acrylate. PVC is chemically inert, making rigid PVC ideal for piping. Plasticized PVC is used in garden hoses, imitation leather, raincoats, and electric plugs.

The Carbide and Carbon Chemicals Corporation developed vinyon filaments and fibers and licensed the American Viscose Corporation to produce them in 1939. Vinyon is a co-polymer of 88% vinyl chloride and 12% vinyl acetate. It was the first plastic fiber produced on a large scale in the United States. The co-polymer is dissolved in acetone and forced through spinnerets to form fibers. The fibers are subjected to hot air to evaporate the solvents and are then stretched in a process similar to cold drawing. The stretching increases the strength of the fiber but lowers its elasticity. The fiber does not take dyes and becomes sticky if heated to over 67°C. At 77°C, garments made of the fiber will shrink. Further research led to the development of the fiber Vinyon N, a co-polymer of vinyl chloride and acrylonitrile patented in 1947.

Dates of commercialization of important vinyl polymers are as follows: PVC (1927; B. F. Goodrich), PS (1930; I. G. Farben), PMMA (1936; Rohm and Haas), PE (1939; ICI), and PTFE (1946; Du Pont).

17.2 APPLICATIONS OF VINYL POLYMER

17.2.1 Polypropylene

Polypropylene (PP) is a linear hydrocarbon polymer, which is expressed as C_nH_{2n}. PP, like polyethylene and polybutene, is a polyolefin or saturated polymer.

PP is one of those most versatile polymers available with applications both as a plastic and as a fiber, in virtually all the plastics end-use markets. Production of PP takes place by slurry, solution, or gas-phase process, in which the propylene monomer is subjected to heat and pressure in the presence of a catalyst system. Polymerization is achieved at a relatively low temperature and pressure, and the product yielded is translucent but readily colored. Differences in catalyst and production conditions can be used to alter the properties of the plastic PE.

Most commercial PP has a level of crystallinity intermediate between that of low-density (LDPE) and high-density PE (HDPE); its Young's modulus is also intermediate. Although it is less tough than LDPE, it is much less brittle than HDPE. This allows PP to be used as a replacement for engineering plastics, such as acetonitrile butadiene styrene (ABS). PP has very good resistance to fatigue. Very thin sheets of PP are used as a dielectric within certain high-performance pulse and low-loss RF capacitors. Giulio Natta produced the first PP resin in Spain, although commercial production began in 1957 and PP usage has displayed strong growth since then. The versatility of the polymer (i.e., the ability to adapt to a wide range of fabrication methods and applications) has sustained growth rates, enabling PP to challenge the market share of a host of alternative materials in a plethora of applications.

Three types of PPs are currently available. Based on the polymerization conditions and the incorporation of ethylene co-polymer, PP can be differentiated into a PP homopolymer, PP block co-polymer, and PP random co-polymer.

1. *Homopolymers (PP-H)* — A general-purpose grade can be used in a variety of different applications.
2. *Block co-polymers (PP-B)* — Incorporating 5–15% ethylene into PP has much improved impact resistance, extending to temperature below 20°C. Their toughness can be further enhanced by the addition of impact modifiers — traditionally elastomers in a blending process.
3. *Random co-polymers (PP-R)* — These co-polymers incorporate co-monomer units arranged randomly (as distinct from discrete blocks) along the PP long-chain molecule. Such polymers, which typically contain 1–7% ethylene, are selected where a lower melting point, more flexibility, and enhanced clarity are advantageous.

17.2.1.1 Applications

The property profiles of the different types of PP also make them suitable for different types of applications. PP-H is normally used in industrial pressure pipe systems and for soil and wastewater pipes and fittings, based mainly on its good mechanical properties and excellent chemical resistance. PP-R is a well-proven material in domestic pressure piping systems for hot and cold water due to its high internal pressure resistance and long lifetime at elevated temperatures. For buried sewerage and drainage applications, wastewater systems, and ducting pipe systems, PP-B is the obvious choice. This is mainly due to its high stiffness, high impact strength, especially at low temperatures, good long-term properties, and excellent chemical resistance. Based on these properties, PP-B is commonly used for other non-pressure systems, particularly for soil and wastewater piping systems and for cable protection systems. PP can be processed by virtually all thermoplastic-processing methods. Most typically, PP products are manufactured by extrusion blow molding, injection molding, and general-purpose extrusion. Expanded PP (EPP) may be molded in a specialized process.

17.2.1.1.1 Flexible Packaging

PP is one of the leading materials used for film extrusion and has, in recent years, benefited versus cellophane, metals, or paper because of its superior puncture resistance, low sealing threshold, and competitive price. PP film is available either as cast film or bi-axially orientated PP (BOPP). The film market may be divided into three main sectors:

1. Food and Confectioneries
2. Tobacco
3. Clothing

17.2.1.1.2 Rigid Packaging

The food and confectioneries sector is the largest of the film markets, with usage ranging from confectioneries to crisps and biscuits. Tobacco products represent a significant market for PP (second largest after food and confectioneries). Rigid packaging subdivides into a multitude of packaging applications from caps and closures to pallets. Reusable and collapsible/stackable crates are a great application for PP, providing ease of transport (both full and empty) and allowing simple, safe, and efficient storage of products; they are ideal for just-in-time (JIT) storage solutions. Consequently, supermarkets are beginning to revert to PP use, and similar products are finding application in the automotive supply chain. Caps and closures manufactured of PP have benefited from growth in the PET bottle market, particularly for mineral water containment and that of edible oil. PP is blow molded to produce bottles for the packaging of a various range of products including condiments, detergent, and toiletries. PP thin-walled containers (e.g., yogurt cups) are also common. PP competes with PS in this field, offering a cheaper material option (processing costs can, however, offset this benefit). PP is a semi-crystalline product and consequently has a narrower processing window than PS; it also tends to display higher shrinkage. Modern thermoforming machinery is capable (with two sets of tools) of processing either PP or PS. Consequently, the future infiltration of PP is very much dependent on price fluctuations.

17.2.1.1.3 Automotive

In the automotive sector, PP is utilized as a monomaterial solution for automotive interiors. The monomaterial dashboard is becoming increasingly achievable. PP film cushioning, film skins, powder slush molding, and even blow-molded parts with integral PP textile covers are emerging. Bumpers, cladding, and exterior trim are also available manufactured from PP. PP developed for such applications provides a low coefficient of linear thermal expansion and specific gravity, high chemical resistance, and good weatherability, processability, and impact/stiffness balance. Improvements with color-at-the-press and pre-colored PP have also reduced or eliminated the need for painting in some applications.

17.2.1.1.4 Industrial

PP is used to manufacture a range of sheet, pipe, compounding, and returnable transport packaging (RTP). With the exception of RTP, where injection molding is used, extrusion dominates the conversion process used for these products. Some PP is utilized by the construction sector — most notably domestic drainage pipes.

17.2.1.1.5 PP Fiber

PP fiber is used in a host of applications including tape, strapping, bulk continuous filament, staple fibers, spunbound, and continuous filament.

17.2.2 Polyvinyl Chloride

Polyvinyl chloride (IUPAC Polychloroethene) (PVC) is a widely used plastic. It is commercially manufactured by polymerizing vinyl chloride monomer. Commercial scale production of PVC began in 1931. Whereas rigid PVC was commercialized in the 1950s, PVC is a naturally rigid plastic that can be made flexible by adding plasticizers. The degree of flexibility depends upon the content of the plasticizers. PVC is also available in liquid form, which is known as plastisols or organosols. PVC is manufactured by suspension, emulsion, bulk or mass, and solution polymerization methods. Worldwide, it is estimated that 70% of PVC is manufactured by the suspension method, 20% by emulsion, 9% by bulk, and 1% by solution.

```
 H     H                  H    H    H    H
  \   /                   |    |    |    |
  C == C    -->  ...  -- C -- C -- C -- C -- ...
  /   \                   |    |    |    |
 Cl    H                  Cl   H    Cl   H
Vinyl Chloride Monomer        Polyvinyl Chloride Polymer
```

In terms of revenue generated, PVC is one of the most valuable products of the chemical industry. Globally, over 50% of PVC manufactured is used in construction. As a building material, PVC is cheap and easy to assemble. In recent years, PVC has been replacing traditional building materials such as wood, concrete, and clay in many areas. Despite appearing to be an ideal building material, concerns have been raised about the costs of PVC to the natural environment and human health. Table 17.1 provides a brief overview of the applications of PVC, and Figure 17.1 gives a brief description of the consumption of PVC in different areas of application.

17.2.2.1 Applications

17.2.2.1.1 General

It is not an easy task to ascertain the accurate information about the amounts of PVC used in different applications. However, a rough indication is all that is required for present intentions toward the use of PVC. Actually, the main concern in reporting the application arises from the wide range of applications for which PVC is used and the wide range of different compositions involved. Table 17.1 provides possibly as good a guide as can be given of the amounts used in some selected applications all around the world [1].

TABLE 17.1
Brief View of Applications for PVC

Bottle	Mineral water, juice, oil bottle, general food, and perfume bottles
Belt	Industrial
Film	Agriculture, packaging, toy, and household products
Sheet	Tape, packaging, wallpaper, and accessories
Leather	Clothing, furniture, handbags, and automobiles
Hose	Garden hose and tubes
Wire coating	Insulation
Injection molding	Shoes and sandals
Tile	Wall and floor
Plate	Industrial and building materials

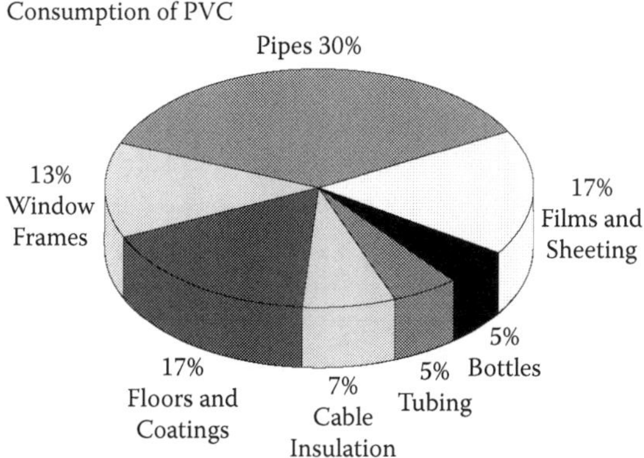

FIGURE 17.1 Consumption chart of PVC.

17.2.2.1.2 Pipework and Similar Products

By far, the largest use for vinyl is in construction applications; 60% of all vinyl is converted into products that are used in residential and commercial building. Vinyl construction products typically are selected for their durability, ease of installation, easy maintenance, and appeal to consumers. In many applications, vinyl has replaced traditional materials such as wood, copper, and aluminum. Pipe applications constitute the major use for vinyl within the construction industry and range from half-inch pipe used in household plumbing to 36-inch or larger mains used in municipal water systems. Vinyl plumbing pipe resists corrosion, sediment buildup, and harsh water conditions. It is easier to install than other types of piping and, because it typically lasts longer, can mean significant cost savings over the life of the system. About 3.3 billion pounds of PVC resin are used annually to make piping products. Two other growing uses for vinyl piping or tubing are in fire sprinkler systems and as nonmetallic electrical conduit. In sprinkler systems, chlorinated PVC (CPVC) piping helps make these important life safety systems easier to install and more affordable. In conduit, vinyl resists short-circuiting and corrosion, and does not conduct electricity, which adds an extra measure of safety to electrical systems. Other popular construction industry uses for vinyl include house siding, window frames, gutters and downspouts, resilient flooring, and wall coverings. In both new construction and remodeling, vinyl combines durability, low maintenance, and good looks, making it a preferred choice for value-conscious architects, builders, and homeowners [2].

17.2.2.1.3 Electrical Applications

17.2.2.1.3.1 Cables and Wires

The single most important component of a wire or cable is its insulation. The selection of insulation is governed by a number of factors: stability and long life; resistance to sunlight (ultraviolet [UV]); dielectric properties; resistance to ionization and corona; resistance to high temperature; resistance to moisture; mechanical strength; and flexibility. No single insulation is ideal in every one of these conditions. It is necessary, therefore, to select a cable with the type of insulation that most fully meets the requirements of the particular installation involved.

Vinyl was first used in electrical applications more than 50 years ago as a replacement for rubber insulation. Today, vinyl commands nearly half of the market for electrical applications such as wire insulation and sheathing. This is because of vinyl's reliable durability and outstanding safety record. Apart from the use in non-metallic conduit, vinyl is used in a number of other electrical applications, ranging from simple house wire to complex telecommunications circuitry. Presently, emerging applications for vinyls include coaxial cable and fiber optic cable. The use of vinyl is also found in a number of electrical components, such as boxes and connectors. Wiring in appliances, cords on power tools, in appliances, and automotive systems are all based on vinyl insulation to preserve electrical power and vinyl covering or jacketing to protect against chemicals, abrasions, cuts, and moisture. Vinyl's inherent flame-retardancy suggests its use in important electrical applications. Vinyl resists flame spread and ignition. It burns at a higher temperature than other materials and will usually not burn once the flame source is removed. These are characteristics that often make vinyl the material of choice in construction and transportation applications. Presently, flexible vinyl is the the predominant material used in wire and cable jacketing and insulation, commanding approximately 40% of a 1.4-billion-pound market. Flexible PVC is also used to make blood storage bags (Figure 17.2), blood tubing used during hemodialysis, endotracheal tubes, intravenous solution dispensing sets, as well as for drug product storage and packaging. The nature of the material means that blood can be stored safely in a lightweight bag for longer periods of time. It can be easily processed into a wide range of flexibilities and is easy to handle during installation. If properly formulated, flexible PVC is non-inflammable and does not harden and crack upon aging. Many different formulations have been reported [3].

Presently, power cable consists of three types of products, depending upon the intensity of power. The low-tension sector requires cables up to 1.1 KV and

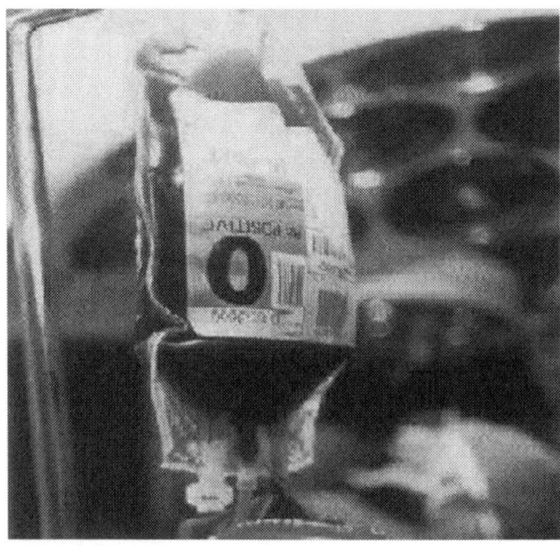

FIGURE 17.2 Use of flexible PVC in blood storage.

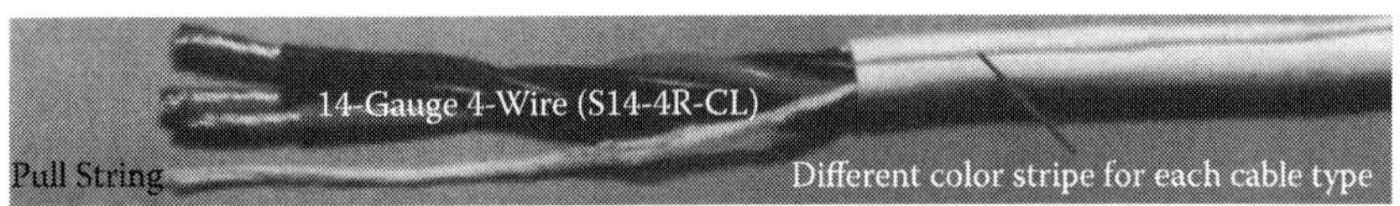

FIGURE 17.3 Spherical wire.

predominantly uses PVC insulation and jacketing. The medium-tension power cable between 1.1 KV and 11 KV uses PE as well as PVC. The high-tension sector is beyond 11 KV. This sector uses PE exclusively [4]. The low-tension cables are used for tertiary distribution of power, mainly from substation to buildings. They also include housing wires. This sector contains many other products for myriad applications. The medium tension cable generally uses silane crosslinkable PE compounds, whereas the high-tension cables use continuous vulcanization cables made from PE (LDPE).

Despite its limitations, PVC is still adequate for a wide range of low voltage applications and other wiring requirements, and is used in many different types of cable. The three forms of commonly used PVC cable are:

1. Circular (i.e., speaker cable) in which a conducting core (internal wire) is jacked with a single layer of PVC or a number of conducting wires are insulated, twisted together, and then covered by an outer PVC jacket for installation, as depicted in Figure 17.3; this outer sheathing protects the cable during installation and makes the cable last longer.
2. Twisted, in which singly insulated conducting cores are simply twisted together.
3. Flat, in which each of the conducting cores (internal wires) are placed in parallel side by side in a single insulated or in separate insulating layers held together by an outer PVC sheath.

17.2.2.1.3.2 Conduit, Terminal Boxes

In recent times, the use of extruded unplasticized PVC has been widely exploited as conduit for electric wiring both in rigid continuous tubular and flexible spirally wound form. Junction boxes and similar fittings are injection molded from unplasticized PVC. In such applications, the main attraction of PVC is due to its freedom from corrosion, although the ease of installation and handling are also important.

17.2.2.1.3.3 Battery Separators

Porous sheets of PVC are normally used for the separators of an accumulator's batteries and are produced by a variety of methods. PVC separators have a very low electrical resistance, which reduces internal loss, saving on electric energy and improving battery performance. Their high porosity ensures easy diffusion of electrolyte and movement of ions, thus guaranteeing battery performance even at high discharge rates. Being completely non-reactive to acids, active metals, and emitted gases, PVC separators, which combine the outstanding characteristics of low electric resistance, chemical cleanliness, higher porosity, low pore size, and superior corrosive resistance with a minimum level of oxidizable organics, are very usable for automobile, traction, stationary, train lighting, and all other lead acid batteries. In addition, a different kind of sheet separator is also produced by other methods, such as extraction of particulate fillers of appropriate size, which is then compounded in before formation of the sheet by extrusion or calendering [5].

17.2.2.1.4 Packaging

17.2.2.1.4.1 General

Performance of a different type has helped vinyl gain a growing share of the packaging market in films, wraps, containers, and bottles. One of the widespread and oldest uses for PVC is as supported and unsupported flexible sheeting or leather cloth for handbags, wallets, shopping bags, suitcases, sport bags, etc., where the materials score on the ground of attractiveness, wear behavior, and low cost. This is, strictly speaking, a packaging application. However, we are more concerned with conventional packaging of commodities for transport, storage, and display. Even under this definition, another application for PVC must be considered, which one would not usually think of as packaging, namely covers for books and wallets for photographic transparencies. These are usually fabricated by HF sealing of calendered plasticized sheeting. The specific permeability behavior of PVC is attractive for a number of packaging applications, and the slow development in this field must be attributed to an unnatural fear of the difficulties of processing PVC arising from its relative instability.

17.2.2.1.4.2 Toxicity

In a fire, all organic materials, including PVC, produce carbon monoxide, which is toxic, odorless, and potentially lethal. Burning PVC releases carbon monoxide, carbon dioxide, hydrogen chloride, and water vapor. Hydrogen chloride is an irritant, but not a narcotic, so its presence in a developing fire situation can serve as a warning while still at levels well below those of dangerous toxicity.

Hydrogen chloride gas, when released from burning PVC, rapidly reacts with water vapor to form corrosive hydrochloric acid. However, a number of independent European studies have confirmed that structural damage in buildings, such as concrete affected by fire, is due to high temperatures in excess of 1000°C instead of as a result of acid attack. Dioxins and furans can also be produced in accidental fires when wood, paper, and many other materials (e.g., PVC) are burned. Practical measurements have indicated that no dioxins and furans are produced when PVC is involved in accidental fires [2].

As has been correctly described elsewhere, anything can be harmful in unfavorable circumstances, and it is erroneous to preclude the possibility of any hazard whatsoever. In practice, PVC is accepted as non-toxic for all reasonable purposes. Unfortunately, the necessary ancillary requirement for low toxicity or non-toxicity is required.

17.2.2.1.5 Medical and Surgical Applications

PVC has been widely used for surgery, pharmaceuticals, drug delivery, and medical packaging. Some products include blood bags, medical containers, fluid bags, tubing, heart and lung bypass sets, masks, gloves, bottles and jars, drainage systems, and ducting. The reason to use PVC in the medical sector is its safety, chemical stability, biocompatibility, chemical resistance, and low cost. In addition, it is usable inside the body and can be easily sterilized.

PVC has not been used in many medical and surgical applications. The main uses have been as a stimulant for external parts of the body in plastic surgery and in artificial limbs, and as tubing in blood transfusion and saline apparatus. The latter two applications arise because of a demand from the medical profession for high clarity to permit ready observation of contents. Clearly, the compositions need to be non-toxic, and considerable controversy exists as to whether PVC should be used at all.

PVC is the most widely used thermoplastic material in medical devices due to its:

Safety: Before medical devices can be used, all the components must be fully understood from a toxicological point of view. Consequently, all the materials used to make such components have to be thoroughly tested and assessed in the European Union before being accepted. Experience based on all available knowledge from international environmental and health-care authorities indicates that PVC is safe. It is the best material existing today, which optimizes all performance and safety requirements at lowest cost.

Chemical stability: Material used in medical applications must be capable of accepting or conveying a variety of liquids without undergoing any significant changes in composition or properties.

Biocompatibility: Whenever plastics are used in direct contact with the patient's tissue or blood, a high degree of compatibility is essential between the tissue/blood and the material. The significance of this property increases with time over which plastic is in contact with the tissue or blood. PVC is characterized by high biocompatibility, and this can be increased further by appropriate surface modification.

Chemical clarity and transparency: Because of its physical properties, products made from PVC can be formulated with excellent transparency to allow for continual monitoring of fluid flow. If color-coded application is needed, virtually any color of the rainbow is available.

Flexibility, durability, and dependability: Not only does PVC offer the flexibility necessary for applications such as blood bags and intravenous (IV) containers, but can also be relied upon for its strength and durability, even under changing temperatures and conditions.

Sterilizability: The absence of sources of infection is a fundamental requirement in medical product applications. PVC products can be easily sterilized using such methods as steam, radiation, or ethylene oxide.

Compatibility: PVC is compatible with virtually all pharmaceutical products in healthcare facilities today. It also has excellent water and chemical resistance, helping to keep solutions sterile.

Resistance to chemical stress cracking: PVC's resilience helps ensure that medical products consistently function during extended use in demanding applications. PVC can easily be extruded to make IV tubing, thermoformed to make "blister" packaging, or blow molded to make hollow rigid containers. This versatility is a major reason why PVC is the material of choice for medical product and packaging designers.

Low cost: The use of PVC plays a big role in containing rising healthcare costs. With PVC accounting for over 25% of medical plastics currently in use, a switch to an alternative could cost the healthcare community hundreds of millions of euros. No less important for the wide variety of applications of PVC are its printability, its transparency or translucency as required by the application, its low tendency to form micro-voids (significant for gloves), and its gloss. These qualities help to promote the safety of patients and medical staff, while also limiting healthcare costs. Indeed, PVC is the best material that meets the performance, safety, and cost criteria for a wide variety of today's medical applications, especially those intended for single use. As a result, more than 25% of all plastic-based disposable medical devices used in hospitals are made from PVC. They have been a major contributor to improving medical safety by reducing the risk of life-threatening infections.

In addition to its specific healthcare benefits, PVC's very versatile properties mean that it can be used in a broad range of other applications. For example, within a hospital, it may be used in water and drainage pipes and in fire-resistant cabling in electrical and telecommunications equipment.

17.2.2.1.6 Toys and Sporting Goods

Play-balls of all sizes that are rotationally cast from PVC plastisols have made a tremendous impact on the toy and sporting goods markets since they were introduced about 10 years ago. They can readily be made in different colors, are easy to inflate, and are cheaper than many other toys. They are also rotationally cast or slush molded from plastisols.

17.2.2.1.7 Transport

A large proportion of plasticized PVC is used in cars and other transport applications. PVC coating is used in a variety of ways for the interior upholstery, interior panels, trim, under-the-car abrasion coatings, and folding hoods. Clear PVC sheeting is used as a window pane in the latter. Cables jacketed with PVC insulation and sheeting are used for wiring the electrical systems. Various kinds of fittings including crash pads, armrests, and other padded fittings are rotationally cast from PVC plastisols and filled with polyurethane foams. Fittings for gear levers, such as sleeves and gaiters, are also cast from plastisols, whereas protective jacketing for plugs and coils is applied by dip-coating. Rigid sheeting is used for internal lining and cladding of vans (e.g, transport of meat), where the chemical resistance, low thermal conductivity, mechanical toughness, and high specific heat are of value. Glass-fiber-reinforced polyester supported unplasticized sheeting is used for tanker trucks [2].

17.2.2.1.8 Agriculture Applications

Polymers have been used in agriculture and horticulture since the middle of the last century. The growing use of plastics in agriculture has enabled farmers to increase their crop production, while restricting the pressure on our environment and quality of our food. Today's plasticulture (i.e., use of plastics in agriculture) results in increased yields, earlier harvests, less reliance on herbicides and pesticides, better protection of our food products, and more efficient water conservation. Agricultural plastics have a vital role to play. Plastics-based agricultural systems provide effective solutions to crop growing in many ways: In arid regions, for example, plastics piping/drainage systems can cut irrigation costs by one- to two-thirds, while as much as doubling crop yield.

The following are some typical applications:

- Growing pots and propagation trays
- Boxes, cases, and crates for crop collecting, handling, and transport
- Fixed and portable tanks for storage and bulk transport of liquids
- Packaging containers and applicators for farm chemicals, fertilizers, insecticides, and herbicides
- Spare parts and accessories for farm equipment and machinery
- Components, such as fittings and spray cones, for irrigation systems

Very large quantities of rigid and flexible pipe and tubing, extruded in PVC and PE, are used in agriculture for applications such as water supply, irrigation systems, soil drainage, and greenhouse heating pipes (Figure 17.4). They are non-corrodible, resistant to chemicals and light, and are easy to install and move. Extruded PE netting is used for shading and protecting crops against birds, rodents, and deer. Another important application is for soil consolidation, where the netting holds the soil in place (especially on sloping banks) until bushes or trees have developed sufficient root structures to stabilize it. Heavier-gauge netting is often used for lightweight fencing for animal farming. Extruded monofilament in PE and PP is widely used for rope and twine, including baler twine and string for agriculture and forage bales.

Clear PVC, PMMA, glass-reinforced polyester, and polycarbonate are used widely for lightweight glazing in applications such as greenhouse covering. Rigid silos and all kinds of containers for culture can be made of these rigid, non-corrodible plastics sheets.

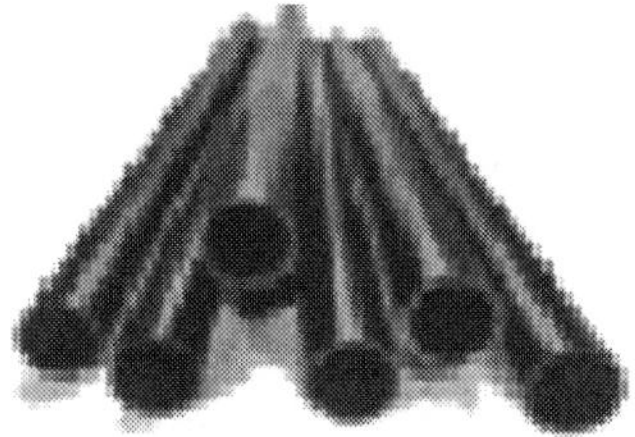

FIGURE 17.4 Pipes made from PVC.

17.2.2.1.9 Furnishings, Lighting, and Associated Fittings

17.2.2.1.9.1 Floor Coverings

The use of vinyl/asbestos floor tiles has been established for many years [6,7]. Their bright colors, good wear properties, heat and chemical resistance, and ease of cleaning made them a better candidate over previous forms of tiles. The increasing use of vinyl in recent years has prompted large-scale production of the product and, therefore, a large number of plants producing vinyl floor tiles by extrusion followed by a two-roll polishing calender have been introduced. Presently, their uses are limited to jaspe, marble, or spatter types. Anything more elaborate would probably be inhibited by increased cost.

Flexible vinyl flooring, either in tile or continuous form, is quieter than vinyl/asbestos, and has better wear resistance, though it tends to be indented slightly more. Another more recently introduced form includes laminates of PVC sheeting with mechanically foamed PVC.

17.2.2.1.9.2 Curtains and Fittings

Although it appears to be difficult for people to accept PVC curtains for general use, calendered, plasticized PVC sheeting can be functional materials for curtains in damp situations (i.e., kitchens, bathrooms, showers). Modern methods of printing permit the production of PVC curtain material in quite complicated multi-color patterns, and the curtains can be very attractive as well as functional.

Many types of curtain rails are now made by the extrusion of unplasticized PVC, and many of the associated fittings are molded from the same material. Despite the common implications that curtain fittings are made of nylon, the latter material is usually confined to the hooks and runners, and one or two of the brackets. Here, PVC scores over brass, etc., by virtue of its freedom from corrosion, the ease with which it can be fabricated into quite complicated forms, its quietness, and its bright color.

Extruded sections of the unplasticized PVC is used for making venetian blinds. Freedom from corrosion, self-coloring, and quietness in use made them advantageous over metals. Its application lies in the use of louvered, calendered, unplasticized PVC films in external window blinds [8].

17.2.2.1.9.3 Wall Coverings

The use of vinyl covering has been prompted because vinyl wall coverings are so easy to clean. They also make it easy to remove sources of known allergens, such as dust and pet dander, thereby improving indoor air quality. In fact, vinyl wall coverings offer significant advantages over competing materials. Vinyl products are extremely durable and compatible with cleaning agents, so that bacteria and other disease-causing microorganisms can be readily removed or disinfected.

The use of PVC falls into three distinct classes, namely unplasticized sheeting as wall panels, lightweight leather cloth, and wallpapers containing clear co-polymer. PVC coated wallpapers offer durable, washable wall coverings. They are long lasting and are available in a wide range of colors, patterns, and textures.

17.2.2.1.9.4 Lighting Fittings

The only major use of PVCs in lighting fittings is as reflectors for fluorescent tubes made from opaque, white unplasticized sheeting. Nevertheless, because this is a satisfactory application, no reason exists as to why translucent materials should not be used for diffusers and globes where the temperatures reached are not too high.

17.2.2.1.9.5 Furniture

PVC leather cloth is used extensively for upholstery, and this application accounts for a good proportion of paste resin manufacture. Co-polymer coating and other forms of laminate are beginning to find application in furniture construction, and shrink-fit extruded tubing from PVC is used to a limited extent. A much publicized but controversial use for PVC in furniture is in inflatable chairs and the like, which are fabricated from plasticized PVC sheeting.

17.2.2.1.10 Clothing

The earliest domestic use of PVC was in the form of calendered plasticized sheeting fabricated into plastic raincoats. Although many of these were unsatisfactory, mainly as a result of a proneness to tearing, excessive stiffening in cold weather, and incorrect formulation or inadequate processing, there is no doubt that this is a good application for the material if properly performed. A few years ago, the use of PVC in clothing increased dramatically due to the introduction of attractive, functional vinyl fabrics and a fashion craze for vinyl clothing. In addition, PVC is widely used for foul weather and industrial protective clothing such as fisherrmen's oilskins and aprons. A major use for PVC is in footwear. Various forms of PVC coated fabric are used for uppers and top-pieces for shoes, boots, and sandals, whereas unit soles, direct molded soles, and molded soles and upper units are now in widespread use. The expanding use of PVC in these applications has been dependent not only on the properties and price of the materials, but also on the development of machines particularly suited to economical production of a variety of types of footwear. For soles and heels, the wear behavior is obviously important. The hardness/softness of the composition is of prime importance, and the bulk softness should generally be between 60–90, with the optimum usually around 70–75. Given proper formulation and processing, durability can be at least as good as and often appreciably better than that of previously established materials such as leather and rubber.

17.2.2.1.11 Communications

17.2.2.1.11.1 Telecommunications

The major use of PVC in telecommunications is as insulation and sheathing for wiring. Unplasticized PVC seems to be an adequate material for telephone instruments.

17.2.2.1.12 Recorded Sound

Unplasticized PVC as a rigid sheet material is used to produce both gramophone records and recording tapes. They are produced almost entirely by compression molding from low molecular weight co-polymers. In the former, PVC seems to provide the best available combination of processing behavior, playing performance, and cost. The co-polymers are used mainly to attain molding cycle times comparable with those required for the earlier records based on shellac. PVC exhibits wear behavior superior to that of other common polymers, although the brilliance of sound reproduction may not be quite as good as with PMMA.

Recording tapes must, among other things, have high strength and modulus in the longitudinal direction. Therefore, they are produced by calendering and stretching unplasticized PVC of relatively high molecular weight — usually by the "Luviterm" process. In strength and modulus, PVC tapes are somewhat inferior to PE terephthalate, but they are appreciably cheaper.

17.2.3 Polyvinyl Acetate

Polyvinyl acetate (PVAc) is a rubbery synthetic polymer. It is prepared by polymerization of vinyl acetate. Partial or complete hydrolysis of the polymer is used to prepare PVA. It was discovered in Germany by Dr. Fritz Klatte in 1912. PVAc is sold as an emulsion in water and as an adhesive for porous materials, particularly wood. It is the most commonly used wood glue, both as "white glue" and the yellow "carpenter's glue;" the former also used extensively to glue other materials such as paper and cloth. It is made by free-radical polymerization of the monomer vinyl acetate.

$$\underset{\text{Vinyl Acetate}}{CH_2{=}CH{-}O{-}C({=}O){-}CH_3} \xrightarrow[\text{Vinyl Polymerization}]{\text{Free Radical}} \underset{\text{Poly(vinyl) Acetate}}{\left[CH_2{-}CH(O{-}C({=}O){-}CH_3)\right]_n}$$

17.2.3.1 Application of PVAc

PVAc is a synthetic polymer and a member of the vinyl ester family. PVAc emulsion adhesives first gained market share by replacing hide glues in the 1940s. Today, vinyl acetate adhesives are the most widely used adhesives on the market, but vinyl acetate emulsions are also heavily used in paints, textile sizing, and non-woven binders. Vinyl acetate emulsion adhesives can be broadly classified as homopolymer or co-polymer.

17.2.3.1.1 Homopolymer Emulsion (PVAc)

PVAc homopolymers were the first PVAc emulsions developed. These polymers are hard and brittle with high molecular weight, high tensile strength, and rapid speed of set. Today, these homopolymers are still the "workhorses" because of their excellent adhesion to a wide variety of "polar" substrates, particularly cellulosic-based substrates such as paper and wood.

PVAc adhesives have the following characteristics:

1. Fast speed of set
2. High strength
3. Excellent adhesion to cellulosic substrates, ceramics, concrete, and glass

4. High molecular weight
5. Cost effective

17.2.3.1.2 Co-polymer Emulsion

Compared with PVAc homopolymers, vinyl acetate co-polymer emulsions offer the advantage of being able to bond difficult-to-bond substrates such as plastic films, coated papers, and metal surfaces. Vinyl acetate co-polymer emulsions are internally flexibilized with a co-monomer such as ethylene (VAE) or an acrylate (VAA). The increased polymer flexibility provides for increased polymer mobility and, therefore, better adhesion.

Applications for PVAc co-polymers include:

1. Textiles
2. Construction
3. Packaging
4. Graphic arts

One of the newer outlets for vinyl acetate [9], and one that has great potential, is in the form of co-polymers with ethylene vinyl acetate (EVA). These are now being produced commercially in the United States, the United Kingdom, and Germany. Applications include molding compositions, floor polishes, heat-sealing compounds, hot-melting adhesives, and water-impermeable coating for paper, often as blends with paraffin wax or chlorinated paraffin wax. The molding and extricable grades of EVA contain approximately 15–20% vinyl acetate co-polymerized with ethylene [10]. They are said to be easily processed in a number of ways, such as injection molding, vacuum forming, cable and wire extrusion, hot-dip coating, and hot-melt adhesive and coating.

More recently, EVA co-polymer dispersions have become available [11]. These contain a lower concentration of ethylene, mainly in the range 10–15%. They find application in paper, textile, and paint.

17.2.3.1.3 Dispersion Polymers

The PVA dispersions are, by far, the major form in which PVAc is used in the market. The main benefit of using the polymer in this form arises from the fact that the dispersions are generally of low viscosity and can be easily handled at a much higher polymer content than solutions. PVAc dispersions are not flammable and do not belong among health-endangering substances such as poisons and other substances harmful to health.

Presently, the use of PVA dispersion has entered into a variety of applications ranging from emulsion to adhesives and has such diverse uses as wallpaper sealers, wood veneering, paper coating, gluing impregnation and finishing of textiles, flooring composition, concrete additives, and as binders in the manufacturing of leather goods. A brief view of some of the uses follows next.

17.2.3.1.3.1 Emulsion Paints

The largest use for PVA dispersions is in the manufacture of emulsion paints. These paints have gained widespread popularity because of their ease of application, lack of odor, and rapid drying time. However, their ease of application contradicts the very complex technology in their manufacture [12].

17.2.3.1.3.2 Concrete and Plaster Additives

The addition of PVA dispersions to cement mortars, concrete, and plaster compositions is widely used to improve the keying of these materials to substrates. Proportions in the region of 5% of the gauging water are used for enabling (e.g., thinner cement coatings to be applied to floors without the risk of failure through poor bonding and cracking). The addition of PVA also improves the cohesion and tensile strength of the composition. As a further aid to bonding, the substrate is often given a coating of the emulsion immediately before the application on the top surface. For this type of work, plasticized homopolymer dispersions containing from 10–20% dibutyl phthalate, based on the total solid content, are usually used.

17.2.3.1.3.3 Flooring Compositions

PVA dispersion based compositions, when combined with sand and other filler media, can be used to construct continuous flooring on appropriate substrates. A hard surface may be laid on a resilient core by applying layers of differing compositions. The lower layers may contain a larger proportion of highly plasticized polymer than the top layer, which in turn will need to be harder to resist abrasion and indentation.

17.2.3.1.3.4 Bonding Fibrous Compositions

Bonding of materials which are fibrous or porous can be achieved by PVA dispersion. Cork, asbestos, cotton waste, flax, wood flour, and scrap leather represent such materials. The PVA is usually added to a slurry of the fibrous material with which it is thoroughly mixed and then precipitated onto the fibers by the addition of a polyvalent salt such as aluminum sulfate. The slurry is then filtered and the coated fibers compressed and dried in heated molds, fusing the polymer with the mass of fibers. The manufacture of shoe insoles from scrap leather is an important example of this process.

17.2.3.1.3.5 Application in the Paper Industry

PVA dispersions have two distinctly separate applications in the paper industry. They may be added to the pulp at the beater stage of manufacture, or they may be used to coat the finished product. Polymer impregnated paper of additional strength is produced if the polymer dispersion is incorporated at the beater stage and in its ensuant precipitation.

Dispersions are applied to finished paper by means of transfer rollers or doctor blades and are dried by passing through hot air ovens. If a high gloss is required, the coated surface is passed over hot calendering rolls. An important outlet for PVA dispersions is wallpaper coating to render the paper smear- and oil-resistant, and to enable the paper to be wiped clean with a damp cloth. A low plasticizer level is employed so that paper does not block on the roll or retain dust when hung. Wallpapers are not calendered, but a semi-gloss finish is achieved through the use of fine particle size dispersion. Improved properties are sometimes claimed with the use of an internally plasticized co-polymer dispersion.

Another application of polymer dispersion in the paper industry is in binding the clay coating that is used to produce a high-quality, dense, smooth surface. Traditionally, starch and casein have been developed for this application. PVA dispersions may also be used to laminate the aluminum foil, and then they are bonded to the aluminum foil by passing both over a hot calender roll to fuse the polymer to metal surface.

17.2.3.1.3.6 Application in the Textile Industry

In the textile industry, PVA dispersions are gradually replacing conventional materials, such as starch and dextrines. They are used for stiffening interlinings in clothing manufacture and are sold as semi-permanent "starch" preparation on the retail market. PVA based sizings are applied to the underside of carpets in the manufacturing stage to increase their stiffness and improve their handling characteristics.

A newer application of co-polymers of vinyl acetate is the bonding of non-woven fabrics such as cotton, nylon, and polyester [13]. PVA dispersions form the basis of many glass-fiber sizes, together with wetting agents, textile lubricants, and Volan (a chromic chloride complex of methacrylic acid) or a silane, such as vinyl triacetoxysilane, to improve the surface properties of the glass fiber for subsequent resin lamination.

With the aid of polymer dispersions, new fields are being found for the use of textile materials A striking example is the invasion of the packaging field by non-woven rayon tapes for strapping. These tapes are made of continuous strands bound together with a PVA emulsion to give a strong product that is competitive with steel and wire tapes and, moreover, is easily dispersed after use.

17.2.3.1.3.7 General Adhesive Applications

PVA dispersions form excellent adhesives for wood, paper, or any porous or fibrous material. Non-porous surfaces cannot be easily glued because the formation of a bond involves loss of water, which is a very slow process if the water only can escape from the glue edges. The amount of plasticizer added to a vinyl acetate homopolymer emulsion used in adhesives is usually only the minimum amount necessary to obtain good fusion between particles. This is particularly true for an adhesive joint that is subject to stress (e.g., in wood joinery work). In this way, the effect of creep is kept to a minimum and the joint strength is at maximum.

Some PVA dispersions have high viscosity and good wet tack, which makes them suitable adhesives for application in high-speed packaging machinery. Though PVA-based adhesives are expensive in comparison to the traditionally used gums and starches in the paper and board industry, the advantage of higher production rates associated with PVA-based adhesives makes them a profitable choice.

PVA emulsions compounded with fillers and extenders to the correct workable consistency are widely used as adhesives for ceramic tiles. These generally contain about 70% whiting or gypsum and 30% polymer, based on the weight of the dried solids. In use, surplus materials can be swabbed off with water before it has dried. Therefore, the compound is much more pleasant to handle, particularly in confined areas, than adhesives containing organic solvents.

17.2.4 Polymethylmethacrylate

17.2.4.1 Introduction

Polymethylmethacrylate (PMMA) is a clear plastic. Commercial grade PMMA is an amorphous polymer of moderate T_g (105°C), light high transparency, and good resistance to acid and environmental deterioration. It is commercially polymerized by free-radical initiators, such as peroxides and azo compounds in suspension or in bulk (e.g., cast polymerization), for sheet and molding compounds or for more specialized applications such as for contact lenses. PMMA may also be polymerized anionically at low temperatures to give highly isotactic (T_g= 45°C, T_m = 160°C) or highly syndiotactic (T_g = 115°C; T_m = 200°C) polymers. PMMA has major applications in the automotive industry (e.g., rear lamps, profiles, and light fixtures), as acrylic sheets for bathtubs, advertisement signs, and lighting fixtures, as well as for composite materials in kitchen sinks, basins, and bathrooms fixtures.

PMMA is more than just plastic and paint. Often, lubricating oils and hydraulic fluids tend to become very viscous and even gummy when they become very cold. This makes it difficult to operate heavy equipment in very cold weather. However, when a small amount of PMMA is dissolved in these oils and fluids, it does not become viscous in the cold, and machines can be operated at temperatures as low as –100°C.

17.2.4.2 Applications

The ability to tailor the hard, stable methacrylates alone or in combination with acrylates to fit specific application requirements, such as outstanding clarity and dimensional stability, and the unusual chemical and light stability of this class of materials has spurred its growth.

17.2.4.2.1 Glazing Materials

The most common plastics used in light-transmitting applications are acrylics, polycarbonate (PC), PVC, and glass-fiber-reinforced polyester (GRP). The largest use of acrylics, especially PMMA, by far, is as a glazing, lighting, or decorative material. PMMA is used for glazing in industrial plants, schools, and other institutional buildings where a high breakage rate (usually caused by vandalism) makes the use of glass costly. Hazardous locations are glazed with PMMA sheets to meet the requirements of safety glazing legislation. PMMA glazing has long been used in military and commercial aircraft. Tinted acrylic sheets can be used to reduce solar heat gain through windows. PMMA sheets, one of the most weather-resistant plastics, are produced worldwide under several proprietary names (e.g., Lucite, Plexiglas, Perspex, and Oroglas).

Acrylic plastics used for light transmission applications are made by the cell casting process. In this method, catalyzed methyl methacrylate monomer syrup is cast between plate glass cells and cured by heating it up to 120°C (248°F). Cell cast sheet is considered to provide the best overall properties of all the forms of acrylic sheet, especially optical clarity. Tabbed and colored sheets are used in decorative applications such as window mosaics, side glazing, patterned windows, and color-coordinated structures. These screens are also used for solar control in sun screens, providing temperature and comfort regulation, while reducing air conditioning and heating costs [14–16].

Because of its impact strength, PMMA is used, for instance, in the lenses of automobile running lights. The spectator protection in ice stadiums is made of PMMA, as are the largest windows and aquariums in the world. The material is used to produce laserdiscs and, sometimes, for DVDs, but the more expensive polycarbonate (also used for CDs) has better properties when exposed to moisture.

17.2.4.2.2 Oil Additives

Long-chain PMMAs are used as additives to improve the performance of internal combustion engines (e.g., lubricating oils and hydraulic fluids). For easy engine starting, it is desirable that the viscosity of lubricating oil be low when cold. As the temperature of the oil increases, the viscosity must be maintained at some acceptable level. Long-chain PMMAs add little viscosity to oil when it is cold, but increase the viscosity of the oil as the temperature is increased. The proper balance of composition and molecular weight of the polymer allows for the formulation of oils of controlled and constant properties.

17.2.4.2.3 Biomedical Applications

The first use of PMMA as a dental device was for the fabrication of complete denture bases. Its qualities of biocompatibility, reliability, relative ease of manipulation, and low toxicity were soon seized upon and incorporated by many different medical specialties. PMMA has been used for (1) bone cements; (2) contact and intraocular lenses; (3) screw fixation in bone; (4) filler for bone cavities and skull defects; and (5) vertebrae stabilization in osteoporotic patients. Although numerous new alloplastic materials show promise, the versatility and reliability of PMMA cause it to remain a popular and frequently used material. In orthopaedics, PMMA bone cement is used to affix implants and to remodel lost bone [17]. It is supplied as a powder with liquid methyl methacrylate (MMA); when mixed together, these yield dough-like cement that gradually hardens in the body [18]. Surgeons can judge the curing of the PMMA bone cement by the smell of MMA in the patient's breath. Although PMMA is biologically compatible, MMA is considered to be an irritant and a possible carcinogen. PMMA has also been linked to cardiopulmonary events in the operating room due to hypotension. Bone cement acts like a grout and not so much like a glue in arthroplasty. Although sticky, it primarily fills the spaces between the prosthesis and the bone, preventing motion. It has a Young's modulus between cancellous bone and cortical bone. Thus, it is a load-sharing entity in the body that does not cause bone resorption.

Methylacrylates are very often used to prepare both soft and hard contact lenses. The detail procedure and chemistry of the contact lenses is reported in Timmer [19], and the methods for making polymeric contact lenses and polymer composition are explained in detail in U.S. Patents 3,951,528; 3,947,401; and 4,239,513 [20–22].

17.2.4.2.4 Optical Applications

Methacrylates are used in the preparation of light-focusing plastic fibers through heat drawing [23] and for low-attenuation optical fibers [24]. The ability to act as a light conduit has led to the use of these materials in medical and scientific applications. Kaetsu et al. and Okubu et al. described methods for the preparation of Fresnel lenses and Fresnel lens films [25,26]. Compositions and methods for the industrial production of cast plastic eyeglass lenses are given in U.S. Patent 4,146,696 [27].

17.2.4.2.5 Other Applications

In semiconductor research and industry, PMMA aids as a resistor in the electron beam lithography process. A solution consisting of the polymer in a solvent is used to spin-coat silicon wafers with a thin film. Patterns on this can be made by an electron beam (using an electron microscope), deep UV light (shorter wavelength than the standard photolithography process), or X-rays. Exposure to these creates chain scission or crosslinking within the PMMA, allowing for the selective removal of exposed areas by a chemical developer. PMMA's advantage lies in that it allows for extremely high-resolution (nanoscale) patterns to be made. It is an invaluable tool in nanotechnology.

Methacrylates are used for the preparation of cultured marble plastic sanitary fixtures [28] and thermoformed bathtubs [29]. The role of high-performance methacrylates in the toy industry in meeting the requirements of the child protection and Toy Safety Act in 1970 is discussed in MacBride [30,31]. Slate impregnated with MMA monomer, followed by polymerization of the monomer, improves the water resistance of the slate [32].

17.2.5 Polyvinyl Alcohol

Polyvinyl alcohol (PVOH) is a water-soluble resin that is produced by the hydrolysis of PVAc, which is made by the polymerization of vinyl acetate monomer. It is classified into two main groups: fully hydrolyzed and partially hydrolyzed grades.

$$\text{--}CH_2\text{-}\underset{\underset{\underset{O}{\|}}{OCCH_3}}{CH}\text{--} + CH_3OH \longrightarrow \text{--}CH_2\text{—}\underset{OH}{CH}\text{--} + CH_3\underset{O}{\overset{}{\underset{\|}{C}}}OCH_3$$

PVOH is a water-soluble synthetic polymer with excellent film-forming, emulsifying, and adhesive properties. This versatile polymer offers outstanding resistance to oil, grease, and solvents, plus high tensile strength, flexibility, and a high oxygen barrier. Suitable applications for PVOH are largely determined by its properties. The basic properties of PVOH depend on its degree of polymerization, degree of hydrolysis, and distribution of hydroxyl groups.

17.2.5.1 Applications

PVOH is generally used in the textile industry for "sizing" cotton fibers: a process where applied chemicals confer strength to the fiber and protect it during the weaving process. PVOH is used widely as an emulsifier and protective colloid in the emulsion polymerization of vinyl acetate and ethylene/vinyl acetate to form high-solid emulsions with outstanding stability and excellent adhesive properties. It is also effective in preparing stable PS and styrene/butadiene co-polymer lattices, and in bead or suspension polymerization of vinyl monomers including vinyl acetate, styrene, and vinyl chloride. It can be used to control particle size and particle size distribution. Polyvinyl alcohols function both as non-ionic emulsifiers and protective colloids in the preparation of oil-in-water type emulsions and dispersions.

The largest application sector for PVOH in the United States is for the manufacture of polyvinyl butyral (PVB). It is a pliable, tough thermoplastic used as the interlayer in laminated safety glass. PVB is formulated to give controlled adhesion to glass and optimum impact performance in laminated safety glass for automotive, security, and general glazing applications. PVB has been used in automotive windshields since 1938, and is now available for use in laminated automotive side window applications for improved security. PVB-laminated glass is also extremely versatile in residential and commercial applications for skylights, atriums, partitions, curtain walls, doors, and even roofs. Glass laminated with PVB will not shatter; even if the glass is broken, the opening is not penetrated because glass fragments adhere to the interlayer.

17.2.5.1.1 Textiles

On a worldwide basis, textile warp sizing is one of the largest applications for PVOH. It imparts high weaving efficiency at low amounts of "add-on" when compared with starch, and it has low biological demand (BOD) compared with many other sizings. Combined with the ability to weave efficiently with low add-on, this results in a low BOD and chemical oxygen demand (COD) in the de-size stream to a finishing mill's wastewater treatment plant or publicly owned treatment works. PVOH can be used alone or in combination with starches and additives, depending on the yarn and weave and on individual mill preference. As a single-component size, PVOH is effective at low add-on for use with the entire spectrum of spun yarns being woven today, including natural, synthetic, and blend yarns such as polyester/cotton spun blends. The low add-on at which PVOH can be applied as a single component size is particularly advantageous for tight fabric constructions that are difficult to weave. In combinations with starch, PVOH improves the strength of the size film and provides the required adhesion to synthetic yarns. Its use is also developed as a finishing resin to impart stiffness to a fabric. It is then used to modify the hand of textile finishes based on thermosetting resins such as Urea formaldehyde and Melamine formaldehyde resins. It is also used as good textile warp size for filament and spun yarn [33].

17.2.5.1.2 Paper Coating

PVOH is an exceptionally powerful pigment binder for paper and paper board coatings, pigmented coatings, in grease proof and other specialty coatings. Its use has also been developed as a pigment binder in clay coating to replace other water soluble binders such as starch, casein, or soy protein [34–37]. The use of PVOH as a pigment binder has been retarded by its lack of water resistance, even though it is economically competitive with casein.

17.2.5.1.3 Adhesives

The unique combination of properties inherent in PVOH has resulted in its use in a wide variety of industrial adhesives. Water solubility — combined with high tensile strength, flexibility, and tack or tackiness — makes it one of the most useful synthetic polymers in the adhesives industry. PVOH is widely used in industrial adhesives for paper and paperboard, and in general-purpose adhesives for bonding paper,

TABLE 17.2 Formulations for PVOH-Based Additives for General Packaging in India

Composition	Parts
PVA Emulsions	100
Plasticizer	10–15
Dextrin	0–100
Clay Filler	0–30
Preservative	0–2
Stabilizer	0–2
Wetting Agent	0–0.2
Secondary water	0–100
Defoamer	0–2
Deodorant	0–1

textiles, leather, wood, and porous ceramic surfaces. Table 17.2 describes the formulations for PVOH-based additives for general packaging in India.

Completely hydrolyzed grades are used in quick-setting water resistant adhesives including paper-laminating adhesives for use in the manufacture of solid fiberboard, spiral wound tubes, cores and drums, and laminated specialties. In many of these applications, PVOH is combined with extenders such as clay. PVOH is also used as a modifier for adhesives based on resin emulsions, particularly PVAc. In these applications, PVOH functions as a protective colloid, emulsifier, thickener, and film former. The higher molecular weight grades of PVOH permit the use of lower adhesive solids by contributing to higher solution viscosity; they also provide excellent mechanical stability properties.

17.2.5.1.3.1 Water-Resistant Adhesives for Paper

PVOH alone, and combined with extenders, clays, and insolubilizers, is used extensively in the preparation of high wet-strength adhesives for paper. Combinations of PVOH with starch or clay are especially effective and economical. When maximum water resistance is required, fully hydrolyzed PVOH are usually selected. Solutions of these grades tend to form a gel structure on long storage. Therefore, where shelf life of the adhesive is important, the special gel-resistant grades are used in manufacturing water-resistant solid fiberboard, for laminating paper, and fabricating paper bags.

17.2.5.1.3.2 Remoistenable Adhesives

The partially hydrolyzed grades of PVOH are usually selected for this application because they are more sensitive to cold water than the completely hydrolyzed grades. When PVOH is used as a remoistenable adhesive, there are fewer tendencies for the coated paper to curl, and problems due to blocking or sticking at high humidities are eliminated.

17.2.5.1.3.3 Binder Adhesives

PVOH-based adhesives are also used as binders for non-woven fabrics, cementations, building products, and ceramics. For most applications where PVOH is used as a binder, excellent results can be obtained by using as little as a 3–5% solution.

17.2.5.1.4 Use as a Colloid Stabilizer

High surface-activity of the PVOH solutions leads to its uses in stabilizing various kinds of hydrogels. The surface tensions of PVOH solutions are largely dictated by the molecular weight and hydrolysis level of PVOH [38]. Molecular weight is measure of the chain length and hydrolysis level is measure of the mole% hydroxyl functionality on the polymer. Generally, the PVOH grades having a degree of hydrolysis between 87 and 89 are used as surface active agents. The main use of PVOH as a colloid stabilizer is in polymer dispersions [39]. In particular, it is one of the most widely used polymers for the colloidal stabilization of vinyl-acetate-based emulsions [40]. PVAc homopolymer and co-polymer emulsions prepared in the presence of PVOH have a number of different applications including adhesives, paints, and textile finishes. Of these, the largest use of PVOH as a colloidal stabilizer is in the production of vinyl-acetate-based emulsions for adhesives. In addition to vinyl acetate, PVOH is also used as a stabilizer for the suspension polymerization of vinyl chloride and the dispersion polymerization of styrene. Usually about 5% of the colloid is used on the weight of the polymer.

17.2.6 Polyvinyl Acetal

Polyvinyl acetal is a complex polymer resulting from the condensation reaction of acetaldehyde and butyl aldehyde with PVA. The aldehyde reacts with adjacent hydroxyl groups on PVA to form acetal rings. The resultant polymer is essentially a co-polymer of acetal rings and unreacted vinyl alcohol. Thus, this polymer exhibits a co-monomer sequence microstructure. It has been reported that acetalization of PVAc is a reversible reaction, and that the distribution of the hydroxyl groups in the polymer is changed by the reaction conditions.

OH OH —+RnCHO→ O O Rn

Polyvinyl Alcohol Polyvinyl Acetal

Synthesis scheme for poly (vinyl –acetal).

When the aldehyde in this reaction is formaldehyde, the product is polyvinyl formal. If the aldehyde in the acetalization reaction is butyraldehyde, then the product is PVB. Polyvinyl formal, PVAc, and PVB have achieved particular commercial importance. The largest application sector for PVAc is the production of safety glass in automotive construction. For architectural uses, plasticized PVB films are used as an intermediate layer in glazing units. The silane-modified PVAcs also have very good suitability for laminated safety glass and glass composites, as well as high-performance safety glass and glazing films, because it is possible to achieve higher tensile stress and further improvement in adhesion to glass. The silane-modified PVAcs have a silicon content of from 0.002–10% by weight, preferably from 0.005–5% by weight, more preferably from 0.01–3% by weight, and most preferably from 0.02–1% by weight, based on the total weight of the silane-modified polyvinyl acetal. Silane-modified PVAcs, where the silicon content is from 0.1–10% by weight, have a high content of free silanol groups or of hydrolyzable alkoxysilane groups or of hydrolyzable alkoxysiloxane groups, and can therefore be crosslinked by using the crosslinking catalysts that are usually used for silanol groups, alkoxysilane groups, or alkoxysiloxane groups.

17.2.6.1 Applications

17.2.6.1.1 Adhesives

The modified polyvinyl acetals are especially capable of being used in the printing ink industry for binders with very good adhesion to various kinds of flexible polymeric films, and for rendering printing inks, which, once applied, have very strong bonding to the substrate and are therefore very difficult to remove from the printed substrate. The excellent adhesion of the silane-modified polyvinyl acetals, particularly PVBs or mixed polyvinyl acetals, make these particularly suitable for use in printing ink formulations. Silane-modified polyvinyl acetals also find their use as binders in the ceramic industry (e.g., binders for green ceramics and ceramic or metal powder in powder injection molding).

The major use of PVB is for an interlayer in automobile safety glass. W.O. Herrmann was the first to propose this use in 1931 [41]. PVB is rubbery and tough and is used primarily in plasticized form as the inner layer and binder for safety glass. When it is mixed with other natural resins, synthetic resins, and drying oil, it can better the resistance to impact flexibility and bond ability of thermoset resins [42–44]. As an adhesive for heat sealing PVB can be mixed with natural resins, synthetic resins, and plasticizing agents and used as an adhesive for heat sealing.

Polyvinyl formal is the hardest of the group; it is used mainly in adhesive, primer, and wire-coating formulations, especially when blended with a phenolic resin. The combination of poly (vinyl formal)–phenolic resin has been developed for use in aircraft manufacturing because of the high-shear-strength values at temperatures up to 250°C, coupled with high peel strength at low temperatures. Adhesive bonds have been durable in many varied climates [45, 46], and the fatigue strength is very high [47]. The polyvinyl formal–phenolic systems are also used in honeycomb construction, brake linings, printed circuits, curtain walls [48], and other structural adhesives [49]. Polyvinyl formal is being used increasingly as a binder in magnetic tapes as well as a surface coating to improve durability.

17.2.6.1.2 Coatings

The first and most extensive application for polyvinyl formal resins is in electrical insulation for magnet wire; this still remains its major application [50]. Polyvinyl formal is combined with other ingredients, such as phenolic resins, to produce blends known as "wire enamels." These compositions are coated onto copper or aluminum wire and cured in ovens at elevated temperatures to give crosslinked film coatings with outstanding electrical, physical, and chemical properties. An advantage of the wire enamels based on Vinylac is the consistent uniformity of the coatings whenever the enamels are cured. This has made the type of insulation based on Vinylac a leader in the magnet wire coating field. Polyvinyl formal can also be used in combination with other resins, such as epoxy and melamine–formaldehyde resins.

17.2.7 Polystyrene

Polystyrene (PS) is a polymer made from the monomer styrene, a liquid hydrocarbon that is commercially manufactured from petroleum. At room temperature, PS is normally a solid thermoplastic, but can be melted at higher temperatures for molding or extrusion, and then re-solidified. Styrene is an aromatic monomer, and PS is an aromatic polymer. The chemical makeup of PS is a long-chain hydrocarbon with every other carbon connected to a phenyl group (an aromatic ring similar to benzene).

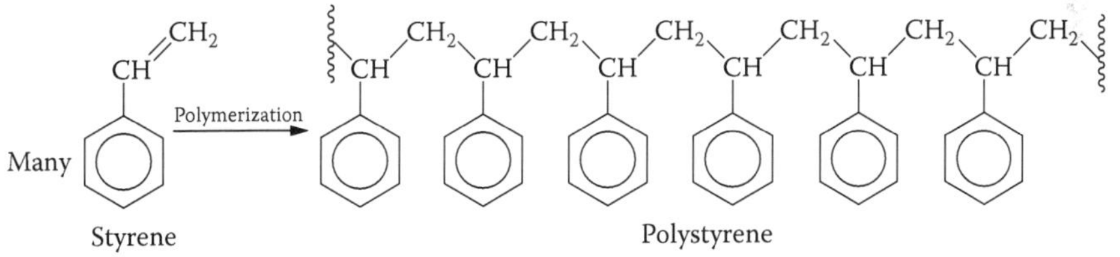

17.2.7.1 Applications

PS is used for the production of a number of applications. However, its major application is as a protective packaging for consumer electronic products and white goods. Its excellent thermal insulation and mechanical protection properties make it ideal to package fish and other foodstuffs. PS also has applications in horticulture as seed trays. The outstanding shock absorbency of expanded PS packaging ensures the protection of a broad range of products. Moreover, its compression resistance means that PS is ideal for stackable packaging goods. When safety is paramount, PS comes into its own. It is used in the manufacture of children's car seats and cycling helmets, where its protective qualities, strength, and shock-absorbency are vital.

17.2.7.1.1 Food Service and Packaging

Many temperature-sensitive pharmaceutical and medical products use PS because of its excellent thermal insulation properties. Strongly relied upon in the food distribution industry, PS is ideal for long distance shipment of perishable foods. PS is highly resistant to heat flow, and the cellular structure of molded PS is essentially impermeable to water. Expanded PS is also used extensively to keep food cold without the need for additional refrigeration (e.g., at fish markets), thus saving energy.

Styrene-based packaging is used to protect valuable electrical and electronic goods and can be reused or recycled where recycling facilities are available in specific areas. One European company is turning rigid television packaging into loose fill material for a second packaging use; Sony reuses PS trays up to 10 times for transporting parts between the Far East and Europe; and Sanyo turns PS packaging into new electrical components.

17.2.7.1.2 Construction

As the importance of energy conservation grows, the need for highly effective insulation materials also increases. PS insulation provides the long-term energy efficiency now demanded by the construction industry. Heating and cooling accounts for 50–70% of the energy costs for the average home. The stable thermal performance of these insulation materials can result in significantly lower heating and cooling costs — savings that really add up over the life of the structure. In addition, their design flexibility means that their benefits can be used in every part of a building — from the foundation, to the walls, to the roof. In addition to reducing energy costs, reducing energy use helps conserve non-renewable fuel supplies. By using more energy-efficient materials and products in construction, the use of fuel and energy is decreased, which translates into reduced air pollution. The energy used to produce PS foam insulation for a typical house is regained after only 1 year through the energy saved. Over a 30-year period, this can add up to 40–60 times the energy used in production and a corresponding reduction of emissions of CO_2 by 10–40 times that used in its production. Comparing different types of insulation material, expanded polystyrene (EPS) insulation has roughly half the economic and environmental cost of mineral wool.

17.2.7.1.3 Office, Commercial, and Industry

These days, when you move into someone else's old office, it is likely that you will feel like you have moved into a brand new one. The furniture and equipment bear little trace of use. That is because most of the office furnishings and equipment you have inherited are made from styrenic plastics — materials that have become indispensable in our working environment. Whether for your wood-finished desktop, computer monitor, letter trays, or the coat hanger behind your door, the properties of styrenic plastics products are equal to the task. In neutral shades or brightly colored, transparent, or opaque, they are not simply pleasing to the eye or touch — they have important qualities that go far beyond mere aesthetics. PS, acrylonitrile-butadiene-styrene (ABS), or styrene-acrylonitrile (SAN) are among the products that make computer casings and monitors strong and heat-resistant; staplers that are resilient enough to withstand repeated impacts; jewel cases that are a cost-effective solution to protect valuable CD-ROMs; and your desktop scratch-resistant despite years of use. The tailored properties and easy processing of styrenic plastics, combined with their remarkable cost-effectiveness, make them an ideal material for the intensively used equipment present in our modern offices.

17.2.7.1.4 Medical and Personal Hygiene

Many vaccines must be maintained at low temperatures. This is easy enough when a refrigerator or freezer is handy, but it is a big challenge when the task is to transport a batch of vaccines to another continent in tropical climates. Thanks to a specially expanded PS package, vaccines are being transported safely from Africa to Latin America. Each shipping unit holds 400 freeze-dried 3-ml vaccines, together with 4 kg (10 pounds) of dry ice. The thermal insulation properties of expanded PS maintain the required temperature inside the sealed package for 72 hours.

17.2.7.1.5 School and Child Care

When it comes to selecting the best food service packaging products, many educational food service systems recognize that PS packaging delivers. No other food service packaging material provides PS's unique combination of performance, economic, and environmental benefits. PS food service packaging is extremely strong yet lightweight, provides excellent insulation, enhances food service sanitation, and contributes to protecting public health. It is less expensive than many other food service packaging options, and has less environmental impact during its manufacture and transport than paperboard food service packaging.

17.2.7.1.6 A Safer Car Ride for Children

Car and booster seats made from styrenic plastic components that meet crash-test standards have helped reduce the death rate from motor vehicle occupant-related injuries among children age 14 and under. Children may not be overly concerned about their own safety during their day-to-day activities, but they certainly know what they like to play with and why. Some styrene-based materials are chosen for their rigidity and are used in toys and bicycle helmets, or the desktops on which children do their homework.

17.2.7.1.7 Marine and Pontoon

In construction of marinas or in many other marine applications, the buoyancy of PS can support more weight than alternative materials. A cubic meter of PS with a density of only 16 kg/m^3 has a buoyancy of 984 kg! Empty enclosures can be punctured and lose buoyancy; a fill of PS prevents sinking as long as the PS stays in place.

17.2.7.1.8 In the Home

Styrenic plastics are used in a broad variety of kitchen appliances, both as a primary material of construction as well as for component use. Styrenics are also used in products that are much more pleasurable than the infamous alarm clock: Writing a letter, drawing, painting, and relying on sturdy luggage for your travels also depend on styrenic plastics. Listening to music, watching television, playing video tapes, or filming the family also depend on styrenic plastics: Consumer electronics is a rapidly growing market with a wide array of end uses where the versatility of styrenic plastics can be fully exploited.

1. Building insulation
    - Resistance to water and water vapor penetration
    - Resistance to attack by fungi and bacteria, and does not support insect or pest life.
    - Self-supporting and lightweight
    - Superior R-values per 25 mm (1") of insulation

2. Roof and ceiling insulation
   Whether a flat roof or a pitched roof, a home or an administration building, factories, workshops or warehouses, or a roof on a garden shed, PS is always involved because of its outstanding insulation. It offers economical answers as an insulating system.

17.2.7.1.8.1 Fire Hazard

PS is classified according to DIN4102 as a "B3" product, meaning highly flammable or "easily ignited." Consequently, though it is an efficient insulator at low temperatures, it is prohibited from being used in any exposed installations in building construction. It must be concealed behind drywall, sheet metal, or concrete. Foamed plastic materials have been accidentally ignited and have caused huge fires and losses. Examples include the Düsseldorf airport, the Channel tunnel, where PS was inside a railcar and caught fire, and the Browns Ferry nuclear plant, where fire went through a fire-retardant, reached the foamed plastic underneath, and went inside a firestop.

17.2.8 Poly Vinyl Ether

Commercial uses have developed for several polyvinyl ethers in which R is methyl ethyl and isobutyl. The vinyl alkyl ether monomers are produced from acetylene and the corresponding alcohols, and the polymerization is usually conducted by cationic initiation using Friedel–Craft-type catalysts.

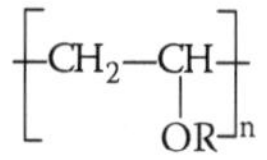

$$\left[-CH_2-\underset{\displaystyle OR}{\underset{|}{CH}}- \right]_n$$

17.2.8.1 Applications

Poly (vinyl ether) must be stabilized by the addition of antioxidants or UV radiation absorbers. They exhibit outstanding tack characteristics and are, therefore, widely used in pressure-sensitive adhesives. They are often modified by the addition of hydrocarbon resins, rosin esters, phenolic resins, or acrylic polymers.

The homopolymers can be cured by radiation or peroxides to produce interesting vulcanizates. Co-polymers with small amounts of vulcanizable groups, such as dienes, ally vinyl ethers, 2-chloroethyl vinyl ether, etc., have also been studied [51]. The vinyl ether can also be the minor component of a co-polymeric elastomer.

17.2.8.1.1 Homopolymers

Among the commercially important homopolymers are those with methyl, ethyl, or isobutyl groups.

17.2.8.1.1.1 Poly (Methyl Vinyl Ether)

This homopolymer is soluble in water at room temperature, but solutions become gradually hazy as the temperature is increased [52]. Eventually, the polymer precipitates; the temperature at which this takes place is known as the "cloud point." The cloud point can be raised (up to 100°C) by the addition of a water-miscible solvent or a surfactant such as the sodium salt of the sulfated adduct of ethylene oxide to nonylphenol.

It has extensive uses in adhesives and coatings, as a non-migrating plasticizer, and as a tackifier. It improves adhesions to metals, glass, plastics, and other surfaces within a broad range of free energies. Because of its hydrophilic character, monomolecular water layers on substrates do not interfere with its adhesion characteristics. Poly (methyl vinyl ether) is a modifier for acrylic pressure-sensitive adhesives for use in tapes, labels, and decals. It can be used as a hot-melt adhesive, as a pigment-wetting agent, and as a plasticizer for printing inks. The polymer also has uses as a textile sizing and as a stabilizer in emulsion polymerization. One particular application has been as a heat sensitizer in the manufacture of rubber-latex dipped goods.

17.2.8.1.1.2 Poly (Isobutyl Vinyl Ether)

This polymer is used either in the dry state or as a solution in an organic solvent. It has excellent adhesion to plastics, metals, and coated surfaces. It also has applications as an adhesion promoter and plasticizer in pressure-sensitive tapes and labels, as well as in various adhesives compositions and surface coatings. The polymer is also useful as a viscosity-index improver for lubricants and as a tackifier for elastomers.

17.2.9 Poly (Vinyl Cinnamate)

Poly (vinyl cinnamate) is conveniently made by the Schotten–Baumann reaction using PVA in sodium and potassium hydroxide solution and cinnamoyl chloride in methyl ethyl ketone. The simplest photodimerizable resist is poly (vinyl cinnamate), made by esterification of PVA with cinnamoyl chloride. Upon exposure, the cinnamate groups can dimerize to yield truxillate or truxinate.

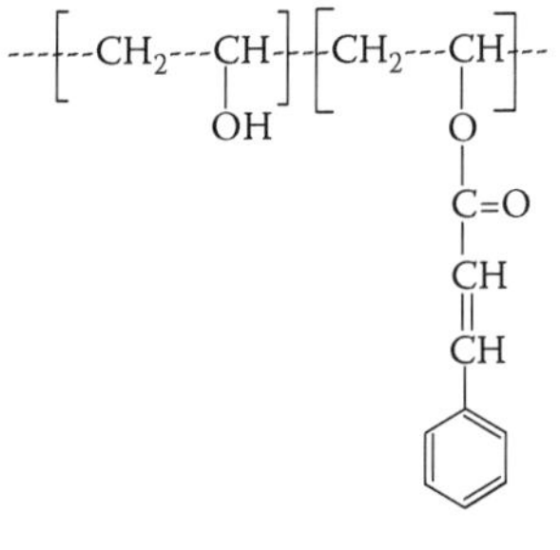

$$---\left[--CH_2---\underset{OH}{CH}-\right]-\left[-CH_2---\underset{\underset{\underset{\underset{C_6H_5}{|}}{\underset{CH}{\|}}}{\underset{CH}{|}}}{\underset{C=O}{\underset{|}{\underset{O}{|}}}}}{CH}\right]--$$

Many cinnamate resins are available, including derivatives of PVA, cellulose, starch, and epoxy resins. It is the epoxy resins that have found the most applications in lithographic materials.

17.2.9.1 Applications

Because cinnamate resins are water-insoluble, a suitable solvent, such as trichloroethylene, is used to achieve the solubility differential for image development. An emulsion of the solvent dispersed in an aqueous phase of gum arabic and phosphoric acid is generally used. The light-exposed regions of the coating are rendered insoluble due to crosslinking and form the printing image.

Poly (vinyl cinnamate) itself is only weakly absorbent above 320 mm. Its photoresponse is generally of the order of a tenth of that in dichromated colloids, but the rate can be accelerated by the use of photosensitizers, such as nitroamines (increase of the order of 100 times), quinones (increase of the order of 200 times for specific ones), and aromatic amino ketones (increase of the order of 300 times for specific ones). A commonly used aromatic ketone is 4,4′-bis (dimethylamino)-benzophenone, also known as Michler's ketone.

Kodak photoresist, which is based upon the poly (vinyl cinnamate) system, is particularly suitable for printed circuit manufacture. It also finds application in some invert halftone photogravure process and photolithographic plates. The property of superior adhesion to metal of cinnamic esters of particular epoxy resins has resulted in the preferred use of these resins for plate making. Very few of the cinnamate-type resists have, however, been used in micro-applications.

17.2.10 Polytetrafluoroethylene

Polytetrafluoroethylene (PTFE) is a vinyl polymer, and its structure, if not its behavior, is similar to PE. PTFE is made from the monomer tetrafluoroethylene by free-radical vinyl polymerization. It was discovered by Roy J. Plunkett (1910–1994) of DuPont in 1938 and introduced as a commercial product in 1946. It is generally known to the public by DuPont's brand name Teflon.

$$\mathrm{F_2C{=}CF_2} \xrightarrow[\text{Vinyl Polymerization}]{\text{Free Radical}} \left[\mathrm{-CF_2-CF_2-}\right]_n$$

Tetrafluoroethylene Polytetrafluoroethylene

PTFE has the lowest coefficient of friction (against polished steel) of any known solid material. It is used as a non-stick coating for pans and other cookware. PTFE is very non-reactive, so is often used in containers and pipework for reactive chemicals. Its melting point is 327°C, but its properties degrade above 260°C.

Other polymers with similar composition are known with the Teflon name: fluorinated ethylene-propylene (FEP) and perfluoroalkoxy polymer resin (PFA). They retain the useful properties of PTFE of low friction and non-reactivity, but are more easily formable. FEP is softer than PTFE and melts at 260°C; it is highly transparent and resistant to sunlight.

It is estimated that the world market for fluoropolymers is between 80,000 and 90,000 tons per year. Although fluoropolymers represent only about 0.1% of all plastics, their use tends to increase at a steady rate because of their outstanding performance characteristics. It is estimated that the world fluoropolymer market could be at 100,000 tons by 2005. PTFE occupies 70% of the total demand for fluoropolymers. The United States accounts for the consumption of 40% of all fluoropolymers. Japan consumes 10,000 tons of fluoropolymers per annum. The world consumption of PTFE is growing steadily as new end uses are developed.

TABLE 17.3
Consumption of PTFE in Various Applications

Resin type	Chemical (%)	Electrical (%)	Mechanical (%)
Fine powder	18	82	—
Granular	33	14	53
Dispersion	30	39	31
Overall by application	25	50	25

17.2.10.1 Applications

Electrical applications consume half of the PTFE produced, and mechanical and chemical applications share equally in the other half. Table 17.3 describes the consumption of PTFE in various applications.

17.2.10.1.1 Automobile Applications

Due to its low friction, PTFE is used for applications where the sliding action of parts is needed: bearings, bushings, gears, slides plates, etc. In these applications, it performs significantly better than nylon and acetal; it is comparable with ultra-high molecular weight PE (UHMWPE), although UHMWPE is more resistant to wear than Teflon. For these applications, versions of Teflon with mineral oil or molybdenum disulfide embedded as additional lubricants in its matrix are being manufactured.

Bearings made of pure and filled PTFE are selected for their low friction, good wear resistance, and excellent chemical resistance from cryogenic temperatures to 260°C. Glass-filled PTFE can be cut or stamped into bearing pads and will resist all weather-related degradation, while remaining an inert interface between disparate construction materials, such as steel and concrete. Granular resin, a large portion of which is supplied as mechanical grade skived tape, is used in over 95% of the bearings fabricated. Significant quantities of granular resins go into making pressure-sensitive tape. The tape is used in moderate temperature, low-friction applications, such as food processing, packaging and heat-sealing equipment, and conveyor surfaces. In automotive applications, seals of PTFE are used on most new power steering units. In aircraft applications, most planes have hydraulic systems containing seals of PTFE. Seals and piston rings are made almost entirely from granular and filled granular resins. Powdered PTFE is used in pyrotechnic compositions as an oxidizer, together with powdered metals such as aluminum and magnesium. Upon ignition, these mixtures form carbonaceous soot and the corresponding metal fluoride and release large amounts of heat. Thus, they are used as infrared decoy flares and igniters for solid fuel rocket propellants [1]. Dispersion resins are used in anti-stick applications.

17.2.10.1.2 Chemical Applications

In the chemical processing and petrochemical sectors, PTFE is used for vessel linings, seals, spacers, gaskets, well-drilling parts, and washers, because PTFE is chemically inert and resistant to corrosion. PTFE is used in a variety of laboratory applications in the form of tubing, piping, containers, and vessels due to its resistance to chemicals and the absence of contaminants attaching to the surface of PTFE products. Because of its chemical inertness, PTFE cannot be crosslinked like an elastomer. Therefore, it has no "memory," and it is subject to creep (also known as cold flow and compression set). This can be both good and bad. A little bit of creep allows PTFE seals to conform to mating surfaces better than most other plastic seals. Too much creep, however, and the seal is compromised. Compounding fillers are used to control unwanted creep, as well as to improve wear, friction, and other properties.

17.2.10.1.3 Electrical Applications

PTFE is one of the best insulators known. In thin sections, it will insulate to 500 volts per mil. Certain grades of PTFE have even greater dielectric strength. It is frequently used in wire and cable wrap, and to separate conductive surfaces in capacitors. Thick-walled, close-tolerance extruded tubing is the PTFE shape of choice where machining or drilling long lengths to close tolerances is impossible. Multi-hole tubing can be extruded. PTFE can be machined into standoff insulators and many different types of high-voltage encapsulation devices for electrical components.

PTFE has excellent dielectric properties. This is especially true at high radio frequencies, making it eminently suitable for use as an insulator in cables and connector assemblies and as a material for printed circuit boards used at microwave frequencies. Combined with its high melting temperature, this makes it the material of choice as a high-performance substitute for the weaker and more meltable PE, which is commonly used in low-cost applications. Its extremely high bulk resistivity makes it an ideal material for fabricating long life, useful devices that are the electrostatic analogues of magnets.

17.2.10.1.4 Semi-Conductor Industry

PTFE is inert, and its operating temperature range is from −177 to 288°C. When made to ultra-pure standards, it is the material of choice for various items used in chip manufacturing, including encapsulation devices for quartz heaters, and the like.

17.2.10.1.5 Food, Beverage, and Pharmaceutical Industries

Virgin PTFE is approved by the Food and Drug Administration for use in the food, beverage, cosmetics, and pharmaceutical industries. Thin film and sheets make an inert, non-toxic slide surface without microscopic depressions where microbes can grow. Conveyor components — profiles, guide rails, and slides — can withstand high temperatures inside baking and drying ovens and other heated segments of the food, cosmetics, or pharmaceuticals manufacturing processes.

17.2.10.1.6 PTFE Compounds

Various fillers can be blended with the PTFE base resin to enhance certain properties (e.g., glass fiber, glass bead, carbon, graphite, molybdenum disulfide, bronze, etc.). PTFE does not melt; it cannot be molded into complex shapes, but must be machined. PTFE is easily machined using standard mechanical woodworking and stamping equipment and tooling. Most shapes are sold slightly oversized for easy trimming and machining to exact sizes.

17.2.10.1.7 Miscellaneous Applications

The uses and applications of PTFE have grown enormously over the past 68 years. Probably one of the most famous uses for PTFE is as a non-stick coating found on cookware (again, under the DuPont Teflon® trademark). However, PTFE has been used in the U.S. space program, as well as in the semiconductor, medical, chemical, automotive, electrical, and petrochemical industries. Plastomer Technologies of Newtown, Pennsylvania, has produced various PTFE shapes, including molded sheets, rods, and cylinders, as well as PTFE skived sheets and films. Other capabilities of Plastomer Technologies include custom machining of PTFE parts to a client's custom requirements, and a wide range of sealing solutions including envelope gaskets, spacers, bottle caps, and liners. As noted previously, PTFE is used in a wide range of industries. Virgin PTFE had been approved by the FDA for use in the pharmaceutical, beverage, food, and cosmetics industries in the form of conveyor components, slides, guide rails, along with other parts used in ovens and other heated systems. PTFE can also be found in the semiconductor sector; it used as an insulator in the production of discrete components such as capacitors and in the chip manufacturing process.

Among many other industrial applications, PTFE is used to coat certain types of hardened, armor-piercing bullets. This reduces the amount of wear on the firearm's rifling. These are often mistakenly referred to as "cop-killer bullets" because of PTFE's supposed ability to ease a bullet's passage through body armor. Any armor-piercing effect is, however, purely a function of the bullet's velocity and rigidity instead of a property of PTFE.

17.2.11 Polyvinyl Pyrrolidone

PVP (polyvinyl pyrrolidone, povidone, or polyvidone) is a water-soluble polymer made from the monomer N-vinyl pyrrolidone.

vinyl-pyrrolidone → poly(vinyl-pyrrolidone)

PVP is crosslinkable to a water insoluble, swellable material either in the course of vinyl pyrrolidone polymerization, by addition of an appropriate multifunctional co-monomer, or by post-reaction, typically through hydrogen abstraction chemistry. PVP is a commonly used inactive ingredient in the preparation of pharmaceutical products. It serves as a dry and damp binding substance in the production of granules and tablets, as well as a thickening agent and solutizer. Crosslinked PVP is also used as a highly active explosive agent and as an accelerating agent for disintegration of solid medications. PVP is a white hygroscopic powder with a weak characteristic odor. In contrast with most polymers it is readily soluble both in water as well as in a large number of organic solvents such as alcohols, amines, acids, chlorinated hydrocarbons, amides, and lactams.

Povidone is the generic name for PVP in pharmaceutical grade. It is the soluble homopolymer of N-vinyl-2-pyrrolidone. As a food and pharmaceutical product additive, it appears in the form identified as K-30. PVP can be plasticized with water and most common organic plasticizers.

17.2.11.1 Applications

17.2.11.1.1 Cosmetics

PVA products are chemically and biologically inert. They are very stable in a wide temperature and pH range, which is due to the PE polymer backbone. No hazard risk exists in case of body contact. The polyvinyl pyrrolidone (PVP-K) series can be used as a film-forming agent, viscosity-enhancement agent, lubricator, and adhesive. They are the key component of hair spray, mousse, gel, lotion, and solution, hair-dying reagent, and shampoo in hair-care products. They can be used as assistant ingredients in skin-care products, eye makeup, lipstick, deodorant, sunscreen, and dentifrice.

17.2.11.1.2 Pharmaceuticals

Povidone K30 is a new and excellent pharmaceutical excipient. It is mainly used as a binder for tablets, dissolving assistant for injection, flow assistant for capsules, dispersant for liquid medicine and stain, stabilizer for enzymes and heat-sensitive drugs, co-precipitant for poorly soluble drugs, and lubricator and anti-toxic assistant for eye drugs. PVPs work as excipients in more than 100 drugs. The complexing ability of PVP has been used to improve the release properties of drugs and anesthetics in topical preparations. Its properties as a suspending agent and detoxificant make for its use in injectable preparations of antibiotics, hormones, and analgesics.

PVP–iodine complex is a useful germicide with greatly reduced toxicity to mammals. It is marketed in gargles, tinctures, and ointments.

17.2.11.1.3 Textiles and Dyes

PVP forms complexes with many dyes and also has surfactant properties that make it useful for fiber dyeing and pigment dispersion. It has been used to improve the dye receptivity of hydrophobic fibers such as polyacrylonitrile or PP. Many dyes become more water soluble in the presence of PVP; the same property makes it useful for stripping cloth or paper. PVP is a suspending agent for titanium dioxide and organic pigments. It promotes better gloss because of its film-forming ability and is used in links and polishes.

17.2.11.1.4 Adhesives

Polyvinyl pyrrolidone (PVP) exhibits numerous essential applications for adhesives, which support its use in highly demanding applications.

Glue sticks can be a demanding application if you just simply focus on safety aspects. PVP is not only a key component for good tack and adhesion, but it is also extensively used in this application because it is non-toxic and it is chemically and biologically inert. Glue sticks are very often used by children, and in addition to not causing any harm if swallowed, PVP can help a great deal to clean any spillage on their outfits because it is also soluble in water.

Wetness indicators are increasingly used in diapers to help with visual check. Adding PVA 64 to hot-melts used for diaper assembly has become a true asset in this field due to its capability to carry the moisture to the wetness indicator. When the diaper becomes wet, PVA 64 carries the humidity to the wetness indicator, which changes in color, indicating that the diaper needs to be changed. Because it is bio-compatible and non-toxic, PVA 64 is perfectly suited for body contact.

Industrial hot-melts should ideally be formulated to have a low melting point to reduce the time and energy required for applying them, and should form a continuous film when spread over the substrates to be bonded. PVA 64 is the product of choice for formulating hot-melts: It provides good tack when processing the adhesive, as well as durable adhesion to the assembly. It reduces the melting temperature and is thermally stable over a wide range of temperatures, and it improves the formation of the adhesive film on the substrates.

PVA 64 is particularly suitable for adhesives formulated for bonding to glass, metal, paper, and most plastic surfaces. Because PVA is soluble in water, it is particularly suited for the production of water-soluble adhesives such as those uses for remoistenable envelopes.

PVP becomes a hydrogel, which is particularly adapted to skin adhesive applications. Once on the skin, it makes the pad easy to remove with no harm. Hydrogels can be combined with inorganic salts to make conductive gels that are applied to pads used for electrosurgery. Moreover, it has neither chemical nor biological impact on the human body and the environment.

PVP is a transparent compound that does not alter in color with time. As such, PVP products will not affect the color stability or the aesthetics of the product.

Co-polymer of vinylpyrrolidone with vinylacetate:

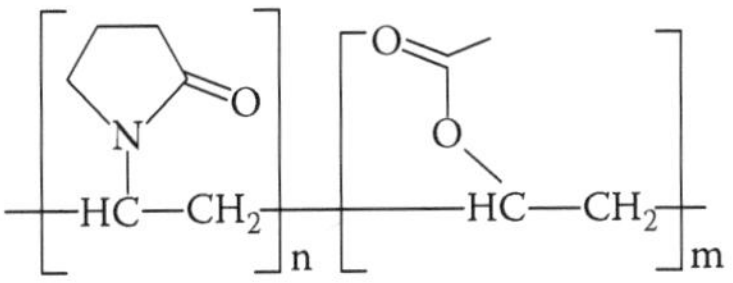

O N O O HC—CH_2 n HC—CH_2 m

Co-polymers with different proportions of vinylpyrrolidone to vinyl acetate can be offered in powder or alcohol solution product. The aqueous product could be cloudy. The clarity of the solution depends on the proportion of vinylpyrrolidone to vinyl acetate. The solution is slightly acidic but not electrolytic.

17.2.11.1.4.1 Applications

17.2.11.1.4.1.1 Cosmetics

PVP/VAc series products play a key role as film-formers and hair-fixing agents. The hydrophobe decreases with the increase in the proportion of vinyl acetate in the molecular structure. This property is extremely useful for PVP/VAc as hair sprays and setting lotions. If being used with PVP K30, the effect would be enhanced.

17.2.11.1.4.1.2 Other Utilities

1. Remoistenable adhesive and other adhesives for paper
2. Viscosity-enhancing agent and protecting colloid for printing ink
3. Binder in coatings (paints) and protecting colloid for printing ink
4. Binder in coatings (paints) and stabilizer for the dispersion of dyeing
5. Film coating agent in pharmaceutical industry

17.2.11.1.4.1.3 Miscellaneous Uses

The monomer is carcinogenic and is extremely toxic to aquatic life. However, the polymer PVP in its pure form is so safe that not only is it edible by humans, it was used as a blood plasma expander for trauma victims after the first half of the 20th century. It is used as a binder in many pharmaceutical tablets; because it is completely inert to humans, it simply passes through the body. PVP added to iodine forms a complex; in solution, it is known under the trade name Betadine. PVP binds to polar molecules exceptionally well, owing to its polarity. This has led to its application in coatings for photo-quality inkjet papers and transparencies, as well as in inks for inkjet printers. PVP is also used in personal care products, such as shampoos and toothpastes, in paints, and in adhesives that must be moistened, such as postage stamps and envelopes. It has also been used in contact lens solutions and in steel-quenching solutions. PVP is the basis of the early formulas for hair sprays and hair gels, and still continues to be a component of some. PVP is an effective suspending agent, either as a primary protective colloid or as a secondary dispersant in polymerization. It is used in paper as a pigment dispersant. In paper coating, it is used as a leveling agent and also to improve the dye receptivity. It is also used as an assistant in the realm of paint, plastics and resins, adhesives, glass fiber, film, ink, TV tubes, detergent, biocide, tabulating, and printing.

17.2.12 Polyethylene

Ethylene is one of the highly used petrochemicals. It may be polymerized by a variety of techniques to produce products as diverse as low-molecular-weight waxes to highly crystalline, high-molecular-weight PE (HDPE). The very first commercialized polyolefin (1939) was a low-crystallinity, low-density PE (LDPE) produced in 1939 by ICI in England. The majority (65%) of LDPE produced in the United States is used as thin film for packaging, whereas the remaining production finds its use in wire and cable insulation, coatings, and injection-molded products.

LDPE is a branched polymer. When no branching exists, it is called linear PE (Figures 17.5 and 17.6), or HDPE. Linear PE is much stronger than branched PE, but branched PE is cheaper and easier to make.

FIGURE 17.5 A molecule of linear polyethylene, or HDPE.

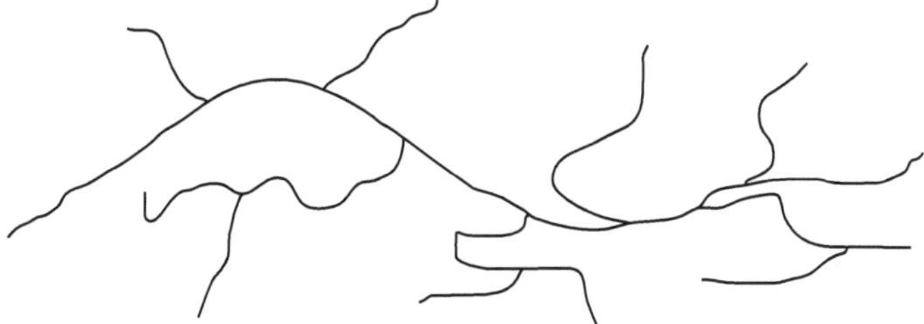

FIGURE 17.6 A molecule of branched polyethylene, LDPE.

Branched PE is often made by free-radical polymerization. Linear PE is made by a more complicated procedure called Ziegler–Natta. UHMWPE is made using metallocene catalysis polymerization.

PE is cheap, flexible, durable, and chemically resistant. LDPE is used to make films and packaging materials, including plastic bags, whereas HDPE is used more often to make containers, plumbing, and automotive fittings. Although PE has low resistance to chemical attack, it was found later that a PE container could be made much more robust by exposing it to fluorine gas, which modified the surface layer of the container into the much tougher "polyfluoroethylene."

17.2.12.1 Classification of PE

UHMWPE is PE with a molecular weight numbering in the millions, usually between 3.1 and 5.67 million. The high molecular weight results in less efficient packing of the chains into the crystal structure as evidenced by densities less than HDPE (e.g., 0.935–0.930). The high molecular weight results in a very tough material. UHMWPE can be made through any catalyst technology, although Ziegler catalysts are most common. Because of its outstanding toughness, cut, wear, and excellent chemical resistance, UHWMPE is used in a wide diversity of applications. These include can- and bottle-handling machine parts, moving parts on weaving machines, bearings, gears, artificial joints, edge protection on ice rinks, and butchers' chopping boards. It has even replaced Kevlar in new bulletproof vests.

HDPE is defined by a density of greater or equal to 0.941 g/cc. HDPE has a low degree of branching and thus stronger intermolecular forces and tensile strength. HDPE can be produced by chromium/silica catalysts, Ziegler–Natta catalysts, or metallocene catalysts. The lack of branching is ensured by an appropriate choice of catalyst (e.g., chromium catalysts or Ziegler–Natta catalysts and reaction conditions). HDPE is used in products and packaging such as milk jugs, detergent bottles, margarine tubs, and garbage containers.

PEX is a medium- to high-density PE that contains crosslink bonds introduced into the polymer structure, changing the thermoplast into an elastomer. (The acronym for medium-density polyethylene is MDPE.) The high-temperature properties of the polymer are improved, its flow is reduced, and its chemical resistance is enhanced. PEX is used in some potable water plumbing systems, as tubes made of the material can be expanded to fit over a metal nipple, and it will slowly return to its original shape, forming a permanent, water-tight connection.

MDPE is defined by a density range of 0.926–0.940 g/cc. MDPE can be produced by chromium/silica catalysts, Ziegler–Natta catalysts, or metallocene catalysts. MDPE has good shock and drop resistance properties. It also is less notch sensitive than HDPE, and stress cracking resistance is better than HDPE. MDPE is typically used in gas pipes and fittings, sacks, shrink film, packaging film, carrier bags, and screw closures.

LLDPE is defined by a density range of 0.915–0.925 g/cc. It is a substantially linear polymer, with significant numbers of short branches, commonly made by colpolymerization of ethylene with short-chain alpha-olefin. LLDPE has higher tensile strength than LDPE. Exhibits higher impact and puncture resistance than LDPE. Lower thickness (gauge) films can be blown compared with LDPE, with better environmental stress cracking resistance compared with LDPE but they are not as easy to process. LLDPE is used in packaging, particularly film for bags and sheets. Lower thickness (gauge) may be used compared with LDPE to make cable covering, toys, lids, buckets and containers, and pipe. Although other applications are available, LDPE is used predominantly in film applications due to its toughness, flexibility, and relative transparency.

LDPE is defined by a density range of 0.910–0.940 g/cc. LDPE has a high degree of short and long-chain branching, which means that the chains do not pack into the crystal structure as well. It therefore has less strong intermolecular forces because the instantaneous-dipole induced-dipole attraction is less. This results in a lower tensile strength and increased ductility. LDPE is created by free-radical polymerization. The high degree of branches with long chains gives molten LDPE unique and desirable flow properties. LDPE is used for both rigid containers and plastic film applications such as plastic bags and film wrap.

VLDPE is defined by a density range of 0.880–0.915 g/cc. It is a substantially linear polymer, with high levels of short chain branches, commonly made by co-polymerization of ethylene with short-chain alpha-olefins (e.g., 1-butene, 1-hexene, and 1-octene). VLDPE is most commonly produced using metallocene catalysts due to the greater co-monomer incorporation exhibited by these catalysts. VLDPEs are used for hose and tubing, ice and frozen food bags, food packaging, and stretch wrap, as well as impact modifiers when blended with other polymers.

HDPE is also widely used in fireworks. In tubes of varying length (depending on the size of the ordnance), HDPE is used as a replacement for the supplied cardboard mortar for two primary reasons: it is much safer than the supplied cardboard tubes, and because if a shell were to malfunction and explode inside an HDPE tube, the tube will not shatter.

Recently, much research activity has focused on the nature and distribution of long-chain branches in PE. In HDPE, a relatively small number of these branches (perhaps 1 in 100 or 1000 branches per backbone carbon) can significantly affect the rheological properties of the polymer.

17.2.12.1.1 Applications

PE shows its true strength when exposed to high temperatures and oxidizing chemicals. It resists flex cracking and abrasion as well as the damaging effects of weather,

UV/ozone, heat, and chemicals. Easy pigmentation and color stability, along with low moisture absorption, complement its use in single-ply roofing systems, automotive and industrial hose and timing belts, and wire and cable jacketing and insulation. Equally important, it has demonstrated long life in harsh environments, as illustrated by liners and floating covers for water and sewage containment.

PE foam is a strong, resilient, closed-cell foam. It is used for shock absorbing, vibration dampening, loose fill, and its most common use: cushioning products in packaging applications. PE is lightweight, shatter proof, flexible, cost-effective, and impervious to mildew, mold, rot, and bacteria.

17.2.12.1.1.1 Packaging

PE foam is a very resilient foam used extensively in packaging. It is available in a number of different densities, sizes, colors, and sheets to meet the needs of the application. Applications include material handling, packaging, or electronic protection. HDPE is used for many packaging applications because it provides excellent moisture barrier properties and chemical resistance. However, HDPE, similar to all types of PE, is limited to those food-packaging applications that do not require an oxygen or CO_2 barrier. In film form, HDPE is used in snack food packages and cereal box liners; in blow-molded bottle form, for milk and non-carbonated beverage bottles; and in injection-molded tub form, for packaging margarine, whipped toppings, and deli foods. Because HDPE has good chemical resistance, it is used for packaging many household as well as industrial chemicals such as detergents, bleach, and acids. General uses of HDPE include injection-molded beverage cases, bread trays, as well as films for grocery sacks and bottles for beverages and household chemicals.

17.2.12.1.1.1.1 Aerospace

Because of the closed-cell, lightweight nature of PE foam it is a very natural choice for the aerospace industry. PE foam offers superior resistance to vibration, compression, and moisture. It also has excellent resistance against UV and ozone and maintains its physical properties at both high and low temperatures. One of the most likely places to find closed-cell foam in an aircraft would be sandwiched between bulkheads. It offers thermal insulation because of the foam's closed-cell construction as well as sound dampening from the surrounding noise of the aircraft.

17.2.12.1.1.1.2 Flotation

Closed-cell foam floats well because of the structure of the foam itself. The cells of the foam are closed meaning they are not interlocking as in open-cell foam structure. Because the foam cells are closed, this gives the foam good buoyancy in that the foam itself cannot become saturated with liquids, causing it to sink. Uses for PE in floatation applications include life jackets and other personal floatation devices. The less dense the material, the more buoyancy it provides. Closed-cell foam is also used in watercraft in much the same manner it is used in the aerospace industry: It is placed between the inner and outer hulls of boats ranging in size from small personal watercraft, such as kayaks, to watercraft as large as speed boats and military attack craft.

17.2.12.1.1.1.3 Appliances

Closed-cell foam insulation is used in HVAC products as thermal insulation. Closed-cell foam insulation offers many benefits over fibrous insulation materials such as fiberglass. Because the cells of PE are closed and the external surface is durable, the foam resists dirt and moisture accumulation. This has the effect of reducing potential fungal or biological pathogen growth. The surface of closed-cell foam is also puncture resistant when compared with other fibrous insulation materials. In the event the surface of the foam is punctured or torn, the properties of PE closed-cell foam remain intact, providing an insulating material that remains resistant to moisture accumulation. Foam PE tapes are manufactured from mainly closed-cell crosslinked foam. These tapes exhibit high strength and excellent weather and heat stability. PE tapes and tape components are used for economical seals, cushioning, and vibration control pads between plastic-to-plastic components in many automotive applications. Other uses include pads and shapes for point of sale display mounts, and general self-adhesive cushion pads for mounting and securing. PE tapes are also recommended for use in the glass sector for Georgian Bar fixing and security glazing applications.

17.2.12.1.1.1.4 Agriculture

Agriculture applications for underground storage tanks range from replacing the old concrete tank by the windmill to pre-treating water before it is released into a sensitive fish farm lagoon. A large volume of stored and treated water can be used to flush and sterilize below hog and cattle pens with elevated floors. PE OcTanks were designed into a methane powered cogeneration system as anaerobic digesters that convert animal waste into combustible gas in a Colorado hog feeding operation. OcTanks can also be used to digest dead poultry for safe disposal.

The agricultural uses for underground storage tanks are vast and varied across the country. Contrast raising aquatic plants and fish or crustaceans in Georgia to hog and cattle production in the Midwest. The most active agricultural underground tank applications have traditionally been in the areas of waste treatment and odor control.

17.2.12.1.1.1.5 Wire and Cable

PE's excellent properties make it easy to use in many kinds of telecom and power applications. Its main characteristics are low dielectric loss, high dielectric strength, chemical inertness, low moisture up-take, and ease of extrusion. Wire and cable insulation and jacketing are produced by extruding the PE through a cross-head and delivering the molten polymer onto the bare conductor (insulation) or the assembled insulated wires (jacketing). PE can be used in its thermoplastic form for telecom insulation and for jacketing. For power applications, crosslinked PE is mainly used because of its improved thermomechanical resistance. Crosslinking can be achieved by use of silane or peroxide crosslinking agents. In certain compounds, additives are also used to improve resistance to aging and to protect against copper catalyzed degradation.

LDPE is the largest single segment of the worldwide plastics market. Demand for this popular material continues to grow (3% per year in the Asia Pacific region) because of its proven success in applications such as film for high-clarity and food packaging, shrink wrap, bags, and agricultural uses; lids and containers; and adhesives. LDPE is predominantly used in film applications due to its toughness, flexibility, and transparency. LDPE has a low melting point, making it popular for use in applications where heat-sealing is necessary. Typically, LDPE is used to manufacture flexible films, such as those used for dry cleaned garment bags, and to produce bags. LDPE is also used to manufacture some flexible lids and bottles, and it is widely used in wire and cable applications for its stable electrical properties and processing characteristics.

17.2.12.1.1.1.6 Tarps

PE tarps or tarpaulins are rip-stop, lightweight plastic tarps with excellent wear, low-moisture, and stain-resistant properties. This makes them a superior choice for many uses over traditional canvas tarps. PE tarps can be used for so many applications that the list is virtually inexhaustible. Homeowner uses range from covering woodpiles, barbecues, lawn equipment, and boats, to camping uses such as providing clean ground cover, a shade canopy, or tent fly. Swimming pool covers are also made from PE that is characterized by its ability to float. Commercially, huge PE tarps are used to cover tennis courts and baseball fields in the rain to keep the ground from getting too wet to play. They are also used to protect gym floors and basketball courts. Truckers use PE tarps for covering trailers, pallets, hay, and other payloads. Special heavy-duty insulated tarps have a polyester fill for keeping out heat and cold.

17.3 BLENDS OF VINYL POLYMERS FOR SPECIALTY APPLICATIONS

17.3.1 Conductive Composites from EPDM-EVA-Carbon Black

In conductive polymer composites at normal temperature and normal concentration of filler, two dominant mechanisms of DC conduction are observed: tunneling and electron hopping. Modern 11-KV cable insulation must operate at considerable electrical stresses. Because of this, the design includes semi-conductive polymer layers, which are applied against the conductors of high-voltage cables. This has the effect of redistributing the electrical stresses and reduces the possibility of electrical breakdown of PE insulation. Conductive polymer layers made of EVA co-polymers are in greater use today. The presence of polar groups in polymer components affects the conductive network formation. While studying the electrical conductivity of carbon-black-filled EVA co-polymer, having more than 30 wt% vinyl acetate, a sharp break in the plot of conductivity versus carbon content cannot be observed, and conductivity increases continuously with increasing vinyl acetate content [53]. Due to the presence of polar groups in vinyl acetate component, the carbon particles are well dispersed in the vinyl-acetate-rich matrices. This section discusses the effect of ENB content on EPDM and VA content on EVA, as well as the blending sequence on the conductivity of composites based on EPDM-EVA-carbon black following the blending formulation shown in the Table 17.4.

TABLE 17.4
Blending Formulations

Blend Numbers	M	N	O	P	Q	R	S	T	U
Royalene-535	–	–	–	–	100	50	50	–	50
Royalene-501	100	–	50	50	–	–	–	50	–
Levaprene-450	–	100	50	50	–	50	50	–	–
Elvax-460	–	–	–	–	–	–	–	50	50
Conducting Carbon black	20	20	20	20	20	20	20	20	20
Dicumyl peroxide (DCP)	1.5	1.5	1.5	1.5	1.5	1.5	1.5	1.5	1.5

17.3.1.1 Conductivity and Its Temperature Dependence

The variation of resistivity with temperature was depicted in Figures 17.2 and 17.3 for low- and high-ENB-containing EPDMs, respectively. High-ENB-containing EPDM exhibits higher conductivity at ambient temperatures. From the figures, it is clear that two types of conductivity occur, depending on the temperature and VA content of EVA. However, the ENB content does not appear to change the mode of conduction over the temperature range studied. For both types of EPDM and elastomeric (high VA content) EVA blends, resistivity increases sharply up to 58°C and then increases slowly. Figure 17.2 reveals that pre-blending followed by black addition gives rise to higher conductivity for low-ENB-containing EPDM; however, blending of black, master batches shows higher conductivity up to 53°C in the case of high-ENB-containing EPDM. High conductivity of the composites having low VA beyond 44°C may be attributed to the structural rearrangement of the carbon black.

It is clear that two types of temperature coefficient of conduction exist, depending on the VA content of EVA, which can be well explained by the Arrhenius type of plot illustrated in Figures 17.4 and 17.5. As observed, an abrupt change occurs in the nature of the electrical conduction at approximately 55°C. The transition temperature does not appear to change with the ENB content of the EPDM. However, the transition temperature shifts toward the lower temperature side as VA content decreases. At this transition point, the temperature coefficient of conductivity changes. For the composites having high VA content EVA, the conductivity has a negative temperature coefficient up to 60°C and thus a metallic type of conductivity is observed. However, composites with low VA content EVA conductivity have a positive temperature coefficient up to 50°C, thus producing the semi-conducting characteristics.

From the linear relationship of the conductivity against 1/T, the activation energies of conduction were calculated and shown in Table 17.5. For the pure EPDM, ENB content does not appear to affect the activation energy of conduction. High-VA-containing EVA has lower activation energy of conduction. For low-ENB-containing EPDM-EVA blend, the pre-blending followed by the black addition (Blend O) has higher activation energy of conduction than the blending of master batches (Blend P). However, the reverse is observed in the case of high-ENB-containing EPDM where black master

TABLE 17.5
Activation Energies and Conductivities of the Blends

Blend Numbers	M	N	O	P	Q	R	S	T	U
ΔE (conduction) [eV]	1.50	1.03	1.46	1.28	1.55	0.95	1.29	0.64	.43
ΔE (curing) [k cals/mole]	32.30	35.20	27.0	30.00	30.00	28.00	26.10	16.70	25.00
Conductivity (δ) at 30°C × 10^4 [mho-Cm^{-1}]	5.52	2.89	46.00	4.30	34.00	5.20	22.00	.0035	.0014

batches (Blend O) followed by blending gives rise to higher activation energy of conduction. Very low activation energy of conduction is observed for low-VA-containing EVA blends (Blends T and U). It is worth mentioning that higher activation energy of conduction imparts high conductivity at ambient conditions (Table 17.5) when compared with similar blending techniques. From Table 17.3, it can be assumed that a correlation exists between the activation energy of curing and that of conduction. High activation energy of curing is associated with low activation energy of conduction for a similar system irrespective of blending sequences. As observed by Norman [54], the complex nature of conduction in plastic/rubber and their blends is the combined effect of viscosity and carbon black structure that has been retained in the vulcanizate and their transformation. The conduction in carbon black composites is achieved by means of electron jumping across the gap between one carbon black aggregate to the other, depending on the gap width [55]. In the case of elastomers and plastics, the band gap theory of conduction, as is used in the usual semi-conductor, may not be applicable. Here, the conduction is mainly a simple, interaggregate conduction through electron tunneling and electron hopping [56]. According to Sumita et al. [53], the conductivity of plastics–carbon black composites is through coagulate carbon particles, whereas in elastomer black composite, more uniform all-around distribution of black is responsible for conduction. In the system of low-VA-containing EVA/EPDM blend, the increase in conductivity (up to 50°C) may be due to an increase in interfacial energy that facilitates the agglomeration of carbon black. High VA content EVA/EPDM blends the increased polar components of surface tension, enhancing the bonding of conduction carbon black and facilitating its more even distribution in the polymer matrix. This increases the availability of carriers to a high level at ambient temperature. With increasing temperature, the scattering phenomena due to collision between short-range electrons probably gives rise to metallic-type conduction.

17.3.2 Heat Shrinkable and Flame-Retardant Polymer Blends Based on EVA

The use of poly (ethylene vinyl acetate) (EVA) in blends and composites is most important from the technological point of view. Because of their good low-temperature flexibility, permeability, and good impact strength, they are of interest as a stretched film for packaging and for cling-wrap purposes. They are in demand as cable insulation due to good resistance to stress cracking and the fact that the polymer may be easily crosslinked [57]. The other characteristics, which have led EVAs to their widespread use are low cost, easy processibility, excellent chemical resistance, excellent electrical insulation properties, toughness, and flexibility even at low temperatures, freedom from odor and toxicity, reasonable clarity of thin films, and sufficiently low permeability to water vapor. The most important features of EVA for which it has been selected here is its lower response temperature of shape memory effect [58–60], good shrinkability [60], good flow behavior [61] (important for a heat-shrinkable product), and low heat conductivity [62] (important for flame-retardancy). The following section discusses the development of EVA blends with various suitable elastomers, with reference to their heat shrinkability and flame-retardancy for commercial application.

17.3.3 Blends of EVA and Polyacrylic Rubber (AR-801) with Reference to Their Shrinkability

17.3.3.1 Effect of Cure Time and Rubber Content on Shrinkability

The variation of shrinkage with cure time at constant (50%), at different temperatures is presented in Figure 17.7. The figure illustrates that with increasing cure time, the shrinkability increases. The polymer chains in the stretched condition exist in four states, and they could be classified into crystalline and amorphous states of coiled and extended chains [63–65]. Only the extended state could provide a driving force for the contraction of the stretched sample [63]. Driving retraction force is provided by comparatively highly oriented amorphous materials. Shrinkage, a

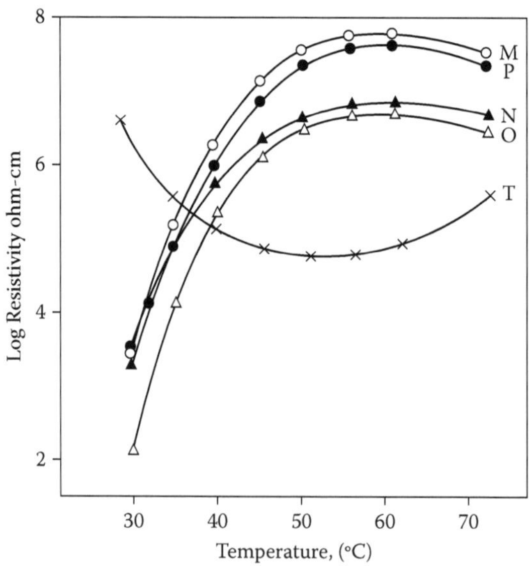

FIGURE 17.7 Variation of resistivity with temperature.

collective property, depends on the concentration and orientation of the oriented amorphous materials [64] and the memory points that depend upon the extent of curing [59, 66]. As greater cure time produces greater amount of crosslinks, so does a greater amount of memory before the stretching process is stored [59, 66]. Thus, with an increase in cure time, recovery increases (i.e., higher shrinkage is obtained owing to a greater number of memory points).

Again, Figure 17.7 shows that with increasing elastomer content, the shrinkage increases at a fixed cure time (20 min). The higher the elastomer content, the higher the amorphous phase will be, which is crosslinked here. Therefore, during stretching, the concentration and extent of orientation amorphous materials will be more. In addition, the number of memory points will be more, because only the elastomer phase is crosslinked here. Thus, it will be more likely to revert to a randomly coiled statistically probable conformation, thus increasing shrinkage.

17.3.3.2 Effect of Temperature on Shrinkage

It can be observed from Figures 17.7 and 17.8 that high temperature (HT) stretched samples show higher shrinkage than room temperature (RT) stretched samples. According to Capaccio and Ward [67], two factors are involved for shrinkage of RT stretched (below T_m of EVA) sample. One factor is the contraction of the amorphous chain, and the other factor is the reorganization of crystallization phase. In the second case, deformation occurs in the crystal, when lamellae of crystals could slip and orient. If the crystal deformation is less, then the contribution of the second factor to the recovery will be less and the first factor only will control the shrinkage. Moreover, in the case of the RT stretched sample, the tendency of the amorphous chain to contract is less compared with HT stretched samples. As for the latter, the orientation

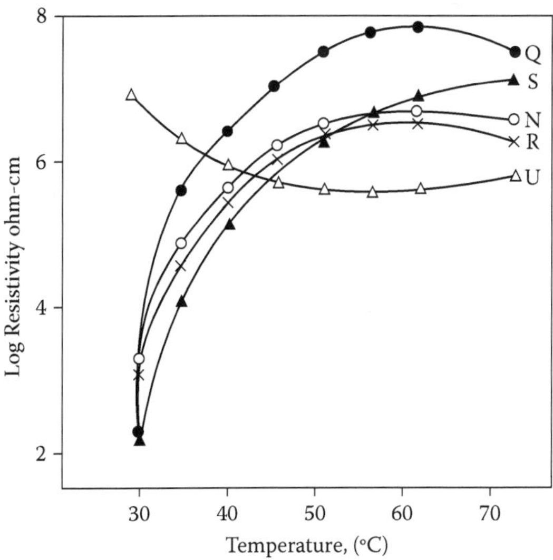

FIGURE 17.8 Variation of resistivity with temperature.

of the material is more due to presence of molten plastic phase with more freedom to flow. Again, the shrinkage of blend stretched at 150°C gets an edge over 100°C.

The total linear deformation of the oriented sample may be described by three rheological components:

1. An instantaneous elastic component, E_1, caused by bond deformation or bond stretching, which is completely recoverable when the stress is released (Hookian behavior).
2. A viscoelastic component, E_2, caused by uncoiling, which is frozen-in to structure when cooled.
3. A plastic (viscous) component, E_3, caused by molecular slipping over each other, which is non-recoverable.

The viscoelastic component can serve as a measure of degree of recoverable orientation (i.e., thermoelastic shrinkage) [68].

At 150°C, the amount of viscoelastic deformation is higher than that at 100°C, so the recovery due to frozen-in internal stress is found to be greater at 150°C [69]. On the other hand, the shrinkage at 180°C, which is a substantial fraction of the initial elastic deformation, is transformed into plastic deformation [70]. At 180°C, due to higher viscous flow, E_3 contributes to a greater extent, compared with E_2, to the total linear deformation. Consequently, the recovery is lower.

17.3.4 EVA Co-Polymer and Polyurethane (PU) Blends with Reference to Their Shrinkability

17.3.4.1 Effect of Cure Time and Elastomer Content on Shrinkability of the Blends

The variation of shrinkage with cure time constant (50%), at different temperatures is shown in Figure 17.8. The figure illustrates that with increasing cure time, the shrinkability decreases. The polymer chains in the stretched condition exits in four states, and they could be classified into crystalline and amorphous states of coiled and extended chains [63–65]. Only the extended state could provide a driving force for the contraction of the stretched sample [63]. Driving retraction force is provided by comparatively highly oriented amorphous materials. Shrinkage, a collective property, depends on the concentration and orientation of the oriented amorphous materials [64]. In this case, both phases (rubber and plastics) were crosslinked. This occurred in coiled state and a greater cure time produced greater amount of crosslinks. This crosslinking in the coiled state is not in favor of orientation during stretching. Therefore, concentration and orientation of the oriented amorphous material will be less for higher cure time, which results in less recovery (i.e., lower shrinkage).

Figure 17.9 illustrates that with an increase in elastomer content, the shrinkage increased to a certain level (EVA/PU = 50/50); then again, with an increase in elastomer content, this property goes down. The more the plastic phase in the blend, the greater is the crystallinity. Consequently, a greater restriction by the crystallized state to the extended chain cores will occur to revert to a randomly coiled conformation with greater entropy. Naturally, with an increase in the elastomer content, the

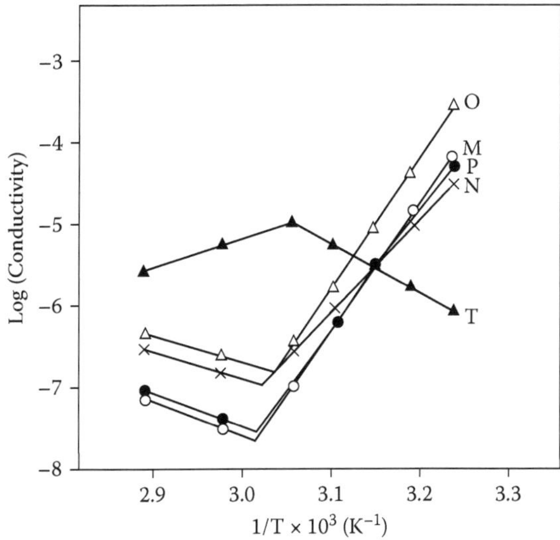

FIGURE 17.9 Plot of Log (conductivity) against 1/T.

extended chains were less restricted and undertook a transition process to a conformation with larger entropy. Actually, coiling took place as a collective process. The driving force for retraction was provided by comparatively highly oriented amorphous materials. A higher PU elastomer means a higher amorphous phase. During stretching, the concentration and orientation of the oriented amorphous materials were more, so the tendency of the extended state to revert to a randomly coiled conformation, thus increasing shrinkage, was greater. Beyond 50% PU blend, though the amount of elastomer was greater still, the shrinkage was lower. The shrinkage force came from the relaxation of internal stress that was frozen-in; internal stress is decreased and a lower shrinkage is obtained.

17.3.4.2 Effect of Temperature on Shrinkage

It is observed from Figures 17.7 and 17.8 that high temperature (HT) stretched samples show higher shrinkage than room temperature (RT) stretched samples. According to Capaccio and Ward [67], two factors are involved for shrinkage of RT stretched (below T_m of EVA) sample. One factor is the contraction of the amorphous chain and the other factor is the reorganization of crystallization phase. In the second case deformation occurs in the crystal, when lamellae of crystals could slip and orient. If the crystal deformation is less, then the contribution of the second factor to the recovery will be less, and the first factor only will control the shrinkage. Moreover, in the case of an RT stretched sample, the tendency of amorphous chain to contract is less compared with HT stretched samples. As for the latter, the orientation of the material is more due to the presence of molten plastic phase with more freedom to flow. Again, the shrinkage of blend stretched at 150°C gets an edge over 100°C. At 150°C, it had an edge over 100°C. At 150°C, the amount of viscoelastic deformation was higher than that at 100°C. Therefore, the recovery due to frozen-in internal stress was found to be greater. The shrinkage at 180°C was appreciably lower than that at 150°C. At a high temperature (180°C), a substantial fraction of the initial elastic deformation was transformed into plastic deformation. At 180°C, the desired alignment component of the deformation, E_2, will be a smaller proportion of the total deformation, mainly because of viscous flow, E_3, which increases as the temperature is increased, resulting in low recovery at 180°C. Tie molecules may also play a role here. During stretching, tie molecules that contact crystal blocks on different fibrils (from microfibrillar model) extend enormously. Those tie molecules are partly responsible for shrinkage. The effect of pulling some chain sections partially out of the crystal block in which the tie molecules are anchored should be remembered. Such pulling during annealing increases the contour length. As a result, the ratio between the counter length and end-to-end distance of the tie molecule is increased, allowing relaxation of initially stretched tie molecules. That may be because of the dropping of retractive force; thus, a lower amount of shrinkage will be obtained.

17.3.5 Conductive Composites of s-PS/Carbon Nanofibers

17.3.5.1 Electrical Resistivity

The volume resistivity was measured on a compression-molded square slab (length 1 cm, thickness 3 mm and breadth 1 cm). A Keithley 230 programmable voltage source test fixture was used to measure volume resistance. This equipment allows resistivity measurement up to 10^{18}. The pressed samples were put into the four-point conductivity apparatus with 500 V through the samples. The current and voltage were measured, and using the following equation, the resistivity in ohm cm was calculated:

where

R = resistance
V = voltage
q = resistivity
ω = width of sample
t = thickness of sample
L = length between inner probes

Figure 17.10 shows a very gradual decrease in the resistivity with increase in the fiber content from 1.1–3.1 phr (parts per hundred parts). The resistivity has reduced from 10^{12} to 1.28×10^{11} Ω cm at 1.1 phr. Then, resistivity is further decreased to 3.18×10^9 Ω cm for 2.1 phr, and then finally drops to 10^9 Ω cm at 3.1 phr. Percolation effects were observed at 2.1 phr of loading, which means the transport of electrons at this loading is maximum.

$$q = \frac{V \times t \times \omega}{I \times L}$$

With the addition of 1 and 5 wt% HDPE, Wu et al. [71] found the percolation threshold of PMMA/CNF composites was reduced from 8.0–4.0 phr. In our present

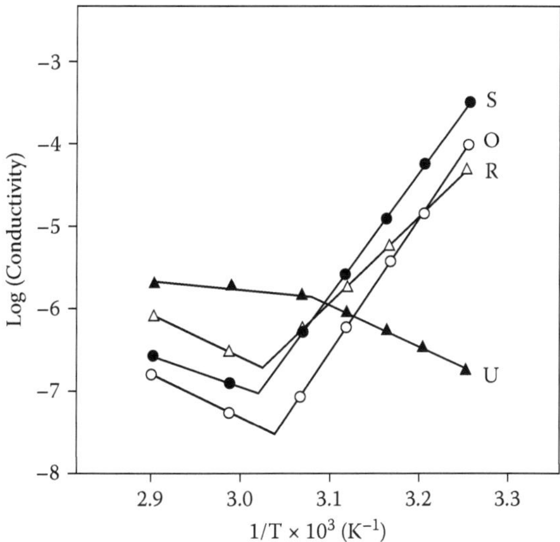

FIGURE 17.10 Plot of Log (conductivity) against 1/T.

work, LCP was added to the s-PS /CNF composites, with the thought that LCP composites more readily form an interconnected conducting network than other thermoplastics, as reported by Lozano et al. [72]; however, the results could not meet our expectations. The decrease in resistivity in the case of s-PS/CNF/LCP to 7.28×10^9 Ω cm, and 6.12×10^8 Ω cm at 1.1 and 2.1 phr for constant loading of LCP, is comparably more than s-PS/CNF at same loading. The reason for this decrease may be due to the network-forming tendency of carbon nanofibers in the presence of LCP, as presented in Figure 17.11. Nevertheless, this small change in resistivity is attributed to the interfacial tension between s-PS and LCP. One composite sample was prepared using modified CNFs (HNO_3/H_2SO4(1:3) at 90°C for 10 min) with fiber loading at 3.1 phr, which showed a decrease in the resistivity to a value of 5.76×10^8 Ω cm, which is slightly lower than the unmodified CNFs composite of the same loading (10^9 Ω cm). This insignificant decrease in the resistivity may be the result of two factors: (1) wrapping the CNFs with insulating polymer due to improved adhesion on modification, which reduces the effectiveness of electrical transport from fiber to fiber, and (2) formation of an oxygen-rich surface layer, which also served as an insulating coating.

17.3.5.2 Temperature Dependence of Electrical Conductivity ($\sigma - T$)

Figure 17.12 is the plot of Log σ versus 1/T. It can be observed that conductivity decreases with an increase in temperature. This indicates the metallic nature of the composites. This kind of behavior arises due to two possible reasons: (1) an increase of the gap between fibers due to thermal expansion of the matrix with temperature; and (2) a scattering phenomenon, which arises due to the collision of the emitted electron with crystallites and impurities present in the system, thus carrier mobility transfer becomes poorer [73–75]. However, it can also be observed that in the vicinity of melting point of samples, the resistivity drops (i.e., a negative temperature

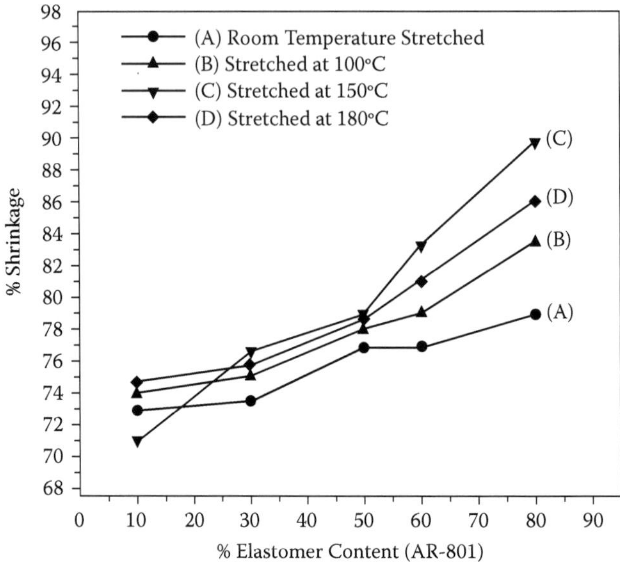

FIGURE 17.11 Variation of % shrinkage with cure time at constant elastomer content (50% AR-801).

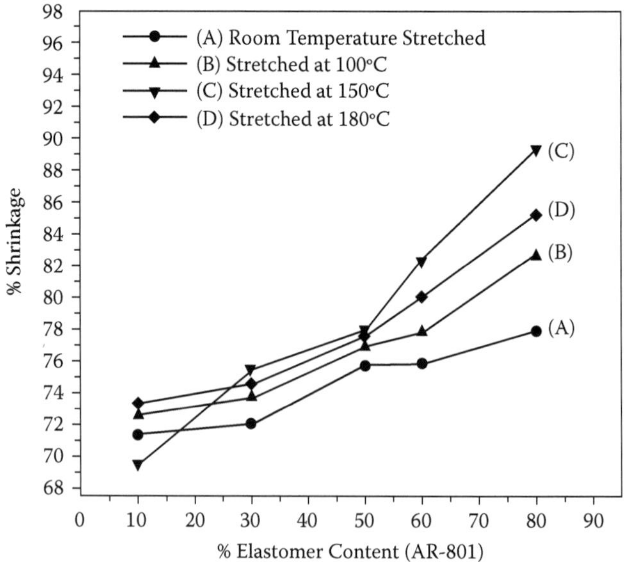

FIGURE 17.12 Variation of % shrinkage with elastomer content at constant cure time (20 min).

TABLE 17.6
Activation Energy Data for s-PS/CNF and s-PS/CNF/LCP Composites

Composites	A	B	C	Modified CNFs at 3.1 phr
E_a activation in (E_v)	.71	.61	.57	.52

s-PS = syndiotactic polystyrene; CNF = carbon nanofibers; LCP = liquid crystalline polymer.

coefficient [NTC] effect appears or material starts to exhibit semiconducting behavior). The activation energy was calculated using the Arrhenius equation:

$$\sigma = \sigma_p \exp - [E_a/(K_B T)]$$

where
σ_p is the pre-exponential factor
E_a is activation energy in eV
K_B is the Boltzmann constant
T is temperature measured in Kelvin

The estimated values of E_a from the plot of Log σ versus 1/T are given in Table 17.6. A low value of activation energy for 3.1 phr CNFs composite, as compared with 1.1 and 2.1 phr, may be attributed to the high conductivity of composite or that less energy is required for the electron to jump from the valance band to the conduction band. However, this value decreases to .52 for modified CNFs composite at 3.1 phr loading.

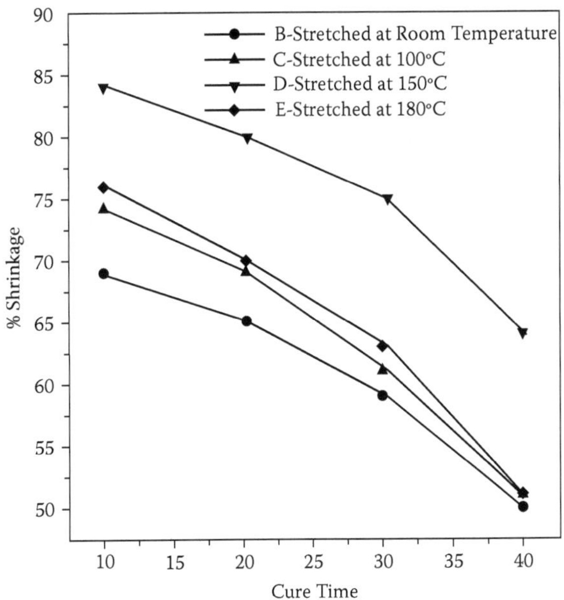

FIGURE 17.13 Variation of shrinkability with cure time: B stretched at room temperature; C stretched at 100°C; D stretched at 150°C; E stretched at 180°C.

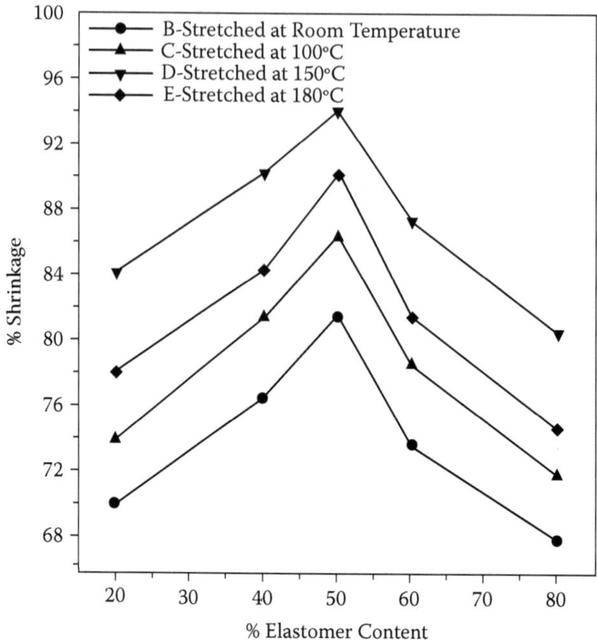

FIGURE 17.14 Variation of shrinkability with elastomer content: B stretched at room temperature; C stretched at 100°C; D stretched at 150°C; E stretched at 180°C.

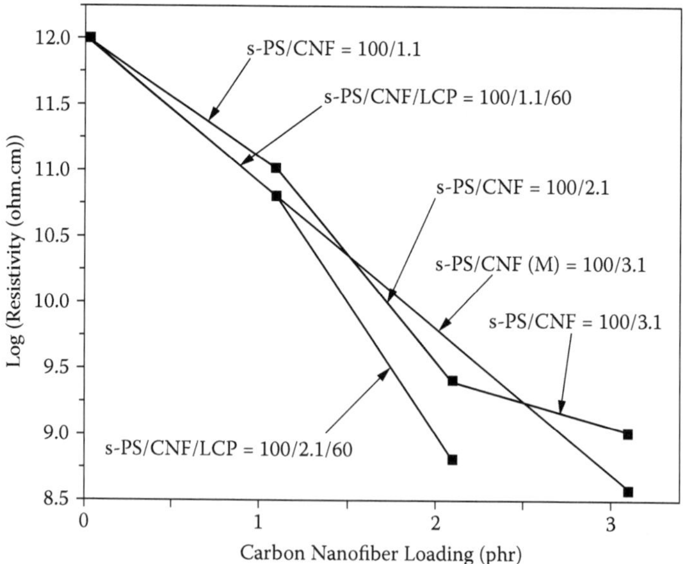

FIGURE 17.15 Shows the electrical resistivity of s-SPS/CNF and s-PS/CNF/LCP (LCP constant) composites at different loading conditions (1.1 phr, 2.1 phr, and 3.1 phr) at room temperature.

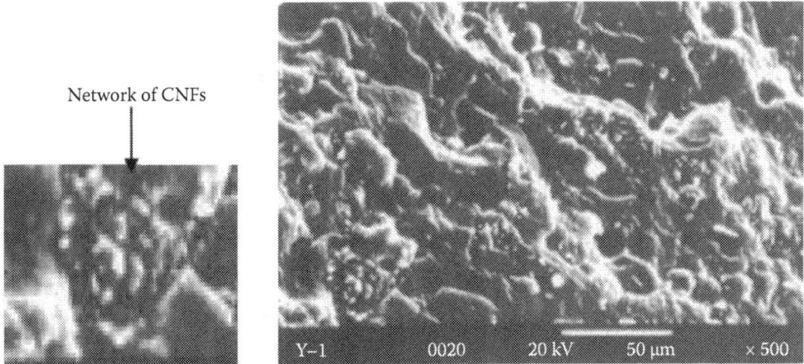

FIGURE 17.16 SEM images of s-PS/CNF/LCP (100/2.1/60) composites.

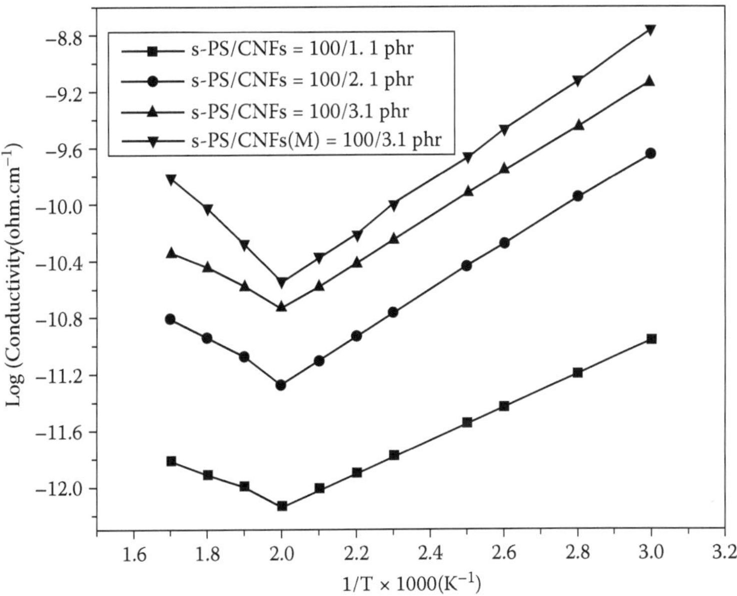

FIGURE 17.17 Dependence of electrical conductivity of s-PS/CNFs composites on temperature (K).

REFERENCES

1. Ehlers, J.F. and Goldstein, K.R., *Kolloidzeitschrift*, 118, 137, 29, 1950.
2. Matthews, G., Liewellyn, I., and Williams, H., *Vinyl and Allied Polymers, Vol. 2,* 1972.
3. Hartmann, A., *Kolloidzeitschrift*, 142, 123, 31, 1955.
4. Booth, D.H., Hollingsworth, P.M., and Lythoge, W.H., *I.E.E. Conf. Rep. Series No. 6*, 1963.
5. Hecker, A. and Cohen, S., *SPE Papers (Quebec)*, 36, 38, 1964.
6. Matthan, J., *Rapra Res.*, 165, 39, 1968.
7. Daltan, W.K., *Br. Plast.*, 35, 136, 50, 1962.
8. ASTM D573–53; D1203-55.52.
9. Ger. Patent 1,126,613, Fardenfabriken Bayer A.G. (42).
10. *Chem. Eng. News*, 42, 25, 49, 1964.
11. *Chem. Wkly.*, 96, 97, 53, 1965.
12. Heiberger, P., *Am. Paint J.*, 39, 48–51, 48, 1955.
13. Adelman, R.L., Allen, G.G., and Sinclair, H.K., *Ind. Eng. Chem. Product Res. Devel.*, 2, 108, 1963.
14. Sheer, A.E., *SPE J.*, 28, 24, 1972.
15. Gambino, H. J. Jr., *Security You Can See Through*, PL-1228, Rohm and Haas Co., Philadelphia.
16. *Mod. Plast.*, 5, 52, 1975.
17. *Chem. Week*, 47, 1973.
18. Masuhara, E. et al., *Jpn. Kokai*, 74, 57, 054, 1974 (to Mochida Pharmaceuticals).
19. Timmer, W., *Chem. Technol.*, 9, 175, 1979.
20. U.S. Patent 3,951,528 (Apr. 20, 1976), Leeds, H.R. (to Pnt Structures, Inc.).
21. U.S. Patent 3,947,401 (March 30, 1976), Stamberger (to Union Optics Corp.).
22. U.S. Patent 4,239,513 (Feb. 13, 1979), Tanaka, K. et al. (to Toyo Contact Lens Co., Ltd.).
23. Ohtsuka, Y. and Hantanaka, Y., *Appl. Phys. Lett.*, 29, 735, 1976.
24. U.S. Patent 4,138,194 (Feb. 6, 1979), Beasley, J.K., Beckerbauer, R., Sehleinitiz, and Wilson, F.C. (to E.I. du Pont de Nemours & Co., Inc.).
25. Kaetsu, I., Yoshida, K., and Okubo, H., *J. Appl. Polym. Sci.*, 24,1515, 1979.
26. Okubo, H., Yoshida, K., and Ketsu, I., *Int. J. Appl. Radiat. Isot.*, 30, 209, 1979.
27. U.S. Patent 4,146,696 (Mar. 17, 1979), Bond, H.M., Torgersen, D.L., and Ring, C.C. (to Buckee-Mears Co.).
28. Wood, A.S., *Mod. Plast.*, 52, 40, 1975.
29. *Mod. Plast.*, 49, 44, 1972.
30. MacBride, R.R., *Mod. Plast.*, 49, 40, 1972.
31. MacBride, R.R., *Mod. Plast.*, 56, 47, 1974.
32. Pyun, H.C., Cho, B.R., and Kwon, S.K., *J. Korean Nucl. Soc.*, 7, 9, 1975.
33. U.S. Patent 2,876,136 (1959), Ford, F.M. (to J. Bancroft & Sons).
34. Colgan, G.P. and Latimer, J.J., *Tappi*, 47(7), 146A, 1964.
35. Colgan, G.P. and Latimer, J.J., *TAPPI*, 44, 818, 1961.
36. Beeman, R.H. and Beardwood, B.A., *TAPPI*, 46, 135 1963.
37. Colgan, G.P. and Plante, P., Poly (vinyl alcohol) in paper coating additives, *TAPPI Monograph No. 25*, Technical Association of the Pulp and Paper Industry, New York, 1963, p. 121.
38. Nakano, S.C. and Motoyama, T., *Kobunshi Kagaku*, 21, 300, 1964.
39. Coker, J.N., *Ind. Eng. Chem.*, 49, 382 1957.
40. Lindemann, M.K., The mechanism of vinyl acetate polymerization, in *Vinyl Polymerization*, Vol. I, Part I, in G.E. Ham, Ed., Marcel Dekker, Inc., New York, 1967, p. 288.

41. German Patent 690,332 (1940), Herrmann, W.O. (to Chemische Forschung Gesellschaft m.b.H.).
42. Reid, E.W., U.S. Patent 2,120,628, 1937 (to Union Carbide).
43. Ryan, J.D., U.S. Patent 2,232,806, 1937 (to Libbey-Owens-Ford Glass Co.).
44. Weihe, A., *Kuunststoffe*, 31, 52, 1941.
45. The tropical durability of metal adhesives, *R.A.F. Tech. Note No. Chem.*, 1349, 1959.
46. Eichner, H.W., Environmental exposure of adhesive-bonded metal lapjoints, *W.A.D.C. Tech. Rep. 59*, Part 1, 564, 1960.
47. Technical Note No. 144, Ciba (A.R.L.) Ltd., 1954.
48. Wu, G., Asai, S., and Sumita, M.A., Self-assembled electric conductive network in short carbon fiber filled poly(methyl methacrylate) composites with selective adsorption of polyethylene, *Macromolecules*, 32, 3534–3536, 1999.
49. U.S. Patent 2,920,990, Been, J.L. and Grover, M. M. (to Rubber and Asbestos Corp.), 1960.
50. Fitzhugh, A. et al., *J. Electrochem. Soc.*, 100, 351, 1953.
51. Lai, J., in *Polymer Chemistry of Synthetic Elastomers*, Part 1, Kennedy, J., Kennedy, P., and Tornqvist, E., Eds., Interscience Publishers, New York, 1986, pp. 331–376.
52. Based on information supplied by Lane, F.B., GAF Corp., New York.
53. Sumita, M. et al., *Colloid Polym. Sci.*, 264, 212, 1986.
54. Norman, H.R., *Conductive Rubber and Plastics*, Elsevier, 1970.
55. Polley, H.M. and Boonstra, B.B.S.T, *Rubber Chem. Technol.*, 30, 170, 1957.
56. Medalla, A., *Rubber Chem. Technol.*, 59, 432, 1986.
57. Brydson, J.A., *Plastic Materials*, Butterworth, London, 1982.
58. Huskic, M. and Sebenik, A., *Polym. Int.*, 31, 41, 1993.
59. Suresh, K. and Pandya, M.V., *J. Appl. Polym. Sci.*, 64, 823, 1997.
60. Fengkui, L. et al., *J. Appl. Polym. Sci.*, 71, 1063, 1999.
61. Arsac, A. et al., *J. Appl. Polym. Sci.*, 74, 11, 2625, 1999.
62. Weil, E.D., *Additivity, Synergism and Antiginism in Flame Retardance*, Plastic Institute of America, New York, 1971.
63. Pakula, T. and Trznadel, M., *Polymer*, 26, 1014, 1985.
64. Wang, M., and Zhang, L.J., *Polym. Sci, Polym. Phys. (Part B)*, 37, 101, 1999.
65. Decchndia, F. et al., *Polym. Sci. Polym. Physics.*, 20, 1175, 1982.
66. Charlesby, A., *Proc. R. Soc. London.*, 215, 187, 1952.
67. Capaccio, G. and Ward, I.M., *Polym. Sci.*, 46, 260, 1982.
68. Ohma, J. et al., *Polym. Blends Composites*, 5, 69, 1981.
69. Ram, A., et al. *Int. J. Polym. Mater.*, 6, 57, 1977.
70. Dechndia, F., Russo, R., and Vittoria, V., *J. Polym. Sci. Polym. Physics.*, 20, 1175, 1982.
71. Wu, G., Asai, S., and Sumita, M.A., Self-assembled electric conductive network in short carbon fiber filled poly(methyl methacrylate) composites with selective adsorption of polyethylene, *Macromolecules*, 32, 3534–3536, 1999.
72. Yang, S., Lozano, K., Lomeli, A., Foltz, H.D., and Jones, R. Electromagnetic interference shielding effectiveness of carbon nanofiber/LCP composites. *Compos. Part A: Appl. Sci. Manuf.* 36:691–697, 2005.
73. Brian, P. et al., Nucleation of polypropylene crystallization by single-walled carbon nanotubes, *J. Phys. Chem. B.*, 106, 5852–5858, 2002.
74. Sandler, K.W. et al., Haffer MSPS. Comparative study of melt spun Polyamide12 fibers reinforced with carbon nanotubes and nanofibers, *Polymer*, 45, 2001–2015, 2004.
75. Jeong, S., Changchun, Z., and Lee, L.J., Synthesis of polystyrene — carbon nanofibers and nanocomposite foams, *Polymer*, 46, 5218–5224, 2005.

18 Recycling of Vinyl Polymers

Mir Mohammad A. Nikje

CONTENTS

FIGURE 18.1 Polymeric product wastes.

18.1 INTRODUCTION

Because disposal of post-consumer polymers is being further constrained by legislation and escalating costs, considerable demand exists for alternatives to disposal or land filling. Among the alternatives available are source reduction, re-use, recycling, and recovery of the inherent energy value through waste-to-energy incineration and processed fuel applications. Each of these options potentially reduces waste and conserves natural resources. In some countries, especially in the United States, plastics recycling was investigated in the late 1980s. Municipal solid wastes (MSW) can be divided into two categories. The first category is based on product type, and the other is based on material type (Figure 18.1). The latter are durable and non-durable goods, containers and packaging, food wastes, yard trimmings, and inorganic wastes. In the other words, material-type categories are paper, paperboards, glass, leather, rubber, plastics, textiles, wood, food, metals, miscellaneous inorganic wastes, and other waste.

Products in the U.S. MSW stream are containers and packaging (33%), non-durable goods (27%), durable goods (16%), yard trimmings (13%), food wastes (10%), and miscellaneous wastes (1%). Similarly, in 1998, in the same way, materials streams as MSW included paper and paper board (38%), yard trimmings (13%), food wastes (10%), plastics (10%), glass (6%), steel (6%), wood (5%), textiles (4%), rubber and leather (3%), other materials (2%), non-ferrous metals (1%), miscellaneous inorganic wastes (1%), and aluminum wastes (1%).

Public pressure for plastics recycling waned in the last half of the 1990s in the United States, although signs indicate that it is increasing in the early 2000s. Today, vinyl polymers are extremely applied in plastic manufacturing. Now, much of the polymer material used in short-term packaging applications ends up occupying valuable landfill space (Tables 18.1 and 18.2). Both synthetic and naturally occurring polymers, which need high energy to be produced, will generally last in these sites for a long period of time because they do not get the necessary exposure to ultraviolet (UV) light and microbes to degrade. Here, they are taking up space and none of

TABLE 18.1
Energy Use for Polymer Production in Europe, 2000

Monomer Type	Energy(GJ/t Product)	Tons CO2(fossil/t Product)
LDPE/LLDPE	78	1.8
HDPE	80	1.7
PP	111	3.4
PVC	57	2.0
PS	87	2.6
Others	360	14.2

Source: Association of Plastics Manufacturers in Europe.

the energy put into making them is being reclaimed. Reclaiming the energy stored in the polymers can be done through incineration, but this can cause environmental damage by release of toxic gases into the atmosphere.

Recycling is a viable alternative in getting back some of this energy in the case of some polymers. Raw materials for polymers are obtained from petroleum, a limited non-renewable resource. Using recycled polymers to replace the petroleum in some cases will help make this resource last longer. As petroleum prices increase, it is becoming more financially valuable to recycle polymers instead of producing them from raw materials. The most commonly recycled materials are PET and HDPE. This is because the original properties of the materials are sufficient to be used for this application without large quantities of additives, which are difficult to remove on recycling. This chapter discusses vinyl polymers recycling. In this chapter, different recycling methods applied for polymers are reviewed, with a focus on vinyl polymers.

The 2004 National Post-consumer Plastics Recycling Report is the 15th annual of the American Plastics Council (APC) plastics recycling study. This study was a cooperative effort between the American Plastics Council (APC), the Association of Post-consumer Plastics Recyclers (APR), and the American Beverage Association (ABA), designed to quantify the amount of and the rate at which post-consumer plastic bottles were recycled in calendar year 2004. Overall, total post-consumer plastic

TABLE 18.2
Annual Plastics Consumption and Recycling Totals (in Tons) in Australia, 2004

Year	Total Recycling	Total Consumption
1997	93,547	1,336,386
2000	167,673	1,530,783
2001	160,854	1,430,874
2002	157,345	1,496,387
2003	189,385	1,521,394

Source: National Plastics Recycling Survey, 2004.

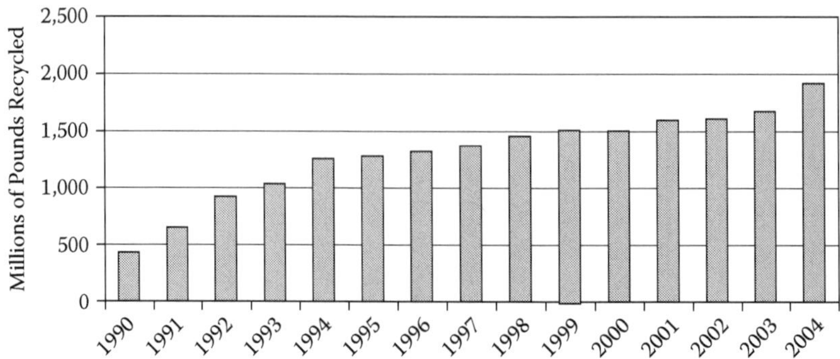

FIGURE 18.2 Total chart of plastic recycling rate in the United States. (From American Plastics Council, 2004. With permission.)

bottle recycling was increased 247 million pounds in 2004 to an all-time high of 1915 million pounds. Conventional (or mechanical) recycling includes flaking or granulating, washing, decontamination, and re-pelletizing of recovered polymer products so they may be fabricated into new, useful, and marketable products. Mechanical recycling can be economically viable, as with the recycling of HDPE milk jugs (Figure 18.2). The process does have some drawbacks. Among them is the requirement for a relatively clean source of post-consumer plastics, the need for efficient separation technology to obtain generically pure resin types, some current end-use market limitations, and often, a labor-intensive process.

It is useful to distinguish three different forms of recycling: closed-loop recycling, open loop recycling, and down-cycling, which can be defined as follows:

1. **Closed-loop recycling** is a recycling process in which a waste material is used for the same purpose as the original purpose or for another purpose requiring at least as severe properties as the previous application so that, after one or several uses, this material can be used over again for the original purpose.
2. **Open-loop recycling** is a recycling process in which a waste material is used for another purpose than the original purpose and will never be used over again for the original purpose.
3. **Down-cycling** is a recycling process in which a (fraction of a) material from a used product is used to make a product that does not require as severe properties as the previous one. These definitions presuppose that we are dealing with a thing that has had some useful life and was turned into waste. These definitions do not appear to cover by-products, given that they have not had a previous use. It thus appears that both re-use and recycling refer to things that have had a purpose and, for some reason, ceased to be used for that purpose. To avoid becoming waste, it is then either re-used or recycled. A rapidly evolving group of advanced recycling technologies, known as depolymerization to monomers, involves collecting polymeric products, sorting by resin type, and then depolymerizing them back into their basic building blocks or monomers. The recovered monomers are then used to produce new resins of the same type.

FIGURE 18.3 Unsorted municipal wastes.

Some problems occur with recycling polymers, the most important of which are:

1. Cost of collection, separation, and sorting of wastes (see Figure 18.3).
2. Unavailability of a stream of clean, homogeneous material and a suitable market for products.
3. In cases where many additives are included in the polymer, the energy required to purify it is greater than that required to produce the plastics from crude oil.
4. Thermosets are difficult to recycle because of the chemical cross-links formed in them. Tables 18.3 and 18.4 list consumption and waste statistics of vinyl polymers in Europe.

TABLE 18.3
Consumption of Vinyl Plastic by Resin (Western Europe)

Polymers	Recycled Polymers (%)	Virgin Polymers (%)	Total (Kilotons)
LDPE/LLDPE	10	90	7121
PP	1	99	5524
PVC	0	100	5243
HDPE	7	93	4837
PS	2	98	2278
EPS	4	96	788

Source: Taylor Nelson Sofres for APME — 2000 Data.

TABLE 18.4
Waste Vinyl Plastics Generation by Resin Type (Western Europe)

Polymer	%
LDPE/LLDPE	32.9
PP	24.3
HDPE	21.5
PVC	11.3
PS	10.0

Source: Taylor Sofres Nelson – 2000 Data.

Economics is a major factor in determining the success or failure of recycling for all materials. Recycling actually occurs when, and only when, consumers purchase recyclable materials that have been collected, sorted, processed, and remanufactured into new products. Recyclable materials separated from garbage should not be viewed as waste, but as a raw material or feedstock for industries to use in making new products. The ultimate success of recycling depends on stable, reliable markets for these materials. Without markets to purchase the collected and separated recyclables, recycling does not happen, with the unfortunate result that these materials often must be disposed of in landfills or waste-to-energy plants. Feedstock recycling (or thermal depolymerization) recycles polymeric products that cannot easily be broken down into their pure generic resin types, or have some level of contamination. A typical feedstock recycling process operates in an oxygen-free environment to prevent the plastics from burning, and results in the recovery of liquid feedstock. These can then be used in place of virgin oil for the production of new plastic resins, fibers, and other valuable petroleum derivatives [1].

Pyrolytic gasification processes usually require harsh conditions, such as high temperatures or catalysts, and produce an olefin-rich hydrocarbon gas or a synthesis gas product. Thermal (or steam) cracking of plastics at elevated temperatures will produce the monomer in good yields but requires a sophisticated distillation to separate and purify the olefins. Synthesis gas, a mixture of carbon monoxide and hydrogen, is produced by the partial oxidation of polymers using high temperatures and a controlled amount of oxygen. The products can be used to synthesize higher value products, such as methyl-*t*-butyl-ether, methanol, and acetic acid. Several pyrolysis liquefaction processes also produce high yields of olefins; when operated at elevated temperatures over 40%, olefin yields have been achieved. Advanced recycling processes to produce either monomers or feed stocks have been demonstrated on a commercial scale. Technical issues seem surmountable, but economic and political hurdles remain [2].

The recovery of monomers or oil from waste polymers by a depolymerization process is called tertiary recycling. Reprocessing scrap as part of a product production is defined as primary recycling, whereas melt recycling is considered secondary recycling, and burning with energy recovery is considered quaternary recycling. Two types of tertiary recycling are used: chemical and thermal. Depolymerization of the plastic by chemical means is called solvolysis, and the process produces a monomer or oligomers. The decomposition of polymers by heat is called thermolysis. If the process is done in the absence of air, it is called pyrolysis. If it is done with a controlled amount of oxygen, it is called gasification. Pyrolysis will produce a liquid fraction, which is a synthetic crude oil, and should be suitable as a refinery feedstock. The non-condensable fraction created during pyrolysis is normally used to provide process heat and any excess is flared.

Gasification of plastic takes place at a higher temperature than pyrolysis and with controlled oxygen addition. The result is a *syngas* that is composed primarily of carbon monoxide and hydrogen. As a mixture, the syngas is valued only as a fuel. However, if the gases are separated, the carbon monoxide and the hydrogen are valued as chemical intermediates, which can have 2–3 times the fuel value of the mixture. A third form of thermolysis is hydrogenation, where the plastic is depolymerized by heat and exposed to an excess of hydrogen at a pressure of over 100 atmospheres. The cracking and hydrogenation are complementary, with the cracking reaction being endothermic and the hydrogenation reaction being exothermic. The surplus of heat normally encountered is handled by using cold hydrogen as a quench for this reaction. Hydro treating can remove many heteroatoms. The resultant product is usually a liquid fuel like gasoline or diesel fuel (Table 18.5).

Thermolysis is a much more versatile and forgiving technology for tertiary recycling than solvolysis. It can handle mixed polymer waste streams along with some level of non-plastic contaminants. Solvolysis requires a relatively pure polymer stream and has little tolerance for contaminants; therefore, the raw material preparation costs are larger.

The most commonly discussed processes for the thermolysis of waste plastic include hydrogenation, gasifiers, fluidized beds, kilns/retorts, and degradative extrusion [3, 4]. Another procedure for recycling or decomposing waste polymers involves placing the post-consumer plastic in a diluent, such as hot oil, with a free-radical precursor, such as PVC or PU, at a low temperature. The thermal decomposition (or pyrolysis) reaction lasts for 1 hour at 375°C and useable products are recovered [5]. Pure or mixed, clean or contaminated plastic wastes can be cracked (i.e., broken down) into

TABLE 18.5
Total Recycling (in Tons) and Recycling Rates of Vinyl Polymers in Australia, 2003

Polymer	Consumption	Domestic Reprocessing	Export for Reprocessing	Total Recycling	Recycling Rate
HDPE	268,070	38,417	23,471	61,888	23.1%
PVC	221,286	7675	1174	8849	4.0%
LDPE/LLDPE	301,848	30,849	5868	36,716	12.2%
PP	240,068	20788	2934	23,721	9.9%
PS	47,800	2422	0	2422	5.1%

Source: PACIA, Australia, 2004.

smaller molecules in low-pressure plasmas at low substrate temperatures and then recycled [6, 7].

Recycling economics depends on finding the recyclates' most valuable form: resin or energy [1]. Biodegradation of general non-medical materials can be divided into microbial and chemical degradation. Chemical degradation includes wind and rain erosion, oxidation, photo degradation, acid/base water, and thermal degradation [8]. One way to deal with the problem of polymer wastes is to make polymers degradable; however, this appears to eliminate the greatest asset of these materials, namely their durability. It also wastes the time, effort, and energy put into making materials in the first place. Water pollution stems from toxic substances leaching out of landfilled materials, and it is a growing problem for the nation's 6000 landfills. So on the surface, degradability is a poor second choice to recyclability, given that degradability does not necessarily make materials disappear; it may only make them physically or chemically smaller. However, not all synthetic materials are recoverable or even worth recovering.

One method that can initiate the degradation process is to make a polymer *photo labile* so that it begins degrading when exposed to the UV region of sunlight. The UV-sensitive carbonyl joints break when struck by UV light. The polymer has the same physical properties as "pure" LDPE but begins falling apart during exposure to sunlight. Photosensitive additives, such as organometallic compounds, can also make a polymer fall apart by initiating a chemical chain reaction on initial exposure to sunlight, breaking open any polyolefin chain. This continues in the absence of sunlight, so that the polymer will fall apart even if it is later buried in a landfill. Some synthetic polymers can serve as food for microorganisms that live in soil and water. ICI Americas has developed a thermoplastic resin, called PHBV random co-polymer, that is stable in air and sunlight but falls apart when it is exposed to bacteria in soil, water, or a sewage-treatment plant. The polymer is actually made by certain soil bacteria when fed a diet supplemented with the monomer 3-hydroxyvalerate. The resulting thermoplastic is a highly crystalline, stiff material that can be processed into film or blown into bottles. Incorporating biodegradable filler, such as starch, into the polymer formulation is also possible. When the polymer is buried, microbes detect the starch and release enzymes that convert the starch into simple sugars that the microbes then absorb as food. This loss of filler causes the polymer product to disintegrate into smaller pieces of polymer [9].

Several more continuous and automated abrasive techniques have been investigated using large flakes of coated plastics in an effort to identify a dry coating removal technique, but so far, all have proven unsatisfactory. High-temperature steam shows promise and is being investigated further [10]. Figure 18.4 shows a product of recycled polyethylene.

It is the recycling process that should be controlled and that the process itself is tested against demanding standards to ensure that it is capable of meeting certain criteria. Because of the individual nature of each recycling process and the interdependencies of each stage, however, it would not be possible to stipulate a single set of operational conditions that should be met for the recycling operation itself. To ensure the safety of recycled materials, each process would have to be considered individually on its merits. Thus, because it is the process itself that should be

FIGURE 18.4 Recycled polyethylene surfaces.

considered, different recyclers using the same process would not need to be considered as separate cases. Additionally, each type of plastic has unique properties that must be considered in relation to the recycling operations and to the end use. Thus, any assessment must also take into account the process that is being applied and the nature of the plastics used. Plastics are known to interact with organic chemicals (i.e., most chemicals) according to their type specific diffusion and sorption behavior. This behavior is a critical parameter that determines the degree of uptake of misused chemicals and thus the potential risk of food contamination. The diffusivity of vinyl plastics generally increases in the sequence:

rigid PVC<polyester<<PS<PP/HDPE<LDPE

18.2 POLYOLEFINS

18.2.1 High-Density Polyethylene

As we know, high-density polyethylene (HDPE) is a linear polymer with the chemical composition of polymethylene, $(CH_2)_n$, and is defined as a product of ethylene polymerization with a density of 0.94 g cm^{-3} or higher (Figure 18.5). HDPE is the largest constituent of municipal wastes. According to the structure, one of the chain ends in an HDPE molecule is a methyl group; the other chain end can be either a

```
Ethylene      ┌────┐
Monomer       │H  H│ H  H  H  H  H  H
    \         │|  |│ |  |  |  |  |  |
     \      ··│C–C │–C–C–C–C–C–C···
      \       │|  |│ |  |  |  |  |  |
       \______│H  H│ H  H  H  H  H  H
              └────┘
```

FIGURE 18.5 Schematic structure of HDPE.

methyl group or a double bond (usually the vinyl group). The number of branches in HDPE resins is low, at most 5 to 10 branches per 1000 carbon atoms in the chain. Even ethylene homo-polymers produced with some transition-metal-based catalysts are slightly branched; they contain 0.5–3 branches per 1000 carbon atoms. Most of these branches are short, and their presence is often related to traces of α-olefins in ethylene. The branching degree is one of the important structural features of HDPE resins [11].

A major reason for the success of HDPE as a packaging material has been its recyclability (see Figure 18.6). While trying to recycle, HDPE containers are separated into two streams for reprocessing:

FIGURE 18.6 Recycling sign of HDPE.

1. Opaque or "*natural*" HDPE, identified as milk bottles and juice bottles.
2. Colored HDPE containers.

Natural HDPE's primary market continued to be new polyethylene bottles for a myriad of non-food applications (i.e., detergent, motor oil, household cleaners, etc.). Pigmented HDPE found increased usage in the production of polyethylene pipe and a wide range of lawn and garden products such as edging, flower, and shrub pots. Plastic lumber consumes a broad range of raw materials due to its diverse nature (recycled bottles, film and mixed rigid containers, plus wide-spec virgin resin). The potential for plastic lumber product applications (both 100% plastic profiles and composite materials composed of plastic/fiber blends) could be huge, because historical demand for residential decking boards apparently continues to grow along with the market for rails, fence posts, and other non-structural applications. The potential for structural applications such as railroad ties, marine pilings, and even small bridges also may become greater as product development and test installations continue to be successful.

Reprocessors have developed strong markets using the "*natural*" HDPE bottles. Many manufacturers limit their processes to manufacture specific products, due to the variation of polymers used in the manufacture of labels and caps. The cap and label polymers create levels of contamination that effectively restrict the opportunities to provide a consistent and standard-quality product that competes with virgin resin. This leaves the recycled product as always being identified as a secondary resin to virgin. Recycled resin also competes in the market place with off-spec or lower priced virgin resin. Colored HDPE containers are being recycled in what could be identified as developing markets. Markets are generally limited due to the impact of color on the final material but can be used in applications such as irrigation piping (Figure 18.7), and compost bins, etc.

18.2.1.1 HDPE Separation Process

The most common method used to separate HDPE during the reprocessing process is by flotation during a wash process. Elutriation is another method used to remove labels or lightweight accessories. It is not suitable for removing high-density material.

FIGURE 18.7 Irrigation pipes made of recycled HDPE. (From The Waste & Resources Action Program, Oxon., United Kingdom. With permission.)

If any contaminants are manufactured from products with a specific gravity (SG) greater than water (the specific gravity of water being 1.0), the polymer will sink, thus enabling the opportunity to separate the polymers by flotation during the washing process. As mentioned before, HDPE is lighter than water. Most plastics have specific gravities in the range of 0.9–1.5 (Figure 18.8).

To obtain an exact specific gravity requires a set of laboratory scales but an order of magnitude can be reached using a simple test procedure. Add a couple of drops of detergent to a beaker of water. The detergent helps to overcome the effects of the surface tension. Drop in a small piece of the plastic you wish to test into the beaker. If it floats, its specific gravity is less than 1.0. If it sinks, the specific gravity is greater than water. Magnets are used to separate any metal contamination from the polymer (Figure 18.9).

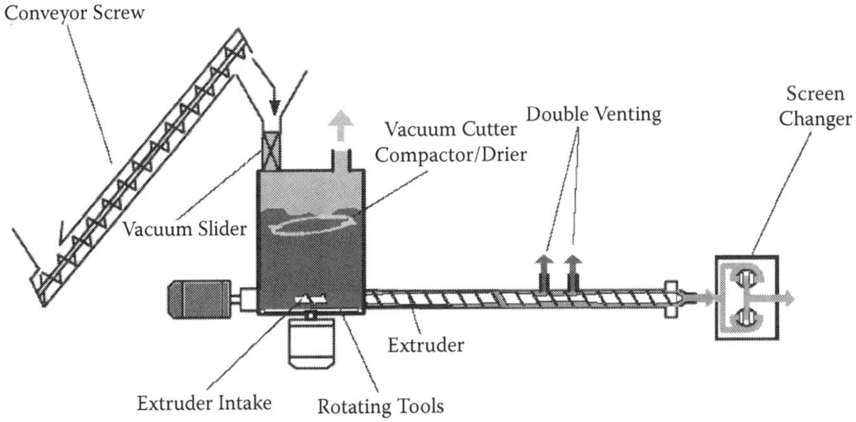

FIGURE 18.8 Process flow for super-clean decontamination of HDPE wastes.

FIGURE 18.9 HDPE flakes separated from metal contamination by magnet. (From R. W. Beck, Inc., Seattle, WA, 2005. With permission.)

18.2.1.2 Recovery and Recycling Process

HDPE milk jugs can be recycled by a cold-water wash step to remove bacteria-generated odor and in the separation stage utilizing a three-compartment sump to separate HDPE using water as the medium (Figure 18.10), and two further stages using heavier media for separation of PVC cap liners and aluminum [12,13]. HDPE crystallizes as densely packed morphological structures (spherulites) under typical conditions from the melt. Spherulites are small spherical objects (usually from 1 to 10 μm) composed of even smaller structural subunits. Polyethylene is semi-crystalline, but introducing an alkene co-monomer in the side chains reduces crystallinity. This in turn has a dramatic effect on polymer performance, which improves significantly as the side chain branch length increases up to hexene, and becomes less significant with octene and longer chains. It is by manipulating this side branching that companies have made various grades of PE suitable for different applications. Thermal degradation of polyethylene gives rise to a continuous spectrum of saturated and

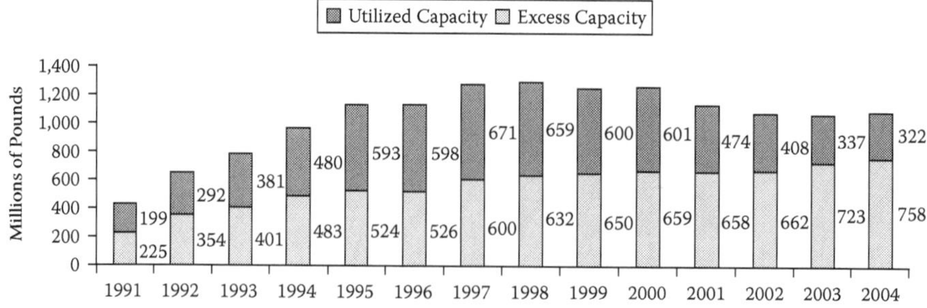

FIGURE 18.10 HDPE bottle wash capacity in the United States. (From R. W. Beck, Inc., Seattle, WA, 2005. With permission.)

unsaturated hydrocarbons from C_2–C_{90}, with lower temperatures favoring larger fragments [14]. The mechanical properties of the recyclates are, in many cases, better than is often assumed. However, the effect of multiple processing as well as of contaminants on the mechanical properties is material dependent. For the optimum utilization of the material properties of recyclates, the use-specific detection of the material parameters (e.g., the detection of the recyclability of structural parts) is necessary [15]. Experimental investigations have indicated that the extrudability (expressed as screw torque, mass intensity of flow, melt flow index, and Barus effect) of recycled polyethylene bags contaminated with mineral fertilizers (e.g., urea, NH_4NO_3, nitro-chalk, super phosphates) depends on the amount and type of fertilizer but not on the particle size of the fertilizer or moisture in the recycled sample. The screw torque and mass intensity of flow decreased and the Barus effect increased with increasing amount of fertilizer contaminant, with the changes being more pronounced at low contaminant concentration. The melt index of recycled polyethylene decreased in the presence of urea, NH_2, NO_3, and super phosphates but increased in the presence of nitro-chalk, apparently due to polyethylene degradation [16].

Melted waste plastic materials such as HDPE, in which quantities of oil have been entrapped, can be mixed with a given dose of a neutralizing agent (e.g., calcium hydroxide) at temperatures of 220–300°C to neutralize the contaminant, thus allowing the bulk polymer to be recycled and to be re-molded into other plastic products, or disposed of in a landfill [17]. Demand for recycled HDPE exceeds supply and much of the recycled resin goes into applications where physical properties are not critical, such as plastic lumber and selected injection molding applications. Recycled HDPE is also being considered in "higher-end" applications such as blow molding, where the material is utilized as a layer in a co-extruded container. Experimental evaluations of post-consumer recycled HDPE is performed using two homo-polymer samples and two co-polymer samples provided by one plastic recycler/converter, with another one homo-polymer sample and one co-polymer sample provided by a different plastic recycler/converter. Results indicate that post-consumer recycled HDPE exhibits adequate processability and a balance of physical properties adequate for a number of non-critical applications in blow molding, summarized as follows:

1. None of the recycled samples exhibits the color of virgin homo-polymer.
2. No evidence of melt fracture is encountered during molding from the recycled plastics.
3. The melt index shifts and swelling indicate that either shear modification or some other form of degradation occurs before or during recycling.
4. The odor and contamination levels in the recycled polymers still need to be reduced to successfully use the recycled ethylene polymers in blow molding.

Further research to rationalize the observed losses in certain physical properties of the recycled resins is still in process [18]. It has been further shown experimentally that the type of recycled plastic used in multi-layer containers, whether co-polymer or homo-polymer, has a significant influence on environmental stress crack resistance (ESCR) performance (i.e., the co-polymers usually outperform the homo-polymers). This fact can be reversed if the co-polymer is more contaminated than the homo-polymer, as the co-polymer is more difficult to clean. The concentration of recycled plastic in the middle layer has minimum effect on the container performance. The three-layer bottle is equivalent in ESCR performance to the virgin monolayer bottle, and the bottles show a slight improvement in ESCR with a thicker inner wall [19]. One of the most important advantages of recycled HDPE is its consistent melt flow index (MFI) and density in most of the recycling plants.

When the branching degree in HDPE increases, its crystallinity and the thickness of its crystalline lamellae decrease. This change brings about significant alterations in the mechanical properties of HDPE; two of the most strongly affected are tensile strength and tensile elongation. HDPE with increased degree of branching are softer and more elastic. An increase in the branching degree from 2–10 per 1000 carbon atoms results in a decrease of the resin tensile strength but a large increase in tensile elongation. Highly oriented HDPE is approximately 10 times stronger than non-oriented polymer because the mechanical strength of a polymer is determined by the number of inter-crystalline links: the tie chains anchored in adjacent crystallites and binding them together. Because these links are few, inter-crystalline boundaries are the weakest elements of the polymer structure. However, because the process of polymer stretching and the dismantling of its original morphological elements are accompanied by a significant increase in the number of inter-crystalline chains, polymer strength thus increases greatly. Similarly, orientation significantly increases polymer rigidity; thus, the elastic modulus of highly oriented HDPE filaments is increased about six times [20, 21].

It has been found that low molecular weight rolled HDPE possess high modulus and yield stress in the roll direction and show brittle fracture in the direction perpendicular to the roll direction. It is suggested that samples with high molecular weight possess more entanglements among the tie chains connecting the lamellar blocks [22].

HDPE is a saturated linear hydrocarbon and for this reason exhibits very low chemical reactivity. The most reactive parts of HDPE molecules are the double bonds at chain ends and tertiary C–H bonds at branching points in polymer chains. Figure 18.11 presents some of the HDPE recyclate's applications.

Because its reactivity to most chemicals is reduced by high crystallinity and low permeability, HDPE does not react with organic acids or most inorganic acids such as HCl and HF. Concentrated solutions of H_2SO_4 (>70%) at elevated temperatures slowly react with HDPE with the formation of sulfur derivatives. HDPE can be nitrated at room temperature with concentrated HNO_3 (approx 50%) and its mixtures with H_2SO_4. Under more severe conditions, at 100–150°C, these acids decompose the polymer and produce mixtures of organic acids. HDPE is also stable in alkaline solutions of any concentration as well as in solutions of all salts, including oxidizing agents such as $KMnO_4$ and $K_2Cr_2O_7$. At room temperature, HDPE is not soluble in any known solvent, but at a temperature above 80–100°C, most HDPE resins dissolve in some aromatic, aliphatic, and halogenated hydrocarbons [11, 21, 23].

Studies involving virgin HDPE/recycled HDPE composites found that the most affected mechanical property was the elongation at break, which decreased with increasing amounts of recycled HDPE. Recycled HDPE was obtained from post-consumer cycle of milk bottles. However, generally it has been found that recycled

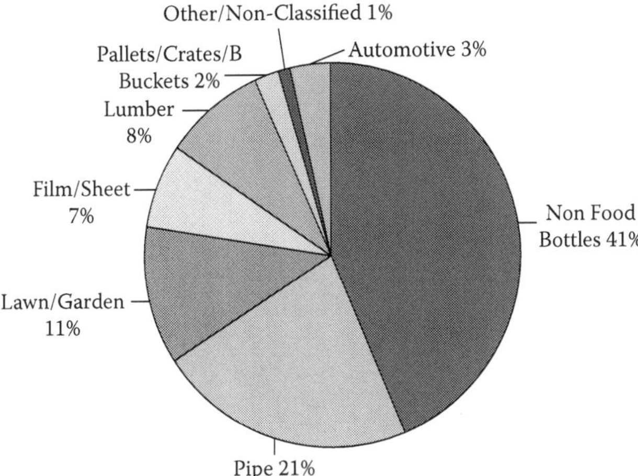

FIGURE 18.11 Application rate of recycled HDPE.

HDPE is a material with useful properties not largely different from those of virgin resin and thus could be used, at an appropriate concentration in virgin HDPE, for different applications [24].

The recycling of homogeneous HDPE from containers for liquids has been found to give rise to materials having mechanical properties that are strongly dependent on the reprocessing apparatus and the processing conditions.

The thermomechanical degradation (Figure 18.12) during processing gives rise to different modifications of the structure depending on the temperature, residence

FIGURE 18.12 Twin screw equipment for HDPE.

time, and applied stress. Generally, if the reprocessing operations are performed in apparatus with low residence time, the mechanical and rheological properties of the raw materials are only slightly influenced by the recycling operations. Significant degradation phenomena and reduction of some mechanical properties are observed on increasing the number of recycling steps in apparatus with larger residence times. By adding antioxidant agents, the polymer maintains the initial properties even after several recycling cycles. The competition between formation of chain branching and chain scission is considered to be responsible for this behavior [25]. The thermal diffusivity of HDPE has been studied over a wide range of temperatures (25–100°C) by melting powdered HDPE in a cylindrical mould at several pressures and recording the temperature profiles at several radial positions. The thermal conductivity of a packed bed of HDPE powder was found to increase with pressure because of the decreased porosity [26].

18.2.1.3 Application of the HDPE Recyclate

As mentioned before, recycled HDPE is used in several areas. One of them is to blend the HDPE recyclate with virgin resin in production of large molded containers [27]. Blow-molded bottles for household and industrial chemicals, such as shampoo, detergents, and bleach, are a major application field for HDPE recyclate [28]. Union Carbide used HDPE recycled as the middle layer of multi-layer blow-molded containers for food contact applications, permitted by the FDA [29]. The mechanical properties of recycled HDPE have introduced it as a part of the raw material in the manufacture of pallets, large injection moldings, sports equipment, kayaks, and plastic films [30–35]. Recycled HDPE is used in film applications while being 40% blended with LLDPE and LEDPE. Recycled HDPE is also modifiable by rubber crumb, which increases the impact strength of HDPE recyclate and decreases its flexural modulus (Figure 18.13) [36–38].

18.2.1.4 New Ideas in HDPE Recycling

Applying artificial weathering to evaluate the light stability of post-consumer HDPE material recycled from bottle crates during the recycling by remelting-restabilization technique has been investigated, and tensile impact strength is monitored during the artificial weathering exposure to study the effect of the re-stabilization. Results illustrate that the restabilization is mandatory for improving the light stability of the post-consumer crate material, ensuring its reusability in the original application [39]. In 2002, a milk bottle recycling project was launched in Northern Ireland [40]. Milk bottles were collected by a deposit system and were sent to a bottle-to-bottle recycling process. As a result of the deposit collection system, the recycled HDPE was completely under source control and had been used in its prior application only for packaging of fresh milk. The collected material was recycled first by a conventional-washing-based recycling process and then further deep-cleansed using a super-clean process. The recycled material was re-used in new milk bottles at 20–30% content without a functional barrier. The intended application was for fresh milk with short storage times under refrigerated conditions. Studying the post-consumer contamination HDPE milk bottles to design a bottle-to-bottle recycling process demonstrates

FIGURE 18.13 Playground devices made of recycled HDPE according to the hygienic requirements.

that predominant contaminants in hot-washed flake samples are unsaturated oligomers, which can be also be found in virgin HDPE pellet samples used for milk bottle production [41]. In addition, the flavor compound limonene, the degradation product of antioxidant additives di-*tert*-butylphenol and low amounts of saturated oligomers, is found in higher concentrations in the post-consumer samples in comparison with virgin HDPE. However, the overall concentrations in post-consumer recycled samples are similar to or lower than concentration ranges in comparison with virgin HDPE. Contamination with other HDPE untypical compounds was rare and is in most cases related to non-milk bottles, which are <2.1% of the input material of the recycling process. The maximum concentration found in one sample of 1 g is estimated as 130 mg kg^{-1}, which corresponds to a contamination of 5200–6500 mg kg^{-1} in the individual bottle. The recycling process is based on an efficient sorting process, a hot washing of the ground bottles, and a further deep-cleaning of the flakes with high temperatures and vacuum. Based on the fact that the contamination levels of post-consumer flake samples are similar to virgin HDPE and on the high cleaning efficiency of the super-clean recycling process, especially for highly volatile compounds, the recycling process investigated has been reported as suitable for recycled post-consumer HDPE bottles for direct food-contact applications. However, handpicking after automatically sorting has been recommended to decrease the amount of non-milk bottles. The conclusions for suitability are valid, providing the migration testing of recyclate contains milk bottles up to 100% and that both shelf-life testing and sensorial testing of the products are successful. The addition of different percentages of recycled material to raw material for injection molding processes implies a change in the rheological behavior of the material, affecting both the way of filling the moulds and the conditions of the process [42]. An HDPE blend

```
       H   H   H   H   H   H   H   H   H   H
       |   |   |   |   |   |   |   |   |   |
~~~~—C—C—C—C—C—C—C—C—C—C—~~~~
       |   |   |   |   |   |   |   |   |   |
       H   H   H   H   H   H   H   H   H   H
```

FIGURE 18.14 Schematic structure of LDPE.

of 100% recycled material increases viscosity up to 10% in regard to raw HDPE, with the recycled pellet size described previously, when tests are performed in a capillary rheometer. This tendency is also observed during the first tests that are being performed with a spiral mould in injection machines.

Recycled HDPE filled with rice hulls obtained from agricultural waste can provide a low cost material for the production of various profiles. These highly filled recycles show a large increase in dynamic shear properties compared with the pure polymer. The changes in properties due to filler may contribute to the large amplitude extrudate tearing observed. This gross tearing poses a significant challenge to industry in extruding these materials [43].

18.2.2 Low-Density Polyethylene

Low-density polyethylene (LDPE) is a thermoplastic polymer belonging to the olefin family. Its properties are influenced by the degree of chain branching within the molecules (Figure 18.14). Polymerizing ethylene under high pressure and temperature produces LDPE. Similar to HDPE, LDPE has a wide range of uses because of its low cost, processability, and high resistance to impact, chemicals, and electricity.

Over one-fourth of the U.S. plastics packaging and disposable market is LDPE products. It is used almost exclusively for packaging and other films (including bags, such as trash bags or other household bags) and coatings. Recovery of LDPE products is generally limited to transport packaging (shrink and stretch films). These are recycled into products such as builders' film (damp proofing), rubbish bags, and agricultural films (see Figure 18.15). Shrink film is the most common form of LDPE collected for recycling. Stretch film has been recycled successfully by several companies in the United States, usually blending a small percentage (10–20%) with other films, to minimize processing difficulties.

FIGURE 18.15 Recycling sign of LDPE.

Most of the polyethylene films consumed are based on a blend of LDPE and linear-LDPE (LLDPE), which makes the film own the properties as desired in its specified application (Figure 18.16).

The techniques used for separation of different kinds of plastics are based on differences in density, shape, color, physicochemical properties, and solubility. The solubility-based processes include stages of dissolving a series of incompatible polymers in a common solvent at various temperatures or in different solvents, so that one polymer is separated each time. These processes differ in the method employed to recover the polymer after the dissolution stage. It can be recovered

FIGURE 18.16 Waste LDPE film bales.

either by rapid evaporation of the solvent, or by adding a proper "anti-solvent" that makes the polymer precipitate, like in the selective dissolution/precipitation (SDP) method. As laboratory experiments indicated, satisfactory separation of the three-component mixture (LDPE/HDPE/PP) is achieved by using the xylene/*iso*-propanol system and dissolving each polymer at different temperatures. All polymers were recovered in small grains [44].

18.2.2.1 LDPE Recycling Methods

Similar to other polymers, several methods are used to recycle LDPE, and accordingly, they have been classified into the following categories: primary (re-melting), secondary (mechanical recycling) (Figure 18.17), tertiary (chemical or thermal recycling), and quaternary (incineration). Recently, much attention has been paid to thermolysis and catalytic polymer degradation techniques as methods of producing various fuel fractions from wastes [45, 46]. In particular, LDPE has been targeted as a potential feedstock for fuel producing technologies [47, 48]. Interest in developing value-added products, such as synthetic lubricants via LDPE thermal degradation, is also growing. The development of value-added recycling technologies is highly desirable because it would increase the economic incentive to recycle polymers. In the first step, LDPE is heated, under a nitrogen atmosphere, to temperatures in the range of 400–450°C. At these temperatures, volatile thermolysis products (PE oil) are produced, which are then condensed and collected. In the second step, the aforementioned oil is hydrogenated at 30–90°C to produce a diesel-type liquid fuel. LDPE oil can also be produced via thermolysis of PE at 420–460°C under an inert, nitrogen atmosphere. The thermolysis characteristics of clean, unprocessed LDPE are essentially the same as unwashed, scrap LDPE packaging. The thermolysis may be performed at higher reaction temperatures in the range 410–440°C with only a small increase in the rate of formation of 1-alkenes over *n*-alkanes. Synthetic diesel fuel produced by hydrogenation of LDPE oil has greatly enhanced properties

FIGURE 18.17 LDPE pelletizer machine. (From CarolMac Co., USA. With permission.)

compared with conventional diesel fuel, but the process requires refinement in terms of producing diesel fuel with lower cloud and pour points [49, 50].

By catalytic degradation of LDPE wastes, the required energy might be decreased because the degradation temperature of wastes could be decreased with catalysts. The zeolites and clay minerals are the most commonly used catalysts. They have effects also on the structure of products, so when the target product is the diesel fuel, isomerization catalysts might be not advantageous. The effects of the additive content in waste LDPE on the degradation is of interest. The cracking of different additive containing waste LDPE has been investigated in the presence of different catalysts and in the absence of catalyst at two temperatures. The temperature and catalysts had significant effect on yields of the liquid products. Using catalysts, the yields are higher and significant change in the product composition can be observed (olefin double bond isomerization, higher olefin content). Cracking temperature affects the yields and to a lesser degree the product composition. The catalyst activity decreases with degradation time, and the activation energy of cracking could be decreased by using catalyst (Figure 18.18) [50–55].

Pillared clays are able to completely convert polyethylene in gaseous and liquid hydrocarbons, showing low coking levels. The selectivity and yield to liquid hydrocarbons are high, as the mild acidity of pillared clays avoided excessive cracking to small molecules. Regenerated catalyst samples show practically identical levels of conversion and selectivity with fresh pillared clay samples.

Furthermore, they produce hydrocarbons with practically the same distribution as the fresh samples, confirming that pillared clays can be completely regenerated. Both high yield to liquid products and re-generability make pillared clays potential catalysts for an industrial process of catalytic cracking of LDPE waste [56].

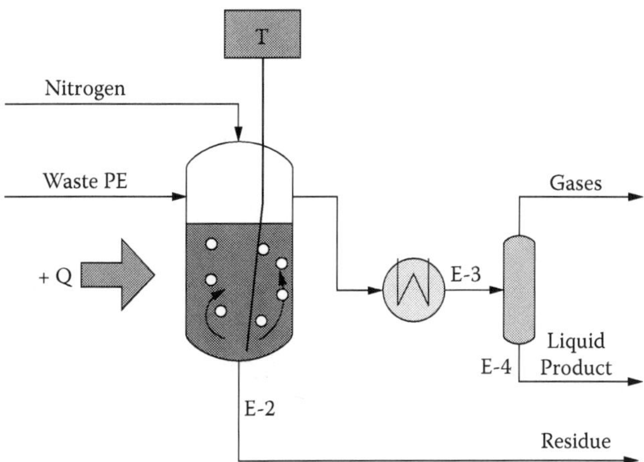

FIGURE 18.18 Schematic structure of catalytic degradation reactor for LDPE.

Many recycling methods are used for LDPE recycling — all trying to reprocess the LDPE thermally, but contamination is always the most important problem, the same as with other polymers. Tackified films by polyisobutylene-based materials are one of those contaminants. Comprehensive and careful sorting is required to ensure removal of contaminants. Other plastics, dirt, adhesives, labels, color pigment, and product residues need to be removed as much as possible to ensure a good-quality recyclate. The more contaminants present the lower the mechanical properties of the material.

Polymer with low molecular weight would be oxidized producing waxes, aldehydes, acids, ketones, etc. [57–59]. Also, experiments with LDPE showed that high processing temperature and high residence times strongly enhance the degradation processes and reduce the mechanical properties, in particular the elongation at break. Greater thermomechanical degradation, better homogenization, and better dispersion of the non-polymeric impurities are responsible for this behavior.

It was also found that introducing additives, like antioxidants (phosphite stabilizer), inert fillers (CaCO3), and impact modifiers (calcium silicate), improves mechanical properties (especially elastic modulus and elongation at break) approaching those of virgin polyethylene. The separation and cleaning steps that would be performed for LDPE are the same as HDPE recycling. Reprocessing of post-consumer recyclate (exposing it to shear and heat stress) produces LDPE that has lower elongation at break, impact strength and stiffness than required, due to the low remaining level of the active form of the stabilizer. Therefore, post-consumer LDPE cannot be safely processed without addition of fresh stabilizers. It is noticeable that even addition of phosphates alone enhances the stability during subsequent processing and heat aging [60].

LDPE is also recycled in blown-film production using factory scrap material. The amounts of the addition of recycled materials to the virgin polyethylene are established, having in mind the requirements for quality performance of the obtained films [61] (Figure 18.19).

FIGURE 18.19 LDPE film recycling plant.

Heterogeneous polymer wastes which contain LDPE are usually recycled in a single step, producing objects with large dimensions and poor mechanical properties. The "secondary" materials (with plastics waste content up to 50%) show mechanical properties similar, except for the elongation at break, to those of the virgin polyethylene. The addition of moderate amounts of calcium carbonate does not significantly reduce the properties of these materials. Thermal results made of LDPE recycling indicate that mixtures reach crystallinities lower than those found in fresh materials. Results also show that recycled materials do not significantly change the crystallization and the fusion temperature, but do change effectively the crystallization rate. From the mechanical test, measurements of the Young's modulus and of the elongation to break show a small increase with a recycled material content. Instead, the resistance to impact is lowered. Overall it can be concluded that it is possible to use recycled LDPE in an amount up to 50% in the final product. Amounts over 50% are not recommended because the greater unsaturation level can catalyze degradation processes, which shorten the final product's useful life. The use of the impact modifier essentially does not improve the mechanical properties of mixtures and addition of antioxidants, and the exposition of samples to photo-oxidation show unusual behavior [62–64].

Recycled LDPE and LLDPE are used in the production of:

1. Builders film and damp-course concrete and brick lining
2. Garbage bags
3. Retail carry bags
4. Lids and closures
5. Black irrigation pipe

```
     H   H   H   H   H   H   H   H   H   H
     |   |   |   |   |   |   |   |   |   |
~~~—C—C—C—C—C—C—C—C—C—C—~~~
     |   |   |   |   |   |   |   |   |   |
     H  CH3  H  CH3  H  CH3  H  CH3  H  CH3
```

FIGURE 18.20 Schematic structure of PP.

18.2.3 Polypropylene

Polypropylene (PP) is the second most common thermoplastic of the olefin family after PE. In 2000, 23% of thermoplastic consumed in Western Europe was polypropylene. PP has a lower impact strength then PE, but a superior working temperature (enabling containers to be "hot-filled") and tensile strength (Figure 18.20). PP has excellent insulation properties, but is most widely used as fibers and filaments produced by extrusion. The fibers are used in some products such as carpets, wall coverings, and for furniture or vehicles.

About 39% of Western Europe's polypropylene consumption in 2000 was from the packaging sector (Figure 18.21). PP is also used for wire insulation, medical devices, piping, and sheeting. Injection molded products are another significant product group, especially for use as medical supplies which need sterilization through heating or irradiation. Most recycled PP comes from vehicles, including battery cases and car fenders. The main recycled process applied for PP is through re-granulation [65]. Because PP is a common and versatile material, it would be a waste to throw it away. It is likely to go into landfill sites. Why not turn it into something useful? Recycling is the path to form a useful material from what has been used before and is rubbish.

FIGURE 18.21 Recycling sign of PP.

During the reprocessing, stabilized PP is made different in color and smell, but undergoes no significant change in mass and alterations to the polymer chains after repeated extrusion. The application properties of the materials are almost unaffected. PP production scrap is usually reground and returned directly to the process. The quality of the material from old parts, cleaned and sorted according to PP grade, matches that of the production scrap (see Figure 18.22). Assurance of the reliability of material supply is more important than processing technology. It is difficult to maintain consistent quality because many different types and grades of PP are available. PP-homo-polymer is formed when propylene is polymerized and PP-co-polymer is made by first polymerizing the propylene, which is followed by co-polymerization of ethylene and propylene. Most polymers are incompatible with each other. Studies have shown that used polypropylene parts that are gathered and sorted centrally by trained personnel can still contain up to 10% foreign material. It is therefore unavoidable to use a costly automated separation technique [66].

In 2001, the European Commission Directive presented a proposal to amend the directive and set a new recycling target of 20% for plastics (mechanical and chemical) by 2006 [67].

FIGURE 18.22 PP scrap made from packaging bands.

In countries where drink products are packaged in returnable bottles (e.g., Germany and the Netherlands), it is possible to recover the polypropylene bottles at the sales outlet assuming that the consumers remember to put the caps back on the bottles. For technical and hygienic reasons, old PP closures cannot be used in the production of new bottles. The material will be turned into reusable pallets or transport containers. PP recycling is more promising in the automotive industry. This is due to the fact that about 500,000 tons of PP is used in the Western European car industry every year. Car bumpers, air ducts, and instrument panels can be recycled and recovered to produce new car parts. The only drawback is that most recycling grades are black and this limits the applications of PP recyclate. To ensure quality, only 20–30% of old material is used in new car parts. Moreover, by producing single polymer assemblies that do not need dismantling after removal, plastic recovering can become more economical and easier. For example, an all-PP bumper system comprised of a glass mat reinforced beam, an energy absorbing PP particle foam, and PP outer skin is available. The cost effectiveness of using PP recyclate depends on the price of new plastic material. The problem is that the price of new plastics fluctuates and sometime it could fall below the cost for recycling/recovering old plastics [68].

18.2.3.1 Recycling Methods for PP

To establish the viability of recycling PP containers, it is necessary to identify high quantity sources either with a low contamination level, or that could be easily cleaned. Of course this is not always achieved and we have to target the low quantity or highly contaminated sources. Cleaning is the first step as with the other polymers, followed by pelletizing and then re-melting for extrusion, blown injection, or molding (Figure 18.23).

It is possible to recycle PP by dissolution of the polymer in a suitable solvent with subsequent polymer recovery and drying to produce a free-flowing powder. An optimum solvent for the process is selected using thermodynamic, kinetic, and energy factors. It has been demonstrated that ethylene tetrachloride is an appropriate solvent for PP recycling process [69]. For some mechanical methods of recycling, PP is pelletized first (Figure 18.24).

FIGURE 18.23 PP flakes from cleaned wastes.

Rheological properties of the photo-oxidization recycled PP blend with virgin polymer indicate some interactions between polymers at low shear rates. On the other hand, at high shear rates, evidence of incompatibility exists. The nominal tensile strength is almost independent of the composition, whereas the other mechanical properties are similar to those of the degraded material. The unusual behavior of the elongation at break is correlated with crystalline phase segregation, which appears with decreasing molecular weight of the degraded component [70–72]. The main effect of recycling on PP is the lowering of the melt viscosity, which is attributed to molecular weight decrease. Recycled PP exhibits greater crystallization rate, higher crystallinity, and equilibrium melting temperature than those measured for virgin PP, and optical characteristics are affected during the recycling process. It is desirable that molecular weight and molecular weight distribution changes would

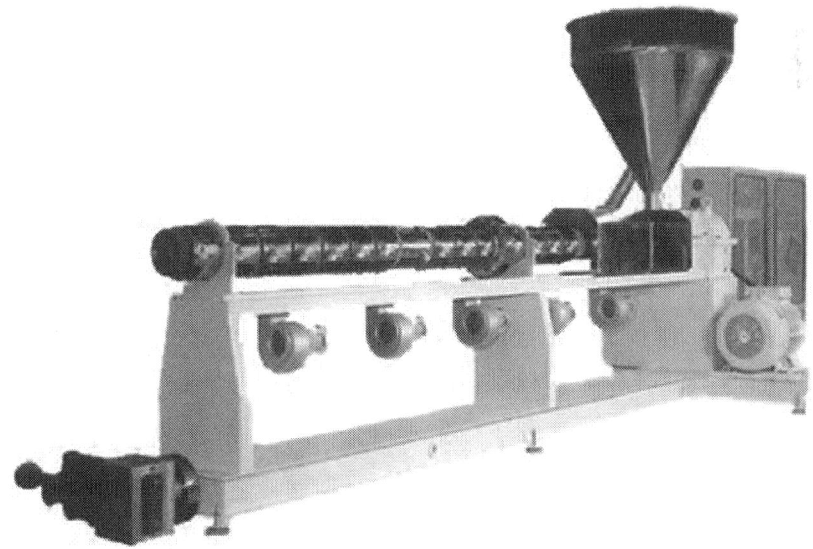

FIGURE 18.24 PP pelletizer machine.

FIGURE 18.25 Cable reels from PP recyclate.

occur during the recycling affecting many characteristics (e.g., gas permeability) [73–77]. PP recycling has always been challenging because the polymer is highly susceptible to thermo-oxidative degradation during extrusion and impurities present in recycled PP tend to degrade even the virgin PP in this process. Thus, the recyclate is stabilized by adding a peroxide decomposer (triphenylphosphite) and a slipping agent (zinc stearate) in contrast to radical scavengers normally used in reprocessing. Radiation processing of PP allows recycling the polymer's wastes either by adding into a fresh and cross-linkable material, or by the reprocessing of aged polymer by a proper exposure to degrade it to lower molecular weight products [78–84]. At any rate, PP recyclate is applied in different industries (Figure 18.25).

18.2.3.2 PP Composite Recycling

PP composites are of widespread application and the potential for recycling them is provided by the dissolution process. In this process, the polymer solution is filtered and new composites are designed. It becomes evident that a significant increase of the tensile modulus and strength occurs as the polymer phase deposited on the fibers increases [85, 86]. As nanotechnology is introduced more and more, nanocomposites are being studied more. PP-based composites are among the most well-known nanocomposites, and PP nanocomposite recycling is a new idea in polymer science and technology [87].

18.2.4 Polyolefin Mixtures Recycling

Recycling of mixed plastic wastes composed of LDPE matrix and PP has been performed by compounding using single-screw or twin-screw extruders. Blends of virgin polymers are prepared to compare mechanical properties of both virgin and regenerated materials. The effect of process parameters and that of different types of compatibilizers is noticeable on recyclate's properties. By adding compatibilizing agents such as ethylene propylene diene monomer (EPDM), elongation at break and impact strength are improved. Twin-screw extrusion of both virgin materials and

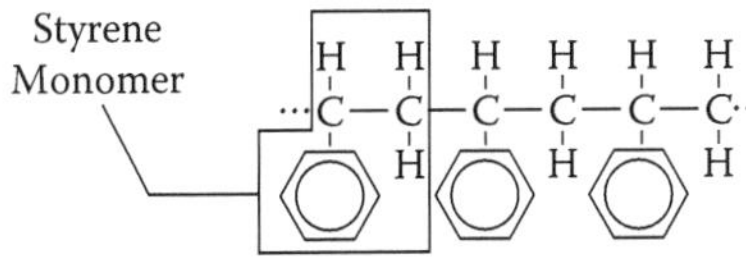

FIGURE 18.26 Schematic structure of PS.

regenerated plastic wastes leads to more homogeneous LDPE/PP blends with better mechanical properties than when using single-screw extrusion. In addition, addition of fibers such as short polyamide fibers in LDPE/PP blends allows the improvement of tensile, flexural, and impact behavior. Thus, the polyamide fibers act as reinforcing agents for polymer matrices in the majority of LDPE. The observed results suggest that a better reinforcing is obtained when affinity between polyamide fibers and LDPE is favored. The thermal decomposition of polyolefins as a recycling route for the production of petrochemical feedstock based on LDPE and PP mixture treating is one of the new ideas. Thermally decomposition product is dissolved in primary heavy naphtha to obtain steam cracking feedstock [88–90].

18.3 POLYSTYRENE

Polystyrene (PS) is a relatively cheap, hard polymer, usually produced by polymerizing styrene monomers (Figure 18.26). High molecular weight PS is used for coatings, whereas lower molecular weight PS grades are used for injection molding. The main weaknesses of PS are that it is brittle, unstable when exposed to UV light, and flammable.

Other forms of PS include expanded polystyrene (EPS), which is produced by using inert volatile solvents as blowing agents, and high-impact PS (HIPS), which is made by incorporating small particles of butadiene rubber. EPS is used mainly as an insulating material in the construction sector, and as an insulator for disposable food containers and protective packaging. The major application for HIPS is fast food packaging. The most abundant form is EPS, although this presents some challenges, mainly due to the fact that the material needs to be densified for transportation, and additives introduced during blowing can be difficult to remove. PS recycling tends to be more limited than other vinyl polymers, because of challenges with collection and processing (see Figure 18.27).

FIGURE 18.27 Recycling sign of PS.

With a large amount of PS in circulation, a significant amount ends up as ground and air pollution (Figure 18.28). All the pollution sources can be categorized into either the industrial or the municipal sources, both types responsible for large amounts of PS pollution. To recycle PS, it must be broken down into its constitutive components. Three general methods of degradation for PS are used: thermal, electromagnetic, and chemical. Thermal energy excites the polymer and various reagents incorporated within the polymer material. The excited reagents are likely to collide and react with the polymer chains at elevated temperatures. The thermal

FIGURE 18.28 PS wastes made from EPS manufacturing.

energy also helps to lower the activation energy requirement, which helps the polymer break down on its own [91].

Burning waste has historically been the method of choice when dealing with wastes that will not biodegrade. This has been researched as a recycling technique for PS. Two processes use thermal energy to recycle PS. The first is simple incineration in a furnace. This involves breaking down the PS at elevated temperatures in an environment containing oxygen. Although this particular process is used widely, adequate measures must be taken to prevent the volatized species from entering the atmosphere. In addition, a method of collection for the monomer and dimer containing oil must be in place. Common returns for incineration are typically around 40%, whereas the optimized incinerator can return nearly 50% with the balance either being released into smokestack soot filters or thermally degraded to other chemical by-products.

Pyrolysis also uses the thermal degradation route, but in an oxygen-deficient environment. This prevents oxidation of the volatized species and gives a monomer/dimer return of around 65%. Again, attention needs to be given to the prevention of air emissions from this process. Using chemical reactions to break PS down into precursors is a relatively new option. Finding a solvent that converts PS into useable material has long been a major obstacle, but recently, the use of catalysts in a thermal degradation process has been developed. This process uses the catalyst to initiate degradation reactions in a pyrolysis process. Most effective of the catalysts are solid bases, but solid acids can also be used less effectively. When in contact with the catalyst at elevated temperatures, PS breaks down. The monomer and dimer return has been as high as 88%. All the previous options result in recovery of oil. This oil is not a pure monomer and it is not a pure dimer. Common contaminants include ethylbenzene, toluene, methylstyrene, and benzene — all of which must be removed from the oil. Although the percentages of these contaminants are minimal,

TABLE 18.6
U.S. Post-Consumer EPS Recycling Rate Summary: 1990–2004 (Millions of Pounds)

Rate Component	1990	1992	1994	1996	1998	2000	2002	2004
Pounds recycled	3.0	20.8	24.2	22.5	19.2	24.9	26.2	25.0
Pounds sold	179	218	238	217	202	206	201	222
Recycling rate	1.7	9.5	10.2	10.4	9.5	12.1	13.0	12.0

Source: Alliance of Foam Packaging Recyclers, 2005.

they must be removed, which adds to the cost and complexity of recycling [92–96]. Table 18.6 lists the recycling rate of EPS in the United States.

18.3.1 Mechanical Recycling of PS

PS is recycled mechanically in four techniques:

1. Post-consumer PS is processed as a partial or direct substitute for virgin polymer, into raw material for the production of loose-fill packing or even for new PS moldings.
2. Re-use in non-foam applications such as plastic stationery products, video and CD cases, plant pots, etc.
3. Extruded applications such as hardwood replacement, garden furniture, windows, and photo frames.
4. Re-use in lightweight concrete and building products (Figure 18.29). It should be mentioned that the HIPS waste is usually densified before recycling (Figure 18.30).

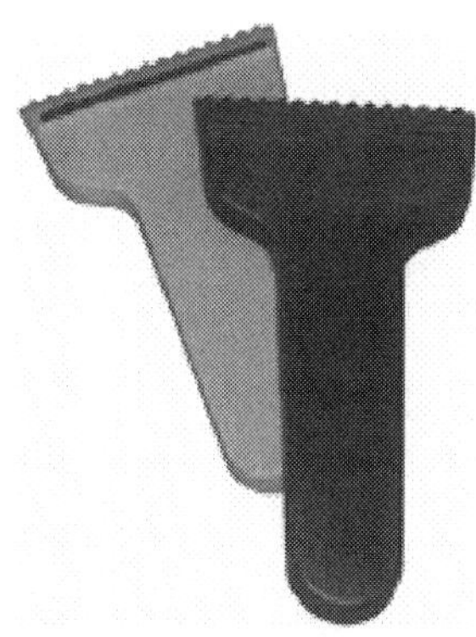

FIGURE 18.29 Product of recycled PS.

FIGURE 18.30 PS scrap densifier. (From MIXACO® Co., Neuenrade, Germany, 2006. With permission.)

Recycling of PS decreases the shear stresses and consequently the viscosity at constant shear rate. Elongation at break of the recyclate would be progressively decreased compared to the virgin polymer, but it has been used in manufacturing of some products that are incompatible with these properties, such as low-weight bricks [97–100].

18.3.2 Other Recycling Methods

Solvent applying methods have been successful for PS recycling. Toluene, tetrahydrofuran (THF), and N-methyl pyrrolidone (NMP) are some solvents that have been applied in recycling processes of post-consumer PS from food packaging and other products. Easier removal of paper and food particles than straightforward melt recovery is one of the advantages. Solvent removal from the PS can be accomplished by azeotropic distillation (for toluene) or dry spinning (for THF). Whereas NMP is removed by extraction, this method is a safe and economical route for large-scale production [101, 102].

In a model solvent technique, PS foam scrap is recovered in the form of small grains. The process mainly comprises dissolution of the waste into benzene or toluene, filtering, dispersion of the solution into water, and subsequent distillation. The alternative solvent/non-solvent systems have been studied on the basis of solution rheology, operating conditions during the recycling procedure, and extents of recovery of PS and solvent. The toluene/water system had no critical influence on the recyclate [103].

d-Limonene is a natural vegetable oil which is extracted from the rinds of citrus fruits and is a good solvent of EPS. Limonene has almost the same solubility as toluene at room temperature. This technique reduces the volume of EPS to about 1/20th of the original. Contracted EPS is recyclable with almost no molecular weight degradation because *d*-limonene acts as an antioxidant of polystyrene during the heating process. This process is performed by a system consisting of an apparatus to dissolve EPS and a recycling plant to separate the limonene solution. The recycling plant can mass reproduce polystyrene with the same mechanical properties as new polystyrene. The recycled polystyrene can be used for packaging [104, 105].

Chemical recycling of waste PS into styrene is mainly by using various kinds of solid acids and bases as catalysts. Solid bases are effective catalysts for the chemical recycling of waste PS because of the high yield of distillates from PS and the high selectivity into styrene in the distillates. Barium oxide appears to be the most effective one. The preparation and the thermal degradation of PS films with dispersed barium oxide powder are the main steps to design a recycling. It is notable that 1 wt% of barium oxide converts about 85% of PS waste to styrene by a simple thermal degradation at 350°C without any addition of other catalytic compounds [106, 107]. Petroleum-based solvents are also newly applied for PS recycling. Using these reagents, waste PS is reduced in volume and gelatinized PS is used as starting material in the process, which consists of distillation followed by thermal cracking and liquefaction (Figure 18.31).

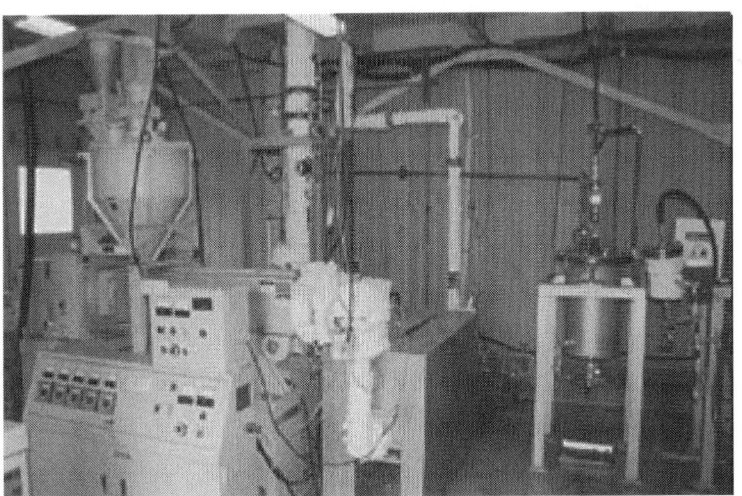

FIGURE 18.31 Plant for converting the PS waste to styrene monomer. (From Toshiba Co., Japan, 2006. With permission.)

Pyrolysis and catalytic pyrolysis are two other processes for PS recycling. The main product from the un-catalyzed pyrolysis of PS is an oily mixture, consisting mostly of styrene and other aromatic hydrocarbons. The generated gases consist of methane, ethane, ethene, propane, propene, and butane. Zeolites are a group of catalysts used in PS pyrolysis. In the presence of catalysts, an increase is observed in the yield of gas and a decrease in the amount of produced oil [108].

PS-based composites are also recycled. Compared with the original composites, the mechanical properties and dimensional stability of the recycled composites would not change significantly even after exposure to extreme conditions. Moreover, the treated composites offer improved properties compared with non-treated and original polymer [109, 110]. New ideas in PS recycling are now forming. One of them is to produce the PS fibers by electrospinning method, which gives small pore size with high surface area fibers [111].

18.4 POLYVINYLCHLORIDE (PVC)

PVC is the second most important polymer worldwide and a large-scale chemical product (Figure 18.32). Thanks to its excellent cost-to-performance ratio and outstanding chemical and environmental resistance, today it is mainly used for long-life applications.

Plants manufacturing PVC began to spring up during the 1930s to meet demand for the versatile material. Just a decade after its conception, PVC was sought for a variety of industrial applications including gaskets and tubing. Joining industries across the nation during the 1940s, PVC manufacturers turned their attention to assisting the war effort. PVC-coated wire was widely used aboard U.S. military ships, replacing wire insulated with rubber. Five companies were making PVC at the last century's midpoint, and innovative uses for PVC continued to be found during the 1950s and 1960s [112].

It is supplied in powdered form by the raw material producers (generally big chemical companies) and comes in a wide range of different types. Compounders or converters then mix processing aids, stabilizers, and other additives into the powder. The PVC is processed into semi-finished and finished products, using either the processing units employed for other thermoplastics or, in most cases, specially modified plants. The following are reasons why PVC enjoys such widespread use:

1. It is the thermoplastic that offers the best price-to-performance ratio.
2. It has a particularly broad property range and can be modified within extensive limits (from hard and rigid to soft and rubber-elastic); one of its attributes is a particularly pronounced resistance to chemicals.
3. It can be processed easily and in a highly versatile manner, and single-sort PVC offers excellent recyclability.

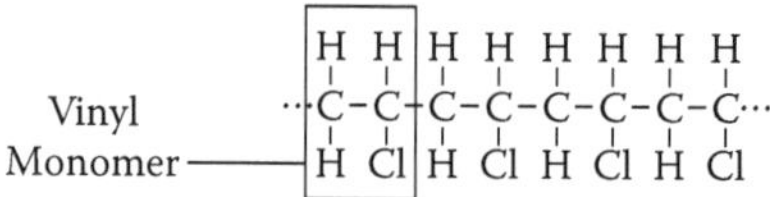

FIGURE 18.32 Schematic structure of PVC.

The excellent cost-to-performance ratio of PVC and the fact that it can be used to achieve a wide variety of tailor-made formulations explain its success in markets as different as building and construction (more than 50% of all PVC sold), cables, the automotive industry, electrical appliances, medical devices, packaging (about 15% of the total), and many other sectors [113].

PVC-based latex has been used on boots, fabric coatings, and inflatable structures, and methods for enhancing PVC's durability have been refined, opening the door to applications in the building trades. PVC piping is transporting water to thousands of homes and industries, aided by improvements in the material's resistance to extreme temperatures.

18.4.1 PVC Recycling

Even after a useful life span of decades, PVC products can be recycled into new applications lasting decades more (see Figure 18.33). A significant amount of PVC scrap, which comes from polymer manufacturers and plastic processors, is collected and recycled. Overall, more than 99% of all manufactured PVC compound ends up in a finished product, due to widespread post-industrial recycling.

FIGURE 18.33 Recycling sign of PVC.

Generically, three principal types of PVC waste are generated:

1. *Production residues*: These arise mostly in the form of cut-offs, in the factory or plant, as the product is made. For many years, such valuable "waste" has been recycled as a matter of good housekeeping, and only a small proportion needs to be disposed of as waste to landfills or by combustion.
2. *Installation waste*: This is made of some sold products, such as flooring, cables, and pipes that have been cut to size during installation. In recent years, the PVC industry has become active in organizing collection systems and in recycling these "waste" products back into new products.
3. *Post-consumer waste*: These are products that have fulfilled their service life and end up in waste streams from different domestic and industry sectors.

Recycling products at the end of their useful lives presents the greatest challenge in terms of economically viable collection, sorting and cleaning. The PVC industry is building on its experience of recycling installation waste and is extending such operations to collect and recycle post-consumer products wherever this is viable. An appropriate collection system is the prerequisite for an effective, long-term program. A variety of systems exist:

1. "Take Back" schemes, including arrangements to collect post-use products, are offered by commercial companies, converters, and local authorities and apply mostly to long-term PVC applications. These are used, for

example, in the construction industry for PVC pipes, windows, profiles, and floorings.
2. "Bring Back" schemes encourage consumers to return used items to collection points (e.g., bottle banks). A major advantage is the pre-sorting of waste by consumers.
3. "Curbside Collection" brings in mostly domestic waste from specific roadside containers placed in front of households. The degree of sorting is usually lower than with the other schemes. A typical example is the yellow bag collection of all types of plastic packaging via the "Dual System" in Germany.

Viable recycling, therefore, depends on many factors, including the following essential requirements:

1. Access through appropriate collection arrangements to a sufficient, steady, and reliable supply of waste materials
2. Technology that makes the recycling economically sensible
3. A constant demand for the recycled products

Several recycling methods are used to convert the waste PVC in to valuable materials. Mechanical recycling of source-separated PVC is a technically relatively simple and common practice. Suitable post-use products are those that are easy to identify and separate from the waste stream or can be kept relatively clean, ending up as a high-quality recyclate for use within the existing range of PVC applications [114–116].

Researchers have found that PVC can easily be separated from other plastics automatically because technology can spot its unique chlorine chemical composition (Figure 18.34). PVC has become one of the most widely used materials, in part because of its cost efficiencies. These efficiencies begin when PVC is produced and continue throughout its lifecycle, encompassing such elements as raw material usage,

FIGURE 18.34 Separated PVC film scraps.

energy used in processing, energy used in distribution and transportation, durability, maintenance requirements, and disposal costs. PVC consistently scores better than other materials in many economic and environmental performance categories.

PVC's beneficial qualities affect every aspect of our lives. Because it will not rust or corrode, PVC is widely used in water pipe to deliver clean water and in sewer pipe to ensure the integrity of wastewater handling systems. PVC pipe installations last up to 50 years, reducing raw material consumption. PVC is also the material of choice for blood bags and tubing, which helps to maintain the world's blood supply and supports critical healthcare procedures such as dialysis. In packaging, PVC helps to keep food safe and fresh during transportation and on store shelves, and provides tamper-resistant packaging for food, pharmaceuticals, and other products. PVC's resistance to high electrical voltage and its ability to bend without cracking make it the leading material for wire and cable insulation. It helps add years to the life of motor vehicles as an under-body coating. PVC's toughness and durability make it the most widely used plastic for building and construction applications such as siding, windows, single-ply roofing membranes, fencing, decking, wall coverings, and flooring. Over 50% of the PVC polymer comes from an inexpensive, renewable resource Selected PVC products are shown to have a distinct lifecycle advantage over competitive materials. PVC has demonstrated its utility and value worldwide. Only 43% of PVC comes from nonrenewable petroleum feed stocks. The balance (57%) comes from salt [117].

In transportation and construction applications, PVC is one of three plastic materials with the lowest energy requirements of the 12 major plastics used. PVC can be safely incinerated in state-of-the-art facilities and its energy recaptured and re-used.

A number of studies, including one by the New York State Energy Research and Development Authority (NYSERDA), have found that the presence or absence of PVC has no effect on the amount of dioxin produced during the incineration process. Instead, incinerator operating conditions (primarily temperature) are the key to controlling dioxin formation. A 1995 study sponsored by the American Society of Mechanical Engineers (ASME), involving the analysis of more than 1900 test results from 169 large-scale commercial incinerator facilities throughout the world, found no relationship between the chlorine content of waste, such as PVC, and dioxin emissions from combustion processes under real-life conditions. Instead, the study stated that the scientific literature is clear that the operating conditions of combustors are the critical factor in dioxin generation. Additionally, in June 1996, the Swedish Environmental Protection Agency declared, "Reducing the quantity of PVC in waste does not reduce the quantity of dioxin in the waste gases." Incinerator scrubbing systems can remove about 99% of the hydrogen chloride (HCl) generated by incinerating PVC plastics and other chlorine-containing compounds and materials. In addition, a study by the Midwest Research Institute determined that wastes containing vinyl plastics are not a significant factor in the cost-effective operation of either medical or municipal solid waste incinerators.

The PVC industry's dioxin emissions are a very small part of overall emissions, constituting less than one-half of 1% of the total emissions to air, water, and land as identified by the EPA. The vinyl industry emits about 12.6 grams of dioxin a year, compared with the EPA's recent estimate of nearly 3000 grams a year from known sources. The EPA also stated that dioxin emissions in the United States decreased by about 80% between 1987 and 1995, primarily due to reductions in air emissions from municipal and medical waste incinerators. Notably, although the amount of dioxin in the environment has decreased sharply over the past 30 years, vinyl chloride monomer (VCM) production has more than tripled. This is the best evidence that the vinyl production chain is a minor contributor to dioxin levels in the environment. Many polymers, including PVC, require additives during the manufacturing process. Individual additives include heat and light stabilizers, colorants, impact modifiers, processing aids, and plasticizers. Additives typically constitute a small part of the overall vinyl formulation and their use is closely regulated by a number of agencies including the U.S. Environmental Protection Agency (EPA), the Food & Drug Administration (FDA), and the National Sanitation Foundation (NSF). All additives used in food and drug applications must have specific regulatory clearance from the FDA.

PVC is produced from vinyl chloride by radical polymerization. The position of the chlorine atom on the hydrocarbon chain leads to high polar secondary valence forces, which are responsible for the relatively high freezing temperature of 80–90°C. These forces, however, also make it possible for a high percentage of low-molecular (or polymeric) liquids that contain similar polar groups (so-called plasticizers) to be bonded into the polymer. This is due to the secondary valence gels that these liquids form with the polymer. The freezing temperature is then lowered — something that is generally exploited to the extent where PVC produced in this way still retains its elastomeric character at room temperature. The polar nature of PVC is also responsible for the fact that only polar liquids can penetrate the PVC and dissolve it. PVC possesses a very high resistance to non-polar substances; in other words, it has excellent chemical resistance. It is also the polar structure of the polymer that permits PVC to take relatively high filler contents — a capability that is even more pronounced in plasticized PVC. This is the reason for the prolonged service life of articles such as floor coverings, because these can contain up to 50% by volume of filler. The drawback to the relatively loose molecular bond between the chlorine atom and the polymer chain is that chlorine atoms can split off during processing, at temperatures of above 140°C or so. This is a problem that was satisfactorily solved many decades ago through the development of additives — so-called stabilizers. These substances, which are mostly metal-based compounds, are mixed in with the polymer during compounding in quantities of about 2%, in conjunction with other aids. They give the products sufficient processing stability and a very long service life, including a sufficient level of light stability. Comparative life cycle analyses often show the excellent environmental performance of PVC, which is preferable to wood and aluminum even without recycling. Increasing recycling rates will give PVC an even greater environmental advantage [118].

18.4.2 Contaminations of Waste PVC

To assess the recycling of PVC, it is necessary to distinguish between the different PVC products and waste types, respectively. The opportunities and limits of recycling are different depending on the product group. To develop a realistic future scenario of PVC recycling, it is also necessary to have a general knowledge of the major factors influencing the recycling quantities. Similar to other polymers, the recycling potentials of PVC are mainly determined by the degree of contamination, which must be accepted for the collected wastes. "Degree of contamination" refers to the degree to which PVC is mixed with other materials when collected and the differences in the composition of the collected PVC material itself. As for the second aspect, it has to be taken into account that the PVC used in products does not consist of pure PVC but of PVC compounds that contain different quantities of additives, such as softeners, filling agents, stabilizers, etc. One major difference in the material composition exists between rigid PVC applications with lower additive contents and soft PVC applications that may contain more than 50% of additives. Even in the same application (e.g., window profiles, pipes, and films), the composition of the PVC material differs between different PVC converters that have their own specific PVC compounds and between different production years, due to technological advances. For example, in cable insulations the content of additives (plasticizers, fillers, and stabilizers) ranges from 50–60% with different mixtures and compounds that are used. The production of high-quality recyclates with defined technical specifications (e.g., strength, elasticity, and color) requires input materials with a defined quality (i.e., pure PVC in terms of the contents of other materials and composition of the PVC compounds). The degree of contamination, which can be achieved for collected PVC wastes, depends mostly on the type of waste in which the PVC products end up and the PVC application (i.e., product group). Concerning the PVC waste types, two major groups must be distinguished:

1. *Pre-consumer* wastes (Figure 18.35) are generated in the production of PVC final and intermediate products (production wastes) and installation wastes from the handling or installation of PVC products. PVC

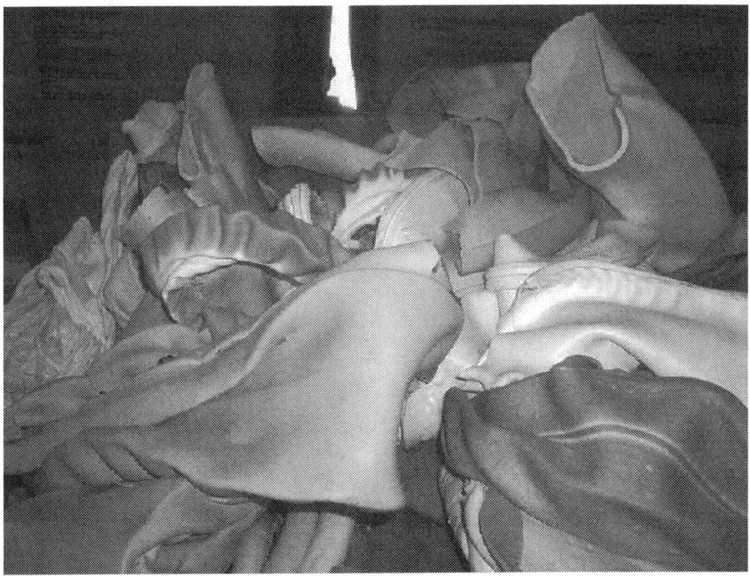

FIGURE 18.35 PVC waste during the product manufacturing.

pre-consumer wastes as a group are comparatively easy to recycle, because they can be collected separately in defined qualities. This is why recycling of PVC pre-consumer wastes are applied mainly in practice.
2. The recycling of *post-consumer* wastes is generally more difficult to realize because they occur in the form of products (e.g., end-of-life products such as pipe, window, and packaging) and thus in more or less mixed waste fractions or as a part of composite materials.

For post-consumer wastes, the different PVC product groups determine, to some extent, in which specific waste flow the PVC occurs. It is also the waste flow (not the material as such) that determines how easily PVC can be separated out as a pure fraction, and it is only the waste flow that can be influenced by waste management measures and policies.

18.4.3 Some Effective Parameters in PVC Recycling

Several factors influence PVC recycling:

1. Legal and organizational factors
   - Recycling regulations
   - Requirements for disposal
   - Voluntary systems/agreements
   - Technical standards and regulations
2. Technical factors
   - Quality versus technical requirements of recyclates
   - Possibility of separation by product types
   - Possibility of separation by material composition
3. Economic Factors
   Net costs (Logistic + Sorting + Processing – Credits for recyclates depending on price of virgin PVC) versus Cost for other waste disposal options
4. Ecological Factors

Possible reductions of environmental impacts in relation to virgin PVC:

- Environmental effects of the recycling processes and products.
- Possible applications of the recyclates.

As mentioned before, a considerable time lag exists between PVC consumption and PVC waste occurring, which is due to the fact that a major part of PVC consumption is processed into long-duration products with lifetimes up to 50 years and more. Nevertheless, a close linkage exists between PVC consumption and PVC waste arising: With some exceptions, all PVC produced will become waste sometime, the only question is when (i.e., waste arising follows PVC consumption with a time-lag). Quite reliable data are available now for the production of PVC and the processors' consumption of PVC. An example of this is shown in Table 18.7. It should be emphasized that these data are from 1999, and the figures are, most likely, much higher today. On the other hand, the amount of post-consumer PVC waste depends

TABLE 18.7
PVC Consumption of Processors in the EU by Product Group (1999)

	PVC Consumption (Compound)	
Product Group	**Kilo Tons**	**%**
Building Products	4250	57%
Packaging	680	9%
Furniture	102	1%
Other household	1346	18%
Electric/electronics	545	7%
Automotive	433	7%
Others	72	1%

Source: European Plastics Converters, Brussels.

primarily on the domestic consumption of the different PVC products. Due to various imports and exports of finished and intermediate products in the product chain, the domestic consumption of PVC in products is much more uncertain than the PVC consumption of the processors.

In the European Union (EU), the biggest consumer of PVC compounds is Germany where about 26% of the total volume is consumed (Table 18.8). Italy, France,

TABLE 18.8
PVC Consumption of Processors in the EU by Country (1999)

	PVC Consumption (Compound)		
	Total (ktons)	**% of Total EU**	**kg per Capita**
Germany	1950	26%	24
Italy	1260	17%	22
France	1140	15%	19
United Kingdom	1070	14%	18
Spain	580	8%	15
The Netherlands	330	5%	22
Belgium and Luxemburg	330	5%	32
Portugal	180	2%	18
Sweden	150	2%	17
Greece	120	2%	11
Austria	100	1%	13
Finland	80	1%	16
Ireland	80	1%	23
Denmark	60	1%	11
EU	7430	100	21

Source: European Plastics Converters, Brussels.

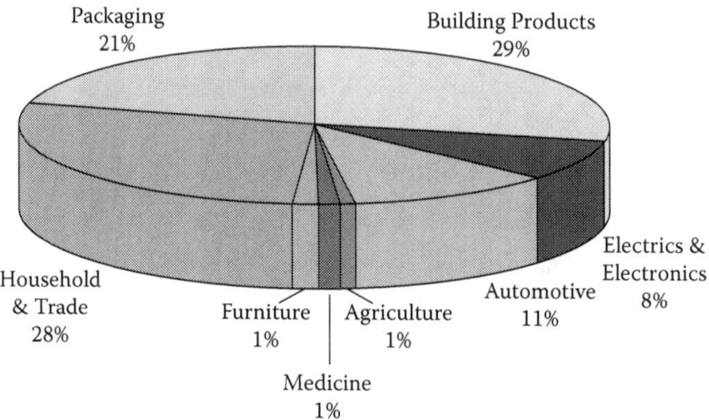

FIGURE 18.36 PVC waste arising in the EU by product group (1999).

and the United Kingdom consume about 15% each (together 45%). Spain consumes less than 10% and all other countries less than 5%.

Due to the fact that the major part of the long-life PVC products are made from rigid PVC, the contribution of flexible products to post-consumer PVC wastes is about 2/3 on a compound-basis (50% on a polymer-basis), whereas flexible PVC amounts to less than 50% (compound-basis) of total PVC consumption [119]. Figures 18.36 and 18.38 show the sources of PVC waste.

18.4.4 Mechanical Recycling of PVC

A major part of the flexible PVC applications are composite products or materials that are difficult to recycle or even cannot be recycled mechanically (e.g. coatings and organosols used for fabric products) (see Figure 18.37). For this reason, the average recycling (mechanical) potential of the PVC waste volume is still lower than it

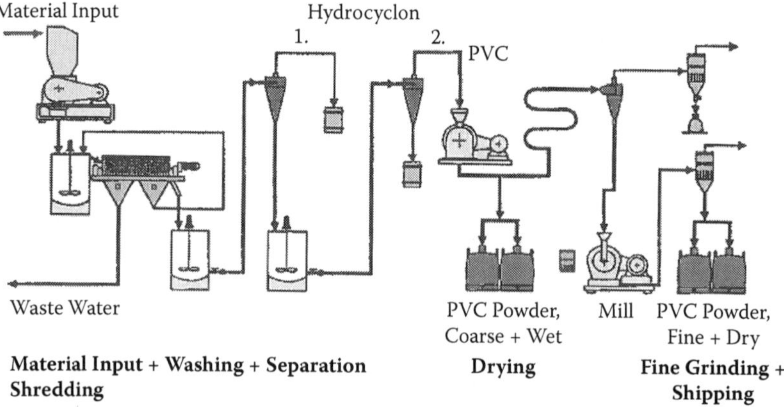

FIGURE 18.37 Plant scheme of mechanical recycling for PVC. (From Hoechst, Inc., Germany, 2006. With permission.)

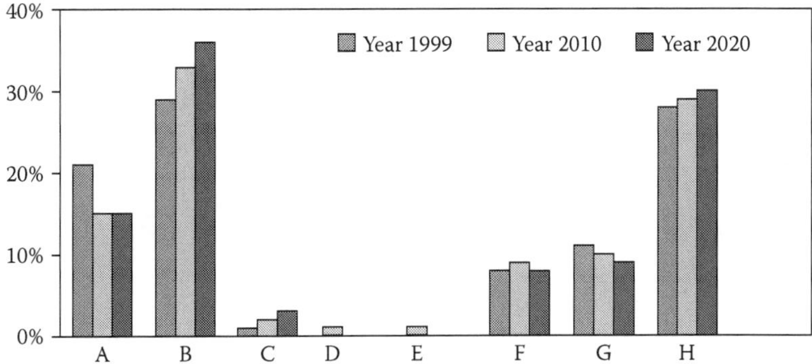

FIGURE 18.38 Sources of PVC post-consumer waste. (A = packaging; B = building products; C = furniture; D = medicine; E = agriculture; F = electronics; G = automotive; H = household and trade.) (From Mechanical Recycling of PVC Wastes, Final Report, 2000. With permission.)

might be in the future when the share of rigid PVC wastes will increase. The PVC waste arising in Germany is about 1000 kilotons (ktons) per year. Of this, 270 ktons are pre-consumer wastes and 730 ktons of post-consumer wastes. Germany is the member state with the highest quantities of recycled PVC. Altogether about 180 kilo tons of PVC are recycled mechanically. Three quarters of this quantity (135 ktons) is pre-consumer waste (120 ktons production wastes and 15 ktons installation cutoffs). The recycling rate for pre-consumer wastes is therefore about 90%. Mechanically recycled post-consumer wastes amount to 45 ktons. This is equivalent to about 8% of the PVC post-consumer waste arising. The major part of the recycled post-consumer wastes (20 ktons) is cable insulation waste. The other important part (15 ktons) is packaging waste that is recovered by the "Duales System Deutschland," the recycling organization responsible to implement the German packaging ordinance. PVC packaging is not recycled separately but together with other plastics in a mixed plastics fraction, which is used in products with low material standards ("down-cycling"). With cable and packaging recycling together, three quarters of the mechanical recycling of post-consumer PVC can be classified as "down-cycling" (35 ktons). The term "recycling system" includes the whole material chain starting with the PVC waste at the place where it arises and ending with the recyclates, which are used for the production of new products (material or energy).

In 2002, EuPR conducted a study on PVC mechanical recyclers in the EU. The aim of this study was to better understand who these independent third-party recyclers are, what their business is, how they recycle post-consumer PVC waste, and whether they are ready to invest in additional capacities. Many European plastics recyclers — around 500 — claim to recycle PVC. The study showed, however, that this is a core business for no more than 30 companies that have the expertise and willingness required dealing with post-consumer PVC waste (washing, grinding, and micronizing).

The main problem with post-consumer waste is to ensure a steady supply of secondary raw material to these recyclers to justify their investments. Recovinyl® was incorporated as a company in 2003. It benefits from the practical experience of major shareholder-recyclers all around Europe. Researches have demonstrated that light concrete made from recycled PVC could be comparable in technical specifications to other similar products. Several are moving through the approval process: obtaining compliance with official technical standards and required regulatory permits; conducting a field test for collection of PVC waste; manufacturing and marketing of the recycled product; and estimating the impact of this process on volume [120]. In mechanical extrude aided recycling, PVC window profiles are recycled in up to nine repeated extrusion processes. The properties, such as impact strength, modulus, Vicat temperature, and thermal stability of recycled window frame profiles from 20- to 25-year-old windows indicate that such recycled PVC is suitable for reprocessing [121–125]. The main reason for reduction in strength, ductility, and other mechanical properties of recycled PVC bottle material separated from the post-consumer waste stream is the presence of impurities, especially PET [126]. Recycled PVC bottle material can be used successfully in calcium–zinc stabilized PVC foam formulations to produce profiles of saleable quality. Increasing amounts of bottle recyclate had no significant effect on gelation time, melt rheology, or plate-out characteristic and gave rise to an improvement on thermal stability. Foam blends can be extruded to produce profiles of good surface finish and low foam density. Up to 100% PVC bottle recyclate would not affect the density, cell structure, or impact properties of co-extruded foam profiles [127–129]. While using recycled PVC in cable insulation, it is necessary to recover copper and PVC from cable forms originating from used motor cars. PVC can be dissolved and separated for re-use in cable and wire insulating [130, 131].

18.4.5 Chemical Recycling of PVC

Besides the mechanical recycling of PVC, several attempts have been made to prepare low-molecular products from PVC by chemical or thermal treatment. Most of the proposed processes use the rather easy dehydrochlorination of PVC either under the influence of heat or alkaline media. PVC is recycled by degradative extrusion, based on the degradation of PVC in an extruder by heat and mechanical energy in the presence of oxygen, steam, and/or catalysts. HCl is the main degradation product which can be used for the synthesis of monomer vinyl chloride, whereas the remaining polymer is not yet completely free of chlorine and has too high a melt viscosity for direct application. Another possibility is the oxidative degradation of PVC by molecular oxygen in aqueous alkaline solution at temperatures between 150 and 260°C with oxygen pressures of 1–10 MPa. The main products are oxalic acid and carbon dioxide, with their yield depending on the reaction conditions and the alkali concentration [132–135].

Many options are available for PVC feed stock recycling for recovery chlorine and organic products. The chlorine is re-used in the form of purified hydrochloric acid or salt and hydrocarbons in the form of energy or as feed stock for the petrochemical industry [136, 137]. A new pilot plant for chemical recycling of PVC is presently installed in Tavaux (France). The process is based on a slag bath gasification developed by Linde AG. PVC waste is converted with oxygen, steam, and sand according to the following formula:

$$2[-CH_2 - CHCl] + O_2 + 2H_2O \rightarrow 2HCl + 4H_2 + 4CO$$

The European Council of Vinyl Manufacturers (ECVM) pronounced a preference for this process for the treatment of PVC-rich waste. They regard the process as robust and economical. The plastic waste as delivered passes a conditioning process in which it is pre-crushed and separated from steel and nonferrous metals before entering the reactor. A pressurized reactor filled with slag is heated up to 1400–1600°C. The slag mainly consists of silicates. PVC, sand, oxygen, and steam are fed into the reactor according to the process conditions. The process is exothermic. Resulting products in the reducing atmosphere are a synthesis gas (CO/H_2) containing HCl and a slag. It is likely that this slag contains most of any metal stabilizers present in the PVC-formulation. HCl is absorbed with water from the synthesis gas. The resulting hydrochloric acid has to be purified from heavy metals, chlorides, and other halogens. Pure HCl gas is produced by distillation of the hydrochloric acid. The HCl-free synthesis gas can be used as feed for chemical processes or as a fuel gas to produce power [138] (Figure 18.39).

BSL Olefinverbund GmbH® has built a plant for the processing of chlorine-containing fluid and solid waste streams. These waste streams originate from all kinds of sources, among others, production waste of BSL and DOW, but also Hg-contaminated sludge from wastewater treatment installations. The goal is to process the waste by thermal treatment and to produce HCl using the energy from the process itself. Tests with mixtures of PVC waste and other waste have been performed. The BSL incineration began in 1999. The plant consists of a pre-treatment of the waste, the thermal treatment and energy recovery, the flue gas purification, the purification of HCl, and a wastewater treatment installation. Natural gas or liquid energy carriers can be added to reach the necessary high temperatures. The waste is incinerated in the rotary kiln and a post-combustion chamber, directly after the rotary kiln, at temperatures of 900–1200°C. During this treatment, HCl is released and recovered. Based on the heat capacity of the waste, halogen content, and potential slag formation, an optimal mixture of wastes is determined. In this way, a continuous production of high-quality HCl can be assured. In addition, the formation of dioxins and

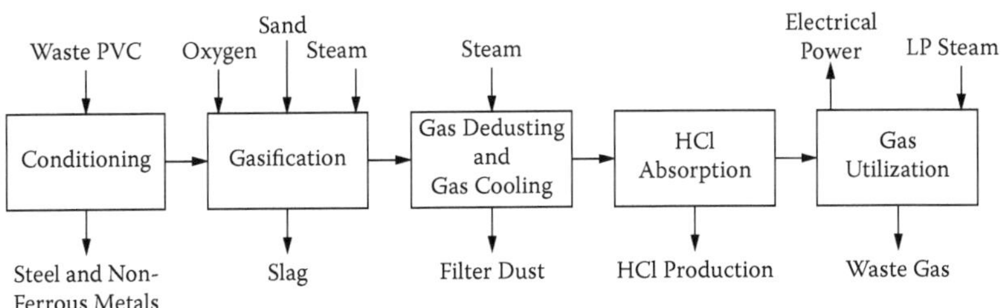

FIGURE 18.39 Pilot plant for chemical recycling of PVC. (From Tavaux®, France. With permission.)

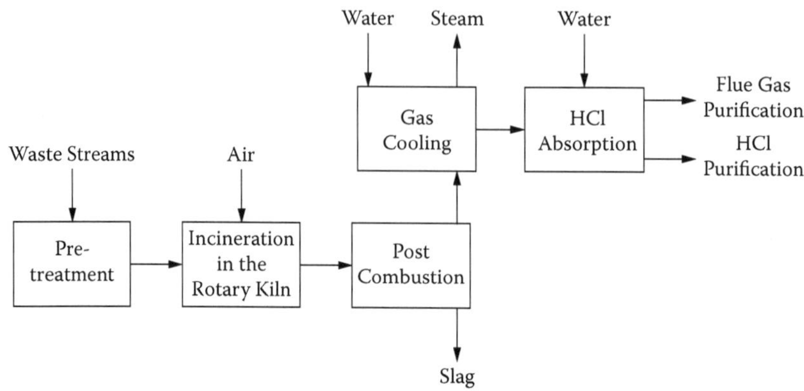

FIGURE 18.40 Schematic representation of BSL gasification process.

furanes can be diminished in this way. The flue gas from the post-combustion is cooled from 1200°C to 230–300°C. In the next step of the process, the flue gas purification, the HCl is absorbed from the flue gas by water. In addition, other impurities are removed from the gas. The raw HCl is then purified to a useful feedstock. The inert products from the incineration are dependent on the chemical composition of the waste. It is likely that the main part of any metals present in a PVC-formulation will end up in this slag [139] (Figure 18.40).

In 1992, Akzo Nobel, as a producer of chlorine and vinyl chloride, began to study a process for feedstock recycling of mixed plastic waste containing PVC. Based on an investigation of all known processes, Akzo Nobel chose in 1994 to use fast pyrolysis technology in a circulating fluid bed reactor system. This technique has been further developed by Battelle for biomass gasification. Akzo Nobel has conducted pilot plant tests (20–30 kg/h) with PVC cable and pipe scrap. Then, experiments on a larger scale (200–400 kg/h) were performed with mixed PVC wastes including artificial leather, roofing, flooring, and packaging material. The results were promising. The process consists of two separate circulating fluid bed reactors at atmospheric pressure: a gasification or fast pyrolysis reactor in which PVC waste is converted at 700–900°C with steam into product gas (fuel gas and HCl) and residual tar, and a combustion reactor that burns the residual tar to provide the heat for gasification. Circulating sand between the gasifier and combustor transfers heat between the two reactors. Both reactors are of the riser type with a very short residence time. This type of reactor allows a high PVC waste throughput. The atmosphere in the gasifier is reducing, avoiding the formation of dioxins.

Depending on the formation of tars, a partial oxidation may be required to convert these tars into gaseous products. The product stream consisting of fuel gas and HCl is quenched to recover HCl. HCl is purified up to specification for oxychlorination. Additives in the waste stream, mainly consisting of minerals and metal stabilizers present in a PVC-formulation, are separated from the flue gas or as a bleed from the circulating sand. The output of the reactor is a synthesis gas with variable composition, which is dependent on the input. If the input contains a lot of PP and PE, relatively a lot of ethylene and propylene will be formed. With proportionally

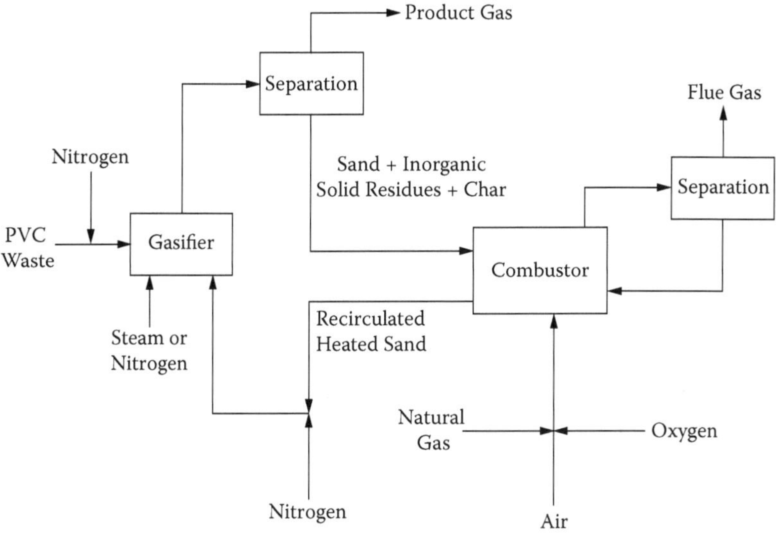

FIGURE 18.41 Schematic plant of PVC recycling from Akzo Nobel process.

more PVC, HCl and CH_4 will be more evident in the product gas. In any case, CO and H_2 will be the main components. In addition, the feed/steam ratio will influence the composition of the gas. If HCl is present in the gas, it will be recovered. From the tests with 100% PVC waste, it appeared that the HCl recovery was higher than 90%, mostly 94–97%. The product from the combustion reactor is fuel gas. Inorganics will be emitted as fly ash from the system [140–142] (Figure 18.41).

Linde KCA in Germany has developed a process to gasify waste materials in a slag bath. The basic technology was developed in the 1950s for gasification of lignite and coal. The process was made suitable to treat PVC waste with the following objectives:

- Maximum possible conversion of the chlorine contained in the PVC into an HCl gas suitable for use in oxichlorination
- Maximum possible conversion of the chemically bound energy of the waste PVC into other forms of energy
- Disposal of the unavoidable waste products of the process in a way complying with environmental regulations

The investigation into the treatment of PVC cable waste started in 1993 on a laboratory scale and was continued in 1995 on a semi-technical scale. At the end of the last decade, a PVC building waste project was performed. In this project, the process was optimized for the treatment of mixed PVC building waste on a semi-technical scale. The chemical and thermal degradation of the PVC waste takes place in a reactor at low pressures (2–3 bar) and moderate temperatures (max. 375°C). In the process, chlorine from the PVC reacts with fillers, forming calcium chloride. Simultaneously, the metal stabilizers that may be present in PVC-waste (Pb, Cd, Zn, and Ba) are converted to metal chloride. This consists of over 60% lead and may be purified and re-used. After completion of the reactions, three main intermediate products are formed: a solid phase product, a liquid product, and a gas phase product. From the gas phase produced in the reactor, hydrogen chloride is collected by absorption in water, and the light gases (mainly carbon dioxide, propane and ethane) are released after incineration. The liquid phase is separated into an organic condensate and an aqueous condensate. Hydrogen chloride solutions are re-used in the downstream separation process. The solid phase is treated in a multi-stage extraction–filtration process. By controlling pH, temperature, as well as the amount of water added, heavy metals are separated from the coke in the filtration and/or evaporation step. Part of the chloride that is not internally re-used finally comes available as calcium chloride from the evaporation step. To minimize the consumption of water, water is recycled between every extraction stage [143–145] (Figure 18.42).

Solvay has developed Vinyloop® according to its collaborations with Ferrari Textiles Techniques (France). This company is specialized in the production of architectural tarpaulin and canvas in PVC/polyester compound. They consider it important that their products be recyclable. The first industrial installation became operational in 2001 at Ferrari Textile Techniques. The process has in fact to be classified as mechanical recycling instead of chemical recycling. The method is based on physical principles, where chemical recycling by definition breaks down a plastic into feedstock; in this process the chemical structure of PVC is unchanged. A pilot plant was planned for 2001. By 2002, the available capacity was 17,000 tons. The process is quite simple in principle. First, the products to be recycled are cut and reduced in size. Then, PVC and its additives are selectively dissolved in a specific solvent such that they become separated from other elements. Finally, PVC is recovered by means of precipitation and dried and is ready for a new life. As indicated, this has to be labeled as mechanical recycling, because the PVC polymer is not broken down into its feed stocks. Yet, unlike classical mechanical recycling processes, where the full

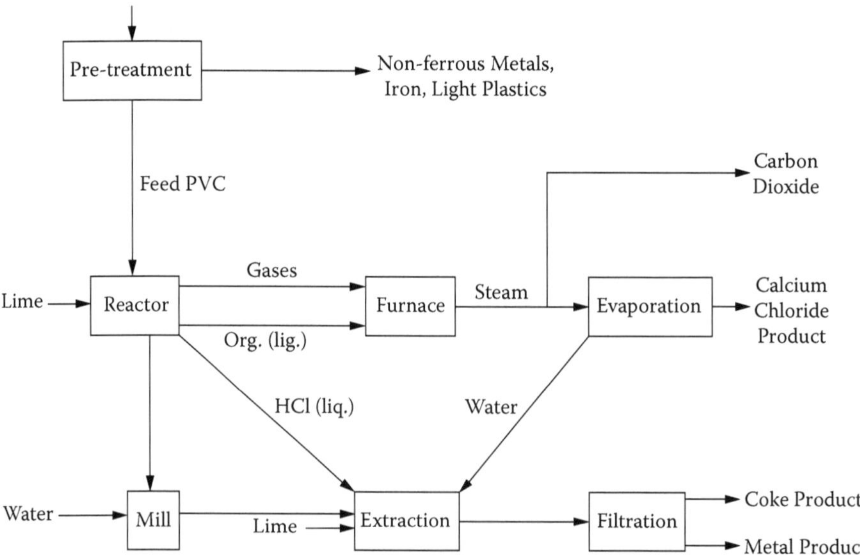

FIGURE 18.42 Schematic plant of PVC recycling. (From Linde KCA, Dresden, Germany, 2001. With permission.)

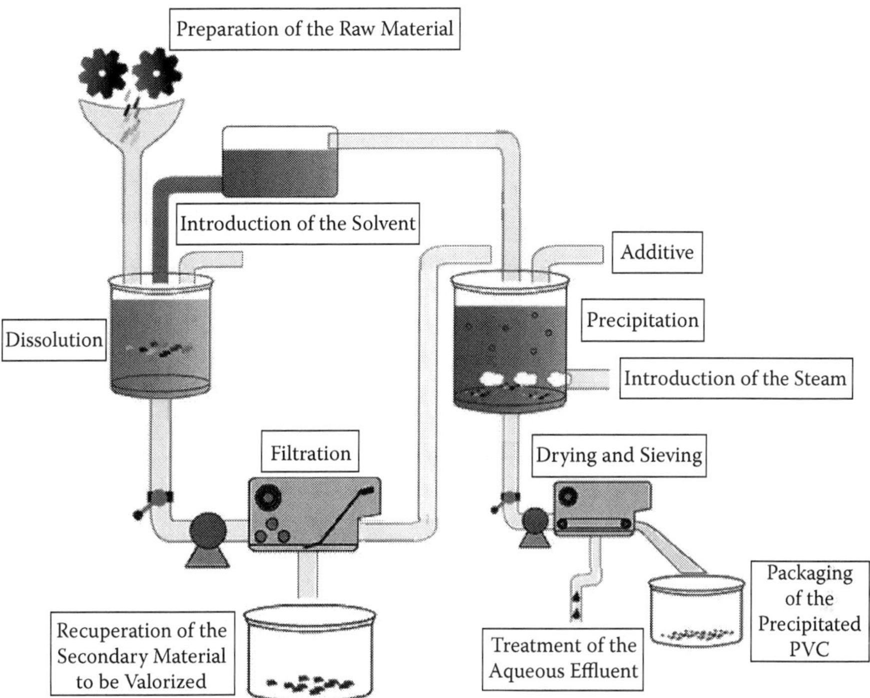

FIGURE 18.43 Schematic plant of PVC recycling by Vinyloop process.

PVC formulation is kept intact, here the components that make up the full formulation are separated. The Vinyloop process (Figure 18.43) is therefore capable of dealing with rather complicated formulations. Solvay claims that the regenerated PVC is comparable in quality to the primary product [146–148].

Whereas chemical recycling of used PVC plays only a minor role, many investigations on the recovery of PVC from scrap are known. Some processes use the dissolution of PVC in organic solvents, such as cyclohexanone, ethyl methyl ketone, or tetrahydrofuran (THF). By selected extraction of PVC from plastic waste, uncontaminated PVC for further application can be obtained. Most of these processes are still in the development phase.

Special interest is given to the separation of PVC and polyesters like polyethylene terephthalate (PET). Glycolysis of PET leads to oligomers that are able to polycondensate with caprolactone. The obtained diols are extended with aliphatic diisocyanates, which, under special conditions, result in polyurethanes that are totally miscible with PVC to give blends with acceptable mechanical properties. The separation of PVC and PET can also be performed by their different mechanical properties or by other automatic sorting devices [149–153]. The basis of such processes is the detection of chlorine in PVC by different physical techniques like X-ray fluorescence or electromagnetic rays. Another possibility for identification of PVC in mixtures with other polymers is offered by the use of marked PVC, which can be prepared by reactive processing of PVC with sodium 2-thionaphtholate or sodium-p-thiocresolate. Such marked PVC can be distinguished from polyolefines by their UV-absorption. The purification of highly polluted PET (with about 6% PVC and 1% polyolefines) is possible by selective dissolution of PVC in THF, ethyl methyl ketone, and in amyl acetate/xylol mixtures for polyolefines [154–157].

Strongly alkaline solutions of sodium hydroxide are able to destroy the hydrophobicity of PET, whereas the hydrophobicity of PVC remains only slightly affected by these solutions. On this basis, a technology involving treatment of PET and PVC particles with alkaline solutions followed by froth floatation of PVC with non-ionic surfactants has been developed and tested. In both steps of this technology, appropriate experimental conditions such as concentration of reagents, temperature, and residence time has been optimized for separation of PVC from PVC/PET mixtures of varying composition. Using this method, 95–100% recovery of PET and PVC can be achieved from a variety of PVC/PET mixtures. PVC/PET mixtures or PVC/PET/PE scraps are also recyclable using electrostatic separation [158, 159]. Recycled PVC is blended with different polymers now. The simplest way is blending of recycled PVC with virgin material. In case of manufacturing oil bottles by these blends, good processing behavior and mechanical properties are observed depending on the amount of the added recyclate. It is difficult to blend PVC with other polymers, because most combinations are incompatible. In some cases, the addition of compatibilizers would help to overcome these problems. In PVC/PS systems, HIPS, which is not good enough as a compatibilizer, can enhance the mechanical properties. Usually, compatibilizers can increase the mechanical properties in general with increasing concentration due to the increase of compatibility. In mixtures of recycled PVC from bottles or tubes with virgin PVC for the manufacturing of new tubes, the particle size and the addition of stabilizers determine the mechanical properties and the homogeneity of the resulting blends. Generally, the recycled PVC did not change the modulus remarkably, whereas the impact strength and the processability in some cases are even improved, but the thermomechanical properties are worse [160–164].

18.4.6 New Aspects in PVC Recycling

PVC has been dehydrochlorinated by microwave oven. Microwave absorbents, such as activated carbon or ferrite, are effective while the decomposition and/or dehydrochlorination yield is proportional to the irradiated microwave energy. In addition, decomposition of PVC mainly accounts for dehydrochlorination, and during microwave irradiation, about 15% of volatilized organic materials are released from the polymer. For commercial PVC resin wastes, more than 90% of dehydrochlorination yield would be obtained by microwave irradiation [165].

Old PVC flooring materials from buildings constructed in the 1970s have been examined in parallel with newly manufactured PVC floorings. Investigating how the important properties of PVC floorings change during their service life, researchers tried to obtain general information regarding the degradation processes in PVC floorings that could influence recycling methods. PVC floorings as

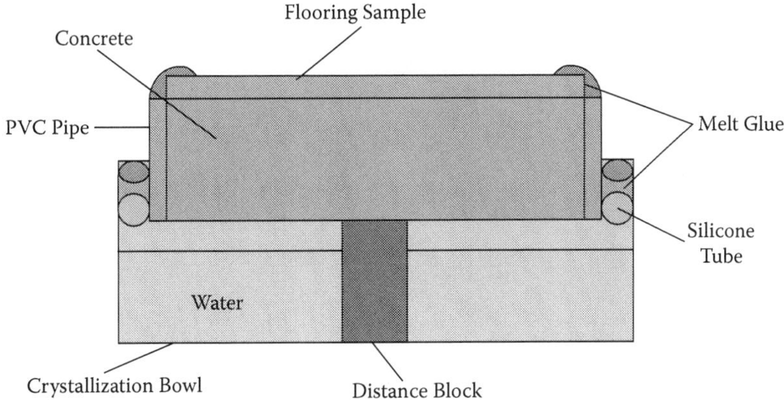

FIGURE 18.44 Experimental set-up for the accelerated aging of PVC flooring on concrete.

plastic waste can be mechanically recycled in the form in which they were recovered without upgrading and without the addition of new plasticizer. The high alkalinity of moist concrete can lead to the decomposition of the plasticizer when PVC flooring is glued onto it. However, the degree of decomposition of plasticizer is very small relative to the mass loss by evaporation and, consequently, should not cause any problems for mechanical recycling. Nevertheless, decomposition products, such as butanol and octanol, can cause indoor environmental problems sometimes designated as "sick-building syndrome." For this reason, gluing directly onto fresh concrete should be avoided. Gluing also makes mechanical recycling less favorable owing to troublesome dismantling and the high degree of contamination from the glue. The heat content in PVC floorings is dependent on the proportions of PVC and plasticizer used. Consequently, changes in the heat content caused by long-term use of PVC floorings should be insignificant [166] (Figure 18.44).

In the attempt to develop an effective process for high quality, high quantity PVC recycling, stepwise process engineering and analytical evaluation of optimization steps are combined. The main development task is to clean up the polymeric macromolecules by removing the quality-interfering substances and by conserving the primary morphology and functionality of the polymers [167, 168]. The performance of blends made from recycled PVC, styrenic acrylonitrile (SAN), and acrylonitrile butadiene styrene (ABS) is improved so that these blends can be used for those applications that must fulfill some requirements with regard to mechanical properties and stability with temperature alterations. Fourier transform infrared spectroscopy (FTIR), differential scanning calorimetry (DSC), Vicat softening temperature (VST), and scanning electron microscopy (SEM) characterize the equipment [169], and the stabilizers are still the most common method for improving the recycled PVC [170]. In addition, the use of waste PVC in blast furnace reduction technology is a new trend for the manufacturing of raw fuel for cement. Raw materials of chemicals from waste PVC and plastisols that contain natural calcium

FIGURE 18.45 "VINYL 2010" program's logo.

carbonate are also made into a paste, which is currently substituted for the virgin PVC [171, 172].

Today, nano technology is one of the most attractive subjects of research all around the world. Nanocomposites are now designed and produced for different applications and PVC-nanocomposites are one of them. Recycled PVC/clay nanocomposites are prepared by melt mixing of recycled PVCs and modified clays. The thermal and mechanical properties of the nanocomposites are improved simultaneously in different clay content composites, and they are being compared to recycled PVC. Specifically, the storage modulus of the nanocomposites with 10 wt% clay loading increases 11 times compared with that of recycled PVC [173].

18.4.7 A Typical Movement for Recycling Called "Vinyl 2010"

Vinyl 2010 is a voluntary initiative by the European vinyl industry to enhance its sustainability profile. Vinyl 2010 (Figure 18.45) is a 10-year plan (initiated in 2000) of commitments, which includes the following:

- Compliance to European Council of Vinyl Manufacturers (ECVM) Charters regarding PVC production emission standards.
- A plan for full replacement of lead stabilizers by 2015, in addition to the replacement of cadmium stabilizers which was achieved in March 2001.
- The recycling in 2010 of 200,000 tons of post-consumer PVC waste; this objective will come in addition to 1999 post-consumer recycling volumes and to any recycling of post-consumer waste as required by the implementation after 1999 of EU Directives on packaging waste, end-of-life vehicles, and waste electronic and electrical equipment.
- The recycling of 50% of the collectable available PVC waste for window profiles, pipes, fittings, and roofing membranes in 2005, and flooring in 2008.
- A research and development program on new recycling and recovery technologies, including feedstock recycling and solvent-based technology.
- The implementation of a social charter signed with the European Mine, Chemical and Energy Worker's Federation (EMCEF) to develop social dialogue, training, health, safety, and environmental standards, including transfer to EU accession countries.

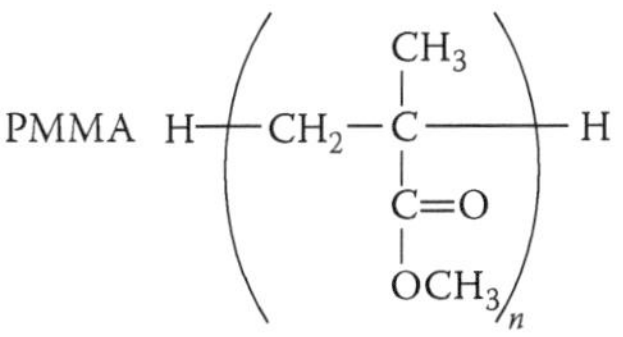

PMMA $$H-\left(CH_2-\overset{\displaystyle CH_3}{\underset{\displaystyle \underset{\displaystyle OCH_3}{\overset{|}{C=O}}}{\overset{|}{\underset{|}{C}}}}\right)_n-H$$

FIGURE 18.46 Schematic structure of PMMA.

- A partnership with local authorities within the Association of Communes and Regions for Recycling (ACRR) for the promotion of best practices and pilot recycling schemes at local level.

The PVC industry will provide a financial support scheme, in particular for new technologies and recycling schemes, allowing up to 250 million Euro of financial contribution over the aforementioned 10-year program.

18.5 POLYMETHYLMETHACRYLATE

Polymethylmethacrylate (PMMA) is a transparent, colorless, thermoplastic polymer that is commonly manufactured as beads for injection molding, and as cast sheet. PMMA products exist in three forms: cast sheet, extruded sheet, and molding compounds [174] (Figure 18.46). The structure of the polymer solid provides PMMA with distinctive optical clarity and the material can be tinted and colored according to requirements.

Its durability and favorable mechanical properties in the form of weather ability and resistance to most aqueous inorganic reagents, high softening point, and high impact strength favor its use in external applications with long life, high quality products. Because of its optical clarity and stability, PMMA has also found uses in the medical field where special grades are used in intraocular lenses, contact lenses, and implants [175]. PMMA for cast sheet used in applications such as sanitary-ware is typically a "high" molecular weight (up to 2,000,000 $gmol^{-1}$). PMMA for molding applications generally has a much lower molecular weight (100,000–200,000 $gmol^{-1}$), contains up to 6% co-monomers, such as ethylacrylate (EA) and methylacrylate (MA), and may contain pigments, coloring materials, and mercaptanes. PMMA is a relatively high-cost plastic and largely maintains its market position as a result of the high functionality associated with its physical properties (see Figure 18.47). The price of polymers is closely related to the energy required for the formation of polymer precursors.

FIGURE 18.47 Recycling sign for polymers that are not categorized, such as PMMA.

The most common production route for PMMA requires the production of a significant number of intermediates (Figure 18.48). The inherent recycling functionality of PMMA places it in a special place in the plastics world. As the simplest indication of recycling attractiveness, PMMA is one of the few plastics that have had a thriving scrap market for many

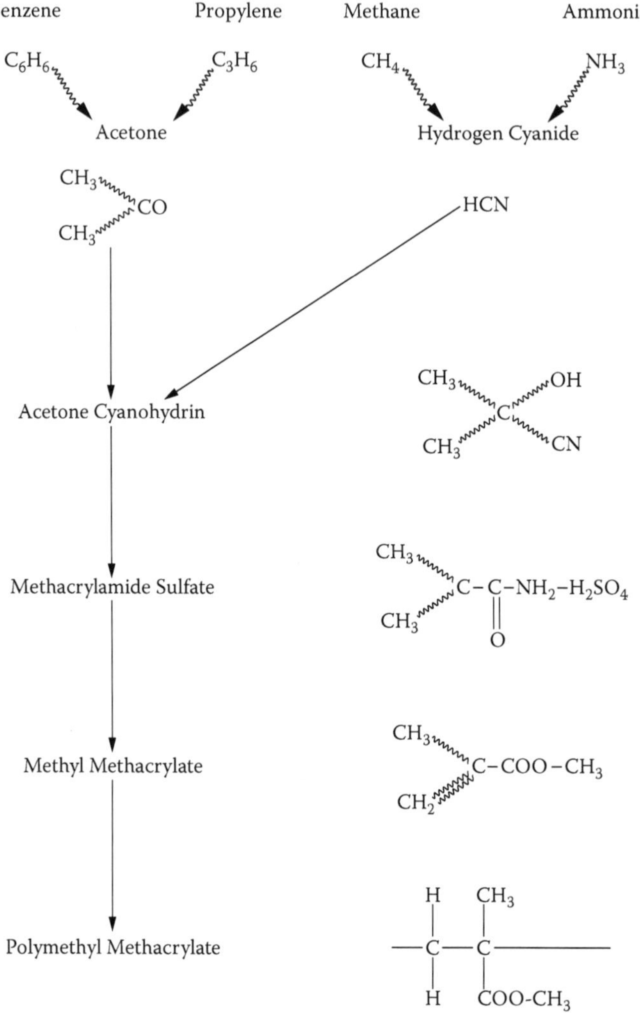

FIGURE 18.48 Schematic chart of PMMA production.

years — approaching 40 years, in fact. Two pathways for recycling are available. Mechanical recycling is common in injection-molded applications, and lower quality scrap sheet and cutting waste can simply be processed into re-grind for re-extrusion. However, the proportion of recyclate that can be put into the process in this manner without altering the special properties of the acrylic is limited to about 15%. In addition, the number of times that mechanical recycling can be performed is restricted. As an alternative, PMMA can be depolymerized or "chemically recycled," a process that can produce a product with essentially virgin quality. Even with relatively crude depolymerization techniques, manufacturers commonly apply up to 30% recycled MMA in colored products and about 10–15% in transparent products. The availability of significant quantities of production rejects and cutoffs at a low price, the energy advantages of recycling, and the fact that recycling plants of moderate size are sufficient for cost-effective operation have made this a viable business.

18.5.1 PMMA Recycling Methods

Most existing acrylic recycling facilities in operation are based on the introduction of chipped scrap to a heat transfer medium (usually molten lead or tin) where depolymerization spontaneously occurs. However, significantly superior technologies are available, even if not widely implemented. For reasons of post-process purification complexity associated with the presence of co-polymers in injection-molded PMMA, the chemical recycling industry operates predominantly on cast PMMA scrap. Unlike the recycling of many other polymers, where industrial actors have recently begun to gather skills and experience related to material and chemical recycling possibilities, the PMMA producer at the center of this work began to work with recycling decades ago (Figure 18.49). The organization has been involved in high grade chemical recycling of internal and external PERSPEX™ production scrap back into the production process at recycled content rates of up to 25% since at least the mid-1960s [176].

PMMA starts to depolymerize into its monomers at 150°C (Figure 18.50). At temperatures between 300°C and 350°C, the reaction is quantitative, the polymer chains are decomposed consecutively, and the formation of the fragments is not statistical.

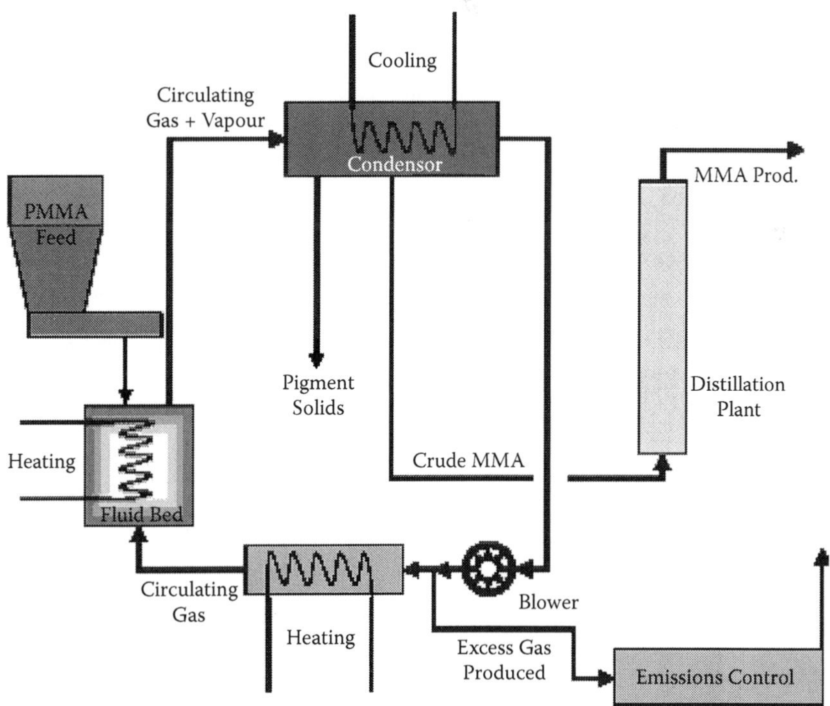

FIGURE 18.49 PMMA recycling plant.

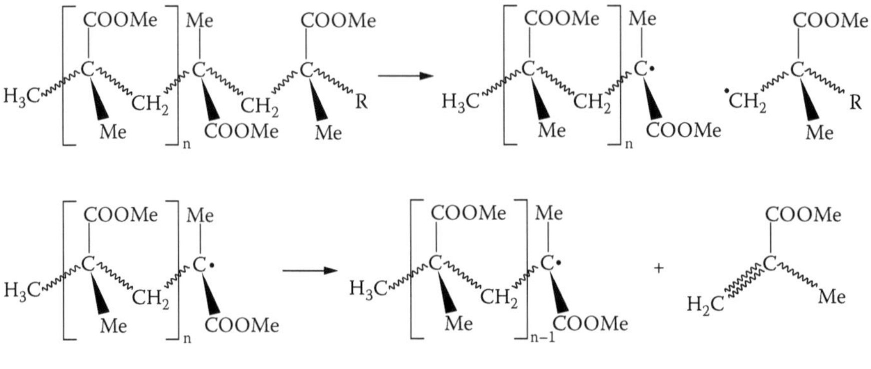

FIGURE 18.50 PMMA recycling mechanism.

In the case of PMMA, the reformation of monomers proceeds in a high yield, as during the pyrolytic degradation tertiary radicals are formed from the quaternary carbon atoms. These are more stable and chemically less reactive than the corresponding secondary and primary radicals. Thus, degradation is the preferential reaction compared with other radical reactions such as recombination. Subsequent polymerization of the purified monomer gives a product that cannot be distinguished from the starting material.

The characteristic of the recyclate made of PMMA mixture with other polymers which is mostly from the automotive industry has been proved to be acceptable for the manufacture of rear light assembly housings [177]. A model solvent technique for the recycling of PMMA decorative sheets is based on dissolution of the sheets in toluene, re-precipitation by n-hexane, washing, and drying. The remaining mixture of toluene and hexane are separated by distillation for further re-use. The polymer is totally recoverable in the form of grains and the solvents yield also is satisfactory. The recyclate is similar to the virgin resin in all properties [178, 179]. During the development of the recycling technology, thermal and oxidative depolymerization of PMMA scrap has been studied under nitrogen and oxygen atmospheres at different heating rates by thermogravimetry (TG) and differential scanning calorimetry (DSC) techniques in the absence and presence of a catalyst. The initial depolymerizaton temperature is around 220°C and final depolymerization temperature around 405°C, whereas under oxygen atmosphere, the temperatures are 220°C and 385°C, respectively [180]. High molecular weight PMMA is also recyclable while not affected in the molecular weight of the monomer [181]. In Japan, the entire Kobe Steel Group, through its unique comprehensive technologies, has made an effort to contribute toward the recycling of waste plastics under the company's "global environment committee" guidelines, in which PMMA is one of the products to be recycled [182]. After the first life cycle of plastics, various recycling processes are available for further utilization of these valuable materials. For an ecologically and economically satisfying solution, the most suitable process has to be chosen. In chemical recycling, polymers are degraded to basic chemical substances that can be re-used in the petrochemical industry. For soiled waste plastics or waste plastics that could not be recycled until now, chemical recycling plays a key role. The pyrolysis of acrylic polymers provides a good example for comparing a fluidized-bed reactor and a tubular reactor on the basis of reactor modeling evaluations. A tubular reactor with internal mass transport is a simplified model for a rotary kiln [183]. Another recycling method for PMMA is depolymerization of post-consumer PMMA back into its starting components. Degradative extrusion in twin screw extruders is used for this process. The major advantages are efficient energy input through shear energy and thermal energy. The process can be operated continuously because residues are performed of the extruder by self-wiping screws. This allows processing contaminated post-consumer PMMA as well as PMMA mixed with other polymers. Furthermore, the process is accelerated by using additives so the mass throughput of the extrusion process is maximized [184, 185].

Recently, as feedstock recycling of polymers has become an attractive option to recycle the increasing amount of plastic wastes and respond to restrictions for waste disposal in many countries, different processes have been investigated, such as degradation of plastics to monomers, pyrolysis into monomers and oil, and gasification into syngas. Pyrolysis of mixed plastic wastes and elastomers is a cost-effective process to recover feed stocks for the petrochemical industry. The Hamburg process [186], which uses an indirectly heated fluidized bed, can be varied to produce mainly monomers, aliphatic hydrocarbons, or aromatics. At temperatures of 450°C, PMMA is depolymerized to more than 98% of the monomer [187]. The MMA-yield is mainly dependent on the residence time of the gas in the reactor and, to a lesser extent, on the operating temperature. At low temperatures, the reaction is kinetically controlled, whereas at high temperatures, heat transfer restricts the overall reaction rate [188].

18.6 COMMERCIAL PARAMETERS IN RECYCLATE MARKETING

Whether a recyclate can be successfully marketed or not will depend on its price, its quality, its availability, and the demand for it. These factors, in turn, are interlinked. The aim must therefore be to achieve a recyclate with precisely defined properties that can be fed into an existing processing cycle — again, with the fewest possible problems — because existing production capacity is essentially tailored to the characteristic properties of virgin PVC:

1. *K-value or viscosity number*: measure of the flow ability of the melt and the possibility of re-melting
2. *Particle size*: measure of conveyability and suitability for transfer to silos and mixing
3. *Grain morphology*: measure of filler absorption capacity and gelation performance
4. *Bulk density*: measure of free flow ability and re-melting output
5. *Thermal stability*: measure of residual stability and post-stabilization
6. *Volatile components*: measure of moisture content and evaporating contents
7. *Cleanliness, fish-eye formation*: measure of soiling
8. Tight tolerances are characteristic of high-grade PVC and are crucial for trouble-free production. Any converter wishing to employ recycled PVC

will thus inevitably require a quality profile that comes as close as possible to that of virgin material. Because this can scarcely be achieved in practice, two options remain open, depending on the quality of the recycled product:

- Direct re-melting in the form of 100% PVC recyclate
- Mixing with virgin PVC, employing compounding techniques

It should, however, be added that the second option is frequently not feasible, or would only be possible after highly elaborate tests.

18.7 RESTABILIZATION OF THE POLYMER RECYCLATE AS AN UPGRADING METHOD

The quality of the recyclate as a raw material intended for reuse is determined by the history of the polymer synthesis, extent of contamination by impurities, primary processing and application, composition (homo-polymer or co-polymer), additives, residual stabilizers such as antioxidants, and processing stabilizers.

Restabilization is the upgrading of recycled material by adjusting heat and light, using stabilizer combinations during processing, and compatibilization of a physical or chemical method enhancing the recovery of polymer during the recycling process, and especially improving the mechanical properties of the recyclate.

As mentioned before, it is very difficult to achieve a 100% pure polymer waste because of the high-cost, time-consuming separation steps. On the other hand, direct recycling of commingled plastics has only a poor chance of producing high-quality materials. Success in the recycling of post-consumer waste is based on enhancing the resistance of the material to deterioration by stabilization [189]. The market demands that recycled plastics meet quality requirements that are comparable with the virgin resins. Prodegrandant effects of structural impurities formed or introduced during the first lifetime of polymers and the loss of stabilizers have to be compensated by proper additives, fitting the type and complexity of the recycle, the source of material, and the next expected application. A constant composition of the recyclate is an important condition for upgrading. Virgin and recycled polymers suffer from analogous degradation mechanisms; therefore, stabilization approaches are valid for both materials. Common principal transformation products of stabilizers are generated from phenolic antioxidants; phosphates from organic phosphites; and oxidation products from sulfides, including sulfur organic acids [190–192].

Reprocessing of polymer waste, exposing it to shear and heat stress, produces polyolefins that have lower elongation at break, impact strength, and stiffness than required, due to the low remaining level of the active form of the stabilizer. Therefore, post-consumer polyolefins cannot be safely processed without addition of fresh stabilizers. Experiments indicate that even addition of phosphates alone enhances the stability during subsequent processing and heat aging. Phosphites such as P-1 or P-2 reduce hydroperoxides to alcohols and hinder their thermo= and photo-initializing effect, besides lowering the influence of the photo-sensitive carbonyl groups, but do not remove them. Phenols as antioxidants deactivate alkylperoxy radicals and guarantee the long-term heat stability of the recyclate during the next lifetime. Phenols and phosphites are used in synergistic combinations [193, 194].

Properties of recycled PS are improved by properly selected antioxidants. PS, when forming part of the post-consumer waste stream, is not a uniform material, but a mixture of polystyrenes differing within broad limits in molecular weight and melt flow parameters, depending on the production technology. Packaging material is the main source of PS used for recycling. Styrenics generally suffer from photo-oxidation and discoloration. Restabilization with low amounts of phenol improves this to some extent [195]. The results obtained with recyclates, based on sorted single plastics, encouraged producers of polymer additives to restabilize mixed household post-consumer polymers.

The feasibility of the restabilization technology was confirmed using polymer wastes consisting of 55–60% polyolefins. Phosphite P-1 (0.1%) with 0.05% of AO-1 makes up the main part of the formulation used to avoid detrimental degradation during processing of the mixture. Light stability is decisive in the use of mixed plastics for outdoor applications, such as noise protection walls.

Thermal degradation of PVC or chemical transformations of sulfur or phosphorus containing antioxidants and halogenated flame-retardants may yield mineral acids such as acids of sulfur or phosphorus and hydrochloric acid. These non-polymeic impurities contaminate the polymers and corrode the processing equipment. Anti-acids are used in reprocessing formulations. They consist mostly of calcium salts of organic acids (e.g., calcium lactate).

Polymers of different structures do not form homogeneous blends. Mixtures of polymers that differ in structure and polarity are compatible with one another only to a very limited extent. They separate in the melt or, at the latest, after cooling in the solid state, because of weak interfaces between the components and local stress concentrations. Therefore, they cannot be mixed in extruders homogeneously to give a molecular dispersion necessary to obtain acceptable material properties. A blend between two polymers is a multi-component system, and the properties are determined by the chemical composition as well as the molecular structure of the individual components and the morphology of the system. Because most plastics are incompatible, a very small amount of dissimilar plastic in a single component waste stream can have detrimental consequences for the mechanical performance of the reprocessed resin. Compatibilization modifies polymer interfaces by reducing the interfacial tension between normally immiscible polymers in the melt during blending, leading to a fine dispersion of one phase in another, increasing the adhesion at phase boundaries, and minimizing phase separation in the solid state. Compatibilizers act as morphology stabilizers and prevent delamination or agglomeration by creating bridges between phases. The compatibilizers are usually block or graft co-polymers, which contain segments that are chemically identical to the blend components, but can also be functionalized polymers that contain reactive side groups. Another possibility is the addition of a free-radical initiator to the blend, which can promote *in situ* formation of graft co-polymers during extrusion. Thermodynamic calculations indicate that efficient compatibilization can be achieved with multi-block co-polymers, potentially for heterogeneous mixed blends [196]. Miscibility of particular segments of the co-polymer in one of the phases of the blend is required. Compatibilizers for blends consisting of mixtures of polyolefins are of major interest for recyclates. Random poly(ethylene-co-propylene) is an effective compatibilizer for LDPE-PP, HDPE-PP, or LLDPE-PP blends. The impact performance of PE-PP is improved by the addition of very low density PE or elastomeric poly (styrene-block-(ethylene-co-butylene-1)-block styrene) tri-block co-polymers (SEBS) [197].

Immiscible blends of HDPE or LDPE with PS are compatibilized with various graft co-polymers, such as PS-graft-PE, PS-graft-EPDM, or block co-polymers, such as SBS tri-blocks, SEBS, and PS-block-polybutadiene. The same block co-polymers are suitable for PP-PS blends. Compatibility of PE with PVC is improved by poly (ethylene-graft-vinyl chloride) or partial chlorinated PE. To compatibilize blends of PE with PET, common for the scrap of beverage bottles, EPDM or SEBS are effective additives. Styrene-butadiene or styrene-(ethylene-co-propylene) block co-polymers are common compatibilizers for the commingled recycle from post-consumer wastes [198–202].

Reactive compatibilizers consist of functional or reactive additives interacting in situ with components of the blend. A polymer chemically identical to one of the components is functionalized to gain a chemical reactivity with the second component. This allows the phases to be held by covalent bonds, making the blend less sensitive to physical stresses. PE, PP, or PS, functionalized (grafted) with maleic anhydride or acrylic acid, poly(ethylene-co-polypropylene), grafted with succinic anhydride or co-polymers of PP with acrylic acid, and styrene with maleic anhydride are examples of effective reactive compatibilizers [203].

However, improperly selected compatibilizers might introduce structures diminishing aging behavior, due to co-oxidation effects; therefore, the restabilization system has to be modified, taking into account the entire polymer blend including the chosen compatibilizer.

The key to the oxidative stability of polymers lies in the control of hydroperoxide formation, which can be accomplished either by inhibiting the radical chain reaction or by deactivating the hydroperoxides. To effectively protect polymers during processing and exploitation, a stabilizer system is required. Antioxidants are recognized as inhibitors of a radical chain reaction involving alkyl and alkylperoxyl radicals as the chain propagating species and hydroperoxides as the indigenous initiators of auto oxidation [204–206]. Antioxidants acting as radical inhibitors are referred to as chain braking (CB) or primary antioxidants, whereas antioxidants acting as peroxide deactivator are called hydroperoxide decomposers (HD) or secondary antioxidants. Phenolic and hindered amines stabilizers (HAS) are the most widely used and effective primary antioxidants. Among the secondary antioxidants, phosphites, phosphonites, and organosulfur compounds can be mentioned. If polymers are intended for outdoor applications, stabilization against photo-oxidative degradation becomes necessary because UV-light, such as heat or mechanical shear, can generate free radicals. UV-stabilizers can operate in several ways. They can act by quenching exited states, by absorbing the UV-light, by decomposing hydroperoxides, or by scavenging radicals [207–211].

REFERENCES

1. Mapleston, P., Three-machine cell makes multi-shell vacuum pipes, *Modern Plastics*, 72(11), 39, 1995
2. Mapleston, P., Auto Recycling Targets Plastics, *Modern Plastics*, 72(5), 48, 1995.
3. Mackey, G., A review of Advanced Recycling Technology, *ACS Symp. Ser.*, 609, 161, 1995.
4. Polaczek, J. and Machowska, Z., Thermal methods of the raw material recycling of plastics wastes, *Polimery* (Poland), 41(2), 69, 1996.
5. Guffey, F.D., Barbour, F.A., Process for waste plastic recycling, World Patent number WO/1994/020590, 1994.
6. Wolpert, V.M., Recycling of plastics (Commodity and Engineering Plastics), International Techno-Economic Report, Vol. 4, Oct.1992.
7. Stabel, U., Wörz, H., Kotkamp, R., Fried, A., Process for recycling of plastics in a steamcracker, European patent number 0713906 A1, 1996.
8. Seymour, R.B. and Carraher, C. E., *Polymer Chemistry: An Introduction,* 4th edition, Marcel Dekker, 1996, 451– 471.
9. Alper, J. and Nelson, G.L., *Polymeric Materials: Chemistry for the Future,* ACS Publication, USA, 1989, 71–83.
10. Grande, J.A., Computer manufacturers make in-roads in use of recyclate, *Modern Plastics*, 72(11), 35–36, 1995.
11. Kirk-Othmer *Encyclopedia of Chemical Technology*, 4th edition, John Wiley & Sons, 17, 724, 1996.
12. Sampson, D.L., Method and apparatus for plastic recycling, World patent number WO/1992/22380, 1992.
13. Scott, G., Biological recycling of polymers, *Polym. Age*, 6, 54–56, 1975.
14. Brandrup, J., *Polymer Handbook*, 3rd Edition, John Wiley & Sons, 1989, 366.
15. Michaeli, W., Dassow, J., Recyclingfähigkeit von verunreinigten, technischen Thermoplasten, *Plastverarbeiter*, 44(8), 34, 1993.
16. Delgrange, J.P., Wilson, T.M.B., BP Fluidized Bed Process for Polyethylene Production, *Polimery* (Warsaw), 1984, 29(10), 402.
17. Maysuzaki, T., Plastic and oil waste processing method, US patent number 5226926, 1993.
18. Nir, M.M., Implications of post-consumer plastic waste, *Plastics Engineering*, 46(10), 21–28, 1990.
19. Sitek, F.A., Restabilization Upgrades Post-Consumer Recyclate, *Modern Plastics*, 70(10), 67–68.
20. Loultcheva, M.K. et al., Recycling of high density polyethylene containers, *Polymer Degradation and Stability*, 57(1), 77, 1997.
21. Kirk-Othmer *Encyclopedia of Chemical Technology*, John Wiley & Sons, 4th edition, 1996, 17, 81.
22. Wang, M.D., Nakanishi, E., and Hibi, S., Effect of Molecular Weight on Rolled High Density Polyethylene. I. Structure, Morphology, and Anisotropic Mechanical Behavior, *Polymer* (UK), 34(13), 2783, 1993.
23. Dontula, N., Campbell, G.A. and Wenskus, J.J., Approach towards moulding parts with constant properties on addition of regrind in *Antec '94 Conference Proceedings II*, 1994, 1783.
24. Pattanakul, C., Selke, S., Lai, C., Miltz, J., Properties of recycled high density polyethylene from milk bottles, *Journal of Applied Polymer Science*, 43(11), 2147, 1991.
25. Haider, N., Consumption and loss of commercial stabilizers in polyethylene exposed to different natural environments, PhD Thesis, Royal Institute of Technology, Stockholm, Sweden, 2000.

26. Woo, M.W., et al., Melting behavior and thermal properties of high density polyethylene, *Polymer Engineering and Science*, 35 (2), 151, 1995.
27. Callari, J.J., New bags are full of trash, *Plastics World*, 1991, 49(12), 40.
28. Anon., Honda, Nissan Trying Out Bumper Recycling, *Plastics News*, October 1991, 3.
29. Anon., *Reuse-Recycle*, 1996, 26, 2.
30. Anon., *Reuse-Recycle*, 1996, 26, 51.
31. Wendrof M.A. , *Proc. TAPPI 1992 laminating and coating conference*, 1992, 247.
32. Smith, M.A., Benham, E.A., Didier, C.M., McDaniel, M.P., Ratzlaff, J., Whitte, W.M., *Proc. Polyehtylene'93*, Zurich, Switzerland, 1993.
33. Edmonson, M.S., Pirtle S.E., *Proc. TAPPI 1992 laminating and coating conference*, 1992, 255.
34. Norwalk, S., *Proc. Recycling plus VII*, Virginia, USA, 1992, 43.
35. Zuch, R.A., Technology for recycling plastic to continue advances, *Paper, Film and Foil Converter,* March 1992.
36. McCarthy, L., Degradable plastics fit best in specialty areas, *Plastics World*, September 1989, 29.
37. Oliphant, K and Baker, W.E., The use of cryogenically ground rubber tires as a filler in polyolefin blends, *Polymer Engineering & Science*, 33(3), 166, 1993.
38. Rajalingam P., Baker, W.E., The role of functional polymers in ground rubber tire-polyethylene composite, *Rubber Chem. Technol.*, 1992, 65, 908.
39. Kartalis, C.N., et al., Mechanical recycling of post-used HDPE crates using the restabilization technique. II: Influence of artificial weathering, *Journal of Applied Polymer Science*, 77(5), 1118, 2000.
40. http://www.greencycle.info
41. Welle, F., Post-consumer contamination in high-density polyethylene (HDPE) milk bottles and the design of a bottle-to-bottle recycling process, *Food Additives & Contaminants*, 22(10), 999, 2005.
42. Javierre, C., Claveria, I., Ponz, L., Aisa, J., Fernandez, A., *Waste Management*, 2006, article in press.
43. Charlton, Z., Vlachopoulos, J. and Suwanda, D., Profile Extrusion of Highly Filled Recycled HDPE, *ANTEC Proceedings*, Orlando Florida, 2914, 2000.
44. Pappa, G., et al., The selective dissolution/precipitation technique for polymer recycling: a pilot unit application, *Resources, Conservation and Recycling,* 34(1), 33, 2001.
45. Shelley, S., Fouhy, K., Moore, S., Plastics reborn, *Chem. Eng.* 1992, 99(7), 30.
46. Ward, M., Companies Offer Chemical Technology to Rescue German Recycling Initiative, *Chemical Week*, 154(16), 24, 1994.
47. Ishihara, Y., et al., Effect of branching of poslyolefin backbone chain on catalytic gasification reaction, *Journal of Applied Polymer Science*, 38(8), 1491, 1989.
48. Horvat, N., Ng, F.T.T., Tertiary polymer recycling: study of polyethylene thermolysis as a first step to synthetic diesel fuel, *Fuel*, 78(4), 459, 1999.
49. McCaffrey, W.C., Kamal, M. R., and Cooper, D. G., Thermolysis of polyethylene, *Polymer Degradation and Stability*, 47(1), 133, 1995.
50. Horvat, N. and Ng, F.T.T., Tertiary polymer recycling: study of polyethylene thermolysis as a first step to synthetic diesel fuel, *Fuel*, 78(4), 459, 1999.
51. Ray, I., et al. Studies on Thermal Degradation Behaviour of EVA/LDPE Blend, *Journal of Elastomers and Plastics (USA)*, 26(2), 168, 1994.
52. Scheirs, J., *Polymer Recycling: Science, Technology & Applications*. Wiley, Canada, 1998.
53. Manos, G., Garforth, A. and Dwyer, J., Catalytic degradation of high-density polyethylene over different zeolitic structures, *Industrial & Engineering Chemistry Research*, 39(5), 1198, 2000.

54. Bate, D.M. and Lehrle, R.S., Kinetic measurements by pyrolysis-gas chromatography, and examples of their use in deducing mechanisms, *Polymer Degradation and Stability*, 53(1), 39, 1996.
55. Pant, D., A new role of alumina in polyethylene degradation: A step towards commercial polyethylene recycling, *Journal of Scientific & Industrial Research*, 64(12), 967, 2005.
56. Manos, G., et al., Tertiary Recycling of Polyethylene to Hydrocarbon Fuel by Catalytic Cracking over Aluminum Pillared, *Energy & Fuels*, 16(2), 485, 2002.
57. Bürkle, D., mandatory recycled plastic content – proc and cons, *Proceeding GLOBEC'96*, Davos, Switzerland, March 18–22, 1996.
58. Schiabello, A., Environmental performance of plastics and stretch film as packaging material, *Proceeding Re-93*, Geneva, Switzerland, 1993.
59. Nam, J.D., A composition methodology for multistage degradation of polymers, *J. Polym. Sci. Polym. Phys. Ed.* 1991, 30, 601.
60. Dintcheva, N.T., Jilov, N. and La Mantia, F.P., Recycling of plastics from packaging, *Polymer Degradation and Stability*, 57(2), 191, 1997.
61. Jelčič, Ž., et al., Polyethylene recycling during processing, *Angewandte Makromolekulare Chemie*, 176(1), 65, 1990.
62. Mehrabzadeh, M., and Farahmand, F., Recycling of commingled plastics waste containing polypropylene, polyethylene, and paper, *Journal of Applied Polymer Science*, 80(13), 2573, 2001.
63. La Martina, F.P., Recycling of Heterogeneous Plastics Wastes. I. Blends With Low-Density Polyethylene, *Polymer Degradation and Stability*, 37(2), 145, 1992.
64. Takuma, K., et al., A Novel Technology for Chemical Recycling of Low-Density Polyethylene by Selective Degradation into Lower Olefins Using H-Borosilicate as a Catalyst, *Chemistry Letters*, 30(4), 288, 2001.
65. Bonte, Y., and Schweda, R., Polypropylene, Kunststoffe Plast Europe, October 2001, v. 91.
66. Seiler, E., *Properties and Applications of Recycled Polypropylene, Recycling and Recovery of Plastics*, Carl Hanser Verlag, 1995.
67. Commission of the European Communities, Proposal for a Directive of the European Parliament and the Council of amending Directive 94/62/EC on packaging and packaging waste, December 2001.
68. Reports on packaging and packaging waste by national competent authorities according to Commission Decision 97/138/EC; written communications from Member States.
69. Drain, K.F., Murphy, W.R. and Otterburn, M.S., Solvents for polypropylene: Their selection for a recycling process, *Conserv. Recycling*, 6(3), 107, 1983.
70. Geetha, R., et al., Photo-oxidative degradation of polyethylene: effect of polymer characteristics on chemical changes and mechanical properties. I: Quenched polyethylene, *Polymer Degradation and Stability*, 19(3), 279, 1987.
71. Valenza, A., La Mantia, F.P., Recycling of polymer waste: Part II—Stress degraded polypropylene. *Polym. Degrad. Stab.*, 1988, 20(1), 63.
72. Albano, C., Rodriguez, A., Recycling of polyolefins. I. Analysis of the mechanical properties of virgin with recycled PP, *Revista De La Facultad De Ingenieria (Venezuela).*, 1988, 13(1), 65.
73. Aurrekoetxea, J., Sarrionandia, M.A. and Urrutibeascoa, I., Effects of recycling on the microstructure and the mechanical properties of isotactic polypropylene, *Journal of Materials Science*, 36(11), 2607, 2001.
74. Brenner, G., Plastics Technology: Recycling of Polypropylene--Making New From Old, *Industrie-Anzeiger (Germany)*, 114(32), 19, 1992.

75. Incarnato, L., et al., Structural modifications induced by recycling of polypropylene, *Polymer Engineering & Science*, 39(9), 1661, 1999.
76. Majumdar, J., et al., Thermal properties of polypropylene post-consumer waste (PP PCW), *Journal of Thermal Analysis and Calorimetry*, 78(3), 849, 2004.
77. Incarnato, L., et al., Influence of recycling and contamination on structure and transport properties of polypropylene, *Journal of Applied Polymer Science*, 89(7), 1768, 2003.
78. Agrawal, A.K., Singh, S.K. and Utreja, A., Effect of hydroperoxide decomposer and slipping agent on recycling of polypropylene, *Journal of Applied Polymer Science*, 92(5), 3247, 2004.
79. Jipa, S., et al., Efectul antioxidant al unor materiale carbonice în polipropilena, *Materiale Plastice*, 39(1), 67, 2002.
80. Burillo, G., Clough, R.L., Czvikovszky, T., Guven, O., Le Moel, A., Liu W., Singh, A., Yang, J., Zaharescu, Y., Polymer recycling: potential application of radiation technology, *Radiat. Phys. Chem.* 2002, 64, 41.
81. Zaharescu, T., Jipa, S. and Giurginca, M., Radiochemical processing of EPDM/NR blends, *J. Macromol. Sci., Pure & Appl. Chem.*, 35, 1093, 1998.
82. Clough, R.L. and Shalaby, S.W., *Radiation Effects on Polymers*, ACS Books, Washington DC, 1991.
83. Jipa, S., et al., Evaluation of the Additivated Isotactic Polypropylene Thermostability by the Chemiluminiscence Method. II. Triazinic Additive, *Materiale Plastice*, 37(2), 63, 2000.
84. Zaharescu, T., Chemical changes in ethylene-propylene elastomers during salt thermal ageing, *Polymer*, 35(17), 3795, 1994.
85. Poulakis, J.G., Varelidis, P.C. and Papaspyrides, C.C., Recycling of polypropylene-based composites, *Advances in Polymer Technology*, 16(4), 313, 1997.
86. Adewole, A.A. and Wolkowicz, M.D., *Handbook of Polypropylene and Polypropylene Composites*. New York, NY: Marcel Dekker, Inc., 2003.
87. Qin, H., et al., Zero-Order Kinetics of the Thermal Degradation of Polypropylene/Clay Nanocomposites, *Journal of Polymer Science Part B: Polymer Physics*, 43(24), 3713, 2005.
88. Hájeková, E., Bajus, M., Recycling of low-density polyethylene and polypropylene via copyrolysis of polyalkene oil/waxes with naphtha: product distribution and coke formation, *J. Anal. Appl. Pyrolysis* 2005, 74, 270.
89. Poulakis, J.G. and Papaspyrides, C.D., The Dissolution/Reprecipitation Technique Applied on High-Density Polyethylene: I. Model Recycling Experiments, *Advances in Polymer Technology*, 14(3), 237, 1995.
90. Bertin, S. and Robin, J.J., Study and characterization of virgin and recycled LDPE/PP blends, *European Polymer Journal*, 38(1), 2255, 2002.
91. Callister, W.D., *Materials Science and Engineering*, John Wiley & Sons, New York, NY, USA, 1991.
92. Zhang, Z., Hirose, T., Nishio, S., Morioka, Y., Azuma, N., Ueno, A., Ohkita, H., Okada, M., Chemical Recycling of Waste Polystyrene into Styrene over Solid Acids and Bases, *Ind. Eng. Chem. Res.* 1995, 34, 4514.
93. Elias, H., *An Introduction to Plastics*, VCH, New York, NY, USA, 1993.
94. Nishizaki, H., et al. *Oil Recovery from Atatic Polypropylene by Fluidized-bed Reactor*, Nippon Kagahu-Kaishi, 1977, 1899.
95. Ide, S., et al., Controlled degradation of polystyrene, *Journal of Applied Polymer Science*, 29(8), 2561, 1984.
96. Sittig, M., *Organic and Polymer Waste Reclaiming Encyclopedia*, NOYES Data Corp., Park Ridge, NJ, USA, 1981.

97. Jamil, F., Some Effects of Recycling High Impact Polystyrene, *Plast. Rubber Process. Appl.*, 9(3), 187, 1988.
98. Kingsbury, T. and Ehrlich, R., Polystyrene recycling: processes, trends, and opportunities, *ANNU RECYCL CONF ARC* , 1998, p. 185–191.
99. Grelle, P.F., and Khennache, O., Recycling structural foam polystyrene: what goes around comes around, The Regional Technical Conference of the Society of Plastics Engineers, Dallas, TX, USA, 11–13 October 1993, p. O1.
100. Koniger, R., Recycling og packaginh foam polystyrene for the production of low-weight bricks, *l'Industria Laterizi*, 6(35), 366, 1995.
101. Rodriguez, F., Recycling polystyrene by dissolution with solvent recovery, The 55th Annual Technical Conference, *ANTEC. Part 3*, Toronto, Canada, 1997, p. 3155.
102. Fujiyoshi, H., et al., Development of expanded polystyrene material recycling system, *Ishikawajima-Harima Giho*, 40(1), 7, 2000.
103. Kampouris, E.M, Papaspyrides, C.D., Lekakou, C.N., A model process for the solvent recycling of polystyrene, *Polymer Engineering and Science*, 1988, 28(8), 534.
104. Noguchi, T., et al., A new recycling system for expanded polystyrene using a natural solvent. Part 1. A new recycling technique, *Packaging Technology and Science,* 11(1), 19, 1998.
105. Noguchi, T., et al., A new recycling system for expanded polystyrene using a natural solvent. Part 2. Development of a prototype production system, *Packaging Technology and Science*, 11(1), 29, 1998.
106. Azuma, N., et al., Chemical recycling and eco-design of polystyrene for better environment, Environment Conscious Materials, Ecomaterials: as held at the 39th Annual Conference of Metallurgists of CIM, Ottawa, Ontario, Canada, 20–23 Aug. 2000, p. 225.
107. Ino, E., Chemical Recycling System of Waste polystyrene Foam DS Integrated Treatment of Gelatinized Resin through Distillation/ Thermal Cracking, *Kogyo Zairyo (Engineering Materials)*, 53(4), 84, 2005.
108. Williams, P.T. and Bagri, R., Hydrocarbon gases and oils from the recycling of polystyrene waste by catalytic pyrolysis, *International Journal of Energy Research*, 2004, 28(1), 31–44.
109. Maldas, D. and Kokta, B.V., Effect of Recycling on the Mechanical Properties of Wood Fiber-Polystyrene Composites. II. Sawdust as a Reinforcing Filler, *Polymer-Plastics Technology and Engineering*, 29(5), 419, 1990.
110. Maldas, D. and Kokta, B.V., Effect of recycling on the mechanical properties of wood fiber-polystyrene composites. Part I: Chemithermomechanical pulp as a reinforcing filler, *Polymer Composites*, 11(2), 77, 1990.
111. Shin, C., A new recycling method for expanded polystyrene, *Packaging Technology and Science*, 18(6), 331, 2005.
112. http://www.azom.com
113. Menges, G., PVC recycling management, *Pure & Appl. Chem.*, 68(9), 1809, 1996.
114. Felger, H.K., *Kunststoff-Handbuch*, Vol. 1 & 2, PVC, Hanser, Munich, Germany 1986.
115. Anon., Product Information from the PVC manufacturers, Brussels.
116. Bittner, Michaeli, Menges, *Die Wiedervenvertung von Kunststoffen*, Carl Hanser Verlag Munich, Vienna, 1995.
117. La Mantia, F.P., Ed., *Recycling of PVC and Mixed Plastics Waste*, ChemTec, Publishing, 1996.
118. La Mantia, F.P., Ed., *Handbook of Plastics Recycling*, Rapra, ChemTech Publishing, UK.
119. Braun, D., Recycling of PVC, *Progress in Polymer Science*, 27(10), 2171, 2002.

120. Mechanical recycling of PVC waste, Final Report, Study for DG XI of the European Commission, 2000.
121. Frey, W., PVC-Kreislaufmodell, *Kunststoffberater* 1994, 39(9), 40.
122. Guterl, M. and Schüle, H., Werkstoffliches Recycling am Beispiel von PVC-Fensterprofilen, *Plastverarbeiter*, 45(10), 78, 1994.
123. Ulutan, S., A recycling assessment of PVC bottles by means of heat impact evaluation on its reprocessing, *J. Appl. Polym. Sci.*, 69(5), 865, 1998.
124. Fehse, B., et al., Possibilities and limits of recycling in the electro/electronics sector, *Plastverarbeiter*, 46(7), 28, 1995.
125. Spaulding, C.H., The recycling of vinyl windows: PVC, glass, and component hardware, *ANTEC '95*. Vol. III, Boston, Massachusetts, USA, 3670, 1995.
126. Arnold, J.C. and Maund, B., The properties of recycled PVC bottle compounds. I. Mechanical performance, *Polym. Eng. Sci.*, 39(7), 1234, 1999.
127. Arnold, J.C. and Maund, B., The properties of recycled PVC bottle compounds. II. Reprocessing stability, *Polym. Eng. Sci.*, 39(7), 1242, 1999.
128. Braun, D. and Kramer, K., Recycling of chalk-filled PVC, *Kunststoffe plast europe (Germany)*, 85(6), 23, 1995.
129. Graham, J., Hendra, P.J. and Mucci, P., Rapid identification of plastics components recovered from scrap automobiles, *Plastics Rubber and Composites Processing and Applications*, 24(2), 55, 1998.
130. Adam, Seine Meinung PVC-Recyclingkonzept fuer Fahrzeugverdrahtungen, *Plastverarbeiter* 1998, 49(2), 23.
131. Heitel, K., Rogner, G., Recycling of PVC cable: materials for cable isolation, *Kunststoffe* 1995, 85(11), 1952.
132. Yoshioka, T., Oxidation of Poly (Vinyl Chloride) Powder by Molecular Oxygen in Alkaline Solutions at High Temperatures, *Nippon Kagaku Kaishi*, 5, 534, 1992.
133. Michaeli, W., Lackner, V., Viscosity reduction and dehydrochlorination of mixed plastics scrap by degradative extrusion., Conf. proc. *ANTEC '94*. San Francisco, Calif. (USA) 1994, 52, 2901.
134. Shin, S.M., Yoshioka, T. and Okuwaki, A., Dehydrochlorination behavior of rigid PVC pellet in NaOH solutions at elevated temperature, *Polymer Degradation and Stability*, 61(2), 349, 1998.
135. Yoshioka, T., et al., Chemical recycling of flexible PVC by oxygen oxidation in NaOH solutions at elevated temperatures, *J. Appl. Polym. Sci.*, 70(1), 129, 1998.
136. Anon. *Kunstst Synth* 1994, 25(12), 28.
137. *Bühl, R.,* Options for poly(vinyl chloride) feedstock recycling. Results of research project on feedstock recycling processes, *Plast. Rubber Compos.*, 1999, 28(3), 131.
138. Information of Arbeitsgemeinschaft PVC und Umwelt, e.v., D-53113 Bonn.
139. Information brochure from BSL Olefinverbund GmbH, 1999.
140. http://www.akzonobel.com/bc/cf1221p1.htm
141. Presentation by: Jaspers, H., Akzo Nobel Chemicals, PVC feedstock recycling, a selection of technologies, 1998.
142. Hoyle, W. and Karsa, D.R., Ed., Chemical aspects of Plastics recycling, Cambridge, The Royal Society of Chemistry, 1997.
143. Information from Linde-KCA-Dresden GmbH bulletin on March 1999.
144. Ullmann's Encyclopedia of Industrial Chemistry, Plastics, Recycling, vol. (A-21), 1992.
145. Leidner, J., *Plastics Waste: recovery of economic value*, Marcel Dekker Inc., New York, 1981.
146. Gebauer, M. and Utzig, J., Das PARAK Verfahren – Alternative zwischen werkstoffichem und rohstofflichem Recycling, *Chemische Technik*, 49 (2), 57, 1997,

147. Utzig, J. and Gebauer, M., A new perspective for plastics recycling to produce paraffin waxes from old plastics materials, R'97, Geneva, 1997.
148. Anon., *Kunstst Recycl. Com. Press* 1997, 4(94), 3.
149. Lusinchi, J.M., et al., Recycling of PET and PVC wastes, *J. Appl. Polym. Sci.,* 69(4), 657, 1994.
150. Nadkarni, V.M., Recycling of Polyester, in *Handbook of Thermoplastic Polyesters,* Ed. Fakirov, S., 2002.
151. Miel, R., Europe progresses on PVC recycling: group sets 2010 goal, sees Vinyloop potential, *Plastics News* (Detroit), 2003, 15(38),9.
152. Thiele, A., Materialrecycling von Thermoplasten über Lösen- Schriftenreihe Kunststoff und Recycling, *Kunststoffberater* 1994, 39(12),12.
153. Anon. *Kunstst Synth* 2000, 47(5), 18.
154. Tersac, G., Purification, par dissolution selective, de déchets d'objets usagés broyés en polychlorure de vinyle, *European Polymer Journal,* 30(2), 221, 1994.
155. Mijangos, C., Hidalgo, M. and Lopez, D., Preparation of Marked Poly (Vinyl Chloride) by Reactive Processing for Identification by UV Devices (Recycling), *Vinyl Technol.,* 16, 162, 1994.
156. Knauf, U., Maurer, A., Holley, W., Wiese, M., Utschick, H., Recycling of PVC/PET composites. Production of pure and homogeneous recyclates, and process monitoring, *Kunststoffe,* 2000, 90(2), 72.
157. Gottesman, R.T., Separation of poly (vinyl chloride) from poly (ethylene terephthalate) and other plastics, *Makromol Chem, Macromol Symp* 1992, 57, 133.
158. Drelich, J., et al., Selective froth flotation of PVC from PVC/PET mixtures for the plastics recycling industry, *Polym. Eng. Sci.,* 38(9), 1378, 1998.
159. Rigo, H.G. and Chandler, A. J., Is there a strong dioxin: chlorine link in commercial scale systems?, *Chemosphere,* 37, 2031, 1998.
160. Lin, H.R. and Lin, C.T., Mechanical properties and morphology of recycled plastic wastes by solution blending, *Polymer Plastics Technology and Engineering,* 38(5), 1031, 1999.
161. Garcia, J.C., Marcilla, A. And Beltran, M., The effect of adding processed PVC on the rheology of PVC plastisols, *Polymer,* 39(11), 2261, 1998.
162. Popovska-Pavlovska, F., Trajkovska, A., Gavrilov, T., Rheological behaviour of VPVC/RPVC Blends, *Makromol Symp,* 2000, 149, 191.
163. Wenguang, M. and La Mantia, F.P., Processing and mechanical properties of recycled PVC and of homopolymer blends with virgin PVC, *J. Appl Polym Sci,* 59(5), 759, 1996.
164. Khunova, V. and Sain, M.M., Optimization of mechanical strength of reinforced composites, *Angew Makromol Chem,* 224(1), 9, 1995.
165. Moriwaki, S., et al., Dehydrochlorination of poly (vinyl chloride) by microwave irradiation, *Applied Thermal Engineering,* 26(7), 745, 2006.
166. Yarahmadi,N., Jakubowicz, I. and Martinsson, L., PVC floorings as post-consumer products for mechanical recycling and energy recovery, *Polymer Degradation and Stability,* 79(3), 439, 2003.
167. *Wissenschaftliche Berichte FZKA,* 7005, D1, 2004.
168. Defosse, M., PVC: industry advances efforts in recycling, recyclate use, *Modern Plastics* (USA), 78(12), 32.
169. Garcia, D., et al., Mechanical Properties of Recycled PVC Blends with Styrenic Polymers, *Journal of Applied Polymer Science,* 101(4), 2464, 2006.
170. Ditta, A.S., et al., A study of the processing characteristics and mechanical properties of multiple recycled rigid PVC, *Journal of Vinyl and Additive Technology,* 10(4), 174, 2004.

171. Sasaki, S., Recycling technology trend of polyvinyl chloride, *Kogyo Zairyo (Engineering Materials, Japan)*, 2001, 49(4), 26.
172. Menke, D., Fiedler, H., Zwah, H., Don't ban PVC: Incinerate and recycle it instead!, *Waste Management & Research,* 2003, 21(2), 172.
173. Yoo, Y., Kim, S.S., Won, J.C., Choi, K.Y., Lee, J.H., Enhancement of the thermal stability, mechanical properties and morphologies of recycled PVC/clay nanocomposites, *Polymer Bulletin,* 2004, 52(5), 373.
174. Markovic, V., Acrylic - Designed for the future, A technical discussion paper from ICI Acrylics, 1997.
175. Boustead, I., Eco-profiles of the plastics and related intermediates, A technical paper for APME and ISOPA, Brussels: Association of Plastics Manufacturers in Europe (APME), 1999.
176. Peck, P., Exploring evolution of industry's responses to high grade recycling from an industrial ecology perspective, vol. 2, Case Studies, 2003.
177. Blass, R., Recycling: Is PMMA a Problem?, *Plastiques Modernes et Elastomeres* (France), 45(7), 63, 1993.
178. Gouli, S., Poulakis, J.G. and Papaspyrides, C. D., Solvent recycling of poly(methyl methacrylate) decorative sheets, *Advances in Polymer Technology,* 13(3), 207, 1994.
179. Papaspyrides, C.D., Gouli, S. and Poulakis, J.G., Recovery of poly(methyl methacrylate) by the dissolution/reprecipitation process: A model study, *Advances in Polymer Technology,* 13(3), 213, 1994.
180. Chandra, R., et al., Recycling of polymethylmethacrylate (PMMA) waste, *Popular Plastics & Packaging* (India), 40(12), 57, 1995.
181. Bigg, D.M., et al., Recycling spent polymethylmethacrylate plastic media blasting-beads, ANTEC '95. Vol. III, Special Areas, Boston, Massachusetts, USA, 1995, p. 3662.
182. Nozue, I., Recycling of plastics and activities at Kobe Steel, Kobe Research and Development (Japan), 47(3), 39, 1997.
183. Sasse, F. and Emig, G., Chemical Recycling of Polymer Materials, *Chemical Engineering & Technology,* 21(10), 777,1998.
184. Michaeli, W. And Breyer, K., Feedstock recycling of polymethyl mathacrylate (PMMA), *Kunststoffe Plast Europe* (Germany), 87(2), 183, 1997.
185. Michaeli, W. And Breyer, K., Feedstock recycling of polymethyl methacrylate (PMMA) by depolymerising in a reactive extrusion process, The 56th Annual Technical Conference, ANTEC. Part 3 (of 3); Atlanta, GA, USA, 1998, p. 2942.
186. Kaminsky, W., Pyrolysis with respect to recycling of polymer, *Die Angew. Makromolekulare Chemie,* 232(1), 151, 1995.
187. Kaminsky, W., Predel, M. and Sadiki, A., Feedstock recycling of polymers by pyrolysis in a fluidised bed, *Polymer Degradation and Stability,* 85(3), 1045, 2004.
188. Smolders, K., Baeyens, J., Thermal degradation of PMMA in fluidised beds, *Waste Management,* 2004, 24(8), 849.
189. Kartalis, C.N., et al., Mechanical recycling of postused high-density polyethylene crates using the restabilization Technique. I. Influence of Reprocessing, *Journal of Applied Polymer Science,* 73, 1775, 1999.
190. Pospíšil, J., Nešpürek, S., Zweifel, H., The role of quinone methides in thermostabilization of hydrocarbon polymers—I. Formation and reactivity of quinone methides, *Polym. Deg. Stab.,* 1996, 54(1), 7.
191. Pospíšil, J., Nešpürek, S., Zweifel, H., The role of quinone methides in thermostabilization of hydrocarbon polymers —II. Properties and activity mechanisms, *Polym. Deg. Stab.,* 1996, 54(1), 15.

192. Joseph, K., Kuriakose, B., Premalatha, C.K., Thomas, S., Melt Rheological Behaviour of Short Sisal Fibre Reinforced Polyethylene Composites, *Plast. Rubber, Compos. Process. Appl.,* 1994, 21, 237.
193. Pospisil, J., Sitek, F.A. and Pfaendner, R., Upgrading of recycled plastics by restabilization: an overview, *Polymer Degradation and Stability,* 48(3), 351, 1995.
194. Vogl, O., Jaycox, G.D., 'Trends in Polymer Science'–Polymer science in the 21st century, *Progress in Polymer Science,* 24(1), 1999, 3.
195. Bonner, J.G. and Hope, P.S., *Polymers Blends and Alloys,* Ed. Folkes M.J. and Hope P.S., Blackie Academic & Professional, 1993.
196. Pospisil, J., Nespurek, S., Highlights in the chemistry and physics of polymer stabilization, *Macrom. Symp.* 1997, 115, 143.
197. Bonelli, C.M.C., Martins, A.F., Mano, E.B., Beatty, C.L., Effect of recycled polypropylene on polypropylene/high-density polyethylene blends, *Journal of Applied Polymer Science,* 80(8), 2001, 1305.
198. Li, T., Topolkaraev, V.A., Hiltner, A., Baer, E., Ji, X.Z., Quirk, R.P., Block-Copolymers as Compatibilizers for Blends of Linear Low-Density Polyethylene and Polystyrene, *Polym. Sci. Polym Phys.,* 1995, 33,667.
199. Welander, M. and Rigdahl, M., Use of an Emulsifying Block Copolymer to Improve Time-Dependent Mechanical Properties of Polyethylene-Polystyrene Blends, *Polymer,* 30(2), 207, 1989.
200. Saleem, M. and Baker, W.E., In situ reactive compatibilization in polymer blends: Effects of functional group concentrations, *Journal of Applied Polymer Science,* 39, 655, 1990.
201. Ajji, A., Morphology and mechanical properties of virgin and recycled polyethylene/polyvinyl chloride blends, *Polym. Eng. Sci.,* 35(1), 64, 1995.
202. Tidjani, A., Comparison of formation of oxidation products during photo-oxidation of linear low density polyethylene under different natural and accelerated weathering conditions, *Polymer Degradation and Stability,* 68(3), 465, 2000.
203. Koning, C., Van Duin, M., Pagnoulle, C., Jerome, R., Strategies for compatibilization of polymer blends, *Prog. Polym. Sci,* 1998, 23,707.
204. Scott, G., Atmospheric oxidation and antioxidants, Elsevier, London, 1965.
205. Grassie, N. and Scott G., *Polymer Degradation and Stabilization,* Cambridge University Press, Cambridge, 1985.
206. Scott., G., *Atmospheric Oxidation and Antioxidants,* Elsevier, London, 1993.
207. Pospisil, J., Transformation of phenolic antioxidants and the role of their products in the lon-term properties, *Adv. Polym. Sci.,* 1980, 36, 70.
208. Pospisil, J., Chemical and Photochemical Behaviour of Phenolic Antioxidants in Polymer Stabilization: a State of the art, Report. I, *Polymer Degradation and Stability,* 40, 217, 1993.
209. Pospisil, J., Chemical and Photochemical Behavior of Phenolic Antioxidants in Polymer Stabilization: a State of the art, Report. II, *Polymer Degradation and Stability,* 39, 103, 1993.
210. Gugumus, F., Re-Evaluation of the Stabilization Mechanisms of Various Light Stabilizer Classes, *Polymer Degradation and Stability,* 39, 117, 1993.
211. Gijsman, P., The mechanism of action of hindered amine stabilizers (HAS) as long-term heat stabilizers, *Polym. Degrad. Stab.,* 1994, 43, 171.

19 Processing of Vinyl Polymers

Chantara T. Ratnam and Hanafi Ismail

CONTENTS

19.1 INTRODUCTION

Vinyl polymers include those polyolefins in which the R subsistent in an olefin monomer (CH_2=CHR) is bonded to the unsaturated carbon through an oxygen atom (vinyl esters, vinyl ethers) or a nitrogen atom (vinyl pyrrolidone, vinyl carbazole). The commonly used commercial vinyl polymers include poly (vinyl acetate), poly (vinyl alcohol), (polyvinyl acetals), poly (vinyl ethers), polyvinyl butyral, polyvinyl fluoride, and polyvinylidene chloride.

The polymer polyvinyl chloride is frequently referred to as vinyl, and is made by polymerization of the monomer vinyl chloride (CH_2=CHCl).

19.1.1 Polyvinyl Chloride (PVC)

Polyvinyl chloride was first discovered in 1838 by Henri Victor Regnault and in 1872 by Eugen Baumann. In the early 20th century, the Russian chemist Ivan Ostromislensky and Fritz Klatte of the German chemical company Griesheim-Elektron both attempted to use PVC in commercial products, but their efforts were not successful as they found difficulties in processing the rigid, brittle polymer. However, PVC became commercial only in 1927, after Waldo Semon of B.F. Goodrich developed the advantage of plasticization. However, development of acrylonitrile rubber (NBR), and in 1942 the discovery of its ability to permanently plasticize PVC spurred rapid penetration of the market [1].

In pure form, PVC is a hard thermoplastic material with a T_g of about 80°C [2]. It is basically an amorphous polymer. Due to the presence of syndiotactic sequences of sufficient length, PVC shows some crystallinity. The crystalline melting point is high (225°) and considerably above normal processing temperatures, which are in the range of 150–200°C [3].

19.1.2 PVC Compounding

PVC has low thermal stability and high melt viscosity; it cannot be processed on its own. Therefore, to produce a useful product other ingredients are added to the PVC resin for the purpose of increasing flexibility, providing adequate heat stability, improving processability, and imparting aesthetic appeal. The compounding ingredients include stabilizers, plasticizers, lubricants, impact modifiers, or processing aids. Materials, such as color pigments, dyes, flame-retardants, fillers, and fungicides, can be added for specific requirements.

19.1.2.1 Heat Stabilizers

The use of thermal stabilizers in PVC formulation is essential to:

- Prevent dehydrochlorination of PVC
- Prevent discoloration of PVC (yellowing, blackening, crosslinking)
- Process PVC at high temperatures, as well as to provide the necessary stabilization required for ultimate application

The choice of a thermal stabilizer is led by the application requirements such as transparency, food contact, and weatherability. Among the common stabilizers are basic lead salts such as basic lead carbonate, tribasic lead sulfate, and dibasic lead sulfate. Organo compounds of other metal used as stabilizers include those of calcium, zinc, and tin, mostly in the form of phenates, octoates, benzoates, and laureates. Barium/cadmium stabilizers are among the oldest PVC stabilizers, but because of serious toxicological and ecological concerns about cadmium and its compounds, they are being successively replaced by other stabilizers, particularly in Europe and Japan. Co-stabilizers are used to improve the efficiency of the stabilizer system and enable it to be tailored to specific requirement. The most frequently used co-stabilizers are antioxidants, polyols, phosphate, β-diketones, and recently zeolites and hydrotalcites.

19.1.2.2 Plasticizers

Plasticizers are low boiling liquids or low molecular weight solids that are added to resins to alter processing and physical properties. They increase resin flexibility, softness, and elongation. They increase low temperature flexibility but decrease hardness. They also reduce processing, temperatures, and melt viscosity in the case of calendering.

Plasticizers fall into two categories based on their solvating power and compatibility with resins: primary and secondary. The primary plasticizers are highly compatible with the resin. Examples of primary plasticizers are Dioctyl phthalate (DOP), Di (n-hexyl; n-octyl; n-decyl) phthalate (linear), and Di-iso decyl phthate (DIDP). The secondary plasticizers have limited compatibility with the resin and are, therefore, used only in conjunction with primary plasticizers to confer some special properties such as:

- Low-temperature flexibility: di-normal octyl decyl adipate (DMODA), dioctyladipate (DOA)
- Flame retardance: Reofas 65 (tri-iso propyl phenyl phosphate)

- Electrical properties: tri-mellitates
- Cost reduction: Cereclor, chlorinated paraffins

19.1.2.3 Fillers

Essentially, fillers are added to formulations to reduce costs, although they may offer other advantages, such as opacity, resistance to blocking, reduced plate-out, and improved dry blending. On the other side, fillers can reduce tensile and tear strength, reduce elongation, cause stress whitening, and reduce low-temperature performance.

The most common fillers used with PVC are calcined clays, and water-ground and precipitated calcium carbonates of particle size around 3 micrometers. Other fillers are silicas and talcs.

19.1.2.4 Lubricants

These materials are of prime importance in PVC processing. They:

- Improve the internal flow characteristics of the compound and reduce the frictional heat build-up (internal lubricants)
- Reduce the tendency for the compound to stick to the process machinery at high temperatures used (external lubricants)
- Improve the surface smoothness of the finished product
- Improve heat stability by lowering internal and/or external friction
- Examples of lubricants are stearic acid, calcium stearate, high melting point waxes, and mineral oil

19.1.2.5 Processing Aids

These may be regarded as low-melt viscosity, compatible solid plasticizers. They are added to lower processing temperature, improve roll release on calenders, reduce plate-out, and promote fusion. They are usually added at concentrations of 5.0%. The most widely used processing aids are acrylic resins, of which acryloid K 120N is an example.

19.1.2.6 Impact Modifiers

These are used in rigid vinyls to improve impact resistance. These are usually acrylic or acrylonitrile butadiene styrene (ABS) polymers used at 10–15 phr levels.

19.1.2.7 Light Stabilizers

Light stabilizers are used for resistance to ultraviolet (UV) radiation. They are used in low concentrations of 0.5–1.5 phr. An example is Tinuvin P, which is produced by Ciba-Geigy.

19.1.2.8 Flame Retardants

PVC is inherently self-extinguishing. However, the plasticizers and additives are not. Therefore, flame-retardants are added. The most widely known one is antimony tri-oxide.

19.1.2.9 Foaming Agents

Foaming agents are chemicals that decompose at predetermined temperatures to produce a certain volume of gas within the molten vinyl and thereby create a foam.

Other important additives used are impact modifiers, biocides, anti-static agents, and pigments.

19.1.3 Rigid PVC

PVC plastics based on the homopolymer (made from one monomer) are of two basic types, rigid or flexible. "Rigid" usually refers to unplasticized PVC, normally containing only polymer, stabilizer, lubricant, and, sometimes, impact modifiers. This term, however, is occasionally extended to include slightly plasticized (up to 20 parts per hundred) products, although these materials should more properly be called "semi-rigid." Other polymers are often added to PVC to improve impact resistance or processing. Products made from rigid PVC compound are hard, tough, and difficult to process, but they have fairly good outdoor stability, superior electrical properties, excellent resistance to moisture and chemicals, and excellent dimensional stability. They are self-extinguishing.

Rigid PVC compounds are used in piping for drains, waste and vent systems, water distribution and irrigation systems, and various building products including house sidings, window sash, building panels, rain gutters, downspouts, flashing, and wall tile.

19.1.4 Flexible PVC

Flexible PVC contains significant amounts of plasticizers (from 20–50 parts per hundred or more) to make it flexible and easy to process. It has lower strength, lower heat resistance, and poorer weathering properties than rigid PVC. Flexible compounds are used in cable and wire insulation, floor and wall coverings, pipe, packaging film, shower curtains, corrugated sheeting, weather stripping, window frames, and decorative wallboard laminates. Flexible PVC is also used increasingly in the automotive industry.

19.1.5 Compounding Process

PVC compounds, which are prepared by blending PVC resin with the additives, are produced in two physical forms:

1. *Granules*: The PVC resin blend is fed to melt processing or extrusion equipment. The molten composition is palletized and cooled, producing PVC compounds in granular form.
2. *Dry Blends*: The PVC resin is blended with the appropriate additives, screened, and packed as a dry powder.

When properly formulated, the PVC compounds can be processed with all the techniques for thermoplastics, and its applications include rigid, elastic, and spongy goods. Other uses include bottles, window frames, pipes, flooring, wallpaper, toys, car seats, guttering, cable insulation, credit cards, and medical products such as blood bags, IV tubing, and much more. However, care should be taken on the metal tooling and dies used in the processes because the HCl formed (even in small amounts) from the PVC in any stage of the heating processes may cause corrosion. Corrosion-resistant coatings and metals are available to minimize the effects of HCl, but these will add cost to the tooling.

19.2 BASIC PROCESSING TECHNIQUES

Vinyl compounds are processed or fabricated into products by two basic methods: melt and liquid processing.

In melt processing, rigid and flexible compounds in solid form are converted into melts by internal (shear) and external heating. This conversion is accomplished in several different types of thermoplastic processing techniques that include injection molding, extrusion, calendering, blow molding, and thermoforming.

19.2.1 Injection Molding

This is the process by which plastic granules are melted, then injected into a mold cavity to create the required component or product shape. Depending on the size and complexity of the product, different pressures are needed to force the liquid plastic into the mold. An injection molding machine consists of three basic parts: the mold plus the clamping and injection units. Figure 19.1 is a plastic injection molding machine.

FIGURE 19.1 Thermoplastic injection molding machine.

The major steps involved in injection molding process include clamping, injection, dwelling, cooling, mold opening, and ejection.

During the injection process, plastic material, usually in the form of pellets, are loaded into a hopper on top of the injection unit. The pellets feed into the cylinder where they are heated until they reach molten form. Within the heating cylinder, a motorized screw mixes the molten pellets and forces them to the end of the cylinder. Once enough material has accumulated in front of the screw, the injection process begins. The molten plastic is inserted into the mold through a sprue, while the screw controls the pressure and speed. *Some injection molding machines use a ram instead of a screw.*

19.2.1.1 Advantages of Injection Molding

- High production rates
- High tolerances are repeatable
- Wide range of materials can be used
- Low labor costs
- Minimal scrap losses
- Little need to finish parts after molding

19.2.1.2 Disadvantages of Injection Molding

- Expensive equipment investment
- Running costs may be high
- Parts must be designed with molding consideration

19.2.2 Extrusion

In extrusion, both single and multiple screw extruders are used to convert vinyl compounds into solid profiles, cellular profiles, pipe, blown film, and flat sheet. Two principle components of an extrusion operation are the extruder and the die. In normal PVC extrusion, PVC granules or pellets and any other materials to be mixed with them are fed into a hopper attached to the extrusion machine. From the hopper, the material falls through a hole in the top of the extruder onto the extrusion screw. This screw, which turns inside the extruder barrel, conveys the PVC forward into a heated region of the barrel where the combination of external heating and heating from friction melts the PVC. The screw moves the molten plastic forward until it exits through a hole in the end of the extruder barrel to which a die has been attached. The die imparts a shape to the molten PVC stream, which is immediately cooled by water to solidify the extrudate, thus retaining the shape created by the die. Auxiliary equipment is used to pull the extrudate away from the extruder at the appropriate rate. Other auxiliary equipment cuts the extrudate to a proper length and packages.

Efficient extruder operation involves interrelated critical factors such as crew speed, feeding rate, temperature control, screw design, and auxiliary equipment. Single-screw extruders tend to only to be used to process rigid PVC in dry blend form for products with thin walls which are made at relatively low weight/hour output. In other areas, dry blends of PVC are usually processed using twin-screw extruders.

This is partly due to the poor thermal stability of PVC. The instability of PVC increases with temperature and shear. PVC also has a low coefficient of thermal conductivity, which makes it more difficult to heat uniformly than many other polymers without intensive mixing and high shear rate. During extrusion, the shear increases by the high melt viscosity of the PVC. Twin-screw machines have much higher surface area of contact among the barrel, the screw, and the polymer than a single-screw machine. This provides much better thermal transfer, minimizing the amount of energy that has to be supplied mechanically via the screw(s). Because the twin-screw extruder is run at low screw speeds, the amount of shear to which the PVC is subjected can be kept to a minimum. Figure 19.2 is an example of a twin-screw extruder, which can be used to process dry blends of PVC.

19.2.3 Calendering

Calendering is used to produce flexible and rigid sheeting in the 2 to 35 mm thickness range. The calender rolls are made of chilled iron and the surface may be either smooth or matte. In this process, molten PVC is compressed in a small gap between two heated cylinders rotating in opposite directions. It is a large-volume processing method, which requires high capital equipment costs compared with extrusion methods. Modern calender trains operate with computerized control systems to ensure optimum quality. In calendering, it is necessary to pre-flux compound on Banbury/roll mill equipment at a rate to match the calender's capacity. A short-barreled extruder downstream of the mixing equipment is often used to screen out any possible foreign contaminant, which might damage the calender rolls.

FIGURE 19.2 A twin screw extruder.

19.2.4 Blow Molding

The blow molding process is used for the production of hollow PVC articles. Basically, the principle of blow molding consists of melting the resin, which forms a hollow tube or "parison" and finally blowing to form the articles. All three stages usually occur simultaneously.

Two types of blow-molding processes are described next.

19.2.4.1 Injection Blow Molding

In this process, a parison is first molded around a core rod in a perform mold and then transferred to a bottle blow-mold cavity. In the cavity, air expands the parison to the shape of the bottle. Injection blow molding is used to produce small bottles and other parts.

19.2.4.2 Extrusion Blow Molding

Extrusion blow molding process is used to produce larger bottles and plastic tanks. In this process, the PVC resin is first extruded as a tube and is then captured by two halves of a mold in a continuous process. A blow pin is inserted, the mold is closed, and air forced into the parison through the blow pin to expand the parison to the form of the mold cavity.

19.2.5 Thermoforming

Thermoforming is a process used to shape thermoplastic sheets and films into discrete parts. In thermoforming, a PVC sheet is heated until it softens. The sheet is then formed to the shape of a mold by the application of external air pressure (i.e., pressure forming which uses air pressure to assist the vacuum) or by pulling vacuum (i.e., vacuum forming, which uses a vacuum to pull it onto a contoured surface) between the sheet and the mold. Thermoforming process also uses the twin sheet forming technique. Twin sheet forming is two pressure or vacuum forming operations occurring simultaneously, which are joined to produce an integrally welded hollow part.

19.2.6 Reinforcing Process

Generally, reinforced plastics or plastics composite materials are made of fibrous reinforcements (usually fiberglass or carbon fibers), which are coated or surrounded by a plastic resin. The material is placed in a mold and solidified, either by thermoplastic or by thermoset molding method. The methods that are commonly employed to process reinforced thermoset plastics can be found elsewhere [4]. Composites with very short fibers tend to have thermoplastics matrices. These very hot fibers are usually blended into the PVC resin by the resin manufacturer that chops the reinforcement. When being extruded or injection molded, the short fibers incorporated in the thermoplastics are able go through small clearances, such as the gap between the extruder screw and the extruder wall or gate that connects the mold cavity with the runner system in the injection mold. Thus, they are processed using all the thermoplastic processing equipment described previously.

19.2.7 Plastisol/Solvent Processing

Some processing methods for PVC do not require melting of the plastic. One of the most common of these is accomplished by adding sufficient solvent to the plastic that the PVC dissolves or becomes suspended in the solvent. The resulting material is called a plastisol or PVC dispersion. The plastisol can be applied to other materials and then dried and fused to form PVC coating or covering. Metal parts, such as racks for dishes, screwdrivers, and other tools, are sprayed with or dipped in plastisol to provide corrosion protection and cushioning.

PVC has the ability to be solvent welded without using heat. Solvent welding employs traditional adhesives and it is much easier to use than the fusion or hot seal method. In this process, the material to be joined, usually PVC pipe and a PVC fitting, are coated to the joining surface with the solvent adhesive. The parts are then allowed to dry.

Other plastics processing techniques, such as compression molding, transfer molding, and reaction injection molding are not widely employed for PVC processing, and they are commonly used to process thermosetting plastics.

19.3 PVC-BASED BLENDS

In unplasticized form, PVC is hard and tough. However, next to polyethylene, PVC is the most abundant thermoplastic material. Due to its low cost and versatility, PVC has received much attention because improvements in processability, heat distortion temperature, impact strength, and service life have been achieved. It is useful to blend PVC with other polymers and elastomers to improve some of its properties.

19.3.1 NBR/PVC Blends

One of the commercially important polymer blends is that of NBR and PVC [5]. A blend of NBR/PVC was the first useful thermoplastic elastomer (TPE) introduced in the market [6]. The unique and useful properties of NBR/PVC blends have been known for over 55 years [7]. In NBR-rich blends, PVC improved the ozone and oil resistance of NBR [8]. Major applications of the NBR-rich NBR/PVC blends are as conveyor belt covers, cable jackets, hose covers and linings, and roller covers [9]. The NBR improves the impact strength of PVC and also acts as a permanent plasticizer to PVC in PVC-rich blends [8]. Some of the major applications of these blends are footwear, cellular thermal insulation, and cellular sporting surfaces [9].

19.3.2 PVC/Polyester Blends

A variety of polyesters are blended with PVC — some as plasticizers for preparation of elastomeric compositions, others as impact modifiers. Polyesters, which are most commonly employed, are oligomeric polyesters of dicarboxylic acids as plasticizers [10]. High molecular weight co-polymers of ethylene and ester monomers such as vinyl acetate are used as plasticizers as well as impact modifiers [11]. A number of publications have reported on the properties of various polyesters as plasticizers for PVC [12].

TABLE 19.1
Blends of PVC

Polymer	Improved Property
ABS, MBS	Impact resistance, hardness, tensile strength, distortion temperature
Acrylics	Impact resistance, transparency, chemical resistance, oil resistance, moldability
Poly(caprolactone)	Plasticization (non-extractable), moldability, impact resistance.
SAN	Low temperature toughness, processability, and dimensional stability.
Poly(urethanes)	Plasticization (non-extractable), elongation, impact resistance, tensile strength, low temperature toughness.
PB	Toughness, weatherability
Poly(dimethyl siloxane)	Processability, heat stability.

19.3.3 PVC/ABS Blends

PVC has poor impact strength. The improvement in this property could be achieved with the addition of elastomeric materials that have a degree of compatibility with PVC [6]. ABS resin is one of the traditional materials that has been used as an impact modifier for PVC because it is thermally stable [13]. PVC/ABS blends can be used in building, automotive, and electrical applications [14]. Khanna and Congdon [15] have developed highly flexible PVC/ABS blends with good impact strength and flame retardant properties.

19.3.4 Polymers Blended with PVC as Processing Aids

Some polymers also act as processing aids. Acrylic and metacrylic polymers, polymethacrylate butadiene styrene (MBS) and ABS terpolymers, and EVA co-polymers are among the most efficient ones [2]. The efficiency of these modifiers as PVC processing aids is related to good mutual polymer–polymer miscibility. The miscibility of EVA co-polymers with PVC depends on the vinyl acetate content of the EVA. The co-polymers with more than 50% vinyl acetate are already miscible with PVC, and they are used mainly as processing aids [16]. The variety of PVC blends and their properties are summarized in Table 19.1 [12–14].

19.4 PVC/ENR BLENDS

Initial studies on PVC/ENR blend system was made by several researchers such as Margaritis and Kalfoglou [17], Ratnam and Nasir [18], Nasir et al. [19], and Varughese and De [20]. Most of these studies investigated the effect of epoxidation levels and blend ratio on the miscibility of the blends. Epoxidized natural rubber with 50% mol epoxidation (ENR50) was found to be mutually miscible with PVC at all compositions [17,20].

19.4.1 Melt Processing

The work by Ratnam and Nasir [18] indicated the need of using suitable mixing conditions to attain optimum blend properties. It has been subsequently reported that mechanical properties of PVC/ENR are greatly influenced by the mixing parameters [21]. The fusion problem associated with the PVC dominant blend was also highlighted by Nasir and Ratnam [22]. They have attributed the continued increase in the Brabender torque until the end of mixing cycle with the incomplete fusion. Similar observations were also reported by George et al. [23] while working on PVC/NBR blends. Nasir et al. [19] have attempted to resolve this problem by modifying the mixing sequence for PVC dominant blends.

19.4.2 Rheological Properties

Rheological properties of PVC/ENR blends with a special reference to the composition dependence and effect of compounding conditions have been studied using a capillary rheometer [24]. The synergism in the apparent shear viscosity with a positive deviation from the logarithmic additive rule was again associated with the blend miscibility. The effect of processing parameters on the rheological behavior of the cured and uncured PVC/ENR blends was also reported by the same authors [25]. For cured blends, it has been shown that the crosslinks play a dominant role at lower shear rates in such a way that their presence impedes flow. At much higher shear rates, flow is controlled by the processing variables, as in the case with the uncured system, and flow improves with increase in temperature. The flow properties of PVC/ENR was found to be pseudoplastic in nature, and the system exhibits elastic phenomena such as die swell and melt fracture.

19.4.3 Miscibility

The miscibility in PVC/ENR system has been reaffirmed through results of dynamic mechanical analysis and morphological studies, which revealed a single T_g and single-phase system, respectively [19]. The synergism in the dynamic storage modulus tensile properties and hardness has supported the earlier findings that PVC/ENR is a miscible system [26, 27]. Hydrogen bonding has been found to be extensively involved in PVC/ENR blend as evidenced from Fourier transform infrared spectroscopy (FTIR). This evidence has been used to justify the nature of the specific interactions that are responsible for miscibility and the improved properties of the blend [26].

19.4.4 Aging

The main problem with the PVC/ENR blends, however, has been the rapid deterioration of physical and mechanical properties at ambient conditions. The unstable nature of these blends has been established by Ishiaku et al. [28, 29] in their studies on thermo-oxidative aging of the PVC/ENR blends. They have attributed the poor aging properties of the blends to two main factors. Principally, the unstable nature of ENR is due to the residual acidity that was inherited from modification of the rubber with a peracid [30, 31]. In addition, the decomposition of PVC liberates HCl, which catalyzes epoxide ring opening reactions [31].

Several attempts have been made to improve the aging behavior of the blend. These include the incorporation of a plasticizer [28], an antioxidant such as

2,2,4,-tri-methyl-1,2-dihydroquinoline (TMQ), a base such as calcium stearate [29], and the addition of a third polymer, nitrile rubber [31]. The addition of a plasticizer [28] and nitrile rubber [32]were found to be effective, particularly in curbing the degradation.

19.4.5 Crosslinking

Blends that undergo thermally induced crosslinking in the absence of any curing agent have been referred to as "self-crosslinkable plastic rubber blends" [33–35]. Ramesh and De [36] reported that an unstabilized PVC/ENR blend forms self-crosslinkable blends during high-temperature molding. Such a blend was found to be immiscible, although they are miscible in uncrosslinked stage [17, 19, 20, 26]. However, addition of a third reactive polymer, such as carboxylated nitrile rubber, made the ternary blend miscible [37]. Subsequently, the same researchers reported that the crosslinking of the miscible PVC/ENR blend during high-temperature molding for prolonged periods has no effect on blend miscibility [38]. The term dynamic vulcanization is commonly used to describe the process of crosslinking the elastomeric component of a thermoplastic elastomer during intensive mixing [39]. Mousa and co-workers [40–42] did studies on dynamically vulcanized PVC/ENR thermo plastic elastomers, which were prepared using a semi-EV vulcanization system. The properties studied include tensile, viscoelastic, swelling index, and thermoxidative aging. The changes in these properties was observed with increasing sulfur concentration from 0–1 phr. The enhancement in property studied has been attributed to the increase in crosslink density with the increase in sulfur loading.

19.5 RADIATION PROCESSING OF PVC/ENR BLENDS

The basis of radiation processing is the ability of the high-energy radiation to produce reactive cations, anions, and free radicals in materials. The role of reactive free radicals, cations, and anions in the production and crosslinking of synthetic polymers is well known [43]. Irradiation processing mainly involves the use of either electron beams from electron accelerators, gamma radiation from Cobalt-60 (Co-60) sources, or UV irradiation on an industrial scale to produce products that are safe, practical, and beneficial [44, 45]. Further details on radiation processing of polymers can be found elsewhere [46].

19.5.1 Sources of Irradiation

19.5.1.1 Cobalt-60 Irradiation

Gamma irradiation from Co-60 gives deep penetration depth due to its high energy. Thus, it is commercially used mainly for the sterilization of medical products. However, a Co-60 irradiator has the disadvantage of producing low dose rates, which allows for more oxygen diffusion, resulting in increased oxidative degradation of polymers. Therefore, a Co-60 source requires an inert atmosphere to induce reactions in polymers.

19.5.1.2 Ultra-Violet Irradiation

Ultra-violet (UV) irradiation from mercury or xenon arc lamps is more widely used. This technique has a practical advantage because it is cheaper and easy to install. However, UV radiation lacks penetration and does not initiate in-depth reactions in thicker samples. Thus, the technique is only employed for the cure of surface coatings.

19.5.1.3 Electron Beam Irradiation

An electron accelerator, also referred to as the electron beam machine, enables irradiation at a high dose rate. Thus, a high product throughput as well as reduced damage caused by oxidative degradation is achieved at the same time. In addition, the absorbed dose rate can be easily controlled and varied, switched on and off quickly with no radiation energy loss during shutdown, and needs no replenishment with time. However, electron beams lack depth of penetration, too.

19.5.2 Radiation Modification of PVC/ENR Blends

Numerous studies were conducted on irradiation crosslinking of poly(vinyl chloride)/epoxidized natural rubber (PVC/ENR), with particular attention to optimization of blending parameters, addition of stabilizers, crosslink enhancers, and antioxidants [47–62]. The PVC/ENR blend was prepared with a Brabender Plasticorder. The blend was irradiated by using a 3 MeV electron beam accelerator with doses ranging from 0–200 kGy. The irradiation was performed at acceleration voltage, beam current, and dose rate of 2 MeV, 2 mA, and 0.78 kGy/sec, respectively. Changes in tensile properties, Shore A hardness, gel fraction, and dynamic mechanical and morphological properties of the blends with irradiation doses were investigated. It was found that PVC/ENR blends crosslink upon electron beam irradiation. Initial studies on the effect of irradiation with doses ranging from 20–100 kGy on various PVC/ENR blend ratio indicated that electron beam irradiation enhances the physical properties of the blend in a relatively mild manner [47, 48]. However, further work by Ratnam and co-workers confirmed that optimum properties could be achieved in the presence of suitable additives [49–62].

19.5.2.1 Stabilization of PVC Phase

An investigation to determine a suitable stabilizer in a PVC/ENR blend for stabilization during blending and irradiation was also published [51, 52]. Among the PVC stabilizers incorporated are tribasic lead sulfate (TBLS), Ca/Zn, Mg/Zn, calcium stearate, and zinc octoate. TBLS was found to be effective in relation to enhancement of blend properties upon irradiation [51, 52]. The observation on the trend shown in Brabender torque-time curves, tensile strength, yellowness index, hardness, and gel fraction has confirmed that tribasic lead sulfate is efficient in stabilizing PVC/ENR blends as well as enhancing the blend properties upon irradiation. Subsequent reports [54, 61, 62] on the effect of lead stabilization of PVC phase

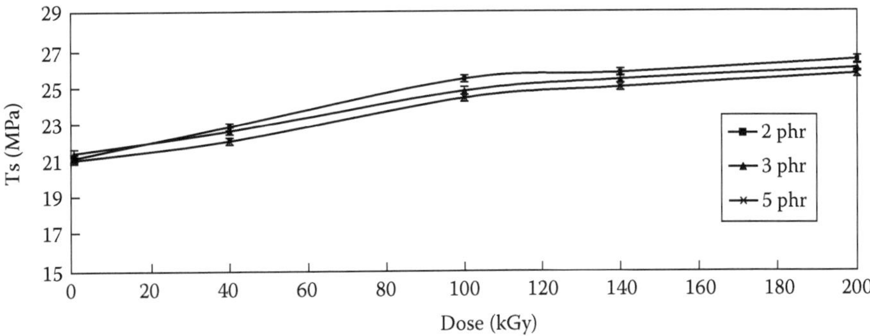

FIGURE 19.3 Effect of irradiation on the Ts of 50/50 PVC/ENR blend at various TBLS levels. (From Ratnam, C.T., *Polym.-Plast. Technol. Eng.*, 41(3), 407, 2002. With permission.)

clearly illustrated the importance of TBLS content in controlling the irradiation-induced crosslinking in PVC/ENR blend. The incorporation of excessive amounts of TBLS generally inhibits the irradiation-induced crosslinking of PVC/ENR blend, although it simultaneously stabilizes the blend against thermal and irradiation-induced degradation. A TBLS content of 2 phr was considered optimum for stabilizing the PVC, with minimum inhibition in irradiation-induced crosslinking. Figure 19.3 depicts the decline in tensile strength of a 50/50 PVC/ENR blend with the increase in TBLS level.

19.5.2.2 Effect of Antioxidants

The effects of three different types of antioxidants — a hindered phenol, a phosphite, and a hindered amine light stabilizer — on the 70/30 PVC/ENR blend were also studied by the same researchers [49]. Results on mechanical properties implied that the hindered phenol is an effective antioxidant for the investigated PVC/ENR blend system. However, further studies [53] on the irradiation processing of PVC/ENR blends in the presence of a 0.25–1.5 phr Irganox 1010 revealed the inhibition of the irradiation-induced crosslinking by the Irganox 1010. Similar reports confirmed that the Irganox 1010 is involved in the stabilization of the blend against irradiation-induced degradation and the 0.5 phr Irganox 1010 is essentially sufficient to provide stability to the blend with a minimum loss in mechanical properties during irradiation. The decline in gel fraction with Irganox 1010 content, as depicted in Figure 19.4 [56], certainly demonstrates the inhibition of the crosslinking by the antioxidant.

19.5.2.3 Effect of Crosslinking Agents

The acceleration of radiation induced crosslinking by the addition of polyfunctional monomers was also reported. The influence of multifunctional acrylates, such as trimethylolpropane trimethacrylate (TMPTA), 1,6-hexaediol diacrylate (HDDA),

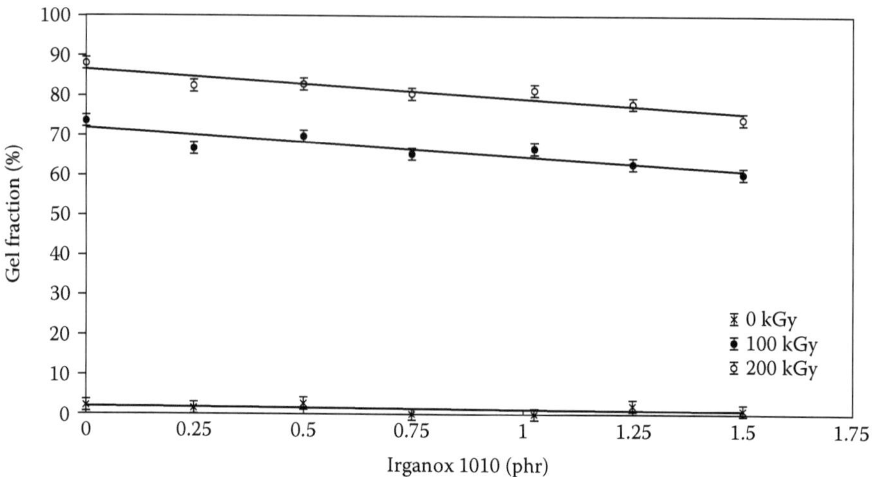

FIGURE 19.4 Effect of Irganox 1010 on the gel fraction of 50/50 PVC/ENR blend. (From Ratnam, C.T. et al., *Polym. Deg. Stab.*, 72(1), 147, 2001. With permission.)

and 2-ethylhexyl acrylate (EHA), as well acrylated polyurethane oligomer, on the 70/30 PVC/ENR blend was investigated. TMPTA was found to render the highest mechanical properties to the blend. Blends containing 3–4 phr TMPTA were found to achieve optimum crosslinking, which in effect caused a maximum in tensile strength at 70 kGy (Figure 19.5). Further addition of TMPTA demonstrated a decline in Ts (at above 40 kGy) due to embrittlement caused by excessive crosslinking and a breakdown of network structure [56, 57].

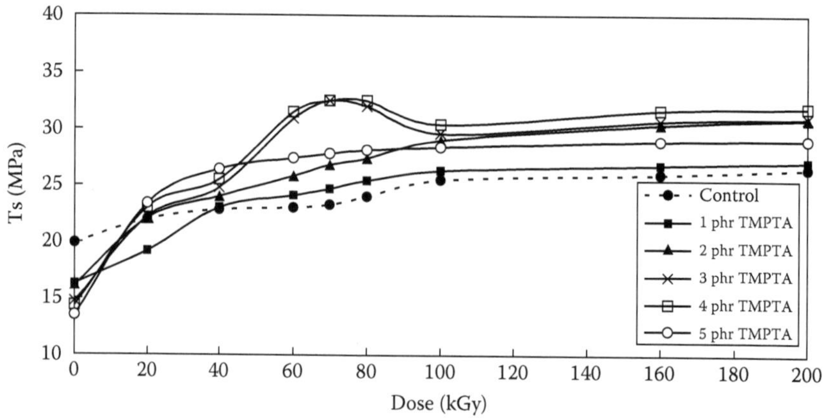

FIGURE 19.5 Effect of TMPTA on the tensile strength of PVC/ENR blend at various irradiation doses. (From Ratnam, C.T. et al., *J. Appl. Polym. Sci.*, 81, 1926, 2001. With permission.)

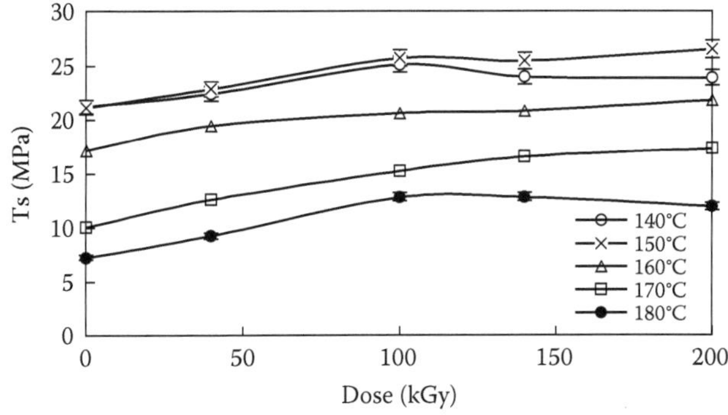

FIGURE 19.6 Effect of irradiation on tensile strength of PVC / ENR blends at various blending temperatures. (From Ratnam, C.T. et al., *Polym.-Plast. Technol. Eng.*, 40(4), 561, 2001. With permission.)

19.5.2.4 Mixing Parameters

To elucidate the effect of mixing parameters on the irradiation crosslinking of the blend, 140–180°C mixing temperatures, 6–30 min total mixing time, and 30–70 rpm rotor speed were utilized [60]. The mixing torque-time curve was correlated with the changes in properties following irradiation. In general, it was observed that the mixing parameters, such as temperature and time, studied in this work are important in maximizing the positive effect of irradiation. Results revealed that a readily compatible blend prepared at 150°C, 50 rpm, and 10 min of minimum mixing time enjoyed maximum benefit from irradiation. However, irradiation was found to impart compatibility to the partially compatible PVC/ENR blends prepared at 140°C. The irradiation-induced degradation was found to be prominent at higher doses for blend that has undergone excessive thermal degradation. Figure 19.6 illustrates the changes in tensile strength of 50/50 PVC/ENR blend with mixing temperature.

19.5.2.5 Dynamic Mechanical Properties

The irradiation-induced crosslinking in PVC/ENR blends were investigated by means of dynamic mechanical analysis [59]. The influence of TMPTA on the dynamic mechanical properties of PVC/ENR blends upon irradiation was also studied. The enhancement in storage modulus (E′) and T_g with irradiation dose indicates the formation of irradiation-induced crosslinks (Figure 19.7). This is further supported by the decrease in tan δ_{max} and loss modulus (E″) peak. The compatibility of the blend was found to improve upon irradiation. A theory proposed by Fox has provided further insight into the irradiation-induced compatibility in the blend as a consequence of crosslinking. Agreement of the results with a theory relating T_g with the distance between crosslinks, M_c, has provided further evidence of irradiation-induced crosslinking. The possible mechanism of crosslinking induced by the irradiation between PVC and ENR was also proposed.

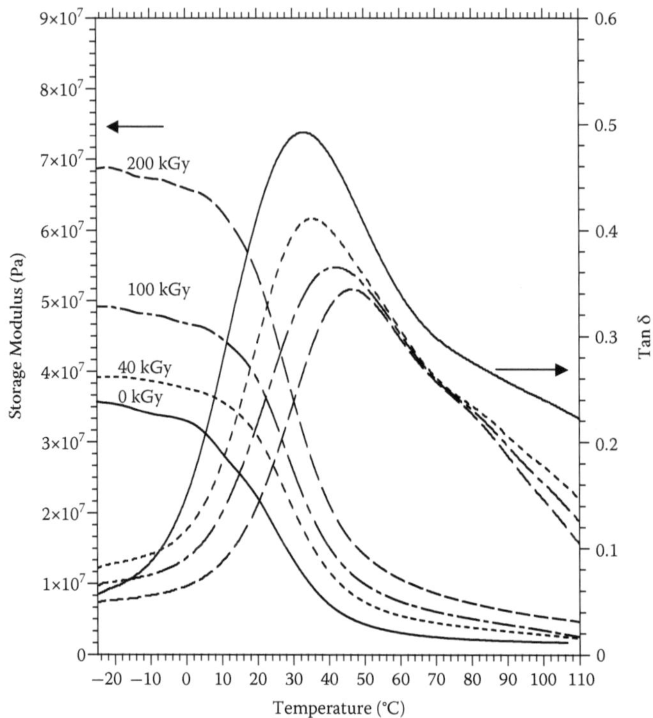

FIGURE 19.7 The effect of irradiation on the temperature dependence tan δ and storage modulus of 50/50 PVC/ENR blend. (From Ratnam, C.T. et al., *Polym. Int.*, 50, 503, 2001. With permission.)

19.6 CONCLUDING REMARKS

PVC resin is the largest volume member of the vinyl family. PVC formulations can be shaped by a variety of techniques and, using very little energy, made into the final product form. Thus, the process versatility of PVC is one of the major factors that contributes to its commercial value. PVC can be made in different forms to permit processing on a wide variety of equipment; each form can be altered further by compounding to achieve particular properties in end products, which range from soft to rigid in nature.

REFERENCES

1. Vinyl Institute. The history of vinyl. http://www.mindfully.org/Plastic/Vinyl-Made.htm.
2. Brydson, J.A., *Plastics Materials*, 4th ed., Butterworths, London, 1982, 291.
3. Toenell, B., Recent development in PVC polymerization, *Polym.- Plast. Technol. Eng.*, 27(1), 1, 1988.

4. Manas, C. and Roy, S.K., *Plastics Technology Handbook,* Marcel Dekker, New York, 1987, 157.
5. Paul, D.R., *Polymer Blends,* Vol. 1, Paul, D.R. and Newman, S. Eds., Academic Press, New York, 1978, Ch. 1.
6. Utracki, L.A. and Favis, B.D., *Handbook of Polymer Science and Technology,* Vol. 4, Cheremisinoff, N.P., Ed., Academic Press, New York, 1989, p. 121.
7. Badum, E., U.S. Patent 2,297,194, 1942.
8. Bhowmick, A.K. and De, S.K., *Thermoplastic Elastomers from Rubber-Plastic Blends,* Eds., De, & Bhowmick, Ellis Horwood, New York, 1990, 13.
9. Pittenger, J.E. and Cohan, C.F., How to use vinyl nitrile rubber resin, *Mod. Plast.*, 25, 81, 1947.
10. Fligor, K.K. and Sumner, J.K., Resinous plasticizer from sebatic acid, *Ind. Eng. Chem.*, 37, 504, 1945.
11. Woo, E.M., Barlow, J.W., and Paul, D.R., Thermodynamic of phase behavior of PVC/ aliphatic polyester blends, *Polymer,* 26, 763, 1985.
12. Ziska, J.J., Barlow, J.W., and Paul, D.R., Miscibility in PVC-polyester blends, *Polymer,* 22, 918, 1981.
13. Utracki, L.A., Melt flow of polymer blend, *Polym. Eng. Sci.*, 23, 602, 1983.
14. Utracki, L.A., Polymer blends and alloys for molding applications, *Polym.-Plast. Technol. Eng.*, 22(1), 27, 1984.
15. Khanna, S.K. and Congdon, W.I., Engineering and molding properties of PVC/NBR and polyester blends, *Polym. Eng. Sci.*, 23, 627, 1983.
16. Stepek, J. and Daoust, H., *Additives for Plastics*, Vol. 5, Springer-Verlag, New York, Ch. 3.
17. Margaritis, A.G. and Kalfoglou, N.K., Miscibility of chlorinated polymers with epoxidised poly (hydrocarbons), ENR/PVC blends, *Polymer,* 28, 502, 1987.
18. Ratnam, C.T. and Nasir, M., *Development in the Plastics and Rubber Product Industries,* Eds., Rajoo, CAPS Enterprise, Kuala Lumpur, 1987, 403.
19. Nasir, M., Ishiaku, U.S., and Mohd Ishak, Z.A., Determination of optimum blending conditions for PVC/ENR blends, *J. Appl. Polym. Sci.*, 47, 951, 1993.
20. Varughese, K.T. and De, P. P., Miscible blends from plasticized PVC and epoxidized natural rubber, *J. Appl. Polym. Sci.*, 37, 2537, 1989.
21. Nasir, M., Ratnam, C.T., and Tee, C. Y., The effect of DOP on the rheological and mechanical properties of PVC/ENR blends, in *Proc. 4th Symp. Malaysian Chem. Engineers*, Kuala Lumpur, 1988.
22. Nasir, M. and Ratnam, C.T., Internal mixer studies on PVC/ENR blends, *J. Appl. Polym. Sci.*, 38, 1219, 1989.
23. George, K.E., Joseph, R., and Francis, D.J., Studies on NBR/PVC blends. *J. Appl. Polym. Sci.*, 32, 2867, 1986.
24. Ishiaku, U.S., Nasir, M., and Mohd Ishak, Z.A., Rheological properties of PVC/ENR blends. Part 1: composition dependence and effect of the compounding conditions, *J. Vinyl Technol.*, 1, 142, 1995.
25. Ishiaku, U.S., Nasir, M. and Mohd Ishak, Z.A., Rheological properties of poly (vinyl chloride)/epoxidized natural rubber blends. Part II: the effect of processing variables, *J. Vinyl Technol.*, 1, 66, 1995.
26. Ishiaku, U.S., Nasir, M., and Mohd Ishak, Z.A., Aspects of miscibility in PVC/ENR blends. Part I: mechanical and morphological properties, *J. Vinyl Technol.*, b16, 219, 1994.
27. Ishiaku, U.S., Nasir, M., and Mohd Ishak, Z.A., Aspects of miscibility in PVC/ ENR blends. Part II: dynamic mechanical properties, *J. Vinyl Technol.*, 16, 226, 1994.

28. Ishiaku, U.S., Poh, B.T., Mohd Ishak, Z.A., and Ng, D., Thermo-oxidative properties of blends of PVC with ENR and NBR rubbers in the presence of an antioxidant and a base, *Polym. Intn.*, 39, 67, 1996.
29. Ishiaku, U.S., Mohd Ishak, Z.A., Ismail, H., and Nasir, M. The effect of DEHP on the thermo-oxidative ageing of PVC/ENR blends, *Polym. Intn.* 41, 327, 1996.
30. Gelling, I.R., Modification of natural rubber latex with peracetic acid, *Rubb. Chem. Technol.*, 58, 86, 1985.
31. Gelling, I.R. and Porter, M., *Natural Rubber Science and Technology,* Roberts, A.D., Ed., Oxford University Press, Oxford, 1988, 359.
32. Ishiaku, U.S., Ismail, H., and Mohd Ishak, Z.A., The effect of mixing time on the rheological, mechanical properties of PVC-ENR blends. *J. Appl. Polym. Sci.*, 73, 75, 1999.
33. Ramesh, P. and De, S.K., Self crosslinkable plastic-rubber blend system based on PVC and carboxylated nitrile rubber, *Polym. Commun.,* 31, 466, 1990.
34. Ramesh, P. and De, S.K., Interfacial crosslinking in an immiscible rubber blend based on polyacrylic acid and polychloroprene, *Polym.*, 33, 3927, 1992.
35. Ramesh, P. and De, S.K., Self-crosslinkable polymer blends based on chlorinated rubber and carboxylated nitrile rubber, *Rubb. Chem. Technol.*, 65, 24, 1992.
36. Ramesh, P. and De, S.K., Self crosslinkable plastic-rubber blend system based on PVC/ENR, *J. Mater. Sci.*, 26, 2846, 1991.
37. Ramesh, P. and De, S.K., Carboxylated nitrile rubber as reactive compatibiliser for self-crosslinkable miscible blends of PVC/ENR, *J. Appl. Polym. Sci.*, 50, 1369, 1993.
38. Ramesh, P. and De, S.K., Evidence of thermally induced chemical reactions in miscible blends of PVC/ENR, *Polymer*, 34, 4893, 1993.
39. Coran, A. Y. and Patel, R., Rubber-thermoplastic compositions. Part I: EPDM-polypropylene thermoplastic vulcanizates, *Rubber Chem. Technol.*, 53, 141, 1980.
40. Mousa, A., Ishiaku, U.S., and Mohd Ishak, Z.A., Oil-resistance studies of dynamically vulcanized PVC/ENR thermoplastic elastomers, *J. Appl. Polym. Sci.,* 69, 1357, 1998.
41. Mousa, A., Ishiaku, U.S., and Mohd Ishak, Z.A., Dynamic vulcanization of PVC/ENR thermoplastic elastomers, part I. Mixing rheology, *Plast. Rubber Comp. Process. Appl.,* 26(8), 331, 1999.
42. Mousa, A., Ishiaku, U.S., and Mohd Ishak, Z. A., Rheological and viscoelastic behaviour of dynamically vulcanized PVC/ENR thermoplastic elastomers, *J. Appl. Polym. Sci.*, 74, 2886, 1999.
43. Allen, G., *Comprehensive Polymer Science*, Vol. 1, Booth, C. and Price, C., Eds., Pergamon Press, Oxford, 1989, Ch. 1.
44. Sakamoto, I. and Mizusawa, K., Industrial application and recent development of electron processing system, *Radiat. Phys. Chem.,* 18, 1341, 1981.
45. Silverman, J., *Radiation Processing of Polymers*, Singh, A. and Silverman, J., Eds., Hanser, Munich 1992, 13.
46. Ratnam, C.T., *Advances in Material Processing*, Vol. 1, Azhari, C.H., Anandastuti, M., Kamal A. Eds., Institute of Material Malaysia, Kuala Lumpur, 2003, 131.
47. Ratnam, C.T. and Zaman, K., Effect of radiation on PVC/ENR blend, in *Proc. Int. Nuclear Conf. 97*, Kuala Lumpur, 1997.
48. Ratnam, C.T. and Zaman, K., Modification of PVC/ENR blends by electron beam irradiation, *Angew. Makromol. Chemie*, 269, 42, 1999.
49. Ratnam, C.T. and Zaman, K., Enhancement of PVC/ENR blend properties by electron beam irradiation: effect of antioxidants. *Polym. Degradation Stability*, 65, 481, 1999.

50. Ratnam, C.T. and Zaman, K., Modification of PVC/ENR blend by electron beam irradiation: effect of crosslinking agents. *Nucl. Instrument Method B*, 152, 335, 1999.
51. Ratnam, C.T. and Zaman, K. Studies on stabilization of PVC/ENR blends by chemiluminescence analysis, *Nucl. Sci. J. Mal.*, 17(1), 37, 1999.
52. Ratnam, C.T. and Zaman, K., Stabilization of PVC/ENR blends, *Polym. Degradation Stability*, 65, 99, 1999.
53. Ratnam, C.T., Irradiation crosslinking of PVC/ENR blend: effect of Irganox 1010 level, *Polym. Int.*, 50, 1132, 2001.
54. Ratnam, C.T., Irradiation crosslinking of PVC/ENR blend: lead stabilization of PVC phase, *Plast., Rubber Composites*, 30(9), 416, 2001.
55. Ratnam, C.T., Nasir, M., and Baharin, A., Irradiation crosslinking of UPVC in the presence of additives, *Polym. Test*, 20(5), 485, 2001.
56. Ratnam, C.T., Nasir, M., Baharin, A., and Zaman, K., Evidence of irradiation-induced crosslinking in miscible blends of PVC/ENR in the presence of TMPTA, *J. Appl. Polym. Sci.*, 81, 1914, 2001.
57. Ratnam, C.T., Nasir, M., Baharin, A., and Zaman, K., Electron beam irradiation of PVC/ENR blends in presence of TMPTA, *J. Appl. Polym. Sci.*, 81, 1926, 2001.
58. Ratnam, C.T., Nasir, M., Baharin, A., and Zaman, K., Electron beam irradiation of PVC/ENR blend in the presence of Irganox 1010, *Polym. Deg. Stab.*, 72(1), 147, 2001.
59. Ratnam, C.T., Nasir, M., Baharin, A., and Zaman, K., Effect of electron beam irradiation on PVC/ENR blend: dynamic mechanical analysis. *Polym. Int.*, 50, 503, 2001.
60. Ratnam, C.T., Nasir, M., Baharin, A., and Zaman, K., Effect of blending parameters on electron beam enhancement of PVC/ENR blends, *Polym.-Plast. Technol. Eng.*, 40(4), 561, 2001.
61. Ratnam, C.T., Irradiation crosslinking of PVC/ENR blend: Effect of stabilizer content and mixing time, *Polymer Testing*, 21(1), 93, 2002.
62. Ratnam, C.T., Irradiation modification of PVC/ENR blend: effect of TBLS content, *Polym.-Plast. Technol. Eng.*, 41(3), 407, 2002.

20 Characterization of Interfaces in Composites Using Micro-Mechanical Techniques

Maya Jacob John, Rajesh D. Anandjiwala, and Sabu Thomas

CONTENTS

20.1 INTRODUCTION

Composite materials offer unique characteristics in properties and can be utilized efficiently in structural applications. The properties offered by composites are, in turn, determined by the constituent materials (i.e., the fiber and the matrix and in many instances, by the fiber–matrix "interface." The term interface is defined as a two-dimensional region between the fiber and matrix having zero thickness (Figure 20.1). The properties of the interface are intermediate between those of fiber and matrix. Matrix molecules can be anchored to the fiber surface by chemical reaction or adsorption, which determine the strength of interfacial adhesion. In certain cases, the interface may be composed of additional constituent as a bonding agent or as an interlayer between the two components of the composite.

As evident from the figure, interface is different from interphase. The word interphase is used as a general term to categorize the polymeric region surrounding a fiber.

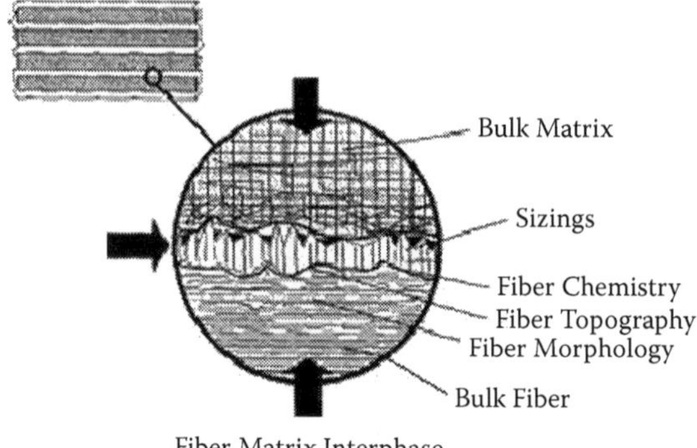

FIGURE 20.1 Schematic representation of interface and interphase in composites.

It consists of polymeric material made from the chemical interaction of sizing or coating on the fiber and the bulk matrix during the curing process. The interphase is also known as the mesophase. Two major issues concerning the interface are its physical dimensions and its role in effecting a bond between fiber and matrix phases.

The fiber/matrix interface in a continuous filament composite transfers an externally applied load to the fibers themselves. Load applied directly to the matrix at the surface of the composite is transferred to the fibers nearest the surface and continues from fiber to fiber via matrix and interface. If the interface is weak, effective load distribution is not achieved and the mechanical properties of the composite are impaired. On the other hand, a strong interface can assure that the composite is able to bear load even after several fibers are broken because the load can be transferred to the intact portions of broken as well as unbroken fibers. A poor interface is also a drawback in situations other than external mechanical loading (e.g., because of differential thermal expansions of fiber and matrix, premature failure can occur at a weak interface when the composite is subjected to thermal stress). Thus, adhesion between fiber and matrix is a major factor in determining the response of the interface and its integrity under stress. Additionally, the interface may be more vulnerable to moisture and solvents than are fiber and matrix.

Several theories[1] that contribute to adhesion have been identified, namely:

1. **Theory of mechanical interlocking**
   Mechanical interactions can occur in several ways (e.g., when a liquid polymer matrix is made to flow on the rough surface of a solid substrate, a "lock and key configuration" results in solidification). Alternatively, differences in thermal expansion coefficients of fiber and matrix may cause compression tightening as temperatures change during processing.
2. **Theory of adsorption interaction**
   According to this theory, primary and secondary forces create adhesion. The most important primary forces are the ionic, covalent, and metallic bonds. The bonds formed by these forces are very strong; their strength is between 60–80 kJ/mol for covalent and 600–1200 kJ/mol for ionic bonds.

The presence of such bonds at the interface is of great importance. The secondary bonds are created by Van der Waals forces (i.e., by dipole–dipole (Keesom), induced dipole (Debye), and dispersion (London) interaction). The strength of these interactions is much lower. Hydrogen bonding and acid–base interactions also have been recognized as important factors. This theory is applied most widely for the description of interaction in particulate filled or reinforced polymers.[2]

3. **Theory of electrostatic interaction**
According to this theory, the polymer and a thin metallic film layer correspond to an electric double layer, and charge transfer between the surfaces creates adhesion. The significance of this theory in particulate filled polymers is very limited.
4. **Theory of interdiffusion**
Adhesion is caused by the mutual diffusion of the molecules of the interacting surfaces. This theory can be applied for the description in polymer blends, but its use is limited when solid surfaces are in contact.

20.2 CHARACTERIZATION OF INTERFACE

The characterization of the interface gives relevant information on interactions between fiber and matrix. In a recent review George et al.[3] have assessed the interface modification and characterization of natural fiber reinforced plastic composites. A critical analysis on the physical and chemical treatment methods that improve fiber–matrix adhesion is presented in the review. Studies of interfacial contributions in wood fiber reinforced polyurethane composites was conducted by Rials et al.[4] The chemical and physical characterization of the interphase has been keenly discussed in an article by Kim and Hodzic.[5] The authors are of the opinion that the gap between physicochemical investigation and bulk material testing is bridged by implementation of novel techniques such as nano-indentation, nano-scratch tests, and atomic force microscopy.

In another interesting study, the interphase in sized glass fiber reinforced poly(butylene terephthalate) composites was extensively investigated by different techniques by Bergeret et al.[6] The authors are of the opinion that the interfacial region of the composite will be affected not only by the coating composition but also by the chemical reactions involved in the vicinity of the fiber and inside the surrounding matrix. Two separate regions were focused upon: the fiber/sizing interphase and the sizing/matrix interphase. The techniques used were, including mechanical and thermomechanical tests, infrared spectroscopy, gel permeation chromatography, carboxyl end group titrations, extraction rate measurements, and viscosity analysis. The results indicated that the adhesion improvement is due to the presence of a short chain-coupling agent and of a polyfunctional additive, which may react both with the coupling agent and the matrix. It was further concluded that it was possible to soften the interphase and consequently increase the composite impact strength.

The role and the importance of interfaces and interphases in multi-component materials have been enumerated by Pukanszky.[7] Recently, researchers have looked into the applicability of different micro-mechanical tests to interface strength characterization and the advantages and disadvantages of stress based and energy based models.[8] In an interesting study,[9] the correlation between post-mortem fracture surface morphology of short fiber reinforced thermoplastics and interfacial adhesion was presented. The authors are of the opinion that fracture surface morphology is dependent on both fiber–matrix interfacial adhesion strength and matrix shear yield strength. Jacob et al.[10] have reviewed the recent advances in characterization of interfaces in biofiber reinforced composites.

The various methods that are available for characterization of the interface are as follows:

1. **Micro-Mechanical Techniques**
   The extent of fiber–matrix interface bonding can be tested by different micro-mechanical tests such as fiber pull-out, micro-debond test, micro-indentation test, and fiber fragmentation test.
2. **Spectroscopic Techniques**
   Electron spectroscopy for chemical analysis/X-ray photoelectron spectroscopy (ESCA), mass spectroscopy, X-ray diffraction study, electron induced vibration spectroscopy, and photoacoustic spectroscopy are successful in polymer surface and interfacial characterization.
3. **Microscopic Techniques**
   Microscopic studies such as optical microscopy, scanning electron microscopy (SEM), transmission electron microscopy (TEM), and atomic force microscopy (AFM) can be used to study the morphological changes on the surface and can predict the strength of mechanical bonding at the interface. The adhesive strength of fiber to various matrices can be determined by AFM studies.
4. **Thermodynamic Methods**
   The frequently used thermodynamic methods for characterization in reinforced polymers are wettability study, inverse gas chromatography measurement, zeta potential measurements, etc. Contact angle measurements have been used to characterize the thermodynamic work of adhesion between solids and liquids as well as surface of solids.

20.2.1 Micro-Mechanical Techniques

Because the interfacial region is inaccessible on a macroscopic level, fine and ingenious investigative methods have to be conceived and developed to provide information about the interface state and quality. In the characterization of the interface, it has been a general rule to use "model composites" where one aspect of the interfacial properties is studied in detail. For the micro-mechanical characterization of the interface properties, techniques like pull-out, fragmentation, and micro-indentation tests are usually used. These test techniques do not simulate a situation that can be found as much in a real composite material but the loading conditions are relatively simple so that the interface properties can be rather easily derived out of these tests.

All these techniques evaluate the bonding strength between the fiber and matrix. Such bond strength values can be used to investigate the dependence of composite performance on the energy absorbing characteristics of the interface and to establish the extent to which the fiber surface treatments can alter bonding.

One of the most common methods is to measure the force required to pull out a fiber embedded in a matrix. This method in addition to its relative simplicity of sample preparation and measurement is expected to give realistic information when one considers the pull-out of fibers from the fracture surfaces of composites. Pull-out tests have been performed on many systems including the steel wire–rubber matrix, the glass fiber–cement matrix, the metal fiber–epoxy matrix, and the glass fiber–polypropylene matrix systems. The pull-out process has various features. The two main aspects are: (1) debonding, which destroys the bond between fiber and matrix, and (2) pull-out, in which a fiber is extracted from a broken matrix against friction. The pull-out test is considered to be a good method of evaluating the interfacial shear load as it can directly measure the interfacial shear strength between the fiber and matrix independent of their properties. The interfacial shear strength is a critical factor that controls the mechanical properties and interlaminate shear strength of composite materials. The fiber pull-out problem has been investigated extensively for purposes of studying the interfacial adhesion and elastic stress transfer between fibers and matrix.[11] A detailed analysis of the pull-out technique is given next.

20.2.1.1 Single-Fiber Pull-Out/Single-Fiber Micro-Droplet Test

Pulling thin fibers out of various matrices has long been considered a desirable way of investigating the basic characteristics of the fiber/matrix interfaces. However, except for the pioneering work of the Russian school of the Institute of Chemical Physics in Moscow, which was reviewed in 1972 by Andreevskaya and Gorbatkina,[12] a number of difficulties were experienced in the 1960s by the few researchers who attempted to prepare specimens (e.g., Norman et al. for glass/polyester systems[13] or Schmidt et al. for carbon/aluminum).[14]

It is relatively easy to work with thick fibers or rods, as demonstrated by De Vekey and Majumdar for glass/cement systems or Kelly and Tyson for tungsten/copper systems. The difficulty one faces with today's high modulus fibers is related to the fineness of the fiber and the corresponding small embedded depth in the matrix, which is required to pull the fiber out instead of breaking it.

20.2.1.1.1 Experimental Techniques

The original Andreevskaya–Gorbatkina method, or "three-fiber method," was used unchanged by Jarvela et al.,[15] but a number of variants have been proposed, including the fiber going through a resin film supported by a tiny hole in a metal sheet or by a rubber stopper with a slit, immersion of the fiber end in a small volume of the resin, etc.

20.2.1.1.2 Specimen Preparation and Testing for Fiber/Resin System

The experimental procedure is presented schematically in Figure 20.2: A drop of liquid resin is spread onto the small aluminum holder (1). A cap with a 0.5-mm central hole containing a piece of fiber (2) is then placed over the holder (3) with the fiber end entering the 10–1000 μm thick resin layer. As many specimens as required can be prepared at one time. All the cylinders are then transferred to an oven and the resin

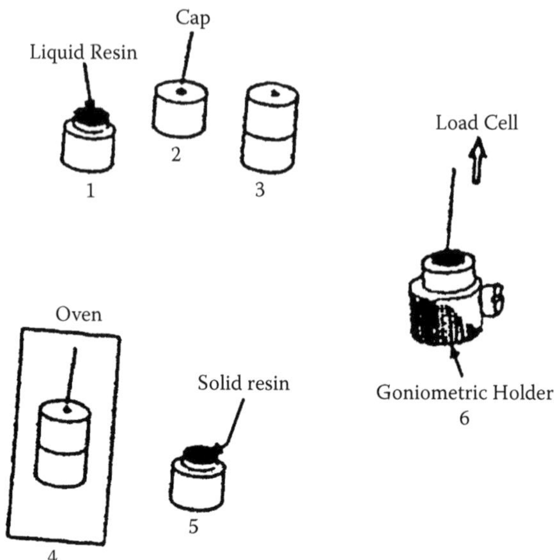

FIGURE 20.2 Schematic illustrations of pull-out test preparation.

properly cured (4) After cooling and removing the cap, each sample is fastened by a screw to a goniometric holder on the crosshead of the tensile machine (5). Tilting is carefully adjusted so that the fiber is aligned with the loading direction (6).

Finally, the crosshead is moved upward to introduce the fiber free end into a 200-μm diameter hole drilled in the brass part on a heating head connected to the load cell. Tin solder is used to fix the fiber. The free length of fiber is adjusted to a few millimeters, and the solder is allowed to cool down. The sample is now ready for the pull-out test. The tensile tester determines the force required to pull out the fiber from the resin.

The fiber is gently smeared with a drop of glycerin to reduce the whipping effect of the debonded fiber when pulling out suddenly from the resin and to allow recovery of the pulled-out fiber so it is unbroken for subsequent examination. In the free part of the fiber, the strain rate is about 10^{-3} s^{-1}, and load is recorded as a function of the displacement throughout the test. After the full extraction has been achieved, the recovered fiber is inspected in the SEM to measure both the fiber diameter and the embedded length, l_e, with l_e given by the distance between the remaining marks of the junction line with the resin (often as a broken wetting cone) and the fiber end. Inspection of the sample also ensures that the fiber does not fail inside the resin button, which would result in an underestimated embedded length.

20.2.1.1.3 Shape of the Force/Displacement Curves

It will be discussed in the following sections that the shape of the force/displacement curves in the pull-out test depends ultimately on the characteristics of the interface and on the dynamics of the test.

For the usual brittle fiber/resin systems, three types of curves are observed in practice. In the first case, for very strongly bonded interfaces, the strain energy stored in the free length of fiber is so high that immediate extraction follows the interface failure. Only the maximum load can be recorded on the curve. If the free length is short enough, the second type of graph can still be obtained. This second type is observed with weakly bonded interfaces: After the interface has failed, the fiber can be extracted progressively in a controlled way and the friction measured up to the final point. The embedded length and the friction load can both be recorded along with the maximum debonding load.

Some authors have observed a third type of curve for Kevlar: one or more peaks are visible in the ascending part of the force/displacement curve. According to these authors, although the first peak corresponds to complete debonding, the latter ones originate in the damage caused by the friction when the non-embedded part of the fiber is pulled up through the matrix. (Kevlar is known to be prone to lateral abrasion.)

From the load/displacement curves, the debonding shear strength is calculated from the equation

$$\tau_d = F_d/D\pi l_e \tag{20.1}$$

where

F_d = Debonding force
D = Fiber diameter, which is measured microscopically
l_e = Embedded length of the fiber

The shear strength of the fiber/resin interface is a key property when investigating the micro-mechanical behavior of composites, because it is a measure of the integrity of the interface. Accurate evaluation of shear strength has, until now, been very difficult, although a keen interest has existed for many years.

Pitkethly and Doble[16] utilized the single fiber pull-out test to characterize the fiber–matrix interface of carbon fiber reinforced epoxy composite. In this study, two fiber–resin combinations were evaluated. In addition to giving a value for maximum interfacial strength, their study yielded information concerning the interfacial shear modulus "G" and interface thickness that is of use in micro-mechanical modeling of the interfacial region. Good agreement also existed between theoretical and experimental values.

The interfacial characterization of flax fiber reinforced thermoplastic composites was performed by Stamboulis et al.[17] They employed the single fiber pull-out technique. Two types of flax fibers were used, namely dew-retted and upgraded Duralin fibers. The Duralin fibers were treated for improved moisture resistance. The interfacial shear strength of dew-retted and upgraded Duralin fibers in LDPE, HDPE, and maleic anhydride modified PP were determined. The force/displacement curves for all the samples were found to be typical of a brittle fracture mode interface behavior. They also did not observe any improvement of interfacial shear strength in the case of upgraded flax fiber reinforced composite.

Another study using a single fiber pull-out was attempted by Van de Velde et al.[18] The authors used dew-retted, hackled long flax treated with propyltrimethoxy silane, phenyl isocyanate, and maleic anhydride polypropylene. The studies revealed that composite prepared with flax fiber treated with MAA-PP exhibited the highest interfacial shear strength. Joseph et al.[19] conducted an interesting study on the comparison of interfacial properties of banana fiber and glass fiber reinforced phenol-formaldehyde composites. They observed that the interfacial shear strength is higher for banana/PF system than for glass/PF system. This was attributed to the hydrophilic nature of cellulose and PF resin. Hydrophilicity of fiber arises from the hydroxyl groups of lignin and cellulose, which can easily form bonds with methylol and phenolic hydroxyl groups of the resole, resulting in a strong interlocking between the two. In the case of glass/PF composites, this kind of bonding is not possible and thus the interfacial shear strength is low. The load/displacement graphs of two systems are given in Figures 20.3 and 20.4. In Figure 20.3, due to the strongly bonded interface in the banana fiber/PF system, the strain energy stored in the free length of the fiber is so high that immediate extraction follows the interface failure. Therefore, the curve indicates only the maximum load. In Figure 20.4, corresponding to the glass/PF system, the first peak corresponds to complete debonding, and the latter ones originate in the damage caused by the friction when the non-embedded part of the fiber is pulled up through the matrix.

The interfacial adhesion of sisal fiber with different thermoset matrices (polyester, epoxy, and phenol formaldehyde) and a thermoplastic resin (LDPE) was investigated by Joseph et al.[20] The authors found that interfacial bonding was maximum in sisal fiber reinforced phenol formaldehyde composites due to chemical bonding between phenol formaldehyde prepolymer and the guaiacyl group of lignocellulose. In the case of sisal-LDPE composites, interfacial bonding is low due to the divergent behavior in polarity of the fiber and matrix.

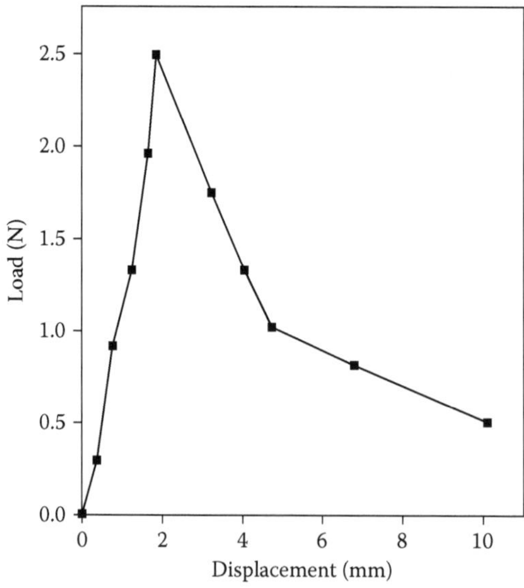

FIGURE 20.3 Load-displacement curve of banana/PF composite.

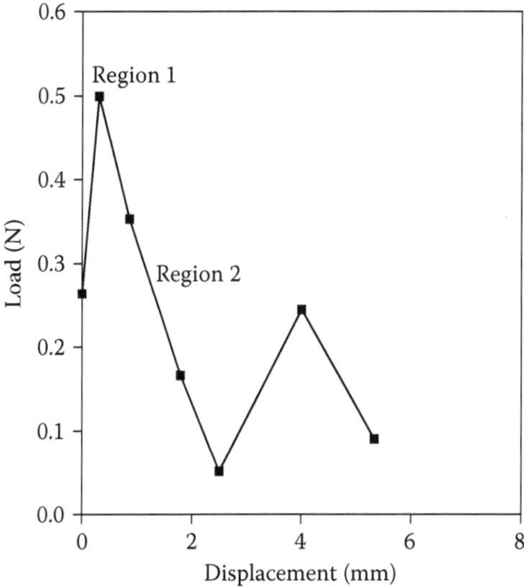

FIGURE 20.4 Load-displacement curve of glass/PF composite.

The interfacial adhesion of flax fiber reinforced polypropylene and polypropylene-ethylene propylene diene terpolymer blends were investigated by Manchado et al.[21] In this study, both the matrices were modified with maleic anhydride. Single fiber pull-out tests were conducted to investigate the extent of interfacial adhesion. They observed that the addition of small proportions of maleic anhydride to the matrices significantly increased the shear strength. The authors were of the opinion that introduction of functional groups in the matrix reduced the interfacial stress concentrations preventing fiber–fiber interactions, which are responsible for premature composite failure. In addition, in the presence of maleic anhydride functional groups the esterfication of flax fibers takes place increasing the surface energy of the fibers to a level closer to that of the matrix. Thus, a better wettability and interfacial adhesion is obtained.

Tanaka et al.[22] studied the effect of water absorption on the interfacial properties of aramid fiber reinforced epoxy composites.

Single-fiber pull-out tests were successfully applied to quantitatively analyze the influence of water absorption on the fiber–matrix interfacial properties. The interfacial strength of aramid/epoxy composite was decreased by 26% after a 7-week immersion time in deionized water at 80°C. The interfacial strength was drastically changed between the 4- and 7-week immersion time and showed a plateau thereafter. *In situ* observations of interfacial crack propagation by a video microscope and an analysis of acoustic emission (AE) signals indicated that the signals obtained during the pull-out process were classified into four types according to fracture modes. AE signals detected a final unstable crack propagation and that fiber breakage had high amplitude.

Pull-out tests were conducted by Sydenstricker et al.[23] in sisal fiber reinforced polyester biocomposites. Sisal fiber was modified by sodium hydroxide and

N-isopropyl-acrylamide solutions. It was observed that all the chemical treatments were effective, and the best results were obtained with the 2% N-isopropyl-acrylamide treatment. A newly developed fiber bundle pull-out technique was developed by Brandsletter et al.[24] to determine the interfacial parameters between carbon fiber bundles and carbon matrix. This test was applied to a bidirectionally layered carbon-fiber-reinforced carbon composite produced on the polymer route with the aid of 8H-satin weave prepreg. The interface between fiber and matrix and the matrix properties were varied by graphitization (a final heat treatment) at three different temperatures. The new test method uses the superposition of external compressive stresses on the inner clamping stresses, which enables the separation of two important interface parameters, the coefficient of friction and the clamping stress. The evaluation procedure also takes into account the real geometry of the pulled-out bundles.

Yuan et al.[25] investigated the effect of air-plasma and argon treatment on interfacial shear strength between sisal fiber and polypropylene by means of single fiber pull-out. They found that optimum treatment parameters were shortest plasma treatment time, medium power level, and medium chamber pressure. Under these conditions, the interfacial shear strength of air plasma treated sisal fibers was found to be higher than argon plasma treated fibers.

The interfacial interaction between ultra-high molecular weight polyethylene fibers and polydicyclopentadiene matrix was analyzed by pull-out test by Devaux et al.[26] Interfacial shear strength was found to reach values of 28 Mpa, confirming strong interactions between the fiber and matrix. The interfacial behavior in carbon fiber reinforced cyanate composites was investigated by Marieta et al.[27] The authors observed that the interfacial shear strengths obtained by this technique were comparable with results of interlaminar shear strength measurements corresponding to woven composite plates.

The interfacial compatibility between sugarcane bagasse fiber and polyester upon chemical modification was investigated by García-Hernández et al.[28] The authors used the fiber pull-out technique to evaluate interfacial shear strength. They observed that interfacial shear strength (IFSS) was higher for treated composites when compared with untreated ones. The interface effects on the mechanical properties and fracture toughness of sisal fiber reinforced vinyl-ester composites were investigated by Li et al.[29] The interfacial shear strength was calculated by single fiber pull-out technique. It was observed that after fiber surface treatments by using different chemical agents, the IFSS was greatly improved. The $KMnO_4$ treatment resulted in the best effect, and the IFSS between permanganate-treated sisal fiber and vinyl-ester resin was almost three times that of the untreated fiber, followed by silane-2 and DCP treatments. For silane-1 treated sisal fibers, IFSS was similar to that of untreated sisal fiber reinforced vinyl-ester (see Figure 20.5).

20.2.1.1.4 Influence of the Embedded Length

When specimens are prepared according to the preceding procedure, various lengths of embedment are obtained. A number of studies emphasize the importance of debonding force F_d, versus embedment length l_e. It is thus necessary to perform several tests with a variety of embedment lengths. A practical limit is dictated by the tensile strength of the fibers at their current length.

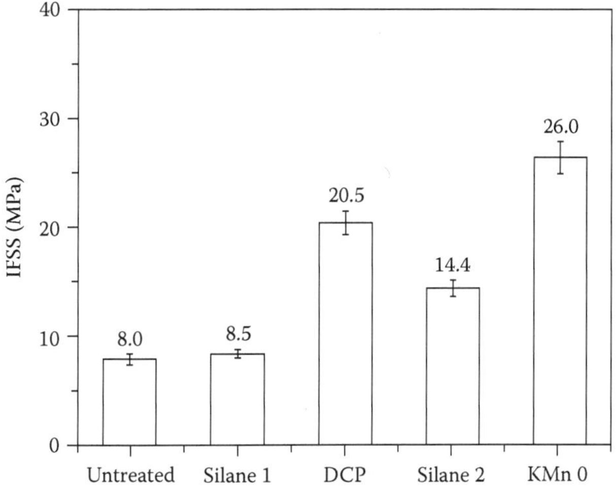

FIGURE 20.5 Variation of interfacial shear strength with chemical modification.

Kim et al.[30] conducted a parametric study on carbon fiber–epoxy matrix composites. The stress field arising in the fiber pull-out test was analyzed by means of finite element method. It was found that within the composite containing a coating layer, the interfacial shear stress is always higher at the fiber/coating interface than at the coating/matrix interface, which is an indication of debond initiation at the former interface in preference to the latter interface.

20.2.1.1.5 Limitations of Single-Fiber Pull-Out

The drawback of the single fiber pull-out test is that it involves only a single fiber. Because the role of neighboring fibers is not taken into account, the thermal stresses and the polymer morphology around the fiber are not the same as in a real composite. Real composites contain multiple fibers and the pull-out fiber is surrounded by a composite medium. In single fiber model composites, the effect of the composite medium surrounding the pull-out fiber has been ignored. Therefore, owing to the influence of the composite medium, surrounding the pull-out fiber, the interfacial debonding process in multi-fiber composites, and the interfacial properties obtained therefrom, would very likely deviate from those of the single-fiber composite pull-out test.

Hampe and Marotzke[31] remedied the previous limitation in an experiment in which pull-out test was conducted with single and multi-fiber samples. Pull-out tests of E-glass fibers in polystyrene and polycarbonate were performed using samples, where the investigated fiber was surrounded by 0–3 other near fibers. The authors found that neighboring fibers can increase the pull-out forces by a factor of three and the interfacial toughness by a factor of four.

Qui and Schwartz[32] developed a method (i.e., single-fiber pull-out from a microcomposite to study the fiber–matrix interface in composites). This method provides a real environment of a fiber pulled out from a real composite. It was observed that the interfacial shear strength would decrease as the fiber volume fraction was increased. In addition, the composite medium was expected to affect the stress transfer in a multi-fiber pull-out test. Thus, research on the multi-fiber composite pull-out problem is critical to understand the stress transfer in real composites. Unfortunately, only little work has been done on these yet unexplored territories.[33] Moreover, real multi-fiber composites are inhomogeneous. The stress transfer between the pull-out fiber and the neighboring matrix is likely to depend on the local fiber volume fraction of the neighboring fibers. This also has not been analyzed yet.

In another study, Fu et al.[34] conducted theoretical analyses of stress transfer in single and multi-fiber composites. A perfectly bonded fiber–matrix interface was assumed. A three-cylinder model, which was used to analyze the stress transfer between parallel fibers in a composite (in an approximate manner) and the effect of interface thickness on the residual thermal stresses in a model composite, was adopted in this study to analyze the multi-fiber pull-out problem. The influence of local fiber fraction and the effects of neighboring and remote fibers on the stress transfer were discussed. When the fibers of the three-cylinder model were short, a fiber length factor for the composite modulus was introduced and its effect on stress transfer was analyzed. The results of the study demonstrated that the neighboring fibers play a role in determining stress transfer, whereas the influence of the remote fibers on the stress transfer is negligible.

Another serious limitation inherent in this procedure arises when very fine fibers are used, as is often the case when the reinforcing elements of the composites are carbon, glass, or aramid fibers. If the required pull-out force exceeds the breaking strength of the fiber, the latter will rupture before pull-out occurs. This restriction can be expressed in terms of a critical embedment length l_c, which can be obtained by defining it in terms of the pull-out requirement

$$l_c = F_p/\pi D\tau \tag{20.2}$$

The expression for fiber tensile breaking stress is

$$F_t = \sigma_f \pi D^2/4 \tag{20.3}$$

where σ_f = ultimate fiber tensile stress at break. In order for pull-out to occur instead of rupture, the applied force cannot exceed F_t; therefore, when we substitute its equivalent from the preceding equation, we obtain a limiting expression for l_c:

$$l_c = D\sigma_f/4\tau \tag{20.4}$$

where

D = diameter of the fiber
τ = interfacial shear strength
σ_f = ultimate fiber tensile stress at break.

An embedded length greater than l_c will result in tensile failure in the portion of the fiber that is not embedded.

Because it is common for the reinforcing fibers to have a diameter of ~ 10 μm and a tensile strength of 2–4 GPa and for the interfacial shear strength to be in the range 10–70 MPa, the embedded length of fiber cannot exceed the range 0.07–1 mm. It is extremely difficult to keep the embedded length down to such small values, and to handle the test specimens and measure shear strengths.

However, a fundamental problem limits the usefulness of any method that requires partial immersion of the fiber in a pool of resin. If the liquid resin appreciably wets the fiber surface, an elevated meniscus will form around the fiber. Unless the liquid is exceedingly dense, the height of the meniscus can be of the order of 1 mm, so that it is virtually impossible to keep the embedded length in the sub-millimeter range under such circumstances. In addition, the formation of this meniscus produces another potential problem. The resin coating is thinner in the meniscus region and may rupture before debonding. As a result, a cone of resin is left behind on the surface of the fiber after pull-out, indicating cohesive failure of the epoxy resin before adhesive debonding at the interface. Although this premature rupture does not invalidate a subsequent shear strength measurement, it does require the inspection of every specimen after debonding to determine the true embedded length. Furthermore, if the two processes occur very closely in time, the possibility exists that the load recorded during pull-out may include the load to fracture the resin cone.

As a result of the preceding problems, another approach has been developed that appears to deal adequately with the limitations of the conventional pull-out method. This method is called the micro-bond technique because it uses only a very small amount of resin for each test in the form of a droplet deposited on the fiber. The micro-bond technique has been performed extensively for purposes of studying the interfacial adhesion and between fibers and matrix.[35]

20.2.1.2 Micro-Bond Test

20.2.1.2.1 Experimental Technique for Thermoset Resins

The procedure involves the deposition of a small amount of resin onto the surface of a fiber in the form of one or more discrete micro-droplets. The droplets form concentrically around the fiber in the shape of ellipsoids and retain their shape after appropriate curing. Once cured, the micro-droplet dimensions and the fiber diameter are measured with the aid of an optical microscope. The embedded length is fixed by the diameter of the micro-droplet along the fiber axis, which is dependent on the amount of resin deposited on the fiber. The practical minimum limit for the embedded length using this technique is 40 μm. Because the liquid drop still retains most of its tendency to minimize its surface area and has very little mass, it forms only a slight meniscus along the fiber. This represents only about 1% of the measured embedded length and therefore can be ignored.

One end of each fiber specimen is glued to a metal tab, which can easily be connected to a load cell. The assembly is then taped to a template and the liquid resin applied to the fiber under a stereo microscope using a fine glass applicator. A small amount of resin on the tip of the applicator is allowed to come into contact with the fiber. Upon retraction of the applicator, some of the resin remains, forming the elliptical droplet around the fiber. Usually, 2 droplets are placed on each fiber, a few millimeters apart, and about 10 fibers are mounted on each template. The collections are cured as required.

The fiber specimens are pulled out of the micro-droplets at a rate of 1 mm min^{-1} using a tensile tester. To grip the droplet (illustrated in Figure 20.6), the micro-vise is mounted on the crosshead of the tester. The plates of the vise are first positioned just above the droplet, and the slit is narrowed until the plates just make contact with the fiber. As the shearing plates continue to move downward, they make contact with the resin and a downward force is exerted on the droplet. The shearing force at the interface is then transferred to the fiber through the fiber–matrix interfacial bond and is recorded by the load cell. When the shearing force exceeds the interfacial bond strength, pull-out occurs and the droplet is displaced downward along the axis of the fiber.

As a general rule, sheared specimens show little or no distortion of the droplet and only occasional traces of debris on the fiber surface. The absence of large resin fragments on the surface of the fiber after pull-out demonstrates that cohesive failure of the matrix does not occur. In practice, two micro-droplets are deposited on each fiber specimen about 5 mm apart. After the lower droplet has been freed in the pull-out test, the slit is opened, the plates are moved up, positioned above the second droplet, and the process is repeated. Thus, two pull-out results are obtained from one fiber specimen. Although more droplets could conceivably be deposited and tested on each fiber, experience has shown that it then becomes difficult to locate and identify droplets and if the fiber itself ruptures, all droplets on it are lost. Therefore, two droplets per fiber are a practical loading.

Craven et al.[36] evaluated the interface of *Bombyx mori* silkworm silk–epoxy composite by micro-bond test. Cured resin droplets were sheared from the silk fibers by the knife-edged jaws of a micro-vice. The micro-bond test has been used in this work due to its suitability for fibers that can carry only low loads. Silk is a strong fiber, but its individual fibers are fine and therefore have a limited load bearing capacity. The microbond technique involves depositing a micro-droplet of resin onto a fiber. After the resin has cured, it is sheared from the fiber by two parallel plates attached to a micro-vice. They observed that the mean interfacial shear strength of silk/epoxy composite is 15 MPa (= 2 MPa). It was found that silk fibers could offer useful reinforcement to an epoxy system due to their high tensile strength and extensibility.

In another study, Luo and Netravali[37] characterized the interfacial properties of green composites made from pineapple fibers and poly (hydroxy butyrate-co-valerate) resin. The IFSS value was found to be very low. The IFSS depends mainly on two factors: mechanical locking and chemical bonding. In the case of pineapple fiber–PHBV resin, H-bonding is possible between ester group on the resin and –OH group on the fiber. However, because of the hydrophobicity of methyl and ethyl pendant groups in the resin, H-bonding probability is low. As a result, IFSS is mainly attributed to high surface irregularity of pineapple fibers and the resulting mechanical interaction. The high viscosity of the resin also seemed to preclude much mechanical bonding. Another interesting study on the IFSS of green composites was conducted by Lodha and Netravali,[38] in which the authors prepared green composites comprised of ramie fiber and soy protein isolate.

Joly et al.[39] studied the physical chemistry of the interface in PP/cellulosic fiber composites. In this study, two treatments were performed on the fibers to decrease the hydrophilic nature of their surface and to make them more compatible with the hydrophobic matrix, by grafting hydrocarbon chains onto a certain number of hydroxyl groups of cellulose fiber. The non-treated composites showed varying interfacial shear stress, whereas the treated composites appeared to have constant debonding forces regardless of the fiber diameter and the embedded surface.

The relationship between glass fiber/polyester and glass fiber/epoxy interface and interlaminar properties of marine composites was investigated by Baley et al.[40] using the micro-debond technique. The influence of cure cycle on bulk properties was also studied. The authors also observed an increase in interfacial shear strength and interlaminar shear strength with cure temperature.

In an innovative approach, Zhandarov and Mader[41] used an indirect estimation of the maximum force recorded in pull-out and micro-bond techniques to calculate the interfacial strength. This method uses the relationship between maximum force and embedded length to determine critical energy release rate and interfacial friction in debonded regions. The interfacial parameters were determined for several fiber–polymer systems, and the results were found to be in good agreement with a similar stress-based approach.

The effect of atmospheric pressure He/air plasma treatment on interfacial shear strength of aramid fiber reinforced epoxy composites was investigated by Hwang et al.[42] The authors used a micro-bond test to calculate IFSS. They observed that plasma treatment for 60 sec increased the interfacial shear strength by 104% over that of untreated. In a similar study Qiu et al.[43] investigated the effect of atmospheric pressure He + oxygen plasma on interfacial shear strength of ultra high modulus polyethylene fibers. They observed that the treated polyethylene fibers had a 65–104% increase in interfacial shear strength over that of control sample.

A problem that is associated with the micro-bond technique is that the maximum debonding force value is influenced by interfacial friction in already debonded regions, and, therefore, these parameters are not purely "adhesional" but depend, in an intricate way, on interfacial adhesion and friction. As a result a new method was developed by Zhandarov et al.[44] using the relationship between maximum force and embedded length to measure fiber–matrix interfacial adhesion and friction. The micro-mechanical tests were modeled for three types of specimen geometries (cylindrical specimens, spherical droplets, and matrix hemispheres in the pull-out test) with different levels of residual thermal stresses and interfacial friction. An interesting result was that the "ultimate IFSS" was not always equal to the "local" bond strength.

20.2.1.2.2 Limitations of the Micro-Bond Test

The specimen preparation for the micro-droplet test whereby a single fiber is pulled out of a small droplet of resin suffers from several difficulties. For instance, the shape of the droplet affects the reliability of the data. Symmetric, round droplets are easier to test and analyze than droplets with flat surfaces, produced when the specimens solidify on a flat substrate. Also the size of the droplet is critical. If the length of the droplet exceeds a critical value, the fiber will fracture before debonding and pull-out. An additional complication with some thermoset materials is that the anticipated curing characteristics may not manifest themselves in a droplet of small size, and thus comparison on a micro-structural level between micro- and macro-specimens may not be possible. Another defect is that this test is not applicable to matrices that are soft.

The previously discussed limitations have been dealt with in a work by Mak et al.[45] In this study, a method was found to form symmetric, regular droplets of 25–35 μm in diameter on glass and carbon fibers with varying surface treatment. The resins used consisted of two epoxies toughened with thermoplastic polyether sulfone (PES). A specially prepared platinum disc was used instead of a variable micro-vice as the test aperture. Thermoplastic reinforced epoxy resin blends were chosen, as the phase separation characteristics of such blends are known to be complex and to vary directly with curing temperature. The shear strength values were calculated from the slope of a plot of the pull-out load versus the embedded length for each system, and experimental results did not have significant variation. This demonstrated that the improvements made to the specimen preparation process were successful.

20.2.1.3 Micro-Indentation Test/Micro-Compression Test

The micro-indentation test was initially developed for fiber reinforced ceramics but has been extended to the other fiber–matrix systems. It is also known as the fiber push-out test. This is the only single-fiber test, which is able to analyze actual composite specimens. The use of a real composite allows a more realistic simulation of thermal stresses, polymer morphology, and the influence of neighboring fibers. Nevertheless, the more realistic testing conditions of the micro-indentation method make it an attractive new technique for many researchers.[47] In this method, a compression force is applied on a single fiber in a well-prepared specimen of a real composite. Desaeger and Verpoest[48] performed the micro-indentation test on different kinds of fiber reinforced polymer composites (carbon and glass fibers embedded in thermoplastic and thermoset matrices.) The compression loads on the fiber will induce, due to difference in elastic properties between the fiber and matrix, high shear stresses at the interface. These shear stresses cause de-cohesion at the interface. Thin slice push-out test has emerged as the de facto standard in evaluating the interfacial properties of metallic and intermetallic matrix composites. Chandra and Ghonem[49] performed thin-slice push-out tests on silicon carbide fiber reinforced titanium matrix composites at various test temperatures with different processing conditions. The authors observed that reaction growth is due to transformation of both coating and matrix. Push-out tests conducted on heat treated specimens revealed that effect of thermal exposure on interfacial properties is predominantly due to exposure temperature. Time of exposure was found to have a secondary role.

In another interesting study Drissi Habte et al.[50] utilized micro-indentation technique to characterize *in situ* mechanical properties and interfacial behavior of Hi-Nicalon/BN/silicon nitride ceramic matrix composites. The study mainly revealed the wide range of application this technique has in analyzing interfacial properties of ceramic matrix composites.

The fracture toughness of glass fiber/epoxy interface using slice compression test (SCT) was evaluated by Tsay et al.[51] The slice compression test is used to evaluate inter-facial properties by loading/unloading a specimen of the epoxy resin containing a single SiO_2 glass fiber between two plates where one has low modulus and the other has high modulus. The interfacial debonding was monitored by using a microscope and a video camera. Energy release rate was calculated by finite element analysis. From the in-situ observation, the authors found that the interface fracture initiates when the radial stress around the fiber changes from compression to tension due to the Poisson's effect. The length of the crack was found to be proportional to the stress as the load increases. It was also found analytically that the energy release rates remain constant once the interface fractures, independent of the initial crack length.

20.2.1.3.1 Experimental Technique

All the micro-indentation tests are usually performed with the aid of an apparatus called the Nano-indenter. The apparatus can record compression loads from a few μN up to a few mN, which can be applied on a single fiber in a well prepared real composite specimen. The equipment has two parts: an optical microscope linked to a camera and a video recorder and a head with the indenter mechanism. The force is applied with a rigid indenter, which is pushed down on a cross-sectional area of the fiber in a thin composite plate. For the test, the specimen is fixed on the table and the places where the indentations have to occur are chosen and memorized by the computer. Afterward, the specimen is moved automatically under the indenter head and the indentations are performed. From the indentation results, the interfacial shear load can be calculated.

20.2.1.3.2 Test Procedure

The test procedure adopted with the Nano-indenter consists of the following steps:

1. Perfect polishing of the specimens
2. Placing of the specimen in the machine one night before the start of the test
3. Calibrations of the distance between the microscope and the indenter head mechanism
4. Selection of a set of fibers having approximately the same diameter and the same environment
5. Video recording of the fibers before the indentations
6. Selection of the different indentation procedures, which have to be applied on the different fibers
7. Fiber indentations
8. Observations of the tested fibers and comparison with the fiber state before the test
9. Noting whether decohesion occurs or not for the different maximum forces selected

The polishing step is crucial for good evaluation of the interface properties. As the observation of decohesion is done by using an optical microscope, the tested surfaces have to be as homogenous as possible. The interface strength can be derived from the decohesion load from shear lag analysis. A first approximation of the interface shear strength out of the results of the micro-mechanical tests is derived from formulae based on shear–lag analysis. The interface debonding shear strength is equal to:

$$\tau_{deb} = nF_{deb}/2\,\pi r^2 \tag{20.5}$$

where

$$n^2 = 2\,E_m/E_f\,(1 + \nu_m)\ln(2\pi/\sqrt{3}\,V_f) \tag{20.6}$$

and

F_{deb} = decohesion force
r = fiber radius
E_f, E_m = Young's modulus of fiber and matrix, respectively
ν_m = Poisson ratio of matrix
V_f = local fiber volume fraction

The main advantage of the micro-indentation test is that it uses a section of a real composite, so that it can readily be used to determine the effect of adjacent fibers. In addition, frictional shear stress can be measured when thin composite sections are used. *In situ* interfacial characterization is also possible. The shear lag analysis approach, though simplified, used for the evaluation of interfacial properties gives reliable results.

20.2.1.3.3 Limitations of the Micro-Indentation Test

One major disadvantage of the technique is the subjectivity in the determination of the load needed to obtain the decohesion. Another disadvantage linked to the use of the Nano-indenter, as the test equipment, is that very weak or strong interfaces cannot be evaluated. The first one is due to polishing limitation and decohesion detection procedure. The second one is due to the maximum load available with the test equipment.

The other limitations of this method are that because the method relies on numerical methods to obtain interfacial shear strength, the result is critically dependent on the assumptions used to describe the stress field around the fiber while under pressure. Accurate interpretation of data from tests in which the fibers are not aligned vertically is another problem. This is because the relative frictional sliding resistance is dependent on the fiber alignment. The fibers must be selected carefully to avoid fiber bundle effects within the sliced composite, as this would also contribute to scatter in the results. Furthermore, this test cannot be used for polymer fibers such as Kevlar. Random errors of up to ± 30% have also been observed in this method.

20.2.1.4 Fragmentation Test/Single-Fiber Composite Test

In the single-fiber composite test (SFC), a single fiber is embedded in a polymer and broken into small pieces. This is the most realistic test from the point of view of the interfacial pressure. The fiber is neither pushed nor pulled directly, and so fiber Poisson effects are similar to that occurring in a fiber composite. Unlike the other methods, it produces only one result for the interfacial shear stress, which is the average for the many fragments produced. The use of Raman spectral lines has made it possible to estimate fiber strains and thus the interfacial shear stress can be estimated directly from the fiber Young's modulus. This makes the test much more powerful. The results obtained with these techniques have been used to assess the efficacy of fiber surface preparation techniques.

20.2.1.4.1 Experimental Techniques

In this test, a single fiber is embedded in a dumbbell shaped tensile coupon, which in turn is subjected to a tensile load. Depending on the level of fiber–matrix adhesion, tensile forces are transferred from the matrix to the fiber, which causes the tensile stresses along the fiber to build up. The tensile stresses are high around the center region of the fiber beyond the transfer lengths of the system. As the loading process continues, the fiber fractures within this center region. When the length of the fiber fragments have reduced to a critical value, no further breakage will occur but the interface will fail in shear. This final fiber fragment length is referred to as the fiber critical length (l_c). It is commonly accepted that this final fiber fragment length is a good indicator of the ability of the interface to transmit loads between the two constituents and a measure of (l_c/d) can be used as an indicator of the fiber–matrix strength. The IFSS is calculated according to the equation developed by Kelly and Tyson

$$\tau = \sigma_f \, d/2\, l_c \tag{20.7}$$

$$\tau = [\sigma_f/2\alpha].K \tag{20.8}$$

where d is the diameter of the fiber, l_c is the average critical length of the fiber, σ_f is the fiber tensile strength at a gauge length equal to the mean fragment length, and K is a coefficient that depends on the variation of the fragment length.[52] If the fragment length varies between $l_c/2$ and l_c, K = 0.75 can be taken as a mean value.

The SFC test originally proposed by Kelly and Tyson for metal–matrix composites provides abundant statistical information from only a few specimens, as well as the interfacial failure modes and the IFSS.

Because the SFC technique considers only an interface around a single fiber, it cannot match well with the interfacial properties of real composites in which the fibers interact with each other. A composite specimen with a small number of fibers was, therefore, developed to model the composites more practically. This composite, called the multi-fiber composite (MFC), is generally prepared with equal inter-fiber separation. This is referred to as the regular MFC. Thus, the MFC test has been used for evaluating the interfacial properties with the fiber to fiber interaction as a function of the inter-fiber distance.

In a recent study,[53] a gradual MFC is proposed, which is characterized by gradually changing the inter-fiber spacing. In the regular MFC with small inter-fiber spacing, the fiber to fiber interaction is strong. On the other hand, in the SFC, no interaction occurs between fibers. It was expected that, therefore, in the gradual MFC test both fragmentation properties obtained in the SFC and the regular MFC tests might appear at the same time. The results showed that as the inter-fiber distance increased, the aspect ratio of broken fibers decreased, whereas the fiber–matrix interfacial shear strength increased. When the reciprocal of the inter-fiber distance was taken, both the aspect ratio and interfacial shear strength were found to show a saturated value. This means that the gradual MFC indicates an upper bound in aspect ratio and a lower bound in interfacial shear strength, whereas the SFC composite indicates a lower bound in aspect ratio and an upper bound in interfacial shear strength. It was found that this fragmentation test could be a new method for composite evaluation, because reducing the difference between these two bounds is effective for composite strengthening.

A new methodology for the prediction of stress transfer in the single fiber fragmentation test, known as the plasticity effect model, has been used by Baker and Jones.[54] A new data reduction technique, known as the cumulative stress transfer function (CSTF technique), which takes the different damage events observed during the fragmentation into account, was observed to obtain a measure of fiber–matrix adhesion from the fragmentation test. The effect of carbon fiber surface treatment on the interface of micro-composite properties was studied using fiber–CSTF technique. It was found that the CSTF technique could predict the fiber–matrix adhesion in a single fiber fragmentation test more accurately than the conventional technique.

Tripathy et al.[55] used the single fiber fragmentation test for determining the critical fiber length and interfacial shear strength of jute fiber reinforced epoxy composites. Four kinds of jute fibers, namely untreated jute fiber, sliver jute filament, bleached jute fiber, and mercerized jute fibers were used. The bleached jute fiber was found to have maximum interfacial adhesion with the epoxy resin.

Zafeiropoulos et al.[56] characterized the interface in flax fiber reinforced polypropylene composites by means of a single fiber fragmentation test. Flax fiber was modified by means of two surface treatments: acetylation and stearation. The authors observed that acetylation improved the stress transfer efficiency at the interface. Stearic acid treatment was also found to improve the stress transfer efficiency but only for lower times. They also found that stearation for longer durations deteriorated the interface. This was attributed to decrease of fiber strength upon treatment of longer duration and the fact that excess of stearic acid acted more as a lubricant than as a compatabilizer.

In an interesting study, Valadez-Gonzalez et al.[57] reported that a maximum quantity of silane coupling agent can be deposited onto henequen fibers. The authors analyzed the tensile and micro-mechanical properties of the composite material. They observed that the interaction between the fiber and matrix was stronger when the fiber surface was physically modified and combined with silane coupling agent. The interfacial shear strength, which was determined from a single fiber fragmentation test, was observed as increasing, reaching a maximum at the intermediate concentrations of silane, and then decreasing with an increase in silane concentration.

An innovative technique was developed by Cabral-Fonseca et al.[58] to determine interfacial properties of two opaque glass fiber/polypropylene systems via fragmentation tests on single filament model composites. Fragmentation tests require the fiber inside the composite to be completely aligned in the loading direction. Because polypropylene is non-transparent, it is not possible to ensure this condition. Therefore, a novel technique was developed to determine the inclination of the filaments embedded in the composite.

The extent of fiber–matrix interfacial adhesion is traditionally evaluated by means of stress based parameter. Recently, the concept of interfacial energy parameter has come into play. Zhou et al.[59] presented an energy balance scheme proposed for the analysis of initial interface debonding, which occurs at fiber breaks during a fragmentation test. The effects of thermal residual stress in the fiber and of friction in the debonded area were incorporated in the energy balance model. An interesting advantage of the energy balance approach to fragmentation test is the fact that it is possible to quantify the interface adhesion of composite systems that never reach saturation. The authors also presented extensive single fiber fragmentation data regarding interface crack initiation regime using sized and unsized E-glass fibers embedded in epoxy matrix.

The interfacial properties and micro-failure degradation mechanisms of bioabsorbable fiber reinforced poly-L-lactide (PLLA) composites were studied by Park et al.,[60] using a fragmentation technique and a non-destructive acoustic emission (AE) technique. The interfacial shear strength between bioactive glass fiber and PLLA was found to be much higher than chitosan fiber/PLLA systems. In the case of bioactive glass fiber, AE energies in tensile fracture were much higher than those after degradation under both tensile and compressive tests. Similar experiments to measure interfacial shear strength were conducted by the same authors[61] for plasma-treated biodegradable poly(p-dioxanone) fiber reinforced poly(l-lactide) composites.

The interfacial shear strength of electrodeposited carbon fiber reinforced epoxy composite was determined and compared using fragmentation and acoustic emission techniques by Park et al.[62] In electrodeposited treated cases, the number of fiber fractures measured by both conventional optical microscopic and acoustic emission methods was more than those of untreated cases. The signal number measured by acoustic emission was also smaller. This was attributed to low carbon fiber fracture signals and the damping effect of ductile epoxy modulus.

20.2.1.4.2 Limitations of the Fragmentation Test

In the SFC, the fibers are broken into fragments by straining the polymer to about $3\varepsilon_{fu}$, where ε_{fu} is the fiber breaking strain. This produces fragments ranging from l_c to $2l_c$, where $2l_c$ = fiber critical length.

The mean interfacial shear stress required to cause this is,

$$\tau_{iu} = \sigma_{fu}/2S_c \tag{20.9}$$

$$S_c = l_c/r \tag{20.10}$$

where

S_c = critical fiber aspect
$2r$ = fiber diameter

The mean fiber fragment length l_m is approximately $3l_c/2$, so a mean interfacial strength can be estimated from l_m. The dependence of mean interfacial shear strength on mean fragment length has long been known to be incorrect because the IFSS has an elastic as well as a supposed plastic component, and the preceding equation is not generally applicable to many practical composite systems. Again, the measurement of l_m is not an easy task because l_m can be exceedingly small.

The failure strain of the matrix must be much larger than the failure strain of the fiber to promote multi-fragmentation of the fiber. This requires the use of matrices, which can undergo large deformations. Consequently, commercial resins utilized in actual composite systems, which typically have low strains to failure cannot be used for this test. Therefore, the interfacial shear strength determined is not directly applicable to the actual composite system.

Another problem is that the embedding matrix can inhibit fiber fracture, which initiates from surface flaws. The magnitude of this effect is dependent on the type of matrix or embedding resin. Therefore, it is important to utilize the actual fiber and resin of a given composite system to determine the interfacial shear strength.

Although the energetics of interface failure for fragmentation[63] may be analyzed quite simply, much confusion exists in the energy dissipation process. In this method, uncertainties will always exist regarding how to partition between interfacial fracture and other dissipative processes such as vibrations.

Another aspect is that friction plays an important role in the debonding process and this is governed by two additional unknowns (i.e., the coefficient of friction, μ, and the pressure across the interface, P). Although some progress has been made with this, values for debonding rely rather heavily on the correctness of the assumptions about μ and P. Thus, according to Piggott,[64] fiber fragmentation as a method of measuring an interface property does not produce clear-cut results.

20.2.2 Comparison of Different Techniques

Piggott[65] used four methods: the single fiber pull-out, micro-debond, micro-compression, and the fragmentation test for the measurement of the failure processes in the carbon-epoxy interface. It was observed that the data obtained from the pull-out tests and fragmentation tests are dependent upon friction. In the case of the micro-debond test, the acquisition of basic data was not straightforward owing to the non-cylindrical shape of the bead. The micro-compression test had the same difficulty because data reduction comes from a finite element analysis, which is apparently based on elastic effects, so that friction is not allowed for. It was seen that the fragmentation tests were roughly one-half of the pull-out results.

The differences between the results with the different methods are due, at least in part, to the different balances between friction and debonding stress. The lower results for fragmentation suggest that friction is playing a major role, and that the fiber fragments could perhaps be nearly fully debonded. In the case of the micro-compression test, results were even lower than the fragmentation values.

Pull-out and push-out measurements were performed on glass fibers in an epoxy resin to determine the dependence of bond strength on test temperature and on fiber surface treatment by Mader et al.[66] A comparative analysis of the two techniques was performed to elucidate elementary processes of polymer-fiber debonding and to determine energy values for adhesional bonds. Differences in bond strength values for pull-out and push-out tests were attributed to failure mechanisms that were either interface-controlled or matrix-controlled. The evidence for the different failure mechanisms characteristic of the two test techniques was provided by an estimation of failure parameters, such as the activation energy for debonding.

In another study, Valadez-Gonzalez et al.[67] used both the single fiber fragmentation test and pull-out to characterize the interface of henequen fiber reinforced HDPE composites. Slight discrepancies were observed between the results obtained from the two micro-mechanical techniques for fibers that were subjected to a pre-impregnation process. From the pull-out experiment, higher interfacial shear strength values were obtained than those for the single fiber fragmentation test. Such discrepancies were attributed to the differences in the experimental configurations in arrangements and mechanical behavior of the fiber, which was impregnated with the matrix. The results obtained from the SFC test was in better agreement with the effective mechanical properties.

Interfacial bond strength between epoxy resin and glass fiber was studied by Mader et al.[68] using the pull-out and push-out techniques. For untreated fibers, these micro-mechanical tests gave similar values of the local interfacial shear strength and critical energy release rate. In the case of treated fibers, both tests showed considerable increase in bond strength. For the modified fibers, however, the pull-out test gave greater values of both interfacial parameters than the push-out test — a result attributed to the different modes of interfacial bonding.

Marshall et al.[69] used the micro-debond and fragmentation tests to measure the properties of interface of carbon/epoxy composites. They reported close agreement for the results obtained from the two techniques.

The fiber–matrix interface was characterized by a new parameter known as the adhesional pressure by Pisanova et al.[70] Adhesional pressure is due to molecular forces acting across the interface. This allows relating it to the work of adhesion, which is performed against normal interfacial forces. The authors investigated the relation between fundamental adhesion (work of adhesion by means of wetting) and practical adhesion (from micro-mechanical tests). Adhesional pressure was calculated using an algorithm based on variation of mechanical analysis of the microbond test. The work of adhesion showed a linear relationship to adhesional pressure values. The advantage of this technique is that it can characterize adhesion for real conditions of the composite formation including irreversible adhesion.

These authors have also studied the acid–base interactions and covalent bond formation at the interface and its contribution to the work of adhesion and adhesion strength.[71]

An experimental study was conducted by Afaghi-Khatabi and Mai[72] to examine the influence of cyclic loading on the interfacial properties of carbon fiber/epoxy resin composites. Two composite material systems having the same fibers and epoxy matrix, but with different fiber surface treatments, oxidized-sized and untreated, respectively, were used in this study. The existence of different interphases in these materials was studied using dynamic mechanical analysis (DMA) techniques. Fatigue tests were conducted at various load levels and then by using DMA and C-scan techniques, the fiber/matrix interfacial degradation was characterized. The results indicated that cyclic fatigue loading did affect the fiber/matrix interfacial properties of the composite laminates. Specimens with oxidized/sized fibers showed less interfacial degradation compared with laminates with untreated fibers.

In an innovative study, the characterization of fiber–matrix interphase was performed by dynamic interphase loading apparatus (DILA). Foley et al.[73] used a piezoelectric transducer that allowed the input of a variety of displacements and loading rates (quasi-static to 50 mm/s) to the indenter. Transient force and displacement values, recorded during the test, were then used to determine the average shear strength and energy absorbed during debonding and frictional sliding during the micro-debonding process. An E-glass/vinyl ester composite was tested under single micro-debonding as well as fatigue loading. Test results showed that the strength and energy-absorbing capability of the interphase was sensitive to loading rate.

In an interesting study by Pothan and Thomas,[74] dynamic mechanical analysis has been used to analyze the interfacial adhesion in banana fiber reinforced polyester composites. The authors observed an additional peak in their tan delta curve and have reported it to be due to an additional polymer layer, which probably is the interphase. The authors also observed an increase of storage modulus with treatment. Storage modulus is found to be directly proportional to the interfacial strength of composite. In addition, the T_g was found to shift to higher temperatures for treated composites indicative of good interfacial adhesion.

The characterization of jute–epoxy composites by means of fatigue loadings was performed by Gassan and Dietz.[75] The loss energy was demonstrated to be a sensitive tool to characterize the nature of fiber–matrix adhesion. The loss energy for composites with poor adhesion between fiber and matrix resulted in significantly higher amounts of consumed energy during a single stress-strain loop than those for composites containing well-bonded fibers.

Despite the enormous amount of micro-mechanical techniques available for interfacial characterization of composites, none of them are industry-friendly because they are viewed as time-consuming and inefficient. In a novel approach, Thomason[76] made use of deriving values for the interfacial shear strength and a fiber orientation factor from a simple combination of the composite tensile stress–strain curve and the fiber length distribution. Despite the recent wealth of activity in the development of micro-mechanical test techniques, little follow-up on this older technique has taken place. In this study, the author explored this analysis by its application to injection molded glass–fiber reinforced thermoplastic composites produced using three matrices (polypropylene, polyamide 6,6, and polybutyleneterephthalate) and containing different levels of glass fiber. The author also described how this analysis can be extended to obtain another important micromechanics parameter: the fiber stress at composite failure. This technique was considered cost effective and less labor intensive. In addition, the data were obtained from "real" composites.

As an extension of the experiment, the author[77] also looked into the mechanical property and fiber length distributions of glass fiber reinforced polypropylene containing different levels of glass fiber, and this data was used as input for measuring micro-mechanical parameters. The author is of the opinion that this method deserves further investigation as a screening tool in composite system development programs (Figure 20.6).

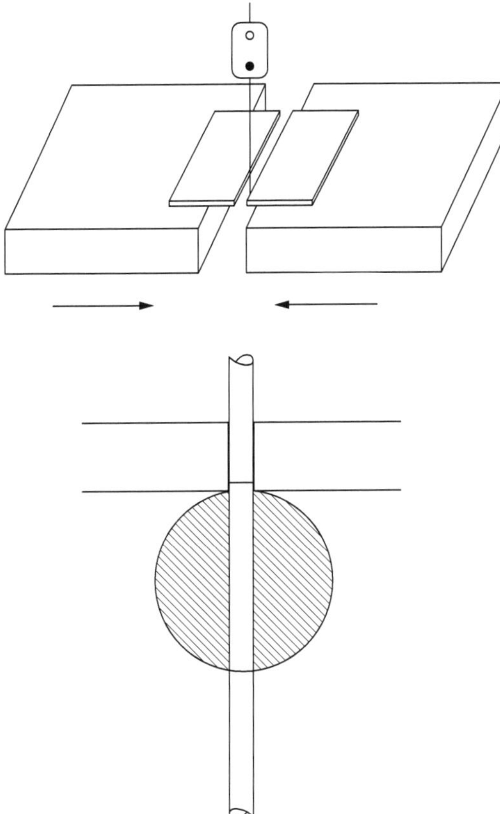

FIGURE 20.6 Arrangement for shear debonding (top) and enlarged schematic of a resin droplet on a fiber under the shearing plates (bottom).

20.3 CONCLUSIONS

The mechanical properties of fiber reinforced composites are dependent upon the stability of interfacial region. Thus, the characterization of the interface is of great importance. All the micro-mechanical techniques evaluate the bonding strength between the fiber and matrix. The pull-out test is considered to be a good method of evaluating the interfacial shear load as it can directly measure the interfacial shear strength between the fiber and matrix, independent of their properties. The interfacial shear strength is a critical factor that controls the mechanical properties and interlaminate shear strength of composite materials. Of all the techniques, micro-indentation test is the only one to be performed on real composite specimens.

REFERENCES

1. Pukanszky, B. and Fekete, E., *Adv. Polym. Sci.,* 1999, 139.
2. Schreiber, H.P. in *The Interfacial Interactions in Polymeric Composites,* Akovali, G., Ed., Kluwer, Amsterdam, p. 21.

3. George, J., Sreekala, M.S., and Thomas, S., *Polym. Eng. Sci.,* 41, 9, 1471–1485, 2001.
4. Rials, T.G., Wolcott, M.P., and Nassar, J.M., *J. Appl. Polym. Sci.,* 80, 546–555, 2001.
5. Kim, J.K., and Hodzic, A., *J. Adhesion,* 79, 4, 383–414, 2003.
6. Bergeret, A., Bozec, M.P., Quantin, J.-C., Crespy, A., Gasca, J.-P., and Arpin M., *Polym. Composites,* 25, 1, 12–25, 2004.
7. Pukanszky, B., *Eur. Polym. J.,* 41, 4, 645, 2005.
8. Zhandarov, S. and Mader E., *Comp. Sci. Tech.,* 65, 1, 149, 2005.
9. Fu, S.Y., Lauke, B., Zhang, Y.H., and Mai, Y.-W., *Composites, Part A,* 36, 7, 987, 2005.
10. Jacob, M., Joseph, S., Pothan, L., and Thomas, S., *Comp. Interf.,* 12, 1–2, 95, 2005.
11. Quek, M.Y. and Yue, C.Y. *J. Mater. Sci.,* 32, 5457–5465, 1997.
12. Andreevskaya, G.D. and Gorbatkina, Y.A., Adhesion of polymeric binders to glass fiber, *Ind. Eng. Chem., Prod. Res. Dev.,* 11, 24–26, 1972.
13. Norman, R.H., James. D.I., and Gale, G.M., *The Chemical Engineer* (Trans. Inst. Chem. Eng.), 182, 243—248, 1964.
14. Schmidt, F.J., Gess, I.J., Essola, C.H., and Buschow, A.G., Proc. 15th Nat. SAMPE Symp. and Exhibition, North Hollywood, 1969, pp. 117–124.
15. Jarvela, P., Latinken, K.W., Purola, J., and Tormala, P., *Int. J. Adhesion Adhesives,* 3,141–147, 1983.
16. Pitkethly, M.J. and Doble, J.B., *Composites,* 21, 389–395, 1990.
17. Stamboulis, A., Baillie C., and Schulz E., *Angew. Makromol. Chemie,* 272, 117–120, 1999.
18. VandeVelde, K. and Kiekens, P., J. *Thermoplastic Composite Mater.,* 14, 3, 244–260, 2001.
19. Joseph, S., Sreekala, M.S., Oommen, Z., Koshy, P., and Thomas, S., *Composites Sci. Technol.,* 62, 1857–1868, 2002.
20. Joseph, K., Varghese, S., Kalaprasad, G., Thomas, S., Prasannakumari, L., Koshy, P., and Pavithran, C., *Eur. Polym. J.,* 32, 10, 1243–1250, 1996.
21. Manchado, M.A.L., Arroyo, M., Biagiotti, J., and Kenny, J.M., *J. Appl. Polym. Sci.,* 90, 2170–2178, 2003.
22. Tanaka, K., Minoshima, K., Grela, W., and Kenjiro, K., *Comp. Sci. Technol.,* 62, 2169–2177, 2002.
23. Sydenstricker, T.H.D., Mochnaz, S., and Amico, S.C., *Polym. Testing,* 22, 4, 375–380, 2003.
24. Brandsletter, J., Peterlik, H., Kromp, K., and Weiss, R., *Composites Sci. Technol.,* 63, 5, 653–660, 2003.
25. Yuan. X.W., Jayaraman, K., and Bhattacharya, D., *J. Adhesion Sci. Technol.,* 16, 6, 703–727, 2002.
26. Devaux, E., Caza, C., Recher, G., and Bierlarski, R., *Polym. Test.,* 21, 4, 457–462, 2002.
27. Marieta, C., Schulz, E., and Mondragon, I., *Composites Sci. Technol.,* 62, 2, 299–309, 2002.
28. García-Hernández, E., Licea-Claveríe, A., Zizumbo, A., Alvarez-Castillo, A., and Herrera-Franco, P.J., *Polym. Composites,* 25, 2, 134–145, 2004.
29. Li, Y., Mai, L-W., and Ye, L., *Comp. Interf.,* 12, 1-2, 141, 2005.
30. Kim, J.-K., Lu, S., and Mai, Y.-W., *J. Mater. Sci.,* 29, 554–561, 1994.
31. Hampe, A. and Marotzke, C., *J. Adhesion,* 78, 2, 167–187, 2002.
32. Qui, Y. and Schwartz, P., *J. Adhesion Sci. Technol.,* 5, 741–763, 1991.
33. Kim, J.K., Zhou, L.M., Bryan, S.J., and Mai Y.-W., *Composites,* 25, 470–475, 1994.
34. Fu, S.Y., Yue, C.Y., Hu, X., and Mai, Y.-W., *Composites Sci. Tech.,* 60, 569–579, 2000.

35. Liu, L., Huang, Y.D., Zhang, Z.Q., and Jiang, N.J., *J. Appl. Polymer Sci.,* 81, 2764–2768, 2001.
36. Craven, J.P., Cripps, R., and Viney, C., *Composites: Part A,* 31, 653–660, 2000.
37. Luo, S. and Netravali, A.K., *J. Mater. Sci.,* 34, 3709–3719, 1999.
38. Lodha, P. and Netravali, A.N., *J. Mater. Sci.,* 37, 17, 3657–3665, 2002.
39. Joly, C., Gauthier, R., and Chabert, B., *Composites Sci. Tech.,* 56, 761–765, 1996.
40. Baley, C., Grohens, Y., Busnel, F., and Davies, P., *Appl. Composite Mater.,* 11(2), 77–98, 2004.
41. Zhandarov, S. and Mader, E., *J. Adhesion Sci. Technol.,* 17, 7, 967–980, 2003.
42. Hwang, Y.J., Zhang, C., Jarrad, B., Stedeford, K., Tsai, J., Park, Y.C., and McCord M., *J. Adhesion Sci. Technol.,* 17, 6, 847–860, 2003.
43. Qiu, Y., Hwang, YJ., Zhang, C., Bures, B.L., and McCord M., *J. Adhesion Sci. Technol.,* 16, 4, 449–457, 2002.
44. Zhandarov, S.F., Mader, E., and Yurkevich, O.R., *J. Adhesion Sci. Technol.,* 16, 9, 1171–1200, 2002.
45. Mak, M., Lowe, A., Jar, B., and Stachurski, Z.J., *Mater. Sci. Lett.,* 17, 645–647, 1998.
46. Marshall, D.B. and Oliver, W.C., *Mater Sci. Eng.,* A126, 95–103, 1990.
47. Chen, E.J.H. and Young, J.C., *Composites Sci. Tech.,* 42, 189–206, 1991.
48. Desaeger, M. and Verpoest, I., *Composites Sci. Tech.,* 42, 215–226, 1993.
49. Chandra, N. and Ghonem, H., *Composites: Part A,* 32, 575–584, 2001.
50. Drissi Habte, M., Natano, K., and Suzuki, K., *Composites: Part A,* 30, 471–475, 1999.
51. Tsay, K.N., Toge, K., and Kawada, H., *Adv. Composite Mater.,* 11, 1, 1–9, 2002.
52. Wimolkiatiask, A.S. and Bell, J.P., *Polym. Composites,* 10, 162–172, 1989.
53. Joung, M., Park, J.-N.K., and Koichi, G., *Composites,* 60, 439–450, 2000.
54. Baker, A.A. and Jones, F.R., *Plast., Rubber Composites,* 30, 1, 2001.
55. Tripathy, S.S., DiLandro, L., Fontanelli, D., Marchetti, A., and Leata, G., *J. Appl. Polym. Sci.,* 75, 1585–1596, 2000.
56. Zafeiropoulos, N.E., Baillie, C.A., and Hodgkinson, J.M., *Composites: Part A,* 1185–1190, 2002.
57. Valadez-Gonzalez, A., Cervantes-Uc, J.M., and Herrera-Franco, P.J., *Composites: Part B,* 39, 3, 309–320, 1999.
58. Cabral-Fonseca, S., Paiva, M.C., Nunes, J.P., and Bernardo, C.A., *Polym. Test.,* 22, 907–913, 2003.
59. Zhou, X.F., Nairn, J.A., and Wagner, H.D., *Composites: Part A,* 1387–1400, 1999.
60. Park, J.M., Kim, D.S., and Kim, S.R., *Composites Sci. Tech.,* 63, 3-4, 403–419, 2003.
61. Park, J.M., Kim, D.S., and Kim, S.R., *Composites Sci. Tech.,* 64, 6, 847–860, 2004.
62. Park, J.M., Kong, J.-W., Kim, J.-W., and Yoon, D.-J., *Composites Sci. Tech.,* 64, 7–8, 983–999, 2004.
63. Piggott, M.R., *Composites Sci. Tech.,* 57, 853–885,1997.
64. Piggott, M.R., *Composites Sci. Tech.,* 57, 965–974, 1997.
65. Piggott, M.R., *Composites Sci. Tech.,* 42, 57–76, 1991.
66. Mader, E., Zhandarov, S., Gao, S.L., Zhou, X.F., Nutt, S.R., and Zhandarov, S.J. *Adhesion,* 78, 7, 547–569, 2002.
67. Valadez-Gonzalez, A., Cervantes-Uc, J.M., Olayo, R., and Herrera-Franco, P.J., *Composites: Part B,* 30, 309–320,1999.
68. Mader, E., Zhou, X.F., Pisanova, E., Zhandarov, S., and Nutt, S.R., *Advanced Composites Lett.,* 9, 203–209, 2000.
69. Marshall, P.I., Attwood, D., and Healy, M.J., *Composites:* 25:7, 1994.
70. Pisanova, E., Zhandarov, S., and Mader, E., *Composites: Part A,* 425–434, 2001.

71. Pisanova, E. and Mader, E., *J. Adhesion Sci. Technol.,* 14, 415–436, 2000.
72. Afaghi-Khatabi, A. and Mai, Y.-W., *Composites: Part A,* 33, 1585–1592, 2002.
73. Foley, M.E, Obaid, A.A., Huang, X., Tanoglu, M., Bogettic, T.A., McKnight, S.H., and Gillespie, J.W. Jr., *Composites: Part A,* 33, 1345–1348, 2002.
74. Pothan, L.A. and Thomas, S., *Composites Sci. Technol.,* 63, 1231–1240, 2003.
75. Gassan, J. and Dietz, T., *Composite Interfaces,* 10, 2-3, 287–296, 2003.
76. Thomason, J.L., *Composites: Part A,* 33, 1283–1288, 2002.
77. Thomason, J.L., *Composites Sci. Technol.,* 62, 10–11, 1455–1468, 2002.

Part V

Parameters

21 Data and Structures

Yusuf Yagci and Munmaya K. Mishra

TABLE 21.1
Comparison of Polymerization Systems

Type	Advantages	Disadvantages
Homogeneous		
Bulk—batch	Simple equipment	May require solution and subsequent precipitation for purification and/or fabrication May require reduction to usable particle sizes Heat control important Broad molecular-weight distribution
Bulk—continuous	Easier heat control Narrower molecular-weight distribution	Requires reactant recycling May require solution and subsequent ppt. for purification and/or fabrication Requires more complex equipment May require reduction to usable particle size
Solution	Easy agitation May allow longer chains to be formed Easy heat control	Requires some agitation Requires solvent removal and recycling Requires polymer recovery Solvent chain transfer may be harmful (i.e., reaction with solvent)
Heterogeneous		
Emulsion	Easy heat control Easy agitation Latex may be directly usable High polymerization rates possible Molecular-weight control possible Usable, small-particle size possible Usable in producing tacky, soft, and solid products	Polymer may require additional cleanup and purification Difficult to eliminate entrenched coagulants, emulsifiers, surfactants, etc. Often requires rapid agitation

(Continued)

TABLE 21.1 (CONTINUED)
Comparison of Polymerization Systems

Type	Advantages	Disadvantages
Precipitation	Molecular weight and molecular-weight distribution controllable by control of polymerization environment	May require solution and reprecipitation of product to remove unwanted material Precipitation may act to limit molecular-weight disallowing formation of ultrahigh-molecular-weight products
Suspension	Easy agitation Higher-purity product when compared to emulsion	Sensitive to agitation Particle size difficult to control

Source: Data compiled from R. B. Seymour and C. E. Carraher Jr., *Polymer Chemistry*, Marcel Dekker, Inc., New York, 1992.

TABLE 21.2
Physical Properties of Some Common Vinyl Monomers

Monomers	Formula	Molecular Weight	Density at 20°C	Boiling Point (°C)	Melting Point (°C)	Solubility[a]
Acrylaldehyde	CH_2:CHCHO	56.06	0.8410	52.5	–87	W, A, E
Acrylamide	CH_2:$CHCONH_2$	71.08	1.122^{30}	125^{25}	84.8	W, A, E, C
Acrylic acid	CH_2:$CHCO_2H$	72.06	1.0511	141.3	12.3	
Acrylonitrile	CH_2:CHCN	53.06	0.8060	77.5	–83	W
Butyl acrylate	CH_2:$CHCO_2C_4H_9$	128.17	0.8986	69^{50}	–64.6	A, E
Butyl methacrylate	CH_2:C $(CH_3)CO_2C_4H_9$	142.20	0.8936	165	–76	A, E
p-Chloro styrene	ClC_6H_4CH:CH_2	138.6	1.0868	192	–15.9	A, E, Bz
Ethyl acrylate	CH_2:$CHCO_2C_2H_5$	100.12	0.924	99	–75	
Methacrylic acid	CH_2:$C(CH_3)CO_2H$	86.09	1.0153	60^{12}	16	W, A, E
Methacrylonitrile	CH_2:$C(CH_3)CN$		0.7998	90.3	–36	
Methyl acrylate	CH_2:$CHCO_2CH_3$	86.09	0.9535	79.6	–75	A, E
Methyl methacrylate	CH_2:CH_2: $(CH_3)CO_2CH_3$	100.12	0.9440	100	–48	A, E
α-Methyl styrene	$C_6H_5C(CH_3)$:CH_2		0.9165^{10}	163.4	–23.2	
Styrene	C_6H_5CH:CH_2	104.15	0.9060	145.2	–30.6	A, E
Vinyl acetate	CH_3CO_2CH:CH_2	86.09	0.9317	72.5	–93.2	A, Bz
Vinyl butyl ether	CH_2:$CHOC_4H_9$	100.16	0.7888	93.8	–112	A, E
Vinyl chloride	CH_2:CHCl	62.5	0.9106	–13.37	–153.79	A, E
Vinyl ether (divinyl ether)	CH_2:CHOCH:CH_2	70.09	0.773	39	–101	
Vinyl ethyl ether	C_2H_5OCH:CH_2	72.11	0.7589	35	–115	A

TABLE 21.2 (CONTINUED)
Physical Properties of Some Common Vinyl Monomers

Monomers	Formula	Molecular Weight	Density at 20°C	Boiling Point (°C)	Melting Point (°C)	Solubility[a]
Vinyl fluoride	CH_2:CHF	46.04	—	−72.2	−161	
Vinyl methyl ether	CH_3OCH:CH_2	58.08	0.7500	12	−122	A, E

[a] W = water; E = ether; A = alcohol; C = chloroform; Bz = benzene.

Source: Portions of the data were compiled from J. B. Bandrup and E. H. Immergut, Eds., *Polymer Handbook*, 3rd ed., Wiley-Interscience, New York, 1989.

TABLE 21.3
Chain Transfer Constants of Solvents to Different Monomers in Free-Radical Chain Polymerization

Monomer	Solvent	Temperature (°C)	$Cs \times 10^4$	References[a]
Acrylamide	Methanol	30	0.13	1
	Water	40	5.8	2
Acrylic acid, ethyl ester	Cyclohexane	50	28.9	3
		60	0.48	3
	Hexane	50	0.524	3
		60	0.593	3
	Toluene	50	0.611	4
		60	0.929	4
Acrylic acid, methyl ester	Benzene	80	0.326	5
	Toluene	60	2.7	6, 7
		80	1.775	5
	CCl_4	40	1.0	8
Acrylonitrile	Acetone	50	1.7	9
	Benzene	60	2.46	10
	CCl_4	60	0.85	11
	Chloroform	60	5.64	10
	Toluene	50	1.153	12
		60	2.632	12
Methacrylic acid, ethyl ester	Benzene	80	0.081	13
	CCl_4	60	0.901	13
	Chloroform	60	0.703	13
	Ethyl acetate	80	0.919	13
	Heptane	80	0.865	13
	Toluene	80	0.436	13
Methyl methacrylate	Acetone	60	0.195	14
		80	0.225	15

(Continued)

TABLE 21.3 (CONTINUED)
Chain Transfer Constants of Solvents to Different Monomers in Free-Radical Chain Polymerization

Monomer	Solvent	Temperature (°C)	Cs × 10^4	References[a]
	Benzene	50	0.036	16, 17
	CCl_4	60	0.040	14
		50	0.82	18
		60	0.925	14
		80	2.393	15
	Chloroform	60	0.454	14
	Chloroform	80	1.124	14
	Cyclohexane	60	12.0	19
		80	0.1	15
	p-Dioxane	80	0.222	15
	Ethyl acetate	60	0.100	20
	Heptane	50	1.8	21
	Methanol	60	0.2	22
		80	0.33	22
	Toluene	60	0.170	6, 14
	Water	60	0	23
Styrene	Acetone	60	4.1	24
	Benzene	50	0.01	17
		60	0.018	25
		70	5.5	26
	Butyl alcohol	50	6.5	27
		60	0.06	28
	t-Butyl alcohol	50	6.6	27
		60	0.22	29
	Carbon tetrachloride	60	84	30
		80	133	31
	Chloroform	60	0.5	28, 32
		80	0.916	33
	p-Dioxane	60	2.75	24
	Ethyl acetate	60	15.5	34
		75	6.67	26
	Hexane	100	0.9	35
	Methanol	60	0.296	36
		80	1.10	22
	Toluene	60	0.105	37
		80	0.15	38
		100	0.53	39
	Water	60	0.006	36
Vinyl acetate	Acetone	60	1.5	19
		70	25.6	40

TABLE 21.3 (CONTINUED)
Chain Transfer Constants of Solvents to Different Monomers in Free-Radical Chain Polymerization

Monomer	Solvent	Temperature (°C)	Cs × 10⁴	References[a]
	Benzene	60	1.07	41
		70	5.27	42
		75	1.4	19
	Butyl alcohol	60	20.0	43
		70	29.1	42
	sec Butyl alcohol	60	31.74	44
		70	6.21	42
	Carbon tetrachloride	60	800	45
		70	2023	46
	Chloroform	60	130	43
	Cyclohexane	60	6.59	44
	p-Dioxane	60	20	19
		70	49.1	42
	Ethyl acetate	50	2.9	47
		60	1.07	44
		70	7.8	42
	Ethyl alcohol	60	25	19
		70	26.3	42
	Heptane	50	17.0	47
	Methanol	60	2.26	43, 48
		70	5.5	49
	Toluene	50	12.0	50
		60	17.8	51
		70	21.1	51
	Xylene	50	14.9	50
		70	278	42
Vinyl chloride	CCl_4	60	280	52
	Chloroform	60	290	52

[a] References:

1. V. F. Gromov, A. V. Matveeva, P. M. Khomikovskii, and A. D. Abkin, *Vysokomol. Soedin., Ser. A.*, 9, 1444 (1967).
2. C. Kwang-Fu, *Kobunski Kagaku*, 29, 233 (1972).
3. P. V. T. Raghuram and U. S. Nandi, *J. Polym. Sci., A-1*, 8, 3079 (1970).
4. M. Raetzsch and I. Zschach, *Plastic Kaut.*, 21(5), 345 (1974).
5. J. N. Sen, U. S. Nandi, and S. R. Palit, *J. Indian Chem. Soc.*, 40, 729 (1963).
6. C. H. Bamford, A. D. Jenkins, and R. Johnston, *Trans. Faraday Soc.*, 55, 418 (1959).
7. C. H. Bamford and E. F. T. White, *Trans. Faraday Soc.*, 52, 716 (1956).
8. I. Sakurada, K. Nom, and Y. Ofuji, *Kobunski Kagaku*, 20, 481 (1963); Chem. Abstr., 63, 8487c (1965).

(*Continued*)

TABLE 21.3 (CONTINUED)
Chain Transfer Constants of Solvents to Different Monomers in Free-Radical Chain Polymerization

9. J. Ulbricht and B. Sandner, *Faserforsch. Textiltech*, 17, 208 (1966).
10. S. K. Das, S. R. Chatterje, and S. R. Palit, *Proc. R. Soc. Ser. A*, 227, 252 (1955).
11. V. A. Dinaburg and A. A. Vansheidt, *Zh. Obshch. Khim.*, 24, 840 (1954).
12. N. T. Srinivasan and M. Santappa, *Makromol. Chem.*, 26, 80 (1958).
13. S. R. Chatterjee, S. N. Khanna, and S. R. Palit, *J. Indian Chem. Soc.*, 41, 622 (1964).
14. R. N. Chadha, J. S. Shukla, and G. S. Misra, *Trans. Faraday Soc.*, 53, 240 (1957).
15. S. Basu, J. N. Sen, and S. R. Palit, *Proc. Roy Soc., Ser. A*, 202 485 (1950).
16. G. Henrici-Olive, S. Olive, and G. V. Schulz, *Makromol. Chem.*, 23, 207 (1957).
17. G. V. Schulz, G. Henrici, and S. Olive, *Z. Elektrochem.*, 60, 296 (1956).
18. J. Pavlinec and E. Borsig, *J. Polym. Sci., Polym. Chem. Ed.*, 19(9), 2305 (1981).
19. J. T. Clarke, R. O. Howard, and W. H. Stockmayer, *Makromol. Chem.*, 44/46, 427 (1961).
20. N. G. Saha, U. S. Nandi, and S. R. Palit, *J. Chem. Soc.*, 427 (1956).
21. M. Lazar and J. Pavlinec, *Chem. Zvesti*, 15, 428 (1961); *Chem. Abstr.*, 55, 22896c (1961).
22. B. R. Bhattacharyya and U. S. Nandi, *Makromol. Chem.*, 149, 231 (1971).
23. U. S. Nandi, P. Ghosh, and S. R. Palit, *Nature*, 195, 1197 (1962).
24. S. L. Kapur, *J. Polym. Sci.*, 11, 399 (1953).
25. R. A. Gregg and F. R. Mayo, *Discuss. Faraday Soc.*, 2, 328 (1947).
26. M. R. Gopalan and M. Santhappa, *J. Polym. Sci.*, 25, 333 (1957).
27. H. J. Dietrich and M. A. Raymond, *J. Macromol. Sci. Chem.*, A6, 191 (1972).
28. R. A. Gregg and F. R. Mayo, *J. Am. Chem. Soc.*, 70, 2373 (1948).
29. G. C. Bhaduri and U. S. Nandi, *Makromol. Chem.*, 128, 183 (1969).
30. C. Walling and J. J. Pellon, *J. Am. Chem. Soc.*, 79, 4776 (1957).
31. K. Katagiri, K. Uno, and S. Okamura, *J. Polym. Sci.*, 17, 142 (1955).
32. F. M. Lewis and F. R. Mayo, *J. Am. Chem. Soc.*, 76, 457 (1954).
33. G. S. Misra and R. N. Chadha, *Makromol. Chem.*, 23, 134 (1957).
34. J. E. Glass and N. L. Zutty, *J. Polym. Sci., A-1*, 4, 1223 (1966).
35. A. P. Titov and I. A. Livshits, *Zh. Obshch. Khim.*, 29, 1605 (1959).
36. R. B. Seymour, J. M. Sosa, and V. J. Patel, *J. Paint Technol.*, 43(563), 45 (1971).
37. Y. Mori, K. Sato, and Y. Minoura, *Kogyo Kagaku Zasshi*, 61, 462 (1958); *Chem. Abstr.*, 55, 4021f (1961).
38. B. M. E. van der Hoff, *J. Polym. Sci.*, 48, 175 (1960).
39. F. R. Mayo, *J. Am. Chem. Soc.*, 65, 2324 (1943).
40. G. P. Scott, C. C. Soong, W.-S. Huang, and J. L. Reynolds, *J. Org. Chem.*, 29, 83 (1964).
41. M. Morton and I. Piirma, *J. Polym. Sci.*, A1, 3043 (1963).
42. A. A. Vansheidt and G. Khardi, *Acta Chim. Acad. Sci. Huang.*, 20, 381 (1959); *Chem. Abstr.*, 54, 11552f (1960).
43. S. R. Palit and S. K. Das, *Proc. R. Soc., Ser. A*, 226, 82 (1954).
44. T. Asahara and T. Makishima, *Kogyo Kagaku Zasshi*, 69, 2173 (1966).
45. A. A. Vansheidt and G. Khardi, *Acta Chim. Acad. Sci. Hung.*, 20, 261 (1959); *Chem. Abstr.*, 54, 6180b (1960).
46. G. Henrici-Olive and S. Olive, *Fortschr. Hochpolymer. Forsch*, 2, 496 (1961).
47. M. Lazar, J. Pavlinec, and Z. Manasek, *Collect. Czech. Chem. Commun.*, 26, 1380 (1961).
48. M. Matsumoto and M. Maeda, *J. Polym. Sci.*, 17, 438 (1955).
49. I. Sakurada, Y. Sakaguchi, and K. Hashimoto, *Kobunshi Kagaku*, 19, 593 (1962); *Chem. Abstr.*, 61, 16159D (1964).

TABLE 21.3 (CONTINUED)
Chain Transfer Constants of Solvents to Different Monomers in Free-Radical Chain Polymerization

50. Kh. S. Bagdasar'ian and Z. A. Sinitsina, *J. Polym. Sci.*, 52, 31 (1961).
51. C. F. Thompson, W. S. Port, and L. P. Wittnaurer, *J. Am. Chem. Soc.*, 81, 2552 (1959).
52. B. A. Englin, T. A. Onishchenko, and R. Kh. Freidlina, *Izv. Akad. Nauk SSSR, Ser. Khim.*, 12, 2542 (1971).

Source: Portions of the data were compiled from J. B. Bandrup and E. H. Immergut, Eds., *Polymer Handbook*, 3rd ed., Wiley-Interscience, New York, 1989.

TABLE 21.4
Typical Free-Radical Chain-Copolymerization Reactivity Ratios of Some Selected Monomers at Different Temperatures

Monomer 1	Monomer 2	r_1	r_2	r_1r_2	Temperature (°C)
Acrylamide	Acrylic acid	1.38	0.36	0.5	60
	Acrylonitrile	1.3	0.8	1.04	60
	Methyl acrylate	1.30	0.05	0.07	60
	Vinylidene chloride	4.9	0.15	0.74	60
Acrylic acid	Acrylamide	0.36	1.38	0.5	60
	Acrylonitrile	1.15	0.35	0.40	50
	Methyl methacrylate	0.33	2.17	0.72	50
	Styrene	0.25	0.15	0.04	60
	Vinyl acetate	2.0	0.1	0.2	70
Acrylonitrile	Acrylic acid	0.35	1.15	0.40	50
	Acrylamide	0.8	1.3	1.04	60
	Butadiene	0.02	0.3	0.006	40
	Butadiene	0.25	0.33	0.08	60
	t-Butyl vinyl ether	0.14	0.003	0.0004	60
	Ethyl acrylate	1.17	0.67	0.78	50
	Ethyl vinyl ether	0.7	0.03	0.021	80
	Methacrylic acid	0.09	2.5		70
	Methyl acrylate	1.5	0.84	1.26	
	Methyl methacrylate	0.15	1.22	0.183	80
	α-Methyl styrene	0.06	0.1		75
	Methyl vinyl ketone	0.61	1.78	1.0858	60
	Styrene	0.04	0.4	0.016	60
	Vinyl acetate	4.2	0.05	0.21	50
	Vinyl chloride	2.7	0.04	0.11	60
	Vinylidene chloride	0.91	0.37	0.34	60
	2-Vinylpyridine	0.113	0.47	0.05	60
	4-Vinylpyridine	0.11	0.41	0.045	60

(Continued)

TABLE 21.4 (CONTINUED)
Typical Free-Radical Chain-Copolymerization Reactivity Ratios of Some Selected Monomers at Different Temperatures

Monomer 1	Monomer 2	r_1	r_2	r_1r_2	Temperature (°C)
Allyl acetate	Methyl methacrylate	0	23	0	60
	Styrene	0	90	0	60
	Vinyl acetate	0.7	1.0	0.7	60
Butadiene	Acrylonitrile	0.3	0.02	0.006	40
	Isoprene	0.75	0.85	0.64	5
	Methacrylonitrile	0.36	0.04	0.014	5
	Methyl methacrylate	0.75	0.25	0.188	90
	Styrene	1.35	0.58	0.78	50
	Vinyl chloride	8.8	0.035	0.31	50
	2-Vinylpyridine	0.94	0.9	0.85	50
Methacrylonitrile	Methyl methacrylate	0.65	0.67	0.44	60
	Styrene	0.32	0.39	0.12	60
	Vinyl acetate	12	0.01	0.12	70
Methacrylic acid	Acrylonitrile	2.5	0.09	0.23	70
	Butadiene	0.53	0.20	0.11	50
	Styrene	0.7	0.15	0.11	60
	Vinyl acetate	20	0.01	0.2	70
	Vinyl chloride	36	0.03	1.08	50
	2-Vinylpyridine	0.58	1.55	0.9	70
Methyl acrylate	Acrylamide	0.05	1.3	0.07	60
	Acrylonitrile	0.67	1.26	0.84	60
	Ethyl vinyl ether	3.3	0	0	60
	Methyl methacrylate	0.50	1.91	0.96	130
	Styrene	0.20	0.75	0.15	60
	Vinyl acetate	9	0.1	0.9	60
	Vinyl chloride	4	0.06	0.24	45
	2-Vinylpyridine	0.20	2.03	0.41	60
	4-Vinylpyridine	0.22	1.7	0.37	60
Methyl methacrylate	Methyl acrylate	1.91	0.5	0.96	130
	Methacrylonitrile	0.67	0.65	0.44	60
	Styrene	0.46	0.52	0.24	60
	α-Methyl styrene	0.5	0.14	0.07	60
	Vinyl acetate	20	0.02	0.4	60
	Vinyl chloride	10	0.1	1.0	68
	Vinylidene chloride	2.53	0.24	0.61	60
α-Methyl styrene	Acrylonitrile	0.1	0.06	0.01	75
	Methyl methacrylate	0.14	0.5	0.07	60
	Styrene	0.3	1.3	0.39	60

TABLE 21.4 (CONTINUED)
Typical Free-Radical Chain-Copolymerization Reactivity Ratios of Some Selected Monomers at Different Temperatures

Monomer 1	Monomer 2	r_1	r_2	r_1r_2	Temperature (°C)
Methyl vinyl ketone	Styrene	0.35	0.29	0.10	60
	Vinyl acetate	7.0	0.05	0.35	70
	Vinyl chloride	8.3	0.1	0.83	70
	Vinylidene chloride	1.8	0.55	0.99	70
Styrene	*p*-Chlorostyrene	0.74	1.03	0.76	60
	Methacrylonitrile	0.39	0.32	0.12	60
	Methyl acrylate	0.75	0.20	0.15	60
	Methyl methacrylate	0.52	0.46	0.24	60
	α-Methyl styrene	1.3	0.3	0.39	60
	Methyl vinyl ketone	0.29	0.35	0.10	60
	Ethyl vinyl ether	80	0	0	80
	p-Methoxy styrene	1.16	0.82	0.95	60
	Methacrylic acid	0.15	0.7	0.11	60
	Vinyl acetate	55	0.01	0.55	60
	Vinyl chloride	17	0.02	0.34	60
	Vinylidine chloride	1.85	0.09	0.17	60
	2-Vinylpyridine	0.55	1.14	0.63	60
	N-Vinylpyrrolidone	24.2	0.08	1.94	60
Vinyl acetate	Ethyl vinyl ether	3.0	0	3.0	60
	Methacrylic acid	0.01	20	0.2	70
	Methacrylonitrile	0.01	12	0.12	70
	Vinyl chloride	0.23	1.68	0.39	60
	Methyl acrylate	0.1	9	0.9	60
	Methyl methacrylate	0.02	20	0.4	60
	Vinylidene chloride	0.1	6.0	0.6	68
	Methyl vinyl ketone	0.05	7.0	0.35	70
	Styrene	0.01	55.0	0.55	60
	Vinyl laurate	1.4	0.7	0.98	60
Vinyl chloride	Methyl acrylate	0.3	3.2	096	60
	Methacrylic acid	0.1	8.3	0.83	70
	Vinylidene chloride	0.05	15.7	0.79	50
	Methyl vinyl ketone	0.08	24.2	1.94	60
	Styrene				
N-Vinylpyrrolidone	Styrene				

Source: Portions of the data were compiled from J. B. Bandrup and E. H. Immergut, Eds., *Polymer Handbook*, 3rd ed., Wiley-Interscience, New York, 1989.

TABLE 21.5
Rate Constants of Some Common Initiators for Vinyl Polymerization

Initiators Solvent	Temperature (°C)	K_d (sec^{-1}) × 10^8	E_a (kcal mol^{-1})	References[a]
		Azo Compounds		
2,2′-Azo-bis-isobutyronitrile				
Benzene	40	54.4	30.7	1
	50	264		1
	60	915		2
	100	152,000		2
Cyclohexane	82	14,300		3, 4
Methyl methacrylate	50	97		5
	70	3,100		6
Styrene	50	297	30.5	7
	70	4,720		7
Toluene	60	903		8
	70	4,000	29.0	9
	100	160,000		9
Ethyl acetate	40	47	30.7	1
	60	936		1
Xylene	50	200		4
	80	15,300	31.3	10
Phenyl-azo-diphenylmethane				
Declain	125	3,440	34	11
Phenyl-azo-triphenylmethane				
Benzene	25	429	26.8	12
	50	13,700		13
Cyclohexane	25	422	24.5	12
	50	9,900		12
Decane	60	57,200		13
Heptane	60	66,000		13
Hexane	60	76,000		13
Octane	60	64,100		13
Toluene	50	17,300		13
	45	8,480		14
		Peroxides		
tert-Butyl peroxide				
Benzene	80	7.81	34	15
	100	88	35	16, 17
	120	1,100	35.3	18
	130	3,220		18
Cyclohexane	95	24.8	40.8	19
	120	630		18
	130	2,590		18

TABLE 21.5 (CONTINUED)
Rate Constants of Some Common Initiators for Vinyl Polymerization

Initiators Solvent	Temperature (°C)	K_d (sec^{-1}) × 10^8	E_a (kcal mol^{-1})	References[a]
Decane	80	1.39		20
	110	201		20
	130	2,480		20
	140	6,660		17
Heptane	80	1.44		20
	110	219		20
Hexane	80	1.64		20
	110	217		20
Octane	80	1.48		20
	110	219		20
Toluene	100	68.2		21
	125	1,600		22
Cumyl peroxide				
Benzene	115	2,050	38	17
Dedecane	128	8,750		23
Acetyl peroxide				
Benzene	50	110	32	17
	60	500		24
	70	2,390		25
Cyclohexane	55	210	31.4	26, 27
	75	3,600		26, 27
Decane	80	6,850		28
Heptane	80	7,720		28
n-Hexane	60	340		29
n-Octane	60	290		29
	80	7,340		28
Toluene	55	270	32	26, 27
	60	500	31	24
	75	4,700		26, 27
	85	15,900		26, 27
Benzoyl peroxide				
Benzene	30	4.8	27.8	30
	60	200	29.7	31
	70	1,170	32	32
	75	1,660	29.7	33
Cyclohexane	80	7,720		34
Decane	80	2,530		20
n-Heptane	80	2,710		20
Hexane	80	2,850		20

(Continued)

TABLE 21.5 (CONTINUED)
Rate Constants of Some Common Initiators for Vinyl Polymerization

Initiators Solvent	Temperature (°C)	K_d (sec^{-1}) × 10^8	E_a (kcal mol^{-1})	References[a]
Styrene	50	70	30.5	35
	60	270		35
	70	990		35
Toluene	30	4.94	28.8	30
	49	60	29.6	1
	60	224		8
	70	1,100		1
Lauroyl peroxide				
Benzene	30	25.6		36
	40	49.1		37
	50	219		37
	60	1,510	30.4	16, 17
	70	5,580		16, 17
Ethyl acetate	40	60.3		37
	50	270		37
	70	3,990		37
tert-Butyl hydroperoxide				
Benzene	130	30	32.9	17
	154	429	40.8	38
	160	660		17
Cyclohexane	100	12		39
Dodecane	86	132	30.7	40
Heptane	172	14,100		21
n-Octane	150	800	39	41
	160	2,500		41
	170	6,900		41
	180	1,820		41
Toluene	100	5.7		39
tert-Butyl peracetate				
Benzene	85	218	36.3	16, 17
	100	1,540		16, 17
Decane	100	1,500	32	17
Hexane	130	50,800		42
Octane	100	2,070		20
tert-Butyl perbenzoate				
Benzene	100	1,070	34.7	16, 17
	110	3,500	34.5	43
	130	33,000		43
Decane	100	1,400	32.0	17
	130	35,600		17
Heptane	115	7,210		20
Octane	115	7,060		20

TABLE 21.5 (CONTINUED)
Rate Constants of Some Common Initiators for Vinyl Polymerization

Initiators Solvent	Temperature (°C)	K_d (sec^{-1}) × 10^8	E_a (kcal mol^{-1})	References[a]
Xylene	119	10,900	33.8	43
	130	34,200		43
tert-Butyl peroctate				
Benzene	70	1,400	31	17
	100	45,500		17
Decane	70	690	31	17
	100	26,400		17
Potassium persulfate				
0.1 *M* NaOH	50	950	33.5	44
	70	2,330		44
	90	35,000		44
Water (pH 3)	50	166		44
Water	40	1,650,000	19.9	45
	50	378,000	29	45
	60	2,180,000		45
	70	5,010,000		45
	80	5,780		46

[a] References
1. C. E. H. Bawn and S. F. Mellish, *Trans. Faraday Soc.*, 47, 1216 (1951).
2. J. P. Van Hook and A. V. Tobolsky, *J. Am. Chem. Soc.*, 80, 779 (1958).
3. L. M. Arnett, *J. Am. Chem. Soc.*, 74, 2027 (1952).
4. T. W. Koenig and J. C. Martin, *J. Org. Chem.*, 29, 1520 (1964).
5. M. B. Lachinov, V. P. Zubov, and V. A. Kabanov, *J. Polym. Sci., Polym. Chem. Ed.*, 15, 1777 (1977).
6. C. H. Bamford, R. Denyer, and J. Hobbs, *Polymer*, 8, 493 (1967).
7. J. W. Breitenbach and A. Schindler, *Monatsh. Chem.*, 83, 724 (1952).
8. S. G. Ng and K. K. Che, *J. Polym. Sci., Polym. Chem. Ed.*, 20, 409 (1982).
9. M. Talât-Erben and S. Bywater, *J. Am. Chem. Soc.*, 77, 3712 (1955).
10. F. M. Lewis and M. S. Matheson, *J. Am. Chem. Soc.*, 71, 747 (1949).
11. S. G. Cohen and C. H. Wang, *J. Am. Chem. Soc.*, 77, 3628 (1955).
12. M. G. Alder and J. E. Leffler, *J. Am. Chem. Soc.*, 76, 1425 (1954).
13. R. C. Neuman and G. D. Lockyer, *J. Am. Chem. Soc.*, 105, 3982 (1983).
14. S. G. Cohen, F. Cohen, and C. H. Wang, *J. Org. Chem.*, 28, 1479 (1963).
15. J. K. Allen and J. C. Bevington, *Proc. R. Soc. (London)*, A 262, 271 (1961).
16. O. L. Mageli, S. D. Butaka, and D. J. Bolton, *Evaluation of Organic Peroxides from Half-Life Data*, Wallace & Tiernan, Lucidol Division, Bulletin 30.30.
17. Anon., *Evaluation of Organic Peroxides from Half-Life Data*, Technical Bulletin, Lucidol Division, Pennwalt (no date).
18. E. S. Huyser and R. M. Van Scoy, *J. Org. Chem.*, 33, 3524 (1968).
19. C. Walling and D. Bristol, *J. Org. Chem.*, 36, 733 (1971).
20. W. A. Pryor, E. H. Morkved, and H. T. Bickley, *J. Org. Chem.*, 37, 1999 (1972).
21. R. Hiatt, T. Mill, K. C. Irwin, and J. K. Castleman, *J. Org. Chem.*, 33, 1421 (1968).

(*Continued*)

TABLE 21.5 (CONTINUED)
Rate Constants of Some Common Initiators for Vinyl Polymerization

22. E. S. Huyser and C. J. Bredeweg, *J. Am. Chem. Soc.*, 84, 2401 (1964).
23. M. S. Kharasch, A. Fono, and W. Nudenberg, *J. Org. Chem.*, 16, 105 (1951).
24. M. W. Thomas and M. T. O'Shaughnessy, *J. Polym. Sci.*, 11, 455 (1953).
25. A. I. Lowell and J. R. Price, *J. Polym. Sci.*, 43, 1 (1960).
26. M. Levy, M. Steinberg, and M. Szwarc, *J. Am. Chem. Soc.*, 76, 5978 (1954).
27. R. D. Schuetz and J. L. Shea, *J. Org. Chem.*, 30, 844 (1965).
28. W. A. Pryor and K. Smith, *J. Am. Chem. Soc.*, 92, 5403 (1970).
29. W. Braun, L. Rajbenbach, and F. R. Eirich, *J. Phys. Chem.*, 66, 1591 (1962).
30. W. E. Cass, *J. Am. Chem. Soc.*, 68, 1976 (1946).
31. J. C. Bevington and J. Toole, *J. Polym. Sci.*, 28, 413 (1958).
32. B. Barnett and W. E. Vaughan, *J. Phys. Chem.*, 51, 926 (1947).
33. A. Conix and G. Smets, *J. Polym. Sci.*, 10, 525 (1953).
34. B. Barnett and W. E. Vaughan, *J. Phys. Chem.*, 51, 942 (1947).
35. S. Molnar, *J. Polym. Sci.*, A-1, 10, 2245 (1972).
36. W. E. Cass, *J. Am. Chem. Soc.*, 72, 4915 (1950).
37. C. E. H. Bawn and R. G. Halford, *Trans. Faraday Soc.*, 51, 780 (1955).
38. R. R. Hiatt and W. M. J. Strachan, *J. Org. Chem.*, 28, 1893 (1963).
39. R. Hiatt and K. C. Irwin *J. Org. Chem.*, 33, 1436 (1968).
40. B. K. Morse, *J. Am. Chem. Soc.*, 79, 3375 (1957).
41. E. R. Bell, J. H. Raley, F. F. Rust, F. H. Seubold, and W. E. Vaughan, *Discuss. Faraday Soc.*, 10, 242 (1951).
42. T. Koenig, J. Huntington, and R. Cruthoff, *J. Am. Chem. Soc.*, 92, 5413 (1970).
43. A. T. Blomquist and A. F. Ferris, *J. Am. Chem. Soc.*, 73, 3412 (1951).
44. I. M. Kolthoff and I. K. Miller, *J. Am. Chem. Soc.*, 73, 3055 (1951).
45. J. K. Rasmussen, S. M. Heilmann, P. E. Toren, A. V. Pocius, and T. A. Kotnour, *J. Am. Chem. Soc.*, 105, 6845 (1983).
46. P. D. Bartlett and K. Nozaki, *J. Polym. Sci.*, 3, 216 (1948).

Source: Portions of the data were compiled from J. B. Bandrup and E. H. Immergut, Eds., *Polymer Handbook*, 3rd ed., Wiley-Interscience, New York, 1989.

TABLE 21.6
Glass Transition Temperature of Some Common Vinyl Polymers

Polymer	T_g (°C)	T_m (°C)
Poly(acrylic acid)	106	
Poly(butyl acrylate)	–54	
Poly(*t*-butyl acrylate)	43–107	
Poly(isobutyl acrylate)	–24	
Poly(acrylamide)	165	
Poly(butyl methacrylate)	13–35	
Poly(methacrylic acid)	228	
Atactic	105	200
Isotactic	38	
Syndiotactic	99	
Poly(vinyl fluoride)	–41	200
Poly(vinyl methyl ether)	–31	
Poly(vinyl alcohol)	85	
Poly(acrylonitrile)	125	
Poly(vinyl chloride)	81	
Poly(vinyl acetate)	32	
Poly(4-bromostyrene)	118	
Poly(4-chlorostyrene)	110	
Poly(α-methyl styrene)	20	
Poly(4-methyl styrene)	97	
Poly(styrene)-isotactic and atactic	100	240

Source: Portions of the data were compiled from J. B. Bandrup and E. H. Immergut, Eds., *Polymer Handbook*, 3rd ed., Wiley-Interscience, New York, 1989.

TABLE 21.7
Structures of Some Common Vinyl Polymers

Acrylonitrile-butadiene-styrene terpolymer (ABS)

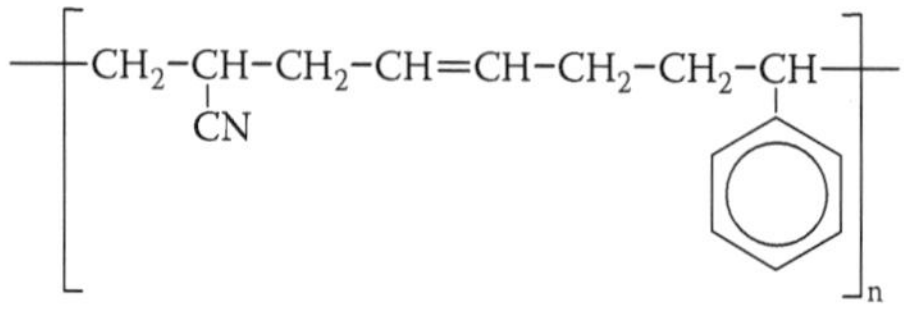

$$\left[-CH_2-\underset{CN}{CH}-CH_2-CH{=}CH-CH_2-CH_2-\underset{C_6H_5}{CH}- \right]_n$$

Butyl rubber

$$\left[-CH_2-\underset{CH_3}{\overset{CH_3}{C}}-CH_2CH{=}\underset{CH_3}{C}CH_2- \right]_n$$

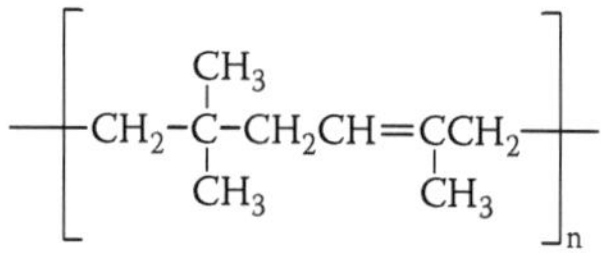

Ethylene-methacrylic acid copolymers (Ionomers)

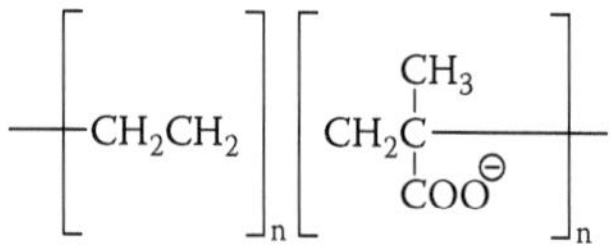

$$\left[-CH_2CH_2 \right]_n \left[CH_2\underset{COO^{\ominus}}{\overset{CH_3}{C}}- \right]_n$$

Nitrile rubber (NBR)

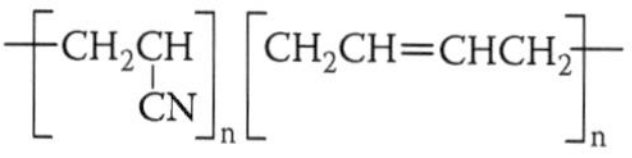

$$\left[-CH_2\underset{CN}{CH} \right]_n \left[CH_2CH{=}CHCH_2- \right]_n$$

Polyacrolein

$$\left[-\underset{HC=CH_2}{CH}-O- \right]_n$$

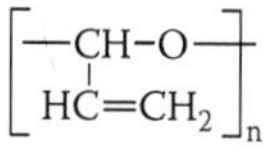

Polyacrylamide

$$\left[-CH_2-\underset{CONH_2}{CH}- \right]_n$$

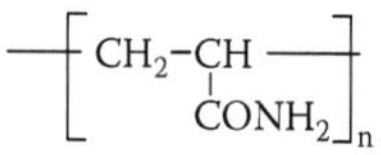

TABLE 21.7 (CONTINUED)
Structures of Some Common Vinyl Polymers

Poly(acrylamide oxime)

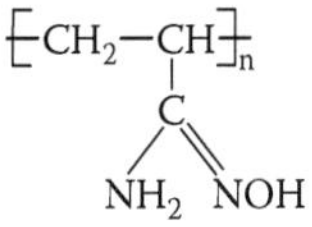

$$\left[CH_2-CH(C(=NOH)NH_2) \right]_n$$

Poly(acrylic anhydride)

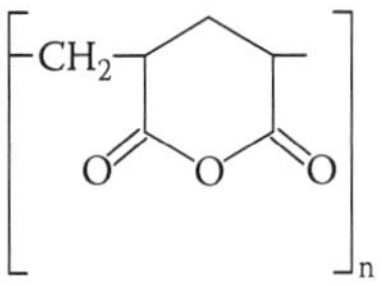

$$\left[CH_2-\text{(glutaric anhydride ring)} \right]_n$$

Polyacrylonitrile

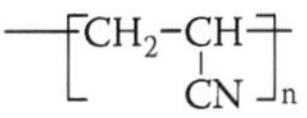

$$\left[CH_2-CH(CN) \right]_n$$

Poly(methyl acrylate)

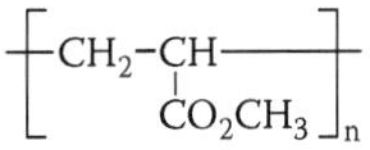

$$\left[CH_2-CH(CO_2CH_3) \right]_n$$

Poly(methyl methacrylates) (PMMA)

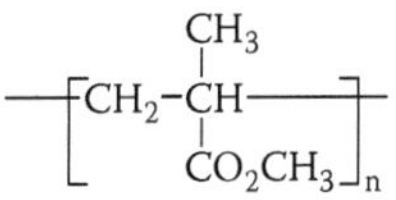

$$\left[CH_2-CH(CH_3)(CO_2CH_3) \right]_n$$

Poly(methyl vinyl ketone)

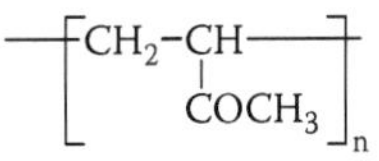

$$\left[CH_2-CH(COCH_3) \right]_n$$

(*Continued*)

TABLE 21.7 (CONTINUED)
Structures of Some Common Vinyl Polymers

Polystyrene (PS)

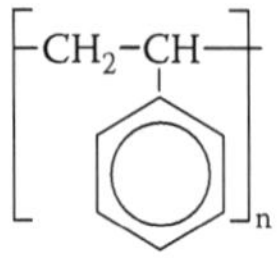

$$\left[CH_2-\underset{\underset{C_6H_5}{|}}{CH} \right]_n$$

Poly(vinyl acetate)(PVAc)

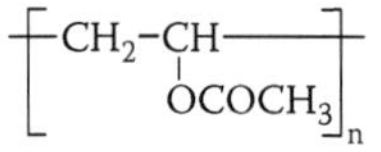

$$\left[CH_2-\underset{\underset{OCOCH_3}{|}}{CH} \right]_n$$

Poly(vinyl alcohol) (PVA)

$$\left[CH_2-\underset{\underset{OH}{|}}{CH} \right]_n$$

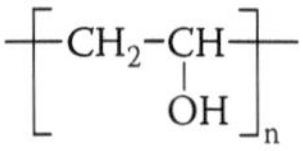

Poly(vinyl *t*-butyl ether)

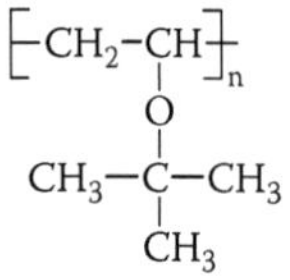

$$\left[CH_2-\underset{\underset{\underset{CH_3-\underset{\underset{CH_3}{|}}{C}-CH_3}{|}}{O}}{CH} \right]_n$$

Poly(vinyl butyral) (PVB)

$$\left[CH_2-CH-CH_2-CH \right]_n \text{ (two O atoms bridged by } CH(CH_2)_2CH_3 \text{)}$$

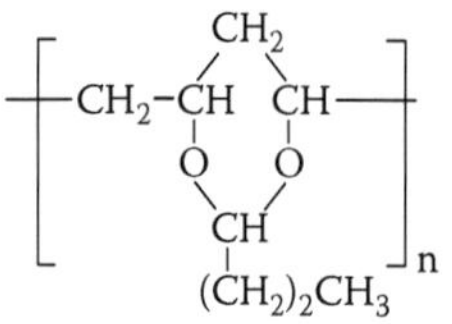

TABLE 21.7 (CONTINUED)
Structures of Some Common Vinyl Polymers

Poly(vinyl butyrate)

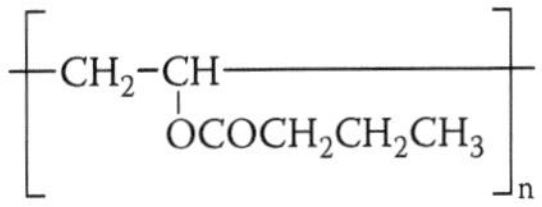

Poly(vinyl carbazole)

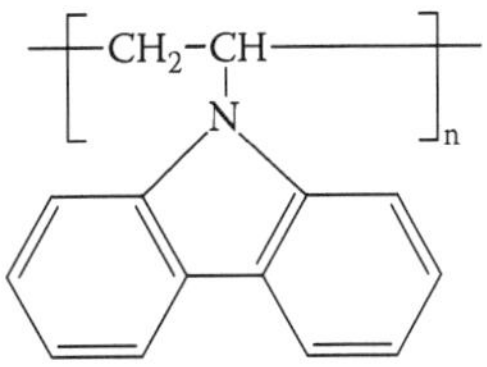

Poly(vinyl chloride) (PVC)

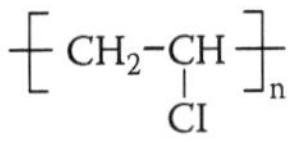

Poly(vinyl chloride-co vinyl acetate)

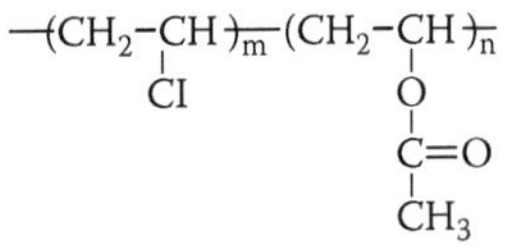

Poly(vinyl formal) (PVF)

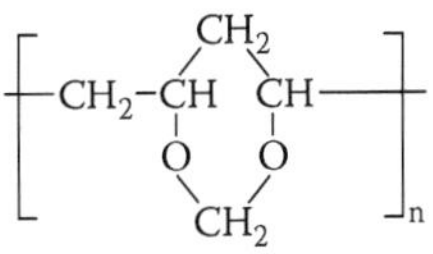

Poly(vinylidene chloride)

(Continued)

TABLE 21.7 (CONTINUED)
Structures of Some Common Vinyl Polymers

Poly(vinyl isobutyl ether)

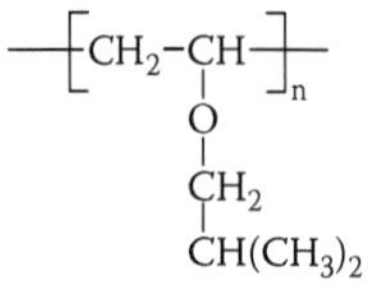

$$-\!\!\left[CH_2-\underset{\underset{\underset{\underset{CH(CH_3)_2}{|}}{CH_2}}{|}}{\underset{O}{|}}{CH}\right]_n\!\!-$$

Poly(vinyl pyridine)

$$\left[CH_2-CH(C_5H_4N)\right]_n$$

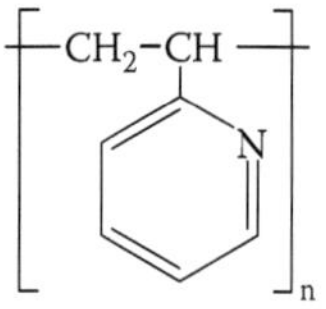

Poly(vinyl pyrrolidone)

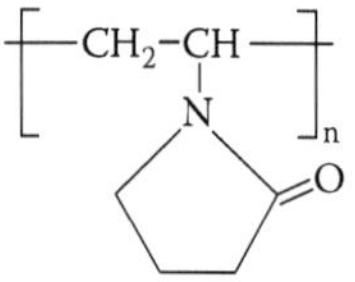

$$\left[CH_2-CH(NC_4H_6O)\right]_n$$

Styrene-acrylonitrile copolymer (SAN)

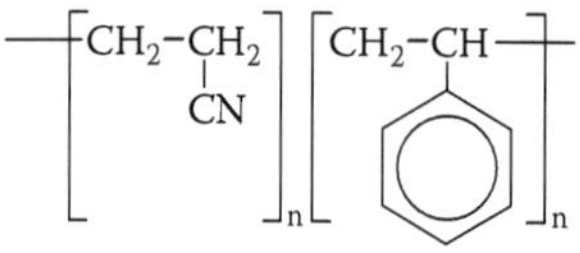

$$-\!\!\left[CH_2-\underset{CN}{\underset{|}{CH_2}}\right]_n\left[CH_2-CH(C_6H_5)\right]_n\!\!-$$

Styrene-butadiene rubber (SBR)

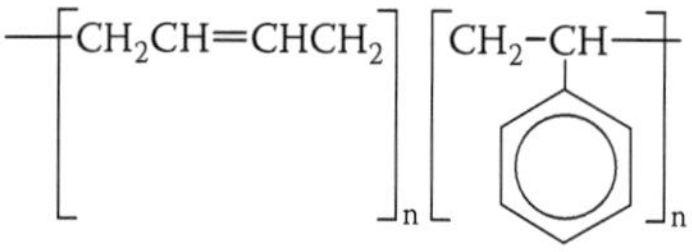

$$-\!\!\left[CH_2CH{=}CHCH_2\right]_n\left[CH_2-CH(C_6H_5)\right]_n\!\!-$$

TABLE 21.8
Trade/Brand Name and Manufacturer of Some Selected Vinyl Polymers

Trade or Brand Name	Product	Manufacturer
Abafil	Reinforced ABS	Rexall Chemical Co.
Absafil	ABS polymers	Fiberfil
Abson	ABS polymers	B. F. Goodrich Chemical Co.
Acralen	Styrene–butadiene latex	Farbenfabriken Bayer AG
Acronal	Polyalkyl vinyl ether	General Aniline Film Corp.
Acrilan	Polyacrylonitrile	Chemstrand Co.
Acrylan-Rubber	Butyl acrylate–acrylonitrile copolymer	Monomer Corp.
Acrylite	Poly(methyl methacrylate)	American Cyanamid Co.
Afcoryl	ABS polymers	Pechiney-Saint-Gobain
Argil	Styrene copolymer monofilament	Shawinigan Chemicals, Ltd.; also Polymer Corp.
Bexone F	Poly(vinyl formal)	British Xylonite
Benvic	Poly(vinyl chloride)	Solvay & Cie S. A.
Bexphane	Polypropylene	Bakelite Xylonite Ltd.
Blendex	ABS resin	Borg-Warner Corp.
Bolta Flex	Vinyl sheeting and film	General Tire & Rubber Co.
Butacite	Poly(vinyl acetal) resins	E. I. du Pont de Nemours Co., Inc.
Butakon	Butadiene copolymers	Imperial Chemical Industries, Ltd.
Butaprene	Styrene–butadiene elastomers	Firestone Tire & Rubber Co.
Butarez CTL	Telechelic butadiene polymer	Phillips Petroleum Co.
Buton	Butadiene–styrene resin	Enjay Chemical Co.
Bu-Tuf	Polybutene	Petrotex Chemical Corp.
Butvar	Poly(vinyl butyral) resin	Shawinigan Resins Corp.
Carina	Poly(vinyl chloride)	Shell Chemical Co. Ltd.
Carinex	Polystyrene	Shell Chemical Co. Ltd.
Celatron	Polystyrene	Celanese Plastics Co.
Cellofoam	Polystyrene foam board	United States Mineral Products Co.
Cerex	Styrene copolymer	Monsanto Chemical Co.
Cobex	Poly(vinyl chloride)	Bakelite Xylonite Ltd.
Cordo	PVC foam and films	Ferro Corp.
Corvic	Vinyl polymers	Imperial Chemical Industries Ltd.
Courlene	Polyethylene (fiber)	Courtaulds
Covol	Poly(vinyl alcohol)	Corn Products Co.
Creslan	Acrylonitrile–acrylic ester copolymers	American Cyanamid Co.
Crystalex	Acrylic resin	Rohm & Haas Co.
Cycolac	Acrylonitrile–butadiene–styrene copolymer	Borg-Warner Corp.
Daran	Poly(vinylidene chloride) emulsion coatings	W. R. Grace & Co.

(Continued)

TABLE 21.8 (CONTINUED)
Trade/Brand Name and Manufacturer of Some Selected Vinyl Polymers

Trade or Brand Name	Product	Manufacturer
Darex	Styrene copolymer resin	W. R. Grace & Co.
Darvan	Poly(vinyldene cyanide)	Celanese Corp. of America
Darvic	Poly(vinyl chloride)	Imperial Chemical Industries, Ltd.
Degalan	Poly(methyl methacrylate	Degussa
Diakon	Poly(methyl methacrylate)	Imperial Chemical Industries Ltd.
Dralon	Polyacrylonitrile fiber	Farbenfabriken Bayer AG
Dynel	Vinyl chloride–acrylonitrile copolymers	Union Carbide Corp.
Dylel	ABS copolymer	Sinclair-Koppers Co., Inc.
Dylene	Polystyrene resins	ARCO Polymer, Inc.
Dylite	Expandable polystyrene	Sinclair-Koppers Co., Inc.
Ecavyl	Poly(vinyl chloride)	Kuhlmann
Elvacet	Poly(vinyl acetate) emulsion	E. I. du Pont de Nemours & Co., Inc.
Elvacite	Acrylic resins	E. I. du Pont de Nemours & Co., Inc.
Elvanol	Poly(vinyl alcohol) resins	E. I. du Pont de Nemours & Co., Inc.
Elvax	Poly(ethylene–co-vinyl acetate)	E. I. du Pont de Nemours & Co., Inc.
Elvic	Poly(vinyl chloride)	Solvay
Evenglo	Polystyrene	Sinclair-Koppers Co., Inc.
Exon	Poly(vinyl chloride)	Firestone Plastics
Flovic	Poly(vinyl acetate)	Imperial Chemical Industries, Ltd.
Fluorel	Poly(vinylidene fluoride)	Minnesota Mining and Mfg. Co.
Foamex	Poly(vinyl formal)	General Electric Co.
Formex	Poly(vinyl acetal)	General Electric Co.
Formvar	Poly(vinyl formal)	Shawinigan Resins Corp.
Fostacryl	Poly(styrene–co-acrylonitrile)	Foster Grant Co.
Fostalene	Plastic	Foster Grant Co.
Fostarene	Polystyrene	Foster Grant Co.
FPC	PVC resins compound	Firestone Tire & Rubber Co.
Gelvatex	Poly(vinyl acetate) emulsions	Shawinigan Resins Corp.
Gelvatol	Poly(vinyl alcohol)	Shawinigan Resins Corp.
Geon	Poly(vinyl chloride)	B. F. Goodrich Chemical Co.
Heveaplus	Copolymer of methyl methacrylate and rubber	Generic name
Hi-Blen	ABS polymers	Japanese Geon Co.
Hostyren	Polystyrene	Hoechst
Hycar	Butadiene acrylonitrile copolymer	B. F. Goodrich Chemical Co.
Implex	Acrylic resins	Rohm & Hass Co.
Kralac	ABS resins	Uniroyal, Inc.
Kralastic	ABS	Uniroyal, Inc.
Koroseal	Poly(vinyl chloride)	B. F. Goodrich Chemical Co.
Kralon	High-impact styrene and ABS resins	Uniroyal, Inc.

TABLE 21.8 (CONTINUED)
Trade/Brand Name and Manufacturer of Some Selected Vinyl Polymers

Trade or Brand Name	Product	Manufacturer
Krene	Plasticized vinyl film	Union Carbide Corp.
K-Resin	Butadiene–styrene copolymer	Phillips Petroleum Co.
Kurlon	Poly(vinyl alcohol) fibers	
Kydex	Acrylic-poly(vinyl chloride) sheet	Rohm & Hass Co.
Kynar	Poly(vinylidene fluoride)	Pennwalt Chemicals Corp.
Lemac	Poly(vinyl acetate)	Borden Chemical Co.
Lemol	Poly(vinyl alcohol)	Borden Chemical Co.
Levapren	Ethylene–vinylacetate copolymers	Farbenfabriken Bayer AG
Lucite	Poly(methyl methacrylate) and copolymers	E. I. du Pont de Nemours & Co., Inc.
Lustrex	Polystyrene	Monsanto Chemical Co.
Lutonal	Poly(vinyl ethers)	Badische Anilin & Soda-Fabrik AG
Lutrex	Poly(vinyl acetate)	Foster Grant Co.
Luvican	Poly(vinyl carbazole)	Badische Anilin & Soda-Fabrik AG
Marvinol	Poly(vinyl chloride)	Uniroyal Inc.
Mipolam	Poly(vinyl chloride)	Dynamit Nobel
Mowilith	Poly(vinyl acetate)	Farbwerke Hoechst AG
Mowtol	Poly(vinyl alcohol)	Farbwerke Hoechst AG
Mowital	Poly(vinyl butyral)	Farbwerke Hoechst AG
Nalgon	Plasticized poly(vinyl chloride)	Nalge Co.
Nipeon	Poly(vinyl chloride)	Japanese Geon Co.
Nipoflex	Ethylene–vinyl acetate copolymer	Toyo Soda Mfg. Co.
Noan	Styrene–methyl methacrylate copolymer	Richardson Corp.
Novodur	ABS polymers	Farbenfabriken Bayer AG
Opalon	Poly(vinyl chloride)	Monsanto Chemical Co.
Oppanol C	Poly(vinyl isobutylether)	Badische Anilin & Soda-Fabrik AG
Orlon	Acrylic fiber	E. I. du Pont de Nemours & Co., Inc.
Paracryl	Butadiene–acrylonitrile copolymer	U.S. Rubber Co.
Pee Vee Cee	Rigid poly(vinyl chloride)	ESB Corp.
Pelaspan	Expandable polystyrene	Dow Chemical Co.
Perspex	Acrylic resins	Imperial Chemical Industries Ltd.
Pevalon	Poly(vinyl alcohol)	May and Baker Ltd.
Philprene	Styrene–butadiene rubber	Phillips Petroleum Co.
Plexiglas	Acrylic sheets	Rohm & Haas Co.
Plexigum	Acrylate and methacrylate resins	Rohm & Haas Co.
Plioflex	Poly(vinyl chloride)	Goodyear Tire & Rubber Co.
Pliovic	Poly(vinyl chloride)	Goodyear Tire & Rubber Co.
Polysizer	Poly(vinyl alcohol)	Showa Highpolymer Co.
Polyviol	Poly(vinyl alcohol)	Wacker Chemie GmbH
Ravinil	Poly(vinyl chloride)	ANIC, S.P.A.

(Continued)

TABLE 21.8 (CONTINUED)
Trade/Brand Name and Manufacturer of Some Selected Vinyl Polymers

Trade or Brand Name	Product	Manufacturer
Resistoflex	Poly(vinyl alcohol)	Resistoflex Corp.
Restirolo	Polystyrene	Societa Italiana Resine
Rhoplex	Acrylic emulsions	Rohm & Haas Co.
Rucon	Poly(vinyl chloride)	Hooker Chemical Corp.
Saflex	Poly(vinyl butyral)	Monsanto Co.
Saran	Poly(vinylidene chloride)	Dow Chemical Co.
Solvar	Poly(vinyl acetate)	Shawinigan Resins Corp.
Solvic	Poly(vinyl chloride)	Solvay & Cie
S-polymers	Butadiene–styrene copolymer	Esso Labs
Staflex	Vinyl plasticizers	Reichhold Chemical, Inc.
Starex	Poly(vinyl acetate)	International Latex & Chemical Corp.
Stymer	Styrene copolymer	Monsanto Co.
Styrocel	Polystyrene (expandable)	Styrene Products Ltd.
Styrofoam	Extruded expanded polystyrene; foam	Dow Chemical Co.
Styron	Polystyrene	Dow Chemical Co.
Sullvac	Acrylonitrile–butadiene–styrene copolymer	O'Sullivan Rubber Corp.
Tedlar	Poly(vinyl fluorocarbon) resins	E. I. du Pont de Nemours & Co., Inc.
Terluran	ABS polymers	Badische Anilin & Soda-Fabrik AG
Texicote	Poly(vinyl acetate)	Scott Bader Co.
Trosiplast	Poly(vinyl chloride)	Dynamit Nobel AG
Trulon	Poly(vinyl chloride) resin	Olin Corp.
Tybrene	ABS polymers	Dow Chemical Co.
Tygon	Vinyl copolymer	U.S. Stoneware Co.
Tyril	Styrene–acrylonitrile copolymer	Dow Chemical Co.
Ultron	Vinyl film	Monsanto Co.
Ultryl	Poly(vinyl chloride)	Phillips Petroleum Co.
Uscolite	ABS copolymer	U.S. Rubber Co.
Vestolit	Poly(vinyl chloride)	Chemische Werke Huls AG
Vestyron	Polystyren	Chemische Werke Huls AG
Viclan	Poly(vinylidene chloride)	Imperial Chemical Industries, Inc.
Vinac	Poly(vinyl acetate) emulsions	Air Reduction Co.
Vinapas	Poly(vinyl acetate)	Wacker Chemie GmbH
Vinidur	Poly(vinyl chloride)	BASF Corp.
Vinoflex	Poly(vinyl chloride)	BASF Corp.
Vinol	Poly(vinyl alcohol)	Air Reduction Co.
Vinylite	Poly(vinyl chloride-co-vinyl acetate)	Union Carbide Corp.
Vinyon	Poly(vinyl chloride-co-acrylonitrile)	Union Carbide Corp.

TABLE 21.8 (CONTINUED)
Trade/Brand Name and Manufacturer of Some Selected Vinyl Polymers

Trade or Brand Name	Product	Manufacturer
Vipla	Poly(vinyl chloride)	Montecatini Edison, S.p.A.
Vybak	Poly(vinyl chloride)	Bakelite Xylonite Ltd.
Vygen	Poly(vinyl chloride)	General Tire & Rubber Co.
Vynex	Rigid vinyl sheeting	Nixon-Baldwin Chemicals, Inc.
Vyram	Rigid poly(vinyl chloride)	Monsanto Co.
Welvic	Poly(vinyl chloride)	Imperial Chemical Industries, Inc.

Source: Portions of the data were compiled from R. B. Seymour and C. E. Carraher Jr., *Polymer Chemistry*, Marcel Dekker Inc., New York, 1992.

TABLE 21.9
Industrially Important Additional Polymers

Name	Repeating Unit	Typical Properties	Typical Uses
Polyacrylonitrile (including acrylic fibers	$-\!(CH_2-CH(CN))\!-$	High strength; good stiffness; tough; abrasion resistant; resilient; good flex life; relatively good resistance to moisture and stains, chemicals, insects, and fungi; good weatherability	Carpeting, sweaters, skirts, socks, slacks, baby garments
Poly(vinyl acetate)	$-\!(CH_2-CH(O-C(=O)-CH_3))\!-$	Water sensitive with respect to physical properties such as adhesion and strength; generally good weatherability; fair adhesion	Lower molecular weight used in chewing gum, intermediate in production of poly(vinyl alcohol), water-based emulsion paints
Poly(vinyl alcohol)	$-\!(CH_2-CH(O-H))\!-$	Water soluble; unstable in acidic or basic aqueous systems; fair adhesion	Thickening agent for various suspension and emulsion systems, packaging film, wet-strength adhesive
Poly(vinyl butyral)	$-\!(CH_2-CH-CH_2-CH)\!-$, with the two CH carbons linked through O–CH(C$_3$H$_7$)–O	Good adhesion to glass; tough; good stability to sunlight; good clarity; insensitive to moisture	Automotive safety glass as the interlayer
Poly(vinyl chloride) and poly(vinylidene chloride) (called "the vinyls" or "vinyl resin")	$-\!(CH_2-CH(Cl))\!-$	Relatively unstable to heat and light; fire resistant; resistant to chemicals, insects, fungi; resistant to moisture	Calendered products such as film sheets and floor coverings; shower curtains, food covers, rainwear, handbags, coated fabrics, insulation for electrical cable and wire, phonograph records

Polytetrafluoroethylene (Teflon)	$-\!\!(CF_2-CF_2)\!\!-$	Insoluble in most solvents; chemically inert; low dielectric loss; high dielectric strength; uniquely nonadhesive; low friction properties; constant electrical and mechanical properties from 27°C to about 250°C; high impact strength; not hard; outstanding mechanical properties	Coatings for frying pans, etc.; wire and cable insulation; insulation for motors. transformers, generators; gaskets; pump and valve packings; nonlubricated bearings
Polyethylene (low-density, branched)	$-\!\!(CH_2-CH_2)\!\!-$	Dependent on molecular weight, branching; molecular-weight distribution, etc.; good toughness and pliability over a wide temperature range; outstanding electrical properties; good transparency in thin films; inert chemically; resistant to acids and bases; ages on exposure to light and oxygen; low density; flexible without plasticizer; resilient; high tear strength; moisture resistant	Films; sheeting used in bags, pouches, produce wrapping, textile materials, frozen foods, etc.; drapes, table cloths; covers for construction ponds, greenhouses, trash can liners, etc.; electrical wire and cable insulator; coating of foils, papers, other films; squeeze bottles
Polyethylene (high-density, linear)		Most of the differences in properties between branched and linear concerns the high crystallinity of the latter; linear polyethylene has a high T_g, T_m, softening range, greater hardness, and tensile strength	Bottles, housewares, toys, films, sheets, extrusion coating, pipes, conduit, wire and cable insulation
Polypropylene	$-\!\!(CH_2-CH(CH_3))\!\!-$	Lightest major plastic; its high crystallinity imparts to it high tensile strength stiffness, and hardness, good gloss, high resistance to marring; high softening range permits polymer to be sterilized; good electrical properties, chemical inertness, moisture resistance	Filament: rope, webbing, cordage; carpeting; injection-molding applications in appliance, small housewares, and automotive fields
Polyisoprene (*cis*-1, 4-polyisoprene)	$-\!\!(CH_2-C(CH_3)=CH-CH_2)\!\!-$	Structure closely resembling that of natural rubber; properties similar to those of natural rubber	Replacement of natural rubber; often preferred because of its greater uniformity and cleanliness

(*Continued*)

TABLE 21.9 (CONTINUED)
Industrially Important Additional Polymers

Name	Repeating Unit	Typical Properties	Typical Uses
SBR (styrene–butadiene rubber)	Random copolymer	Generally slightly poorer physical properties than those of natural rubber	Tire treads for cars, but inferior to natural rubber with respect to heat buildup and resilience, thus not used for truck tires; belting; molded goods, gum, flooring, rubber shoe soles, electrical insulation, hoses
Butyl rubber (copolymer of isobutylene with small amounts of isoprene added to permit vulcanization)	Amorphous isoprene—largely 1,4 isomer	Good chemical inertness; low gas permeability; high viscoelastic response to stresses; less sensitive to oxidative aging than most other elastomers; better ozone resistance than natural rubber; good solvent resistance	About 70–60% used for inner tubes for tires
Polychloroprene (Neoprene)	Mostly 1,4 isomer	Outstanding oil and chemical resistance; high tensile strength; outstanding resistance to oxidative degradation and aging; good ozone and weathering resistance; dynamic properties are equal to or greater than those of most synthetic rubber and only slightly inferior to those of natural rubber	Can replace natural rubber in most applications: gloves, coated fabrics, cable and wire coatings, hoses, belts, shoe heels, solid tires
Polystyrene	$-\!\!\left(\mathrm{CH_2-CH(C_6H_5)}\right)\!\!-$	Clear; easily colored; easily fabricated; transparent; fair mechanical and thermal properties; good resistance to acids, bases, oxidizing, and reducing agents; readily attacked by many organic solvents; good electrical insulator	Used for the production of ion-exchange resins, heat- and impact-resistant copolymers, ABS resins, etc.; foams, plastic optical components, lighting fixtures, housewares, toys, packaging appliances, home furnishings
Poly(methyl methacrylate)	$-\!\!\left(\mathrm{CH_2-CH(C{=}O)(OCH_3)}\right)\!\!-$	Clear transparent, colorless; good weatherability; good impact strength; resistant to dilute basic and acidic solutions; easily colored; good mechanical and thermal properties; good fabricability; poor abrasion resistance compared with glass	Available in cast sheets, rods, tubes, and molding and extrusion compositions; applications where light transmission is needed, such as tail- and signal-light lenses, dials, medallions, brush backs, jewelry, signs, lenses, skylight “glass”

Source: Portions of the data were compiled from R. B. Seymour and C. E. Carraher Jr., *Polymer Chemistry*, Marcel Dekker, Inc., New York, 1992.

Index

A

D

E

F

G

H

I

K

L

M

N

O

P

Q

R

S

T

V

W

X

Z